The Origin

and Early

Diversification

of

Land Plants

Smithsonian Series in Comparative Evolutionary Biology
Douglas H. Erwin, Smithsonian Institution
V. A. Funk, Smithsonian Institution

The intent of this series is to publish innovative studies in the field of comparative evolutionary biology, especially by authors willing to introduce new ideas or to challenge or expand views now accepted. Within this context, and with some preference toward the organismic level, a diversity of viewpoints is sought.

ALSO IN THE SERIES

Parascript
Parasites and the Language of Evolution
Daniel R. Brooks and Deborah A. McLennan

The Development and Evolution of Butterfly Wing Patterns
H. Frederick Nijhout

Hawaiian Biogeography
Evolution on a Hot Spot Archipelago
Edited by Warren L. Wagner and V. A. Funk

The Origin and Early Diversification of *Land Plants*

A Cladistic Study

Paul Kenrick
Swedish Museum of Natural History

Peter R. Crane
The Field Museum and the University of Chicago

SMITHSONIAN INSTITUTION PRESS
Washington and London

COPY EDITOR: Gail K. Schmitt
PRODUCTION EDITOR: Deborah L. Sanders
DESIGNER: Janice Wheeler

Library of Congress Cataloging-in-Publication Data

Kenrick, Paul.
The origin and early diversification of land plants : a cladistic study / Paul Kenrick and Peter R. Crane.
p. cm. — (Smithsonian series in comparative evolutionary biology)
Includes bibliographical references (p.) and indexes
ISBN 1-56098-730-8 (cloth : alk. paper).
— ISBN 1-56098-729-4 (paper : alk. paper)
1. Plants—Evolution. 2. Plants—Phylogeny. 3. Cladistic analysis. 4. Paleobotany. I. Crane, Peter R. II. Title. III. Series.
QK980.K44 1997
581.3′8—dc21 97-24710

British Library Cataloguing-in-Publication Data available

Manufactured in the United States of America
06 05 04 03 02 01 00 99 98 97 5 4 3 2 1

∞ The paper used in this publication meets the minimum requirements of the American National Standard for Information Sciences—Permanence of Paper for Printed Library Material ANSI Z39.48-1984.

To the botanists and paleobotanists whose data and interpretations provided the substance and inspiration for this book

Contents

Preface

This work began while P.K. was a post-doctoral research associate supported by the Department of Geology at the Field Museum, Chicago. Additional support was provided by the National Science Foundation of the United States (grant DE 90-20237), the Swedish Natural Science Research Council (grants G-GU 9381-309; G-GF 9381-310; B-AA/BU 10728-300), and the Scientific Exchange Program of the Swedish Royal Academy of Sciences. Significant progress in completing this book was made while P.R.C. was supported by a grant from the Mellon Foundation as a Senior Mellon Fellow at the Laboratory of Molecular Systematics at the Smithsonian Institution working with E. A. Zimmer, and while P.K. was employed in the Department of Palaeobotany at the Swedish Museum of Natural History, Stockholm. P.R.C. also acknowledges the support of the Department of the Geophysical Sciences at the University of Chicago.

We are grateful to R. A. Chapman, K. Renzaglia, R. Brown, E. M. Friis, B. Lemmon, P. Herendeen, E. Taylor, T. N. Taylor, W. A. DiMichele, C. F. Delwiche, M. J. Donoghue, P. G. Gensel, L. Graham, G. Mueller, V. A. Albert, E. A. Zimmer, and W. E. Bemis for helpful discussions and numerous constructive comments. In particular we thank D. W. Stevenson and especially R. M. Bateman for their thorough reviews of the complete manuscript. We are also indebted to W. A. DiMichele, R. M. Bateman, P. G. Gensel, B. Mishler, L. Lewis, V. A. Albert, W. H. Wagner, K. M. Pryer, A. R. Smith, J. E. Skog, and G. W. Rothwell for access to unpublished manuscripts. We thank C. Shute and the Keeper of Paleontology, the Natural History Museum, London, and H. Kerp, H. Hass, and the late W. Remy, Abt. Paläobotanik, University of Münster, for permitting us to examine and illustrate fossil specimens held in their collections. We are also grateful to K. Khullar for help in preparing the manuscript. We would like to express special thanks to the copy editor, Gail Schmitt, for the many improvements that she made to the manuscript.

P.K. gratefully acknowledges the support of Dianne Edwards and a United Kingdom Natural Environment Research Council Grant (GR3/5069) during earlier work that provided some important data for Chapters 4 and 5, including Figures 5.18, 5.20, 5.21, 5.22, and 5.23. Most of the reconstructions were drawn by Pollyanna Lidmark. Jony Eriksson drafted Figures 5.1, 5.2, 5.4, 7.28, and 7.29. Figure 4.25 was drawn by Clara Simpson.

Finally we thank our wives, Anne and Elinor, for their patience, understanding, and good humor during the long process of completing this work.

The Origin and Early Diversification of *Land Plants*

1 Introduction

The origin and early diversification of land plants *(embryophytes)* and the related evolutionary transition from green algae to bryophytes and pteridophytes have long been recognized as pivotal events in plant evolution. Their significance for understanding plant diversity is comparable to that of the Cambrian "explosion" for investigating the diversity of Metazoa. From a systematic perspective, the evolution of a land flora marks the origin of many of the major groups of green plants, and a clear understanding of relationships among these early groups is critical to a full appreciation of large-scale phylogenetic patterns in the plant kingdom. From a morphological perspective, the emergence of land plants was a period of unparalleled innovation. Embryophyte morphology evolved rapidly under terrestrial conditions—from the relative simplicity of the green algae to the complex morphologies that had become characteristic of land plants by the end of the Devonian. From a physiological perspective, the drastic changes inherent in the transition from aquatic to terrestrial existence had profound biological consequences, and many of the morphological, physiological, and biochemical innovations of land plants can be interpreted as adaptations necessary for survival in a gaseous, rather than an aquatic, medium. From an ecological perspective, the evolutionary patterns and processes underlying the origin of a land flora are central to understanding the initial assembly of terrestrial ecosystems and therefore the evolution of those groups of animals that are characteristic of terrestrial environments (e.g., insects and tetrapods).

Approaches to investigating land plant origins span several disciplines, including those dealing directly with the fossil record—such as paleobotany, palynology, geology, and stratigraphy—as well as neobotanical studies of comparative plant morphology, plant development, and molecular biology. Over the last century, studies in these areas have accumulated a massive body of data from both living and fossil plants, but the information has remained scattered and has not yet been used fully in clarifying the early evolution of terrestrial plants. Different data sets provide diverse and complementary perspectives, but it is clear to us that a complete and maximally useful hypothesis of land plant origins should be based on an evaluation of all of the relevant information.

The aim of this book is to begin to synthesize the data that are currently available for resolving phylogenetic, and hence evolutionary, patterns among "basal" groups of land plants. Our analysis uses the important theoretical developments embodied in a cladistic approach and focuses on comparative morphology. This study is the first detailed cladistic treatment to evaluate both neobotanical and paleobotanical data together across all major basal lineages of embryophytes and tracheophytes. The results of our morphological study are compared and evaluated against other neobotanical evidence, in particular against the growing body of molecular sequence data. Our approach is to determine the distribution of features (characters) across taxa and to combine the emergent patterns into a coherent, parsimonious, and explicit cladistic hypothesis of relationships among major plant groups. Such a hypothesis is a prerequisite for useful discussion of land plant diversification, including a detailed understanding of the evolutionary changes in physiological, developmental, and ecological capabilities accompanying the origin of land plants and the early development of terrestrial ecosystems. This phylogenetic framework is also necessary for the rational design of research strategies in these and many other areas of plant science.

BACKGROUND

Early Ideas on the Phylogeny of Land Plants

The origin of land plants has persisted as a central problem in evolutionary botany for more than a century, and during this period, approaches to phylogenetic reconstruction have undergone dramatic changes. The history of ideas on land plant phylogeny strongly reflects shifting perspectives on phylogenetic reconstruction in botanical science as a whole.

The earliest contributions to clarifying the phylogeny of land plants focused on evidence from the comparative morphology of extant groups and were heavily influenced by Hofmeister's (1869) classic research on the unity and diversity of plant life cycles. The discovery of remarkable similarities in the structures and processes associated with sexual reproduction in bryophytes, ferns, and certain seed plants (Hofmeister 1869) led to widespread acceptance of land plant monophyly, and the recognition of motile male gametes in bryophytes and pteridophytes pointed to an aquatic ancestry. Based on similarities in photosynthetic pigments, cell structure, and spermatozoid morphology, it was also widely accepted that land plants probably originated from organisms at a level of organization comparable to that of green algae (Bower 1890, 1908; Campbell 1895; Celakovsky 1874; Haeckel 1868, 1894; Lignier 1903, 1908; Pringsheim 1878).

The most explicit early phylogenetic scheme for land plants is that of Haeckel (1868), which represents relationships among the major pre-Darwinian classes of plants as an unambiguous, dichotomously branching tree (Figure 1.1). Haeckel based this tree on morphological similarities (e.g., archegonia, sporangia, tracheids, seeds), which were interpreted as evidence of shared common ancestry. In current cladistic terms, Haeckel viewed land plants (Muscinae, Filicinae, Phanerogamae), vascular plants (Filicinae, Phanerogamae), seed plants (Phanerogamae), and angiosperms as monophyletic groups, whereas he interpreted bryophytes (Muscinae), pteridophytes (Filicinae), and gymnosperms (Gymnospermae) as paraphyletic. Liverworts (Hepaticae) were placed in a basal position among embryophytes, and mosses (Frondosae) and vascular plants were viewed as sister taxa. Haeckel envisaged the origin and first major radiation of land plants as taking place rapidly in the early Devonian (Figure 1.1).

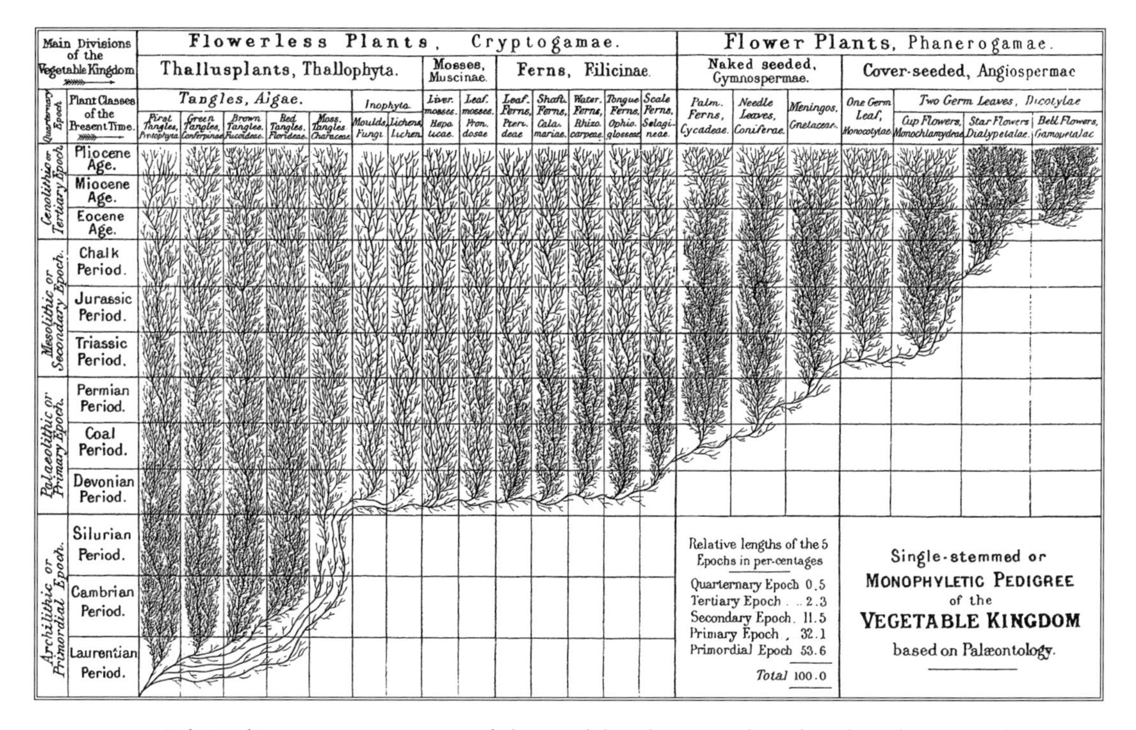

FIGURE 1.1. Relationships among major groups of plants and their divergence through geological time according to Haeckel 1876, Figure 5.

Notwithstanding the inferred monophyly of land plants, the remote position of bryophytes relative to vascular plants was generally acknowledged by early workers based on the conspicuous differences in development of the gametophyte and sporophyte, and these differences also led to practical problems in comparing the two groups. Because of the paucity of comparative data and an almost nonexistent fossil record, early ideas on the relationships between bryophytes and vascular plants were influenced strongly by two competing theories on the evolution of plant life cycles.

The Antithetic Theory, developed by Celakovsky (1874) and Bower (1890), proposed that the green algal ancestor of land plants had a haplobiontic life cycle with an anisomorphic alternation of generations in which the gametophyte was multicellular and the sporophyte was unicellular (zygote only) (Figure 1.2). According to this hypothesis, the simple sporophyte of bryophytes was intermediate between the unicellular condition of certain green algae and the highly differentiated, multicellular sporophytes of vascular plants. In contrast, the Homologous Theory of Pringsheim (1878), Potonié (1899), Hallier (1902), and Tansley (1907) proposed that the green algal ancestor of land plants had a diplobiontic life cycle with an isomorphic alternation of generations, in which both the gametophyte and sporophyte were well differentiated and multicellular (Figure 1.2). According to this hypothesis, the simple sporophyte of bryophytes had been reduced from more complex green algal ancestors. Proponents of both theories supported their ideas by pointing to "algae" with appropriate life cycles, but the antithetic theory received broader acceptance between 1890 and 1920 because green algae, the most likely "ancestral group," were thought then to be exclusively haplobiontic. The simplicity of the sporophyte in all bryophytes was also used as an argument against reduction from more complex forms (Bower 1890; Church 1919; Fritsch 1916).

Further clarification of relationships among pteridophytes and seed plants came during the latter part of the nineteenth century through comparative studies of extant groups (e.g., Bower 1890,

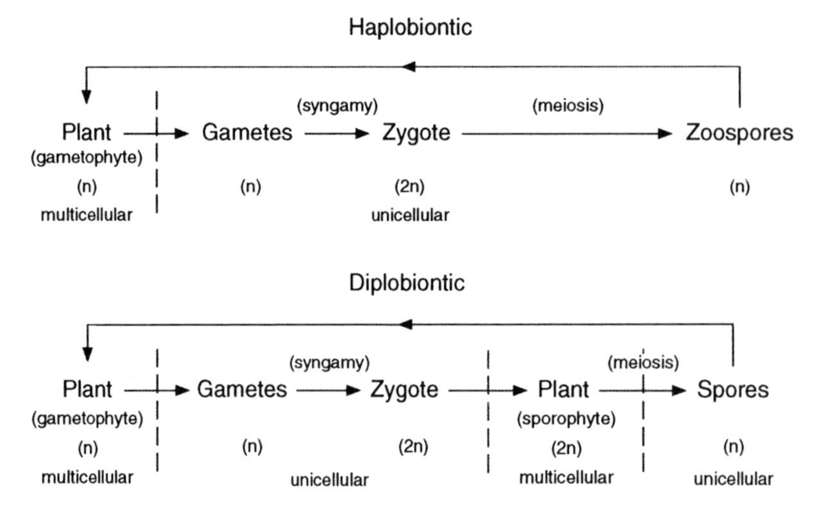

FIGURE 1.2. Main features of haplobiontic and diplobiontic life cycles. Multicellular and unicellular phases are separated by a vertical broken line. Diagrams modified after Kenrick 1994, Figure 1. *Haplobiontic* (Gr. *haploos,* "single" + Gr. *bioo,* "to live") life cycles are characterized by a *single* multicellular generation: in plants with haploid, haplobiontic life cycles, only the gametophyte is multicellular. This type of life cycle is typical of many green algae, including the Charophyceae. *Diplobiontic* (Gr. *diplos,* "double" + Gr. *bioo*) life cycles are characterized by *two* multicellular generations. The gametophyte phase produces gametes, whereas the sporophyte phase produces spores. Alternation of generations occurs between multicellular haploid and diploid phases. Diplobiontic life cycles occur in some green algae (e.g., some Ulvophyceae) and in all land plants. Current systematic evidence favors an origin of diplobiontic life cycles in land plants from haplobiontic life cycles in closely related charophycean algae via the evolution of a multicellular sporophyte phase.

1894, 1904; De Bary 1877; Goebel 1887; E. C. Jeffrey 1902), detailed anatomical descriptions of Paleozoic fossils (e.g., Renault 1879, 1888; Williamson 1871, 1872a,b, 1881, 1887), and renewed synthetic approaches in numerous texts (e.g., Campbell 1895; De Saporta and Marion 1881; Engler and Prantl 1898–1900; Goebel 1887; Potonié 1899; Renault 1879, 1888; Sachs 1868; Scott 1900; Solms-Laubach 1891; Van Tieghem 1891; Warming 1891). Initially, the ferns, or megaphyllous pteridophytes, were considered to be the group from which seed plants and microphyllous pteridophytes *(Equisetum* and lycopsids, which then included *Psilotum)* evolved. For example, in Haeckel's tree (Figure 1.1), ferns ("Leaf Ferns, Pterideae") are placed among basal vascular plants, whereas lycopsids ("Scale Ferns, Selagineae") are in a more derived position as a sister group to seed plants ("Flower Plants, Phanerogamae"). The basal position of ferns in early phylogenetic schemes was a holdover from their intermediate position in classifications that predated Darwin and Hofmeister. This position for ferns was supported by similarities between Hymenophyllaceae (filmy ferns) and mosses, including the filmy character of the leaf, filamentous prothallus, projecting sexual organs, a single well-defined apical cell, and the absence of roots (Bower 1891). Although widely accepted (e.g., Campbell 1895; Hallier 1902; Potonié 1899), the basal position of ferns was challenged around the turn of the century by evidence on the comparative morphology and life cycles of extant groups (Bower 1891), as well as by accumulated data from the Paleozoic fossil record that blurred the traditionally accepted distinctions between extant pteridophytes and gymnosperms.

Paleobotanical studies by Williamson (1871, 1872a,b, 1881, 1887) and Renault (1879, 1888) demonstrated that cambial activity was not unique to seed plants and had probably evolved independently in several Paleozoic pteridophyte groups, including fernlike taxa, lycopsids, and sphenopsids. This discovery, followed by the recognition that much Carboniferous fernlike foliage had actually been produced by a newly recognized group of fossil seed plants (Cycadofilicales of Jeffrey [1902] and Potonié [1899]; pteridosperms of Oliver and Scott [1904]), reduced the perceived distinctiveness of seed plants and pointed to a close relationship with ferns. This conclusion was also consistent with results emerging from comparative studies of extant plants. Based on detailed anatomical investigations of extant pteridophytes and gymnosperms, Jeffrey (1902) concluded that ferns and seed plants both possessed a significant anatomical similarity termed the *leaf gap,* and he argued that this, along with megaphyllous leaves and abaxial sporangia, was evidence that ferns were more closely related to seed plants than to either lycopsids or *Equisetum.*

Further evidence on the relationships of ferns followed from ontogenetic investigations of pteridophytes. Studies of sporangium ontogeny led to the development of the concepts of the eusporangium and leptosporangium (see Chapter 3), as well as to recognition of the eusporangium as the general condition in vascular plants (Bower 1890; Goebel 1887). The distinctive leptosporangium is confined to a single group containing all extant ferns (except Marattiales and Ophioglossales), and on this basis, Goebel argued that leptosporangiate ferns were unlikely to be ancestral to other vascular plants. This idea received further support from Bower's (1891) observation that the eusporangium significantly predates the leptosporangium in the fossil record. Building on these results, Scott (1900, 513–519) emphasized the Marattiales in his summary of the similarities between extant ferns and cycads and noted a general resemblance in the large compound leaves and frequent circinate vernation (especially in the cycad *Stangeria,* which was originally placed in the fern genus *Lomaria*). Scott also highlighted the similarities between the arrangement of pollen sacs on the sporophylls of cycads and in the sori of certain ferns, as well as the combination of fernlike and cycadlike characters in Paleozoic pteridosperms ("seed ferns") such as the medullosans.

The pioneering comparative studies of extant and fossil taxa of the late nineteenth century culminated in Bower's seminal treatise, *The Origin of a Land Flora* (1908), which developed several key phylogenetic conclusions, including (1) the close phylogenetic relationship between charophycean algae and embryophytes; (2) the marked differences between lycopsids and other vascular plants; (3) the broadly intermediate position of bryophytes with respect to green algae and tracheophytes (Antithetic Theory); and, (4) the relatively primitive characteristics of Ophioglossales and Marattiales with respect to other ferns. At the same time however, Bower explicitly drew attention to the considerable morphological discontinuities that existed between green algae and bryophytes, and also between bryophytes and pteridophytes. In doing so he anticipated a role for the fossil record but thought it unlikely that informative fossil evidence would be forthcoming.

The Influence of Paleobotanical Data on Interpretations of Early Land Plant Phylogeny

The first good evidence of early fossil land plants (Dawson 1859) preceded *The Origin of a Land Flora* by almost fifty years. In a series of pioneering papers (1861, 1862, 1870, 1871), Dawson described several early Devonian plants from Canada (Gaspé Bay) and Scotland and interpreted these early fossils as primitive vascular plants showing a degree of morphological simplicity unknown in extant groups. His work was controversial (see, for example, Solms-Laubach 1891), but the prevailing skepticism was not universal, and Lignier (1903, 1908), for example, utilized Dawson's paleobotanical discoveries in developing his ideas on the relationships between Sphenophyllales and Equisetales (Lignier 1903) and in discussing the evolution of microphylls and megaphylls (Lignier 1908).

Subsequent reports of fossils plants from early Devonian rocks in Bohemia (Potonié and Bernard 1904) and Norway (Halle 1916a,b; Nathorst 1913, 1915) enhanced the credibility of Dawson's results, and the discovery of unequivocal, exceptionally well preserved Devonian plant fossils from the Rhynie Chert, Scotland (Kidston and Lang 1917, 1920a,b, 1921), dispelled any further doubts. The Rhynie Chert plants provided the first detailed evidence on the morphology and anatomy of primitive pteridophytes (Bower 1920).

The documentation of morphologically simple vascular plants from the Rhynie Chert and other Devonian localities partially bridged the morphological discontinuity between bryophytes and tracheophytes that had been highlighted by Bower (1908) and had a lasting influence on phylogenetic ideas and the classification of land plants (Chapter 4). Most significantly, these discoveries also added impetus to the collection and description of Devonian fossil plants and, together with a relative decline in comparative work on extant taxa, shifted research on the early diversification of land plants to an almost exclusive focus on the fossil record.

During the 1950s, 1960s, and 1970s, studies of Devonian macrofloras and palynofloras increased rapidly, but evolutionary interpretations of the accumulating data were based mainly on inferences from stratigraphic patterns. An important summary was provided in the *Traité de paléobotanique* (Andrews, Arnold, et al. 1970; Boureau 1964; Boureau and Doubinger 1975; Boureau, Jovet-Ast, et al. 1967), and the paleobotanical-stratigraphic approach culminated in a series of influential papers by Banks (1968, 1970, 1975b, 1980) and by Chaloner (1970; Chaloner and Sheerin 1979). Several important reviews and much new paleobotanical data have been published subsequently (e.g., Edwards and Edwards 1986; Gensel and Andrews 1984; Gray 1985; Knoll, Niklas, et al. 1984; Knoll, Niklas, and Tiffney 1979; Selden and Edwards 1989; Stewart 1983; Stewart and Rothwell 1993; T. N. Taylor 1981, 1988a; T. N. Taylor and E. L. Taylor 1993). These have continued to refine our knowledge of stratigraphic patterns and extinct botanical diversity but have had less impact on our understanding of relationships among major plant groups.

Approaches to Reconstructing the Phylogeny of Early Land Plants

Despite pioneering attempts to integrate data from both extant and fossil plants into an inclusive evolutionary scheme (Bower 1935; Campbell 1940; G. M. Smith 1938; Zimmermann 1930, 1938), interest in reconstructing phylogenetic relationships declined rapidly in the two decades following the Rhynie Chert discoveries. In part, this decline reflected reduced research in the face of rapid developments in other areas of botanical science during the first half of this century (e.g., cytology, plant physiology, biochemistry; H. H. Thomas 1932, 1951), but there was also increasing dissatisfaction with what came to be regarded as mere phylogenetic speculation (e.g., Lang 1915; Lotsy 1916; Tansley 1923). In hindsight this view can be seen to have arisen through the difficulties of dealing with crucial theoretical and methodological issues.

A key factor contributing to the decline of phylogenetic studies seems to have been increasing awareness of "homoplastic or parallel development in the Plant Kingdom" (Wardlaw 1952, 96; see also, for example, Bower 1923a, 1935; Lang 1915). The term *homoplasy,* which denotes similarity due to convergent or parallel evolution, was coined by Lankester (1870) at about the same time as paleobotanical discoveries were demonstrating the independent origin of secondary xylem in different groups of land plants (Renault 1879, 1888; Williamson 1871, 1872a,b, 1881). Growing recognition of convergent and parallel evolution probably contributed to the development of the many polyphyletic views on the evolution of land plants (e.g., Arber 1921; Axelrod 1952, 1959; Campbell 1895; Church 1919; De Saporta and Marion 1881; Fritsch 1916; D. H. Scott 1920, 1924; Seward and Ford 1906). It also highlighted the absence of an appropriate methodological framework in which to develop and choose among competing phylogenetic hypotheses and thereby distinguish homology and homoplasy. This methodological impasse undoubtedly discouraged expression of phylogenetic ideas. For example, Lang (1915, 2) commented that "the inter-relationships of plants look more like a bundle of sticks" than the tree developed by Haeckel (1868), and Harris, Millington, and Miller (1974, 85) succinctly summarized the situation when they commented that the "building" of phylogenetic trees is "no longer held in esteem, for we know it can be done in many different ways."

Many phylogenetic ideas were also weakened by undue emphasis on hypothetical transformation series or stratigraphic patterns. For example, a priori acceptance of one of the two competing theories on the evolution of the life cycle in early land plants—the Homologous and Antithetic Theories (Bower 1894, 1908; Campbell 1895; Hallier 1902; Potonié 1899) generated quite different predictions about the phylogenetic relationships of bryophytes. Similarly, Bower's (1894, 1908) early ideas on the evolution of the pteridophyte strobilus (through sterilization and septation of a mosslike sporangium) strongly influenced his interpretation of lycopsids as a basal group in vascular plants. Stratigraphic evidence on the absence of leptosporangiate ferns in the early Paleozoic was seen as inconsistent with the supposed basal phylogenetic position of this group (Bower 1891; Scott 1900), and the relatively sudden stratigraphic appearance of early land plants in the fossil record has been used frequently as an argument for polyphylesis (e.g., Andrews and Alt 1956; Axelrod 1959; Leclercq 1954).

Over the last three decades, the development of phylogenetic systematics—*cladistics* (Hennig 1965, 1966)—has fostered intensive discussion of the philosophical and methodological bases for reconstructing phylogenetic patterns and has clarified crucial theoretical and methodological issues that blocked progress with plant phylogeny in the early decades of this century. Computer software for rapid numerical cladistic analyses of large or complex data sets, as well as advances in molecular biology that have made possible the rapid amplification and sequencing of RNA and DNA from a variety of organisms, have also reactivated interest in phylogenetic reconstruction.

The basic insight of cladistics is that systematic groups should be defined only on the basis of characters they alone possess (Figure 1.3). At a given hierarchical level, the concept of overall similarity *(phenetics)* is separated into a generalized, nondefinitional component *(symplesiomorphy)* and a less generalized, definitional component *(synapomorphy)*. Where different characters indicate conflicting groups (and therefore conflicting hierarchical patterns), the principle of *parsimony* is applied, and the simplest of competing patterns is preferred (Figure 1.4). The "true" synapomorphies recognized in this way are *homologies* (taxic homologies), and the problems of determining polarity, synapomorphy, and homology can therefore be viewed as synonymous (Hill and Crane 1982; Nelson and Platnick 1981; Patterson 1982; Rieppel 1988).

The result of a cladistic analysis is therefore merely a pattern of nested synapomorphies that defines relationships in a relative way and, like any other hierarchical pattern, can be summarized in a treelike diagram. There is no necessary relation to the theory of evolution and no necessary time axis. Although cladograms can thus be largely independent of evolutionary preconceptions, most systematists wish to interpret them in phylogenetic terms. Indeed, for many systematists the compatibility between cladistic reasoning and a simple model of character change during phylogeny is an important reason for preferring a cladistic rather than a phenetic approach.

Interpreted in phylogenetic terms, a hierarchical pattern of synapomorphies can be viewed as a nested series of evolutionary novelties, and the principle of parsimony as a criterion for preferring phylogenetic hypotheses that minimize homoplasy (parallelism, convergence, reversal). Groups defined by synapomorphies *(clades)* can be interpreted as including a common ancestor and all of its descendants *(monophyletic)* (Figure 1.5). In contrast, *paraphyletic* groups *(grades)* are definable only by a combination of plesiomorphic and apomorphic features and can be viewed as containing some, but not all, of the descendants of a common ancestor (Figure 1.5; see Ax 1987; Brooks

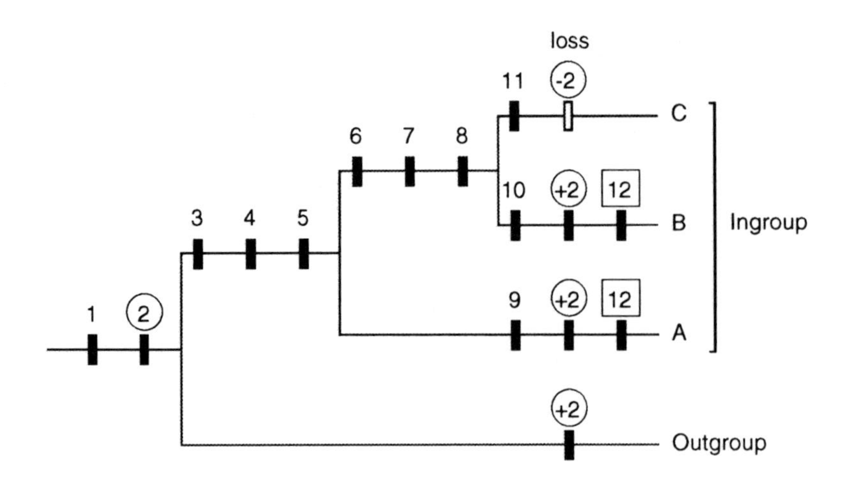

FIGURE 1.3. Similarities and differences analyzed for phylogenetic information content for three ingroup taxa: A, B, C. Numbers refer to characters 1–12. *Uninformative similarities: (1)* Symplesiomorphy—uninformative general similarity shared by all three ingroup taxa plus outgroup. *(3, 4, 5)* Symplesiomorphy (or holapomorphy)—uninformative general similarity shared by all three ingroup taxa but *not* present in outgroup. This is a synapomorphy of the ingroup. *Uninformative differences: (9, 10, 11)* Autapomorphy—uninformative difference unique to each of the three ingroup taxa. These may be synapomorphies of the terminal taxa. *Misleading similarities: (2)* Plesiomorphy—uninformative similarity between two (A and B) of the three ingroup taxa (states of character 2 shown on branches: + = present, – = absent). Character 2 is applicable at a more general level (below the outgroup) and also at a more restricted level (defining clade C through loss), but is uninformative with respect to the relationships of the three ingroup taxa. *(12)* Homoplasy—uninformative similarity between two (A and B) of the three ingroup taxa. Character 12 is a shared derived character supporting a clade A + B (not shown) but conflicting with characters 6, 7, and 8, which support the clade B + C. The principle of parsimony favors the clade B + C over A + B. Note that in this example character 12 can be interpreted as convergence in A and B or as a holapomorphy of the clade A + B + C that has been lost in C. *Informative similarities: (6, 7, 8)* Synapomorphy—the only informative character(s) with respect to relationships among the three ingroup taxa. The *consistency index* (CI) of a data set is a measure of the amount of homoplasy (Farris 1989; Kluge and Farris 1969). The consistency index of a character is the ratio of the minimum number of possible changes over the amount of actual change. For a binary character (e.g., character 2) with two states (present or absent), the minimum number of changes is one. The actual number of changes on the tree is two: one gain at the base of the tree and one loss in taxon C. Thus, CI of character 2 = 0.5. The consistency index for each character (characters 1–12) is calculated and summed for the tree and divided by the total number of characters (CI = 11/12 = 0.92) (see Wiley, Siegel-Causey, et al. 1991 for further explanation). Conventionally, the CI is calculated after removing uninformative characters (e.g., autapomorphies and invariant characters of the ingroup) and in this case would be CI = 4/5 = 0.8. Note that characters scored in a data matrix as "?," for whatever reason, complicate this situation and may artificially inflate the consistency index. *Character optimization* identifies the most parsimonious distribution of characters on a tree (Farris 1970). However, sometimes two or more equally parsimonious optimizations are possible for a single tree. For example, it is equally parsimonious to interpret character 12 as a parallelism in taxa A and B or as a synapomorphy of clade (A, B, C) and a reversal in taxon C. ACCTRAN optimization "*acc*elerates" *tran*sformations and would favor a reversal in taxon C. DELTRAN optimization "*del*ecerates" *tran*sformations and favors parallelisms in A and B.

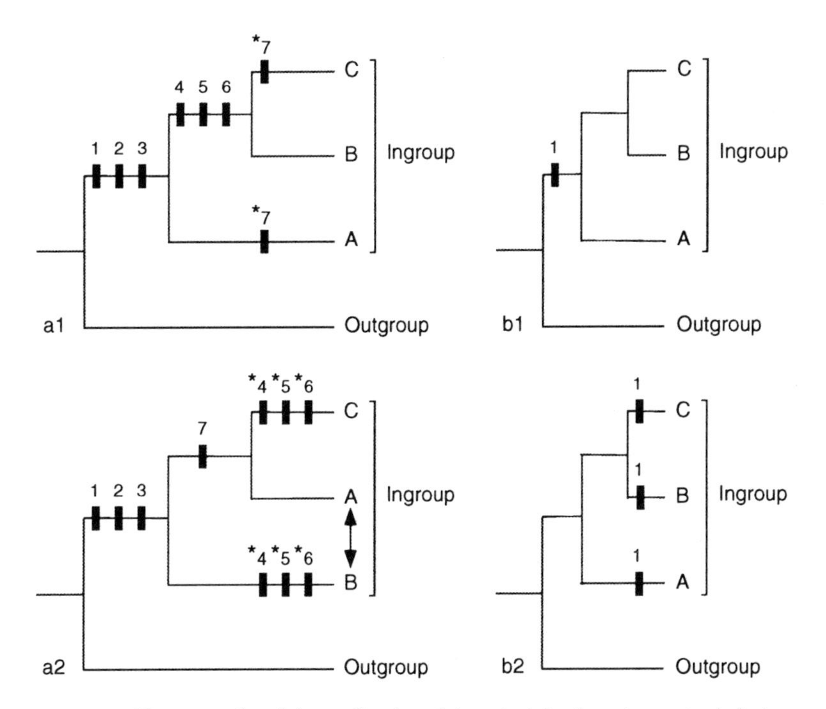

FIGURE 1.4. Two examples of the application of the principle of parsimony in cladistic analysis. (**a**) Group recognition. The relationship between taxa A, B, and C is more parsimonious in cladogram a1 than in a2. In cladogram a1, only character 7 is homoplasic (indicated by asterisk); in cladogram a2, characters 4, 5, and 6 are homoplasic. Cladogram a1 is more parsimonious because it involves the least homoplasy (i.e., the least number of hypotheses of evolutionary novelty and the strongest character support for one of the three possible pairs of taxa A, B, C). (**b**) Character origination. The relationships of taxa A, B, and C (established on the basis of other characters, not shown) is the same in b1 and b2, and character 1 is a synapomorphy of the ingroup. The hypothesis that character 1 has only one evolutionary origin (b1) is more parsimonious than proposing three independent origins (b2).

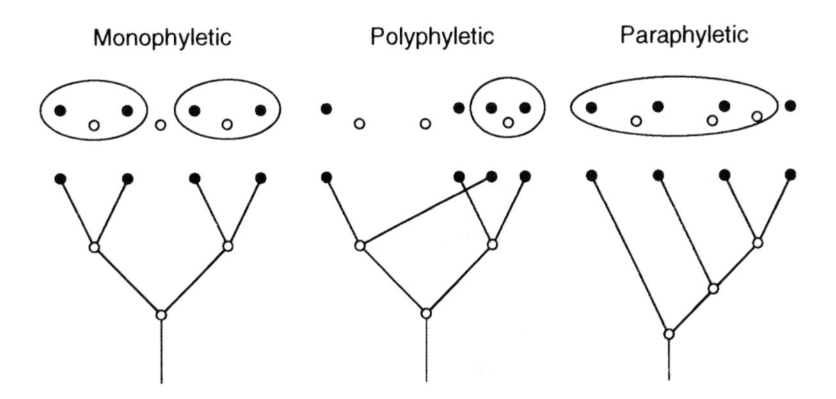

FIGURE 1.5. Terminology of groups and definitions. Closed circle = extant group; open circle = ancestral species. *Monophyletic group (clade):* an ancestral species and all of its descendants. *Polyphyletic group:* a group that does not contain an ancestral species common to all taxa (i.e., the ancestor is placed in another taxon). *Paraphyletic group (grade):* an ancestral species and some but not all of its descendants. Diagram modified after Hennig 1966, Figure 18.

and McLennan 1991; Forey, Humphries, et al. 1992; Smith 1994; Wiley 1981; Wiley, Siegel-Causey, et al. 1991 for a more complete discussion of systematic theory and cladistic techniques).

An important feature of phylogenetic systematics is that it preserves a direct relationship between the recognition of groups and the distribution of characters. It thus provides a methodological framework in which competing hypotheses of relationship can be compared critically at the level of individual characters, more or less independently of ideas on the nature or course of the evolutionary process.

Paleobotanical Data in Systematic Studies

Recent advances in systematic theory have done much to clarify the role of paleontological data in systematic studies (Donoghue, Doyle, et al. 1989; Doyle and Donoghue 1992, 1993; Gauthier, Kluge, and Rowe 1988; Huelsenbeck 1994; Patterson 1982; Smith 1994). Paleontological data have a variety of applications (e.g., biostratigraphy, paleoecology, paleogeography), but in a cladistic context, neither fossils nor their stratigraphic distribution have a special role in phylogenetic reconstruction (see, for example, Nelson and Platnick 1981; Patterson 1982; Smith 1994; contra: Axelrod 1952; Davis and Heywood 1973; Fisher 1992; Hughes 1976, 1994). In cladistic studies, cladograms are generated from morphological data, and evidence from the fossil record simply provides important additional data on the diversity of extinct land plants. Stratigraphic data become significant a posteriori and provide both a temporal context and a temporal calibration. Numerous examples show that fossil evidence contributes most effectively to resolving relationships among ancient groups that have become isolated through extinction. Such isolated groups are problematic both in morphological and in molecular data sets because evolutionary change following cladogenesis combined with extinction of "intermediate" taxa frequently obscures evidence of homology, thus creating problems in recognizing relationships. The major groups of land plants and green algae provide clear examples of this problem.

The importance of paleobotanical data in systematic studies of basal land plants is that they allow increased sampling of morphological diversity in the lycopsid, sphenopsid, fern, seed plant, and vascular plant stem groups. The inclusion of stem group fossils (e.g., *Aglaophyton, Archaeopteris,* zosterophylls) with character combinations not seen among extant taxa helps to clarify homologies and provides important additional morphological data for the recognition of parsimonious phylogenetic patterns. Although comparative data from extant groups such as green algae, bryophytes, or lycopsids is equally as relevant to the study of embryophyte phylogeny as data from well-preserved early fossils, the fossil record is probably the only means of increasing taxon sampling in problematic stem groups. In this context, detailed comparative studies of information-rich fossils and whole-plant reconstructions based on well-preserved material are of maximum systematic interest. Studies of this kind provide most of the paleobotanical information synthesized in subsequent chapters of this book. Paleobotanical research aimed at extending stratigraphic or geographic ranges has less relevance to systematic problems than do careful descriptions and comparative analyses of characters.

THE ORGANIZATION OF THIS BOOK

In this book we apply the techniques of phylogenetic systematics in attempting to resolve relationships that were established during the early diversification of land plants approximately 450–350 Ma (mega-annum, or millions of years before present). The approach is fundamentally similar to that used in attempting to clarify relationships among major groups of seed plants (e.g., Crane 1985; Doyle and Donoghue 1986, 1992; Nixon, Crepet, et al. 1994) and is guided by the same basic philosophy of treating living and fossil taxa together as part of a single comparative analysis using cladistic methods. The group terminology and conven-

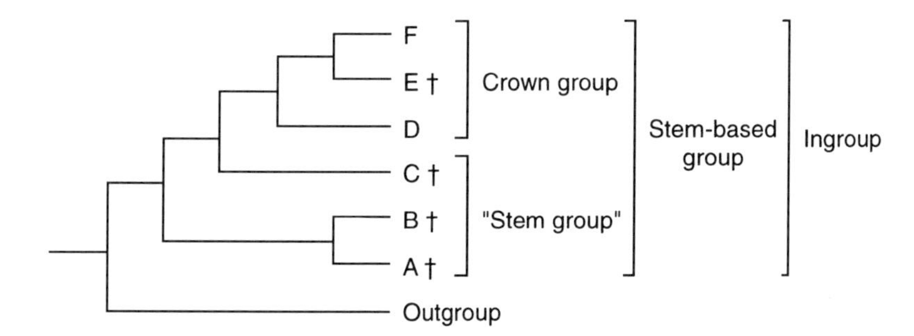

FIGURE 1.6. Terminology, definitions, and conventions used in this book. † = extinct; quotation marks = paraphyletic group. *Ingroup:* taxa under study. *Outgroup:* closely related taxon that is not part of the ingroup and is used for comparative purposes. *Sister group:* the most closely related clade to any other clade in an analysis; for example, A is a sister group to B and vice versa, and (A + B) is a sister group to (C + D + E + F). *Crown group:* the clade containing all the extant taxa; for example, (D + E + F). Note that the crown group may also contain extinct groups. *Stem-based group:* the clade containing the crown group plus one or more extinct sister groups. *Stem group:* the stem-based group minus the crown group. The stem group contains only extinct taxa and is paraphyletic by definition.

tions that we employ are summarized in Figure 1.6. One objective of this book is to develop phylogenetic hypotheses that have predictive value for future research and that are sufficiently robust to withstand the addition of new data from both living and fossil plants. An equally important goal is to render current ideas on early land plant evolution more explicit and to place them in a theoretical and empirical framework that facilitates discussion and that highlights critical characters and taxa for future study. We believe that this renewed synthetic effort is not only necessary but is also timely with respect to recent advances in the systematics of green algae (Chapter 2) and bryophytes (Chapter 3), emerging new data on early fossil land plants (Chapter 4), and significant new work in other areas of land plant phylogenetics (Chapters 2, 3, and 7). It is also clear that thoroughly developed phylogenetic hypotheses based on morphological data will be crucial for comparison with the rapidly emerging phylogenetic results based on molecular sequence data.

Our focus is on the origin and evolution of embryophytes and tracheophytes, and we attempt to integrate the existing paleobotanical data primarily with knowledge of extant liverworts, hornworts, mosses, lycopsids, sphenopsids, ferns, and Psilotaceae. Our work extends previous phylogenetic studies of some these groups (e.g., Mishler and Churchill 1984, 1985b; Mishler, Lewis, et al. 1994) and complements other recent high-level cladistic studies of land plants that have focused on integrating neobotanical and paleobotanical evidence in seed plants, basal angiosperms, and certain subgroups of lycopsids (Bateman 1992; Bateman, DiMichele, and Willard 1992; Crane 1985; Doyle and Donoghue 1986, 1992; Doyle, Donoghue, and Zimmer 1994; Nixon, Crepet, et al. 1994; Rothwell and Serbet 1994).

In Chapter 2 we provide a brief review, in cladistic terms, of the major monophyletic groups that are recognized in current kingdom-level classifications. In subsequent chapters we provide detailed treatments of embryophytes (Chapter 3), polysporangiophytes, tracheophytes (vascular plants), and euphyllophytes (Chapter 4), zosterophylls (Chapter 5), and lycopsids (Chapter 6), focusing in particular on relationships among basal members of these groups. In these chapters we also compare our results with the phylogenetic ideas developed by earlier authors. Finally (Chapter 7), we evaluate the implications of our results in terms of the

TABLE 1.1

Cladistic Classification of Green Plants

INFRAKINGDOM EMBRYOBIOTES

Superdivision Marchantiomorpha

- Division Marchantiophyta
  - Class Marchantiopsida
    - Order Sphaerocarpales
    - Order Monocleales *incertae sedis*
    - Order Marchantiales *incertae sedis*
    - Order Calobryales
    - Order "Metzgeriales"
    - Order Jungermanniales

Superdivision Anthoceromorpha

- Division Anthocerophyta
  - Class Anthocerotopsida

Superdivision Bryomorpha

- Division Bryophyta
  - Class Bryopsida
    - SUBCLASS SPHAGNIDAE
    - SUBCLASS ANDREAEIDAE
      - Order Takakiales
      - Order Andreaeales
      - Order Andreaeobryales
    - SUBCLASS TETRAPHIDAE
    - SUBCLASS POLYTRICHIDAE
    - SUBCLASS BUXBAUMIIDAE
    - SUBCLASS BRYIDAE

Superdivision Polysporangiomorpha

- Plesion Horneophytopsida
- Plesion *Aglaophyton major*
- Division Tracheophyta
- PLESION RHYNIOPSIDA
- SUBDIVISION LYCOPHYTINA
  - Plesion *Zosterophyllum myretonianum incertae sedis*
  - Class Lycopsida
    - Plesion Drepanophycales
    - Order "Lycopodiales"
    - Plesion Protolepidodendrales
    - Order Selaginellales
    - Order Isoetales
  - Plesion Zosterophyllopsida
    - Plesion *Zosterophyllum divaricatum*
    - Plesion Sawdoniales (families *sedis mutabilis*)
      - Plesion Sawdoniaceae
      - Plesion Barinophytaceae
      - Plesion "Gosslingiaceae"
      - Plesion Hsuaceae
- SUBDIVISION EUPHYLLOPHYTINA
- **Plesion *Eophyllophyton bellum***
- **Plesion *Psilophyton dawsonii***
- **Infradivision Moniliformopses (classes *sedis mutabilis*)**
  - Plesion "Cladoxylopsida" (subclasses *sedis mutabilis*)
    - PLESION "CLADOXYLIIDAE"
    - PLESION STAUROPTERIDAE
    - PLESION ZYGOPTERIDAE
  - Class Equisetopsida
  - Class Filicopsida (subclasses *sedis mutabilis*)
    - SUBCLASS OPHIOGLOSSIDAE
    - SUBCLASS PSILOTIDAE
    - SUBCLASS MARATTIIDAE
    - SUBCLASS POLYPODIIDAE
- **Infradivision Radiatopses**
- Plesion *Pertica varia*
- Supercohort Lignophytia (cohorts *sedis mutabilis*)
  - PLESION "ANEUROPHYTALES"
  - PLESION "ARCHAEOPTERIDALES"
  - PLESION "PROTOPITYALES"
  - COHORT SPERMATOPHYTATA

Notes: See discussion on classification in Chapter 7 for additional details and the more complete higher-level classification of green plants in Table 7.1. Quotation marks = paraphyletic group.

TABLE 1.2
Informal Names of Monophyletic Higher Taxa and Corresponding Linnean Taxa

Informal name	Corresponding Linnean taxon (or taxa)
Embryophytes	Embryobiotes
Land plants	Embryobiotes
Liverworts	Marchantiomorpha, Marchantiophyta, Marchantiopsida
Hornworts	Anthoceromorpha, Anthocerophyta, Anthocerotopsida
Mosses	Bryomorpha, Bryophyta, Bryopsida
Stomatophytes	Anthoceromorpha + Bryomorpha + Polysporangiomorpha
Polysporangiophytes	Polysporangiomorpha
Tracheophytes	Tracheophyta
Vascular plants	Tracheophyta
Eutracheophytes	Lycophytina + Euphyllophytina
Lycophytes	Lycophytina
Lycopsids	Lycopsida
Rhizomorphic lycopsids	Isoetales
Arborescent lycopsids	Isoetales
Fern–*Equisetum*–seed plant clade	Euphyllophytina
Sphenopsids	Equisetopsida
Ferns	Filicopsida
Leptosporangiate ferns	Polypodiidae
Lignophytes	Lignophytia
Seed plants	Spermatophytata

evolution of particular characters and discuss their significance for previous ideas on the developmental, functional, and adaptive basis of early land plant evolution.

A brief review of current information on all taxa considered in our analyses, together with information on other important plants, is given in Appendix 1. Data matrixes for our analyses are included in Appendixes 2–5. Throughout this book we adopted a relatively conservative approach to homology decisions in order to minimize the extent to which our treatment of particular characters predisposes our analyses toward certain results. This approach allows us to consider possible transformational homologies a posteriori in more detail in Chapter 7.

TABLE 1.3
Problematic Groups

Taxon	Status
Green algae	Paraphyletic
Charophycean algae	Paraphyletic
Bryophytes	Paraphyletic
Protracheophytes	Paraphyletic
Rhyniophytina *sensu* Banks	Paraphyletic, possibly polyphyletic
Zosterophyllophytina *sensu* Banks	Paraphyletic
Trimerophytina *sensu* Banks	Paraphyletic
Pteridophytes	Paraphyletic
Progymnosperms	Paraphyletic
Pteridosperms	Paraphyletic or polyphyletic
Gymnosperms	Paraphyletic

Note: For further details, see Table 7.4.

2 Plants in the Hierarchy of Life

Higher-level classifications of life clearly illustrate the distinction between monophyletic and paraphyletic groups. The earliest classifications of living organisms (e.g., Linné 1753) recognized the differences between plants and animals, but by the mid-nineteenth century, it became evident that neither fungi nor protozoans could be accommodated easily within this basic dichotomy. Hogg (1860) instituted the term *Protoctista* for "organisms neither animal nor plant," and Haeckel (1866) subsequently recognized animals, plants and "protists" as the three main branches in his "tree of life" (Figure 2.1). Chatton (1925) first defined the distinction between what later became known as prokaryotes and eukaryotes, and this convention was incorporated by Copeland (1956) as part of the rationale for a four-kingdom scheme that recognized Monera (prokaryotic organisms), Protoctista (protists, higher fungi, algae except chlorophytes), Animalia, and Plantae. In 1959, Whittaker emphasized the distinctiveness of the fungi (based largely on their characteristic extracellular mode of nutrition and distinctive chitinous cell walls) and their probable origin from unicellular, colorless, flagellate ancestors. Subsequently, five groups were formally recognized at the kingdom level (Table 2.1, Figure 2.2; Margulis 1974; Whittaker 1969; Whittaker and Margulis 1978). In cladistic terms, at least two of these five kingdoms are probably artificial (or paraphyletic; see below), but the five-kingdom approach has been widely adopted with only minor modification (e.g., Margulis and Schwartz 1988; Raven, Evert, and Eichhorn 1992). Other schemes have been proposed (e.g., Cavalier-Smith 1978, 1981, 1987), but none has received such widespread acceptance.

THE FIVE KINGDOMS IN A PHYLOGENETIC CONTEXT

The five kingdoms (Figure 2.2) can also be visualized as a nested set of groups (Figure 2.3). Advances

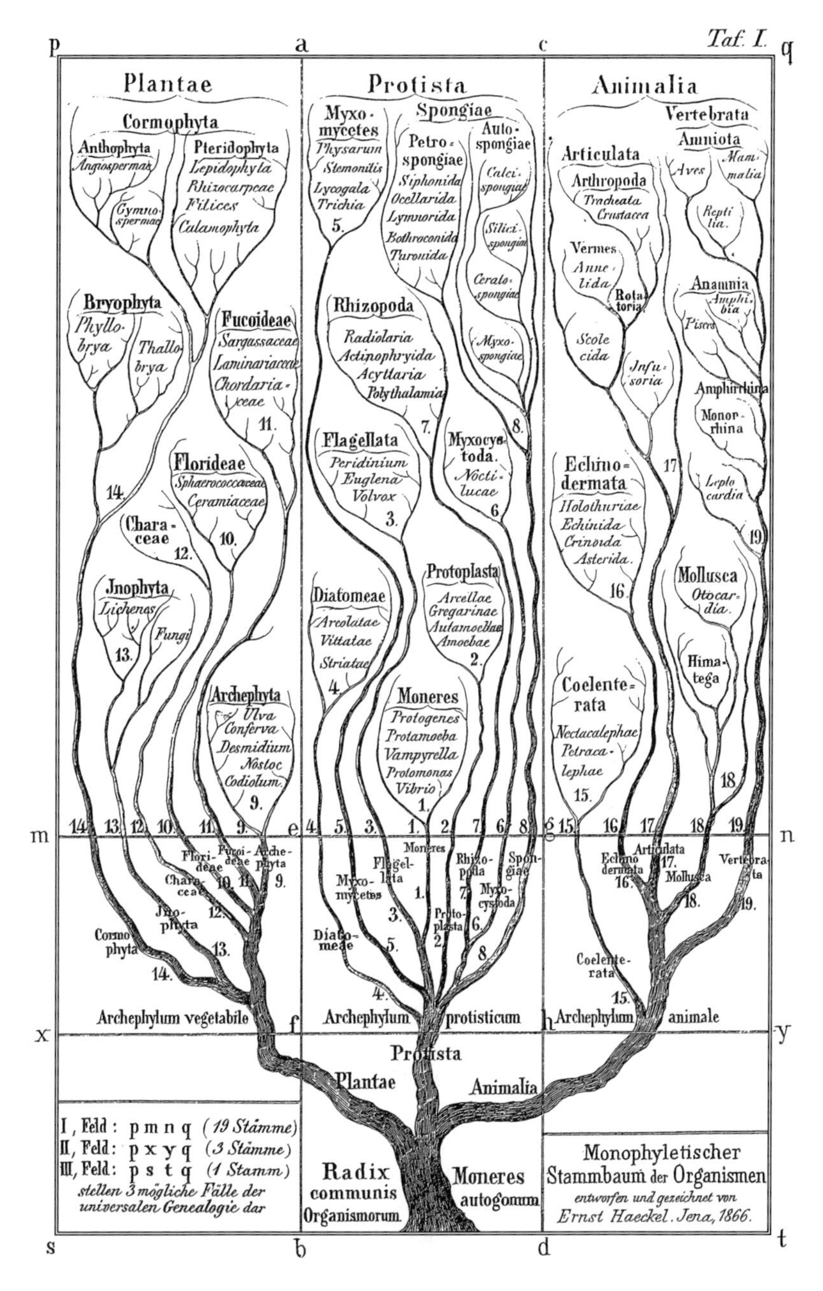

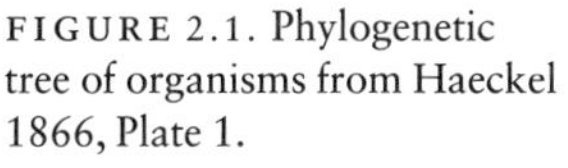
FIGURE 2.1. Phylogenetic tree of organisms from Haeckel 1866, Plate 1.

in molecular biology emphasize the fundamental unity of life, with defining features (synapomorphies) that include the possession of self-replicating nucleic acids and other basic biochemical and structural characteristics (e.g., the bilayered lipid cell membrane) (Gogarten 1995). Eukaryotes may be defined by the possession of a membrane-bound nucleus, in which the nucleic acids are organized into chromosomes, and also by the possession of "a dynamic cytoskeletal and membrane system" that have the capability of engulfing external materials and bringing them into the cell (Knoll 1992, 622; Margulis and Schwartz 1988). Most eukaryotes are also fundamentally chimeric, their plastids and other organelles having arisen via many endosymbiotic events throughout the course of evolution.

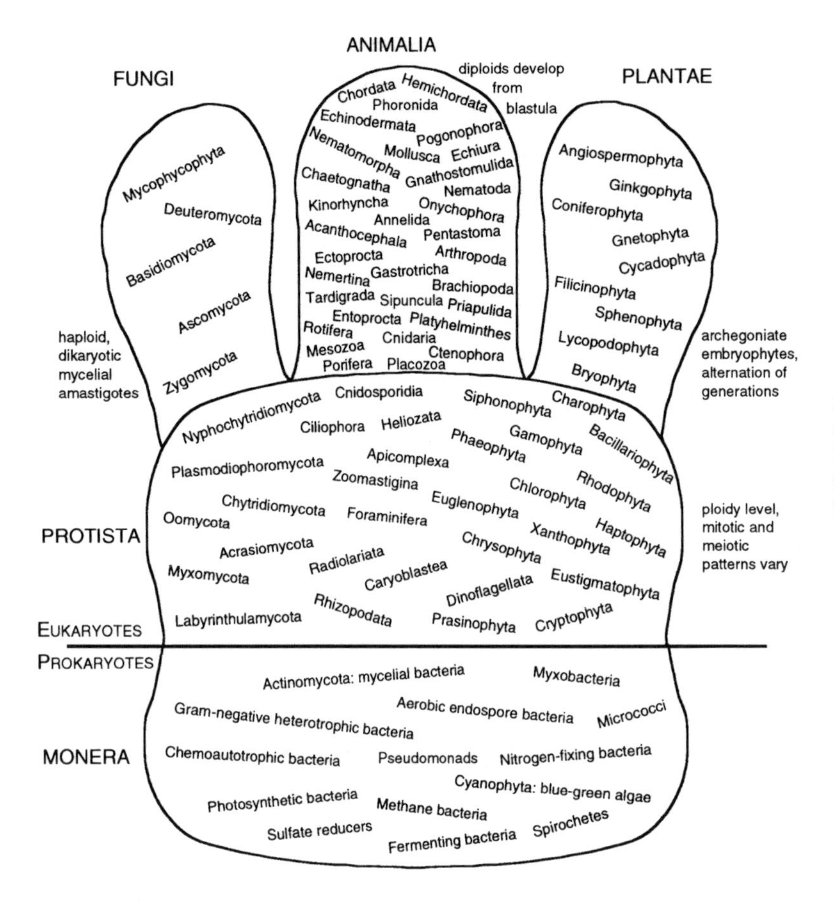

FIGURE 2.2. The five kingdoms (Monera, Protista, Fungi, Animalia, Plantae) and their relationships. Redrawn from Margulis 1974, Figure 7.

TABLE 2.1

The Five-Kingdom Classifications of Whittaker and Margulis

Kingdom	Whittaker 1969	Margulis 1974
Monera	Bacteria, blue-green algae	All prokaryotes: bacteria, blue-green algae, mycelial bacteria, gliding bacteria, etc.
Protista	Protozoa, chrysophytes	All eukaryotic algae: green, yellow green, red and brown, and golden-yellow; all protozoa; flagellated fungi; slime molds; and slime net molds
Plantae	Green algae, brown algae, red algae, bryophytes, tracheophytes	Metaphyta: all green plants developing from embryos
Fungi	Slime molds, fungi	Amastigomycota: conjugation fungi, sac fungi (molds), club fungi (mushrooms), yeasts
Animalia	Multicellular animals	Metazoa: all animals developing from blastulas

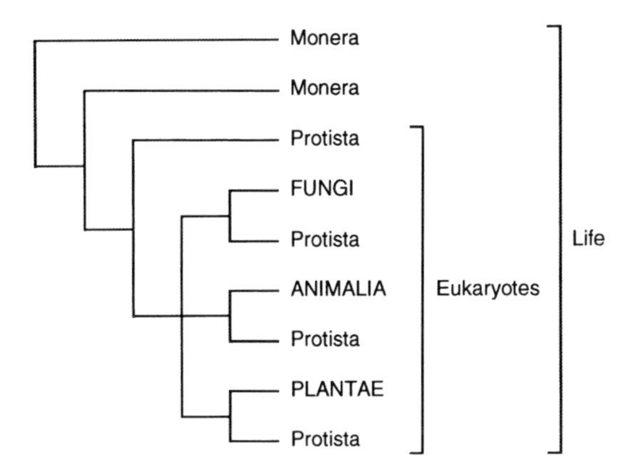

FIGURE 2.3. Cladistic representation of the five-kingdom classification (Monera, Protista, Fungi, Animalia, Plantae), emphasizing the paraphyly of Monera and Protista. Five monophyletic groups can be recognized (life, eukaryotes, Fungi, Animalia, Plantae).

Current hypotheses suggest that the Metazoa (animals, including Porifera) are defined by their characteristic spermatozoid development (which results in the formation of tiny polar bodies), the presence of collagen, and the differentiation of somatic cells (Ax 1989). Fungi (Chytridomycetes, Zygomycetes, Ascomycetes, Basidiomycetes) are defined by their characteristic extracellular enzymatic digestion, chitinous cell walls, and distinctive α-aminoadipic acid–lysine pathway (Barr 1992; Kendrick 1985; Margulis and Schwartz 1988). Within the fungi, the chytrids are basal, and the Zygomycetes are the sister group to the Ascomycetes and Basidiomycetes.

The fifth clade in the five-kingdom scheme is the plants, and in this book we follow Copeland's original concept, which includes the green algae (Chlorophyta *sensu lato*) within the plant kingdom (see below). Recognition of these five groups (life, eukaryotes, animals, fungi, plants) emphasizes that two of the traditionally recognized five kingdoms, Monera and Protista (Protoctista) (Figure 2.3), can be defined only by exclusion. The apparent unity of the Monera arises because they lack the characteristic structural features of eukaryotes (Woese, Kandler, and Wheelis 1990), whereas protists (protoctists) are effectively the "residue" of eukaryotes after all organisms with the distinctive attributes of animals, fungi, or plants have been removed from consideration.

Within the limits of current sampling and analytic techniques, the identification of eukaryotes, animals, fungi, and plants as putative clades is broadly supported by analyses of nucleotide sequence data from both 16S (small subunit) (Sogin, Edman, and Elwood 1989) and 28S (large subunit) ribosomal RNA (rRNA) (Perasso, Baroin, et al. 1989). However, among the Monera, analyses of rRNA sequences and other data provide evidence for the recognition of two distinct groups: the true Bacteria (including Cyanobacteria–blue-green algae) and the Archaebacteria (Archaea), which include the thermoacidophiles, halobacteria, and methanogens (Figure 2.4).

The recognition of these two substantial and putatively monophyletic groups (*domains:* Woese 1989)—Bacteria and Archaea (comprising the Crenarchaeota [extreme thermophiles] and Euryarchaeota [methanogens, extreme halophiles, and their relatives])—substantially undermines the systematic homogeneity of prokaryotes and suggests that there are three main branches to the tree of life (Figure 2.4, Table 2.2; Woese, Kandler, and Wheelis 1990). Cladistic results (Larsen et al. 1993; Gogarten 1995; Sogin, Gunderson, et al. 1989; Sogin, Edman, and Elwood 1989), which indicate that Archaea are more closely related to Eucarya (eukaryotes) than either is to Bacteria, further emphasize the paraphyletic status of the Monera (Figures 2.3 and 2.4; Tables 2.1 and 2.2). There is also the suggestion that the Archaea themselves may be paraphyletic, with the Crenarchaeota (eocytes) more closely related to eukaryotes than to the Euryarchaeota (Rivera and Lake 1992).

Current estimates suggest that there are more than 100,000 species at the protist grade (Margulis, Corliss, et al. 1990), and at this level of evolution, endosymbiotic and lateral transfers of genes substantially complicate analyses of phylogenetic patterns (e.g., Gogarten 1995; Delwiche and Palmer 1996). The relationships of many groups remain uncertain, but some securely identified patterns of higher-level relationships are beginning to emerge based on ultrastructural, biochemical, and sequence data (e.g., Lipscomb 1989; Patterson and Sogin 1993; Perasso, Baroin, et al. 1989; Ragan

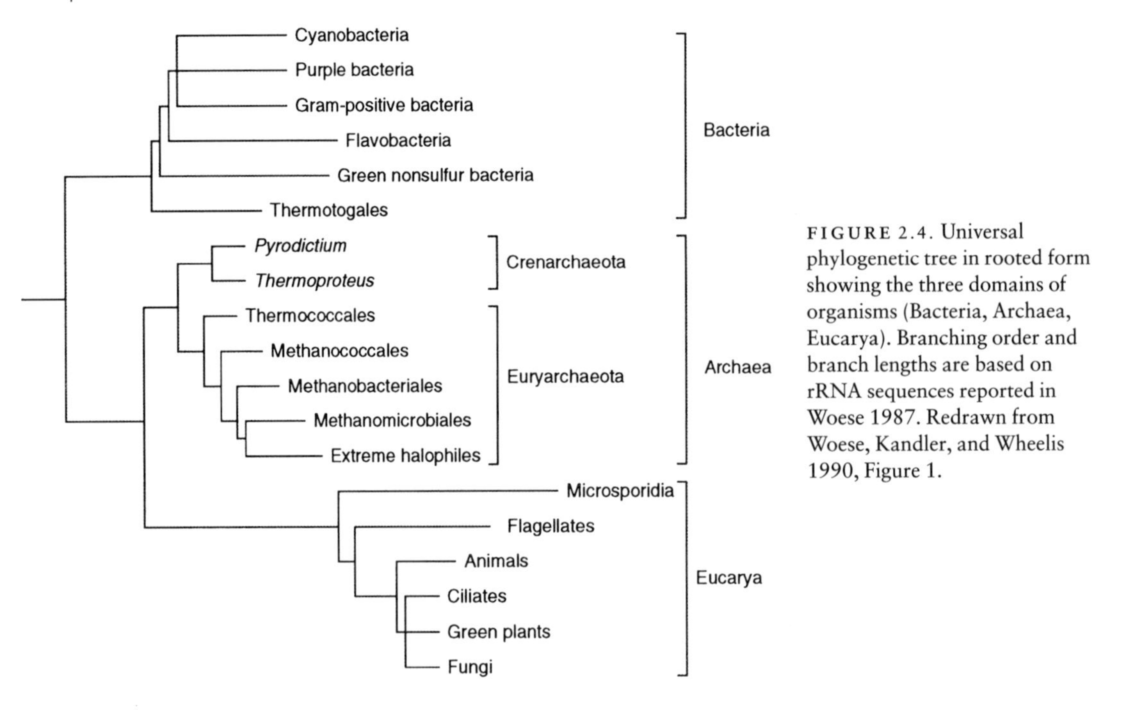

FIGURE 2.4. Universal phylogenetic tree in rooted form showing the three domains of organisms (Bacteria, Archaea, Eucarya). Branching order and branch lengths are based on rRNA sequences reported in Woese 1987. Redrawn from Woese, Kandler, and Wheelis 1990, Figure 1.

1989; Sogin, Edman, and Elwood 1989). Eukaryotes are characterized by the possession of a nucleus, and those taxa currently thought to be basal include a variety of anaerobic parasites of animals that lack mitochondria (e.g., *Giardia*). Other lineages that diverge relatively early (e.g., euglenids and entamoebas) generally possess mitochondria, and subsequently there is a group of abruptly diverging lineages (i.e., lineages that are very similar in their 16S rRNA sequences), which form the "crown" of the eukaryotic tree (Figures 2.5 and 2.6; Knoll 1992; Wainright, Hinkle, et al. 1993).

In addition to the monophyly of green plants (which is well supported; see below), other potential higher-level monophyletic groups identified to date within the eukaryote crown group include

TABLE 2.2
Two Recent Organismal Classifications at the Domain Level

Mayr 1990	Woese, Kandler, and Wheelis 1990
Domain "Prokaryota" ("Monera")	
Subdomain Eubacteria	Domain Bacteria
[no kingdoms given]	[no kingdoms given]
Subdomain Archaebacteria	Domain Archaea
Kingdom Euryarchaeota	Kingdom Euryarchaeota
Kingdom Crenarchaeota	Kingdom Crenarchaeota
Domain Eukaryota	Domain Eucarya
Subdomain "Protista"	[no kingdoms given]
Subdomain Metabionta	
Kingdom Metaphyta (plants)	
Kingdom Fungi	
Kingdom Metazoa (animals)	

Notes: Alignment across page denotes correspondence between the two systems. See also Woese 1991 and Mayr 1991. Quotation marks = paraphyletic group.

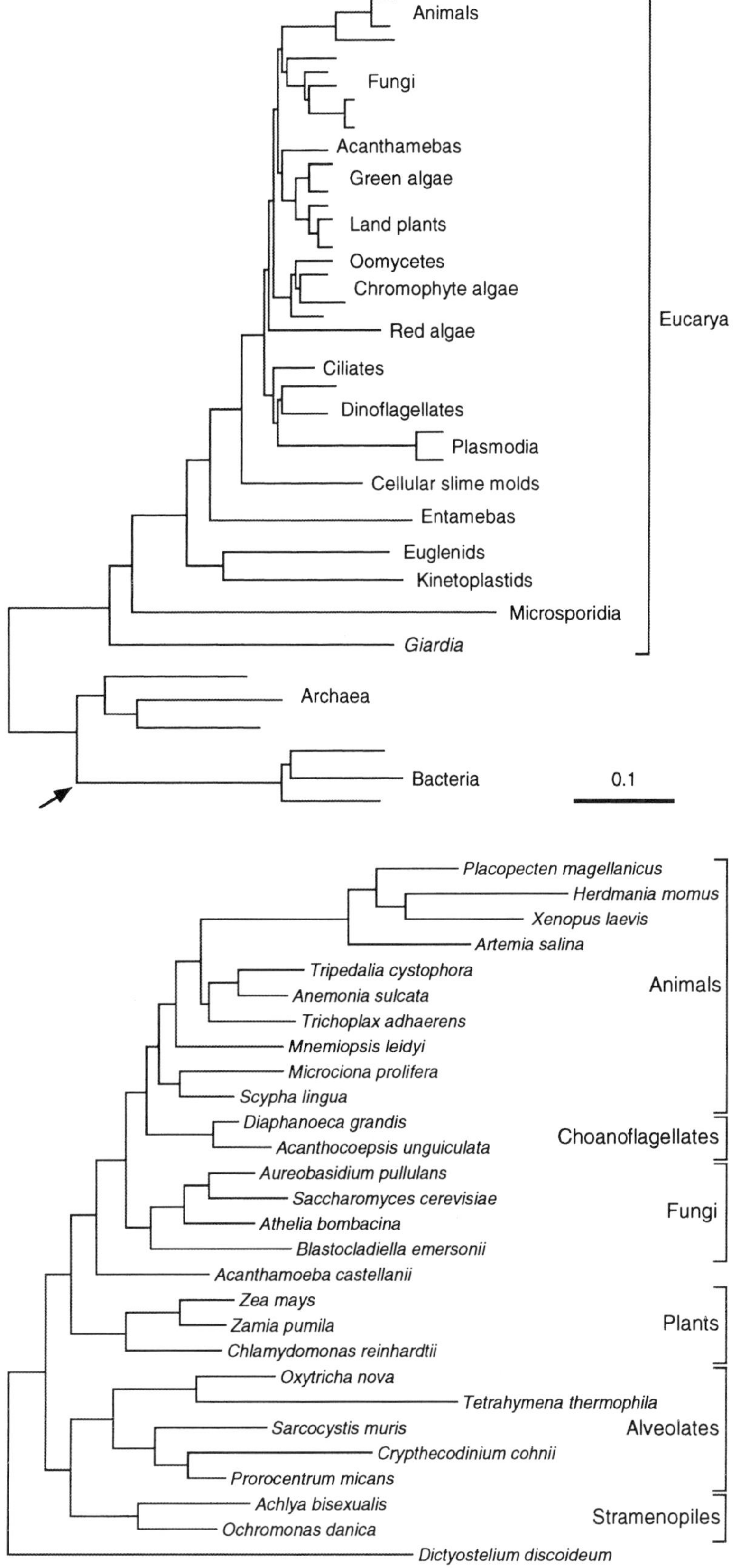

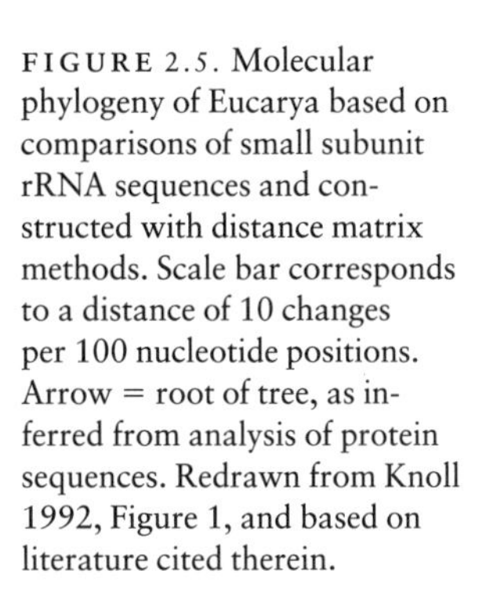

FIGURE 2.5. Molecular phylogeny of Eucarya based on comparisons of small subunit rRNA sequences and constructed with distance matrix methods. Scale bar corresponds to a distance of 10 changes per 100 nucleotide positions. Arrow = root of tree, as inferred from analysis of protein sequences. Redrawn from Knoll 1992, Figure 1, and based on literature cited therein.

FIGURE 2.6. Relationships among animals, fungi, plants, and other eukaryotic groups inferred from complete 16S-like rRNAs. The topology is based on computer-assisted alignment of 16S-like rRNA and Maximum Likelihood methods. The tree is rooted on *Dictyostelium discoideum*. Redrawn from Wainright, Hinkle, et al. 1993, Figure 1.

the stramenopiles (diatoms, brown algae, xanthophytes, chrysophytes, oomycetes, and labyrinthulids: Wainright, Hinkle, et al. 1993) and the dinozoans or alveolates (dinoflagellates, ciliates, and apicomplexans: Baroin, Perasso, et al. 1988; Patterson and Sogin 1993; Sogin, Edman, and Elwood 1989; Wainright, Hinkle, et al. 1993; Wolters 1991; Figure 2.6). In general, most analyses suggest that red algae (Rhodophyta) diverged relatively early in the "crown" of the eukaryote tree (Sogin, Edman, and Elwood 1989; Wolters 1991) and that the earliest diverging eukaryotes include facultative aerobes that are parasitic or symbiotic within animal hosts (e.g., *Giardia,* microsporidians, *Trichomonas:* Sogin, Edman, and Elwood 1989).

An important result of several recent studies, which were based on complete 16S-like rRNA sequence data (Wainright, Hinkle, et al. 1993) and structural rearrangements in 25 protein sequences (Baldauf and Palmer 1993), is support for earlier suggestions that animals (Metazoa) and fungi are sister taxa, to the exclusion of plants (Chlorobionta). There is also evidence to suggest that fungi and Metazoa may have been derived from a flagellated protist similar to extant choanoflagellates (Wainright, Hinkle, et al. 1993), as originally suggested by Cavalier-Smith (1987; see also Figure 2.6). In analyses based on small subunit rRNA sequence data, the choanoflagellates are resolved as the basal lineage of the animal clade, followed by the divergence of sponges defining the base of the Metazoa *sensu stricto* (Figure 2.6; Wainright, Hinkle, et al. 1993).

One recent summary of phylogenetic relationships among major groups of organisms is presented in simplified form in Figure 2.5. As discussed by Knoll (1992), the temporal patterns implicit in this and similar schemes are broadly consistent with the first recorded appearance of these groups in the fossil record. Early indirect evidence of life is provided by ferric minerals (Banded Iron Formations) in some of the oldest known sedimentary rocks (c. 3,800 Ma, Isua Group, Greenland), which are thought to have been produced through reactions with temporary, localized excesses of biologically generated oxygen (Towe 1994). The earliest direct fossil evidence of organisms included at the Monera level is provided by stromatolites (c. 3,500 Ma, Warrawoona Group of Western Australia; Awramik 1992) and microbial filaments of presumed cyanobacterial (oscillatoriacean) origin (c. 3,465 Ma, Apex Basalt of Western Australia; Gogarten 1995; Schopf 1993).

The earliest evidence of probable eukaryotes is provided by macroscopic, spirally coiled, carbonaceous fossils similar to the enigmatic fossil *Grypania* from rocks dated at about 2,100 Ma (Negaunee Iron Formation, Michigan; Han and Runnegar 1992). In younger rocks (c. 1,800–1,900 Ma), large spheroidal acritarchs from the Chuanlinggou Formation, China, are also interpreted as eukaryotes. The existence of eukaryotes at this time is further supported by the presence of steranes in rocks of the Barney Creek Formation, northern Australia, dated at about 1,690 Ma. Steranes are degradation products of the sterols that are characteristic of the membranes of eukaryotes (Knoll 1992). Current paleontological and biogeochemical data therefore suggest that eukaryotes were components of late Paleoproterozoic ecosystems from around 2,100 Ma, although their diversity, up to around the beginning of the Neoproterozoic (c. 1,000 Ma), was apparently limited.

Throughout the Neoproterozoic Era the diversity of eukaryotes increased substantially and includes a few taxa for which relationships to extant lineages can be suggested (Knoll 1992). Most spectacularly, silicified carbonates from Somerset Island, northern Canada, have yielded an unequivocal bangiophyte red alga from rocks dated at 1,260–950 Ma (Butterfield, Knoll, and Swett 1990; Knoll 1992). Subsequently, the Neoproterozoic record of macroscopic and microscopic eukaryotes increases dramatically (Knoll and Walter 1992; Knoll 1992). Structurally simple, but presumed multicellular, ediacaran "animals" are first recorded at about 570 Ma, whereas diverse and less equivocal Metazoa are recorded at around the beginning of the Phanerozoic (c. 540 Ma) (Knoll and Walter 1992). Multicellular algal remains that are morphologically similar to green algae, such as *Cladophora* and *Coelastrum,* are recorded from the

Svanbergfjellet Formation of Spitsbergen (Butterfield, Knoll, and Swett 1988; Knoll 1992), which is dated at approximately 750 Ma. Chemical biomarkers that are diagnostic of green algae (Summans and Walter 1990), along with putative prasinophyte microfossils, are also common during the Neoproterozoic (1,000–540 Ma) (Knoll 1992).

THE PLANT KINGDOM (CHLOROBIOTA)

The possession of diverse ultrastructural and biochemical features clearly indicates that the green algae are more closely related to embryophytes than to any other group of eukaryotes, as was proposed in the nineteenth century by Pringsheim (1878), as well as by Bower (1908) and others. Therefore, following Copeland (1956), we regard the plant kingdom as encompassing both groups. This definition contrasts with the much less inclusive circumscription of plants adopted in many widely used texts and also with that accepted in some five-kingdom classifications (e.g., Margulis 1974, 1996; Margulis and Schwartz 1988) in which the plant kingdom is synonymous with land plants (embryophytes) (Table 2.1). Although green algae are often placed among the protists (e.g., Margulis 1974, 1996; Margulis, Corliss, et al. 1990; Margulis and Schwartz 1988), the morphological, ultrastructural, and biochemical features that they share with land plants (embryophytes) provide overwhelming evidence of a close phylogenetic relationship (see below). In view of these profound shared features, as well as the current uncertainties over the closest protist relatives of green plants (see below), we place the protist-plant boundary below the level of green algae.

All green plants (Chlorobiota) possess a suite of distinctive characters that set them apart from other organisms, and several major clades can be recognized within the group (Figure 2.7). The possession of the photosynthetic pigments chlorophyll *a* and *b*, the storage of α-1,4 glucan (starch) in the chloroplast, and the stellate structure that links the nine pairs of microtubules in the base of the flagella (undulipodia) of motile cells are unique among eukaryotes (Bremer 1985; Mattox and Stewart 1984; Melkonian 1990). In addition, the characteristic chloroplasts of green plants are surrounded by only two unit membranes and have internal thylakoid membranes organized in pairs or stacks *(grana)*. Structures traditionally interpreted as chloroplasts in the distantly related photosynthetic euglenoids are surrounded by three membranes, perhaps suggesting an independent origin by endosymbiosis of a unicellular green alga (see Walne and Kivic 1990, 283). Chloroplasts in the Chlorobiota were almost certainly derived by endosymbiosis of a photosynthetic prokaryote, with many of the prior functions of the prokaryote genome subsequently having been lost or transferred to the nucleus of the host cell (see, for example, Baldauf, Manhart, and Palmer 1990; McFadden and Gilson 1995). A cyanobacterial origin for all plastids is supported by recent evidence from DNA sequences of the *tufA* gene (Delwiche, Kuhsel, and Palmer 1995).

Whereas the circumscription of green plants is straightforward and the origin of the chloroplast is clearly via endosymbiosis from a cyanobacteria-like prokaryote, identifying the close relatives of the original host cell is more problematic. Phylogenetic analyses of rRNA sequence data place certain amoeboid taxa such as *Acanthamoeba* (Cedergren, Gray, et al. 1988; Clark and Cross 1988; Gunderson, Elwood, et al. 1987; Sogin, Edman, and Elwood 1989) or *Mastigamoeba* (Wolters 1991) in a sister group relation to green plants, but this potential relationship requires further study (Patterson and Sogin 1993).

Within the Chlorobiota the green algae are clearly a grade of organization comprising those green plants that are excluded from the embryophytes (Figure 2.7). As currently circumscribed, the green algae include approximately 500 genera and more than 16,000 species (Melkonian 1990). They are a manifestly paraphyletic group, and there are strong morphological, ultrastructural, and molecular data indicating that certain taxa (e.g., Charophyceae) are more closely related to embryophytes than to other taxa of green algae (e.g., Chlorophyceae, Ulvophyceae) (Figure 2.7).

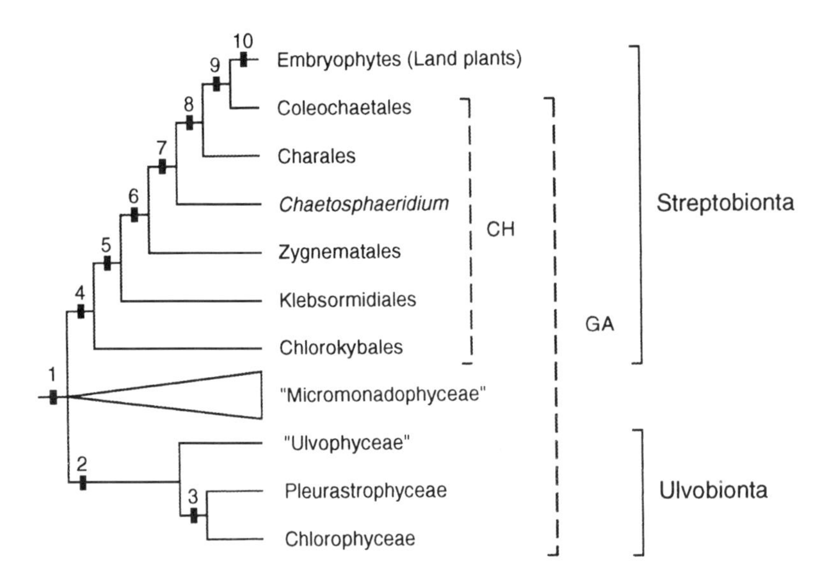

FIGURE 2.7. Cladistic relationships among Chlorobiota (green plants), based on results of Bremer (1985), Graham, Delwiche, and Mishler (1991), Kenrick (1994), Mattox and Stewart (1984), Mishler and Churchill (1985b), and Mishler, Lewis, et al. (1994). GA = green algae; CH = Charophyceae; quotation marks = paraphyletic group. The Micromonadophyceae are probably not monophyletic and their relationships are currently poorly understood. Uncertainty still exists over the branching order of some groups (see references above). Numbers refer to characters supporting these nodes: *(1)* presence of starches (α-1,4 glucans) consisting of amylose and amylopectin, a "stellate" structure in the flagellar transition region of motile cells, several chloroplast features (two membranes, thylakoids in pairs or stacks, loss of endoplasmic reticulum: Melkonian 1990), transfer from chloroplast to nucleus of the small subunit ribulose bisphosphate carboxylase gene (Douglas and Durnford 1989; Newman and Cattolico 1990), presence of chlorophyll *b*, loss of phycobilins, and ribosomal RNA (rRNA) sequence data (Chapman and Buchheim 1991); *(2)* unique ultrastructural characters of the flagellum; *(3)* distinctive features of cell division involving a phycoplast; *(4)* ultrastructural features of the flagellum and cell division, including a distinctive multilayered structure (MLS) associated with the cytoskeletal microtubules in swimming sperm, dispersion of the nuclear envelope during mitosis, loss of rhizoplast and eyespot in motile cells, similar flagella insertion, other similarities in cytoskeletal microtubule organization, and intron data in the tRNA Ala and Ile chloroplast genes (Manhart and Palmer 1990); *(5)* filamentous growth; *(6)* phragmoplast during cell division, polyphenolics, and a sexual growth response; *(7)* plasmodesmata, sheathed hairs, branching, apical growth, oogamy; *(8)* egg retention, enlargement of zygote; *(9)* polyphenolics induced by sex, zygote retained through sporogenesis; *(10)* archegonia plus embryo (possibly linked), antheridia, cuticle, sporangium (spore mass encapsulated within a sterile, cellular wall), spore walls containing sporopollenin (Graham 1990), details of spermatozoid ultrastructure (i.e., orientation of MLS and the presence of bicentriolar centrosomes: Graham and Repavich 1989), and details of cell division (i.e., cortical microtubule arrays and preprophase microtubule bands: Brown and Lemmon 1990a). Monophyly of embryophytes is also consistent with analyses of conserved portions of the chloroplast DNA (see Graham, Delwiche, and Mishler 1991), the position of the *tufA* gene in chloroplast and nuclear DNA (Baldauf, Manhart, and Palmer 1990), and analyses of rRNA sequence data (Chapman and Buchheim 1991; Waters, Buchheim, et al. 1992).

Morphologically the green algae exhibit considerable diversity, ranging from unicellular flagellates to complex multicellular organisms that may take filamentous (e.g., *Spirogyra*), colonial (e.g., *Volvox*), parenchymatous (e.g., *Chara*), or other forms. The more complex levels of organization appear to have been derived independently and repeatedly in different green algal clades (Mattox and Stewart 1984). Morphological diversity among the green algae is paralleled by exceptional ecological amplitude that encompasses not only marine, brackish, and freshwater aquatic settings, but also extends to fully subaerial, subterranean, epiphytic, epizoic, endophytic, and endozoic habitats (Melkonian 1990).

The green algae were first separated from other algal groups by Harvey (1836) as the Chlorospermae, and Blackman (1900) was the first to propose the phylogenetic divergence of the group from ancestral unicellular flagellates (Melkonian 1990; Round 1984). Subsequently, Pascher (1914, 1931) recognized the main lines of morphological divergence from unicellular flagellates, and many of his groups remain embodied in modern classifications, for example, the Volvocinae, Tetrasporinae, Protococcineae, Ulotrichineae, Siphonineae, and Siphonocladineae (Melkonian 1990). Fritsch (1935) incorporated most of these groups into his widely used nine-order system (see Round 1984 for historical review). The modern renewal of interest in the systematics of the green algae developed largely from the application of transmission electron microscope (TEM) techniques, particularly to clarify the structure of the flagella apparatus and details of cell division and cytokinesis (Pickett-Heaps 1969, 1975; Pickett-Heaps and Marchant 1972). These accumulated ultrastructural data, supplemented by information from biosynthetic pathways (Kremer 1980), and more recently by molecular sequences (Baldauf, Manhart, and Palmer 1990; Chapman and Buchheim 1991; Manhart and Palmer 1990; Rausch, Larsen, and Schmitt 1989), form the basis for modern hypotheses of relationship at the green algal grade.

A significant early result from TEM studies of green algae was the recognition of a group of motile chlorophytes whose cell bodies and flagella were covered by nonmineralized organic scales (Manton and Parke 1960). These plants are now recognized as the *prasinophytes,* or Micromonadophyceae, a group of uncertain status that may be one of the most basal lineages among green plants (Melkonian 1990). More extensive application of ultrastructural studies led to the recognition of two major groups among the green algae, the Chlorophyceae *sensu lato* and the Charophyceae, based on features of cell division, flagella insertion, and other characters (Moestrup 1974; Pickett-Heaps 1975; Pickett-Heaps and Marchant 1972; Stewart and Mattox 1975). Somewhat later the Chlorophyceae were subdivided into the Chlorophyceae *sensu stricto* and the Ulvophyceae (Stewart and Mattox 1978). Using implicitly cladistic argumentation, Stewart and Mattox proposed that the charophycean algae, and in particular, the Charales and Coleochaetales, were more closely related to land plants (embryophytes) than they were to other taxa at the green algal level (Figure 2.7). These conclusions broadly supported the suggestions of Pringsheim (1878), Bower (1890), and C. Jeffrey (1962) of a close relationship between charophytes and land plants. Bower (1908), in particular, raised the possibility that *Coleochaete*-like plants may have been involved in the origin of the elaborated sporophyte generation of embryophytes and that bryophytes were, in some sense, intermediate between green algae and the "higher" embryophytes.

The fossil record of green algae significantly predates that of land plants and is dominated by larger calcium carbonate–depositing calcareous groups and the decay-resistant stages of microscopic unicellular, colonial, or filamentous forms (Tappan 1980; T. N. Taylor and E. L. Taylor 1993). The earliest putative green algae resemble extant unicellular and filamentous species and come from the Neoproterozoic of Australia (Bitter Springs Formation, 900 Ma; Schopf 1968; Schopf and Blacic 1971) and Spitsbergen (Svanbergfjellet Formation, 700–800 Ma; Butterfield, Knoll, and Swett 1988; Knoll 1992). Some of the Spitsbergen fossils resemble *Coelastrum* and *Cladophora.* If this similarity is correctly interpreted, such fossils would indicate that many major lineages of green algae were already differentiated by around the middle of the Neoproterozoic Era. Other microscopic Precambrian fossils include the cysts of Micromonadophyceae (prasinophytes: Moczyolowska 1991) and colonial forms such as *Botryococcus* (Chlorococcales, Chlorophyceae: Tappan 1980). The calcareous Dasycladales (Ulvophyceae) appear in the mid-Cambrian (Berger and Kaever 1992; Nitecki and Toomey 1979; Riding 1991; Riding and Voronova 1985; Tappan 1980), and the Codiaceae (Ulvophyceae) appear in the lower Paleozoic (Elliott 1984).

STREPTOBIONTA

The first cladistic analysis of relationships among the major groups of green plants (Parenti 1980) supported a sister group relationship between charophytes and embryophytes, but more detailed documentation and more rigorous cladistic analyses of relationships among green algae and basal embryophytes were presented by Bremer (1985), Mishler and Churchill (1985b), and Sluiman (1985) (Chapter 3). All of these studies yielded broadly similar results with respect to the phylogenetic position of land plants (Figure 2.7). An excellent up-to-date review of the origin of land plants and their relationship to charophycean algae is provided by Graham (1993). The group comprising charophycean algae and embryophytes has been named the streptophyte clade (Streptobionta).

Using a relatively limited suite of primarily ultrastructural characters, Sluiman (1985) followed Stewart and Mattox (1975) in emphasizing the cladistic distinction between green plants with a cruciate flagellar apparatus (Chlorophyceae, Pleurastrophyceae, Ulvophyceae: Mattox and Stewart 1984) and those with a noncruciate flagellar apparatus. Sluiman also recognized several nested putative synapomorphies linking the Charophyceae *sensu lato* (including the Zygnematales) with embryophytes. The Charales and Coleochaetales were recognized as the algal groups most closely related to land plants, with the Charales as the most immediate sister group. Analyses by Bremer (1985) and Mishler and Churchill (1985b) considered a larger suite of features but produced broadly similar results. Differences from Sluiman's conclusions include a close relationship between the Pleurastrophyceae and Chlorophyceae (as suggested by Mattox and Stewart 1984), as well as details of resolution within the charophyte grade. *Chlorokybus* was resolved as basal (Sluiman placed *Chlorokybus* closer to *Klebsormidium*), and the positions of *Coleochaete* and the Charales were reversed such that *Coleochaete* was placed as the embryophyte sister taxon (Figure 2.7).

The most complete recent synthesis of ultrastructural data bearing on the question of charophycean-embryophyte relationships is that of Graham, Delwiche, and Mishler (1991). The most parsimonious cladogram derived from this analysis provides additional character support for the phylogenetic patterns proposed earlier by Bremer (1985) and Mishler and Churchill (1985b). *Coleochaete* is resolved as paraphyletic, with thalloid or parenchymatous species such as *C. orbicularis* placed as most closely related to embryophytes (tree length = 26 steps in an analysis of 21 characters). Trees that were one step longer also supported the paraphyly of *Coleochaete,* but whereas trees that were two steps longer continued to support the monophyly of the Charales-*Chaetosphaeridium*-*Coleochaete*-embryophyte clade, relationships among these groups were unresolved.

These results are broadly consistent with the phylogenetic implications of recent analyses of rRNA nucleotide sequences (Chapman and Buchheim 1991) and chloroplast protein-synthesis genes (Baldauf, Manhart, and Palmer 1990; Manhart and Palmer 1990). Analyses of Manhart and Palmer (1990) support the *Coleochaete*-Charales-embryophyte group, to the exclusion of *Spirogyra* and other green algae, based on their shared possession of two chloroplast tRNA introns. Baldauf, Manhart, and Palmer (1990) show that the chloroplast *tufA* gene present in the Ulvobionta (Chlorophyta *sensu* Bremer 1985) has been transferred to the nucleus in some but not all of the Charophyceae. The ribosomal RNA gene sequences of Chapman and Buchheim (1991) support the notion that the Charophyceae are a grade closely related to embryophytes. Preliminary results place the Charales as the sister taxon to the embryophytes (Chapman and Buchheim 1991, Figure 14), but because *Coleochaete* and *Klebsormidium* are somewhat surprisingly placed within a paraphyletic Zygnematales, further analyses are needed to provide more robust rRNA support for the hypothesized charalean-embryophyte sister group pattern. An excellent recent review of molecular and other evidence pertaining to the phylogeny of green algae is provided by McCourt (1995).

The earliest unequivocal characean remains are based on calcified oogonia *(gyrogonites)* from the late Silurian (Pridoli), and the fossil record suggests that the Charales underwent their first major diversification during the Silurian and Devonian Periods (Feist and Grambast-Fessard 1991). Based on gyrogonite morphology, three major groups can be recognized within the Charales: The Sycidales comprise forms having gyrogonites composed of vertical cells and are first found during the Silurian. The Trochiliscales comprise forms having gyrogonites composed of dextrally spiraling cells and range from the Devonian to the Carboniferous. The third group, the Charales, includes the extant taxa and is defined by the presence of gyrogonites with sinistrally spiraling cells. Within the Charales a preliminary cladistic analysis (Martín-Closas and Schudack 1991) recognizes a major subclade (Quinquespiralia) that is defined by the possession of oosporangia with five sinistrally spiraling cells and includes the post-Paleozoic families Porocharaceae, Clavatoraceae, Raskyellaceae, and Characeae. The earliest charalean gyrogonites are from the middle Devonian, whereas the first record of the Quinquespiralia is from the Permian. The Quinquespiralia underwent a major diversification during the Mesozoic, leading to the differentiation of the six extant genera. Relationships among extant Characeae are reviewed by McCourt, Karol, et al. (1996).

Excluding calcified gyrogonites, the early fossil record of charophytes is poor, but several groups can be traced to the middle Paleozoic. Unicellular fossils resembling extant members of the Zygnematales have been recorded from middle Devonian cherts (Baschnagel 1966). Cellular, disk-shaped compressions known as *Parka decipiens* resemble the extant charophycean alga *Coleochaete orbicularis* in overall morphology (Niklas 1976). These fossils first appear in the upper Silurian and are common in lower Devonian freshwater facies. The affinities of *Parka* are still equivocal because the ploidy level of the thallus and associated spore masses are uncertain, and many morphological and anatomical details are not well understood (Hemsley 1990, 1994b; T. N. Taylor 1988a; T. N. Taylor and E. L. Taylor 1993). In contrast, the presence of the distinctive vegetative remains of *Ni-*

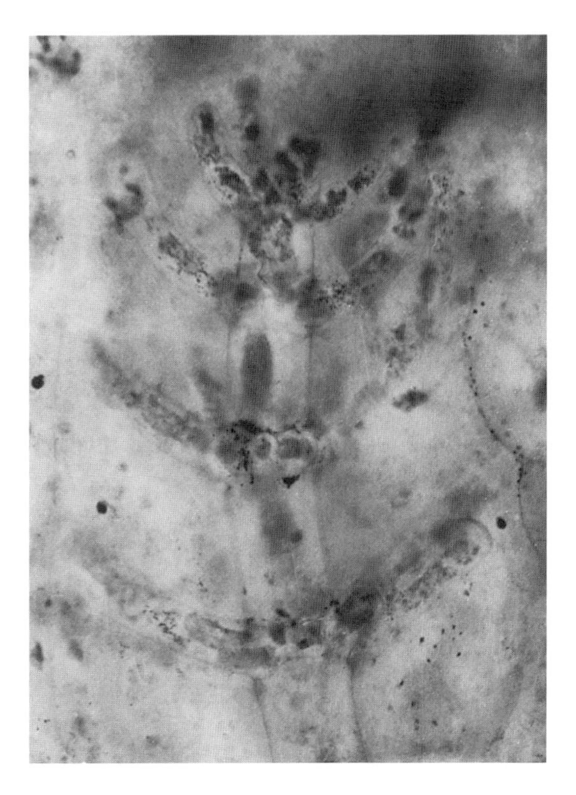

FIGURE 2.8. *Palaeonitella cranii,* a charophyte from the lower Devonian Rhynie Chert, Scotland: septate axis showing nodal organization with four whorls of branches. Scale bar = 75 μm. Reproduced from Taylor, Hass, and Remy 1992, Figure 1 (Remy Collection, slide 2261, Abt. Paläobotanik, University of Münster) with the permission of H. Hass.

tella-like organisms in the lower Devonian Rhynie Chert (D. S. Edwards and Lyon 1983; Taylor, Hass, and Remy 1992; Taylor, Remy, and Hass 1992) provides clear evidence of essentially "modern" charophytes and establishes unequivocally that the differentiation of several charophyte groups had already occurred by the time of the Silurian-Devonian boundary (Figure 2.8). Vegetative and reproductive remains are also known from the late Devonian of South Africa (Gess and Hiller 1995).

3 Embryobiota

The Embryobiota are terrestrial plants that possess several distinctive features, some of which are clearly adaptations to the land environment. Features characteristic of embryophytes include a diplobiontic alternation of generations, desiccation-resistant haploid spores, a cuticle, and distinctive female and male sexual organs *(archegonia* and *antheridia)*. Traditional classifications recognize two major embryophyte groups: bryophytes and tracheophytes.

Bryophytes are small, structurally simple land plants in which the gametophyte phase of the life cycle is larger, and usually more complex, than the corresponding sporophyte. In all bryophytes the sporophyte is unbranched (but see Bower 1935, Figure 440), is retained on the gametophyte, and consists only of a foot embedded in gametophyte tissue and a short stalk *(seta)* bearing a single sporangium *(capsule)*. As enumerated by Schuster (1966, 138), the shared features of bryophytes include a heteromorphic life cycle in which the sporophyte plays a "secondary role" and a *lack* of specialized vascular tissues. Three major groups of bryophytes are recognized by most authors: the Marchantiopsida (liverworts), the Anthocerotopsida (hornworts), and the Bryopsida (mosses) (Doyle 1970; Parihar 1977; Raven, Evert, and Eichhorn 1992; Schofield 1985; Schuster 1966; Smith 1955; Watson 1971).

In contrast to bryophytes, tracheophytes have a conspicuous, complex, independent sporophyte generation and a small, inconspicuous gametophyte, which is clearly reduced in derived groups. In seed plants the male gametophyte consists of only a few cells, and the female gametophyte is retained on, and nourished by, the sporophyte. Current phylogenetic thinking recognizes two major extant lineages of tracheophytes—a lycopsid lineage and an *Equisetum*–fern–seed plant lineage—along with several problematic early fossil groups, including rhyniophytes, zosterophylls, and trimerophytes (e.g., Banks 1970; Bremer, Humphries, et al. 1987; Meyen 1987; Raven, Evert, and Eichhorn 1992).

Extant diversity varies greatly among the major embryophyte groups, and it is clear from the fossil record that much extinction has occurred in the history of many lineages. Liverworts and mosses are diverse and geographically widespread and include approximately 8,000 and 10,000 species, respectively (Schofield 1985). Hornworts, comprising some 400 species, are the least diverse of the bryophyte groups (Schofield 1985). For many years, hornworts were classified with liverworts but are now usually treated separately (Schuster 1966, 1979, 1984c, 1992). Compared to hornworts, and in the context of other major groups of land plants, the species diversity of mosses and liverworts is remarkable and comparable to that of ferns. Liverworts are especially morphologically and ecologically diverse and include epiphytic, rupestral, and ground-dwelling forms (Schuster 1966). Among tracheophytes, in contrast, several major groups are much less diverse.

Current systematic studies indicate that bryophytes are the extant paraphyletic residue of non-tracheophyte embryophytes. The three bryophyte groups (liverworts, hornworts, and mosses) and the tracheophytes usually are treated as individually monophyletic, either explicitly (Mishler and Churchill 1984, 1985b; Mishler, Lewis, et al. 1994; Parenti 1980) or implicitly, but a diphyletic origin of tracheophytes that involves the independent evolution of lycopsids is considered likely by some authors (Banks 1970; Raven, Evert, and Eichhorn 1992). Based on recent cladistic analyses mosses, hornworts, or both, appear more closely related to vascular plants than to liverworts (Kenrick and Crane 1991; Mishler and Churchill 1984, 1985b; Mishler, Lewis, et al. 1994).

SYSTEMATICS

Historical Background

The discovery by Hofmeister (1869) of fundamental similarities in the life cycles of the major land plant groups had a profound effect on phylogenetic thinking in botany. All embryophytes have complex diplobiontic life cycles with an alternation of multicellular diploid and haploid phases (Figure 1.2), but there are also significant morphological differences between the phases, as well as marked differences among the groups. An even greater variety of life cycles, spanning a range of haplobiontic and diplobiontic types, occur in the algae (John 1994). Early ideas on land plant evolution were strongly influenced by the more conspicuous aspects of these algae and embryophyte life histories.

One influential hypothesis, the Homologous Theory (see Chapter 1), held that the putative algal ancestor of land plants had a diplobiontic life cycle, similar to that in embryophytes, in which both haploid and diploid phases were isomorphic, morphologically complex, free-living organisms (Figures 1.2 and 3.1; Church 1919; Fritsch 1916, 1945; Hallier 1902; Potonié 1899; Pringsheim 1878; Tansley 1907; Zimmermann 1930, 1952). Proponents of this view focused on the life cycles of green, brown, and red algae and argued that some of the more complex features of embryophytes were already present in their algal ancestors, including a heterotrichous growth form (i.e., a branching system consisting of prostrate and upright axes), apical growth, dichotomous branching, and primitive sporangia (Figures 3.1 and 3.2; Fritsch 1945; Zimmermann 1952). According to this hypothesis, the sporophyte generation of vascular plants was further elaborated, and the gametophyte generation further reduced, from the ancestral condition. In contrast, the sporophyte generation of bryophytes was thought to have evolved by reduction from a more complex, branched structure, and consequently, liverworts, hornworts, and mosses were not considered to be phylogenetically intermediate between green algae and vascular plants. Because retention of the sporophyte on the gametophyte was not a feature of the hypothetical algal ancestor, nutritional dependence was seen as an evolutionary novelty that characterized the bryophytes. According to this interpretation, bryophytes are a natural evolutionary group representing an early lineage independent from other land plants.

An alternative hypothesis, the Antithetic Theory (see Chapter 1), developed by Bower (1890, 1908, 1935) and based on the earlier views of Celakovsky (1874), held that land plants evolved from filamentous terrestrial or freshwater green algae that had a haplobiontic (gametophyte-dominated) life cycle (Figure 1.2). Proponents of this hypothesis

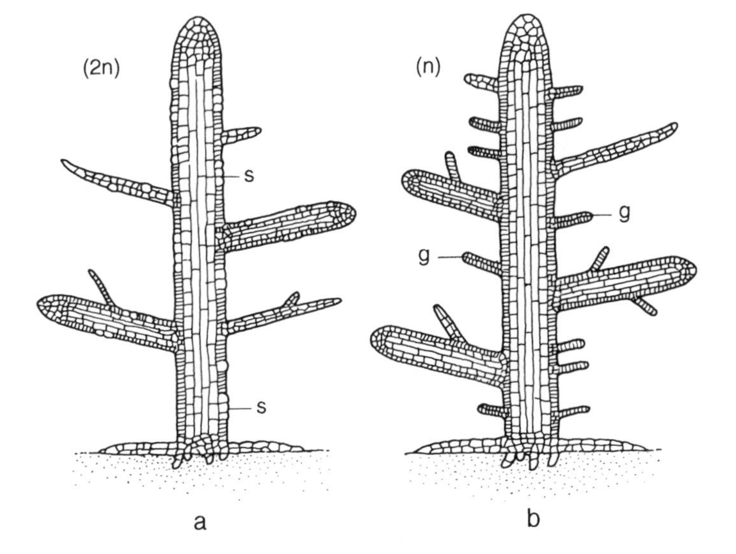

FIGURE 3.1. Hypothetical early land plant with an isomorphic, diplobiontic life cycle (redrawn from Fritsch 1945, Figures 1 and 2). Each of the two generations comprises a branched, axial erect system with central conducting cells attached to a basal prostrate system. (**a**) Diploid (sporophyte, 2n) phase with superficial sporangia *(s)*. (**b**) Haploid (gametophyte, n) with superficial gametangia *(g)*.

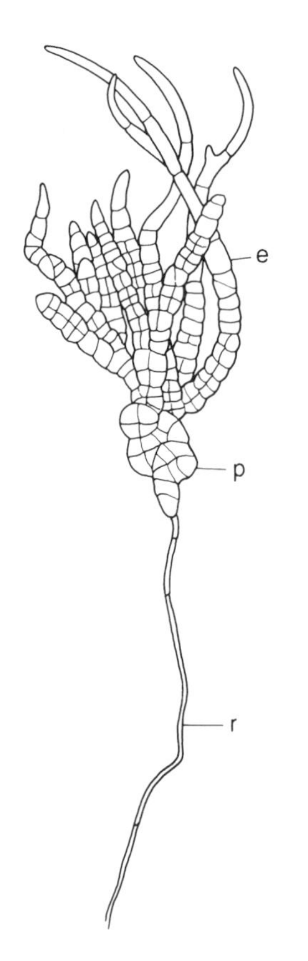

FIGURE 3.2. *Fritschiella tuberosa.* Habit of small mature plant showing two distinct parts to the thallus, which is characteristic of the heterotrichous habit: prostrate system *(p)* with rhizoids *(r),* and erect system *(e)* of branched filaments (×300). Redrawn from Fritsch 1945, Figure 5.

envisaged several critical stages involving the evolution of *oogamy* (large, nonmotile female gametes) followed by the retention of the egg cell, and later the retention of the zygote, within a specially modified structure—the archegonium. In light of this hypothesis, the sporophyte generation of land plants evolved by progressive sterilization and corresponding "lengthening" of sporophyte ontogeny (Bower 1904). Multicellularity evolved by the delay of zygotic meiosis and the insertion of mitotic cell divisions into the diploid phase of the life cycle (Figure 1.2). The sporophyte generation was "intercalated" into the life cycle during the ecological transition from aquatic to terrestrial habitats, and the profound morphological differences between the sporophyte and gametophyte in embryophytes were explained as the result of differing evolutionary histories. From this perspective, bryophytes were seen to be broadly intermediate between green algae and tracheophytes.

Modern systematic studies, based on an impressive variety of biochemical and ultrastructural information in extant plants (see Chapter 2), favor the origin of embryophytes from charophycean green algae that are closely related to *Coleochaete* (Figure 3.3) or to the Charales (Figure 3.4; Bremer 1985; Graham 1993; Graham, Delwiche, and Mishler 1991; Kenrick and Crane 1991; Mishler and Churchill 1984, 1985b; Mishler, Lewis, et al.

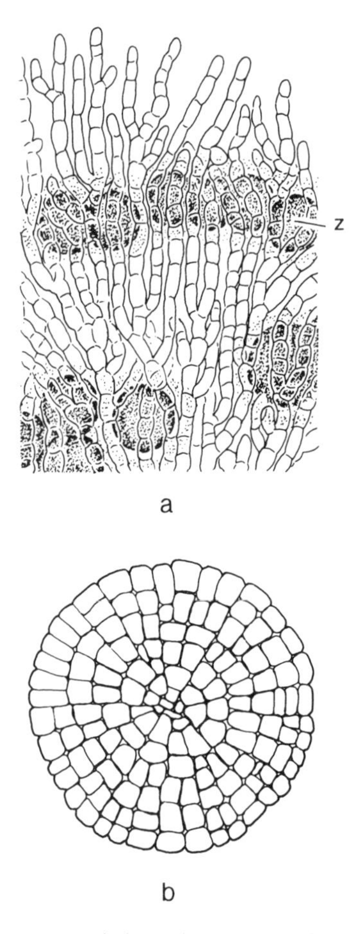

FIGURE 3.3. Morphology of two species of *Coleochaete.* Illustration based on photographs in Graham 1993 (Figures 4.5 and 4.6). (**a**) *C. pulvinata:* part of plant showing branched filamentous morphology typical of the genus and spherical zygotes *(z)* surrounded by other thallus cells (×100). (**b**) *C. orbicularis:* entire plant with disk-shaped, thalloid morphology characteristic of some members of the genus (×160).

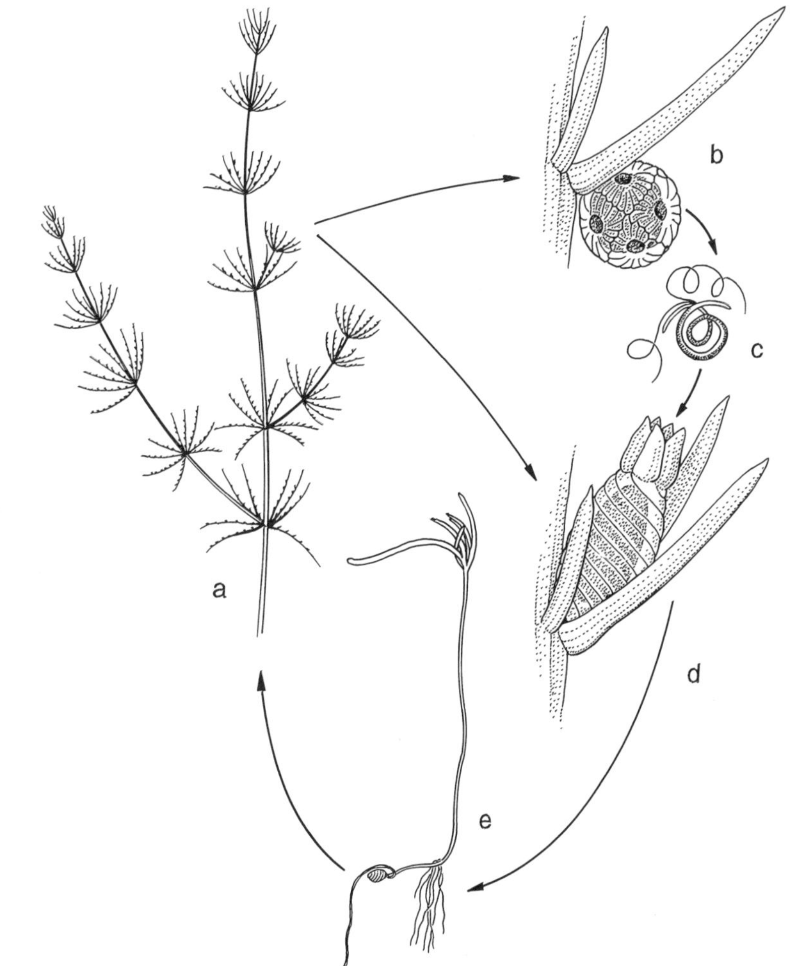

FIGURE 3.4. Life cycle of *Chara* (based on Engler and Prantl 1898–1900, Figure 119). (**a**) Gametophytic plant (×0.7). (**b**) Antheridium (×35). (**c**) Biflagellate, coiled spermatozoid (×500). (**d**) Calcified oogonium (zygote remains in oogonium after fertilization and the whole structure is shed from the gametophytic plant) (×35). (**e**) Young gametophyte that has emerged from a dispersed oogonium (×3). Note that the nonmotile zoospore of *Chara*, which is retained within the oogonium, is atypical of green algae.

1994). Because these green algae are exclusively haplobiontic and oogamous, current evidence strongly supports Bower's Antithetic Theory. Furthermore, this theory is consistent with the simplicity of liverwort, hornwort, and moss sporophytes, both in mature form and ontogeny, and the intimate association of the sporophyte and the gametophyte in basal embryophytes (i.e., bryophytes) follows naturally from their oogamous algal ancestors.

The Phylogenetic Position of Embryophytes

Current thinking on the phylogenetic position of embryophytes can be divided into two broad classes of hypotheses—polyphyletic and monophyletic—that differ principally in the way that comparative data are interpreted. Polyphyletic hypotheses focus on *differences* among major embryophyte groups to suggest independent origins from within the green algae (e.g., Crandall-Stotler 1980, 1984, 1986; Duckett and Renzaglia 1988a; Philipson 1990, 1991; Schuster 1977, 1979, 1981, 1984c). Most modern proponents of embryophyte polyphyly take a moderate view and concentrate on possible independent origins of embryophytes from among charophycean algae, frequently invoking hypothetical extinct charophytes to support these ideas. One recent interpretation views bryophytes, vascular plants, and certain charophycean algae as monophyletic but interprets the multicellular sporophyte and features of the gametophyte generation, such as archegonia and antheridia, as homoplastic (i.e., independently derived) in liverworts, hornworts, mosses, and vascular plants (e.g., Duckett and Renzaglia 1988a; Longton 1990; Philipson 1991). In other words, although the major embryophyte groups are viewed as closely related, their divergence occurred while they were still at the charophyte level of organization, and morphological similarities among the bryophyte classes and vascular plants are interpreted as convergences. In our view, such interpretations are problematic. At a theoretical level, differences (unique features) are uninformative about relationships, and undue emphasis on "dissimilarity" opens the way for highly divergent and unparsimonious phylogenetic interpretations. At an empirical level, it remains to be demonstrated that different subgroups of embryophytes are more closely related to different groups of charophycean algae than they are to each other.

The monophyletic hypothesis of embryophyte phylogeny interprets the many *similarities* of bryophytes and vascular plants as evidence of shared ancestry. Structures such as the archegonium, antheridium, stomates, and sporangium are viewed, either explicitly or implicitly, as homologous. According to this interpretation, these structures were inherited directly from a common embryophyte ancestor, and differences among the groups arose *after* subsequent cladogenesis. Monophyletic hypotheses range from those that interpret bryophytes and vascular plants as divergent evolutionary lines from land plants of a very primitive archegoniate type (Bower 1935; Lignier 1903) to others that envisage vascular plants as having evolved from one of the bryophyte classes, usually the hornworts (Campbell 1940; Smith 1938, 1955). Still other hypotheses consider one or more of the bryophyte classes as having evolved by reduction from more complex ancestors at the rhyniophyte level of organization (Bateman 1996b; Cronquist, Takhtajan, and Zimmermann 1966; Remy and Remy 1980; Robinson 1985; T. N. Taylor 1988a) or even from basal vascular plants (Garbary, Renzaglia, and Duckett 1993; see references in Schuster 1966, 136, and also Schuster's own views on hornworts, 139, 298).

Previous Cladistic Approaches

During the 1980s, several attempts were made to resolve some of the issues outlined above by means of a cladistic approach. The first such analysis of higher-level relationships in land plants incorporated all of the major groups of bryophytes and vascular plants and supported a monophyletic hypothesis of embryophyte origin (Figure 3.5; Parenti 1980). The bryophytes themselves were resolved as

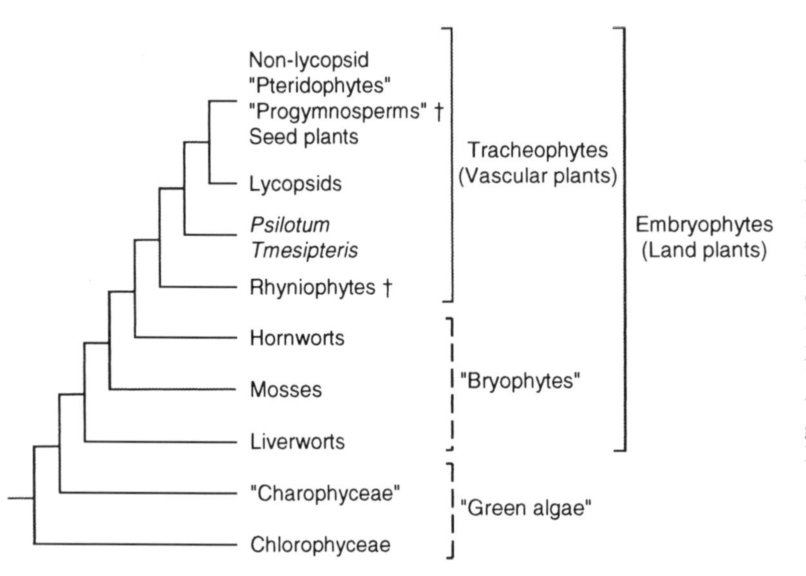

FIGURE 3.5. Cladistic relationships among embryophytes, from Parenti 1980, Figure 4 (nonnumerical analysis): simplification showing major relationships. † = extinct; quotation marks = paraphyletic group. Note the basal position of *Psilotum* and *Tmesipteris* relative to other vascular plants, and the sister group relationship between hornworts and vascular plants.

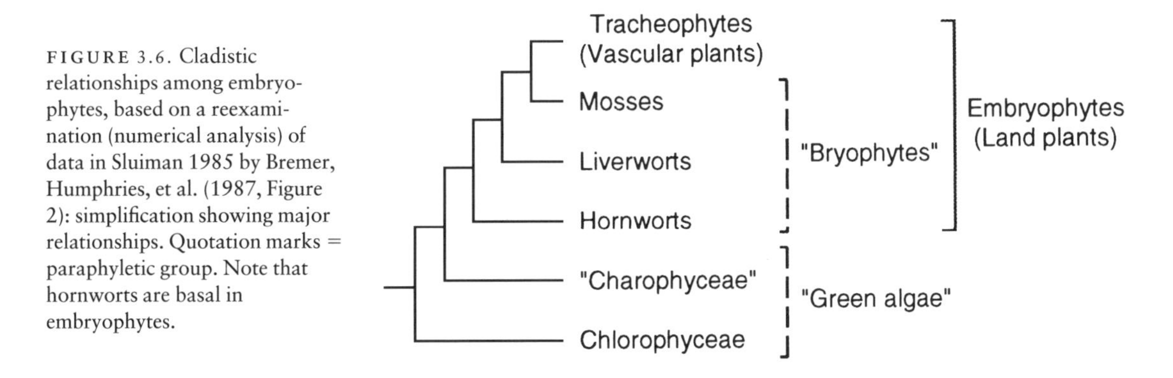

FIGURE 3.6. Cladistic relationships among embryophytes, based on a reexamination (numerical analysis) of data in Sluiman 1985 by Bremer, Humphries, et al. (1987, Figure 2): simplification showing major relationships. Quotation marks = paraphyletic group. Note that hornworts are basal in embryophytes.

paraphyletic; this interpretation supported their broadly intermediate position between the green algae and tracheophytes. Parenti's analysis stimulated discussion, but its impact was reduced by a variety of problems (Bremer and Wanntorp 1981b; Mishler and Churchill 1984; Parenti 1982; Smoot, Jansen, and Taylor 1981; Young and Richardson 1982). Subsequent analyses by Bremer and Wanntorp (1981a,b) and Sluiman (1985) supported embryophyte monophyly but left the relationships among the major groups of bryophytes and the vascular plants unresolved. Reexamination of Sluiman's data by Bremer, Humphries, et al. (1987) resulted in improved resolution (Figure 3.6).

An early analysis by Sluiman (1985) focused on relationships within the green algae, based on ultrastructural and chemical characters. In addressing phylogenetic patterns within land plants, sporophyte and gametophyte morphology were dismissed on the assumption that most medium- to large-scale morphological features are not homologous (Sluiman 1985, 226). According to this interpretation, similarities such as stomates in the sporophytes of mosses, hornworts, and vascular plants were not used as evidence of relationships. In view of the limited data set, the a priori dismissal of some morphological features, and other problems (see Bremer, Humphries, et al. 1987; Theriot 1988), Sluiman's conclusions should be treated with caution.

A more complete and balanced phylogenetic analysis of selected green algae and bryophytes was presented by Mishler and Churchill (1984, 1985b; Figure 3.7). In agreement with Parenti (1980), they interpreted embryophytes and vascular plants as monophyletic and bryophytes as paraphyletic. They also supported the monophyly of the three major groups of bryophytes and considered mosses to be the sister group of the vascular plants. Even though some aspects of this analysis have been criticized (Mishler and Churchill 1985a, 1987; Robinson 1985; Whittemore 1987), these papers made a major contribution to systematizing the distribution of characters at the bryophyte level. Furthermore, although the criticisms highlighted important aspects of characters and char-

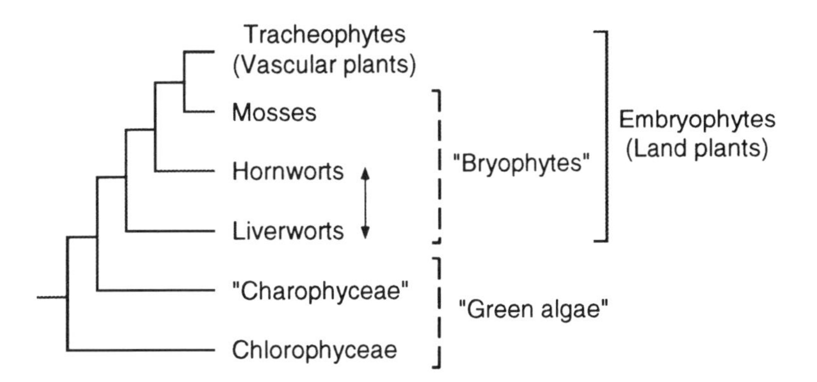

FIGURE 3.7. Cladistic relationships among embryophytes, after Mishler and Churchill 1984, Figure 1 (numerical analysis). Quotation marks = paraphyletic group. Note the paraphyly of bryophytes, the basal position of liverworts, and the sister group relationship between mosses and vascular plants. The position of hornworts and liverworts is the reverse of that shown in Figure 3.6.

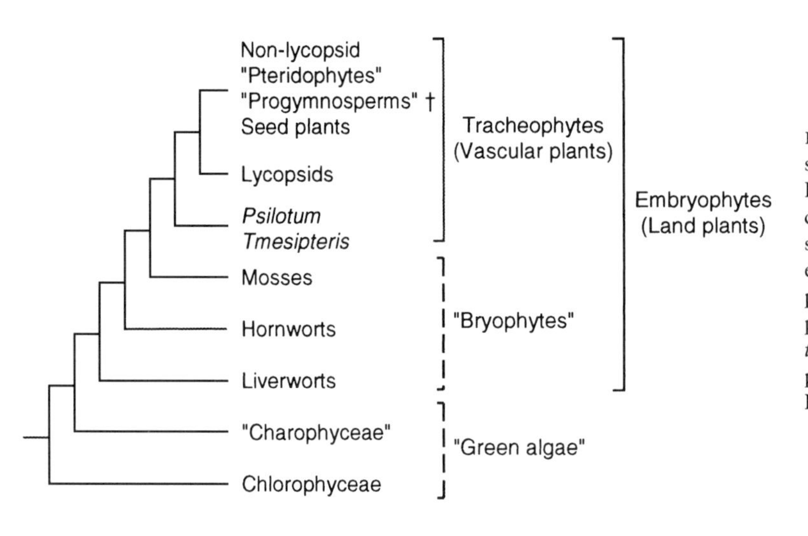

FIGURE 3.8. Cladistic relationships among embryophytes, after Bremer 1985, Figure 2 (numerical analysis): simplification showing major relationships. † = extinct; quotation marks = paraphyletic group. Note the basal position of *Psilotum* and *Tmesipteris* relative to other vascular plants, as in Parenti 1980 (see Figure 3.5 herein).

acter coding, they did not alter substantially the major patterns of relationship. Mishler and Churchill's results were incorporated into the summary of green-plant phylogeny presented by Bremer (1985; Figure 3.8). More recent numerical analyses of morphological data support these conclusions (Mishler, Lewis, et al. 1994) and were slightly modified as part of an analysis of basal embryophytes that included fossil taxa (Figure 3.9; Kenrick and Crane 1991).

Significant new morphological evidence based on ultrastructural aspects of male gametes and gametogenesis have recently been introduced in a cladistic study by Garbary, Renzaglia, and Duckett (1993). The results of this analysis differ significantly from previous ones and resolve the bryo-

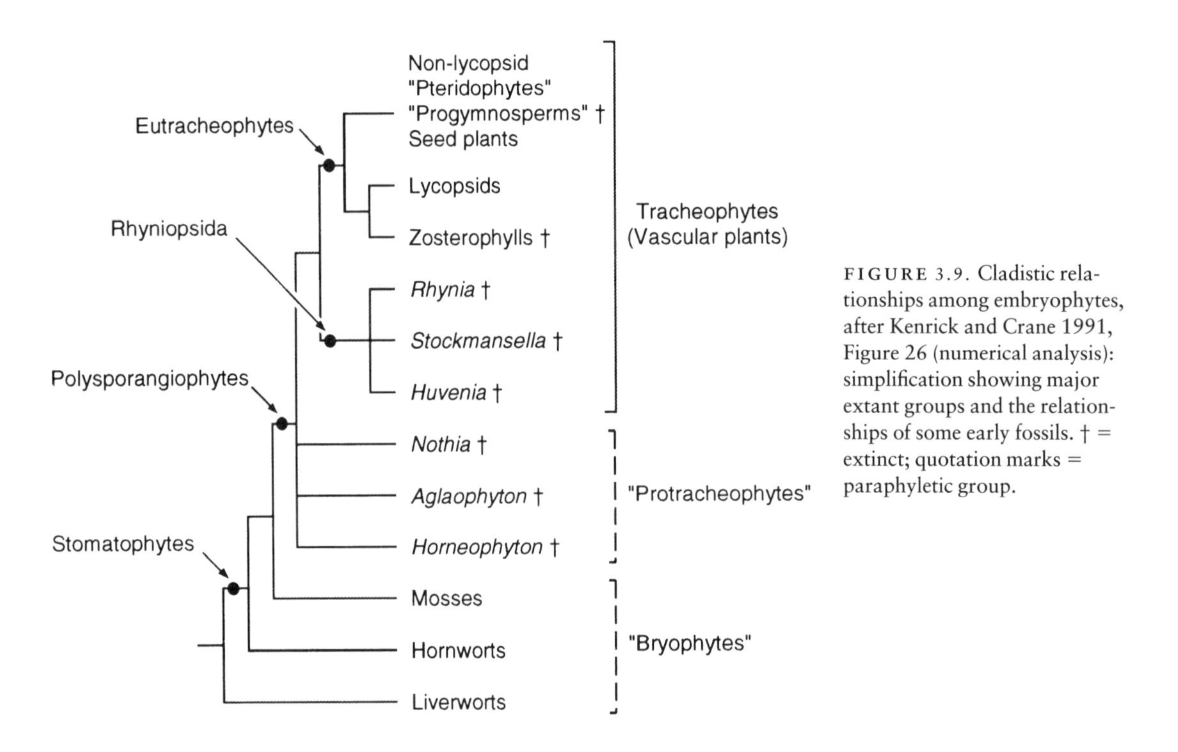

FIGURE 3.9. Cladistic relationships among embryophytes, after Kenrick and Crane 1991, Figure 26 (numerical analysis): simplification showing major extant groups and the relationships of some early fossils. † = extinct; quotation marks = paraphyletic group.

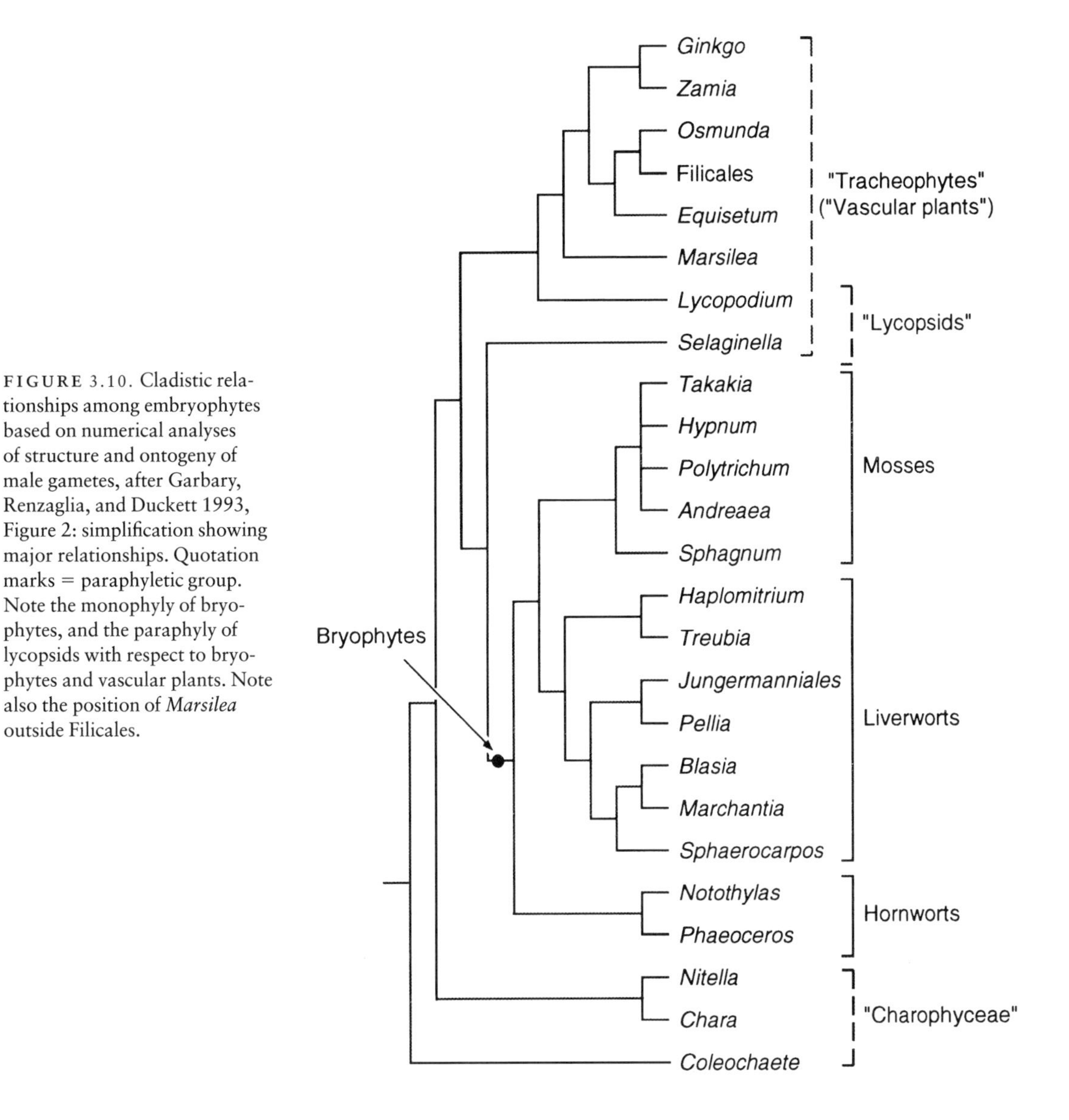

FIGURE 3.10. Cladistic relationships among embryophytes based on numerical analyses of structure and ontogeny of male gametes, after Garbary, Renzaglia, and Duckett 1993, Figure 2: simplification showing major relationships. Quotation marks = paraphyletic group. Note the monophyly of bryophytes, and the paraphyly of lycopsids with respect to bryophytes and vascular plants. Note also the position of *Marsilea* outside Filicales.

phytes as monophyletic, the Lycopodiaceae as the sister group to non-lycopsid vascular plants, and the Selaginellaceae as the sister group to bryophytes (Figure 3.10; Garbary, Renzaglia, and Duckett 1993). This controversial result supports a hypothesis of bryophyte reduction from lycopsid-like ancestors. Garbary, Renzaglia, and Duckett (1993) suggest that differences from previous interpretations may be due to the nature of their data. Many features of the motile apparatus of the male gamete may be structurally or functionally correlated and could thus result in anomalous systematic groupings.

The Contributions of Molecular Data

The earliest applications of molecular data to the phylogeny of bryophytes and vascular plants were based on phenetic analyses of 5S ribosomal RNA sequences (Hori, Lim, and Osawa 1985; Van de Peer, De Baere, et al. 1990). Results from these studies support the monophyly of bryophytes,

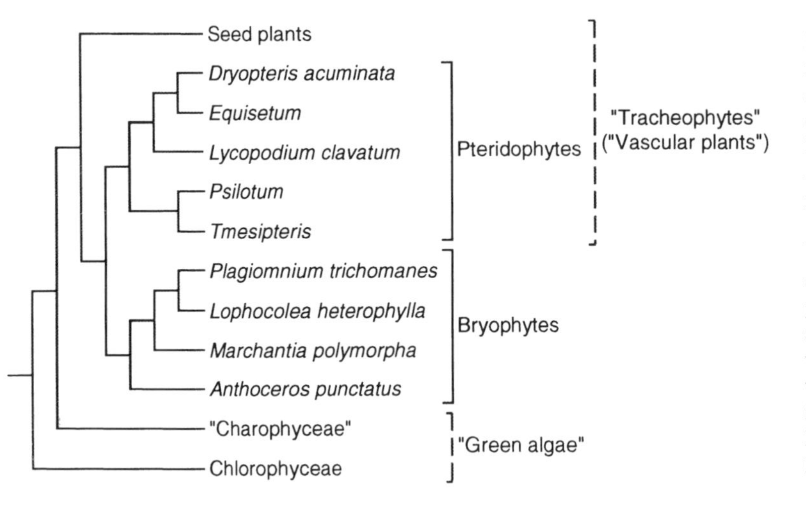

FIGURE 3.11. Phenetic relationships among green algae and embryophytes based on 5S rRNA data, after Hori, Lim, and Osawa 1985, Figure 3: simplification showing major relationships. Quotation marks = paraphyletic group. Note the basal position of seed plants and the monophyly of pteridophytes and bryophytes. Note also that bryophytes are nested within vascular plants, but not with the same relationships as in Garbary, Renzaglia, and Duckett 1993 (see Figure 3.10 herein).

pteridophytes, and gymnosperms and thus conflict with recent cladistic analyses that are based on morphological data. Hori, Lim, and Osawa (1985) proposed a sister group relationship between bryophytes and pteridophytes and suggested that this clade is the sister group to seed plants (Figure 3.11). On the basis of very similar data, Van de Peer, De Baere, et al. (1990) placed pteridophytes as a sister group to seed plants and bryophytes as a sister group to the more inclusive vascular plants (Figure 3.12). Most of the differences between these two analyses may be due to the different methods used to estimate phenetic distance (Van de Peer, De Baere, et al. 1990). Cladistic analysis of the data from Hori, Lim, and Osawa (1985) yields little structure and low consistency indexes (CI = 0.528; Bremer, Humphries, et al. 1987). The 57 equally parsimonious solutions found by Bremer, Humphries, et al. (1987) yielded highly incongruent results, including such bizarre groupings as *Chlamydomonas* (green alga) with *Lemna* (aquatic angiosperm) within angiosperms, and bryophytes as the sister group to seed plants. General concerns have been expressed about using the 5S rRNA molecule to resolve relationships at any taxonomic level because of its short length (about 120 bases, most of which are not informative), the presence of compensating substitutions (i.e., many positions do not vary independently), and the high rate of change in the variable portions (Halanych 1991; Steele, Holsinger, et al. 1991).

Phylogenetic analysis of sequence data from

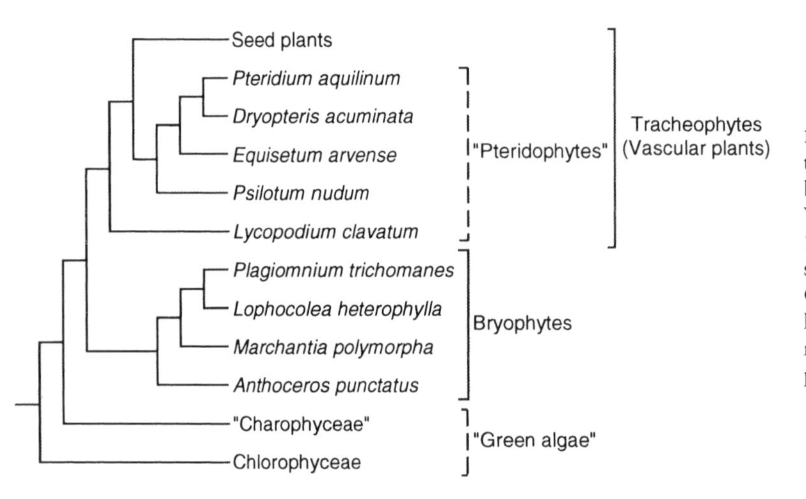

FIGURE 3.12. Phenetic relationships among embryophytes based on 5S rRNA data, after Van de Peer, De Baere, et al. 1990, Figure 2: simplification showing major relationships. Quotation marks = paraphyletic group. Note the sister group relationship between vascular plants and bryophytes.

larger subunits of nuclear-encoded rRNA molecules has proved more useful in resolving relationships of a broad range of organisms and for putatively ancient, as well as relatively recent, lines of divergence (see Chapman and Buchheim 1991; Waters, Buchheim, et al. 1992). The larger subunits provide a greater number of potentially informative sites, as well as relatively conserved and relatively variable regions that may provide resolution at different phylogenetic depths. Waters, Buchheim, et al. (1992) sequenced parts of both the small subunit (18S-like rRNA) and large subunit (26S rRNA) for two hornworts *(Phaeoceros laevis, Notothylas breutelii),* three marchantioid liverworts *(Asterella tenella, Conocephalum conicum, Riccia austinii),* and three mosses *(Atrichum angustatum, Fissidens taxifolius, Plagiomnium cuspidatum),* generating 120 informative characters. These sequences were compared to those of previously sequenced vascular plants *(Equisetum, Zamia, Glycine, Oryza, Zea).* With one exception, the phylogeny recovered by Waters, Buchheim, et al. (1992) supports the earlier conclusions of Parenti (1980) and the more detailed analysis of Mishler and Churchill (1984, 1985b). Using *Klebsormidium flaccidum* and *Coleochaete nitellarum* (Charophyceae) as outgroups, they concluded that embryophytes are monophyletic and that marchantioid liverworts are basal relative to other the taxa sampled (Figure 3.13). Liverworts, hornworts, mosses, and vascular plants are each supported as monophyletic groups, and the bryophytes as a whole are resolved as paraphyletic. Relationships among the four major groups of embryophytes were not strongly supported: solutions one step longer than the most parsimonious result failed to resolve relationships among vascular plants, mosses, hornworts, and liverworts.

In a later study, Mishler, Thrall, et al. (1992) obtained sequences from parts of the chloroplast 16S and 23S rRNA for eleven bryophytes (mosses: *Andreaea, Polytrichum, Sphagnum, Tetraphis, Thuidium;* hornworts: *Anthoceros;* liverworts: *Conocephalum, Marchantia, Metzgeria, Porella, Sphaerocarpos*). *Chlamydomonas* and *Chlorella* were used as green algal outgroups, and two angiosperms *(Zea* and *Nicotiana)* were used to represent vascular plants. Analysis of the combined 16S and 23S data yielded a tree supporting a sister group relationship between mosses and vascular plants and a basal position for liverworts within embryophytes (Figure 3.14). This result is consistent with the previous analysis by Mishler and Churchill

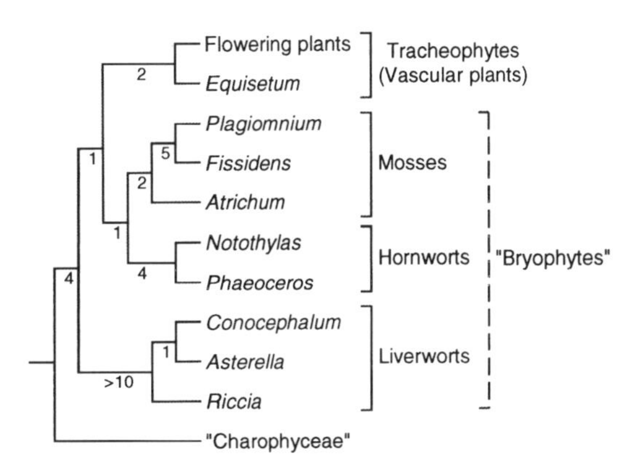

FIGURE 3.13. Cladistic relationships among embryophytes based on 120 informative characters from seven regions in the small subunit (SSU) and large subunit (LSU) rRNAs, after Waters, Buchheim, et al. 1992, Figure 2: simplification showing major relationships. Numbers on the branches are Bremer Support indexes; quotation marks = paraphyletic group. Note the paraphyly of bryophytes to vascular plants, the basal position of liverworts in embryophytes, and the sister group relationship of mosses to hornworts.

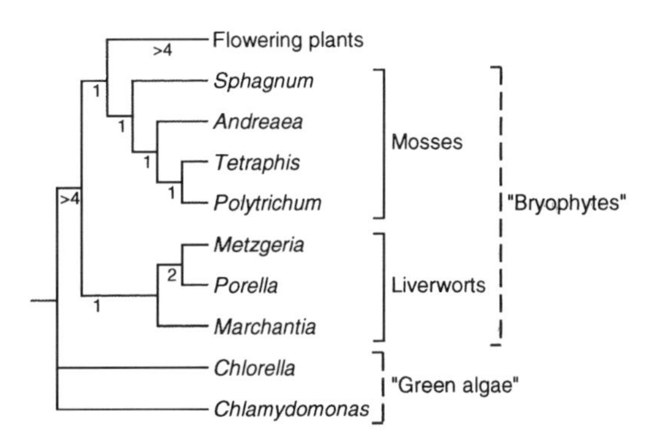

FIGURE 3.14. Cladistic relationships among embryophytes based on analysis of chloroplast encoded 16S and 23S rRNAs, after Mishler, Thrall, et al. 1992, Figure 6: simplification showing major relationships. Numbers on the branches are Bremer Support indexes; quotation marks = paraphyletic group. Note the paraphyly of bryophytes to vascular plants, the basal position of liverworts in embryophytes, and the absence of hornworts from the analysis.

(1984) based on morphology and other features, but the critical hornwort group was not included because 16S data were not available. Bremer Support (Decay Index = DI) was weak on all major branches (DI = 1), except for the strongly supported embryophyte and angiosperm clades (DI ≥ 4). Independent analyses of the 16S and 23S data yielded different results. The 23S data provided an uninformative consensus tree that supported only the embryophyte clade, and the 16S data gave a strange result that divided liverworts and placed marchantioid taxa in a sister group relationship with vascular plants.

More recently, Bopp and Capesius (1996) analyzed relationships among 20 liverworts, the hornwort *Anthoceros,* and 19 mosses, based on 18S rRNA sequences. Both maximum parsimony and neighbor-joining analysis supported liverwort paraphyly with respect to mosses and hornworts. The Marchantiopsida were resolved as monophyletic and as a sister group to a clade comprising the Jungermanniales, hornworts, and mosses. The effect of including vascular plants on the relationships among bryophytes was not explicitly investigated in this analysis.

The rRNA studies have provided some limited additional support for the previous cladistic results of Mishler and Churchill (1984, 1985b), including the monophyly of embryophytes, paraphyly of bryophytes, and the basal position of liverworts. In contrast to the results of Garbary, Renzaglia, and Duckett (1993), which were based on male gametes and gametogenesis, neither the morphological analyses of Mishler and Churchill (1984, 1985b) nor the molecular studies of Mishler, Thrall, et al. (1992) and Waters, Buchheim, et al. (1992) support the monophyly of bryophytes. Both molecular studies offer only weak support for relationships within embryophytes. A sister group relationship between hornworts and mosses suggested by Waters, Buchheim, et al. (1992) contradicts the moss–vascular plant relationship proposed by Mishler and Churchill (1984, 1985b). Both Waters, Buchheim, et al. (1992) and Mishler, Thrall, et al. (1992) acknowledged the preliminary nature of their analyses, particularly the limited systematic sampling, which is currently being addressed (Mishler, Chapman, pers. comm. to P.K. and P.R.C.). Problems are particularly acute within liverworts, mosses, and vascular plants, and monophyly of these three groups cannot be tested rigorously until leafy liverworts, mosses (such as *Sphagnum, Takakia, Andreaea, Andreaeobryum*), and homosporous lycopsids are also included in the analysis.

Early phylogenetic results from chloroplast-encoded *rbcL* DNA sequences conflict strongly with all previous rRNA- and morphology-based studies (Manhart 1994). Sampling included 35 taxa of green algae and land plants, including 6 Charophyceae *(Klebsormidium, Sirogonium, Spirogyra, Nitella, Chara, Coleochaete),* 3 mosses *(Sphagnum, Brotherella, Andreaea),* 2 liverworts *(Bazzania, Marchantia),* and the hornwort *Megaceros.* All of the major vascular plant groups were represented. Separate analyses using different weighting schemes gave different results. An analysis using equal weighting supported embryophyte and seed plant monophyly but resolved vascular plants and bryophytes as polyphyletic (Figure 3.15). *Sphagnum* and *Andreaea* were resolved as paraphyletic to a lycopsid–seed plant clade. Another moss *(Brotherella)* and the liverwort *Bazzania,* together with the hornwort *Megaceros,* were resolved as paraphyletic with respect to a clade consisting of *Equisetum,* Ophioglossaceae, Psilotaceae, and leptosporangiate ferns. *Marchantia* was resolved as a sister group to all other embryophytes. In this *rbcL* analysis, homoplasy levels were very high, as indicated by consistency indexes as low as 0.28.

Morphology, Spermatogenesis, and Molecules: Compared and Combined

The most recent reexamination of relationships among extant green algae and basal land plants by Mishler, Lewis, et al. (1994) used three different data sets based on general morphology (a variety of "traditional" characters), on spermatogenesis, and on molecular sequences. Compatibility was examined by analyzing the data sets separately and in various combinations. A consensus based on all of the data reproduced the earlier findings of Mishler and Churchill (1984, 1985b), despite substan-

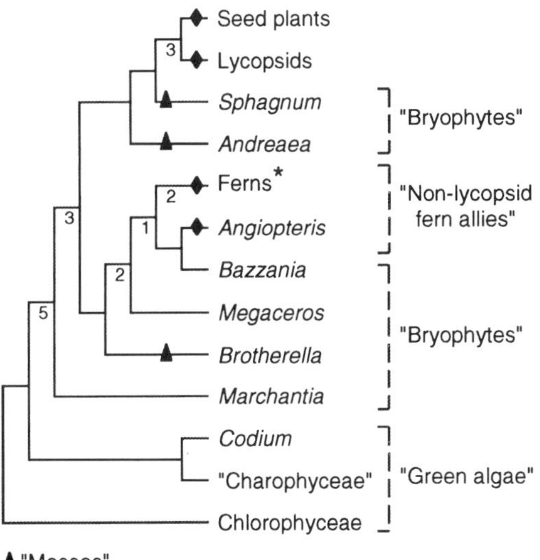

FIGURE 3.15. Cladistic relationships among embryophytes based on analysis of chloroplast encoded *rbcL* DNA, after Manhart 1994, Figure 2: simplification showing major relationships. The tree is based on all three positions having equal weighting and is rooted on four species of green algae. Numbers on the branches are Bremer Support indexes; quotation marks = paraphyletic group. *Ferns in this figure include leptosporangiate ferns, the Ophioglossales, *Psilotum,* and *Equisetum.* Note the polyphyly of bryophytes and vascular plants and the sister group relationship between lycopsids and seed plants.

tial disagreement in some patterns of relationship when the data sets were analyzed individually. When Mishler, Lewis, et al. (1994) reexamined the sperm-spermatogenesis data of Garbary, Renzaglia, and Duckett (1993), the analysis generated a reduced set of 65 characters (there were originally 90) but reproduced the earlier result of bryophytes nested among vascular plants. Morphological characters favored bryophyte paraphyly with respect to vascular plants, but relationships among the major bryophyte clades and tracheophytes were poorly resolved. Comparison with various molecular data sets showed that the *rbcL* sequences gave the most divergent results with respect to relationships among major clades. Two factors that may lead to spurious groupings are (1) extinction in the stem lineages of major clades, which restricts sampling for molecular studies to a small number of divergent extant taxa, and (2) increased changes in particular parts of molecular sequences, which result in overprinting of phylogenetic signals (site saturation) (Albert, Backlund, and Bremer 1994; Albert, Backlund, Bremer, Chase, et al. 1994). The total available evidence (general morphology, male gamete ultrastructure, and molecular sequences) supported the monophyly of embryophytes and resolved bryophytes as paraphyletic to vascular plants. Liverworts were found to be basal in the embryophytes, and mosses were resolved as a sister group to vascular plants (Mishler, Lewis, et al. 1994).

Phylogenetic Questions and Aims of Analysis

In general, current analyses of morphological, ultrastructural, and molecular data point to the monophyly of embryophytes and tracheophytes, the paraphyletic status of the bryophytes, and the basal position of liverworts. The most substantial uncertainties concern the relative phylogenetic positions of mosses, hornworts, and vascular plants, and there is also some disagreement between molecular and more traditional data on these issues. In particular, divergent results based on *rbcL* data underline the need for a rigorous and broadly based morphological analysis of these groups.

In this chapter we will analyze relationships among major extant groups of bryophytes and vascular plants, including one well-known fossil that exhibits a potentially significant combination of characters not present in any extant group. The aim is to clarify phylogenetic relationships *among* liverworts, hornworts, mosses, and basal polysporangiophytes rather than to focus on relationships within these groups. A particularly important objective is to test the monophyly of vascular plants with respect to bryophytes and charophycean algae. Most of the characters used have been employed in previous analyses, although we have made many modifications in character definition and scoring.

CHOICE OF TAXA

Our choice of taxa to include in this analysis of basal embryophytes is a compromise dictated by

TABLE 3.1
Genera Included in the Cladistic Analysis of Basal Embryophytes

Taxon	Exemplars
Polysporangiophytes	*Equisetum, Huperzia, Horneophyton* †
Mosses	*Polytrichum, Sphagnum, Andreaea, Andreaeobryum*
Putative moss	*Takakia*
Hornworts	*Anthoceros, Notothylas*
Liverworts	*Haplomitrium, Sphaerocarpos, Monoclea*
"Charophyceae"	*Coleochaete* *, *Chara* *

Notes: See Appendix 1 for taxon descriptions and Appendix 2 for data matrix including levels of inapplicable and unknown characters. † = extinct. * = outgroup. Quotation marks = paraphyletic group.

the phylogenetic questions that we are seeking to address, the need to keep the analysis within manageable limits, and our desire to include those taxa of maximum interest with respect to phylogenetic issues at this level. The taxa considered are listed in Table 3.1, and brief descriptions and reviews of their features are provided in Appendix 1.

Our analysis uses multiple low-level taxa, mainly genera or species, as representatives (exemplars, placeholders) of major groups. This approach allows tests of monophyly for higher-level taxa and permits potential variability in states within higher-level groups to be assessed accurately and explicitly (Nixon and Davis 1991). It reduces the problems that arise when higher taxa are scored using characters from several different genera or species within the group *(compartmentalization),* and as far as we are aware, has eliminated "composite taxa," which can have a combination of features that does not occur in any known species. Compartmentalization uses large clades as the units of analysis (e.g., Doyle, Donoghue, and Zimmer 1994; Mishler and Churchill 1984, 1985b; Nixon, Crepet, et al. 1994). This approach often simplifies the problems of analysis by reducing the number of taxa to be considered and also reduces character-sampling problems (missing data), which generally increase at the lower levels of the systematic hierarchy. However, the disadvantages of compartmentalization include the following: (1) the monophyly of large taxa is assumed and not tested by the analysis, (2) plesiomorphic character states of large taxa are sometimes difficult to score because of variation within the group, and (3) generalizations about character distributions that are based on inadequate sampling are difficult to justify.

Outgroups

There is strong support from morphological, ultrastructural, biochemical, and molecular data for a pattern of relationships in which the Charophyceae are a grade of organization comprising plants that are more closely related to embryophytes than to other green algae (Bremer 1985; Graham 1993; Graham, Delwiche, and Mishler 1991; see also Chapter 2). We therefore focus on taxa in the Charophyceae as the most reasonable outgroups for analyzing relationships among basal embryophytes. The genus *Coleochaete* is particularly important because several derived morphological and ultrastructural characters link the genus to the embryophytes, perhaps even making it paraphyletic (Graham, Delwiche, and Mishler 1991). This possibility is not pursued here, and we include only the well-studied species *Coleochaete orbicularis* (Figure 3.3). *Chara* is included as a representative of the Charales (Figure 3.4). The third genus of the Charophyceae, *Chaetosphaeridium,* includes only four species (Thompson 1969), all of which are poorly understood, particularly in details of sexual reproduction and ultrastructure of cell division (Graham, Delwiche, and Mishler 1991). For these reasons, the genus is not considered further.

Marchantiopsida

Mishler and Churchill (1985b) resolved the liverworts as a monophyletic group supported by three synapomorphies. Of these, the presence of oil bodies is the best corroborated because oil bodies are known in over 90% of liverwort species and have not been observed in outgroup taxa. The second character, the presence of elaters, is more problematic because these sterile cells could be homologous with the pseudoelaters of hornworts and therefore may be a synapomorphy at a more inclusive systematic level (see character 3.11 and discussion below). The third synapomorphy, lunularic acid, appears to be widespread in liverworts, but to our knowledge the closely related charophycean algae have not been assessed for this feature. Support for liverwort monophyly therefore is not strong, and recent molecular analyses have not yet sampled a sufficiently wide range of taxa to address this question rigorously (Manhart 1994; Mishler, Thrall, et al. 1992; Waters, Buchheim, et al. 1992).

Because of uncertainties over liverwort monophyly, and because relationships within liverworts are poorly resolved, selecting appropriate hepatic taxa to include in our analysis was problematic. Mishler and Churchill (1985b) considered the relationships among the basal taxa Sphaerocarpales, Marchantiales, and Monocleales and a clade comprising Calobryales, Metzgeriales, and Jungermanniales as unresolved. We include representative members of three of these groups (as exemplars for liverworts) in our analysis: *Monoclea* (Monocleales; Figure 3.16), *Haplomitrium* (Calobryales; Figure 3.17), and *Sphaerocarpos* (Sphaerocarpales; Figure 3.18). In future studies, it would also

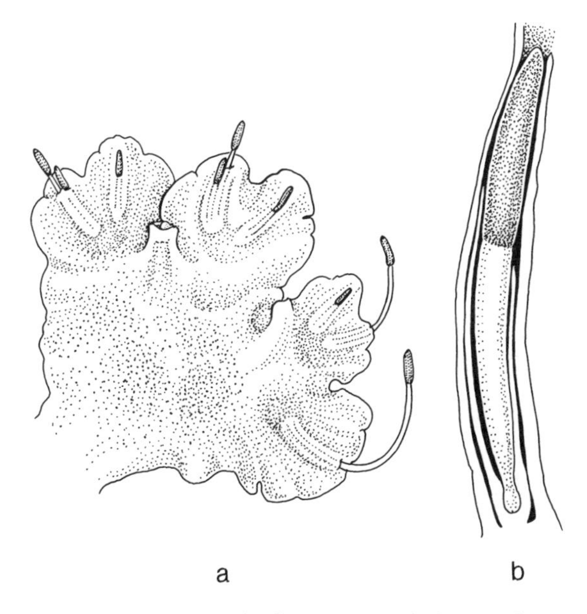

FIGURE 3.16. *Monoclea forsteri* (Monocleales, Marchantiopsida). (**a**) Gametophyte bearing sporophytes (×1). Redrawn and modified from Schuster 1984b, Figure 89.12. (**b**) Immature sporophyte just prior to expansion of seta and emergence from the involucre formed by surrounding gametophytic tissues (×10). Redrawn and modified from Schuster 1984b, Figure 88.4.

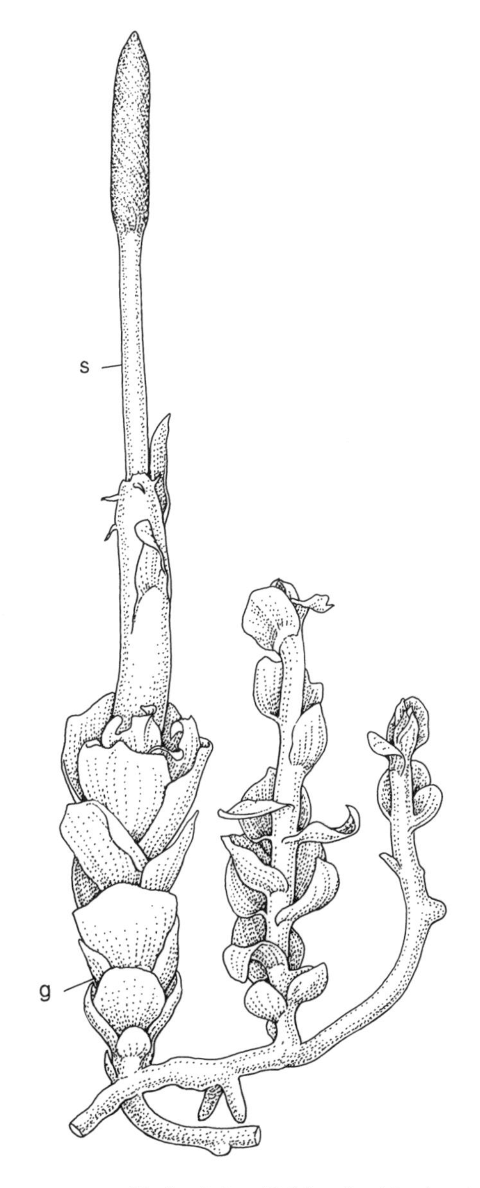

FIGURE 3.17. *Haplomitrium* (Calobryales, Marchantiopsida). Leafy, axial gametophyte *(g)* bearing solitary terminal sporophyte *(s)* (×10). Redrawn from Schuster 1984b, Figure 37.1.

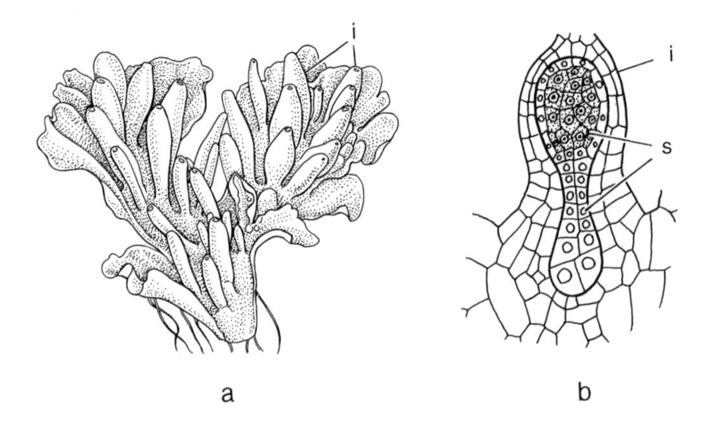

FIGURE 3.18. *Sphaerocarpos* (Sphaerocarpales, Marchantiopsida). (**a**) Female plant of *S. texanus* showing crowded, tubular involucres *(i)*, which contain the immature sporophytes (×10). Redrawn from Schuster 1984b, Figure 86.1. (**b**) Longitudinal section through a young sporophyte *(s)* of *S. cristatus* before emergence from the involucre *(i)*. Differentiation of archesporial tissue *(stippled area)* prior to spore formation (×170). Redrawn from Smith 1938, Figure 25B.

be desirable to include basal Metzgeriales, as well as basal taxa from the two large and probably monophyletic groups Marchantiales and Jungermanniales (Table 3.2).

Anthocerotopsida

The Anthocerotopsida are a small and well-supported monophyletic group (Crandall-Stotler 1981, 1984; Hasegawa 1988; Hässel de Menéndez 1988; Mishler and Churchill 1985b; Renzaglia 1978; Schuster 1984c, 1992). Mishler and Churchill (1985b) cited 11 synapomorphies of the group, and Hässel de Menéndez (1988) identified 17 (Table 3.3). Even though some of these characters may be linked, or may be synapomorphies at a more inclusive level (see character discussion below and Mishler and Churchill 1987; Whittemore 1987), there is strong support for hornwort monophyly, and this is also consistent with results from analyses of molecular data (Mishler, Lewis, et al. 1994; Waters, Buchheim, et al. 1992).

TABLE 3.2
Classification of Marchantiopsida

Hepaticae
"Marchantiidae"
Sphaerocarpales
Monocleales
Marchantiales
Jungermanniidae
"Calobryales"
"Metzgeriales"
Treubiales
Jungermanniales

Source: After Schuster 1984b.

Note: Quotation marks = paraphyletic group.

Five genera are usually recognized within the Anthocerotopsida: *Anthoceros, Dendroceros, Notothylas, Megaceros,* and *Phaeoceros,* but some authors recognize an additional one or more genera (e.g., Hasegawa 1988; Hässel de Menéndez 1988; Renzaglia 1978; Schuster 1984c, 1992). Despite many potentially useful systematic features, relationships among these genera are not well resolved. In part this is because some of the genera are poorly defined, but there is also a lack of comparative data across groups (Mishler and Churchill 1985b). Poorly defined genera include *Phaeoceros* and *Anthoceros,* which may be paraphyletic with respect to either *Megaceros* and *Dendroceros* or to *Notothylas,* depending on where the cladogram is rooted. Both *Megaceros* and *Dendroceros* are also poorly defined (Mishler and Churchill 1987; Whittemore 1987). Despite these difficulties, the most parsimonious result from Mishler and Churchill's cladistic analysis (1985b) resolved *Notothylas* as a sister group to a clade containing the other four genera, and this relationship is consistent with the two-family classification of Schuster (1984c). An alternative, less parsimonious solution noted

TABLE 3.3

Synapomorphies of the Hornwort Clade according to Mishler and Churchill

Character	Generation	Comments
Cuneate apical cell	G	Wedge-shaped cell with four cutting faces (see character 3.46)
Pyrenoid	G (and S?)	Protein aggregate contained within the chloroplast and composed almost exclusively of ribulose 1,5-bisphosphate carboxylase-oxygenase (see character 3.29)
Mucilage cells	G	Occur in the thalli of all hornworts
Mucilage cleft	G	Stomalike pore of mucilage cavity on the ventral face of the thallus (see character 3.27)
Nostoc symbiont	G	*Nostoc* symbionts within the ventral mucilage cavities
Endogenous antheridia	G	Stalked antheridia in chambers within the thallus (see character 3.20)
Embedded archegonium	G	No stalk cell; archegonium located within the thallus surface (see character 3.21)
Spermatozoid ultrastructure	G	Unique features in the male gamete (see summary in Garbary, Renzaglia, and Duckett 1993)
Vertical division of zygote	S	First division of zygote typically vertical as opposed to transverse in most other embryophytes
Pseudoelaters	S	Elongate unicellular forms with a single helical thickening in *Dendroceros* and *Megaceros* to multicellular unthickened branched structures in other genera (see character 3.11)
Columella	S	Columnar mass of sterile tissue that develops within the spore mass of the sporangium of hornworts, mosses, and some early fossil polysporangiophytes (see character 3.10)

Source: Mishler and Churchill 1985a.

Notes: G = gametophyte. S = sporophyte.

by Mishler and Churchill (1985b) and favored in some more traditional schemes interprets the small, simple sporophytes of *Notothylas* as derived and the taxa with larger sporophytes, such as *Dendroceros* and *Megaceros,* as basal. The fully resolved cladogram given by Hässel de Menéndez (1988) also favors this interpretation, but these results are problematic because most nodes do not appear to be supported by characters (most characters appear to be autapomorphic), and the basis for character polarization is not discussed. In view of the uncertainties over relationships in hornworts, our analysis includes the two divergent morphological types represented by *Anthoceros* (Figure 3.19) and *Notothylas* (*sensu* Renzaglia 1978).

Bryopsida

Mosses are treated implicitly as monophyletic in traditional classifications, and this is supported by recent cladistic analyses (Mishler and Churchill 1984; Mishler, Lewis, et al. 1994). However, to allow for the possibility of polyphyly, as well as for alternative character-state trees, we include *Andreaea* (Figure 3.20), *Sphagnum* (Figure 3.21), and *Andreaeobryum,* which are generally placed near the base of the group. In addition, we include *Polytrichum* (Figure 3.22) as an exemplar for the Polytrichales and "higher" (peristomate) mosses (Tetraphidales, Buxbaumiales, Bryales), which comprise a clade defined by the presence of a peri-

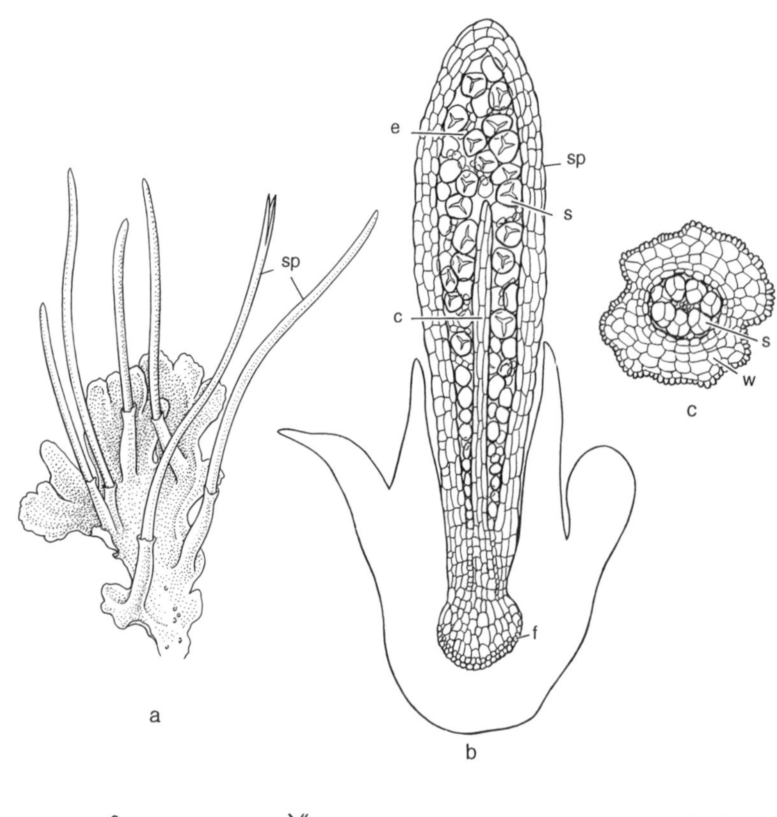

FIGURE 3.19. *Anthoceros (Phaeoceros) laevis* (Anthocerotopsida). (**a**) Part of thalloid gametophyte bearing upright, tubular sporophytes *(sp)* (×1). Redrawn and modified from Schuster 1984c, Figure 2.1. (**b**) Longitudinal section of young sporophyte *(sp)* showing sporangium with spores *(s)*, elaters *(e)*, and columella *(c)* (×50). Sporophyte is attached to gametophyte by bulbous foot *(f)* at base. (**c**) Transverse section through sporangium showing central columella, spore tetrads *(s)*, and outer tissues of sporangium wall *(w)* (×50).

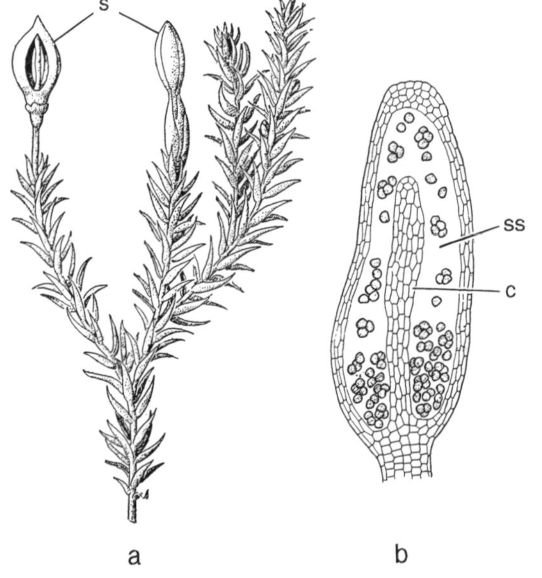

FIGURE 3.20. *Andreaea petrophila* (Bryopsida). Drawings reproduced from Smith 1938, Figure 54A. (**a**) Leafy gametophyte bearing mature and immature sporophytes *(s)* (×2). Note the pseudopodium. (**b**) Semidiagrammatic longitudinal section through sporangium. Spore sac *(ss)* surrounds a central sterile columella *(c)* (×70).

stome, a cylindrical rather than overarching spore mass, and the presence of an operculum (possibly convergent with that of *Sphagnum*) (see Mishler and Churchill 1984 for details). We also include the formerly problematic taxon *Takakia* (Figure 3.23; Crandall-Stotler 1986; Crandall-Stotler and Bozzola 1988; Schuster 1966, 1984a). The recent discovery of antheridia and a sporophyte generation in *Takakia* provide strong evidence for a relationship with mosses (Davison, Smith, and McFarland 1989; Renzaglia, Smith, et al. 1991; Smith and Davison 1993).

Polysporangiophytes

The three polysporangiophyte taxa included in this analysis have been chosen as exemplars to represent the two major lineages of vascular plants that are hypothesized in current systematic treatments: the lycopsids and the fern–seed plants (Chapter 4). This selection of taxa also allows for the possibility that tracheophytes are polyphyletic. The extant

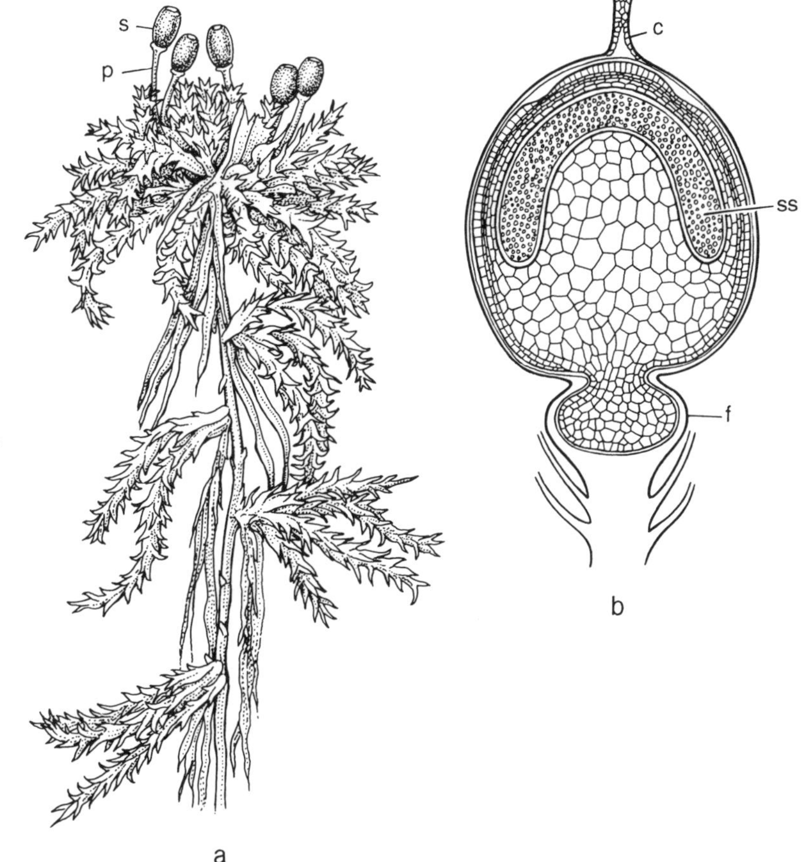

FIGURE 3.21. *Sphagnum squarrosum* (Bryopsida). (**a**) Gametophyte bearing five sporophytes *(s)* on pseudopodia *(p)* (×6). Redrawn from Schofield 1985, Figure 4-3C. (**b**) Longitudinal section through young sporophyte still enclosed within calyptra *(c)* (×40). Note the bulbous foot *(f)* and the convex spore sac *(ss)* with protruding columella. Redrawn from Bower 1935, Figure 43H.

lycopsid *Huperzia* (Figure 3.24) and the extant sphenopsid, *Equisetum* (Figure 3.25), are both included, along with *Horneophyton* (Figure 3.26), which is one of the best-known, early fossil land plants and which was placed near the base of polysporangiophytes in our previous analysis of embryophyte relationships (Kenrick and Crane 1991). The recent description of *Langiophyton,* the probable gametophyte of *Horneophyton,* also facilitates comparison of gametophyte features.

CHARACTER DESCRIPTIONS AND CODING

Characters Scored for Almost All Taxa

3.1 • SPOROPHYTE CELLULARITY The presence of a multicellular sporophyte generation (sporic meiosis) is characteristic of all embryophytes and does not occur in charophycean algae. This feature was formerly incorporated in the "embryo" character of Mishler and Churchill (1984), which has been criticized because it incorporates multicellularity as well as the nutritional and developmental interactions between the gametophyte and sporophyte. These features are no longer considered linked because nutritional interactions between the sporophyte and gametophyte have been observed recently in *Coleochaete* (Graham 1993; Graham, Delwiche, and Mishler 1991). The unique multicellularity of land plant sporophytes is therefore recognized as an independent feature equivalent to the "sporic meiosis" character of Graham, Delwiche, and Mishler (1991) and Graham (1993) and to the "embryo" character of Mishler, Lewis, et al. (1994).

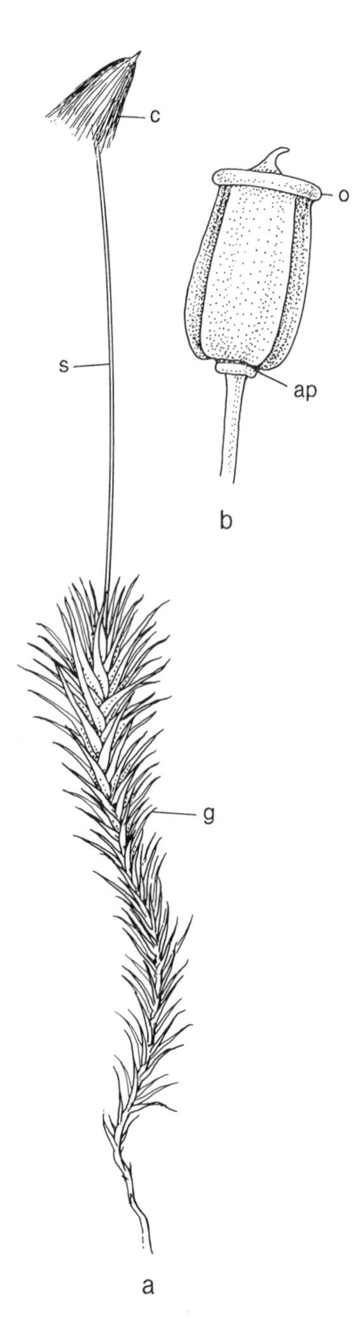

FIGURE 3.22. *Polytrichum commune* (Bryopsida). (**a**) Part of leafy gametophyte *(g)* with attached sporophyte *(s)* (×2). Apex of sporophyte bears a fibrous calyptra *(c)*. (**b**) Sporangium with calyptra removed showing operculum *(o)*, ridged sporangium wall, and basal apophysis *(ap)* (×5).

Cellularity of sporophytes is scored as a simple binary character: 0 = unicellular; 1 = multicellular. There were no inapplicable cases; none were missing data (see Appendix 2).

3.2 • SPOROPHYTE INDEPENDENCE OR DEPENDENCE In vascular plants the sporophyte is initially dependent on the gametophyte (its foot absorbs nutrients from the gametophyte) but develops complete physiological independence during early ontogeny (Bierhorst 1971). In liverworts, hornworts, and mosses, the sporophyte remains in intimate contact with gametophytic tissues during spore production (Ligrone, Duckett, and Renzaglia 1993). Similarly, a connection with gametophytic tissues is retained by the zygote in *Coleochaete* (Graham 1993), and in *Chara,* the zygote is retained within the oogonium until meiosis is complete even though the oogonium is shed after fertilization (Graham 1993).

Sporophyte independence or dependence is scored as a binary character: 0 = dependent; 1 = independent. There were no inapplicable cases; none were missing data (see Appendix 2). See character 4.1 and Appendix 3 for comparison.

3.3 • WELL-DEVELOPED SPORANGIOPHORE The unbranched seta of moss sporophytes and the branched sporangium-bearing axes of vascular plants were scored as homologous structures by Mishler and Churchill (1984) and by Mishler, Lewis, et al. (1994) under the term "aerial sporophyte axis" (Mishler and Churchill 1984, character 29). We support this interpretation because the sporophyte axis is persistent and internally differentiated in both groups, and because outgroup comparison with hornworts, liverworts, and charophytes suggests that the plesiomorphic land plant sporophyte is probably a small, epiphytic, simple sporangium (see below for further discussion).

In common with most mosses, an elongate seta is present in *Takakia* (Smith and Davison 1993) and *Polytrichum* (Parihar 1977), and there is a short, broad seta in *Andreaeobryum* (Murray 1988). Both *Andreaea* (Murray 1988) and *Sphag-*

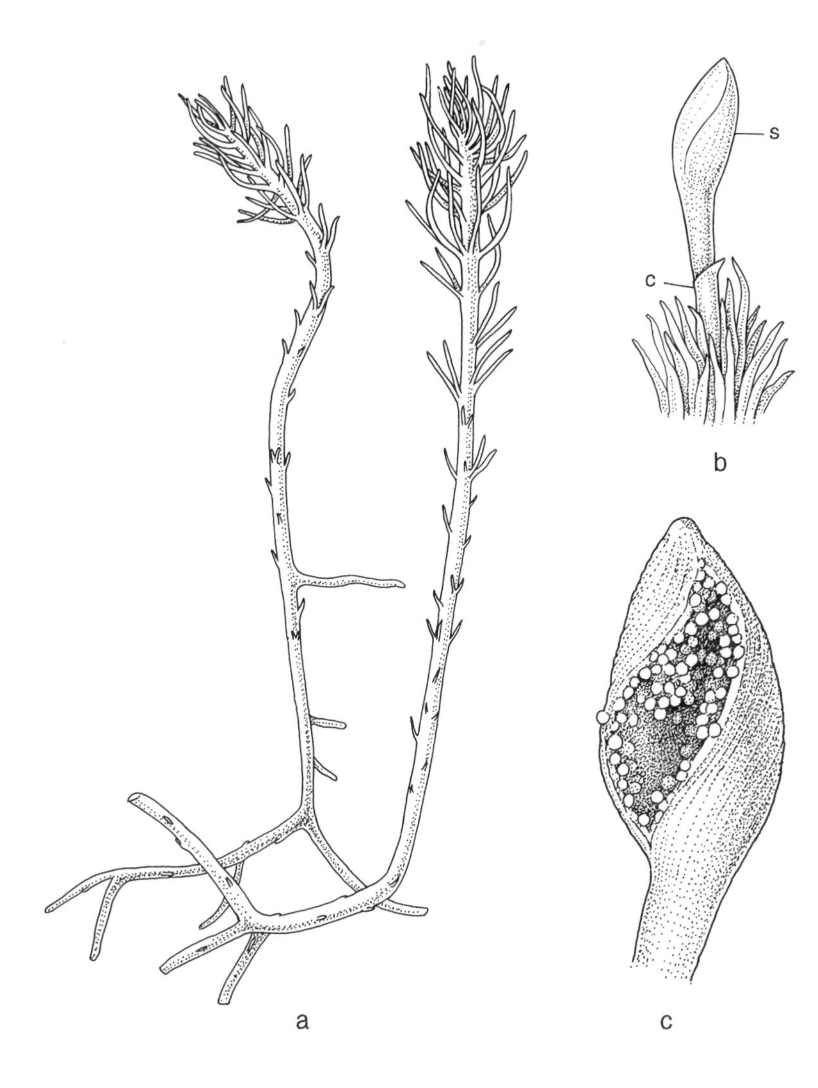

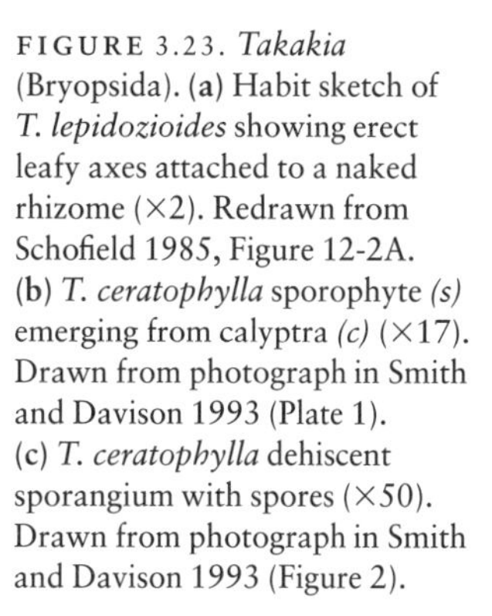

FIGURE 3.23. *Takakia* (Bryopsida). (**a**) Habit sketch of *T. lepidozioides* showing erect leafy axes attached to a naked rhizome (×2). Redrawn from Schofield 1985, Figure 12-2A. (**b**) *T. ceratophylla* sporophyte *(s)* emerging from calyptra *(c)* (×17). Drawn from photograph in Smith and Davison 1993 (Plate 1). (**c**) *T. ceratophylla* dehiscent sporangium with spores (×50). Drawn from photograph in Smith and Davison 1993 (Figure 2).

num (Parihar 1977) are usually described as possessing an undeveloped or suppressed seta, but some authors interpret this condition as a true absence (Watson 1971). To allow for the possibility that the condition in *Sphagnum* and *Andreaea* is plesiomorphic for mosses, or for embryophytes as a whole, we score these taxa as lacking a seta.

Tissue differentiation within the seta of many mosses is quite pronounced and shows similarities to early fossil vascular and nonvascular plants, including the possession of a central conducting strand and a peripheral cortical layer of thick-walled cells (Figures 4.20 and 4.21). In mosses the occurrence of a well-developed seta without a conducting strand has not yet been demonstrated (Hébant 1977). A central conducting strand of unthickened elongate cells, which lose their cytoplasm at maturity, has been documented in *Takakia* (Renzaglia, Smith, et al. 1991; Smith and Davison 1993), *Andreaeobryum* (Murray 1988), and *Polytrichum* (Hébant 1977). The elongated cells of this strand are sometimes intermixed with parenchyma (Hébant 1977; Renzaglia, Smith, et al. 1991). In both *Polytrichum* (Hébant 1977) and *Andreaeobryum* (Murray 1988), there is a peripheral cortical layer (or layers) of thick-walled cells.

Similarities between the conducting strand in the sporophytes of such taxa as *Polytrichum* (Figure 4.20; Hébant 1977, Figure 113) and in those of

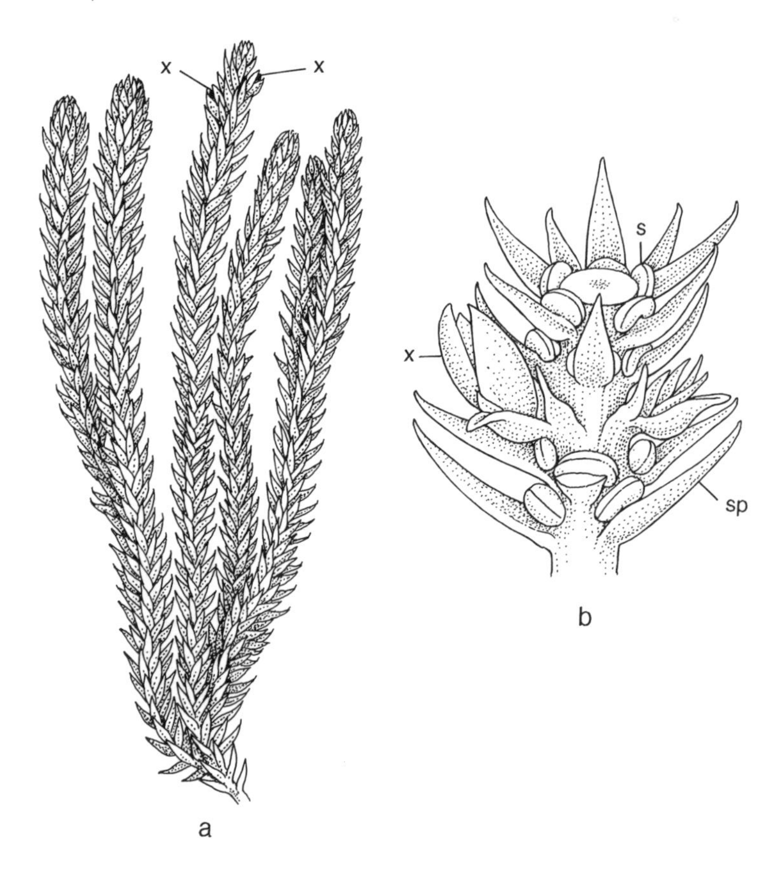

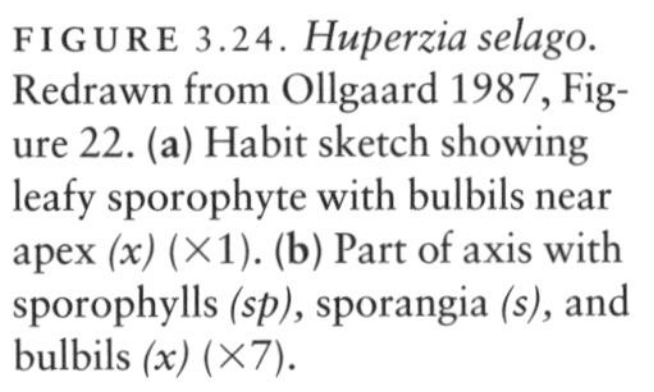

FIGURE 3.24. *Huperzia selago.* Redrawn from Ollgaard 1987, Figure 22. (**a**) Habit sketch showing leafy sporophyte with bulbils near apex *(x)* (×1). (**b**) Part of axis with sporophylls *(sp)*, sporangia *(s)*, and bulbils *(x)* (×7).

such early fossil polysporangiophytes as *Aglaophyton* (Figure 4.21; D. S. Edwards 1986, Figures 37–42) are so striking that the central strand of *Aglaophyton* has been interpreted as hydrome surrounded by a *leptome sheath* (D. S. Edwards 1986). Unbranched axes of the mosslike fossil sporophyte *Sporogonites* show a similar level of cellular differentiation. There is a central region of elongated and unthickened cells, possibly conducting tissue, surrounded by a poorly preserved, presumably parenchymatous, tissue and a layer of thick-walled epidermal cells (Halle 1916a, 1916b, 1936a). No thick-walled cortical tissue has been observed. A similar simple level of differentiation is evident in the branched axes of such early fossil taxa as the nonvascular plants *Aglaophyton* (D. S. Edwards 1986; Kidston and Lang 1917, 1920a) and *Horneophyton* (Eggert 1974; El-Saadawy and Lacey 1979b; Kidston and Lang 1920a; Remy and Hass 1991d) and the vascular plants *Psilophyton dawsonii* (Banks, Leclercq, and Hueber 1975) and *Gosslingia breconensis* (Kenrick and Edwards 1988a).

In hornworts, there is no structural unit comparable to the seta of mosses. The meristematic zone between the capsule and the foot is not treated as homologous with the seta because the cell lines derived from this continually dividing zone ultimately differentiate into capsule tissues: jacket, spores, elaters, and columella (Renzaglia 1978). Similarly, following Schuster (1966), Crandall-Stotler (1980), Mishler and Churchill (1984), and Mishler, Lewis, et al. (1994), we interpret the seta of such liverworts as *Haplomitrium, Monoclea,* and *Sphaerocarpos* as nonhomologous (therefore, independently derived) to the seta of mosses. There is little internal differentiation in the massive setae of these three taxa (Schuster 1984a), and no liverwort sporophyte is known to possess a central strand of conducting tissue (Hébant 1977, 1979). Generally, the seta in liverworts is an exceedingly ephemeral structure that elongates rapidly just

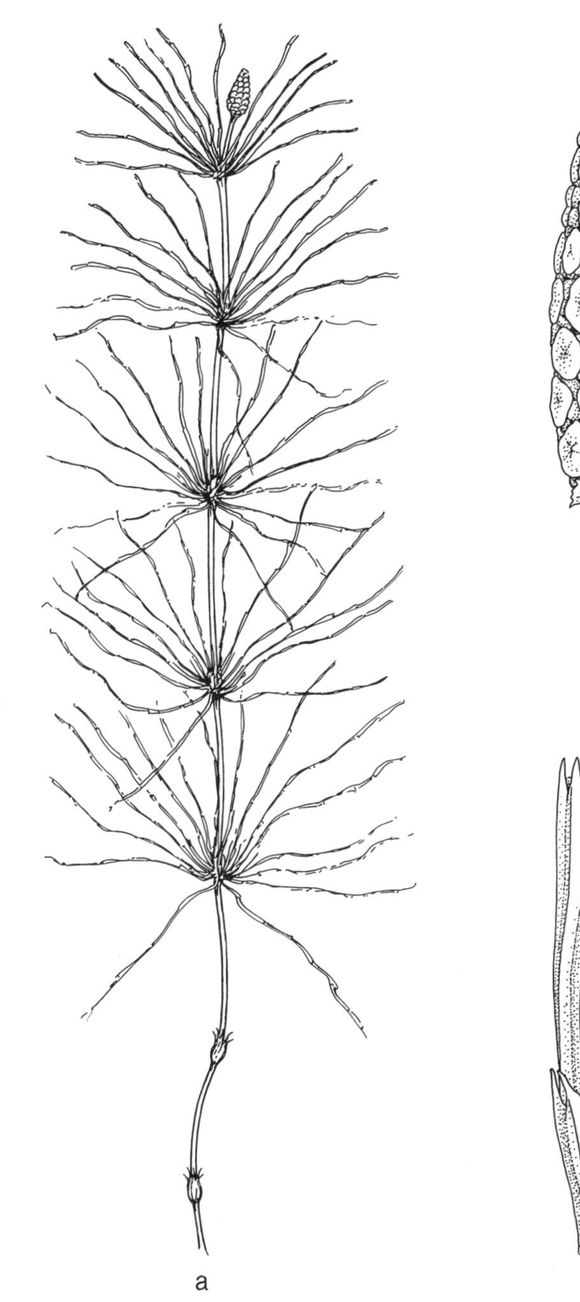

FIGURE 3.25. *Equisetum* cf. *pratense*. (**a**) Habit sketch showing branch with whorls of branches and terminal cone (×1.5). (**b**) Cone with peltate sporangiophores cut away on right to show attachment of sporangia *(s)* (×3). (**c**) Small whorled leaves on part of branch (×8).

prior to, or after, capsule dehiscence (Schuster 1966, 1984a, 1984b, 1984c). A seta is not observable (and is therefore inapplicable) in the unicellular sporophytes of the charophycean outgroups.

The presence or absence of a well-developed sporangiophore, defined as a sporophyte axis with internal differentiation of tissues, is treated as a binary character: 0 = sporangiophore absent; 1 = sporangiophore present. There were two inapplicable cases; none were missing data (see Appendix 2). See character 4.2 for comparison.

3.4 • SPOROPHYTE BRANCHING The presence of a branched sporophyte producing multiple sporan-

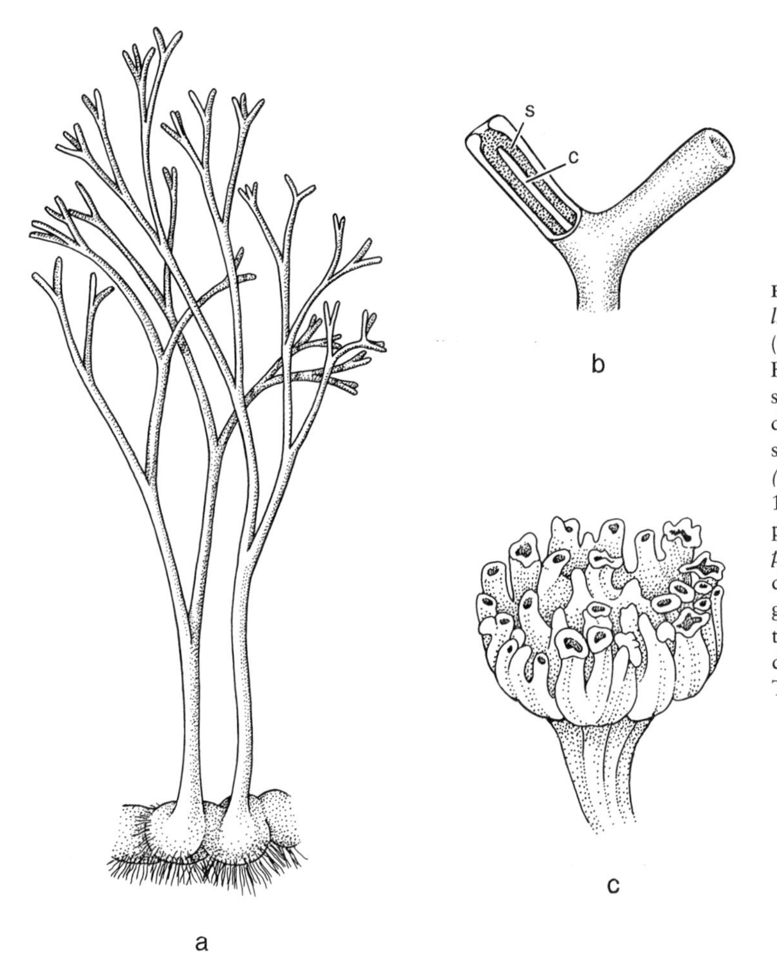

FIGURE 3.26. *Horneophyton lignieri.* (**a**) Reconstruction of habit (×2). Redrawn from Stewart 1983, Figure 9-4. (**b**) Details of branched sporangium. Left branch shows details of interior: spore sac *(s)* surrounds a central sterile columella *(c)* (×4). Redrawn from Stewart 1983, Figure 9-4. (**c**) Putative gametophyte of *Horneophyton (Langiophyton mackiei):* reconstruction of distal axis showing terminal gametangiophore with fingerlike processes that bear archegonia (×5). Redrawn from Remy and Hass 1991d, Text-Figure 9.

gia from the product of a single fertilization is diagnostic of all extant tracheophytes, as well as of several fossil taxa possessing other kinds of conducting cells (e.g., *Aglaophyton, Horneophyton* [Figure 3.26], *Nothia*). The axis bearing the sporangium in mosses, hornworts, and liverworts is unbranched except in a few rare instances in some mosses, such as *Bryum* and *Hypnum,* where a single dichotomy results in two sporangia (Bower 1935, 592, Figure 440). Sporophyte branching is not observable (and is therefore inapplicable) in the unicellular sporophytes of the charophycean outgroups.

The presence or absence of a branched sporophyte is treated as a binary character: 0 = branched sporophyte (monosporangiate) absent, 1 = branched sporophyte (polysporangiate) present. There were two inapplicable cases; none were missing data (see Appendix 2). See character 4.3 for comparison.

3.5 • INTERCALARY MERISTEM In hornwort sporophytes (e.g., *Anthoceros, Notothylas*), as soon as the columella, archesporium, and jacket are differentiated in the primary capsular region of the sporogonium, apical growth ceases and further upward growth is achieved solely through the activity of a meristematic region at the base of the capsule (Figure 3.19; Renzaglia 1978). This upward growth is extensive in *Anthoceros* but ceases early in the development of *Notothylas* sporophytes. The intercalary meristem adds continually to all three regions of the capsule so that progressive differentiation and maturation of tissue can be seen in a single sporophyte. We interpret this characteristic as a uniquely derived feature of hornworts because no comparable meristematic region is known in other fossil or extant land plants. We follow Crandall-Stotler (1980) and Mishler and Churchill (1984, character 24) and interpret the transitory "intercalary meristem" phase in the embryonic sporophyte of mosses as a parallelism. As discussed by Whittemore (1987; see reply by Mishler and Churchill [1987]), the intercalary meristem character is linked to at least two other characters cited by Mishler and Churchill (1985b): nonsynchronous spore production and sporangium length. Both are related directly to the activities of the intercalary meristem and are not used here. The feature of sporangium length could be reformulated in terms of the duration of meristematic activity. The intercalary meristem is not observable (and is therefore inapplicable) in the unicellular sporophytes of the charophycean outgroups.

The presence or absence of an intercalary meristem is treated as a binary character: 0 = intercalary meristem absent; 1 = intercalary meristem present. There were two inapplicable cases; none were missing data (see Appendix 2).

3.6 • FOOT FORM The foot is a specialized mass of tissue that develops early in the ontogeny of embryophytes, forms an intimate connection with the gametophytic tissues, and performs a nutritive function for the young sporophyte (Figure 3.19; Ligrone, Duckett, and Renzaglia 1993). Two distinctive types of foot form are recognized: a bulbous mass of tissue is the general condition in embryophytes, whereas an elongate tapering foot is characteristic of some mosses.

The foot region of liverwort sporophytes usually is more or less bulbous and rarely penetrates gametophytic tissue (Ligrone and Gambardella 1988a,b; Schuster 1984a). Similarly, a relatively massive, more or less bulbous foot is characteristic of hornworts, such as *Anthoceros, Notothylas,* and *Dendroceros* (Ligrone and Renzaglia 1990; Renzaglia 1978), and is the common condition in vascular plants. During early embryology in *Lycopodium clavatum,* four cells of the proembryo adjacent to the suspensor cell differentiate to form a bulbous foot (Bierhorst 1971, 24). Similar foot morphology and ontogeny is known in the Ophioglossaceae, Psilotaceae (Frey, Campbell, and Hilger 1994a,b), Schizaeaceae, Stromatopteridaceae, and Osmundaceae and other leptosporangiate ferns (Bierhorst 1971), but the embryogeny of *Equisetum* is poorly understood. Mosses possess a more complex and more highly organized foot at the base of the sporophyte, which is quite distinct from the foot of liverworts and hornworts. In the Bryidae and Andreaeidae the foot is an elongate tapering structure that penetrates deeply into the conducting strand of the gametophyte stem and has a group of collapsed cells at the extreme tip (Ligrone and Gambardella 1988a). This type of tapering foot has been observed in *Takakia* (Renzaglia, Smith, et al. 1991), *Polytrichum* (Ligrone and Gambardella 1988b), *Andreaea,* and *Andreaeobryum* (Murray 1988). In marked contrast to other moss sporophytes, *Sphagnum* possesses a bulbous foot (Ligrone and Renzaglia 1989). Foot form is not observable (and is therefore inapplicable) in the unicellular sporophytes of the charophycean outgroups and is unknown although potentially observable in the fossil *Horneophyton.*

Foot form is treated as a binary character: 0 = bulbous; 1= tapering. There were two inapplicable cases; one was missing data (see Appendix 2).

3.7 • XYLEM-CELL THICKENINGS Water-conducting cells (xylem elements, tracheids) with differentially thickened cell walls are characteristic of the sporophytes of extant vascular plants but not of bryophytes and green algae. In nearly all extant vascular plants, conspicuous pits characterize the metaxylem cells, whereas annular or helical thickenings are typical of the earliest protoxylem elements to differentiate (Figure 3.27; Bierhorst 1960). Evidence from the fossil record indicates that cells with helical or annular thickenings are more general than pitted cells because many early vascular plant fossils possess tracheids with only helical thickenings (Kenrick and Crane 1991). Well-defined wall thickenings are absent in the early polysporangiophyte fossil *Horneophyton*

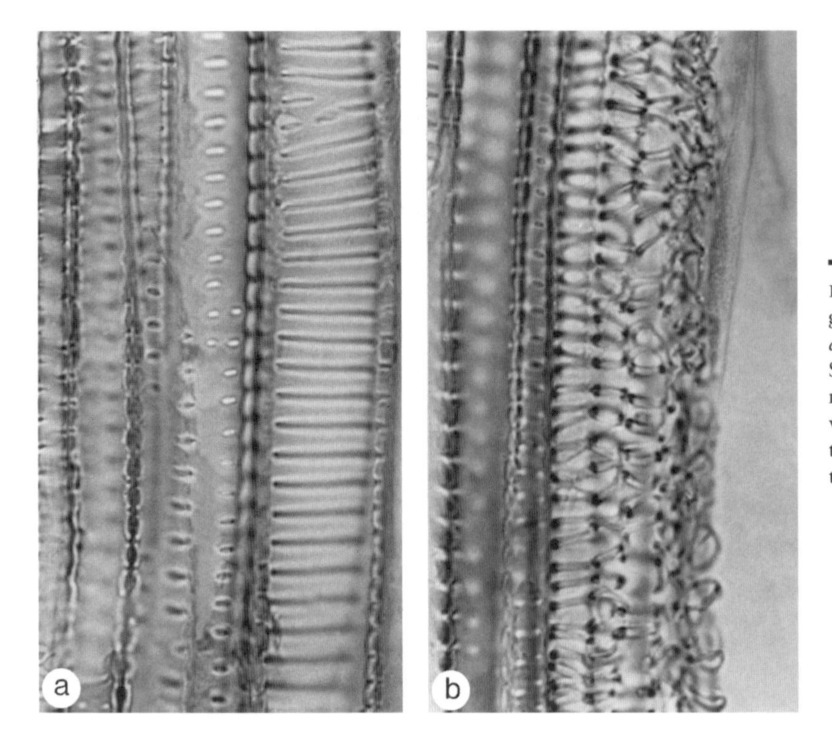

FIGURE 3.27. Light micrographs of tracheids of *Lycopodium volubile* (Lycopodiaceae). Scale bar = 25 μm for both micrographs. (**a**) Metaxylem with scalariform pitting. (**b**) Protoxylem with annular-helical thickenings.

(Remy and Hass 1991d). Xylem cells are not observable (and are therefore inapplicable) in the unicellular sporophytes of the charophycean outgroups.

The presence or absence of differential thickenings in water-conducting cells is treated as a binary character: 0 = simple annular or helical thickenings absent; 1 = simple annular or helical thickenings present. There were two inapplicable cases; none were missing data (see Appendix 2). See character 4.17 for comparison.

3.8 • SPORANGIUM Sporangia in all embryophytes comprise a spore mass surrounded by a sterile outer layer. Ontogenetically, sporangia are characterized by early differentiation of the archesporial developmental pathway from sterile (amphithecial) tissues. The sporangium is generally more or less ovoid or obovoid, and dehiscence by one or more longitudinal sutures is common in liverworts, hornworts, *Andreaea, Andreaeobryum, Takakia,* and the sporangia of many early vascular plant fossils. A more elaborate dehiscence mechanism, involving an operculum and a peristome, characterizes most mosses (Figure 3.28).

In vascular plants, two types of sporangia (eusporangia and leptosporangia) are often recognized (Gifford and Foster 1989). The *eusporangium* is usually large with a relatively thick wall comprising two or more cell layers, one of which may ultimately degenerate. Ontogenetically, the presence of a eusporangium is first recognized when a group of one or more rows of superficial cells begins to divide periclinally. Typical eusporangia occur in lycopsids and many early fossil tracheophytes. The distinctive *leptosporangium,* which is typical of most ferns, is usually small with a wall comprising a single layer of cells and a distinctive annulus that is typically oriented vertically (Bierhorst 1971). The ontogeny of the leptosporangium is first recognized when a single superficial cell begins to divide obliquely or periclinally. We agree with Bierhorst (1971, 266) that the concepts of eusporangium and leptosporangium are often difficult to define clearly and that a continuum of sporangium form exists; however, certain specific

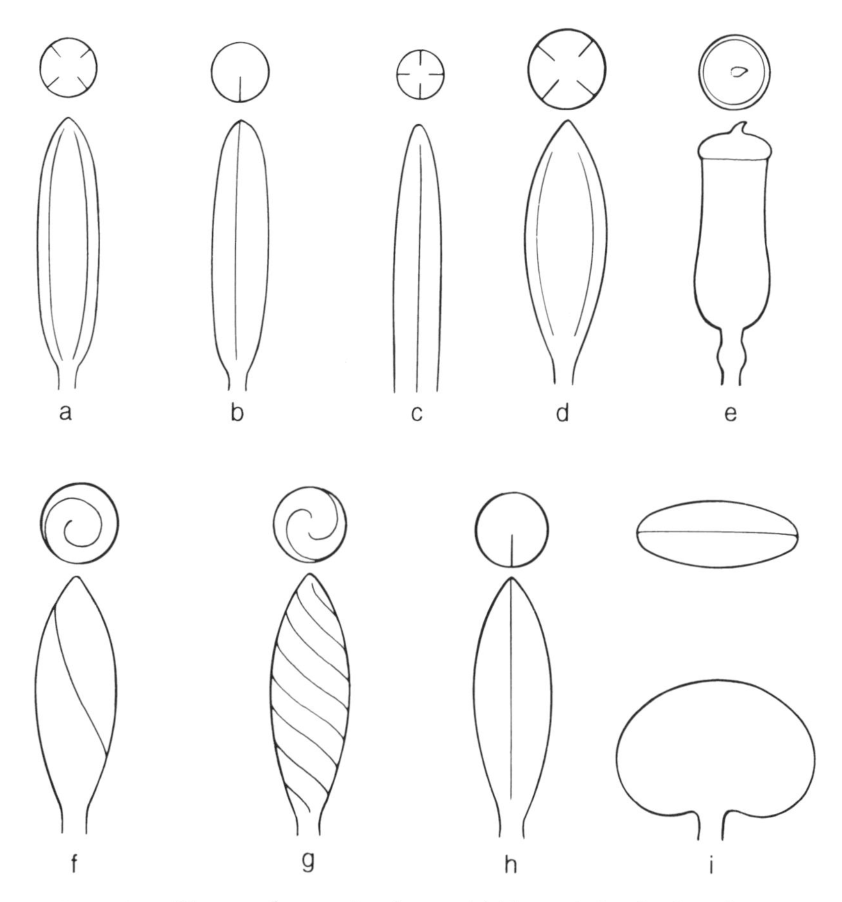

FIGURE 3.28. Diagram of sporangium form and dehiscence in basal embryophytes. Different dehiscence mechanisms in liverworts, hornworts, mosses, and early vascular plants (lateral and apical views). (**a**) Four valves attached at apex typical of some Marchantiopsida. (**b**) Single vertical slit typical of some Marchantiopsida. (**c**) Four valves attached at apex typical of Anthocerotopsida. (**d**) Four valves attached at apex typical of valvate mosses (e.g., *Andreaea:* Bryopsida). (**e**) Operculate dehiscence (e.g., *Polytrichum:* Bryopsida). (**f**) Single helical slit (*Takakia:* Bryopsida). (**g**) Multiple helical slits typical of some stem group vascular plants in Rhyniophytina (e.g., *Huvenia*). (**h**) Single dehiscence line along one side of sporangium (e.g., *Psilophyton:* Euphyllophytina). (**i**) Reniform sporangium with single distal slit typical of basal Lycophytina.

features of sporangium morphology, and perhaps ontogeny, are potentially useful for defining groups within vascular plants (e.g., dehiscence, annulus, sporangium development in leptosporangiate ferns). A sporangium is not observable (and is therefore inapplicable) in the unicellular sporophytes of the charophycean outgroups where meiotic cell division occurs within the zygospore wall.

We score the sporangia (capsules) of liverworts, hornworts, and mosses as provisionally homologous with the sporangia of vascular plants and treat the presence or absence of sporangia as a binary character: 0 = sporangium absent; 1 = sporangium present. There were two inapplicable cases; none were missing data (see Appendix 2).

3.9 • SPORANGIUM DEHISCENCE Mechanisms of sporangium dehiscence in liverworts generally involve longitudinal splitting of the sporangium into four non-twisted valves. The four-valved condition is thought to be common in the group (and perhaps plesiomorphic) because the lines of splitting coincide with the boundaries between the four major quadrants of the embryo that are established very early in ontogeny (Figure 3.28; Schuster 1966, 585; 1984a, 879). In most liverworts, the sporangium dehisces completely from apex to base, but in some cases, dehiscence occurs only for two-thirds to three-quarters the length of the capsule (Schuster 1966, 591). In a few groups of liverworts, there are no well-marked valves and dehiscence is irregular (e.g., the Fossombroniaceae of the Jungermanniidae, many Marchantiales). *Haplomitrium* is basically two- to four-valved (Schuster 1966, 634), but generally only two dehiscence lines are functional (Schuster 1984b, 938, 1042). In *Monoclea* there is only one functional line of dehiscence (Schuster 1984a, 887; 1984b, 1040), and *Sphaerocarpos* is indehiscent (Schuster 1984b, 1038). Mishler and Churchill (1985b) consider dehiscence into two or four (usually four) regular valves as a synapomorphy of their Calobryales-Metzgeriales-Jungermanniales clade, which character is interpreted as derived relative to the Monocleales, Sphaerocarpales, and Marchantiales. The absence of dehiscence, or the presence of fewer dehiscence lines, in related sister groups is considered to be the plesiomorphic condition.

In hornworts, the most common form of dehiscence is by the gradual splitting and twisting of the capsule along two longitudinal sutures with the valves usually remaining attached at the apex (Figure 3.28; Renzaglia 1978; Schuster 1984c). More rarely, three or four suture lines develop in *Phaeoceros,* and in other genera there is sometimes one line or none at all (Renzaglia 1978). *Notothylas* is generally indehiscent but may split along one or two lines. In indehiscent species of *Notothylas,* two lines of dehiscence form but are nonfunctional (Schuster 1984c, 1086).

In mosses, the most common form of dehiscence is along a suture that encircles the apex of the capsule, resulting in a detachable lid or operculum (Figure 3.28). This mechanism is thought to have been lost in cleistocarpous taxa. Mishler and Churchill (1984) interpret the presence of an operculum as a synapomorphy of their peristomate moss clade, but as independently evolved in *Sphagnum* (resolved as basal among mosses in their analysis). In the two putatively plesiomorphic mosses *Andreaea* and *Andreaeobryum,* there are usually four to six main longitudinal sutures, but one to two, or more, shorter, subsidiary sutures also may occur (Figure 3.28). The valves of the capsule usually remain attached at the apex (Murray 1988). In *Takakia,* dehiscence occurs via a single, longitudinal, helical suture that extends from the sporangium apex to base (Smith and Davison 1993) (Figures 3.23 and 3.28).

Among fossil taxa, details of dehiscence in the early mosslike sporophyte *Sporogonites* are equivocal, but certainly there is no well-defined operculum. Halle (1936a) thought that dehiscence probably occurred through longitudinal sutures, and at least five well-marked longitudinal furrows were present in the midregion of the permineralized capsule wall. If correctly interpreted, this mechanism would be most similar to that in *Andreaea* and *Andreaeobryum* (Figure 3.28). Dehiscence in the nonvascular plant *Aglaophyton* is equivocal, and although D. S. Edwards (1986) considered the possibility of oblique longitudinal splitting of the wall (Remy 1978), he concluded that the features observed may have been taphonomically induced. Similar uncertainty exists about the mechanism of dehiscence in *Rhynia* (D. S. Edwards 1980). Sporangia of the early vascular plants *Huvenia* (Hass and Remy 1991) and *Stockmansella* (Fairon-Demaret 1985, 1986b) dehisce through several oblique longitudinal sutures (Figure 3.28), but the valves remain attached at the apex as in the mosses *Andreaea* and *Andreaeobryum.* In *Horneophyton,* no longitudinal sutures were observed, but an apical slit may exist (Eggert 1974; El-Saadawy and Lacey 1979b). Dehiscence in poorly preserved compression fossils is usually difficult to determine, but in some well-preserved species of *Cooksonia* (e.g., *Cooksonia pertonii:* Edwards, Fanning, and Richardson 1986), no obvious dehiscence feature was observed, whereas in *Cooksonia caledon-*

ica (D. Edwards 1970a), dehiscence is clearly similar to that in zosterophylls.

Dehiscence in zosterophylls and lycopsids is characterized by a single suture that passes around the entire distal rim of the sporangium, producing, in most cases, two equal valves (Figure 3.28). In those trimerophytes in which dehiscence has been described, such as *Psilophyton dawsonii* (Banks, Leclercq, and Hueber 1975) and *Psilophyton crenulatum* (Doran 1980), there is one longitudinal suture from apex to base along one side of the sporangium (Figure 3.28). A similar type of dehiscence is characteristic of a group of taxa including *Rhacophyton* (Andrews and Phillips 1968), *Tetraxylopteris* (Bonamo and Banks 1967), and probably also *Pseudosporochnus* (Leclercq and Banks 1962), all of which are probably closely related to the trimerophytes.

Dehiscence of the sporangium of embryophytes is a complex character that deserves more detailed analysis in adequately preserved fossil material (Figure 3.28). However, for the purposes of this analysis, we draw a distinction between the general condition in embryophytes of linear valvate dehiscence, which involves one or more longitudinal slits in the sporangium, and the operculate condition common in many mosses. These two types of dehiscence are treated as a binary character. A sporangium is not observable (and is therefore inapplicable) in the unicellular sporophytes of the charophycean outgroups. We regard dehiscence as poorly understood in the fossil *Horneophyton* and score this plant as unknown.

Sporangium dehiscence is scored as a binary character: 0 = linear; 1 = operculate. There were two inapplicable cases; one was missing data (see Appendix 2). See character 4.29 for comparison.

3.10 • COLUMELLA The *columella* is a columnar mass of sterile tissue that develops within the spore mass of the sporangium of hornworts, mosses, and some early fossil polysporangiophytes (Figures 3.19–3.21, 3.29). Most mosses possess an elongate columella that occupies the center of the capsule and extends to the apex. In *Polytrichum,* it is elongate and square in transverse section (Parihar 1977), whereas in *Andreaea* and *Andreaeobryum,* it is massive, rodlike, and lobed in transverse section (Murray 1988). A columella has been documented in the recently discovered sporophyte of *Takakia* (Smith and Davison 1993). In *Sphagnum* the columella is massive and more or less circular in transverse section (Parihar 1977). In *Sphagnum, Andreaea,* and *Andreaeobryum,* the spore sac is dome-shaped (i.e., continuous over the top of the columella), whereas most other mosses possess the presumed derived condition in which the spore sac is cylindrical (i.e., not continuous over the top of the columella: Mishler and Churchill 1984, characters 45 and 46). In hornworts the columella ranges from very small (sometimes absent) in *Notothylas* to an elongate structure extending the length of the sporangium in *Anthoceros.* In *Megaceros* the columella is massive, showing up to 40 cells in transverse section (Renzaglia 1978). In all hornworts the spore sac is continuous over the apex of the columella (Parihar 1977; Schuster 1984c). The columella is absent in liverworts (Schuster 1966, 1984a,b). The masses of longitudinally aligned, elaterlike cells *(elaterophores)* that accompany free elaters in certain members of the Metzgeriales (Schuster 1966) differ developmentally and morphologically from the columella of mosses and hornworts, and we treat them as nonhomologous structures.

Among the relevant fossils, there is some doubt about whether *Sporogonites* possesses a columella. Halle (1916a, 1936a) described a central sterile area within the spore mass inside the sporangium and inferred the presence of a columella, but this interpretation has been criticized by Arber (1921) because no cells were preserved in the sterile central region. However, given the nature of the mineral in which the plant was preserved (pyrite), partial decay may have resulted in the preservation of only the more robust tissue systems (e.g., spore mass, epidermis), as is the case in many pyritized early land plants (Kenrick and Edwards 1988a). The absence of spores in the mineralized central part of the sporangium is better viewed as evidence favoring the presence of a sterile region and there is no doubt from the illustrations that the spore mass is continuous over this columella region, as in hornworts and putatively basal mosses *(Sphagnum, Andreaea, Andreaeobryum)*.

An unequivocal columella is present in the sporangium of the early polysporangiophyte fossil *Horneophyton* (Figure 3.29; Eggert 1974; El-Saadawy and Lacey 1979b; Kidston and Lang 1920a). In this plant the sporangium is branched with a continuous columella and a spore cavity among the branch levels. At the distal end of the sporangium, the spore mass is continuous over the apex of the columella (Eggert 1974; El-Saadawy and Lacey 1979b). The "sterile pad" of tissue that projects a short distance into the spore cavity in such plants as *Rhynia* (D. S. Edwards 1980) and in some extant homosporous lycopsids (Bower 1894; Mishler and Churchill 1985b) is more difficult to interpret. There is no evidence for a well-marked columella in *Aglaophyton* (D. S. Edwards 1986) or in such early vascular plants as *Psilophyton* (Banks, Leclercq, and Hueber 1975) and *Asteroxylon* (Lyon 1964). A columella is not observable (and is therefore inapplicable) in the unicellular sporophytes of the charophycean outgroups.

The presence or absence of a columella is treated as a binary character: 0 = columella absent; 1 = columella present. There were two inapplicable cases; none were missing data (see Appendix 2). See character 4.32 for comparison.

3.11 • ELATERS AND PSEUDOELATERS *Elaters* and *pseudoelaters* are sterile cells that develop in the archesporium of liverworts and hornworts (Figure 3.30). Elaters are thought to have a nutritive function relating to spore development and may also aid in dehiscence and dispersal. Elaters in

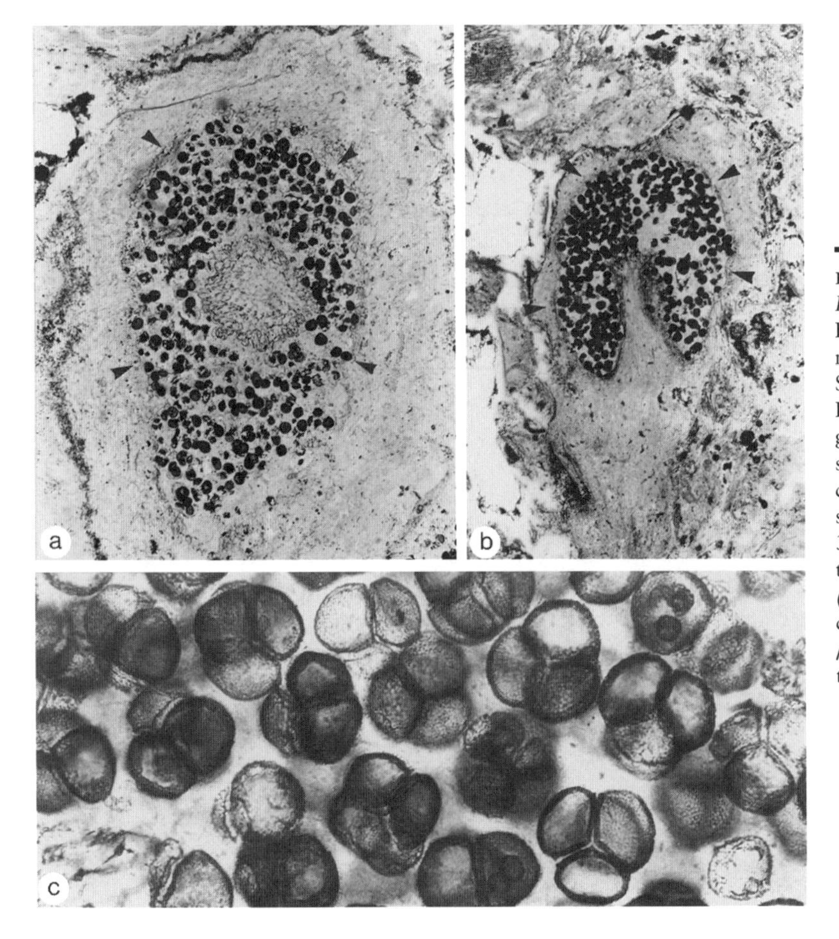

FIGURE 3.29. Sporangium of *Horneophyton* from the lower Devonian Rhynie Chert. Specimen in the collection of the Swedish Museum of Natural History, Stockholm. Photographs by P.K. (**a**) Transverse section showing central sterile columella and surrounding spore sac *(arrows)* (scale bar = 340 μm). (**b**) Longitudinal section showing convex spore sac *(arrows)* overarching central columella (scale bar = 565 μm). (**c**) Spores adhering in tetrads (scale bar = 40 μm).

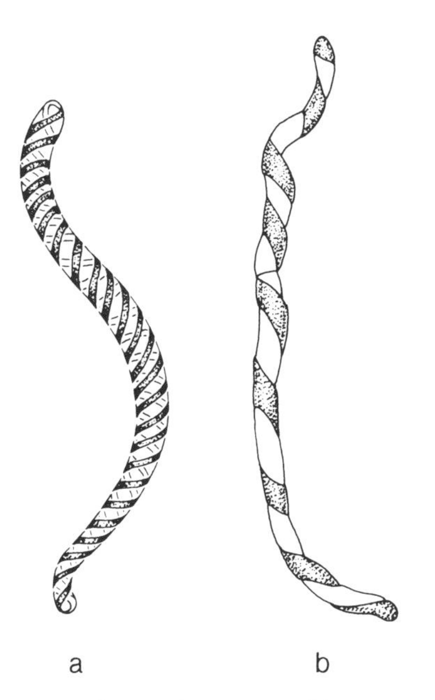

FIGURE 3.30. Elaters in liverworts and hornworts. (**a**) Elater of *Plagiochasma rupestre* (Marchantiopsida) (×330). Redrawn from Schuster 1984b, Figure 90.9. (**b**) Elater of *Dendroceros endiviaefolius* (Anthocerotopsida) (×1,000). Redrawn from Schofield 1985, Figure 19-2E.

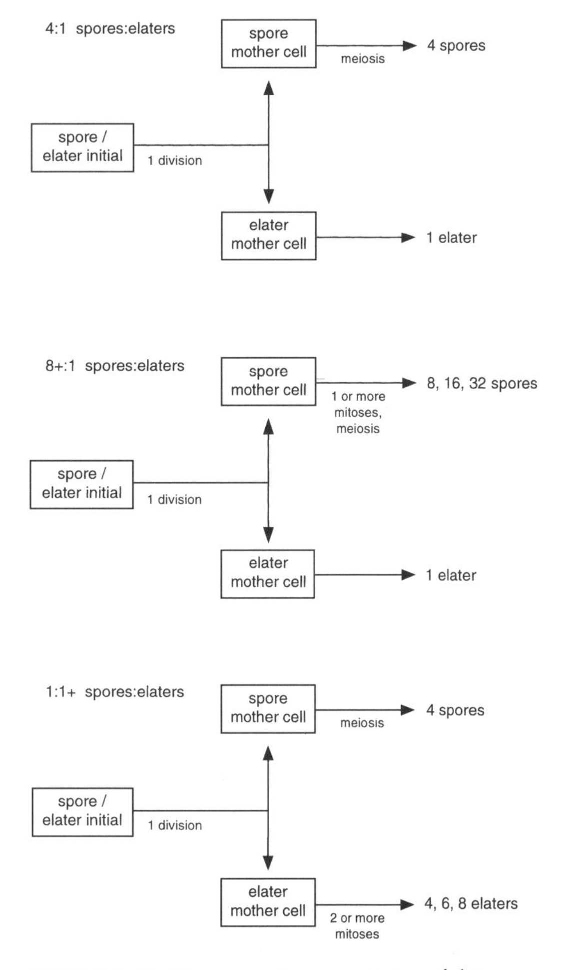

FIGURE 3.31. Diagrammatic representation of elater ontogeny in liverworts and hornworts. Data from Schuster 1966, 168. **Top.** 4:1 ratio of spores to elaters in some Marchantiales (Marchantiidae) and *Dendroceros* (Anthocerotopsida). **Middle.** 8+:1 ratio of spores to elaters in most Marchantiidae and Jungermanniidae. **Bottom.** 1:1+ ratio of spores to elaters in most Anthocerotopsida.

liverworts are one-celled, usually slender, tubular structures. The walls are thin except for one to four helical thickenings (Schuster 1966, 149). In the Ricciaceae the entire archesporium is converted into spores, and elaters are absent. In the Sphaerocarpales, elaters are small, ovoid to spherical, and lack helical thickenings (Schuster 1966, 170). The so-called pseudoelaters of hornworts range from elongate unicellular forms with a single helical thickening in *Dendroceros* and *Megaceros* to multicellular, unthickened branched structures in other genera (Renzaglia 1978, 50).

In both liverworts and hornworts, the production of sterile cells within the spore mass is closely related to the production of spores (Figure 3.31). In the simplest condition, a single archesporial cell divides to give one cell that differentiates into an elater and a second cell (spore mother cell) that undergoes meiosis to produce a tetrad of spores. This division produces a 4:1 ratio of spores to elaters. The same ratio of spores to elaters also occurs in the hornwort *Dendroceros* (Schuster 1966, 169) and in certain marchantialean liverworts (Schuster 1966, 168). The 4:1 ratio was considered the plesiomorphic condition for hornworts and liverworts by Schuster (1966, 170). In most liverworts, the ratio of spores to elaters is increased by the occurrence of one or two mitotic divisions in the daughters of the spore mother cell after the spore-elater division: in *Haplomitrium* the ratio is 12:1, and in *Monoclea* it is 32–36:1 (Schuster 1984a, 888). In hornworts, except for *Dendroceros,* the converse occurs: after the spore-elater division, the spore mother cell undergoes immediate meiosis, while one or more mitotic divisions may take place in the daughters of the elater mother cell; thus the number of elaters may equal or exceed the number of spores. The occurrence of mitotic divisions in the elater daughter cells after the spore-elater division and after the adhesion of these daughter cells into filaments, gives rise to the "septate," multicellular elaters of some hornworts.

The elaters and pseudoelaters of liverworts and hornworts were treated as nonhomologous, independently derived, apomorphic features of those groups by Mishler and Churchill (1984, 1985b) and Mishler, Lewis, et al. (1994) based on "clear differences in development" and morphology. In our view, this interpretation was premature because the morphological differences between the pseudoelaters of some hornworts (absence of thickenings and multicellularity) and the elaters of liverworts are directly related to a small developmental change involving the insertion of one or more mitotic cell divisions in the elater daughter cells after the spore-elater division (Figure 3.31). Furthermore, in the hornwort *Dendroceros*, the 4:1 spore-to-elater ratio characteristic of some liverworts is maintained, and the resulting elaters are also similar in being unicellular with a helical thickening (Figure 3.30). For these reasons we view the differentiation of sterile cells from the cell line leading to spores, which occurs just prior to the production of the spore mother cells, as a significant similarity in position, timing, and development in liverworts and hornworts. Patterns of cell division are identical in some liverworts and hornworts and result in an identical 4:1 spore-to-elater ratio as well as in morphologically similar elaters. The significant differences that do occur in other liverworts and hornworts merely involve the insertion of one or two mitotic cycles—in the daughter cells of the spore mother cell and elater mother cell, respectively—after the spore-elater division and may be apomorphic alterations of a similar basic character in these groups. We provisionally score the elaters in hornworts and liverworts as homologous.

Elaters have not been observed in any mosses or nonvascular or vascular polysporangiophytes. Although derived from sporogenous tissue, the sterile strands of material called *trabeculae* that radiate upwards from the subarchesporial pad in the sporangia of *Isoetes*, and possibly also in the microsporangia of some arborescent lycopsids (Bower 1894), are very different and unlikely to be homologous to elaters on morphological grounds. The spore-trabeculae division occurs much earlier in sporangium ontogeny, and trabeculae are absent in the putatively more basal lycopsids and polysporangiophytes.

The presence or absence of elaters or pseudoelaters is treated as a binary character: 0 = elaters or pseudoelaters absent; 1 = elaters or pseudoelaters present. There were no inapplicable cases; none were missing data (see Appendix 2). See character 4.33 for comparison.

3.12 • ZOOSPORE FLAGELLA Flagellate reproductive cells (asexual zoospores) are probably plesiomorphic for charophycean algae because systematic studies strongly support an origin of this group from unicellular flagellate forms (Bremer 1985; Graham, Delwiche, and Mishler 1991; Mishler and Churchill 1985b). Zoospores are present in the Chlorokybales, *Klebsormidium, Chaetosphaeridium,* and *Coleochaete* and are absent in the Zygnematales, the Charales, and embryophytes. Meiosis is followed directly in the Charales and Zygnematales by growth of new gametophytes without a flagellate stage (Bold and Wynne 1985). In embryophytes, the zoospore homologue is a nonflagellate, sporopollenin-walled, air-borne spore.

The presence or absence of flagella on zoospores is treated as a binary character: 0 = zoospore flagella present; 1 = zoospore flagella absent. There were no inapplicable cases; none were missing data (see Appendix 2).

3.13 • PERINE The *perine* is the normal outer wall layer of the spores of homosporous ferns and lycopsids and is also found in the microspores of the Isoetaceae and Selaginellaceae (Lugardon 1990). A distinct perine layer is present in mosses but is absent in liverworts, hornworts, and *Coleochaete* (Brown and Lemmon 1990b; Graham 1990; Mogensen 1983). It is well established from develop-

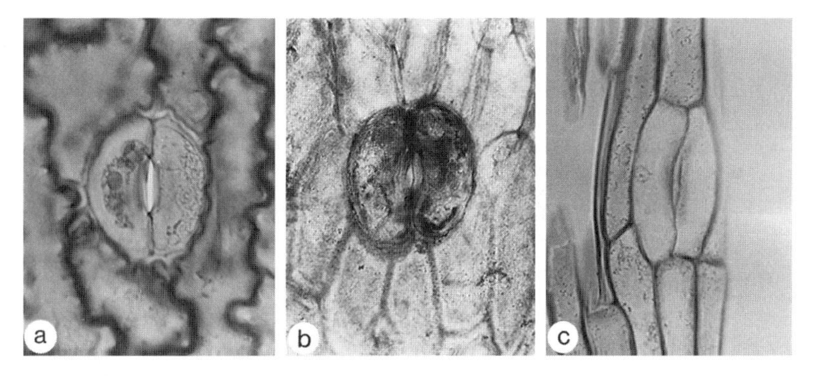

FIGURE 3.32. Stomates in extant Lycopodiaceae, Anthocerotopsida, and extinct protracheophytes. (**a**) *Lycopodium volubile* (Lycopodiaceae) (scale bar = 25 μm). (**b**) Extinct *Aglaophyton (Rhynia) major* (protracheophyte) (scale bar = 60 μm). (**c**) *Anthoceros* sp. (scale bar = 35 μm).

mental studies that the perine of pteridophytes is derived from tapetal material. Little is known about the origin of perine in mosses, although it is probably of extrasporal origin (Brown and Lemmon 1990a; Lugardon 1990). Perine has been identified in such putative basal mosses as *Sphagnum, Andreaeobryum,* and *Andreaea* (Brown and Lemmon 1988), as well as in such potentially more derived taxa as *Polytrichum* (Olesen and Mogensen 1978). In coding this character, we treat the perine of mosses and pteridophytes as potentially homologous, but a more precise definition of how perine deposition is qualitatively different from the deposition of other tapetally derived wall material is desirable. A TEM study of the spores of the fossil *Psilophyton* failed to show a distinctive perine layer but indicated that the remains of an outer ornamented layer could represent a partially developed or very plesiomorphic type of perispore (Gensel and White 1983). This character is unknown (missing) in *Chara,* in the fossil *Horneophyton,* and in *Takakia.*

The presence or absence of a perine is treated as a binary character: 0 = perine absent; 1 = perine present. There were no inapplicable cases; three were missing data (see Appendix 2).

3.14 • CUTICLE Most embryophyte sporophytes possess a cuticle, even though it may be very thin in some bryophytes (Hébant 1977, 13, 86, 110; Lee and Priestley 1924, 530; Paton and Pearce 1957) and possibly absent in some mosses (Watson 1971, 126). The cuticles of vascular plants are composed of biopolyesters and waxes (Kolattukudy 1980), but the chemical composition of functionally similar layers in bryophytes has not been well characterized. Further comparative work on the chemistry and structure of the cuticle in land plants is required. Cuticles are often preserved in the fossil record and have been identified in many polysporangiophyte fossil sporophytes (D. Edwards 1993), including *Psilophyton, Rhynia, Aglaophyton, Nothia* (Edwards, Edwards, and Rayner 1982), *Horneophyton* (Kidston and Lang 1920a), *Sporogonites* (Halle 1936a), and *Protosphagnum* (Jovet-Ast 1967).

In green algae, cuticles are generally regarded as absent, but a specialized surface layer in *Coleochaete* may be functionally similar to the cuticle of land plants. No waxes have been detected in this layer, which is not well characterized chemically (Graham 1993).

The presence or absence of a waxy cuticle is scored as a binary character: 0 = cuticle absent; 1 = cuticle present. There were no inapplicable cases; none were missing data (see Appendix 2).

3.15 • STOMATES A *stomate* consists of two more or less reniform guard cells surrounding a pore that leads to a substomatal chamber (Figure 3.32). Stomates are present in the sporophytes of some hornworts and mosses and of most vascular plants but are completely absent in liverworts. Stomates are structurally quite different from the epidermal pores observed in the gametophytes of some marchantialean liverworts. In *Riccia* and *Ricciocarpus* the pores develop from small vertical apertures that become enlarged during ontogeny as epidermal cells undergo further divisions. The pore may be bordered by four, five, or six cells, and these cells are sometimes further modified through subsequent divisions to form rings of cells, which may be elevated or have thickened walls (Schuster 1984a).

In hornworts, stomates are normally present in *Anthoceros* and *Phaeoceros,* where they are developmentally and structurally identical to those of higher plants (Renzaglia 1978). Stomates are absent or vestigial in the sporophytes of *Notothylas, Dendroceros,* and *Megaceros* (Renzaglia 1978; Schuster 1984c).

Stomates are present in the capsules of most mosses, although there are a significant number of taxa in which stomates are absent (Paton and Pearce 1957). Variation occurs even at the subgeneric level. In *Polytrichum,* there are no stomates in *P. anaum, P. aloides,* and *P. urnigerum,* but *P. alpinum* has a neck immediately below the capsule at the upper end of the seta on which there are over 250 large stomates (Paton and Pearce 1957). In other *Polytrichum* species, there is a groove between the *apophysis* (the swelling at the base of the sporangium) and the body of the sporangium containing long-pored stomates that range in number from about 20 in *P. alpestre* to nearly 200 in *P. formosum.* In some taxa the stomates are confined to the neck region of the sporophyte, and the absence of stomates is sometimes correlated with absence of the neck. Stomates are found in *Sphagnum* (Paton and Pearce 1957) but are absent in the plesiomorphic mosses *Andreaea, Andreaeobryum* (Murray 1988), and *Takakia* (Smith and Davison 1993). Given the variability of this feature across taxa (and even within certain genera), the nonfunctional nature of some stomates, and the small size of the sporophytes, it is likely that the absence of stomates in some mosses and hornworts is due to phylogenetic loss.

No stomates were observed in the Norwegian material of the mosslike fossil *Sporogonites* by Halle (1916a, 1936a), but since the material was examined only in thin sections, stomates would have been difficult to detect and may have been overlooked. They were observed in morphologically similar material from Wales assigned to the same species by Croft and Lang (1942). Stomates have been recorded from many early land plant fossils, including *Aglaophyton* (Kidston and Lang 1917, 1920a), *Cooksonia pertonii* (Edwards, Fanning, and Richardson 1986), *Horneophyton* (El-Saadawy and Lacey 1979b), *Rhynia* (Kidston and Lang 1917, 1920a), *Asteroxylon* (Kidston and Lang 1920b), about nine zosterophylls (see Chapter 5), and also such trimerophytes as *Psilophyton* (Banks, Leclercq, and Hueber 1975). The presence or absence of stomates has not been recorded in many other early fossils because of inadequate preservation.

Stomates have been described in several fossil gametophytes belonging to basal polysporangiophytes from the Rhynie Chert: *Langiophyton, Lyonophyton,* and *Kidstonophyton* (Kenrick 1994; Remy, Gensel, and Hass 1993; Remy and Hass 1991a,b,c,d; Remy and Remy 1980). In many species of hornwort, mucilage-containing clefts occur on the underside of the thallus. These have been variously described as, "mucilage clefts" that "resemble stomates" (Renzaglia 1978, 42), "slime pores or mucilage slits" representing "vestigial stomates" (Parihar 1977, 126), or simply "stomates" that are viewed as homologous with the stomates in the sporophyte generation (Schuster 1984c, 1079). Stomates and any other structures that could potentially be interpreted as stomatal homologues are absent in the gametophytes of mosses (Schofield and Hébant 1984, 639) and liverworts (Schuster 1966). Stomates are not observable (and are therefore inapplicable) in the unicellular sporophytes of the charophycean outgroups.

The presence or absence of stomates is treated as a binary character: 0 = stomates absent; 1 = stomates present. There were two inapplicable cases; none were missing data (see Appendix 2). See character 4.12 for comparison.

3.16 • GAMETOPHYTE FORM Axial gametophytes are found in the majority of basal extant vascular plants including the Lycopodiaceae, Ophioglossaceae, Psilotaceae, Stromatopteridaceae, and various Schizaeaceae, hence this is widely regarded as the plesiomorphic condition (Bierhorst 1971, 78). Gametophytes in some basal ferns, such as the Osmundaceae and Marattiales, are thalloid. Those of *Equisetum* are more difficult to interpret but are generally described as thalloid (Duckett 1973). All hornworts and the liverworts in the Sphaerocarpales, Marchantiales, and Monocleales are characterized by dorsiventral, thalloid gametophytes. Among liverworts, only those of the relatively derived Jungermanniidae (e.g., *Haplomitrium*) are axial and approach radial symmetry (Schuster 1984c). In contrast, the mature gametophytes of all mosses are radial and axial (Mishler and Churchill 1984, 1985b).

Evidence from flora of the lower Devonian Rhynie Chert suggests that plants at the rhyniophyte grade such as *Aglaophyton* and *Horneophyton,* and possibly also plesiomorphic zosterophylls such as *Nothia,* had gametophytes that were axial and therefore most comparable to the gametophytes of mosses and basal extant vascular plants (Kenrick 1994; Remy, Gensel, and Hass 1993). Morphologically complex, radial gametophytes occur also in *Chara,* whereas gametophytes are thalloid or filamentous in *Coleochaete.*

The form of the gametophyte is treated as a binary character: 0 = thalloid gametophyte; 1 = axial gametophyte. There were no inapplicable cases; none were missing data (see Appendix 2).

3.17 • GAMETOPHYTE LEAVES There is a great range in the form of leaflike appendages in liverworts. Ontogenetic and systematic studies suggest that leaves probably evolved several times: once in the Jungermanniales and independently in the Metzgeriales and Calobryales (Crandall-Stotler 1981, 1984; Mishler and Churchill 1985b; Schuster 1966, 1979). In *Haplomitrium,* leaves are noncostate (i.e., lack a midrib), occasionally pluristratose (possessing multiple cell layers) basally, and basically unlobed and lacking teeth (Schuster 1966, 637). *Sphaerocarpos* is described as having two rows of horizontally inserted, non-costate, unistratose "leaves" (Schuster 1984b, 1036), but we follow Mishler and Churchill (1987) and interpret the plant body, including the "leaves," as basically a thalloid structure, as in the closely related form *Monoclea* (Schuster 1984b, 1040).

Leaves in mosses are basically simple in outline although there is great variety in shape. Generally the leaf is unistratose except at the *costa* (midrib), and the leaf apex is acute or acuminate, tapering gradually or abruptly from the body of the lamina (Schofield and Hébant 1984, 630).

The gametophytes of hornworts (Renzaglia 1978), of all extant vascular plants (Bierhorst 1971), and of early fossil polysporangiophytes (Kenrick 1994; Remy, Gensel, and Hass 1993) are leafless. Leaves are also absent in all charophycean algae.

The presence or absence of leaves on the gametophyte is scored as a binary character: 0 = leaves absent; 1 = leaves present. There were no inapplicable cases; none were missing data (see Appendix 2).

3.18 • ARCHEGONIUM Archegonia of similar overall morphology are characteristic of all major embryophyte groups (Smith 1955). In embryophytes, the archegonium is flask-shaped and comprises a sterile outer cellular layer surrounding a basal chamber containing the egg cell (Figure 3.33). The neck of the flask varies greatly in length and in vascular plants usually consists of four cell rows, but more are observed in many mosses (Goebel 1887) and liverworts (Schuster 1984a, 860). At maturity, a central sperm canal leads from the egg to the outer surface of the gametophyte.

In vascular plants and hornworts, the archegonium is sunken within the gametophyte and only the upper part of the neck protrudes from the surface of the plant. Archegonia of similar morphology have recently been observed in early Devonian gametophytes *(Langiophyton)* (380–408 Ma) that probably belong to *Horneophyton lignieri* (Remy and Hass 1991d). In liverworts and mosses, the archegonium is usually sessile or borne on a short, multicellular stalk. In many liverworts, sterile tissue, which takes various forms, surrounds the archegonium. In the Sphaerocarpales the sterile tissue is flask-shaped and surrounds both the antheridium and the archegonium (Schuster 1984a), whereas tubular outgrowths are common in other taxa (e.g., *Monoclea, Pellia, Pallavicinia*). In the

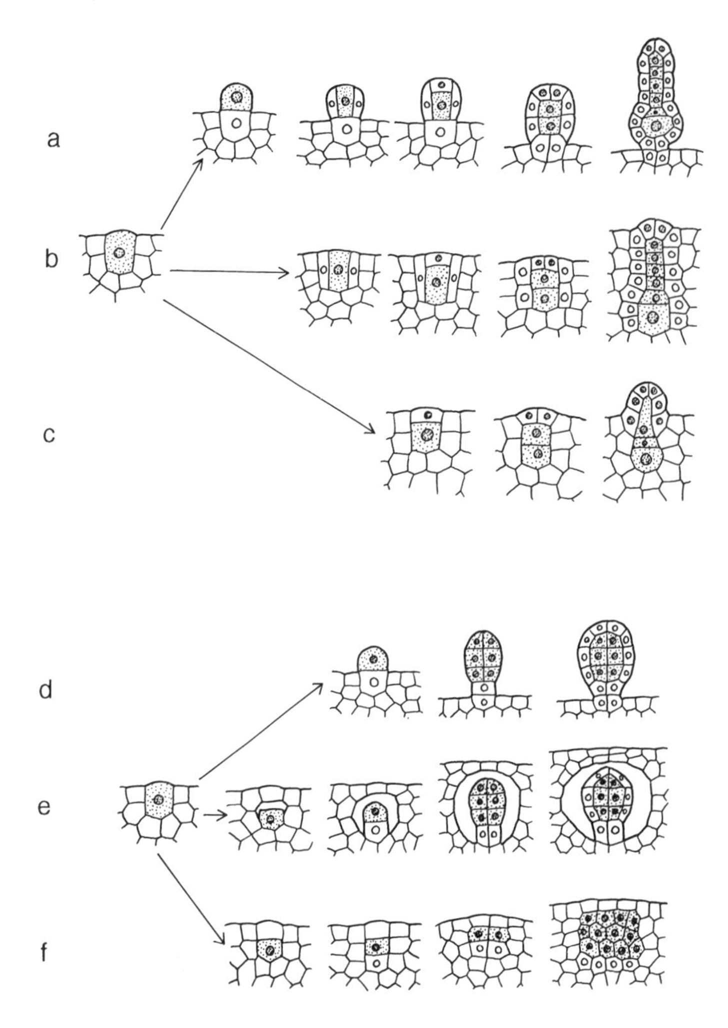

FIGURE 3.33. Simplified diagram of archegonium and antheridium ontogeny in major groups of embryophytes (redrawn from Smith 1938, Figures 68 and 69). Note that many variations in ontogeny exist within these groups; some of these are discussed further in Chapter 7. (**a**) Archegonium of Marchantiopsida. (**b**) Archegonium of Anthocerotopsida. (**c**) Archegonium of Tracheophyta. (**d**) Antheridium of Marchantiopsida. (**e**) Antheridium of Anthocerotopsida. (**f**) Antheridium of Tracheophyta.

Marchantiales the archegonia are sunken and surrounded by dense tufts of scales. Leaflike structures frequently surround the archegonia in the Jungermanniales (Schuster 1984a, 864).

Archegonia develop from a superficial cell in all liverworts and hornworts (Schuster 1984a, 858), mosses (Smith 1955), and vascular plants (Gifford and Foster 1989), and distinctive patterns of cellular cleavage characterize archegonial development in various groups (Figure 3.33).

The oogonia (archegonia analogues) of *Chara* are highly distinctive structures and are the most complex gametangia in green algae (Pickett-Heaps 1975). In *Chara* the cell beneath the archesporial cell divides to produce five sterile cells (jacket cells) that elongate helically around the maturing egg cell to form a sterile sheath that is one cell thick (Grant 1990; Pickett-Heaps 1975). The sterile cells then divide unequally to produce one or two tiers of cells surrounding an apical pore. A calcium carbonate film is secreted by the jacket cells and encases the mature egg cell in a hard protective layer. Calcified oogonia (gyrogonites) of this kind have an extensive fossil record extending to the late Silurian (Pridoli: 414 Ma, Feist and Grambast-Fessard 1991). Structurally, the oogonia of the Charales are quite different from the archegonia of embryophytes. Oogonia in *Coleochaete* are less elaborate and develop from a single, more or less superficial cell. There is no specialized sterile jacket as in *Chara*. Following fertilization, peripheral cells surrounding the young zygote divide, submerging it in the thallus (Graham and Wilcox 1983). Because of these differences, we do not feel justified in scoring the archegonia of embryophytes as homologous with the oogonia of the Charophyceae.

The presence or absence of an archegonium is scored as a binary character: 0 = archegonium absent (oogonium); 1 = archegonium present. There were no inapplicable cases; none were missing data (see Appendix 2).

3.19 • ANTHERIDIUM MORPHOLOGY Antheridia of similar overall morphology are characteristic of all major embryophyte groups. The antheridium comprises a more or less spherical mass of spermatozoids enclosed in a jacket of sterile cells (Figure 3.33).

The antheridia of *Chara* are highly distinctive structures that develop elongate spermatogenous filaments (Pickett-Heaps 1975), which produce biflagellate sperm. Antheridia of *Coleochaete* are smaller and simpler than those of charalean algae. Filamentous species produce sperm in unicellular branches, and thalloid species produce simple, multicellular antheridia with naked spermatogenous tissue (Graham 1993, 186).

Antheridia enclosed in sterile jackets are scored as homologous in all land plants and are treated as fundamentally different to the antheridia with naked spermatogenous tissues in *Coleochaete* and *Chara*. The antheridium of *Horneophyton* has not been observed. The morphology of the antheridium is scored as a binary character: 0 = naked antheridium; 1= antheridium with jacket. There were no inapplicable cases; one was missing data (see Appendix 2).

3.20 • ANTHERIDIUM DEVELOPMENT In all embryophytes, antheridia originate from a single superficial cell, and early antheridium development parallels that of the archegonium in many groups (Figure 3.33; Bierhorst 1971; Gifford and Foster 1989; Schuster 1984a). A unique type of endogenous antheridium development occurs in hornworts and results in one or more fully developed antheridia contained within chambers in the thallus (Schuster 1984a, 859; 1984c, 1082). A periclinal division of a superficial cell gives an inner antheridium initial and a sterile outer "cover" cell. Unlike in other embryophytes, these stalked antheridia differentiate from the inner antheridium initial and within a cavity that develops through divisions in neighboring thallus cells. Interestingly, exogenous antheridium development (which is usual in bryophytes) occasionally occurs in hornworts when antheridium ontogeny begins without the initial periclinal division that produces the cover cell (Lampa 1903). In many liverworts, antheridia develop within cavities or are surrounded by other protective structures (e.g., as in *Monoclea*); however, in these cases the antheridium initial is exogenous, and the protective structures develop by the overgrowth of the surrounding tissues.

In lycopsids, antheridia are embedded individually within the tissues of the gametophyte, whereas antheridia in other vascular plants, such as the Ophioglossaceae, Equisetaceae, Psilotaceae, and leptosporangiate ferns, are sessile on the surface of the gametophyte (Bierhorst 1971). The endogenous antheridia of lycopsids develop from a single surface cell that undergoes a periclinal division. The major difference from antheridium development in hornworts is that in lycopsids the inner cell differentiates only spermatogenous tissues, whereas the outer cell forms the antheridium wall. We do not treat this condition as homologous with cavity development in hornworts, in which several stalked antheridia develop freely within a space in the thallus. The antheridia of *Horneophyton* have not been observed.

The type of antheridium development is treated as a binary character: 0 = exogenous development; 1 = endogenous development. There were no inapplicable cases; one was missing data (see Appendix 2).

3.21 • ARCHEGONIUM POSITION The archegonia of all liverworts and mosses are either superficial and sessile or are borne on short stalks (Bower 1935; Schuster 1984a). In all liverworts the archegonium develops from a protruding superficial cell that divides to form a short filament, which then differentiates into the archegonium (Schuster 1984a, 858). Archegonium development in hornworts and extant vascular plants can be traced to a superficial cell that develops within the thallus so that only the archegonial neck protrudes (Bierhorst 1971; Renzaglia 1978; Schuster 1984a); otherwise, in hornworts, archegonium ontogeny is similar to that in other bryophytes. Archegonia in the early Devonian fossil *Horneophyton lignieri* (Remy and Hass 1991d) are in a similar position to those in hornworts and vascular plants. We treat the position of the female reproductive organs (oogonia) in *Coleochaete* and *Chara* as potentially homologous to that in land plants and score them as basically superficial. In some *Coleochaete* species, submersion of the zygote within the thallus can occur through division of the surrounding cells after fertilization.

Archegonium position is treated as a binary character: 0 = archegonia superficial; 1 = archegonia sunken. There were no inapplicable cases; none were missing data (see Appendix 2).

3.22 • GAMETANGIA DISTRIBUTION Gametangia distribution is complex and variable in embryophytes, and character recognition is problematic because positional criteria are difficult to define across groups with such substantial variations in morphology. Nevertheless, we have attempted to incorporate gametangial distribution as a character by determining the position of the gametangium relative to the apical meristem or apical cell.

In mosses, gametangia are basically terminal, generally on the main stem or on a reduced branch (Schofield 1985). In *Andreaea, Andreaeobryum* (Murray 1988, 317), and *Takakia* (Davison, Smith, and McFarland 1989; Hattori and Mizutani 1958; Murray 1988; Smith and Davison 1993), the archegonia are sometimes displaced laterally by continued apical growth. In *Polytrichum* the gametangia are terminal, borne on an expanded cuplike structure surrounded by leaves; renewed apical growth the following year produces a new length of stem and a new group of terminal gametangia (Schofield 1985). In *Sphagnum,* archegonia terminate the branches, and no further apical growth occurs (Schofield 1985). The antheridia of *Sphagnum* occur in the leaf axils and may extend a short distance along the stem from the apical meristem (Schofield 1985). In hornworts, the gametangia are not terminal but are scattered along the midline of the thallus (Renzaglia 1978, 45; Schuster 1984c, 1992).

Gametangia distribution shows the greatest variability in liverworts. In the Calobryales, the gametangia are axillary to the leaves (Schuster 1966, 539). These "fertile" leaves are clustered near the apex in *Calobryum* (Schuster 1966, 634) but can be more scattered in *Haplomitrium* (Schuster 1966, 637). Archegonia occur terminally, but in the absence of fertilization, apical growth may continue (Schuster 1966, 629). In all leafy Jungermanniales the antheridia are also axillary to leaves and are normally confined to specific regions of the plant (Schuster 1966, 539, 644). The archegonia of the Jungermanniales are strictly terminal on main, lateral, or ventral branches (Schuster 1966, 643). In the Metzgeriales the antheridia and archegonia are superficial and often mixed in acropetal succession along one side of the midrib (Schuster 1966, 539). In such genera as *Pallavicinia, Moerckia,* and *Treubia,* the antheridia are similarly distributed but occur in the axils of small scales. In *Pellia, Noteroclada, Blasia,* and *Verdoornia,* antheridia are dorsal on the thallus and occur in acropetal succession in several irregular lines. In these genera, thallus tissue grows up and around each antheridium, forming a discrete chamber (Schuster 1966, 540). The antheridia of such taxa as the Aneuraceae, Metzgeriaceae, and Hymenophytaceae occur along the midrib of special sexual branches (Schuster 1966, 540), and the gametangia of the Sphaerocarpales occur dorsally on the axial region, with the exception of *Riella,* in which the antheridia are on the dorsal winglike outgrowth of the axis. In the Sphaerocarpales, each gametangium develops its own bottle-shaped involucre (Schuster 1984b, 1036). The gametangia of the Monocleales occur dorsally on the axial region—the antheridia develop acropetally in chambers within the thallus, and the archegonia are covered by a dorsal flap of tissue.

In the recently described gametangiophores of early fossil polysporangiophytes, the gametangia are borne terminally on an expanded disc or cup-shaped head (Kenrick 1994; Remy, Gensel, and

Hass 1993; Remy and Hass 1991b,c,d; Remy and Remy 1980). Similarly, the gametangia in the gametophytes of the Lycopodiaceae are basically terminal on an expanded disk-shaped or convoluted gametangiophore (Bierhorst 1971; Bruce 1976b, 1979; Kenrick 1994). In contrast, in other homosporous vascular plants, such as *Equisetum,* the Schizaeaceae, Psilotaceae, Ophioglossales, and Marattiales, the gametangia are not confined to the terminal regions of the plants (Bierhorst 1971).

In charophycean algae with complex gametangia, such as *Chara,* the oogonia and antheridia are scattered along the axes of the plant. In other taxa, such as the "thalloid" species of *Coleochaete,* the gametes develop directly from filaments or superficial cells.

Gametangia distribution is treated as a binary character: 0 = gametangia widely distributed (nonterminal); 1 = gametangia more or less terminal. There were no inapplicable cases; none were missing data (see Appendix 2).

3.23 • PARAPHYSES *Paraphyses* are sterile, multicellular, generally uniseriate, chlorophyllous hairs found among the gametangia of most mosses, including those of the Andreaeopsida (Murray 1988) and of *Polytrichum* (Parihar 1977). Paraphyses are absent in *Sphagnum, Takakia,* and *Buxbaumia* (Mishler and Churchill 1984; Schofield and Hébant 1984), as well as in hornworts (Schuster 1984c) and also in *Equisetum* (Bierhorst 1971, 94). Multicellular paraphyses are present among the gametangia in some Lycopodiaceae, including *Huperzia selago* (Bower 1935, 265), but apparently are absent in heterosporous lycopsids. Paraphyses or paraphyllia often occur in the axils of male bracts in the Jungermanniales and are probably modified slime papillae (Schuster 1984a, 819). The systematic position of these taxa within liverworts suggests that these structures are not homologous with the paraphyses of mosses and lycopsids.

The presence or absence of paraphyses is difficult to score for many vascular plants with free-living, subterranean gametophytes because of their similarity with the septate rhizoids of such taxa as the Psilotaceae, Stromatopteridaceae, Schizaeaceae, Osmundaceae, and some Ophioglossaceae (Bierhorst 1971, 256). In vascular plants, this morphological similarity extends also to location, because many such "multicellular rhizoids" occur among gametangia. Hairs are also found in association with gametangia on the gametophytes of some leptosporangiate ferns (Stokey 1951). Paraphyses are not present in charophycean algae, nor have they been observed in the early polysporangiate fossil *Horneophyton* (Remy and Hass 1991d).

The presence or absence of paraphyses is treated as a binary character: 0 = paraphyses absent; 1 = paraphyses present. There were no inapplicable cases; none were missing data (see Appendix 2).

3.24 • RHIZOID CELLULARITY Multicellular uniseriate or biseriate gametophytic rhizoids with oblique cross walls are characteristic of mosses and occur in almost all genera (Crundwell 1979), including *Andreaea, Andreaeobryum* (Murray 1988), and *Polytrichum* (Parihar 1977), and also in the protonema of *Sphagnum* (Parihar 1977). Rhizoids *(sensu stricto)* are absent in *Takakia* (Murray 1988, 327), but the mucilage pads on the lower shoots may be potentially homologous structures. Rhizoids are also absent in *Haplomitrium* (Schuster 1966, 634). In hornworts (Renzaglia 1978) and liverworts (Schuster 1984b), rhizoids are exclusively unicellular when they are present.

Unicellular rhizoids are common in the free-living gametophytes of vascular plants and may occur in clusters on the exposed portions of the dominantly endosporic megagametophytes of *Selaginella* (Bierhorst 1971). Rhizoids are septate (multicellular) in the gametophytes of the Psilotaceae, Stromatopteridaceae, Schizaeaceae, some Ophioglossaceae, and some Osmundaceae (Bierhorst 1971). Unicellular rhizoids are present on the lower surface of the gametophytes of *Equisetum* (Bierhorst 1971, 94) and *Huperzia* (Engler and Prantl 1902, 570).

Rhizoids are only preserved in the fossil record under exceptional conditions, such as those that generated the Rhynie Chert. Rhizoids are known to be unicellular in the early fossil polysporangiophytes *Aglaophyton, Horneophyton,* and *Rhy-*

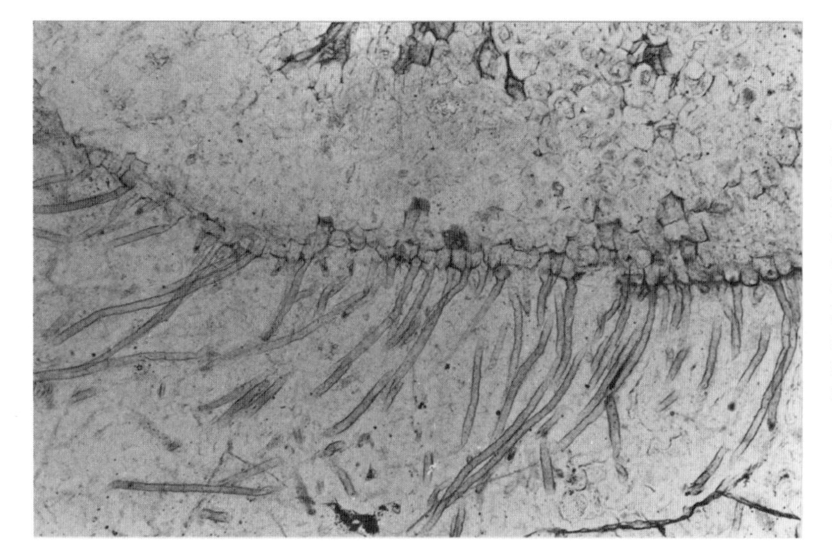

FIGURE 3.34. Unicellular rhizoids emerging from the lower surface of a prostrate axis of a lower Devonian plant (probably *Horneophyton*) in the Rhynie Chert. Scale bar = 240 μm. Specimen from the collection of the Swedish Museum of Natural History, Stockholm. Photograph by P.K.

nia (Kidston and Lang 1920a), but it is unknown whether these are attached to sporophytes or gametophytes (Figure 3.34). Rhizoids appear to be absent in the rhizome of the basal lycopsid *Asteroxylon* from the Rhynie Chert (Kidston and Lang 1920b, 645), but sporophytic, unicellular rhizoids have been observed in the closely related zosterophyll *Serrulacaulis furcatus* (Hueber and Banks 1979).

Potential rhizoid homologues occur in both charophycean outgroups. In the Characeae the entire photosynthetic shoot system is anchored by nongreen, single-celled processes that penetrate the soil (Wood and Imahori 1965). In *Coleochaete* and many other "algae," the unicellular rhizoidlike hairs are thought to deter herbivory or to perform a nutritive function.

The cellularity of rhizoids is treated as a binary character: 0 = unicellular rhizoids; 1 = multicellular rhizoids. There were two inapplicable cases; none were missing data (see Appendix 2).

3.25 • PROTONEMA TYPE The ontogeny of moss gametophytes (except *Sphagnum*) begins with an elaborate persistent, prostrate, branched growth *(protonema)* of elongated cells (Mishler and Churchill 1984; Nehira 1983). In the Andreaeopsida, persistent protonemas with both filamentous and simple, multicellular axial parts occur (Murray 1988), and both appear to be modifications of the basic type of protonema that occurs in other mosses. These modifications perhaps developed in association with the rock-dwelling habit of the Andreaeopsida (Mishler and Churchill 1984). We follow Mishler and Churchill (1984) in treating these two protonemal types as homologous structures. In *Sphagnum,* however, a very different elliptical, thalloid protonema develops from a short filament. The protonema of *Takakia* has not been observed.

In the Anthocerotopsida, protonemas comprise globose masses of cells and develop directly from the germinating spore (Renzaglia 1978). This situation is identical to that in the liverwort *Haplomitrium* and similar to that in many other liverworts (Nehira 1983). In *Sphaerocarpos* the germ tube develops into a filamentous protonema consisting of a few cells; a cylindrical thallus then develops from the apex. No protonemal stage appears to be present in *Monoclea* (Nehira 1983), and an apical cell develops directly.

In many vascular plants, at the stage in the life cycle in which bryophytes produce a protonema, a few early cell divisions produce an undifferentiated globular mass of cells before the development of the apical cell and of the gametophyte thallus (Bierhorst 1971). Globular masses consisting of a few cells have been observed in *Equisetum* (Duckett 1973; Gifford and Foster 1989) and the Lycopodiaceae (Bierhorst 1971).

No protonemal stage is recognizable in *Coleo-*

chaete, but a well-differentiated juvenile thallus of limited growth occurs in *Chara* (Bold and Wynne 1985). The protonema of *Takakia* and the fossil *Horneophyton* are currently unknown.

Despite the wide range of morphologies recognizable during early gametophyte development in embryophytes, the moss protonema is generally sufficiently distinctive to provide an unambiguous character. We provisionally define a binary character that distinguishes the elaborate filamentous protonemas of mosses from the relatively undifferentiated protonemas of other embryophyte groups.

The type of protonema is treated as a binary character: 0 = nonfilamentous or undifferentiated protonema; 1 = elaborate, persistent, and filamentous. There were no inapplicable cases; two were missing data (see Appendix 2).

3.26 • OIL BODIES Membrane-bound intercellular bodies containing diverse oils are present in the gametophytes of more than 90% of liverworts (Schuster 1966, 205). Oil bodies are present in *Haplomitrium* (Schuster 1966, 629), *Monoclea* (Schuster 1984b, 1044), and *Sphaerocarpos* but are not found in green algae, hornworts, mosses, and vascular plants. The moss *Takakia,* placed in the Calobryales by Schuster (1984b), was thought to have oil bodies similar to those in other hepatics. However, a recent TEM study of mature photosynthetic cells showed that liverwort-type oil bodies are absent (Crandall-Stotler 1986). The oil bodies of liverworts appear to be fundamentally different in both ultrastructure and development from the lipid bodies that are found in every other group of land plants as well as in algae (Duckett 1986, 29). Oil bodies are unknown in *Horneophyton.*

The presence or absence of oil bodies is treated as a binary character: 0 = oil bodies absent; 1 = oil bodies present. There were no inapplicable cases; one was missing data (see Appendix 2).

3.27 • MUCILAGE CLEFTS The ventral tissue of the thalli of all genera of the Anthocerotopsida except *Phaeoceros* is characterized by the presence of mucilage cavities (slime clefts or mucilage clefts) that develop acropetally. These cavities open to the exterior via a stomalike pore (Renzaglia 1978; Schuster 1984c). This character was treated as a synapomorphy of the hornwort clade by Mishler and Churchill (1985b, character 4).

The presence or absence of mucilage clefts is scored as a binary character: 0 = mucilage clefts absent; 1 = mucilage clefts present. There were no inapplicable cases; none were missing data (see Appendix 2).

3.28 • PSEUDOPODIUM Both *Andreaea* and *Sphagnum* are unique among embryophytes in possessing a prominent, specialized gametophytic structure on which the sporophyte is exserted—the *pseudopodium.* In *Andreaea* the pseudopodium lengthens rapidly by cell elongation. It is usually naked, but in a few species, it bears unfertilized archegonia or rudimentary leaves. There is no pseudopodium in *Andreaeobryum* (Murray 1988) and *Takakia* (Smith and Davison 1993), although a morphologically similar structure occurs in *Sphagnum* (Parihar 1977; Schofield and Hébant 1984).

The presence or absence of a pseudopodium is treated as a binary character: 0 = pseudopodium absent; 1 = pseudopodium present. There were no inapplicable cases; none were missing data (see Appendix 2).

Characters with Significant Missing Data

3.29 • PYRENOIDS *Pyrenoids* are protein aggregates contained within the chloroplasts of many photoautotrophs. The proteins are composed almost exclusively of ribulose 1,5-bisphosphate carboxylase-oxygenase, which is the principal carbon dioxide–fixing enzyme (Vaughn, Campbell, et al. 1990). Pyrenoids of similar form occur frequently in the green algae, including *Chlamydomonas* and *Chlorella* (Vaughn, Campbell, et al. 1990), in several species of *Coleochaete* (Duckett and Renzaglia 1988b), and in other nongreen "algae" such as *Euglena* (Kiss, Vasconcelos, and Triemer 1986) and *Porphyridium* (McKay and Gibbs 1989). The plastid ultrastructure of hornworts has been intensively studied, and pyrenoids are present in *Phaeoceros, Notothylas, Anthoceros,* and some species of *Dendroceros;* they are absent in *Megaceros* and other *Dendroceros* species (Duckett and Renzaglia

1988b). Pyrenoids are absent in liverworts, mosses, and vascular plants.

This character is problematic because previous general conclusions about its systematic distribution are based on observations in relatively few taxa. In the taxa relevant to this study, the presence of pyrenoids has been confirmed in *Anthoceros, Notothylas* (Vaughn, Campbell, et al. 1990), and *Coleochaete* (Duckett and Renzaglia 1988b). The absence of pyrenoids has been confirmed in *Haplomitrium, Sphaerocarpos, Takakia,* and *Sphagnum* (Duckett and Renzaglia 1988b).

The presence or absence of a pyrenoid is treated as a binary character: 0 = pyrenoids present; 1 = pyrenoids absent. There were no inapplicable cases; eight were missing data (see Appendix 2).

3.30 • BICENTRIOLAR CENTROSOMES Centriolar development in the spermatid mother cells in the antheridium of land plants with biflagellate sperm (i.e., liverworts, hornworts, mosses, Lycopodiaceae, Selaginellaceae) is reported to occur by midpoint separation of a coaxial bicentriolar centrosome followed by rotation of the centrioles to form a parallel pair (Graham and Repavich 1989; Renzaglia and Duckett 1988). In contrast, orthogonal centriolar replication, a process in which a new centriole arises from the side of a parental centriole at its base, is thought to be common among protists and occurs in *Chara, Nitella,* and *Coleochaete* (Graham and Repavich 1989). In land plants with multiflagellate spermatozoids (e.g., *Equisetum*), centrioles originate around the periphery of the blepharoplast (Garbary, Renzaglia, and Duckett 1993). Character scores are based on the data provided by Graham and Repavich (1989), Garbary, Renzaglia, and Duckett (1993) and Renzaglia and Duckett (1991).

This character is problematic because general conclusions about its distribution are founded on observations from relatively few taxa. Centriolar development has not been observed in *Monoclea, Andreaeobryum, Huperzia,* and the fossil *Horneophyton.*

The developmental origin of bicentriolar centrosomes is treated as an unordered multistate character: 0 = orthogonal centriolar replication; 1 = coaxial bicentriolar replication; 2 = peripheral to blepharoplast. There were no inapplicable cases; four were missing data (see Appendix 2).

3.31 • ORIENTATION OF LAMELLAE IN MULTILAYERED STRUCTURES One of the major differences between the multilayered structures (MLS) in the sperm of charophycean algae and embryophytes is the angle made between the long axis of the S_1 microtubules and that of the underlying S_2 lamellae. This angle is reported as 40–45° in embryophytes and as 90° in charophycean algae (Garbary, Renzaglia, and Duckett 1993; Graham and Repavich 1989; Sluiman 1985).

This character is problematic because general conclusions about its distribution are based on observations from relatively few taxa. Character scores are based on the data provided by Sluiman (1983), Garbary, Renzaglia, and Duckett (1993) and Renzaglia and Duckett (1991). The MLS has not been observed in *Monoclea, Andreaeobryum, Huperzia, Anthoceros,* and the fossil *Horneophyton.*

The orientation of the MLS lamellae is treated as a binary character: 0 = angle of 90°; 1 = angle of 40–45°. There were no inapplicable cases; five were missing data (see Appendix 2).

3.32 • PREPROPHASE BANDS IN MITOSIS *Preprophase bands* (PPBs) are a transitory array of microtubules marking the site where the new cell plate will join the parental walls in a mitotically dividing cell. The PPB is a characteristic feature of the cell cycle in higher plants; it has been reported in the three major clades of bryophytes as well as in vascular plants.

This character is problematic because general conclusions about its distribution are based on observations from relatively few taxa. In the groups analyzed here, the PPB has been observed in *Selaginella* and *Isoetes,* some ferns and angiosperms, *Sphagnum, Notothylas, Phaeoceros,* and *Marchantia.* PPBs are absent in green algae, including *Coleochaete* and *Chara* (Brown and Lemmon 1990a).

The presence or absence of PPBs in mitosis is treated as a binary character: 0 = PPBs absent; 1 = PPBs present. There were no inapplicable cases; 11 were missing data (see Appendix 2).

3.33 • LUNULARIC ACID Lunularic acid appears to be restricted to liverworts (Gorham 1977; Suire and Asakawa 1979) and was detected in all 76 liverwort taxa examined by Gorham (1977); these spanned the Marchantiales, Sphaerocarpales *(Sphaerocarpos),* Monocleales *(Monoclea),* Calobryales *(Haplomitrium),* Metzgeriales, and Jungermanniales. However, lunularic acid was absent in the hornworts *Anthoceros* and *Phaeoceros,* as well as in all seven mosses examined in the same study (including *Sphagnum* and *Polytrichum*). Lunularic acid was also absent in three species of the Ulvobionta *(Chlorella, Ulva, Sargassum),* one species each of the Cyanophyceae and Xanthophyceae, two ferns, and two lichens. Similarly, Pryce (1972) found that lunularic acid was not present in *Selaginella, Equisetum,* and *Polypodium.* The absence of lunularic acid in green algae is uncertain because of conflicting results (Gorham 1977; Pryce 1972). To our knowledge, the presence or absence of lunularic acid in *Coleochaete* and *Chara* has not been evaluated.

The presence or absence of lunularic acid is treated as a binary character: 0 = lunularic acid absent; 1 = lunularic acid present. There were no inapplicable cases; eight were missing data (see Appendix 2).

3.34 • D-METHIONINE The use of different metabolic pathways in the metabolism of the L and D isomers of amino acids in plants appears to characterize certain higher-level taxonomic groups (Markham and Porter 1978, 258; Pokorny 1974; Pokorny, Marcenko, and Keglevic 1970; Suire and Asakawa 1979). Pokorny, Marcenko, and Keglevic (1970) and Pokorny (1974) surveyed the metabolic pathways for the L and D isomers of the amino acid methionine in over 130 taxa including 22 angiosperms, 13 gymnosperms, 9 ferns, *Equisetum,* 3 lycopsids, 27 mosses, 10 liverworts, 4 hornworts, 6 green algae, and 35 other species of algae, fungi, lichens, and bacteria. In bacteria, algae, and liverworts, the metabolic pathways of the L and D isomers are qualitatively identical and involve a deamination reaction. In the vascular plants, mosses (except *Mnium*), and hornworts surveyed, the D isomers of methionine are distinguished from its L isomers and are acylated to either *N*-acetyl or *N*-malonyl derivatives. Likewise, the 11 fungi surveyed were able to distinguish D-methionine. Surveyed taxa relevant to this study include *Sphaerocarpos, Anthoceros, Sphagnum, Andreaea, Polytrichum,* and *Lycopodium.* The ability of fungi to distinguish D-methionine should be interpreted as a parallelism or convergence, whereas the inability of the moss *Mnium* to distinguish the L and D isomers seems likely to be a loss. To our knowledge, this metabolic character has not been assayed in *Coleochaete* and *Chara.*

The ability to distinguish D-methionine is treated as a binary character: 0 = L and D isomers of methionine not distinguished; 1 = L and D isomers of methionine distinguished. There were no inapplicable cases; eight were missing data (see Appendix 2).

Other Potentially Relevant Characters

Many other characters were considered potentially relevant to resolving relationships at the general embryophyte level but were not included in this analysis. Some of these characters are discussed below, along with our reasons for not including them.

3.35 • ORIENTATION OF THE FIRST DIVISION OF THE ZYGOTE The orientation of the first division of the zygote with respect to the neck of the archegonium is transverse in most embryophytes. In hornworts, the first division is vertical but in some instances, transverse divisions have been recorded (see Schuster 1984c, 1083). In vascular plants, there are intermediate situations between strictly vertical and strictly horizontal (Bierhorst 1971). In *Equisetum,* both transverse and oblique first divisions have been recorded in the zygote (Gifford and Foster 1989, 196), whereas in lycopsids the first division is transverse in the Lycopodiaceae and Selaginellaceae and oblique in the Isoetaceae (Gifford and Foster 1989). The orientation of the first division (of meiosis in this case) is unobservable in the charophycean outgroups because an archegonium is not present. The vertical first division of the zygote appears to be an autapomorphy of hornworts and is therefore uninformative about relationships among other embryophyte groups. This character was not included in our analysis because the condition is equivocal in many vascular plants.

3.36 • PLACENTAL TRANSFER CELLS The junction between the sporophyte foot and the surrounding tissue of the gametophyte *(vaginula)* is termed the *placenta.* Distinctive cells with internal labyrinthine walls *(transfer cells)* occur on either side, or both sides, of the placenta in bryophytes. In polytrichaceous mosses (Hébant 1977; Ligrone and Gambardella 1988a), *Takakia* (Renzaglia, Smith, et al. 1991) and *Andreaea* (Ligrone and Gambardella 1988a; Murray 1988), transfer cells occur on the sporophyte side of the junction, whereas in the Buxbaumiidae and Eubryideae, transfer cells are found on both sides of the placenta (Ligrone and Gambardella 1988a). *Sphagnum* differs from other mosses that have been examined in that morphologically differentiated transfer cells have not been observed in either gametophyte or sporophyte tissues (Ligrone and Renzaglia 1989). In those hornworts that have been investigated *(Phaeoceros laevis, Anthoceros punctatus, Dendroceros tubercularis),* transfer cells have been found on the gametophyte side of the placenta only (Ligrone and Renzaglia 1990). Ultrastructural details in the placental region of liverworts are available only for some members of the Sphaerocarpales, Marchantiales, and Metzgeriales. In these groups, transfer cells have been observed on both sides of the placenta (Ligrone and Gambardella 1988b). In the green algae, cells with labyrinthine walls very similar to those in the transfer cells of embryophytes have been observed only in the gametophytic tissue surrounding the zygotes of *Coleochaete orbicularis* (Delwiche, Graham, and Thomson 1989; Graham and Wilcox 1983); similar cells were not observed in two other species of *Coleochaete—C. soluta* and *C. scutata.* Transfer cells have been reported in the placenta of some ferns (Gunning and Pate 1969, 1974). The morphology and position of placental cells in embryophytes may provide several useful characters at different hierarchical levels, but more detailed documentation of this feature in a broader range of extant polysporangiophytes is desirable. This information was not used in our analysis because more detailed comparative work is required to identify potential homologies in different embryophyte groups and in green algae.

3.37 • TRIPARTITE LAMELLAE During spore development, the first phase of activity leading to the permanent exine is the production of tripartite lamellae (TPL) external to the spore cytoplasm. The TPL are associated with the initiation of exine formation in the spores of most hepatics, mosses, and pteridophytes, and also in the pollen grains of seed plants (Brown and Lemmon 1986, 1990b). The TPL have not been reported in green algae but occur in all liverworts investigated except *Corsinia* (Brown and Lemmon 1990b). TPL have been identified in *Haplomitrium* (Brown and Lemmon 1986), although *Monoclea* and *Sphaerocarpos* have not yet been investigated with respect to this feature. TPL are absent in the hornworts *Anthoceros, Notothylas,* and *Dendroceros* and also in the basal mosses *Andreaea* and *Andreaeobryum* (Brown and Lemmon 1988). TPL have been observed in *Sphagnum* (Brown and Lemmon 1988) and *Polytrichum* (Olesen and Mogensen 1978) and appear to be characteristic of mosses in general (Brown and Lemmon 1988, 1990b). Among extant pteridophytes, TPL have been recorded in *Lycopodium clavatum* (Lugardon 1990; Uehara and Kurita 1991). This character was not included in our analysis because a greater breadth of sampling is necessary, and the absence of TPL from some hornworts, mosses, and liverworts suggests that the distribution of this character is complex.

3.38 • INTERNAL DIFFERENTIATION Axis anatomy in leafy hepatics is always extremely simple (Schuster 1984a, 776). In erect taxa, the cells of the cortex can be very thick-walled, and in some taxa, relatively large clear and translucent (hyaline) cells may develop in the cortex. In the Sphaerocarpales the gametophyte consists of a simple parenchymatous axis showing no internal differentiation (Schuster 1984a, 823). Similarly, in the Monocleales, there is little internal differentiation in the thallus (Schuster 1984a, 826). In the Marchantiales the thallus has a more complex internal structure: a unistratose (one cell thick) dorsal epidermis overlies chlorophyllose aerenchyma, which in turn overlies a ventral parenchymatous zone with poorly defined epidermis (Schuster 1984a, 827). A central strand of water-conducting cells (*sensu* Hébant 1977, cells that are dead and empty at maturity as a result of lysosomal breakdown of protoplast late in development) does not normally occur in liverworts. Only a few leafy liverworts in the genus *Haplomitrium* and a few thalloid ones in the Metzgeriales possess water-conducting cells, although elongate conducting parenchyma may be more common (Hébant 1977). In *Haplomitrium* and *Moerckia,* water-conducting cells retain the general appearance of parenchyma cells of the cortex, but in *Pallavicinia, Symphogyna,* and *Hymenophyton,* the conducting cells are elongate and have tapering ends (Hébant 1977, 26).

In hornworts, cellular differentiation within the thallus is negligible. An epidermis of photosynthetic cells overlies an undifferentiated parenchymatous interior (Renzaglia 1978; Schuster 1984c).

Anatomically, moss gametophytes generally are more highly differentiated than those of liverworts or hornworts. A typical moss stem comprises an epidermis having a thin cuticle but no stomates, an outer cortex of *stereids* (supporting cells with thickened walls), and an inner cortex of thin-walled conducting parenchyma. In many species there is also a central strand of water-conducting cells (Hébant 1977; Schofield and Hébant 1984). Important variations occur in the relative development of the central strand (particularly in the leptome), in the presence or absence of leaf traces, and in the degree of development of other tissues. In the Andreaeopsida, stems consist of more or less uniform cells with thicker-walled outer cortical cells and a collenchymatous inner cortex (a cortical cylinder forming stereids with age). Water-conducting cells are absent in the center, but conducting parenchyma is present (Hébant 1977; Murray 1988). In *Takakia,* there are one or two thick-walled cortical cells and a central strand of elongate water-conducting cells with transverse or slightly oblique end walls (Hattori and Mizutani 1958; Hébant 1977; Schuster 1984b).

In *Sphagnum,* the outer cortex consists of thin-walled cells or two or three layers of hyaline cells that are dead at maturity. As in the leaves, the hyaline cells are strengthened by various types of helical thickening (Hébant 1977). The cortex surrounds a layer of narrow, thick-walled, elongate cells with deeply pigmented walls. There is a central region of thin-walled elongate parenchymatous cells (Parihar 1977), and large pores or simple pits have been observed in the cell walls of all tissue systems (Baker 1988).

In *Polytrichum,* cortical cells containing starch underlie a poorly defined epidermis. The peripheral cortex is composed of elongate tapering cells that gradually merge into more compact parenchyma. Inside the cortex is a rudimentary pericycle surrounding a zone of *leptoids* (phloemlike cells). To the interior is the *hydrome sheath,* which consists of one or two layers of cells with dark brown, suberized walls. The central hydrome contains water-conducting tissue consisting of a sheath of thin-walled cells and a central cylinder of thick-walled cells. Leaf traces may be observed throughout the axis (Hébant 1977; Tansley and Chick 1901).

Evidence from the fossil record suggests that such early polysporangiophyte sporophytes as *Aglaophyton* and *Horneophyton,* and possibly also such basal zosterophylls as *Nothia,* had gametophytes that were radial and axial and that these gametophytes possessed morphological and anatomical complexity comparable to that of their corresponding sporophytes as well as the gametophytes of extant mosses (Kenrick 1994; Remy, Gensel, and Hass 1993; Remy and Hass 1991b,c,d; Remy and Remy 1980). In contrast, the gametophyte generation in extant lycopsids and ferns is very small and simple, and there is little or no internal anatomical differentiation (Bierhorst 1971; Gifford and Foster 1989). Anatomical complexity in the gametophytes of mosses and early polysporangiophyte fossils requires further comparative study and may provide several additional characters.

3.39 • MUCILAGE HAIRS Nearly all liverworts possess mucilage-producing hairs or papillae that are often initiated in a regular pattern close to the growing apex (Schuster 1966, 506; 1984a, 801). Mucilage papillae are absent in hornworts (Schuster 1984a, 802). Two types of mucilage papillae occur in *Takakia* (Proskauer 1962; Schuster 1966): stalked axillary papillae, which are "closed" and comparable to those of the Jungermanniales (mucilage is secreted through the cell wall), and beaked "open" papillae, which are probably not homologous with those of other hepatics (Schuster 1984a, 802). Stalked, hairlike, axillary mucilage papillae occur in *Andreaea* and are typically mosslike (Murray 1988, 315). In the very young leaves of *Sphagnum,* there is usually one axillary hair that disappears in the mature leaf (Parihar 1977, 173). The mucilage papillae of *Andreaeobryum* are very similar to those of *Takakia* and differ in position and form from those reported in other mosses. There is also a terminal beak or papilla through which mucilage is extruded (Murray 1988, 315). Mucilage hairs were not used as a character in our analysis because the great variety in form, function, and position of these structures in embryophytes has not been surveyed adequately. Homology assignments among the different major groups based on current knowledge would be controversial.

3.40 • SPORANGIUM SHAPE The capsule of liverworts is radially symmetrical and varies in shape from cylindrical through ovoid to spherical (Schuster 1966). Among the Marchantiidae (*sensu* Schuster; Table 3.2), shape varies: ellipsoidal to cylindrical in *Monoclea,* ellipsoidal in *Lunularia,* obovoid in *Marchantia* and *Conocephalum,* and more or less spherical in *Sphaerocarpos, Reboulia,* and *Asterella* (Schuster 1984a). A similar range of variation is observable in the Jungermanniidae (*sensu* Schuster), from spherical types such as *Pellia* to cylindrical types like *Haplomitrium.* Capsule shape is an important systematic character at the family level in liverworts, and the most common type is ovoid with several (often four) straight, approximately vertical lines of dehiscence (Schuster 1966).

The capsule of hornworts is radially symmetrical and cylindrical. Its great length in all taxa except *Notothylas* results from continuous meristematic activity at the capsule base. In *Notothylas,* capsule growth begins normally but ceases early in the growing season (Renzaglia 1978; Schuster 1984c).

The capsule of mosses is radially symmetrical and a great variety of forms is observable in the Tetraphidales, Polytrichales, Buxbaumiales, and Bryales (Watson 1971). Capsules range from basically spherical to cylindrical, having a range of modifications that reflect structures associated with the dehiscence mechanism (e.g., peristomate mosses). In the putatively more basal mosses, *Takakia* (Smith and Davison 1993), *Andreaea,* and *Andreaeobryum* (Murray 1988), the capsule is elliptical to fusiform, but in *Sphagnum* it is spherical (Watson 1971). In the mosslike fossils *Sporogonites* (Halle 1936a) and *Tortilicaulis* (D. Edwards 1979), capsule form is elliptical to fusiform. In the early polysporangiophytes the sporangia are radially symmetrical and commonly more or less fusiform, for example, in *Aglaophyton* (D. S. Edwards 1986) and in such early vascular plants as *Rhynia* (D. S. Edwards 1980; Kidston and Lang 1917, 1920a) and *Psilophyton* (Banks, Leclercq, and Hueber 1975). The sporangia of *Horneophyton* (Eggert 1974; El-Saadawy and Lacey 1979b) and *Caia* (Fanning, Edwards, and Richardson 1990) are unique among embryophytes in that they are branched and otherwise barely distinguishable from sterile axes. A markedly different form of sporangium occurs in lycopsids and zosterophylls—in these groups, sporangia are usually dorsiventrally flattened and approximately reniform. This type of sporangium is typical in the Lycopodiaceae (Bierhorst 1971) and in early fossils such as *Asteroxylon* (Lyon 1964), *Nothia* (El-Saadawy and Lacey 1979a), *Renalia* (Gensel 1976), *Zosterophyllum llanoveranum* (D. Edwards 1969a), and *Trichopherophyton* (Lyon and Edwards 1991; see Chapters 5 and 6).

Sporangium shape is a complex feature in basal

embryophytes that is difficult to define because of an apparently continuous range of form. Sporangium shape was not used as a character at this level in our analysis, but it is useful at lower taxonomic levels (see characters 4.27 and 4.28).

3.41 • SPORANGIUM WALL THICKNESS Sporangia in all members of the Marchantiidae (*sensu* Schuster; Table 3.2) and Calobryales have walls that are one cell thick, whereas the more derived Metzgeriales and Jungermanniales have capsule walls that are from two to ten cell layers thick (Schuster 1984a,b). The capsule wall is usually formed from thin-walled cells, which are stiffened by the development of various types of localized thickenings (Schuster 1966). In the hornworts, the sporangium wall is from four to six cells thick, with the outer layer forming a distinct epidermis, which in *Phaeoceros* and *Anthoceros* contains stomates (Renzaglia 1978; Schuster 1984c). The capsule wall of mosses consists of several layers of cells covered by an epidermis, which may contain either functional or nonfunctional stomates. In *Andreaea* and *Andreaeobryum* the wall contains from five to ten layers of cells (Murray 1988), in *Sphagnum* it contains from four to six layers (Parihar 1977), and in *Polytrichum* the wall is several layers thick. Details of the sporangium wall in the recently discovered sporophyte of *Takakia* remain to be described.

In the mosslike fossil plant *Sporogonites,* the capsule wall is thick and consists of several layers of cells (at least two or three—probably more), including a thick-walled epidermis (Halle 1936a). In such early fossil polysporangiophytes as *Aglaophyton* (D. S. Edwards 1986), *Horneophyton* (Eggert 1974), and *Nothia* (El-Saadawy and Lacey 1979a), and also in such early fossil vascular plants as *Rhynia* (D. S. Edwards 1980) and *Psilophyton* (Banks, Leclercq, and Hueber 1975), the sporangia are clearly several layers thick. Likewise, among extant vascular plants, the sporangia of basal pteridophytes are also several layers thick. In *Huperzia selago,* for example, the sporangium is three or four cell layers thick (Bower 1894), and the thick-walled sporangia of eusporangiate ferns have been extensively documented (e.g., Bierhorst 1971).

In their analysis of relationships within liverworts, Mishler and Churchill (1985b) noted that the polarity of this character (their character 19) was problematic because green algae do not form capsules, and hornworts, mosses, and vascular plants have multistratose sporangium walls. Nevertheless, they argued that the unistratose condition was plesiomorphic in liverworts, based on a functional outgroup comparison within liverworts and on a hypothetical transformation series from an undefined capsule wall to a single-layered wall to a multilayered wall (which they consider a priori the most likely scenario). This interpretation leaves open the possibility that the multistratose sporangium walls in hornworts, mosses, and vascular plants are homologous and implies that they evolved independently of multilayered walls in the Jungermanniales and Metzgeriales.

We did not use sporangium wall thickness as a character at this level in the analysis because it is variable, the discrete states are difficult to define, and the feature has not yet been adequately surveyed.

3.42 • NONSYNCHRONOUS SPORE PRODUCTION The nonsynchronous production of spores was listed as a hornwort synapomorphy by Mishler and Churchill (1987). This character appears to be directly linked to the unique intercalary meristem of hornworts, which continually differentiates sporangium tissues in most taxa (see discussion in Mishler and Churchill 1987; Whittemore 1987).

3.43 • MALE GAMETOGENESIS Ultrastructural features of male gametogenesis have been used to examine relationships among basal land plants and green algae. Garbary, Renzaglia, and Duckett (1993) listed 90 characters, but these were later reduced to 65 in a subsequent study by Mishler, Lewis, et al. (1994). Characters have been compiled based on the development of spermatogenous cells, on the flagellar apparatus, and on cytoskeletal, nuclear, plastidial, and cytoplasmic structure. We recognize that the ultrastructural features of male gametes and gametogenesis have potential importance for systematic studies at this level, but much of this information is currently difficult to assess. First, the raw data on flagellar and cytoskeletal structure are widely scattered in the primary literature, and there has not yet been a review that critically evaluates, systematically examines, and effectively summarizes these features while taking full account of possible difficulties in the interpretation of transmission electron micrographs. We have found parallel problems of "data dispersal" in the paleobotanical literature, which is one reason why the fossil record has been underutilized. Second, it is currently unclear how thoroughly the various ultrastructural and developmental features have been sampled for potential intraspecific and interspecific variability. In other areas of plant science (e.g., palynology), ultrastructural data of this type are often poorly sampled because they are time-consuming and technically difficult to obtain. Similar sampling problems occur with molecular sequences. Third, as recognized by Garbary, Renzaglia, and Duckett (1993) and Mishler, Lewis, et al. (1994), character correlations are of particular concern when large data sets are based on one small aspect of comparative morphology, particularly in systems that may have strong functional constraints, such as the motile apparatus of swimming sperm.

3.44 • CHLOROPLAST ULTRASTRUCTURE Several ultrastructural features of the chloroplast are thought to be characteristic of various embryophyte groups (Duckett and Renzaglia 1988b). As with spermatozoid ultrastructure, these features are problematic because they have been sampled only in a relatively small number of taxa. Furthermore, because of the difficulties in obtaining these data through transmission electron microscopy, character conceptualization and comparison is often very difficult. Channeled stroma or lateral interconnections between chloroplast grana are thought to be absent in liverworts, mosses, and vascular plants but are present in green algae and hornworts (Duckett and Renzaglia 1988b). We chose not to include this character because the availability of data for relevant taxa is currently too sparse.

3.45 • CHEMICAL CHARACTERS Surveys of chemicals such as flavenoids and terpenes have found some correlations with major groups of green plants (Markham and Porter 1978; Mishler and Churchill 1984; Suire and Asakawa 1979). Similarities in terpene structure among liverworts and tracheophytes have been noted (Suire and Asakawa 1979), and although flavonoids are common in land plants, they are rarely observed in green algae (Markham and Porter 1978). The presence of polyphenolics in the cell walls of water-conducting cells was used as a synapomorphy of the moss-tracheophyte clade by Mishler and Churchill (1984). We have not included chemical data in our analysis because we regard current taxon sampling as inadequate for using this feature to address high-level systematic questions among embryophytes.

3.46 • CUNEATE APICAL CELL A cuneate (wedge-shaped) apical cell was treated tentatively as a synapomorphy of hornworts by Mishler and Churchill (1985b), although the apical cell of *Dendroceros* is hemidiscoidal (Crandall-Stotler 1984). This character is problematic because apical cell morphology varies considerably in the gametophytes of embryophytes and is often difficult to define. Cuneate apical cells also occur in some basal thalloid liverworts, such as the Marchantiidae (Crandall-Stotler 1984; Schuster 1984a), which led Mishler and Churchill (1985b) to suggest that this type of morphology may be more general in embryophytes. Because of the considerable variability in apical cell morphology among embryophyte gametophytes, we decided not to use this character in our analysis.

3.47 • AERIAL CALYPTRA In embryophytes the growing sporophyte is generally surrounded by the archegonial remnant *(calyptra, involucre)*, which eventually splits, allowing the sporophyte to protrude. In all mosses except *Sphagnum*, rupturing of the archegonium results in a distinctive calyptra of variable size and shape that forms a sheathing cap, which protects the elongating tip of the sporangium (Schofield and Hébant 1984). This feature was used as a synapomorphy of mosses (except *Sphagnum*) by Mishler and Churchill (1984), but it may be more widely distributed because similar caplike structures have been observed in the hornwort *Anthoceros* (Renzaglia 1978; Schuster 1984c). We have excluded this character from our analysis because of uncertainty with regard to the condition in hornworts.

3.48 • XYLEM AND PHLOEM The hydrome and leptome of mosses were homologized with the xylem and phloem of tracheophytes by Mishler and Churchill (1984) based on the significant similarities in cell wall structure within these tissue systems established by the comparative studies of Hébant (1977) and Scheirer (1980, 1990). Xylem and phloem were thus interpreted by Mishler and Churchill (1984) as synapomorphies of their moss-tracheophyte clade. The water-conducting cells of mosses *(hydroids),* particularly of the Polytrichales, have much in common with the tracheids of vascular plants and the water-conducting cells of early polysporangiophytes (D. Edwards 1993; D. S. Edwards 1986; Hébant 1977; Kenrick and Crane 1991; see also Chapter 4). Similarities include cell elongation, the hydrolysis of oblique end walls; strong acid phosphomonoesterase activity associated with cell differentiation, cell death, and loss of the protoplast at maturity; and the presence of phenolic substances associated with the cell walls (Scheirer 1980, 1990). Both hydroids and tracheids are elongate and lose their protoplasts at maturity. Similarly, the leptoids of mosses strongly resemble the phloem elements of vascular plants in such features as cell elongation, oblique end walls, retention of the nucleus, the presence of many enlarged plasmodesmata, controlled autolysis of the protoplast, and the presence of cell organelles at maturity (Scheirer 1980, 1990).

We acknowledge the similarities discussed above and consider that these general aspects of xylem and phloem may well be homologous in mosses and tracheophytes. However, we chose not to score xylem and phloem in our analysis for two reasons. First, detailed observations on the hydrome and the leptome in mosses and the phloem in tracheophytes are currently based on relatively few taxa. Second, we wanted to avoid overweighting the moss-tracheophyte relationship in our analysis. Our character 3.3 (well-developed sporangiophore) recognizes significant similarities in internal tissue differentiation in the sporophytes of mosses and tracheophytes compared to the undifferentiated sporophytes of hornworts and liverworts. To score xylem and phloem as absent in liverworts and hornworts would potentially overweight this character. Our approach is therefore conservative but risks underestimating the character support for the moss-tracheophyte relationship.

ANALYSIS

Our analysis is based on 14 extant taxa and 1 well-preserved early fossil polysporangiophyte, *Horneophyton* (Table 3.1). The charophycean algae *Coleochaete* and *Chara* were used as outgroups. Relationships were assessed using 34 characters. It should also be noted that several characters do not affect the outcome of the analysis. The presence or absence of a sporangium (character 3.8) was included to emphasize that its distribution is consistent with an assumption of homology in embryophytes and to highlight the difficulty of scoring the outgroup state.

The most problematic aspect of the analysis is that most of the sporophyte characters were found to be inapplicable in the charophycean outgroups because the sporophytes (zygotes) are unicellular. Thus, the presence or absence of characters such as the sporangium or stomates on the sporophyte cannot strictly be scored for charophycean algae. Given that major structural innovations are an important feature of biological evolution, the issue of inapplicable characters deserves more detailed study. Several alternatives have been suggested for circumventing such problems in cladistic analyses (e.g., Maddison 1993), but none of these is entirely satisfactory. We adopted three different approaches to scoring inapplicable characters in the charophycean outgroups: scoring as unknown ("?"), assigning unique outgroup states, and attempting to polarize ingroup states with reference to greater levels of generality (see below).

Scoring outgroup states as unknown does not allow polarization (determination of plesiomorphic and apomorphic conditions) of ingroup states and therefore does not affect tree rooting. The problem with this approach is that it affects character optimization on the tree because, operationally, with the analytical approaches currently available, ingroup states are assigned to these inapplicable outgroup characters (Platnick, Griswold, and Coddington 1991). Assigning unique states to outgroup characters has the advantage of maintaining an operational distinction between inapplicable and missing character states. Furthermore, the inapplicable states do not influence the outcome of the analysis because the outgroup states are unique and do not affect tree rooting. The effect on character optimization is that ingroup states cannot be optimized onto the outgroups. Both approaches produce the same result because neither allows inapplicable outgroup states to be used in character polarization. As a result, tree rooting is based only on characters with known outgroup states. The identical result of these two approaches is referred to as analysis 3.1 (Figure 3.35).

An attempt was also made to polarize characters that were inapplicable in the outgroups through reference to greater levels of generality. With this approach, the absence of stomates was justified as plesiomorphic because they are absent in all other living organisms. Other characters such as the presence of a sporangium and branching are still equivocal because both states occur in other potential outgroups. This approach did not affect ingroup tree topology but rooted the tree in a slightly different place. This analysis is referred to as analysis 3.2.

In all the analyses, trees were generated using the Branch-and-Bound routine of PAUP (Phylogenetic Analysis Using Parsimony, version 3.1.1; Swofford 1990), which finds all the most parsimonious solutions for a given data set. Trees were rooted using outgroup comparison with a minimum of two outgroups, and the effects of the different outgroups were tested together and individually. Consistency indexes given throughout were calculated excluding uninformative characters. Tree stability was tested using Bremer Support (Donoghue, Olmstead, et al. 1992; Källersjö, Farris, et al. 1992), which measures the relative amount of support for each clade. The number of additional steps required before support for a clade collapses was calculated by computing consensus trees starting with the shortest tree and adding one step each time. A Bremer Support value (Decay Index) of 3 indicates

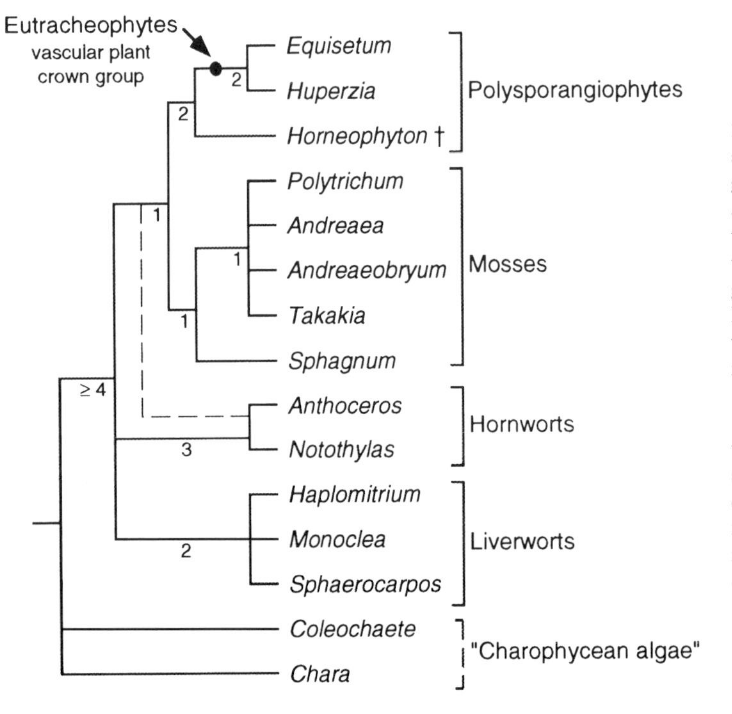

FIGURE 3.35. Cladogram of relationships among major embryophyte groups: semistrict consensus tree from analysis 3.1. Two charophycean algae were designated as outgroups. † = extinct; quotation marks = paraphyletic group. Numbers on the branches are Bremer Support indexes. Analysis 3.1 was a Branch-and-Bound search (by means of PAUP version 3.1.1; Swofford 1990) of 15 taxa and 34 characters (data in Appendix 2), resulting in 54 trees that were 50 steps in length (consistency index excluding uninformative characters = 0.67; rescaled consistency index = 0.54). Result of analysis 3.2 is indicated by broken line (see text for details).

that three additional steps are required before alternative clades can be found. Character state changes are based on ACCTRAN optimization, which favors reversals over parallelisms.

RESULTS

An analysis based on 15 taxa and 34 characters yielded 54 equally parsimonious trees of 50 steps with consistency indexes of 0.67 (analysis 3.1: Figure 3.35). These trees support a monophyletic interpretation of liverworts, hornworts, mosses, and vascular plants, as well as the monophyly of polysporangiophytes and embryophytes. All trees also support bryophyte paraphyly with respect to vascular plants and resolve mosses as the vascular plant sister group in agreement with Mishler and Churchill (1984) and Mishler, Lewis, et al. (1994). There are two major differences between our consensus tree (Figure 3.35) and previous morphology-based analyses. First, our analysis includes one early fossil polysporangiophyte, which is placed consistently in the vascular plant stem group. Second, the relationship among liverworts, hornworts, and the moss-polysporangiophyte clade is not fully resolved. Examination of all 54 trees showed only two alternative positions for hornworts and liverworts. One set of trees supported a sister group relationship based on the presence of elaters (Figure 3.36, Table 3.4). The other set of trees supported a sister group relationship between liverworts and the moss-polysporangiophyte clade based on loss of the pyrenoid in the chloroplast. Surprisingly, no support was found for the hornwort-moss-polysporangiophyte clade suggested by Mishler and Churchill (1984) and by Kenrick and Crane (1991), apparently because of the inability to polarize two critical characters (stomates and D-methionine) as well as inconsistencies and ambiguities in their state distributions. The character state for stomates cannot be observed in the unicellular zygotes of the charophycean algae, and the situation is further complicated by the fact that stomates are absent in some critical basal mosses (*Takakia, Andreaea, Andreaeobryum*). The D-methionine character is more problematic because it has not been observed in many key taxa, including both outgroups. One of the 54 equally parsimonious trees is illustrated in Figure 3.36, and the

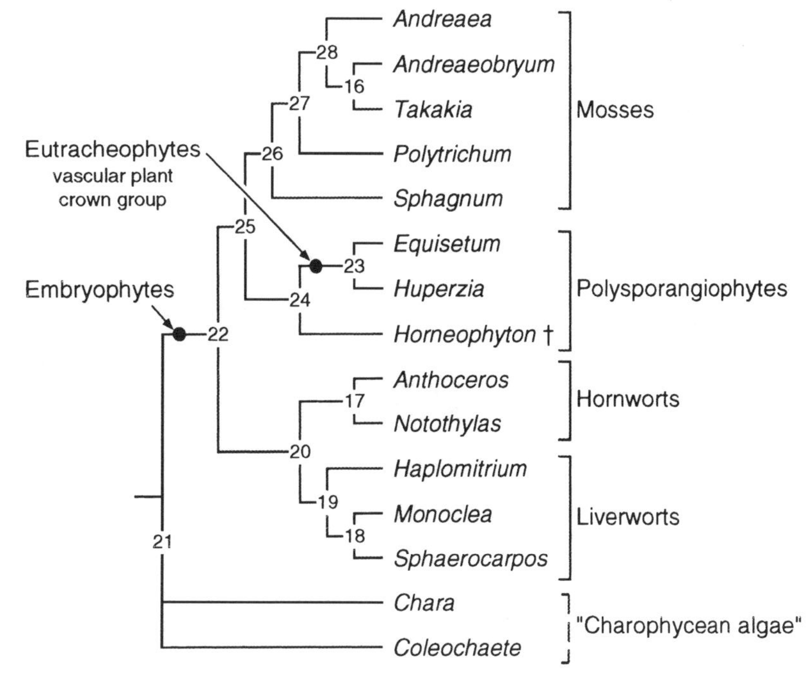

FIGURE 3.36. Character-state changes on one of the 54 most parsimonious trees from analysis 3.1: ACCTRAN optimization. † = extinct; quotation marks = paraphyletic group. Analysis 3.1 was a Branch-and-Bound search (by means of PAUP version 3.1.1; Swofford 1990) of 15 taxa and 34 characters (data in Appendix 2), resulting in 54 trees of length 50 (consistency index excluding uninformative characters = 0.67; rescaled consistency index = 0.54). Numbers in the figure represent internal nodes; see Table 3.4 for character-state changes corresponding to those nodes.

TABLE 3.4

Character-State Changes on Nodes of Tree in Figure 3.36

Beginning node → end node	Character	Change in state	CI
Node 16 → *Takakia*	3.23 Paraphyses	1 → 0	0.333
Node 19 → *Haplomitrium*	3.16 Gametophyte form	0 → 1	0.250
	3.17 Gametophyte leaves	0 → 1	0.500
	3.22 Gametangia distribution	0 → 1	0.333
Node 20 → node 17	3.5 Intercalary meristem	0 → 1	1.000
	3.20 Antheridium development	0 → 1	1.000
	3.21 Archegonium position	0 → 1	0.500
	3.27 Mucilage clefts	0 → 1	1.000
	3.29 Pyrenoids	1 → 0	0.500
Node 20 → node 19	3.10 Columella	1 → 0	0.500
	3.15 Stomates	1 → 0	0.500
	3.26 Oil bodies	0 → 1	1.000
	3.33 Lunularic acid	0 → 1	1.000
	3.34 D-Methionine	1 → 0	1.000
Node 21 → *Chara*	3.16 Gametophyte form	0 → 1	0.250
Node 21 → *Coleochaete*	3.12 Zoospore flagella	1 → 0	1.000
Node 21 → node 22	3.1 Sporophyte cellularity	0 → 1	1.000
	3.14 Cuticle	0 → 1	1.000
	3.18 Archegonium	0 → 1	1.000
	3.19 Antheridium morphology	0 → 1	1.000
	3.29 Pyrenoids	0 → 1	0.500
	3.30 Bicentriolar centrosomes	0 → 1	1.000
	3.31 Orientation of lamellae in multilayered structures	0 → 1	1.000
	3.32 Preprophase bands in mitosis	0 → 1	1.000
Node 22 → node 20	3.11 Elaters and pseudoelaters	0 → 1	1.000
Node 22 → node 25	3.3 Well-developed sporangiophore	0 → 1	0.333
	3.13 Perine	0 → 1	1.000
	3.16 Gametophyte form	0 → 1	0.250
	3.22 Gametangia distribution	0 → 1	0.333
Node 23 → *Huperzia*	3.23 Paraphyses	0 → 1	0.333
Node 23 → *Equisetum*	3.16 Gametophyte form	1 → 0	0.250
	3.22 Gametangia distribution	1 → 0	0.333
Node 24 → node 23	3.7 Xylem-cell thickenings	0 → 1	1.000
	3.10 Columella	1 → 0	0.500
Node 25 → node 24	3.2 Sporophyte independence or dependence	0 → 1	1.000
	3.4 Sporophyte branching	0 → 1	1.000
	3.21 Archegonium position	0 → 1	0.500
	3.30 Bicentriolar centrosomes	1 → 2	1.000
Node 25 → node 26	3.9 Sporangium dehiscence	0 → 1	0.500
	3.17 Gametophyte leaves	0 → 1	0.500
	3.24 Rhizoid cellularity	0 → 1	1.000
Node 26 → node 27	3.6 Foot form	0 → 1	1.000
	3.23 Paraphyses	0 → 1	0.333
	3.25 Protonema type	0 → 1	1.000
Node 26 → *Sphagnum*	3.3 Well-developed sporangiophore	1 → 0	0.333
	3.28 Pseudopodium	0 → 1	0.500
Node 27 → node 28	3.9 Sporangium dehiscence	1 → 0	0.500
	3.15 Stomates	1 → 0	0.500
Node 28 → *Andreaea*	3.3 Well-developed sporangiophore	1 → 0	0.333
	3.28 Pseudopodium	0 → 1	0.500

Notes: ACCTRAN optimization. CI = consistency index of character. Tree no. 24.

character changes at internal nodes are listed in Table 3.4.

Because of inapplicable character states in the outgroups for many of the sporophyte characters, an attempt was made to polarize these characters with reference to greater levels of generality in a separate analysis (analysis 3.2). Polarization was only attempted for characters with unequivocal states in both potential outgroups (characters 3.5–3.7, 3.10, 3.15). An analysis based on 15 taxa and 34 characters yielded 27 equally parsimonious trees of 51 steps with consistency indexes of 0.66 (analysis 3.2: Figure 3.35). The effect of rescoring some of the inapplicable outgroup characters was to resolve a hornwort-moss-polysporangiophyte clade that is fully congruent with the pattern of relationships suggested in previous morphology-based studies (Kenrick and Crane 1991; Mishler and Churchill 1984; Mishler, Lewis, et al. 1994).

Tree Stability

Tree stability was tested using Bremer Support (Decay Index = DI) measurements. Support was strongest for embryophytes (DI ≥ 4) and hornworts (DI = 3) and weakest for monophyly of mosses and the moss-polysporangiophyte clade (DI = 1). Intermediate levels of support were found for liverworts, polysporangiophytes, and eutracheophytes (DI = 2). The second analysis, which reproduced Mishler and Churchill's result (1984), showed uniformly weak support (DI = 1) for all groups except hornworts (DI = 3) and embryophytes (DI ≥ 4).

Three taxa were responsible for 40% of the homoplasy in the data: the liverwort *Haplomitrium* and the mosses *Sphagnum* and *Takakia*. The axial gametophytes, terminal gametangia, and gametophytic leaves of *Haplomitrium* are convergences with similar features in the moss-polysporangiophyte clade (Figure 3.36, Table 3.4). Removal of *Haplomitrium* from analysis 3.1 improved the consistency index (0.71) and reduced the tree length to 47 but did not change the tree topology or improve branch support. A similar effect was observed when *Sphagnum* was removed from analysis 3.1. Removal of *Takakia* from analysis 3.1 improved the consistency index slightly (0.68), increased support for the *Polytrichum*-Andreaeopsida clade, and reduced tree length to 49 but did not change the tree topology.

Removal of the fossil *Horneophyton* had the greatest effect on tree topology. Removing *Horneophyton* from analysis 3.1, and reanalysis based only on extant taxa, removed support for the moss–vascular plant clade and left the relationships among tracheophytes and major groups of bryophytes unresolved. Gametophyte morphology in *Equisetum*, interpreted most parsimoniously as a modification of the basic axial condition in vascular plants in the full analysis, becomes more equivocal with the removal of those polysporangiophytes possessing axial gametophytes. Removing *Horneophyton* from analysis 3.2 modifies tree topology to support a sister group relationship between hornworts and vascular plants and a sister group relationship between mosses and liverworts. However, all these relationships are weakly supported (DI = 1).

Embryophytes

Embryophyte monophyly was the most strongly supported relationship in this analysis (DI ≥ 4). Several tests of embryophyte monophyly were performed by rooting trees on *Chara* and forcing *Coleochaete* into a sister group relationship with various embryophyte clades (e.g., hornworts, polysporangiophytes). Unequivocal embryophyte synapomorphies include multicellular sporophytes, cuticles, archegonia, and antheridia (Figure 3.36, Table 3.4). The sporangium is also a putative synapomorphy of embryophytes, but its state cannot be observed in the outgroups. Other synapomorphies (which are less secure because they are unknown in many taxa) are centriolar development, microtubule orientation in the MLS of motile sperm, the presence of PPBs, and spore walls containing sporopollenin. Several other potential synapomorphies are based on characters with equivocal distributions. Loss of the pyrenoid in the chloroplast could be a synapomorphy of embryophytes (an independent acquisition in hornworts or independent losses in liverworts and the moss-polysporangiophyte clade). Elaters may be a synapomorphy of embryophytes (lost in the moss-polysporangiophyte clade) or may be independent acquisitions in liverworts and hornworts. Analysis 3.2 supports stomates as a synapomorphy of a hornwort-moss-polysporangiophyte clade, but in analysis 3.1 the distributions of character states are equivocal. Stomates could be a synapomorphy of embryophytes (lost in liverworts), or they could be independent acquisitions in hornworts and the moss-polysporangiophyte clade. According to either scenario, stomates are lost in such mosses as *Takakia, Andreaea,* and *Andreaeobryum.*

Stomatophytes

The hornwort–moss–vascular plant clade of Mishler and Churchill (1984) (stomatophytes, *sensu* Kenrick and Crane 1991) was only supported in analysis 3.2. In analysis 3.1, with the conservative outgroup scoring, none of the 54 most parsimonious trees supported this stomatophyte clade. In the second analysis, the hornwort-moss-polysporangiophyte clade is weakly supported (DI = 1) by the presence of stomates (which are most parsimoniously interpreted as losses in some mosses) and a columella in the sporangium (which is lost in extant vascular plants but is present in some early polysporangiophytes). The D-methionine character of Mishler and Churchill (1984) could not be polarized and therefore does not provide support for this group.

The Moss-Polysporangiophyte Clade

The sister group relationship between mosses and vascular plants suggested by Mishler and Churchill (1984) is consistent with recent cladistic analyses that have included early fossil taxa (Kenrick and Crane 1991). A clade comprising mosses and polysporangiophytes (vascular plants and early nonvascular fossils) was supported in our analyses (DI = 2). Unequivocal synapomorphies include axial gametophytes and terminal gametangia (both of which are independently derived in *Haplomitrium* and the Jungermanniales), as well as a perine layer on the spores and persistent and internally differentiated sporophytes (sporangiophores are considered to be lost in *Andreaea* and *Sphagnum*) (Figure 3.36, Table 3.4).

One particularly interesting feature of these analyses is that the inclusion of the fossil *Horneophyton* is essential to resolving the moss-polysporangiophyte clade. Removal of *Horneophyton* from analysis 3.1 results in an unresolved polytomy among the four major embryophyte groups, whereas its removal from analysis 3.2 results in weak support (DI = 1) for a vascular plant–hornwort clade and a moss-liverwort clade. The vascular plant–hornwort clade was supported by the presence of stomates (which are considered to be independently derived in mosses) and of sunken archegonia. The moss-liverwort clade was very weakly supported by the absence of the pyrenoid in both taxa. The loss of the pyrenoid is unlikely to be a synapomorphy at this level because, as far as we are aware, it is also absent in all of the vascular plants that have been examined. Unfortunately, the presence or absence of a pyrenoid has not been determined in *Equisetum* and *Huperzia,* which represent vascular plants in our analysis.

Polysporangiophytes and Vascular Plants

The polysporangiophyte clade comprises all vascular plants as well as early stem group fossils (Kenrick and Crane 1991) and was supported in both analyses (DI = 2). Unequivocal polysporangiophyte synapomorphies include sunken archegonia (which are independently derived in hornworts), an independent alternation of generations, and multiple sporangia (sporophyte branching) (Figure 3.36, Table 3.4). A change from bicentriolar to peripheral centrosome development mapped onto the tree in Figure 3.36 as a synapomorphy of polysporangiophytes. The placement of this change was determined using ACCTRAN optimization and is equivocal because the conditions in *Huperzia* and in the fossil *Horneophyton* are unknown. This change is more likely to be an autapomorphy of *Equisetum* because peripheral centriolar development is characteristic of species with multiflagellate gametes. The eutracheophyte clade is supported on the basis of xylem thickenings and the loss of the columella within the sporangium (DI = 2). The monophyly of eutracheophytes is supported by the presence of xylem cell thickenings and the loss of the columella (Figure 3.36, Table 3.4; see also additional character support in Chapter 4).

Liverworts, Hornworts, and Mosses

The monophyly of the major groups of bryophytes was supported in both analyses, with support for the hornworts being strongest (DI = 3). Monophyly of hornworts is supported by the presence of an intercalary meristem, mucilage clefts, sunken archegonia (convergence with polysporangiophytes), and pyrenoids and by antheridium production within a cavity in the thallus (Figure 3.36, Table 3.4). Monophyly of both mosses and liverworts had only weak support (DI = 1). Monophyly of mosses is supported by the presence of gametophytic leaves and multicellular rhizoids. Operculate dehiscence is a synapomorphy of mosses in Figure 3.36 (Table 3.4), and linear dehiscence in the Andreaeopsida is a reversal. The position of these characters is reversed in other equally parsimonious trees. Our analysis confirmed the problematic nature of *Sphagnum* that was highlighted previously by Mishler and Churchill (1985b). Both *Sphagnum* and *Takakia* are significant sources of homoplasy and are partially responsible for weak support of the moss clade. The placement of *Takakia* in a basal position in the mosses (Garbary, Renzaglia, and Duckett 1993; Smith and Davison 1993), rather than with *Haplomitrium* in the Calobryales (Schuster 1984b), is confirmed primarily by data from the recently described sporophyte.

The monophyly of liverworts is supported by the presence of oil bodies. Other possible synapomorphies, depending on how the characters are optimized onto the tree, include the loss of the columella, stomates, and D-methionine and the presence of lunularic acid (Figure 3.36, Table 3.4). The two other characters cited by Mishler and Churchill (1985b) were found to be equivocal. The distribution of lunularic acid is poorly understood in the ingroup, and the condition in the two charophycean outgroups is unknown. Elaters in hornworts and liverworts were scored as separate characters by Mishler and Churchill (1984) on the basis of developmental differences. We argue that there are good grounds for considering these structures as homologous and have scored them as such in this analysis (see character 3.11). Analysis 3.1 found some trees consistent with this hypothesis (i.e., a liverwort-hornwort clade based on the elaters character). In other trees it was equally parsimonious to consider elaters as a synapomorphy of embryophytes (and as lost in the moss-polysporangiophyte clade) or as independent acquisitions in the liverworts and hornworts.

Experiments

The results obtained in this analysis are broadly consistent with previous cladistic treatments but differ radically from other proposals of relationship among the major land plant groups and green algae. We tested the relative parsimony of some alternative hypotheses of relationship against our data matrix by forcing topological constraints on the trees generated during the analysis.

EMBRYOPHYTE POLYPHYLY Embryophyte polyphyly has been widely discussed, particularly in relation to the major bryophyte groups and vascular plants (Crandall-Stotler 1980, 1984, 1986; Duckett and Renzaglia 1988a; Philipson 1990, 1991; Schuster 1977, 1979, 1981, 1984c). Recently, hornworts have attracted much attention, and several bryologists have argued that this group probably evolved independently from within the green algae or Charophyceae (e.g., Crandall-Stotler 1984; Duckett and Renzaglia 1988a; Schuster 1984c). The argument for an independent origin of hornworts is based on the many unique features of this group, and any similarities to other embryophytes are interpreted as convergences. Our analysis, in common with other cladistic treatments, finds the unique features of hornworts to be autapomorphic and the similarities to other embryophytes to be interpreted most parsimoniously as homologies. If an independent origin of hornworts from within charophycean algae is specified, the resulting trees are at least seven steps less parsimonious than our most parsimonious result (Figure 3.35). This highly unparsimonious solution invokes massive homoplasy with multiple origins of features such as multicellular sporophytes, a bulbous sporophyte foot with transfer cells, a sporangium, a columella, elaters, a cuticle, stomates, an arche gonium, an antheridium, and bicentriolar centrosome replication and similarities in the MLS of motile sperm and PPBs. A similar, highly unparsimonious result (at least seven extra steps) is obtained if bryophyte monophyly is constrained and an independent origin of the group from within charophycean algae enforced.

BRYOPHYTE MONOPHYLY AND REDUCTION FROM VASCULAR PLANTS Several other hypotheses of grouping within land plants were found to be only marginally less parsimonious than our most parsimonious tree. The monophyly of bryophytes and a sister group relationship with vascular plants required only two additional steps. However, the monophyly of bryophytes is not supported by any synapomorphies. Hypotheses involving bryophyte reduction from either rhyniophytes (requiring three additional steps) (Bateman 1996b; Remy and Remy 1980; Robinson 1985; T. N. Taylor 1988a) or vascular plants (requiring more than six additional steps) (Garbary, Renzaglia, and Duckett 1993) were still less parsimonious. Both hypotheses assume homoplastic loss of polysporangiophyte or vascular plant features, losses for which there is no evidence. Removal of *Takakia* from the mosses and inclusion with the Calobryales (i.e., sister to *Haplomitrium*) in liverworts (Schuster 1984b) required an additional four steps. Homoplastic characters include independent acquisition of a sporangiophore (internally differentiated sporophyte axis), a columella, and a tapering mosslike sporophyte foot, as well as the loss of elaters and oil bodies.

DISCUSSION

Paleobotanical data combined with results from cladistic analyses indicate that the liverworts, hornworts, mosses, and vascular plants are the products of ancient cladogenic events that began in the mid-Ordovician and were probably complete by the late Silurian (see Chapter 7). The extant members of these clades exhibit a remarkable range of morphological diversity and have diverged to such an extent that significant problems are encountered in neobotanical systematic studies that compare morphology or molecular sequences. The morphological problems highlighted in our analysis include difficulties in obtaining comparative data and controversy about primary homology assignments because of the morphologically divergent nature of the groups. In particular, critical sporophyte characters are difficult to polarize because the outgroups lack a multicellular sporophyte generation. Furthermore, the morphological divergence of the sporophyte and gametophyte generations between bryophytes and vascular plants has probably resulted in the loss or extensive modification of suites of characters. These problems are compounded by patchy sampling of critical morphological data within groups. Sampling problems are currently most acute in the new morphological data sets based on ultrastructural studies of male gametes and gametogenesis, of cell organelles, and of aspects of mitosis and meiosis.

Recent molecular studies based on the *rbcL* gene and 18S and 26S rRNA sequences have encountered similar problems, which are manifested in low consistency indexes, potential difficulties with "long branch attraction," and incongruence with the data of comparative morphology. Cladistic studies based on *rbcL* sequences currently provide the most comprehensive taxonomic sample for basal land plants but the results are highly incongruent with morphology-based studies (Albert, Backlund, and Bremer 1994; Albert, Backlund, Bremer, Chase, et al. 1994; Manhart 1994). The evolution of the *rbcL* sequence appears to be strongly constrained by its function, with most nucleotide changes being functionally neutral (i.e., comprising substitutions that are silent with respect to amino acid changes or involving function-conserving amino acid replacements) and quasi-ultrametric (i.e., more or less clocklike) (Albert, Backlund, Bremer, Chase, et al. 1994). The combination of quasi-ultrametric change and long internal branches resulting from ancient cladogenic events implies that much of the similarity in *rbcL* sequences among major basal embryophyte groups is spurious (i.e., convergent) and that incongruence with morphology-based studies is caused by the relatively high levels of homoplasy (Albert, Backlund, and Bremer 1994; Albert, Backlund, Bremer, Chase, et al. 1994).

One promising new approach to addressing phylogenetic relationships among basal embryophytes has been to combine data from molecular sequences and comparative morphology into a single analysis based on the "total evidence" (Albert, Backlund, Bremer, Chase, et al. 1994; Mishler, Lewis, et al. 1994). The rationale behind the total-evidence approach is that it provides the most stringent test of homology because all data contribute equally to determining tree topology. Furthermore, the emergent pattern is perhaps more likely to reflect underlying phylogenetic patterns because convergence is unlikely to co-vary in widely different classes of data. Mishler, Lewis, et al. (1994) combined neobotanical data from 18S and 26S rRNA with morphological data sets based on male gamete ultrastructure and "general" comparative morphology. This combined analysis is congruent with studies based on "general" morphology (Mishler and Churchill 1984, 1985b; this study), even though individual analyses of rRNA and ultrastructural data produced different results. Further progress through neobotanical studies requires extending the existing database by increased sampling, in particular by the use of alternative molecular sequences and by the incorporation of data on larger-scale genomic rearrangements.

Because of the ancient and divergent nature of the major basal extant embryophyte groups, paleobotanical data might be expected to play a crucial role in phylogenetic analyses by allowing more sampling along stem lineages. The analyses presented here and in Chapter 4 differ from previous treatments in their explicit inclusion of data from the fossil record. Our results demonstrate that whereas paleobotanical data have a significant impact on phylogenetic studies in vascular plants (see Chapter 4), evidence from the fossil record currently plays a minor role in understanding relationships among charophycean algae and basal embryophytes at the bryophyte level of organization. Our cladistic analysis indicates that most well-documented early macrofossils are more closely related to vascular plants than to any other embryophyte group. One remarkable feature of the fossil record is the absence of unequivocal stem group liverworts, hornworts, or mosses.

Although slightly less well supported, our results are generally congruent with the tree presented by Mishler and Churchill (1984) and Mishler, Lewis, et al. (1994). Similarities include support for the monophyly of embryophytes, liverworts, hornworts, mosses, and vascular plants; for the paraphyletic relationship between bryophytes and vascular plants; and for a sister group relationship between mosses and vascular plants. The early fossil polysporangiophyte *Horneophyton* was found to be essential to resolving the moss–vascular plant clade because of ambiguities introduced through a stringent approach to character scoring and analysis. The main ambiguity in our analysis with respect to liverworts and the moss–vascular plant clade concerns the position of hornworts. This group was placed as a sister group to the moss–vascular plant clade by Mishler and Churchill (1984). The difficulties of resolving the position of hornworts seem to hinge around the divergent morphology of the group and character-sampling problems in basal embryophytes. The discovery of early hornwort stem group fossils could also have a considerable impact (see Chapter 7).

The late appearance of mosses, liverworts, and hornworts in the macrofossil record contrasts strongly with the early abundance of vascular plants and is often cited as stratigraphic evidence favoring the evolution of bryophytes from within vascular plants. We regard this hypothesis as unlikely because experiments with our data set indicate that such a pattern of relationship is more than six steps less parsimonious than the most parsimonious solution, in which tracheophytes "emerge" from plants at the bryophyte grade. Furthermore, we found no characters that would support a stem group relationship between any early fossil vascular plant and any of the main bryophyte clades. The two most plausible alternative hypotheses to bryophyte paraphyly are bryophyte monophyly and a sister group relationship with vascular plants (two steps less parsimonious) and a protracheophyte origin for both vascular plants and bryophytes (three steps less parsimonious). We interpret the late appearance of bryophyte macrofossils and the complete absence of unequivocal evidence for stem group bryophytes as probably caused by a combination of geological, taphonomic, and collector biases (see Chapter 7).

The early fossil record contains several enigmatic forms that may be related to major embryophyte clades or possibly to other groups that were important in early terrestrial ecosystems (e.g., fungi). Currently, these taxa are either too incomplete or problematic to contribute usefully to a numerical cladistic analysis. *Spongiophyton* and *Protosalvinia* are enigmatic, dichotomously branched, thalloid organisms. A possible relationship with land plants has been suggested for *Spongiophyton* based on the presence of a cuticle and spores (Chaloner, Mensah, and Crane 1974; Gensel, Chaloner, and Forbes 1991), but the absence of critical information on structures involved in sexual reproduction has made more precise systematic assignment problematic. Recently, Stein, Harmon, and Hueber (1994) have reinterpreted *Spongiophyton* as a lichen, based on new anatomical information. Internally, *Spongiophyton* comprises an organized network of fungal hyphae strongly resembling a fungal-lichen cortex. Fungal hyphae terminate in pores on the outer surface comparable to cyphellae or pseudocyphellae of lichens, but an unequivocal phycobiont has not yet been documented.

Protosalvinia is a thalloid organism bearing tetrads of large (200 μm), thick-walled spores in terminal depressions on the thallus surface (Niklas and Phillips 1976). Ultrastructural studies suggest that the spores are meiotic, but unlike in other embryophytes, sporopollenin is absent in the spore wall (W. A. Taylor and T. N. Taylor 1987). Many contrasting affinities have been suggested for *Protosalvinia,* ranging from ferns to red or brown algae (review by Hemsley 1994a; T. N. Taylor and E. L. Taylor 1993).

An embryophyte relationship for the enigmatic lower Devonian plant *Orestovia* seems likely (Krassilov 1981; Snigirevskaya and Nadler 1994). *Orestovia* occurs in large quantities as the dominant component of the middle Devonian Barzas Coals in the Kuzneckij Basin, Russia. Small, narrow, microperforate, branched, cuticularized axes bear numerous stomates and tubercles. Internally, these axes comprise a "central strand" of tracheid-like cells and possibly several cuticularized layers. The tracheid-like cells are predominantly helical and poorly preserved, and many appear to be solid rods rather than hollow tubes. No other cellular tissues are preserved except for abundant spores within some axes. Spores are large (150–190 μm) and a range of morphologies have been documented. Spores inside the axes are subtriangular, having conspicuous verrucate ornament on the distal surface and a relatively smooth proximal surface. Spores adhering to the external surface of the axes are more or less echinate. *Orestovia* has been interpreted as a primitive vascular plant (Krassilov 1981), but a hornwort affinity may also be possible. Such an affinity is suggested by the presence of spores within elongate axes; possible reinterpretation of the "central strand" as a columella and the unusual "tracheid-like" cells as elaters also needs to be considered. More recently, Snigirevskaya and Nadler (1994) interpreted most of the preserved material of *Orestovia* as rhizomatous and drew comparisons with extant Marsileaceae. Unequivocal reproductive structures are unknown.

Nematothallus (D. Edwards 1982) and *Cosmochlaina* (D. Edwards 1986) are microscopic cellular cuticular fragments that are sometimes associated with whefts of branched tubes. Some cuticles have pores in the surface (D. Edwards 1986), but apart from their cuticular, cellular nature, there is little additional evidence to support an affinity with land plants (Strother 1993).

A common group of upper Silurian and Devonian fossils of highly equivocal systematic affinity is characterized by a tubular internal organization. These fossils were originally mistaken for fossil wood, but subsequent studies have shown that the tubes are not the tracheids characteristic of vascular plants. The affinities of *Prototaxites, Nematasketum,* and *Pachytheca* to plants are poorly supported and diverse relationships to such groups as brown algae and fungi have been suggested. *Prototaxites* is an axial fossil ranging from a few millimeters to 90 cm in diameter with a complex structure comprising large tubes surrounded by smaller branching filaments (Schmid 1976). A relationship with ascomycete fungi such as *Clavaria* has recently been suggested based on anatomical comparisons (Hueber 1994). *Nematasketum* is common in the upper Silurian and Devonian and also has an axial organization. Internally, *Nematasketum* is composed of an outer layer of parallel, unbranched, smooth-walled tubes and an inner layer of branched filaments, some of which have internal thickenings. *Pachytheca* strongly resembles *Nematasketum* and *Prototaxites* in internal structure but is spherical (Burgess and Edwards 1988; Gerrienne 1990; Niklas 1976; Schweitzer 1983b). These organisms may be the source of the common microscopic tubular fragments often extracted in palynological preparations from the Silurian and early Devonian (Burgess and Edwards 1991; Wellman 1995).

Clear evidence for the presence of charophycean algae first appears in the late Silurian in the form of characteristic gyrogonites typical of the Charales (Feist and Grambast-Fessard 1991). The extensive fossil record of the Charales is based mainly on the calcified gyrogonites, and recent cladistic treatments have focused exclusively on characters obtained from these parts (Martín-Closas and Schudack 1991). Vegetative remains of a *Nitella*-like plant have been described from the lower Devonian Rhynie Chert (Figure 2.8; Edwards and Lyon 1983; Taylor, Remy, and Hass 1992). Other early evidence of charophycean algae comes from oval compression fossils of *Parka decipiens,* which resemble extant *Coleochaete orbicularis* (Figure 3.3; Hemsley 1990; Niklas 1976; T. N. Taylor 1988a; T. N. Taylor and E. L. Taylor 1993). *Parka* has a pseudoparenchymatous morphology and bears small, approximately oval structures that have been interpreted as either haploid spores or diploid zygotes (Hemsley 1990; T. N. Taylor and E. L. Taylor 1993).

4 Polysporangiophytes

Polysporangiophytes are one of the most inclusive groups of land plants and are diagnosed by the presence of axial, branching sporophytes that bear multiple sporangia (Kenrick and Crane 1991). As currently circumscribed, the group includes tracheophytes, as well as less specialized and less well understood extinct nonvascular taxa that lack tracheids and are thus intermediate between bryophytes and vascular plants (a protracheophyte grade *sensu* Kenrick and Crane 1991). Polysporangiophytes diversified through the Silurian and Devonian Periods and have an important and diverse fossil record. Traditional classifications recognize three early extinct groups—the Rhyniophytina, the Trimerophytina, and the Zosterophyllophytina—and several smaller groups of uncertain status, for example, the Barinophytaceae and Sciadophytaceae (Banks 1968, 1975b). The Rhyniophytina are small, structurally simple, leafless vascular plants with terminal sporangia. The Trimerophytina are similar to the Rhyniophytina in many respects but are larger and have more complex branching. The Zosterophyllophytina are also structurally simple, leafless, vascular plants, but they strongly resemble lycopsids in sporangium morphology and in certain anatomical details (see Figures 5.1 and 5.2). Extant polysporangiophytes comprise two major lineages of vascular plants: lycophytes (Lycophytina: Tables 1.1 and 7.1) and an *Equisetum*–fern–seed plant group (Euphyllophytina: Tables 1.1 and 7.1), as well as several problematic taxa such as the Psilotaceae (Bremer, Humphries, et al. 1987; Crane 1990). Resolving phylogenetic patterns in basal polysporangiophytes mainly requires clarifying the relationships among these diverse groups.

Current systematic studies indicate that the extinct Devonian groups Zosterophyllophytina and Trimerophytina are closely related to the extant lycopsid and non-lycopsid tracheophyte lineages, respectively (Banks 1968, 1975b). The Zosterophyllophytina have been treated either as strongly paraphyletic to lycopsids (Crane 1990) or as containing a large monophyletic component that is the sister group to lycopsids (Gensel 1992; Kenrick and Crane 1992). Relationships in the Trimerophytina are not well resolved (Gensel 1992), but a strongly paraphyletic relationship with non-lycopsid tracheophytes has been suggested and is consistent with the assumed "ancestral" status of the group (Crane 1990).

The Rhyniophytina are widely recognized as the most problematic group of early fossil land plants (Banks 1992; Crane 1990; Edwards and Edwards 1986; Gensel 1992; Hass and Remy 1991; Kenrick and Crane 1991; Meyen 1987; Schweitzer 1983b; T. N. Taylor 1988a). Rhyniophytes have been interpreted as "ancestral," either to tracheophytes as a whole (Chaloner and Sheerin 1979), or to the less inclusive Trimerophytina in conjunction with the non-lycopsid tracheophytes (Banks 1970), and thus the group is probably paraphyletic. This chapter evaluates the relationships among basal polysporangiophytes in a cladistic context, focusing mainly on early fossil taxa and particularly on putative rhyniophytes.

A BRIEF HISTORY OF RELEVANT PALEOBOTANICAL DISCOVERIES

Several pioneering studies recognized and described plant fossils from Devonian rocks in the mid-nineteenth century (e.g., Göppert 1852, Miller 1859), but the most significant contributions were made by the Canadian geologist J. W. Dawson during the 1850s and 1860s based on collections made in Gaspé Bay, Canada. Dawson described and reconstructed a very simple plant (*Psilophyton princeps:* Figure 4.1) with dichotomously branching leafless axes bearing terminal fusiform sporangia. Although he recognized that *Psilophyton* was quite unlike any known pteridophyte, he interpreted it as perhaps closely related to extant Marsileaceae (Pilulariae) or Psilotaceae (1859, 1871). Dawson's reconstruction and ideas on botanical affinities were strongly criticized by contemporary botanists, who considered *Psilophyton* to be possibly a fern petiole, a root system or even an alga (Solms-Laubach 1891). Support for his interpretation came eventually from more detailed descriptions of lower and middle Devonian plants from Norway (Nathorst 1913, 1915), including the confirmation of tracheids in the spiny stems of *Psi-*

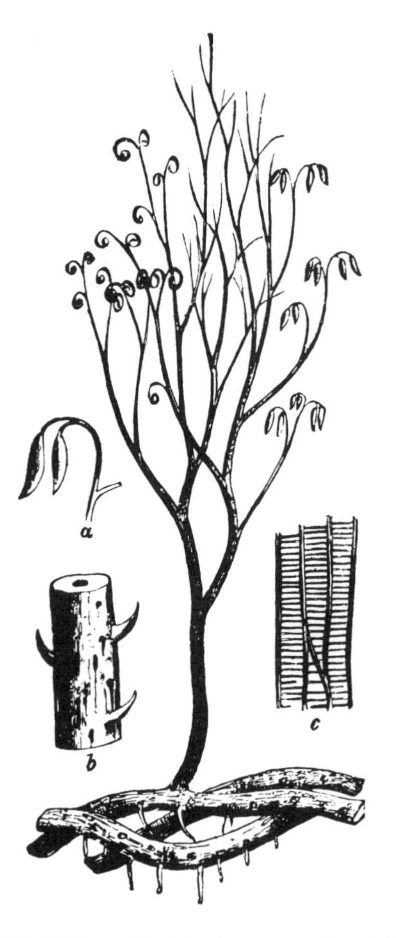

FIGURE 4.1. Dawson's original reconstruction of the early land plant *Psilophyton princeps* based on fossils from Gaspé Bay, Quebec, Canada (reproduced from Dawson 1870, Figure 19): (a) pair of fusiform sporangia; (b) axis with prominent nonvascular spines; (c) tracheids preserved in pyrite. This reconstruction is now known to be a composite of three remotely related taxa (Banks, Leclercq, and Hueber 1975; Hueber 1967, 1971b). Axes with fusiform sporangia belong to the trimerophyte *Psilophyton princeps*. Axes with circinate vernation belong to the zosterophyll *Sawdonia ornata*. Rhizomatous axes probably belong to *Taeniocrada*.

lophyton ornatum, which conclusively established its status as a vascular plant (Halle 1916b). An important review of these and other early paleobotanical studies is given by Arber (1921).

Early descriptions of Devonian plants were based mainly on fragmentary compression fossils, but in 1912, almost complete permineralized plants were discovered in a silicified chert of lower Devonian age near the village of Rhynie, Scotland (Gensel and Andrews 1984; Mackie 1914; Rice 1994). This discovery marked a watershed in the development of paleobotany, and there are many parallels between the Devonian plants of the Rhynie Chert (Chaloner and Macdonald 1980) and the Cambrian animals of the Burgess Shale (Gould 1989) in terms of their influence on ideas of botanical and zoological evolution. The fossil plants from the Rhynie Chert were similar to those described by Dawson, Halle, and others, but their exceptional anatomical preservation (Figure 4.2) provided unequivocal documentation of their land-plant status and pteridophyte affinities (Kidston and Lang 1917, 1920a,b, 1921). These early Devonian plants attracted great interest because they partially filled the phylogenetic hiatus between extant bryophytes and pteridophytes that was emphasized by Bower (1908). In addition, the overall form of the plants approximated the hypothetical intermediates between bryophytes and vascular plants that had been proposed by Lignier (1908), which were based in part on Dawson's *Psilophyton princeps.*

The paleobotanical discoveries outlined above ushered in a period of collection and description of early land plants that continues today. Most research has concentrated on the diverse and informative macrofloras preserved in the Devonian sediments of northern Europe and North America. Most of the known macrofossils are tracheophytes or putative tracheophytes. There are very few descriptions of bryophytes (cf. *Sporogonites:* Appendix 1), although there are a wealth of enigmatic fossils *(Parka, Prototaxites, Protosalvinia, Orestovia, Pinnatiramosus)* that may be relevant at this level but whose relationships are poorly understood. So far, the macrofossil record has contributed little to understanding the earlier transition from green algae to land plants, but the record of dispersed spores provides strong evidence for a diversification of embryophytes, perhaps at the bryophyte grade, from the upper Ordovician onward (Caradoc; c. 450 Ma) (Gray 1985, 1991, 1993; Gray and Shear 1992; W. A. Taylor 1995a). Spore floras also indicate significant modifications of floristic composition through the Silurian (Burgess

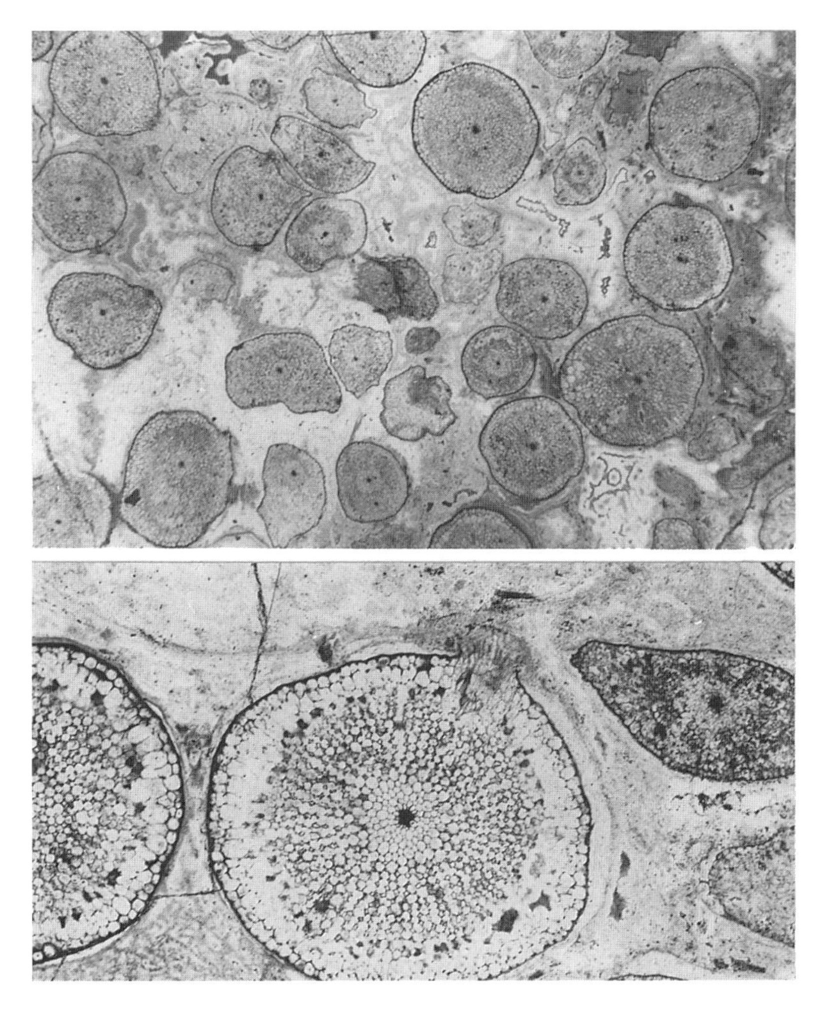

FIGURE 4.2. Thin section of Rhynie Chert showing transverse sections through axes of the early land plant *Rhynia gwynne-vaughanii*. Photographs by P.K. **Top.** Many stems preserved in growth position (×8). Specimen from the Natural History Museum, London, Lang Collection 1077. **Bottom.** Stem showing excellent soft tissue preservation at the cellular level (×25). Specimen from the Natural History Museum, London, Scott Collection 3133.

1991; Burgess and Richardson 1991; Dufka 1995; Gray 1985, 1991; Gray and Boucot 1977; Gray, Massa, and Boucot 1982; Gray and Shear 1992; Richardson and McGregor 1986; Strother 1991; Strother and Traverse 1979; Wellman 1993a,b; Wellman and Richardson 1993).

SYSTEMATICS

The Rise and Fall of the "Psilophyte" Concept

Following the description of fossil plants from the lower Devonian Rhynie Chert (Kidston and Lang 1917, 1920a,b, 1921), most authors recognized a group of plants at the pteridophyte level called the Psilophytales. The Psilophytes were defined as "[a] Class of Pteridophyta characterized by the sporangia being borne at the ends of certain branches of the stem without any relation to leaves or leaf-like organs" (Kidston and Lang 1917, 780). The concept of the Psilophytales was founded on very simple plants such as *Rhynia* (Figure 4.3) and *Psilophyton* (Figure 4.1), but as knowledge of Devonian floras increased, a wider range of taxa were assigned to this group. Phylogenetically, the Psilophytales (Rhyniophyta *sensu* Cronquist, Takhtajan, and Zimmermann [1966]) were considered by many botanists as the ancestral group from which all pteridophytes and possibly seed plants were derived (e.g., Bower 1935; Eames 1936; Høeg 1967; Kräusel 1938; Leclercq 1954; Scott 1920; Smith 1955; Stewart 1964; Zimmermann 1952). This idea and the associated concepts of morphological evolution were developed most extensively

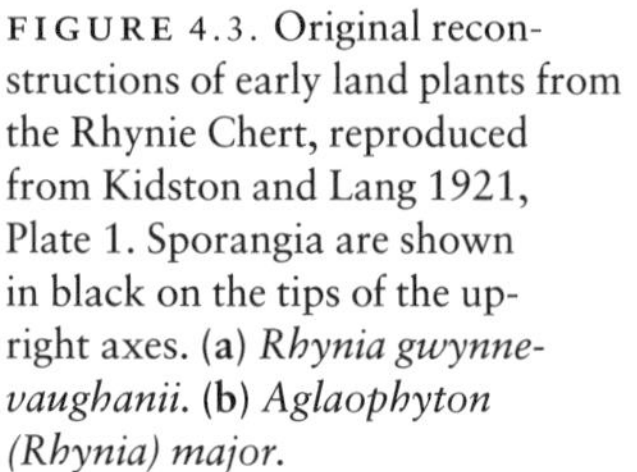
FIGURE 4.3. Original reconstructions of early land plants from the Rhynie Chert, reproduced from Kidston and Lang 1921, Plate 1. Sporangia are shown in black on the tips of the upright axes. (**a**) *Rhynia gwynne-vaughanii.* (**b**) *Aglaophyton (Rhynia) major.*

by Zimmermann in his Telome Theory, which he elaborated in a series of articles (Zimmermann 1930, 1938, 1952, 1965). The potential ancestral status of the Psilophytales was extended further by some authors who thought that even bryophytes may have been derived from the group by "retrograde evolution" (Cronquist, Takhtajan, and Zimmermann 1966; Schuster 1966).

By the 1960s, it was widely recognized that the assignment of newly described fossils to the psilophytes had resulted in a group of taxa with diverse relationships. Although many alternative classifications have been proposed since the work of Kidston and Lang (e.g., Arnold 1947; Eames 1936; Hirmer 1927; Høeg 1967; Kräusel 1938; Walton 1940), the most influential was undoubtedly that of Banks (1968, 1975a,b; Table 4.1). He presented a major synthesis of the existing paleobotanical data on early land plants and introduced a new classification of the Psilophytales that substantially clarified the systematic placement and phylogenetic relationships of many Devonian fossil plants. Banks's classification (1968, 1975b) replaced the group Psilophytales with three distinct subdivisions: the Rhyniophytina, the Trimerophytina, and the Zosterophyllophytina. Interpreted phylogenetically, these groups highlighted an early dichotomy in vascular plant evolution. The Rhyniophytina were considered to be ancestral to the Trimerophytina, a lineage ultimately giving rise to ferns and seed plants. The Zosterophyllophytina were thought to be ancestral to lycopsids, and together, these groups were interpreted as an independently evolved lineage of vascular plants (Banks 1970, 79; Figure 4.4). More recently, others have placed the Rhyniophytina in a still more central position as ancestral to both major lineages of vascular plants (Figure 4.5; Chaloner and Sheerin 1979; Edwards and Edwards 1986; Meyen 1987; Selden and Edwards 1989; Stewart 1983), and perhaps also to bryophytes (Remy 1982; T. N. Taylor 1988a).

New Data, New Methods, New Problems

Despite the utility of Banks's classification, as well as widespread acceptance of its phylogenetic implications, the status of each of the three subdivisions has been challenged recently on theoretical and empirical grounds. In particular, cladistic theory has raised questions concerning group definition and conceptualization. Further problems have emerged through the description of new taxa and through new information on certain critical anatomical features. There is a growing list of "aberrant" taxa that are difficult to accommodate in Banks's tripartite classification (Banks 1992).

RHYNIOPHYTINA In many respects, the Rhyniophytina is the most unsatisfactory and problematic of Banks's three subdivisions. The concept of this group (Banks 1968, 1970, 1975a,b) was based in part on the oldest and simplest taxa formerly assigned to the now redundant Psilophytales. In its original form, the subdivision accommodated "[n]aked (leafless) plants whose aerial branches fork dichotomously, some terminated by sporangia; horizontal rhizomes with rhizoids, roots apparently lacking; slender, solid xylem strands, probably centrarch" (Banks 1975b, 407). The diagnosis was based mainly on three well-preserved plants from the Rhynie Chert: *Rhynia gwynne-vaughanii* (the type species), *Aglaophyton (Rhynia) major,* and *Horneophyton lignieri* (Banks 1975b). Five other less well known Silurian and Devonian genera were included, and another five genera were associated with the subdivision under the heading "questionable Rhyniophytina" (Table 4.2).

The "ancestral" status of the Rhyniophytina implies that the group is paraphyletic, and this characterization precludes unambiguous definition by features that the group alone possesses (Crane 1990; Kenrick and Crane 1991). The inclusion of an increasing number of poorly preserved early fossils, such as *Steganotheca, Salopella, Dutoitea, Eogaspesiea, Yarravia, Hedeia, Hostinella, Aphyllopteris, Eorhynia,* and *Tortilicaulis,* has introduced additional problems and has added no further criteria for group definition. Whereas parts of the diagnosis (e.g., forked axes) refer to unique features (i.e., apomorphies) that provide a straightforward criterion for group membership, others (e.g., absence of leaves) are symplesiomorphies and effectively set an "upper limit" to the Rhyniophytina. The complications that arise are well illustrated by such "problematic" taxa as *Cooksonia caledonica* (D. Edwards 1970a), *Huvenia* (Hass and Remy 1991), *Uskiella* (Shute and Edwards 1989), *Renalia* (Gensel 1976), *Nothia* (El Saadawy and Lacey 1979a), *Sartilmania* (Fairon-Demaret 1986a), *Stockmansella* (Fairon-Demaret 1985, 1986b), and *Hsua* (Li 1982, 1992), which appear to fall between the Rhyniophytina and Banks's other two subdivisions. *Renalia,* for example, is included in the Rhyniophytina by some workers on the basis of plesiomorphic features (e.g., terminal sporangia), and in the Zosterophyllophytina by others on the basis of apomorphic characters (e.g., sporangium form and dehiscence) (Table 4.2).

TABLE 4.1

Three Influential Classifications of "Psilophytales"

Kräusel 1938[1]	Høeg 1967[2]	Banks 1975b[3]
Psilophytales	Psilophyta	Tracheophyta
Rhyniaceae	Psilopsida	Rhyniophytina
Rhynia sensu lato	Rhyniales	Rhyniales
Horneaceae	Rhyniaceae	Rhyniaceae
Horneophyton (=_Hornea_)	_Rhynia sensu lato_	_Rhynia sensu lato_
Sporogonites	_Horneophyton_ (=_Hornea_)	_Horneophyton_ (=_Hornea_)
_Yarravia_*	_Cooksonia_*	_Cooksonia_*
_Cooksonia_*	_Dutoitea_ (=_Dutoitia_)*	_Steganotheca_*
Psilophytaceae	_Hicklingia_	_Salopella_*
Psilophyton	_Taeniocrada_*	_Dutoitea_ (=_Dutoitia_)*
_Dawsonites_*	Questionable Rhyniaceae	_Eogaspesiea_*
_Taeniocrada_*	_Codonophyton_*	Questionable Rhyniophytina
Hicklingia	Sciadophytaceae	_Taeniocrada_*
Zosterophyllaceae	_Sciadophyton_*	_Hicklingia_
Zosterophyllum	Sporogonitaceae	_Nothia_
_Barinophyton_*	_Sporogonites_	_Yarravia_*
_Pectinophyton_*	_Eogaspesiea_*	_Hedeia_*
Rebuchia (=_Bucheria_)*	Zosterophyllaceae	Zosterophyllophytina
Sciadophytaceae	_Zosterophyllum_	Zosterophyllales
_Sciadophyton_*	_Rebuchia_ (=_Bucheria_)*	Zosterophyllaceae
Dutoitea (=_Dutoitia_)*	_Gosslingia_	_Zosterophyllum_
Asteroxylaceae	Hedeiaceae	_Rebuchia_ (=_Bucheria_)*
Asteroxylon	_Hedeia_*	_Sawdonia_
_Thursophyton_*	_Yarravia_*	_Gosslingia_
Hostimella (=_Hostinella_)*	Psilophytales	_Crenaticaulis_
Pseudosporochnaceae	Psilophytaceae	_Bathurstia_*
Pseudosporochnus	_Psilophyton_	Trimerophytina
Drepanophycaceae	_Asteroxylon_	Trimerophytales
_Drepanophycus_*	_Thursophyton_*	Trimerophytaceae
Baragwanathia	_Psilodendrion_*	_Psilophyton_
Haplostigma	_incertae sedis_	_Trimerophyton_
Protolepidodendron	_Psilophytites_*	_Pertica_
_Barrandeina_*	_Hostimella_ (=_Hostinella_)*	_Dawsonites_*
_Duisbergia_*	_Dawsonites_*	_Hostimella_ (=_Hostinella_)*
Protopteridaceae	_Nothia_	_Psilodendrion_*
Rellimia (=_Protopteridium_)	_incertae sedis_	_Psilophytites_*
Aneurophyton	Barinophytales	_incertae sedis_
Cladoxylon	Barinophytaceae	Sciadophytaceae
	_Barinophyton_*	_Sciadophyton_*
	_Pectinophyton_*	Barinophytaceae
	_Protobarinophyton_*	
	_Barinostrobus_*	
	Barrandeinaceae	
	_Barrandeina_*	
	_Bröggeria_*	
	_Barrandeinopsis_*	
	_Duisbergia_*	
	incertae sedis	
	Trimerophyton	
	_Loganophyton_*	

See notes on opposite page.

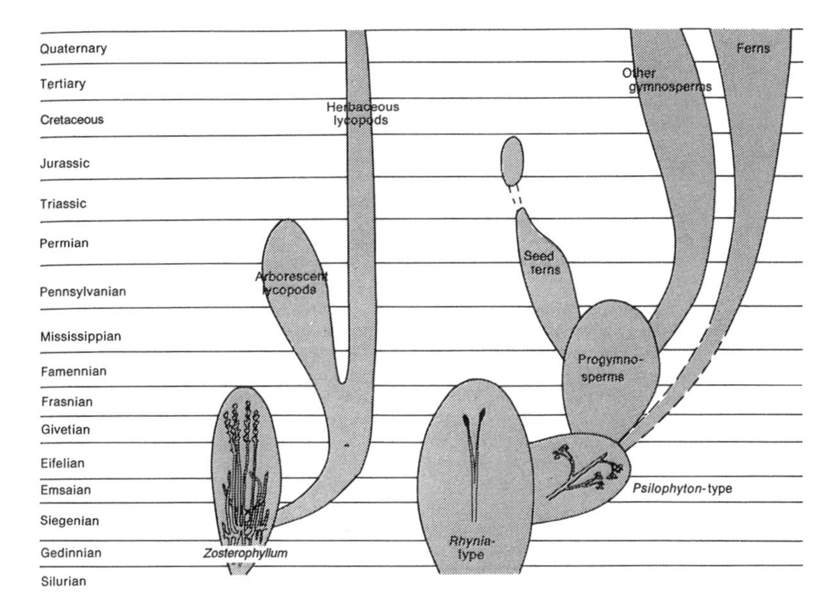

FIGURE 4.4. Phylogenetic relationships among basal vascular plants through geologic time according to Banks 1970, Figure 5-2. Note the hypothesized diphyletic origin of vascular plants from unspecified ancestral groups. Original figure reproduced here with permission.

Another important result from recent studies of the Rhyniophytina is that the systematic position and circumscription of the widely discussed and relatively large genus *Cooksonia* has become increasingly equivocal. The monophyly of *Cooksonia* is doubtful, and until a more restrictive definition is available it is perhaps better viewed as a form genus (see below). Furthermore, despite the frequent citation of this genus as comprising the earliest vascular plants, putative tracheids have been described from only one species of *Cooksonia* (Edwards, Davies, and Axe 1992). For many such plants it remains an open question as to whether the tracheids were merely not preserved or whether the plants were at a protracheophyte level of evolution.

Problems with the circumscription of the Rhyniophytina have been exacerbated by the removal of certain critical, well-preserved taxa based on the results of recent anatomical studies. Since, by definition, the Rhyniophytina are vascular plants, the presence of tracheids is an important criterion for group membership. Clarification of tracheid structure in early land plants shows that whereas some members of the Rhyniophytina possess this key anatomical feature (e.g., *Rhynia gwynne-vaughanii, Renalia*), others have unornamented water-conducting cells (e.g., *Aglaophyton [Rhynia] major:* D. S. Edwards 1986; *Nothia:* El-Saadawy and Lacey 1979a) or lack well-developed tracheids (e.g., *Horneophyton:* Kidston and Lang 1920a; Remy and Hass 1991d). *Aglaophyton* is now known to have a level of anatomical differentiation comparable to that of moss sporophytes (D. S. Edwards 1986; Hébant 1977), whereas *Rhynia gwynne-vaughanii, Sennicaulis, Stockmansella,* and *Huvenia* have been shown to possess highly distinctive tracheids that differ from those

←

Note: * = relatively poorly understood taxa or form genera with significant amounts of missing information.

[1] "Psilophytales" comprise a wide range of taxa from simple plants of the *Rhynia*-type to more complex taxa now known as progymnosperms, lycopsids, and early ferns. The Psilophytales comprise all early, extinct vascular plants.

[2] Changes from Kräusel 1938: The Psilophytales (Psilophyta) are restricted to simpler plants of the *Psilophyton, Rhynia, Zosterophyllum*-type. The Pseudosporochnaceae and Protopteridaceae are removed, and the Drepanophycaceae are transferred to lycopsids. The Barinophytales are transferred from Zosterophyllaceae to *incertae sedis.*

[3] Changes from Høeg 1967: Three new groups are recognized within Tracheophyta for simpler members of the Psilophytales. Other taxa transferred to such groups as lycopsids or progymnosperms. The Barinophytaceae and Sciadophytaceae are retained as *incertae sedis,* as in Høeg 1967.

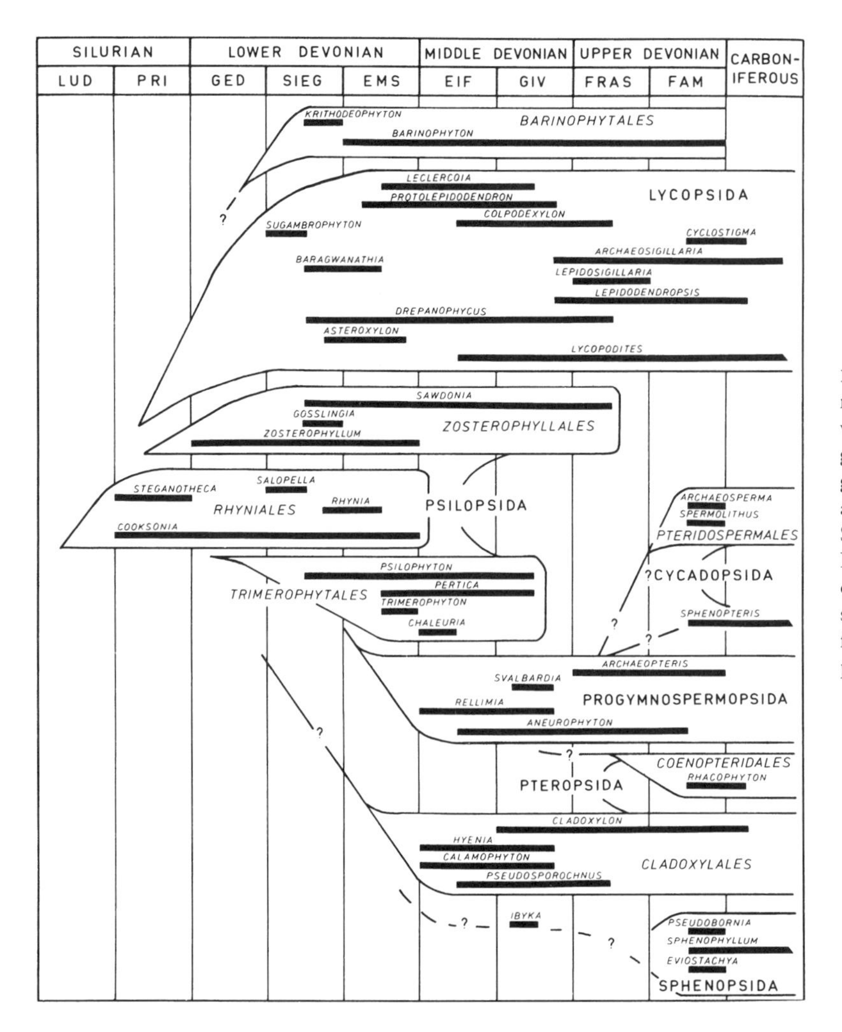

FIGURE 4.5. Phylogenetic relationships among basal vascular plants and stratigraphic distribution of key genera and major groups according to Chaloner and Sheerin 1979, Text-Figure 6. Note the monophyletic origin of vascular plants from unspecified precursors. Original figure reproduced here with permission.

of other vascular plants (Kenrick and Crane 1991). Notwithstanding the possibility of phylogenetic loss, taxa without tracheids have been removed from the group (Table 4.2), and it is now clear that many sporophytic features usually associated with the Rhyniophytina—such as branching and terminal, fusiform sporangia—are general for land plants as a whole and, in fact, define the more inclusive polysporangiophyte clade (Kenrick and Crane 1991).

The new discoveries and problems outlined above have prompted several recent attempts to revise Rhyniophytina, either by removing taxa or by creating loosely associated form-groups (Banks 1992; Edwards and Edwards 1986; Hass and Remy 1991; Meyen 1987; Schweitzer 1983b; T. N. Taylor 1988a). As a result, the Rhyniophytina have slowly dissolved into an increasingly problematic and heterogeneous collection of plants. Examination of four recent reviews (Banks 1992; Edwards and Edwards 1986; Meyen 1987; T. N. Taylor 1988a) shows that the group contains only one species on which all authors agree: the type species, *Rhynia gwynne-vaughanii* (Table 4.2).

Edwards and Edwards (1986), T. N. Taylor (1988a), and Banks (1992) retained the basic concept of Rhyniophytina *sensu* Banks (1975b) but questioned the criteria for membership in the group (Table 4.2). Among other features, all authors require the demonstration of tracheids for unequivocal placement in the subdivision, and in practice, this requirement has reduced the number of taxa included in the Rhyniophytina to two or possibly three. "Nonvascular" plants (e.g., *Aglaophyton [Rhynia] major:* D. S. Edwards 1986) and taxa in which sporangia are not terminal (e.g., *Nothia aphylla:* El-Saadawy and Lacey 1979a) were excluded. Meyen (1987; see Table 4.2) adopted a different approach and formally recognized the similarities between some members of the Rhyniophytina *sensu* Banks and the zosterophylls on the one hand and trimerophytes on the other. Zosterophyll-like rhyniophytes (e.g., *Renalia, Hicklingia, Nothia, Hsua*) were placed within his class Zosterophyllopsida, whereas trimerophyte-like rhyniophytes (e.g., *Rhynia, Cooksonia*) were placed with trimerophytes in an order Rhyniopsida. One rhyniophyte was placed in its own order—Horneophytopsida (Table 4.2).

Two alternative solutions have been suggested for the classification of poorly preserved taxa usually associated with the Rhyniophytina. Edwards and Edwards (1986) and Banks (1992) have an informal grouping, rhyniophytoid, *within* Rhyniophytina for the many taxa in which critical features of stem anatomy have not been described but that conform in general appearance to more informative vascular plant fossils such as *Renalia* or *Rhynia gwynne-vaughanii* (Table 4.2). T. N. Taylor (1988a) proposed an informal grouping *outside* Rhyniophytina (cooksonioids) for the same equivocal rhyniophytoids and other nonvascular taxa (Table 4.2).

Early cladistic treatments have highlighted the problematic nature of the Rhyniophytina that results from the absence of unequivocal defining features (Crane 1989, 1990). In Crane's hypothesis, the "transitional" morphologies of certain rhyniophytes were accounted for by the interpretation that some (e.g., *Renalia*) are more closely related to the zosterophylls and lycopsids, whereas others (e.g., *Rhynia*) are more closely related to the trimerophytes (Figure 4.6). A preliminary numerical analysis by Gensel (1992) left relationships among the Rhyniophytina and the Trimerophytina unresolved but also found *Renalia* to be more closely related to zosterophylls and lycopsids than to other rhyniophytes. Kenrick and Crane (1991) identified a small clade (Rhyniopsida) by placing *Rhynia gwynne-vaughanii, Huvenia kleui,* and *Stockmansella langii* in a sister group relationship to all other vascular plants. This clade was delimited by synapomorphies, whereas other taxa often associated with the Rhyniophytina were found to form grade of organization in the vascular plant stem group (Figure 3.9). These preliminary cladistic analyses underline the potentially diverse relationships of taxa traditionally included in Rhyniophytina *sensu* Banks (1975b), Edwards and Edwards (1986), T. N. Taylor (1988a), and Banks (1992).

TRIMEROPHYTINA The Trimerophytina, as developed by Banks (1968), was based originally on the rather poorly understood genera *Trimerophyton* and *Dawsonites,* but the concept of the group was clarified and extended through the inclusion of *Pertica* and the detailed morphological study of *Psilophyton dawsonii* by Banks, Leclercq and Hueber (1975). In its original form, the subdivision accommodated plants having a "[m]ain axis branched spirally and dichotomously; lateral branches forked either into three units or dichotomously; fertile branches, much-forked, terminated in a mass of paired, fusiform-ellipsoidal sporangia; axes smooth, punctate or spiny; xylem strand large, solid, centrarch, composed of scalariform tracheids" (Banks 1975b, 409). Three other less well known Devonian genera were also included (Table 4.3).

The concept and definition of Trimerophytina has changed little over the last twenty years, although two new species of *Pertica* have been described—*P. varia* (Granoff, Gensel, and Andrews 1976) and *P. dalhousii* (Doran, Gensel, and Andrews 1978)—and several problematic taxa have been associated informally with the group (Table 4.3; Banks 1992). *Psilophyton* is the largest and most diverse genus, and newly described species include *P. charientos* and *P. forbesii* (Gensel 1979), *P. crenulatum* (Doran 1980), *P. coniculum* (Trant

TABLE 4.2
Some Systematic Treatments of the Rhyniophytina

Banks 1975b	Gensel and Andrews 1984[1]	Edwards and Edwards 1986[2]
Rhyniophytina	Rhyniophytina	Rhyniophytina
Rhyniales	Rhyniales	Rhyniales
Rhyniaceae	Rhyniaceae	Rhyniaceae
Rhynia gwynne-vaughanii	*Rhynia gwynne-vaughanii*	*Rhynia gwynne-vaughanii*
Aglaophyton (Rhynia) major	*Aglaophyton (Rhynia) major*	*Taeniocrada decheniana* *
Horneophyton	*Horneophyton*	*Renalia (pro parte)*
Cooksonia *	*Cooksonia* *	*Hostinella (pro parte)* *
Steganotheca *	*Steganotheca* *	*Aphyllopteris (pro parte)* *
Salopella *	*Salopella* *	*Taeniocrada (pro parte)* *
Dutoitea *	*Dutoitea* *	"Rhyniophytoids"
Eogaspesiea *	*Eogaspesiea* *	*Eogaspesiea* *
Questionable Rhyniophytina	Questionable Rhyniophytina	*Cooksonia* *
Taeniocrada *	*Taeniocrada* *	*Steganotheca* *
Hicklingia	*Renalia*	*Salopella/Eorhynia* *
Nothia	*Nothia*	*Hedeia/Yarravia* *
Yarravia *	*Yarravia* *	Questionable Rhyniophytina
Hedeia *	*Hedeia* *	*Dutoitea* *
	Zosterophyllophytina	*Hsua*
	Hicklingia	*Horneophyton*
		incertae sedis
		Hicklingia
		Aglaophyton (Rhynia) major
		Nothia

Note: * = poorly understood taxa with significant amounts of missing information or form genera.

[1] Classification similar to that of Banks 1975b except *Hicklingia* moved to Zosterophyllophytina. The new taxon *Renalia* is problematic because it appears to be intermediate between Rhyniophytina and Zosterophyllophytina.

[2] (1) Rhyniophytina reduced to two well-understood taxa *(Rhynia gwynne-vaughanii* and *Renalia hueberi); Horneophyton* removed. (2) An informal category of "rhyniophytoid" established for plants in which anatomy is unknown. (3) *Hicklingia, Nothia,* and *Aglaophyton* (=*Rhynia major*) moved to *incertae sedis.*

and Gensel 1985), and *P. szaferi* (Zdebska 1986). Variation in branching pattern, and possibly also anatomy, indicates that some taxa such as *Pertica* may have a closer relationship to aneurophytalean progymnosperms than to *Psilophyton* (Gensel 1984; Gensel and Andrews 1984).

Crane (1990) suggested that the Trimerophytina are paraphyletic with respect to non-lycopsid tracheophytes; a hypothesis consistent with the "ancestral" status of the group envisaged by Banks (1968, 1970; Figure 4.4). Many of the features used to define the Trimerophytina appear to be more general in polysporangiophytes and vascular plants (e.g., dichotomous branching, terminal fusiform-ellipsoidal sporangia, solid centrarch xylem). The more restrictive aspects of the definition are clearly present in such groups as progymnosperms (e.g., scalariform tracheids, terminal masses of paired sporangia) and this suggests that these features, too, define more inclusive clades. Furthermore, several putative members of this group are known to have complex primary xylem morphology (Banks 1992; Gensel 1984; Gensel and Andrews 1984; T. N. Taylor and E. L. Taylor

Meyen 1987[3]	Taylor 1988[4]	Banks 1992[5]
Rhyniopsida	Rhyniophyta	Rhyniophytina
Rhyniales	*Rhynia gwynne-vaughanii*	Rhyniales
Rhynia gwynne-vaughanii	*Renalia hueberi*	Rhyniaceae
Cooksonia *	*Horneophyton*	*Cooksonia* *
Steganotheca *	*incertae sedis* "cooksonioids"	*Rhynia gwynne-vaughanii*
Trimerophytales	*Cooksonia* *	*Uskiella* *
[subtaxa not listed]	*Aglaophyton (Rhynia) major*	"Rhyniophytoids"
Horneophytopsida	*Eogaspesiea* *	*Eogaspesiea* *
Horneophyton	*Steganotheca* *	*Steganotheca* *
Zosterophyllopsida	*Tortilicaulis* *	*Salopella* *
Hicklingia	*Salopella* *	*Eorhynia* *
Nothia	*Eorhynia* *	*Hedeia* *
Hsua	*Hedeia* *	*Yarravia* *
Renalia	*Yarravia* *	*Caia* *
incertae sedis		*Dutoitea* *
Aglaophyton (Rhynia) major		*incertae sedis*
Taeniocrada		*Aglaophyton (Rhynia) major*
		Horneophyton
		Taeniocrada decheniana *
		Renalia
		Nothia
		Hicklingia *
		Huia *
		Hsua
		Stachyophyton *

[3](1) Rhyniophytina *sensu* Banks not recognized. Rhyniales reduced to one well-understood taxon. (2) Four taxa, including *Renalia* transferred to Zosterophyllopsida. (3) *Taeniocrada*, etc., moved to *incertae sedis.*

[4](1) Rhyniophytina reduced to three well-understood taxa. (2) All other taxa placed in an informal category of "cooksonioid" *(incertae sedis).*

[5](1) Rhyniophytina reduced to one well-understood taxon *(Rhynia gwynne-vaughanii).* (2) An informal category of "rhyniophytoid" for plants in which anatomy is unknown. (3) *Hicklingia, Nothia,* and *Aglaophyton* (=*Rhynia major*) moved to *incertae sedis.*

1993) that strongly resembles that of aneurophytalean progymnosperms and of other more complex taxa such as *Ibyka* (Skog and Banks 1973).

Phylogenetic Questions and Aims of Analysis

Despite the problematic nature of the Rhyniophytina, the recent accumulation of new data provides the opportunity to begin to resolve the relationships of these and related plants (e.g., Gensel 1992; Hueber 1992; Kenrick and Crane 1991). In this chapter we analyze the relationships among the better-known members of the Rhyniophytina *sensu lato* and major early groups of land plants, including representative zosterophylls, extant lycopsids, trimerophytes, fernlike plants, and progymnosperms. We incorporate new information from the fossil record and focus particularly on relationships among basal tracheophytes and enigmatic nonvascular taxa such as *Aglaophyton.* Specific questions to be addressed include (1) Does current knowledge support the monophyly of vascular plants, or are there at least two independently evolved lines as favored by Banks (1970)? (2) How

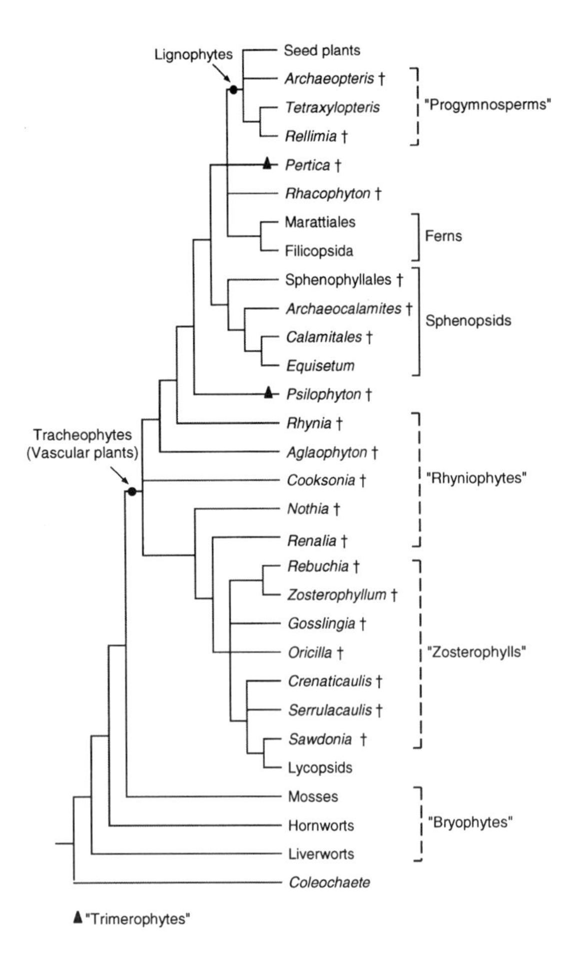

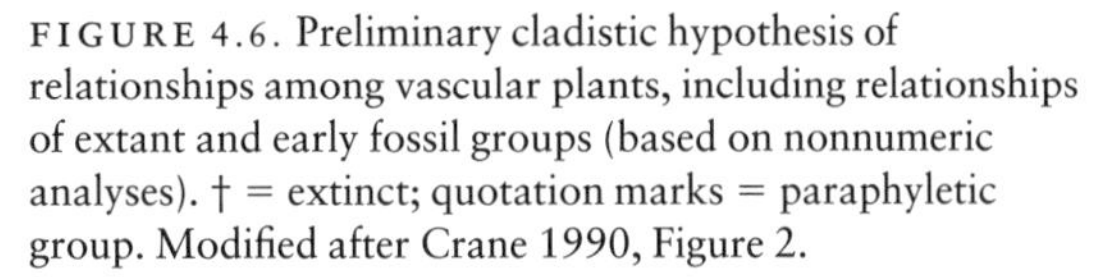
FIGURE 4.6. Preliminary cladistic hypothesis of relationships among vascular plants, including relationships of extant and early fossil groups (based on nonnumeric analyses). † = extinct; quotation marks = paraphyletic group. Modified after Crane 1990, Figure 2.

TABLE 4.3

Some Treatments of the Trimerophytina

Banks 1975b	Gensel and Andrews 1984[1]	Meyen 1987[2]	T. N. Taylor and E. L. Taylor 1993[3]
Trimerophytina	Trimerophytina	Rhyniopsida	Trimerophyta
Trimerophytales	Trimerophytales	Rhyniales	*Psilophyton*
Trimerophytaceae	Trimerophytaceae	*Rhynia gwynne-vaughanii*	*Trimerophyton* *
Psilophyton	*Psilophyton*	*Cooksonia* *	*Pertica*
Trimerophyton *	*Trimerophyton* *	*Steganotheca* *	*Yunia* *
Pertica	*Pertica*	Trimerophytales	*Dawsonites* *
Dawsonites *	*Dawsonites* *	*Psilophyton*	*Hostinella* *
Hostinella *	*Hostinella* *	*Trimerophyton* *	
Psilodendrion *	*Psilodendrion* *	*Pertica*	
Psilophytites *	*Psilophytites* *		

Note: * = poorly understood taxa with significant amounts of missing information or form genera.

[1] No changes from Banks 1975b.

[2] Consideration of better-known taxa; close relationship with Rhyniales explicitly stated.

[3] Addition of one new taxon.

well supported is the recent suggestion (Kenrick and Crane 1991) of a protracheophyte grade of nonvascular polysporangiophytes, or is it more parsimonious to interpret the absence of tracheids in *Aglaophyton, Horneophyton,* and *Nothia* as having evolved by loss? (3) Can the genus *Cooksonia* be supported as a natural (potentially monophyletic) taxon? (4) Can the Rhyniophytina and Trimerophytina be salvaged as useful and phylogenetically meaningful systematic concepts? and (5) What information is required from the fossil record in order to improve the resolution among basal polysporangiophytes and tracheophytes?

CHOICE OF TAXA

A total of 34 taxa were selected to develop a framework of relationships among basal polysporangiophytes and tracheophytes (Table 4.4). The putatively basal liverworts *Haplomitrium* (Figure 3.17) and *Sphaerocarpos* (Figure 3.18) were used as outgroups based on the analyses in Chapter 3. *Anthoceros* (Figure 3.19) was included as an exemplar for hornworts and *Polytrichum* (Figure 3.22) as an exemplar for mosses to allow for the possibility that tracheophytes may be diphyletic: that some may be more closely related to mosses and others more closely related to hornworts. Putative rhyniophytes included were *Horneophyton lignieri* (Figure 3.26), *Aglaophyton major* (Figure 4.7), *Rhynia gwynne-vaughanii* (Figure 4.8), and *Nothia aphylla* (Figure 5.13) from the Rhynie Chert; the *Taeniocrada*-like plants *Huvenia kleui* and *Stockmansella langii;* and seven species of cooksonioid morphology, including *Cooksonia caledonica* (Figure 4.9), *Cooksonia cambrensis, Cooksonia pertonii, Sartilmania jabachensis,* and *Uskiella spargens* (Figure 4.10), as well as the larger and more complex *Renalia hueberi* (Figure 4.11) and *Hsua robusta* (Figure 4.12). Undoubtedly the most problematic fossils in this phase of our analysis were those assigned to *Cooksonia.* The systematics of this difficult genus is considered in more detail below. *Horneophyton* was also included in our previous analysis of basal embryophytes (Chapter 3).

As exemplars for zosterophylls, we included three taxa that represent some of the major forms traditionally assigned to the group: *Sawdonia ornata, Gosslingia breconensis,* and *Zosterophyllum llanoveranum.* As exemplars for lycopsids we included the extant homosporous *Huperzia* (Figure 3.24; see also Chapter 3) and the Rhynie Chert lycopsid *Asteroxylon mackiei.* We included *Psilophyton dawsonii* (Figure 4.13), *Psilophyton crenulatum,* and *Pertica quadrifaria* as exemplars for trimerophytes. Lignophytes were represented by the aneurophytalean progymnosperm *Tetraxylopteris schmidtii* (Figure 4.14). *Rhacophyton ceratangium* (Figure 4.15), *Ibyka amphikoma* (Figure 4.16), and *Pseudosporochnus nodosus* (Figure 4.17) were considered because of their potential significance in understanding the origin of sphenopsids and ferns.

Other problematic taxa considered were *Sporogonites exuberans* (Figure 4.18) (possible bryophyte), *Tortilicaulis offaeus,* and *Caia langii* (Figure 4.19), and the putative trimerophytes *Yunia dichotoma* and *Eophyllophyton bellum.*

Banks 1992[4]
Trimerophytina
Trimerophytales
Trimerophytaceae
Psilophyton
Trimerophyton *
Pertica
Dawsonites *
Hostinella *
"likely Trimerophytina"
Yunia *
"perhaps advanced beyond Trimerophytina"
Oocampsa *
Gothanophyton *
Tursuidea *

[4] Consideration of possibly related taxa under informal headings.

TABLE 4.4
Taxa Examined in the Cladistic Analysis of Basal Polysporangiophytes

Taxon	Exemplars
Hornworts	***Anthoceros***
Mosses	***Polytrichum***
Zosterophylls	***Gosslingia breconensis*** †
	Sawdonia ornata †
	Zosterophyllum llanoveranum †
Lycopsids	***Asteroxylon mackiei*** †
	Huperzia selago
Ferns	*Pseudosporochnus nodosus* †
	Rhacophyton ceratangium †
Trimerophytes	*Pertica quadrifaria* †
	Psilophyton crenulatum †
	Psilophyton dawsonii †
Putative trimerophytes	***Eophyllophyton bellum*** †
	Yunia dichotoma †
Sphenopsids	*Ibyka amphikoma* †
Lignophytes	***Tetraxylopteris schmidtii*** †
Putative rhyniophytes	***Aglaophyton (Rhynia) major*** †
	Cooksonia caledonica †
	Cooksonia cambrensis †
	Cooksonia pertonii †
	Horneophyton lignieri †
	Hsua robusta †
	Huvenia kleui †
	Nothia aphylla †
	Renalia hueberi †
	Rhynia gwynne-vaughanii †
	Sartilmania jabachensis †
	Stockmansella langii †
	Uskiella spargens †
incertae sedis	*Caia langii* †
	Sporogonites exuberans †
	Tortilicaulis offaeus †
Liverworts	***Haplomitrium*** *
	Sphaerocarpos *

Notes: All taxa included in analysis 4.1. Taxa in bold included in analysis 4.2. See Appendix 1 for taxon descriptions and Appendix 3 for data matrix. * = outgroup. † = extinct.

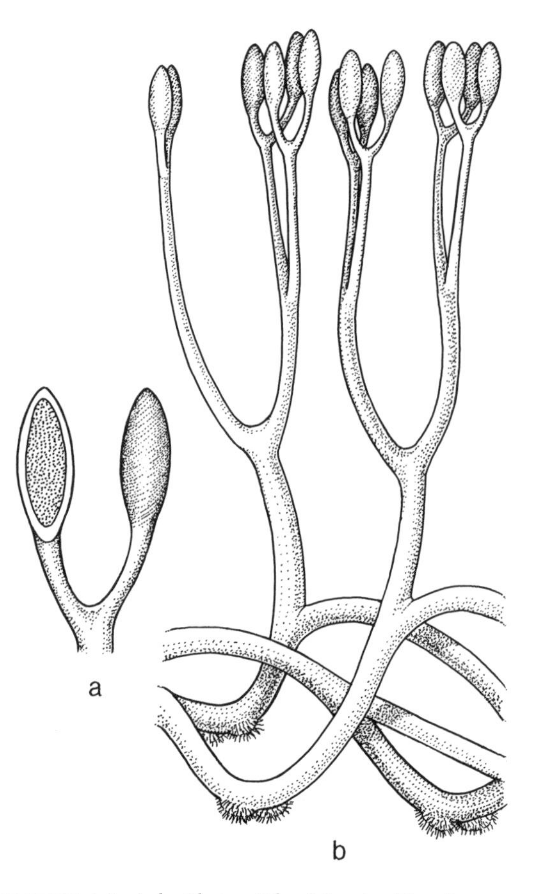

FIGURE 4.7. *Aglaophyton (Rhynia) major.* Drawings based on D. S. Edwards 1986, Figure 57, and Kidston and Lang 1921, Plate 1. (**a**) Details of sporangium, including view of interior; no columella present (×3). (**b**) Reconstruction of habit: terminal, fusiform sporangia; rhizomatous axes with clusters of unicellular rhizoids (×0.8).

Cooksonia

The genus *Cooksonia* was created originally for plants having "[d]ichotomously branched, slender, leafless stems, with terminal sporangia that are short and wide. Epidermis composed of elongated, pointed, thick-walled cells. Central vascular cylinder consisting of annular tracheides [*sic*]" (Lang 1937a, 288). Because the compression fossils on which this definition is based are tiny and dichotomize only once or twice, Lang recognized that they could be either the terminal portions of a larger plant or the almost complete specimens of a very small plant. The absence of associated larger axes from several localities suggested that *Cooksonia* was a small plant, and Lang recognized two species: *C. pertonii* (the type species) and *C. hemisphaerica.*

The generic diagnosis of *Cooksonia* is problematic because it does not provide a clear apomorphic

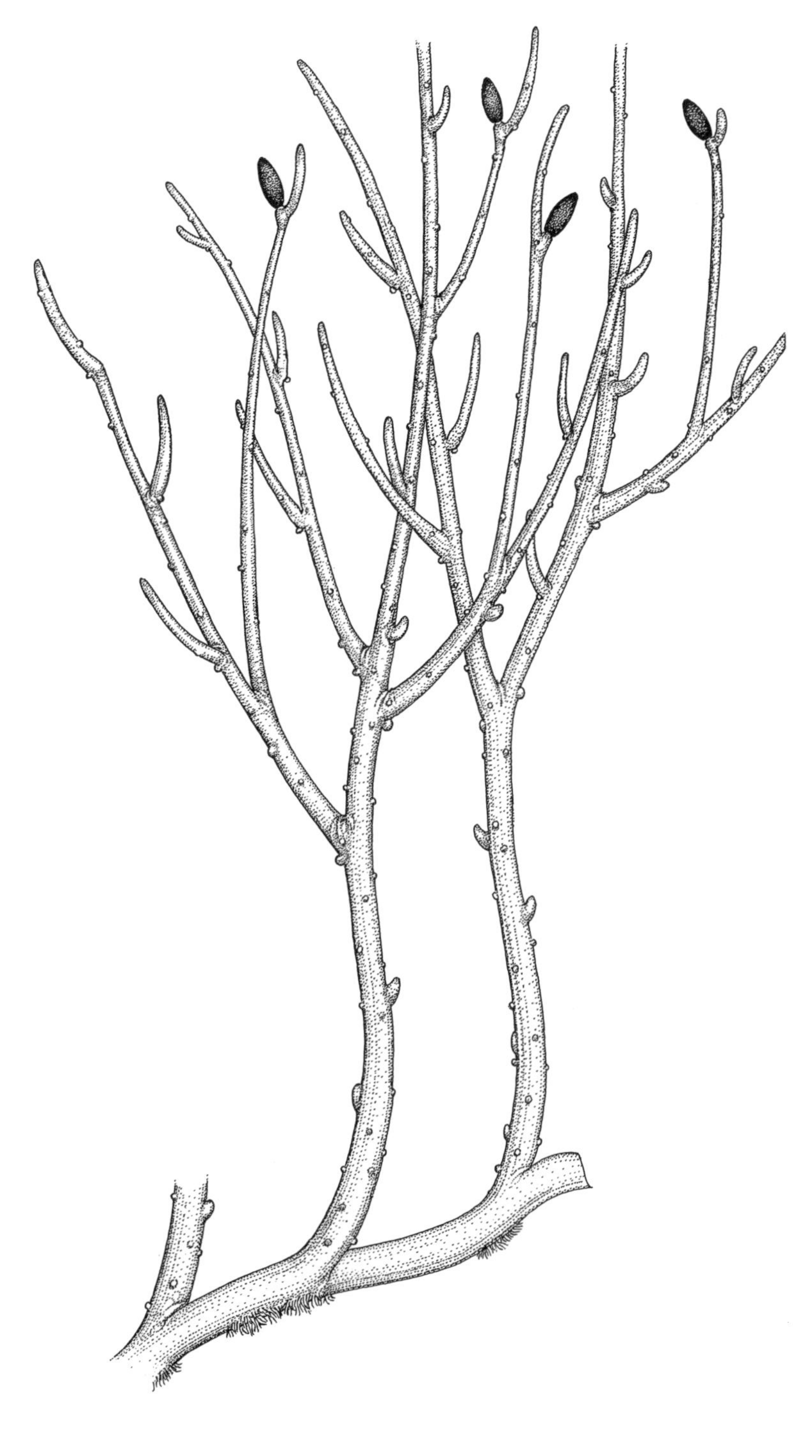

FIGURE 4.8. *Rhynia gwynne-vaughanii.* Reconstruction of habit: fusiform sporangia (heavy stipple) on short lateral branches; numerous adventitious branches and small "hemispherical projections" on axes; rhizomatous axes with clusters of unicellular rhizoids (×2). Redrawn from D. S. Edwards 1980, Figure 1, and Kidston and Lang 1921, Plate 1.

defining feature. Slender, dichotomously branching axes, terminal sporangia, elongate epidermal cells, and possibly tracheids with annular thickenings in some species are general features common to a wide range of taxa at this level of plant evolution (i.e., plesiomorphic). Leaflessness is also plesiomorphic at this level. Short and wide sporangia, the most restrictive part of the diagnosis, encompasses a wide range of sporangium shapes, from hemispherical to reniform, and does not distin-

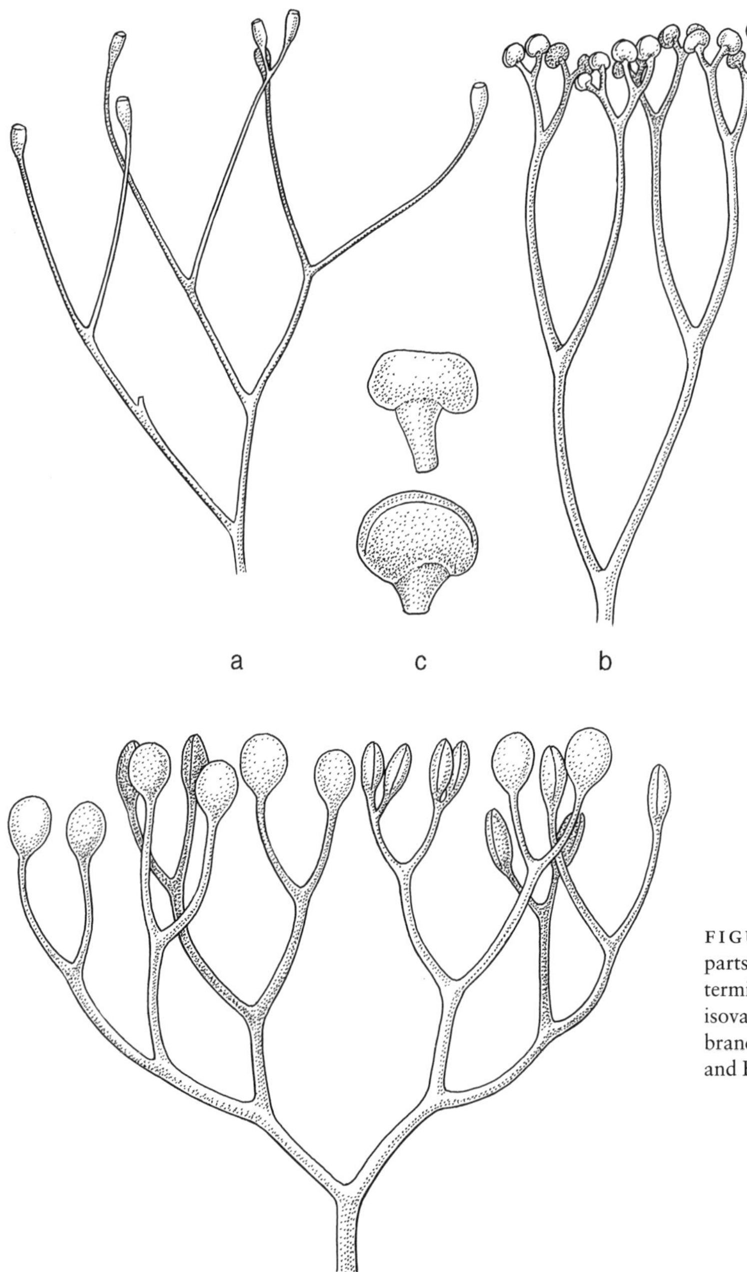

FIGURE 4.9. Reconstructions of distal parts of two rhyniophytes with isotomous branching. Drawings based on D. Edwards 1970a, Text-Figures 1 and 4, and Stewart 1983, Figure 9.2. (**a**) *Steganotheca striata:* terminal ellipsoidal sporangia with truncated apex (×2). (**b**) *Cooksonia caledonica:* terminal reniform sporangia (×1). (**c**) Details of sporangium of *C. caledonica* showing distal dehiscence line on lower specimen (×7).

FIGURE 4.10. Reconstruction of distal parts of *Uskiella spargens* showing terminal spatulate sporangia with distal isovalvate dehiscence on isotomous branching axes (×1). Redrawn from Shute and Edwards 1989, Figure 71.

guish among the variety of dehiscence mechanisms that have been inferred. At the level of basal embryophytes, the generic characteristics listed by Lang are thus plesiomorphic, and this creates obvious difficulties for the definition of *Cooksonia.*

As the morphology and anatomy of certain *Cooksonia* species have become better understood, there has been a tendency to segregate new genera rather than to improve the original definition. Thus, such taxa as *Hsua* (Li 1982), *Steganotheca* (D. Edwards 1970a), and *Uskiella* (Shute and Edwards 1989) were all formerly assigned to *Cook-*

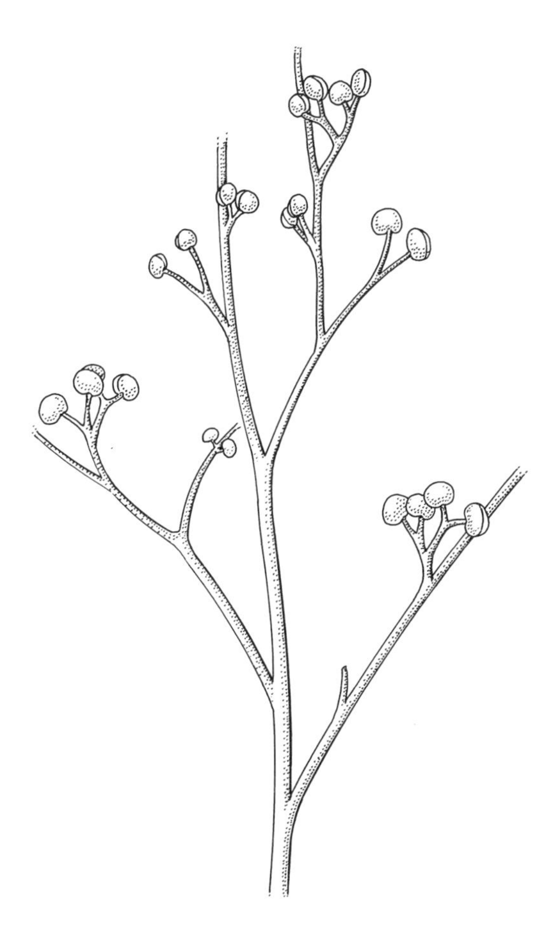

FIGURE 4.11. Reconstruction of distal parts of *Renalia hueberi* showing terminal reniform sporangia on an anisotomous branching system (×1.5). Redrawn from Gensel 1976, Figure 1.

sonia. Furthermore, there are several morphologically similar genera, such as *Pertonella* (Fanning, Edwards, and Richardson 1991) and *Dutoitea* (Høeg 1930; Rayner 1988), that differ from *Cooksonia* only in the presence of small, spinelike enations.

The morphological and anatomical details of most plants currently assigned to *Cooksonia* are very poorly understood. Spores have been obtained from well-preserved material of *C. pertonii*, *C. hemisphaerica,* and *C. caledonica* (D. Edwards 1979; Edwards, Davies, and Axe 1992; Edwards and Fanning 1985; Edwards, Fanning, and Richardson 1986; Fanning, Edwards, and Richardson 1992; Fanning, Richardson, and Edwards 1988; Lang 1937a). Poorly preserved tracheid-like cells (measuring about 28 μ in diameter) with annular thickenings were obtained from a sterile axis associated with *C. hemisphaerica* (Lang 1937a). Lang was convinced that this sterile axis belonged to *C. hemisphaerica,* but apart from the evidence of association, there is no confirmation that the two plant fragments were produced by a single species. Association evidence by itself is not compelling, and more recent investigations of well-preserved material suggest that tracheid-like cells were absent from at least some specimens assigned to *C. pertonii* (Edwards, Fanning, and Richardson 1986). Putative water-conducting cells were de-

FIGURE 4.12. Partial reconstruction of *Hsua robusta* showing reniform sporangia on isotomous lateral branches. Note the large main axes bearing small lateral branchlets and downward pointing branched appendages (subordinate branches, see character 4.6) (×0.25). Redrawn from Li 1982, Figure 1.

FIGURE 4.13. Partial reconstruction of *Psilophyton dawsonii* showing pseudomonopodial system with differentiation of subordinate fertile and vegetative lateral branches. Sporangia are fusiform in terminal clusters (×2). Reproduced, with permission, from Banks, Leclercq, and Hueber 1975, Text-Figure 13.

scribed in *C. pertonii* ssp. *apiculispora* by Edwards, Davies, and Axe (1992). These cells were unusually narrow for tracheids (3.2–11 μm in diameter) but appeared to possess annular or helical thickenings. No other details of the wall structure have been clearly established. Epidermal structures, stomates, and some details of internal anatomy have been described only in *C. pertonii* (Edwards, Davies, and Axe 1992; Edwards, Fanning, and Richardson 1986).

The occurrence of pseudomonopodial branching systems bearing *Cooksonia*-type sporangia (Ananiev and Stepanov 1969; Edwards and Fanning 1985; Gensel 1976; Li 1982) indicates that at least some of the specimens assigned to *Cooksonia* are parts of larger plants. Further problems are also introduced by the recognition that individual species of *Cooksonia* distinguished by small differences in sporangium shape are themselves probably unnatural. For example, Fanning, Richardson, and Edwards (1988, 1991) report three different spore genera obtained from the sporangia of about 40 different compression fossils assignable to *C. pertonii,* the type species. They acknowledge the artificial nature of *C. pertonii* and recognize three subspecies corresponding to the three spore types: *C. pertonii* ssp. *pertonii, C. pertonii* ssp. *synorispora,* and *C. pertonii* ssp. *apiculispora.*

Because of the problems outlined above, we do not attempt to treat *Cooksonia* as monophyletic. For our analysis, we have selected a representative sample of specimens currently assigned to *Cooksonia* and taxa previously assigned to the genus (see Appendix 1 for details). Our analysis of each species is based on specimens from a single locality.

Psilotaceae

Although the Psilotaceae are often hypothesized to be relevant to the early evolution of land plants (e.g., Stewart 1983), they were not included in our analysis (Appendix 1) for several reasons. First, given the problematic nature of their sporophyte and indications of filicalean affinities based on gametophyte morphology, spore structure, and mo-

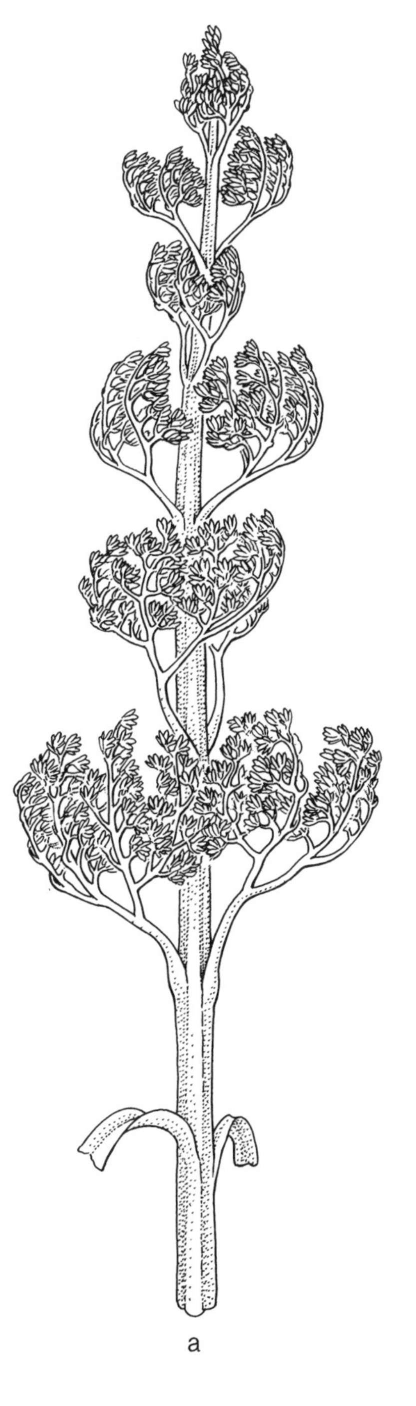

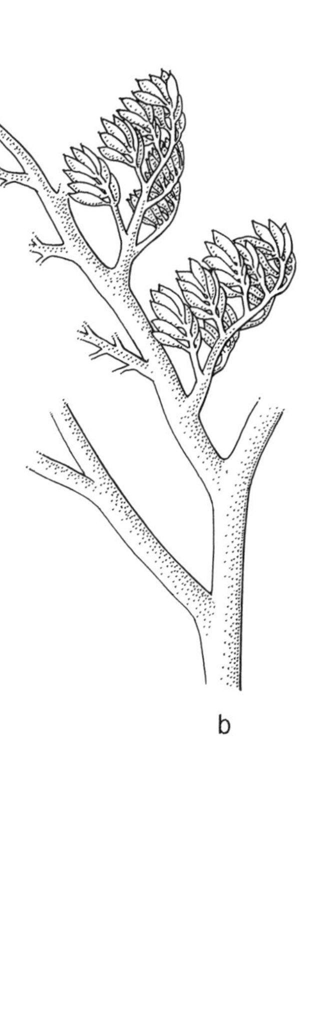

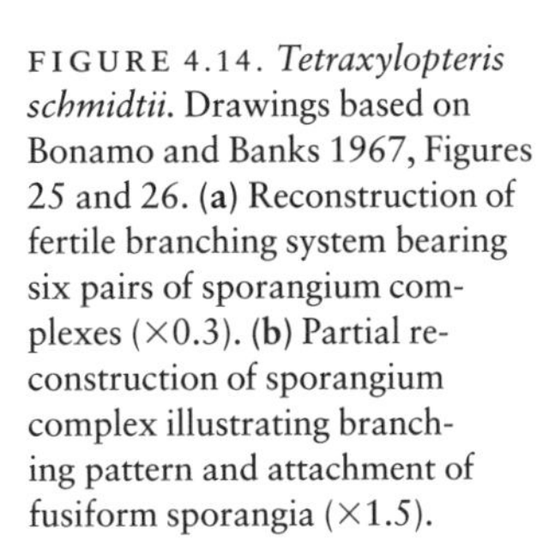

FIGURE 4.14. *Tetraxylopteris schmidtii.* Drawings based on Bonamo and Banks 1967, Figures 25 and 26. (**a**) Reconstruction of fertile branching system bearing six pairs of sporangium complexes (×0.3). (**b**) Partial reconstruction of sporangium complex illustrating branching pattern and attachment of fusiform sporangia (×1.5).

lecular data (see Chapter 7), any critical analysis of the relationships in the family must include the relevant groups of extant ferns. Unfortunately, because filicalean relationships are only currently being resolved, an adequate treatment of this question is beyond the scope of this book. We will briefly consider the relationships of Psilotaceae in Chapter 7.

FIGURE 4.15. *Rhacophyton ceratangium.* Reconstruction of basal part of a fertile frond showing two lateral units on the right, each composed of two pairs of fertile and sterile "pinnae" (×0.8). Two lateral units on the left are completely sterile. The fertile "pinnae" have been replaced by similar sterile *aphlebiae.* Reproduced from Andrews and Phillips 1968 (Text-Figure 4), with the permission of Academic Press.

CHARACTER DESCRIPTIONS AND CODING

Characters Scored for Almost All Taxa

4.1 • SPOROPHYTE INDEPENDENCE OR DEPENDENCE In extant vascular plants, the sporophyte is initially dependent on the gametophyte (absorbing nutrients through the foot) but develops complete physiological independence during early ontogeny (Bierhorst 1971). In liverworts, hornworts, and mosses, the sporophyte remains in intimate contact with gametophytic tissues during spore production (Ligrone, Duckett, and Renzaglia 1993). It is likely that the mature sporophyte attained complete physiological independence in the Rhynie Chert fossils for which in situ growth and rooting or rhizomatous structures have been demonstrated (i.e., *Rhynia, Aglaophyton, Horneophyton, Asteroxylon*). It is also probable that the sporophytes of trimerophytes and their descendants were physiologically independent because rooting structures or rhizomatous axes have been demonstrated in many taxa (e.g., *Psilophyton crenulatum:* Doran 1980; *Pseudosporochnus nodosus:* Leclercq and Banks 1962). Similarly, rhizomatous axes are known in zosterophylls (e.g., *Serrulacaulis furcatus:* Hueber and Banks 1979; *Zosterophyllum myretonianum:* Lele and Walton 1961). The independent status of other early fossil sporophytes is more problematic. Rooting structures are unknown for many rhyniophytes and other early problematic compression fossils. Furthermore, although a gametophytic structure has been described for the early, apparently monosporangiate, taxon *Sporogonites* (Andrews 1958), suggesting that the sporophyte is dependent, we have chosen to score this character as unknown because the material is so poorly preserved. The results of our analysis provide a posteriori support for the dependency of the *Sporogonites* sporophyte.

The independence or dependence of the sporophyte generation is treated as a binary character: 0 = dependent; 1 = independent. There were no inapplicable cases; eight were missing data (see Appendix 3). See character 3.2 for comparison.

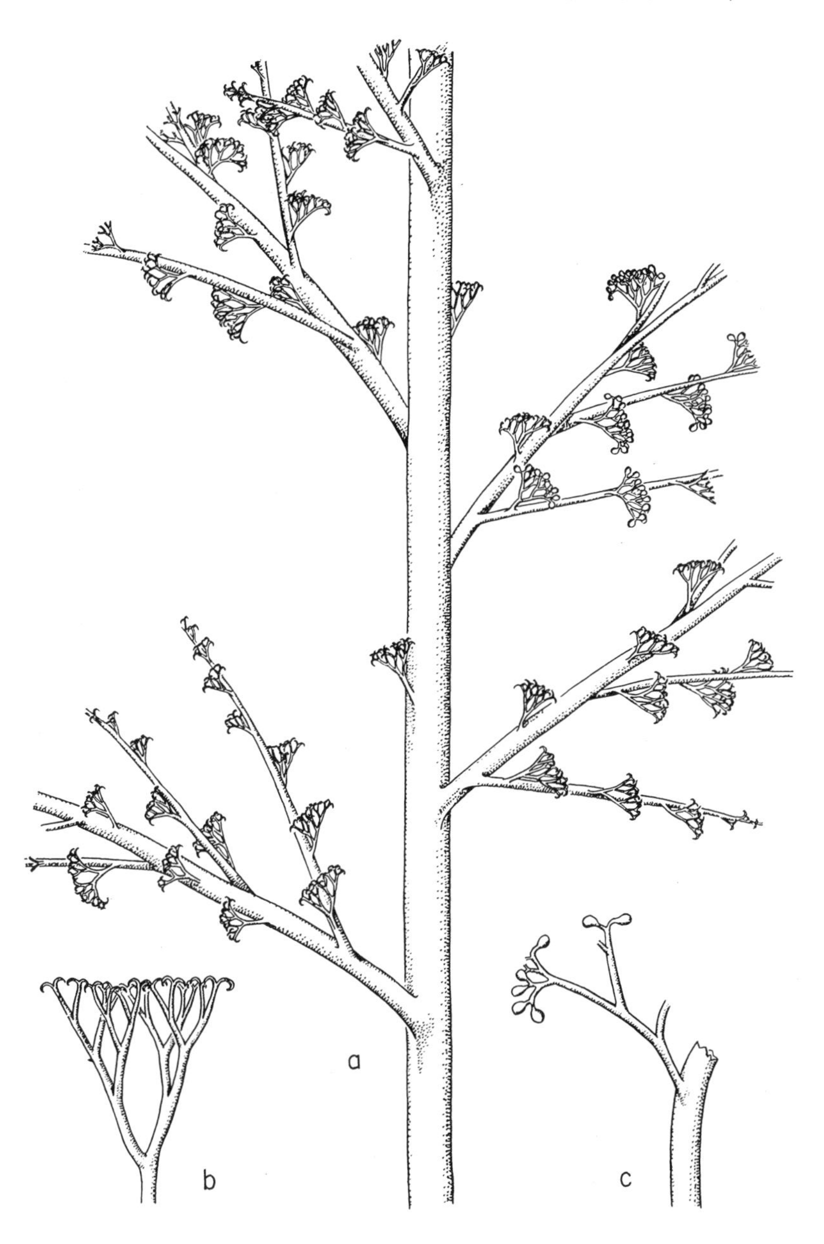

FIGURE 4.16. *Ibyka amphikoma*. Drawings reproduced, with permission, from Skog and Banks 1973, Figure 1. (**a**) Partial reconstruction of plant (×0.2). (**b**) Sterile ultimate appendage (×1). (**c**) Partial reconstruction of fertile ultimate appendage (×1).

4.2 • WELL-DEVELOPED SPORANGIOPHORE In their analysis of extant embryophytes, the unbranched seta of mosses and the branched sporangium-bearing axes of vascular plants were treated as homologous structures by Mishler and Churchill (1984) and termed an *aerial sporophyte axis*. In both groups, the sporophyte axis is persistent and internally differentiated (Figures 4.20 and 4.21). The seta of liverworts was treated as independently derived because of its ephemeral nature and differences in structure and development (e.g., Crandall-Stotler 1980; Mishler and Churchill 1984; Schuster 1966). The short or absent seta in *Sphagnum* and *Andreaea* were interpreted as examples of reduction or loss. Our analysis of relationships in basal embryophytes supports these interpretations (Chapter 3).

The presence or absence of a well-developed sporangiophore (sporophyte axis with internal differentiation) is treated as a binary character:

FIGURE 4.17. *Pseudosporochnus nodosus*. Reconstruction of a terminal branch bearing sterile and fertile ultimate units (×1). Reproduced from Leclercq and Banks 1962 (Text-Figure 3), with the permission of E. Schweizerbart'sche Verlagsbuchhandlung (Nägele u. Obermiller), Johannesstrasse 3A, D-70176 Stuttgart.

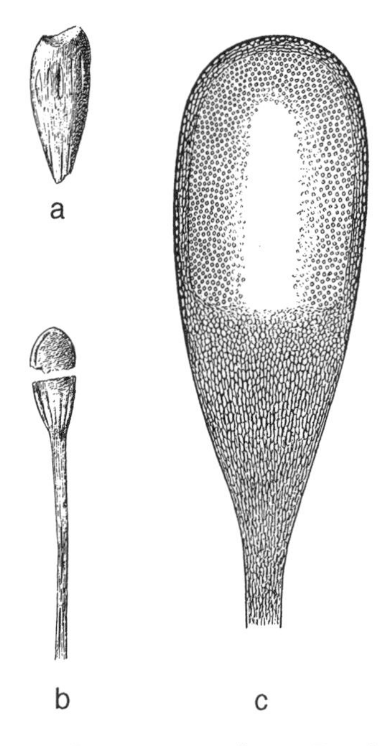

FIGURE 4.18. *Sporogonites exuberans*. Drawings reproduced from Halle 1936a (Figure 1), with permission. (**a**) Sporangium (capsule) fragment showing five short vertical slits (dehiscence features?) on the outer surface (×2). (**b**) Sporangium terminal on slender axis (×2). (**c**) Partial reconstruction of sporangium showing parts of interior and exterior (×7). Spore sac surrounds a central sterile region (columella?).

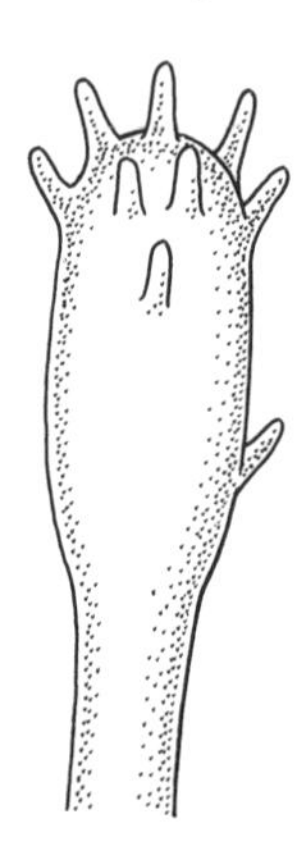

FIGURE 4.19. *Caia langii*. Reconstruction of external form of sporangium with minute sterile protrusions (×15). Redrawn from Fanning, Edwards, and Richardson 1990, Figure 5.

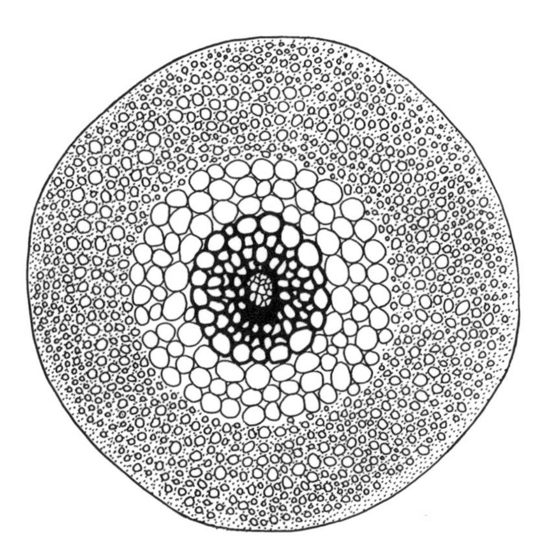

FIGURE 4.20. *Polytrichum juniperinum.* Transverse section of mature seta illustrating internal differentiation of sporophyte axis (×200). *From center to periphery:* central conducting strand of thin-walled hydroids (xylem), leptoids (phloem), inner cortex, outer cortex of distinctive thick-walled cells. Modified after Hébant 1977, Figure 113.

0 = sporangiophore absent; 1= sporangiophore present (differentiated sporophyte axis). There were no inapplicable cases; none were missing data (see Appendix 3). See character 3.3 for comparison.

4.3 • SPOROPHYTE BRANCHING The presence of a branched sporophyte producing multiple sporangia from a single fertilization event is diagnostic of all extant tracheophytes, as well as of several fossil taxa in which other kinds of conducting cells occur (e.g., *Aglaophyton, Horneophyton, Nothia;* Figure 4.21). The axis bearing the sporangium in mosses, hornworts, and liverworts is unbranched, except in a few rare instances in some mosses such as *Bryum* and *Hypnum,* where a single dichotomy may result in two sporangia (Bower 1935).

The presence or absence of a branched sporophyte is treated as a binary character: 0 = branched sporophyte (monosporangiate) absent; 1 = branched sporophyte (polysporangiate) present. There were no inapplicable cases; none were missing data (see Appendix 3). See character 3.4 for comparison.

4.4 • BRANCHING TYPE Polysporangiophytes exhibit several orders of branching. Dichotomy in which the daughter axes are approximately equal in diameter *(isotomous)* is common in at least the distal branches of many early polysporangiophytes. In some taxa, such as *Aglaophyton* (Figure 4.7), *Rhynia* (Figure 4.8), and *Cooksonia* (Figure 4.9), more or less isotomous branching appears to predominate at all levels. In other taxa, dichotomy is *anisotomous* (daughter axes of unequal diameter), and the wider axis overtops narrower laterals *(pseudomonopodial),* giving the appearance of a monopodial system. Anatomically, this pseudomonopodial branching is characterized by complete cellular continuity (Scheckler 1976, 204). It is typical of most pteridophytes, occurring for example, in zosterophylls (e.g., *Gosslingia*) and in such basal euphyllophytes as *Psilophyton* (Figure 4.13), as well as in more derived groups such as the Aneurophytales (Figure 4.14; Scheckler 1976). In monopodial branching systems, smaller branches develop laterally, often in the axils of leaves, and not by apical dichotomy. This type of branching occurs in seed plants and in some ferns.

In contrast to the pseudomonopodial condition, in monopodial branching there is a developmental lag in the lateral branch trace compared to the xylem in the main axis, and this delay can result in cell-pattern differences in the lateral branch trace of the two types of branching (Scheckler 1976). These differences notwithstanding, in most fossils it is difficult to distinguish between monopodial (nonaxillary) and pseudomonopodial branching systems. In plants in which pseudomonopodial or monopodial branching predominates, more or less isotomous dichotomy can occur in the distal regions.

Branching type is treated as a binary character: 0 = more or less isotomous branching predominating; 1 = pseudomonopodial branching. This character is inapplicable in monosporangiate extant taxa as well as in fossils presumed to be monosporangiate. There were five inapplicable cases; none were missing data (see Appendix 3).

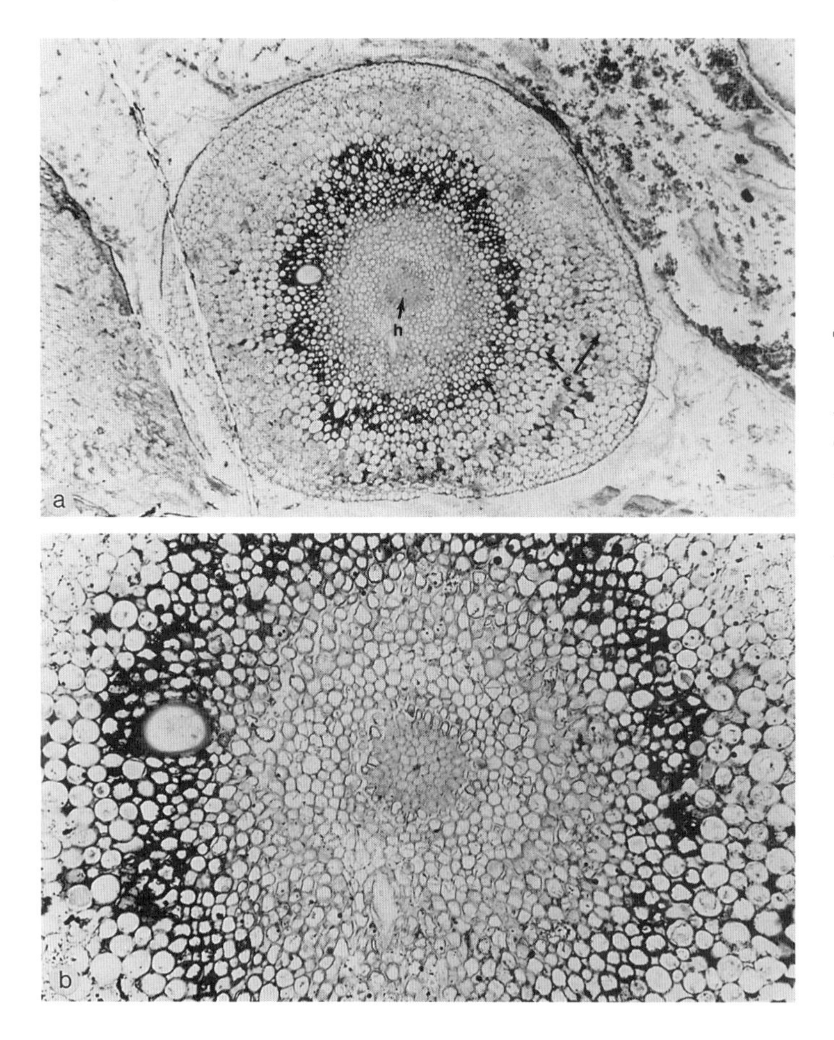

FIGURE 4.21. *Aglaophyton (Rhynia) major.* Transverse section illustrating internal differentiation of stem. Specimen from the collection of the Swedish Museum of Natural History, Stockholm. Photographs by P.K. (**a**) Complete stem, showing hydroids *(h)*, possible leptoids *(l)*, and cortex *(c)* (scale bar = 615 μm). (**b**) Central region of water-conducting cells (scale bar = 205 μm).

4.5 • BRANCHING PATTERN The pattern of branch formation in early fossil land plants encompasses many forms (Figure 4.22) and frequently can be characterized in both compressions and anatomical permineralizations. In many zosterophylls, lateral branches arise alternately on either side of the main axis but in the same plane, thus producing markedly planar branching systems (D. Edwards 1970b; Edwards, Kenrick, and Carluccio 1989; Kenrick and Edwards 1988b; Rayner 1983; Remy, Hass, and Schultka 1986; Remy, Schultka, and Hass 1986). In certain more basal Euphyllophytina (e.g., *Psilophyton*), branch departure is helical (Banks, Leclercq, and Hueber 1975), whereas in others, it is basically four-rowed, opposite, and decussate (tetrastichous) (e.g., *Tetraxylopteris*, Bonamo and Banks 1967; Scheckler and Banks 1971a). *Rhacophyton* has a distinctive quadriseriate form of branching in its basal axes (e.g., *Rhacophyton*, Cornet, Phillips, and Andrews 1976). In many early polysporangiophytes, branching clearly is nonplanar, but no regular pattern is discernible (e.g., *Rhynia, Aglaophyton, Huvenia*). This type of branching is treated provisionally in a general category of "nonplanar," but further investigation is required to more precisely characterize branching in these taxa.

Branching pattern is treated as an unordered multistate character with five character states (Figure 4.22): 0 = no marked branching pattern (absent); 1 = helical branching; 2 = planar alternate branching; 3 = tetrastichous branching; 4 =

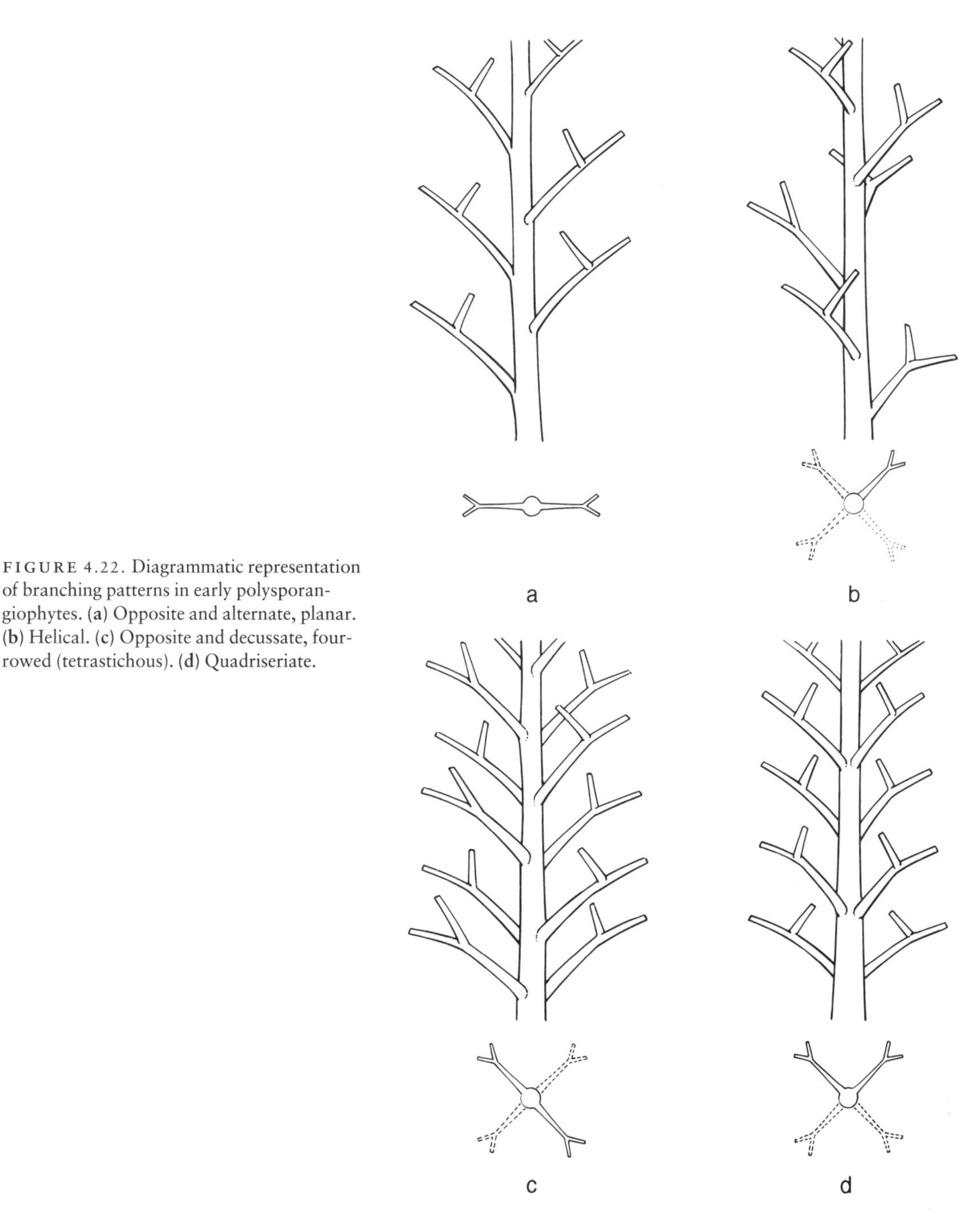

FIGURE 4.22. Diagrammatic representation of branching patterns in early polysporangiophytes. (**a**) Opposite and alternate, planar. (**b**) Helical. (**c**) Opposite and decussate, four-rowed (tetrastichous). (**d**) Quadriseriate.

quadriseriate branching. This character is inapplicable in monosporangiate extant taxa as well as in fossils presumed to be monosporangiate. There were five inapplicable cases; three were missing data (see Appendix 3). See character 5.2 for comparison.

4.6 • SUBORDINATE BRANCHING In addition to the main branching system, small subordinate (often undeveloped) axes are common in many zosterophylls. One common type of branching pattern involves the production of a single, usually small, undeveloped, circinate axis slightly below each dichotomy of the main branching system (Figure 5.15). These small axes are oriented perpendicular to the plane of the main branching system. Because of their orientation and distribution, they are usually observable in compression fossils only as small conical bumps that indicate their point of attachment to the main axis (Figures 5.15 and 5.16). Axes of this type have been termed *axillary tubercles* (D. Edwards 1970b) or *angular organs* (Remy, Hass, and Schultka 1986).

The division of the xylem involving a dichotomy of the main axis and the simultaneous production of a subordinate branch (axillary tubercle) has been observed in *Gosslingia* and *Deheubarthia,* and the pattern of associated vasculature is similar in both plants (see Chapter 5 and Figure 5.17). In *Tarella* and *Anisophyton potoniei* (Remy, Hass, and Schultka 1986), these small circinate branches develop into a dichotomous branching system that is smaller than the main axial system. Similar subordinate branching systems have been shown in *Sawdonia, Hsua,* and *Anisophyton gothanii* (Remy, Schultka, and Hass 1986).

In many zosterophylls, such as *Gosslingia, Deheubarthia, Anisophyton, Crenaticaulis,* and *Thrinkophyton,* the subordinate branch is positioned at, or slightly below, each dichotomy of the main axial system. We treat subordinate axes in this position as homologous with those that occur slightly above the dichotomy of the main axial system (e.g., *Hsua, Sawdonia*).

Subordinate axes comparable to those described above are unknown in homosporous lycopsids. The noncircinate, subordinate branching systems described by Schweitzer (1980c) and Rayner (1984) in *Drepanophycus spinaeformis* were interpreted as roots based on their orientation within the sediment and their attachment to leafless rhizomatous axes. Lateral branch formation in the fossils *Asteroxylon* and bulbil (small leafy branch) formation in extant *Huperzia selago* (Stevenson 1976) are leaflike in origin and do not involve true dichotomy, as is the case with axillary tubercle branching.

In *Psilophyton dawsonii,* one of the branches that are produced by two successive closely spaced dichotomies (termed *double dichotomies* by Banks, Leclercq, and Hueber 1975) often aborts. The result is a superficial similarity to axillary branching in zosterophylls. However, important points of difference in *P. dawsonii* are (1) that any one of the three branches produced may abort, (2) that this type of branching is confined to the lower part of the fertile region, and (3) that the aborted branch is neither circinate or basally oriented.

Subordinate branching is treated as an unordered multistate character with three states: 0 = subordinate branching absent; 1 = more or less axillary branches (Figure 5.16); 2 = branch abortion. Subordinate branching is inapplicable in monosporangiate extant taxa as well as in fossils presumed to be monosporangiate. There were five inapplicable cases; one was missing data (see Appendix 3). See character 5.3 for comparison.

4.7 • *RHYNIA*-TYPE ADVENTITIOUS BRANCHING
A distinctive type of adventitious aerial branching occurs in *Rhynia.* The axes bear small hemispherical projections that give rise to lateral branches in which the vascular strand is not continuous with that of the main axis. Similar branching occurs in *Huvenia* but has not been reported for *Stockmansella* or any other extant or fossil plant considered in this analysis.

The presence or absence of *Rhynia*-type adventitious branching is treated as a binary character: 0 = *Rhynia*-type adventitious branching absent; 1 = *Rhynia*-type adventitious branching present. *Rhynia*-type adventitious branching is inapplicable in monosporangiate extant taxa as well as in fossils presumed to be monosporangiate. There were five inapplicable cases; two were missing data (see Appendix 3).

4.8 • CIRCINATE VERNATION The axes of most zosterophylls exhibit a distinctive coiled apical region similar to circinate vernation in ferns (Figure 5.18). Circinate vernation does not occur in any extant lycopsid and is absent in early lycopsid-like fossils such as *Asteroxylon, Baragwanathia,* and *Drepanophycus.* In addition, circinate vernation has not been observed in basal polysporangiophytes such as *Rhynia, Aglaophyton, Horneo-*

phyton, and *Cooksonia*-like plants, nor in basal Euphyllophytina. Dawson's well-known original reconstruction of *Psilophyton princeps* (Figure 4.1; Dawson 1859, 1870) illustrates a plant apparently with circinate growth, but it was later shown to be based on at least three distinct plants, one of which *(Sawdonia ornata)* is a zosterophyll (Banks, Leclercq, and Hueber 1975; Hueber 1967). The vegetative apexes and terminal sporangia of species of *Psilophyton* (Figure 4.13) and other trimerophytes (*sensu* Banks) are often recurved, but this is not part of an overall circinate pattern of development. Similarly, there is recurvation but no evidence of circinate development in *Tetraxylopteris* (Figure 4.14), *Rhacophyton* (Figure 4.15), *Ibyka* (Figure 4.16), and *Pseudosporochnus* (Figure 4.17).

Circinate vernation is treated as an unordered multistate character with three states: 0 = no circinate vernation or recurvation (absent); 1 = circinate vernation (present); 2 = recurvation. There were no inapplicable cases; none were missing data (see Appendix 3). See character 5.4 for comparison.

4.9 • MULTICELLULAR APPENDAGES Nonvascular multicellular spines or teeth *(enations)* are an easily recognizable feature of many early polysporangiophytes (Figure 5.20). Multicellular spines are common in many zosterophylls, where they may be filiform (Edwards, Kenrick, and Carluccio 1989; Hueber 1971b; Lang 1932; Rayner 1984; Zdebska 1982), expanded at the tip (Hao 1989a; Remy, Hass, and Schultka 1986), or deltoid in shape (Banks and Davis 1969; Hueber 1971a; Hueber and Banks 1979). Morphologically similar spines are found in some rhyniophytes, such as *Huvenia* and *Dutoitea,* and also in some species of *Psilophyton* (Doran 1980; Gensel 1979).

Microphylls of basal homosporous lycopsids generally are somewhat larger than the spines discussed above and are either fully vascularized or clearly influence the differentiation of vascular tissue in the adjacent axis (e.g., *Asteroxylon:* Figure 6.16). We define the *microphyll* as a stem outgrowth that influences the differentiation of axial vascular tissue. This character can be observed in well-preserved compression fossils and permineralizations. In extant homosporous lycopsids, a single vascular strand is found within each leaf. In *Asteroxylon,* elongated cells (procambium?) occur in the leaf, but the differentiation of tracheids stops at the leaf base (Hueber 1992).

Multicellular appendages are treated as an unordered multistate character with three states: 0 = no appendages (absent); 1 = multicellular spines; 2 = microphylls. There were no inapplicable cases; none were missing data (see Appendix 3). See character 5.5 for comparison.

4.10 • DICHOTOMOUS PINNULELIKE APPENDAGES Isotomously branching, nonplanar, small, ultimate appendages are present in many taxa of the Euphyllophytina. These structures are leaflike in several respects but are unwebbed. In the Aneurophytales (Scheckler 1976) and in *Rhacophyton* (Cornet, Phillips, and Andrews 1976), these appendages may be planar or nonplanar. Similar structures are present also in *Ibyka* (Figure 4.16; Skog and Banks 1973), *Pseudosporochnus* (Figure 4.17; Leclercq and Banks 1962), *Eophyllophyton* (Hao and Beck 1993), and *Estinnophyton* (Fairon-Demaret 1979). Comparable small vegetative branchlets occur also in *Psilophyton dawsonii* (Figure 4.13; Banks, Leclercq, and Hueber 1975).

The presence or absence of dichotomous pinnulelike appendages is treated as a binary character: 0 = dichotomous pinnulelike appendages absent; 1 = dichotomous pinnulelike appendages present. This character is inapplicable in monosporangiate extant taxa as well as in fossils presumed to be monosporangiate. There were five inapplicable cases; one was missing data (see Appendix 3).

4.11 • CONICAL SPORANGIUM EMERGENCES Sporangia in the rhyniophytes *Caia* and *Pertonella* (Fanning, Edwards, and Richardson 1991) are distinctive in having small, more or less conical emergences (Figure 4.19). Although they are not shown in most reconstructions (e.g., Figure 3.26), similar emergences have also been documented in *Horneophyton* (Eggert 1974).

The presence or absence of conical sporangium

emergences is treated as a binary character: 0 = conical sporangium emergences absent; 1 = conical sporangium emergences present. There were no inapplicable cases; none were missing data (see Appendix 3).

4.12 • STOMATES The presence or absence of stomates are treated as a binary character: 0 = stomates absent; 1 = stomates present. There were no inapplicable cases; 15 were missing data (see Appendix 3). See character 3.15 for detailed comparisons.

4.13 • STEROME The stems of many early polysporangiophytes have a well-developed peripheral zone *(sterome)* consisting of several layers of thick-walled, decay-resistant cells (Figure 5.21; D. Edwards 1993). This zone is continuous in many zosterophylls, including *Gosslingia, Sawdonia,* and *Crenaticaulis.* No equivalent tissue has been recorded in the basal members of the Lycopodiales, and although thick-walled cortical tissues are present in extant Lycopodiaceae, these tissues are unlignified in the basal members of the clade (e.g., *Huperzia*).

A sterome is present in many tracheophytes and has been recorded in one species of *Cooksonia (C. pertonii)* and in plants of broadly similar morphology, such as *Uskiella,* but is unknown in many taxa in which the anatomy is not preserved. Plants from the Rhynie Chert, such as *Rhynia, Aglaophyton, Horneophyton, Nothia,* and *Asteroxylon,* do not possess a thick-walled, decay-resistant peripheral zone. In *Psilophyton dawsonii,* the sterome is similar to that in zosterophylls (Banks, Leclercq, and Hueber 1975), but in more derived members of the Euphyllophytina, such as *Pseudosporochnus* (Leclercq and Banks 1962; Leclercq and Lele 1968; Stein and Hueber 1989), the aneurophyte *Tetraxylopteris* (Scheckler and Banks 1971a; Scheckler and Banks 1972), and possibly *Ibyka* (Skog and Banks 1973), peripheral decay-resistant tissues form a discontinuous sheath of alternating bundles of fibers and parenchyma *(sparganum cortex).*

A sterome is treated as an unordered multistate character with three states: 0 = no sterome (absent); 1 = continuous sterome several cell layers thick; 2 = discontinuous sterome of fiber bundles. This character is inapplicable in taxa lacking well-developed sporangiophores (e.g., liverworts, hornworts). There were 3 inapplicable cases; 12 were missing data (see Appendix 3).

4.14 • XYLEM STRAND SHAPE The shape of the xylem strand in transverse section exhibits a variety of forms in early polysporangiophytes (Figure 4.23). In most zosterophylls the xylem strand of the main axes ranges from a solid elliptical to a strap-shaped protostele. This has been illustrated in *Gosslingia, Sawdonia, Barinophyton, Crenaticaulis, Deheubarthia, Konioria, Thrinkophyton, Zosterophyllum llanoveranum, Huia,* and possibly *Krithodeophyton.* In some of the smaller species, the xylem strand is only slightly elliptical or terete. In *Zosterophyllum llanoveranum,* the strand is elliptical within or close to the fertile spike, but terete in the region below. In *Z. fertile,* the strand appears also to be terete. In *Nothia,* the strand is slightly elliptical, and in *Huia, Trichopherophyton,* and possibly *Krithodeophyton,* it is elliptical to terete. In *Yunia dichotoma,* the xylem strand is clearly terete (Hao and Beck 1991b).

A solid stellate protostele is typical of the extant Lycopodiaceae (Wardlaw 1924) and of the early fossil lycopsids *Asteroxylon, Baragwanathia,* and *Drepanophycus* (see the discussion of lycopsid characters in Chapter 6). In larger or horizontal axes of some extant species, the stellate strand becomes dissected into a series of anastomosing bands.

The basal members of the Euphyllophytina, such as *Psilophyton,* have a xylem strand that is more or less terete in cross section (Figure 4.24). In the larger axes of *P. dawsonii,* the xylem strand may become weakly three-lobed (Banks, Leclercq, and Hueber 1975). In more derived taxa, such as the Aneurophytales, the xylem may be three-lobed (Stein 1993; Wight 1987) or four-lobed, as in *Tetraxylopteris* (Figure 4.24; Scheckler 1976; Scheckler and Banks 1971a). A predominantly three-lobed xylem strand also characterizes *Ibyka* (Skog and Banks 1973). Wight (1987) and Stein (1993)

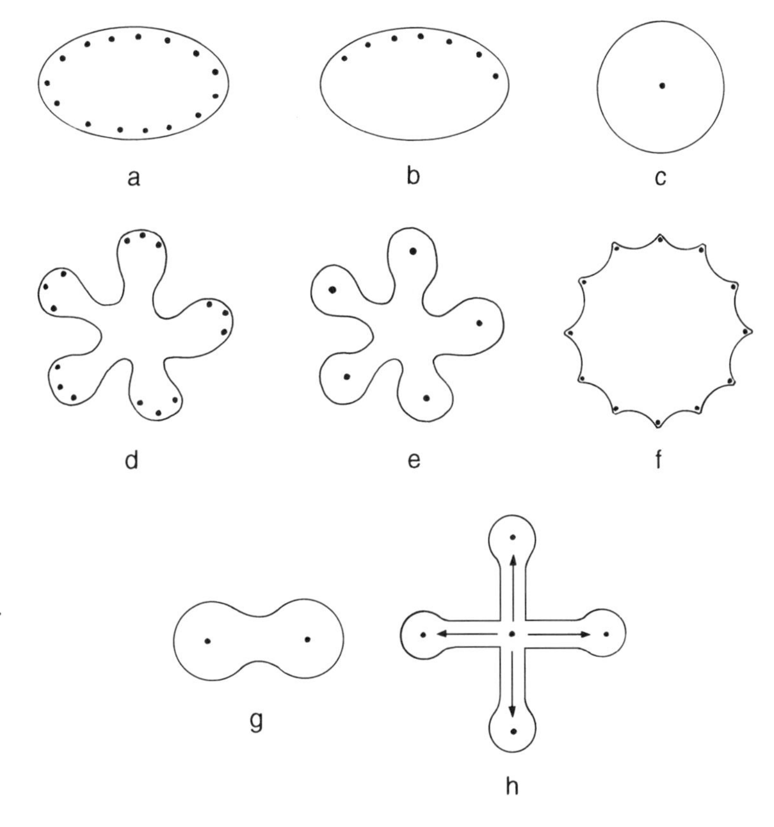

FIGURE 4.23. Diagrammatic representation of primary xylem shape and protoxylem position in early polysporangiophytes. Black dots mark the position of protoxylem elements. (**a**) Zosterophylls: elliptical, exarch. (**b**) Zosterophylls: elliptical, exarch on one side only. (**c**) Basal Euphyllophytina: terete, centrarch. (**d**) Lycopodiaceae and some early fossil lycopsids (e.g., *Asteroxylon*): stellate, exarch. (**e**) Some lycopsids: stellate, weakly mesarch. (**f**) Protolepidodendrales and ontogenetic stages of rhizomorphic lycopsids: ribbed, exarch. (**g**) Branch trace in some early putative ferns: clepsydroid, mesarch. (**h**) Aneurophytales and some trimerophytes: lobed, centrarch.

have shown that in the Aneurophytales, the outline of the xylem is probably a direct result of branch formation and of the close internodal spacing. In *Rhacophyton,* the primary xylem is clepsydroid (Figure 4.23) in cross section (Andrews and Phillips 1968; Dittrich, Matten, and Phillips 1983), and in *Pseudosporochnus,* the axis is polystelic with clepsydroid branch strands (Leclercq and Banks 1962; Leclercq and Lele 1968; Stein and Hueber 1989). In the basal polysporangiophytes and in mosses for which conducting strands are present, the xylem (or its probable homologue in mosses) is approximately terete in cross section.

The shape of the xylem strand is treated as an unordered multistate character with five states (Figure 4.23): 0 = xylem strand approximately terete; 1 = xylem strand elliptical to strap-shaped; 2 = xylem strand stellate; 3 = xylem strand lobed; 4 = xylem strand clepsydroid. This character is inapplicable in taxa lacking well-developed sporangiophores. There were 3 inapplicable cases; 11 were missing data (see Appendix 3). See character 5.7 for comparison.

4.15 • PROTOXYLEM The position of the protoxylem is a distinctive feature in many early polysporangiophytes (Figure 4.23). In most zosterophylls, the protoxylem is interpreted as *exarch* (protoxylem peripheral to metaxylem), based on the presence of cells with small cross-sectional diameters in a peripheral position in the xylem strand. The exact distribution of protoxylem around the edge of the xylem strand has not been investigated in detail and may provide further useful characters. In *Gosslingia,* and possibly in *Deheubarthia* and *Konioria,* the protoxylem appears to be on *one side* only of the xylem strand. An exarch xylem strand has been described in *Gosslingia, Sawdonia, Barinophyton, Crenaticaulis, Konioria, Thrinkophyton, Trichopherophyton,* and possibly also in *Zosterophyllum llanoveranum.*

In the extant Lycopodiaceae, xylem maturation

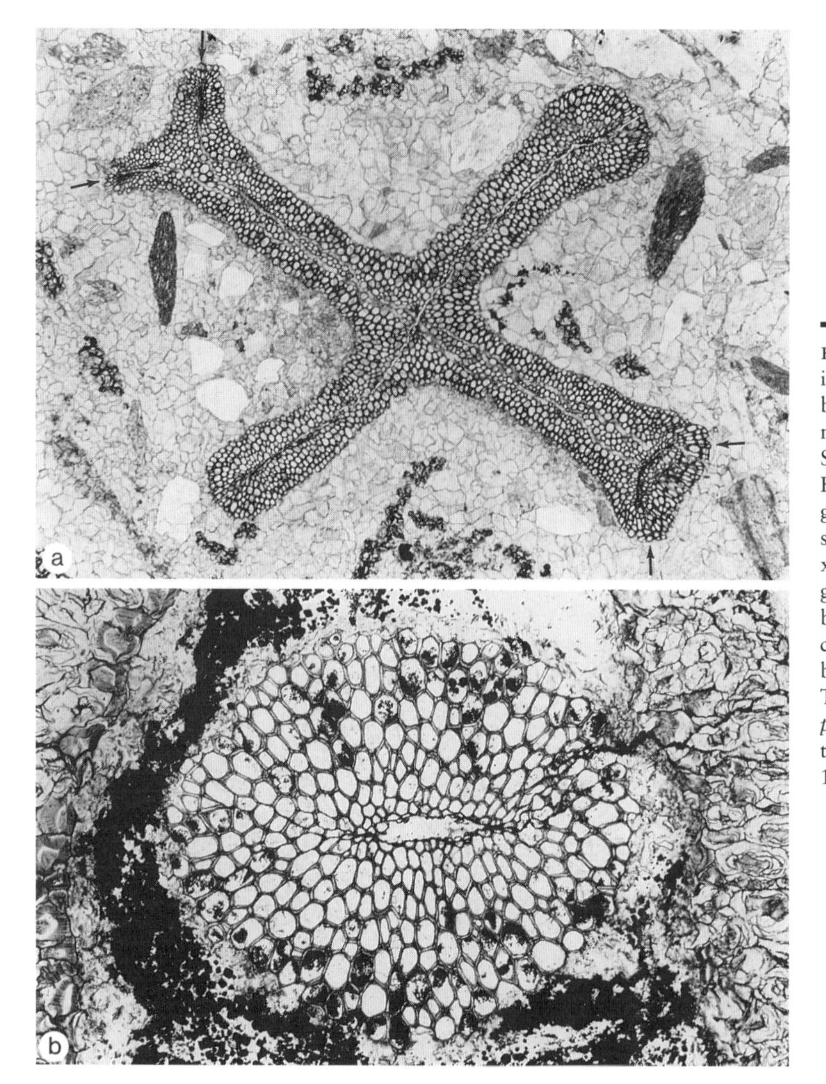

FIGURE 4.24. Xylem shape in transverse section in two basal Euphyllophytina. Specimens from the collection of the Swedish Museum of Natural History, Stockholm. Photographs by P.K. (**a**) Transverse section of four-lobed primary xylem from a probable progymnosperm precursor (scale bar = 370 μm). *Arrows* indicate lobes forming prior to branch trace departure. (**b**) Terete xylem strand of *Psilophyton dawsonii* with centrarch protoxylem (scale bar = 100 μm).

in the stem is almost uniformly exarch, protoxylem being situated at the outer edge of each xylem arm. Only rarely in large stems is slight mesarchy observed (Bierhorst 1971). Xylem maturation in the early fossil lycopsid *Drepanophycus* appears to be exarch (Rayner 1984) or possibly mesarch (Fairon-Demaret 1971). In *Asteroxylon,* both exarch and shallowly mesarch conditions seem to occur (Kidston and Lang 1920b).

Centrarch xylem maturation (protoxylem central and surrounded by metaxylem) is typical in such basal members of the Euphyllophytina as *Psilophyton* (Figure 4.24; Doran 1980; Gensel 1979; Hartman and Banks 1980). Similarly, the Aneurophytales have been interpreted as "fundamentally centrarch" (Wight 1987, 209): the protoxylem strands that are present in the arms of the lobed xylem are all connected to a central strand, and in this respect their development is similar to that in *Psilophyton.* Centrarch maturation of conducting tissues appears to be present in some basal polysporangiophytes such as *Nothia* and *Yunia* (diarch in part). However, in other taxa, which are often described as centrarch (e.g., *Horneophyton, Rhynia*), and in other basal polysporangiophytes such as *Aglaophyton* and *Stockmansella,* the distinction between protoxylem and metaxylem on the basis of cell size is difficult and sometimes impossible to make. This situation suggests that protoxylem formation may not be a feature of some of these groups. Cell patterns that could be interpreted as reflecting protoxylem-metaxylem differentiation are not seen in the water-conducting strands of bryophytes (Hébant 1977).

So-called mesarch xylem maturation is typical in *Rhacophyton* (Dittrich, Matten, and Phillips 1983), *Pseudosporochnus* (Leclercq and Banks 1962; Stein and Hueber 1989), and *Ibyka* (Skog and Banks 1973). In *Pseudosporochnus hueberi,* the position and distribution of the protoxylem varies between the peripheral and central xylem strands, from discrete strands near the outer edge to a more median position (Stein and Hueber 1989).

The absence or presence and type of protoxylem is treated as an unordered multistate character with four states: 0 = protoxylem absent; 1 = protoxylem centrarch; 2 = protoxylem exarch; 3 = protoxylem mesarch. This character is inapplicable in taxa lacking well-developed sporangiophores. There were 3 inapplicable cases; 13 were missing data (see Appendix 3). See character 5.8 for comparison.

4.16–4.19 • TRACHEID CELL WALL We recognize four separate characters in the tracheid cell wall: (1) a thick, lignified, decay-resistant layer (present in most but not all tracheids); (2) helical-annular thickenings; (3) metaxylem pitting; and (4) small pitlets or holes in the lignified wall layer either between the gyres of annular or helical thickenings or within pits (Figures 4.25 and 4.26). The distribution of these features in selected taxa from this analysis is shown in Table 4.5. The four characters of the tracheid cell wall are treated individually in the next four sections.

4.16 • DECAY-RESISTANCE OF THE TRACHEID CELL WALL The tracheid cell walls of most vascular plants typically are relatively thick because of extensive lignification of the cellulose fibers, which confers decay resistance in fossils (Kenrick and Edwards 1988a). Thick-walled, decay-resistant tracheids are typical of fossil euphyllophytes (e.g., *Psilophyton dawsonii*) and of the Lycopodiales (e.g., *Leclercqia, Minarodendron*). The pattern of lignification is somewhat less extensive in the zosterophylls and basal lycopsids (Kenrick and Crane 1991; Kenrick and Edwards 1988a) but still comprises more than half of the cell wall thickness. In all of these cells, the decay-resistant part of the cell wall is a solid, homogenous layer (Figures 4.25 and 4.26).

In *Nothia aphylla,* the water-conducting cells have no thickenings or pitting, yet their walls are of comparable thickness (and contrast in light microscopy) to those of the lycopsid *Asteroxylon* from the Rhynie Chert, and appear considerably more robust than the water-conducting cells of *Rhynia* and *Aglaophyton* from the same locality. In the S-type cells described by Kenrick, Edwards, and Dales (1991), Kenrick and Crane (1991), and Kenrick, Remy, and Crane (1991) in *Sennicaulis hippocrepiformis, Rhynia gwynne-vaughanii, Stockmansella langii, Huvenia kleui,* and *Sciadophyton* sp., the cell wall between thickenings is thinner, and the continuous, decay-resistant inner layer is extremely thin. Decay resistance manifests itself also in an enigmatic heterogeneous "spongy" layer. In many taxa with unthickened water-conducting cells, such as *Aglaophyton* and the polytrichaceous mosses, the cell walls are very thin and there is little or no decay resistance or lignification.

The decay-resistance of the tracheid cell wall is treated as a binary character: 0 = limited or no decay resistance; 1 = extensive decay resistance (lignification) of the cell wall. This character is inapplicable in taxa lacking well-developed sporangiophores. There were three inapplicable cases; eight were missing data (see Appendix 3).

4.17 • TRACHEID THICKENINGS The water-conducting tissue of many early polysporangiophytes consists entirely of tracheids with annular or helical thickenings (Kenrick and Crane 1991), whereas in other extant and fossil lycophytes and in euphyllophytes with various types of pitted metaxylem elements, cells possessing annular or helical thickenings are confined to the early-formed proto-

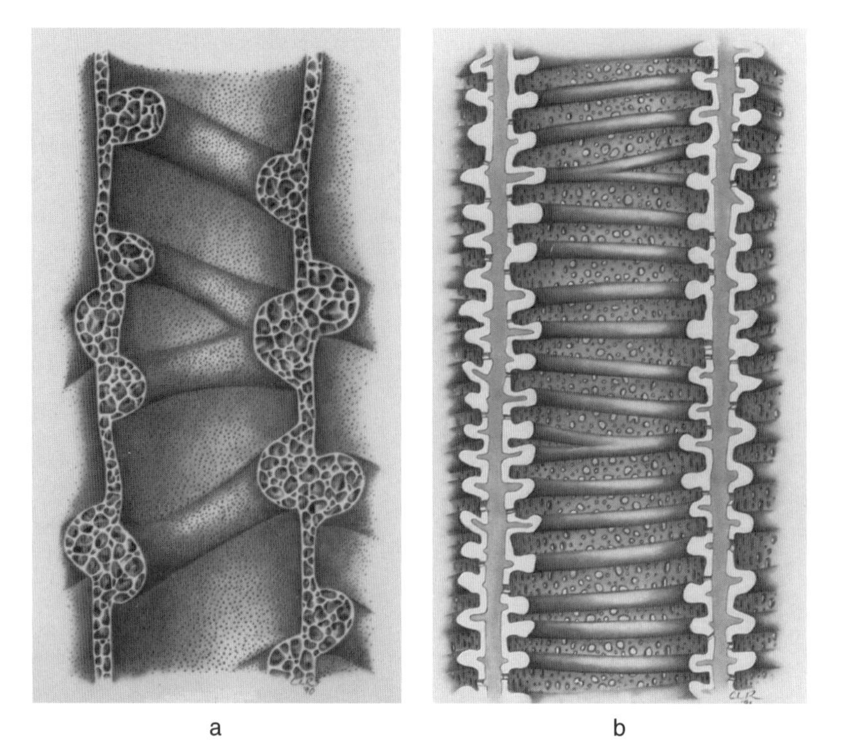

FIGURE 4.25. Reconstructions of tracheids from two lower Devonian plants, (**a**) *Sennicaulis hippocrepiformis,* and (**b**) *Gosslingia breconensis.* Drawings of cut-away longitudinal sections, showing interior of cells, reproduced from Kenrick and Crane 1991, Figures 22 and 23. (**a**) S-type tracheid typical of Rhyniopsida (×800). The two-layered cell wall comprises a very thin decay-resistant inner layer facing cell lumen and a "spongy" outer layer, which is partially mineralized in fossils. Minute plasmodesmata-sized perforations occur over the entire inner surface of the cell wall. (**b**) G-type tracheid typical of many zosterophylls and early lycopsids (×800). The two-layered cell wall comprises a decay-resistant inner layer facing cell lumen and a nonresistant outer layer, which is often heavily mineralized in fossils. Distinct perforations occur in the decay-resistant wall between thickenings.

xylem (Figures 4.25 and 4.26; Bierhorst 1960, 1971). Cells with annular or helical types of wall thickening have been described in all zosterophylls that have been investigated using scanning electron microscopy and in basal fossil members of the Lycopodiales, including *Asteroxylon, Baragwanathia,* and *Drepanophycus* (Kenrick and Crane 1991). Helically thickened cells also typify the xylem strand of such early polysporangiophytes as *Rhynia, Stockmansella,* and *Huvenia* (Kenrick and Crane 1991; Kenrick, Edwards, and Dales 1991; Kenrick, Remy, and Crane 1991).

The presence or absence of helical and annular thickening in tracheids is treated as a binary character: 0 = unthickened cells (absent); 1 = simple annular or helical thickenings in metaxylem, protoxylem, or both (present). This character is inapplicable in taxa lacking well-developed sporangiophores. There were three inapplicable cases; six were missing data (see Appendix 3). See characters 3.7 and 5.9 for comparison.

4.18 • METAXYLEM TRACHEID PITTING Pitting in the cell wall of metaxylem tracheids is found in several plant groups by the Emsian stage of the late lower Devonian (Figure 4.26). Prior to this time, metaxylem cell wall thickening was exclusively annular or helical. Scalariform pitting is typical of

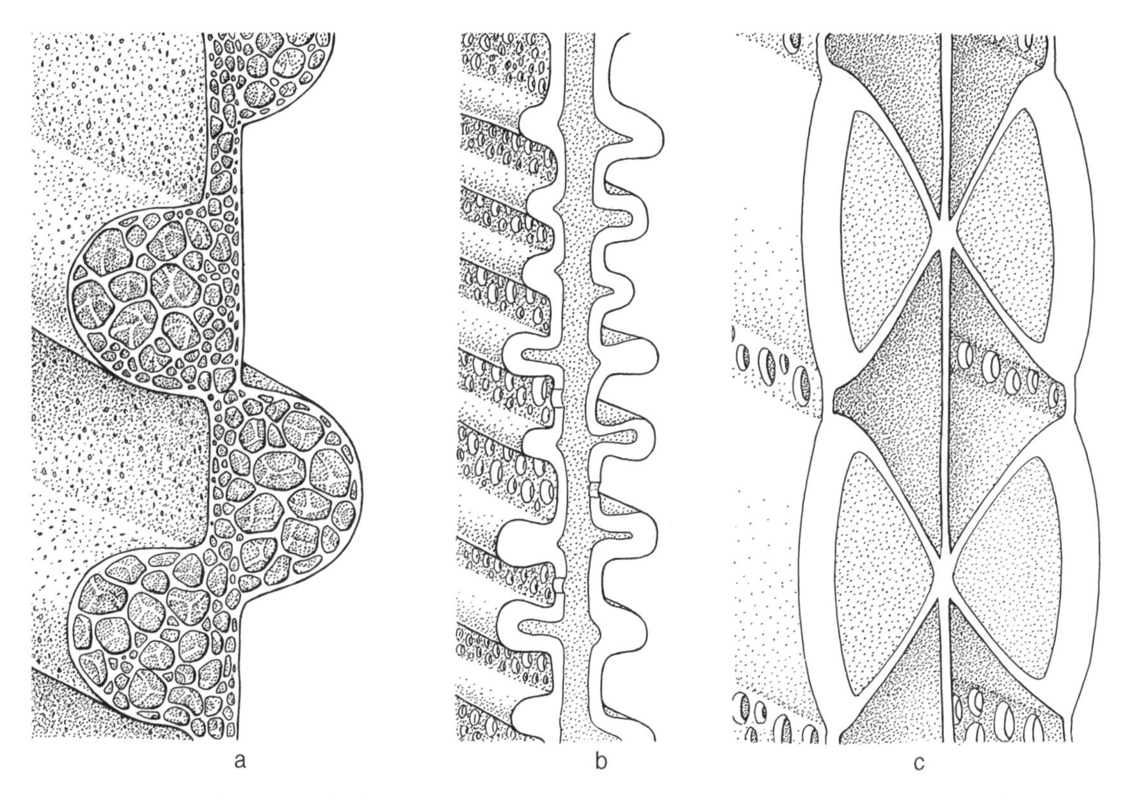

FIGURE 4.26. Details of tracheid cell wall construction in three distinct tracheid types common in early vascular plants. All drawn to the same scale (×2,500). (**a**) S-type cell wall with helical thickenings typical of Rhyniopsida. The two-layered cell wall comprises a very thin decay-resistant inner layer facing cell lumen and a "spongy" outer layer, which is partially mineralized in fossils. Minute plasmodesmata-sized perforations occur over the entire inner surface of the cell wall. (**b**) G-type cell wall with annular-reticulate thickenings typical of many zosterophylls and early lycopsids. The two-layered cell wall comprises a decay-resistant inner layer facing cell lumen *(white)* and a nonresistant outer layer *(stippled)*, which is often heavily mineralized in fossils. (**c**) P-type cell with scalariform pitting typical of plesiomorphic Euphyllophytina (e.g., *Psilophyton dawsonii*). The two-layered cell wall comprises a decay-resistant inner layer facing the cell lumen *(white)*, pit chambers, and a nonresistant inner layer within the scalariform bars, which is often replaced by mineral in fossils. Characteristically, the scalariform pit apertures are covered with a perforate sheet of decay-resistant material.

TABLE 4.5

Distribution of Tracheid Wall Characters among Ten Critical Fossil Genera

Genus	Annular or helical thickening	Thick, inner, decay-resistant layer to cell wall	Pitlets restricted to area between thickenings or within pits	Scalariform or circular pitting
Horneophyton	–	–	–	–
Aglaophyton	–	–	–	–
Rhynia	+	–	–	–
Gosslingia	+	+	+	–
Nothia	–	+	–	–
Asteroxylon	+	+	+	–
Leclercqia	+	+	–	+
Lepidodendron	+	+	+	+
Psilophyton	+	+	+	+
Tetraxylopteris	+	+	?	+

Note: See also character descriptions 4.16–4.19.

many early fossil Lycopodiales, including *Minarodendron* and the arborescent lycopsids of the Carboniferous Period (Cichan, Taylor, and Smoot 1981). Bordered pitting of the scalariform or circular type is also found in this group, including in *Leclercqia* and in extant *Huperzia selago* (Bierhorst 1960). Scalariform bordered pitting predominates in basal fossil euphyllophytes such as *Psilophyton* (Doran 1980; Gensel 1979; Hartman and Banks 1980), *Ibyka* (Skog and Banks 1973; Stein 1982), and *Rhacophyton* (Dittrich, Matten, and Phillips 1983). Both scalariform pits and circular bordered pits are present in *Pseudosporochnus* (Leclercq and Banks 1962; Stein and Hueber 1989) and *Tetraxylopteris* (Scheckler and Banks 1971a).

The presence or absence of pitting in metaxylem tracheids is treated as a binary character: 0 = scalariform or circular bordered pitting absent; 1 = scalariform or circular bordered pitting present. This character is inapplicable in taxa lacking well-developed sporangiophores. There were three inapplicable cases; seven were missing data (see Appendix 3). See character 5.9 for comparison.

4.19 • SIMPLE PITLETS IN THE TRACHEID WALL The presence of irregular pitlike openings *(pitlets)* in the cell wall is a common feature in many early polysporangiophyte taxa (Figures 4.25 and 4.26). Such structures are found between annular and helical bars in zosterophylls and basal lycopsids (Kenrick and Crane 1991), as well as in the pit apertures of more derived fossil lycopsids, such as *Minarodendron,* and *Selaginellites* (Rowe 1988b), and in the arborescent lycopsids of the Carboniferous Period (Cichan, Taylor, and Smoot 1981). Similar structures are found in basal members of the Euphyllophytina, such as *Psilophyton* (Doran 1980; Gensel 1979; Hartman and Banks 1980). This feature appears to have been lost in many later fossils and also in extant groups. However, remnants of these pitlike openings may occur in the early-formed protoxylem of the Lycopodiaceae, Equisetaceae, and Ophioglossaceae (Bierhorst 1960).

The presence or absence of simple pitlets in the tracheid wall is treated as a binary character: 0 = pitlets absent; 1 = pitlets present. This character is inappplicable in taxa lacking well-developed sporangiophores. There were 3 inapplicable cases; 16 were missing data (Appendix 3).

4.20 • ALIGNED XYLEM The radially aligned metaxylem typical of many basal Euphyllophytina and arborescent lycopsids differs in important respects from the more familiar woody tissue derived from a bifacial cambium in seed plants (Cichan 1985a,b, 1986a,b; Cichan and Taylor 1982, 1990; Gensel and Andrews 1984; Leclercq and Bonamo 1971; Rothwell and Pryor 1991).

The cambium of arborescent lycopsids is unifacial and apparently a defining feature of that group. It is very unlikely to be homologous with that in seed plants. This conclusion, based on an analysis of character distribution, is corroborated by detailed histological study that highlights the developmental differences of the unifacial cambium of the arborescent lycopsids from that of the bifacial cambium of seed plants (Cichan 1985a; Rothwell and Pryor 1991).

In the basal euphyllophytes, many taxa exhibit a radial alignment of the metaxylem, but there is often no histological evidence of cambial activity, and ray cells are absent. The developmental processes underlying radial alignment in these instances are unclear (Gensel and Andrews 1984; Leclercq and Bonamo 1971) but may involve radial divisions of incipient cambial cells. Radially aligned xylem, whether or not the product of cambial activity, has been noted in the larger axes of *Psilophyton dawsonii* (Banks, Leclercq, and Hueber 1975), *P. crenulatum* (Doran 1980), the Aneurophytales (Beck and Wight 1988; Scheckler and Banks 1971a), *Rhacophyton* (Andrews and Phillips 1968; Dittrich, Matten, and Phillips 1983), *Pseudosporochnus* (Stein and Hueber 1989, 351), and the cladoxylalean *Rhymokalon* (Scheckler 1975).

The presence or absence of aligned xylem cells is treated as a binary character: 0 = alignment of xylem cells absent; 1 = radial alignment of xylem cells present. This character is inapplicable in taxa lacking well-developed sporangiophores. There were three inapplicable cases; nine were missing data (see Appendix 3).

4.21 • XYLEM RAYS Ray cells are present in some of the more derived members of the Euphyllophytina. Uniseriate and multiseriate rays have been identified in *Tetraxylopteris* (Aneurophytales: Scheckler and Banks 1971a) and uniseriate rays are present in *Rhacophyton* (Dittrich, Matten, and Phillips 1983).

The presence or absence of xylem rays is treated as a binary character: 0 = rays absent; 1 = rays present. This character is inapplicable in taxa lacking well-developed sporangiophores. There were three inapplicable cases; six were missing data (see Appendix 3).

4.22 • SPORANGIOTAXIS There are several readily recognizable patterns of sporangium arrangement among the zosterophylls. The most common one consists of two vertical rows of pedicellate sporangia that may become one row in the basal or distal regions (Figure 5.24). Where two vertical rows are present, they are usually on opposite sides of the axis and in the same plane as the major plane of branching (e.g., *Zosterophyllum* subgenus *Platyzosterophyllum;* Gensel 1982a; Gerrienne 1988; *Crenaticaulis, Thrinkophyton:* Figure 5.11). Another arrangement found in some zosterophylls is dorsiventral rows (Remy, Hass, and Schultka 1986; Remy, Schultka, and Hass 1986). In putatively basal Lycophytina, sporangia are arranged around the axis, possibly in helixes. This pattern is characteristic of such taxa as *Huia,* several species of *Zosterophyllum* subgenus *Zosterophyllum (Z. myretonianum, Z. deciduum), Nothia, Gumuia,* and *Hicklingia.* Helical sporangiotaxy is characteristic also of many extant Lycopodiaceae (Bierhorst 1971; Øllgaard 1987). Among early fossil lycopsids, sporangia appear to be helical in *Baragwanathia,* but in *Drepanophycus spinaeformis* and *Asteroxylon mackiei,* sporangiotaxy is poorly understood; however, the sporangia do not appear to be in one or two vertical rows but rather arranged around the axis.

In taxa with terminal sporangia, such as bryophytes, early polysporangiophytes, and many basal members of the Euphyllophytina, the sporangia cannot be said to exhibit sporangiotaxis. In potentially more derived members of the euphyllophytes, such as *Tetraxylopteris* (Figure 4.14; Bonamo and Banks 1967) and *Rhacophyton* (Figure 4.15, Andrews and Phillips 1968), sporangia appear to be arrayed on one side of a dichotomous, nonplanate, leaflike structure.

Sporangiotaxis is treated as an unordered multistate character with three states (Figure 5.24): 0 = no sporangiotaxis (none); 1 = rowed arrangement; 2 = helical sporangia. This character is inapplicable in taxa that are monosporangiate and those that have terminal sporangia. There were 27 inapplicable cases; none were missing data (see Appendix 3). See character 5.10 for comparison.

4.23 • SPORANGIUM ATTACHMENT The distribution of sporangia on the axes of early polysporangiophytes shows three distinctive patterns. In zosterophylls and lycopsids, sporangia are attached laterally on short stalks (Figures 5.1 and 5.24); in some basal taxa, such as *Zosterophyllum myretonianum, Nothia,* and *Asteroxylon,* these stalks are known to be vascular. In bryophytes, sporangia are terminal on the end of a short axis (seta). Similarly, in basal polysporangiophytes, they are often solitary at the ends of the main branching system, for example in *Aglaophyton* (Figure 4.7), *Cooksonia* (Figure 4.9), and *Uskiella* (Figure 4.10). Sporangia in *Rhynia* and *Stockmansella* are lateral and sessile, whereas in *Huvenia* they terminate short, branched axes.

Sporangium attachment is treated as an unordered multistate character with three states: 0 = sporangia terminal; 1 = sporangia terminal on short, lateral unbranched stalks; 2 = sporangia lateral sessile. There were no inapplicable cases; none were missing data (see Appendix 3). See character 5.18 for comparison.

4.24 • SPECIALIZED FERTILE ZONE In trimerophytes, progymnosperms, and many early fernlike taxa, sporangia are paired in densely branched clusters at the ends of smaller branches (e.g., *Psi-*

lophyton crenulatum: Doran 1980; *Ibyka:* Skog and Banks 1973; Figure 4.16) or arranged more or less on one side of dichotomously branched leaflike structures (e.g., *Tetraxylopteris:* Figure 4.14; *Rhacophyton:* Figure 4.15; Andrews and Phillips 1968; Bonamo and Banks 1967).

The presence or absence of a specialized fertile zone is treated as a binary character: 0 = specialized fertile zone absent; 1 = fertile zone of densely branched sporangium clusters present. This character is inapplicable in monosporangiate taxa. There were six inapplicable cases; none were missing data (see Appendix 3).

4.25 • SPECIALIZED SPORANGIUM-AXIS JUNCTION In most early polysporangiophytes the sporangium-axis junction may be described as nonspecialized: the sporangium is attached directly to the axis and the morphological and anatomical transition between axis and sporangium is continuous. However, in several fossils *(Stockmansella, Huvenia, Rhynia gwynne-vaughanii),* there is a distinctive pad of tissue into which the sporangium is slightly sunken (Figures 4.27 and 4.28). This pad is situated on the end of a short axis or attached laterally on the main axes.

The presence or absence of a specialized sporangium-axis junction is treated as a binary character: 0 = specialized pad of tissue absent; 1 = specialized pad of tissue present. There were no inapplicable cases; none were missing data (see Appendix 3).

4.26 • SPORANGIUM ABSCISSION Sporangium abscission is rare in early polysporangiophytes but appears to be characteristic of several taxa. Sporangium abscission occurs in *Stockmansella* and probably also in *Rhynia gwynne-vaughanii* (Figure 4.28; D. S. Edwards 1980, 1986), and an abscission layer forms at the base of the sporangium in both species. The occurrence of sporangium abscission is unknown in the related taxon *Huvenia.* Sporangium abscission has also been recorded in *Zosterophyllum deciduum* (Figure 5.9; Gerrienne 1988).

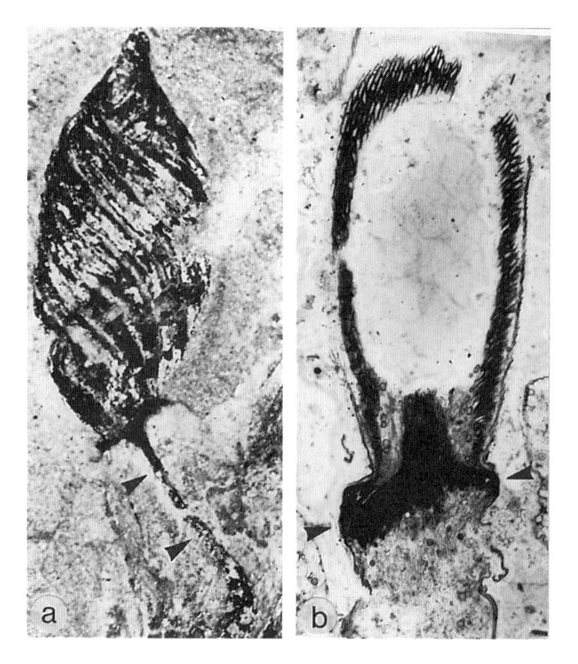

FIGURE 4.27. Sporangia of Rhyniopsida. (**a**) Compression fossil of *Huvenia kleuii* showing sporangium with multiple slit dehiscence attached to an axis (×9). Axis contains the coalified remains of a vascular strand *(arrows).* Reproduced, with permission, from Hass and Remy 1991, Plate 49 and Figure 5 (specimen from the Remy Collection, Abt. Paläobotanik, University of Münster). (**b**) Oblique longitudinal section through silicified sporangium of *Rhynia gwynne-vaughanii* showing orientation of epidermal cells and with the conspicuous dark "pad" of tissue at the base *(arrows)* (×10). Specimen from the Natural History Museum, London, Lang Collection 125. Photograph by P.K.

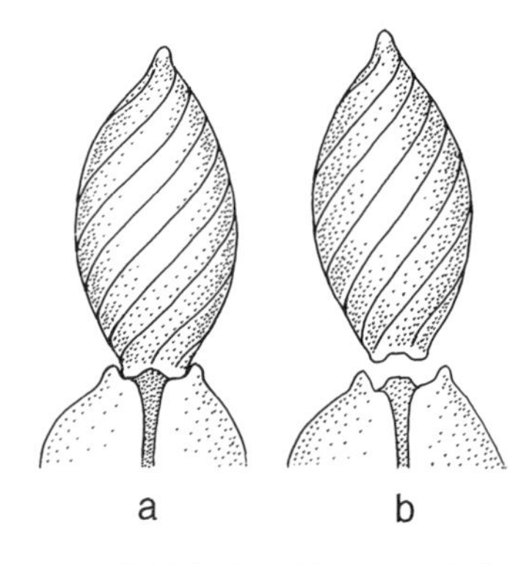

FIGURE 4.28. Multiple slit dehiscence and abscission mechanism in *Huvenia kleuii.* Drawings based on Hass and Remy 1991, Text-Figure 7. (**a**) Sporangium attached to axis (×5). (**b**) Detachment of sporangium leaving a distinctive circular scar with a central prominence (×5).

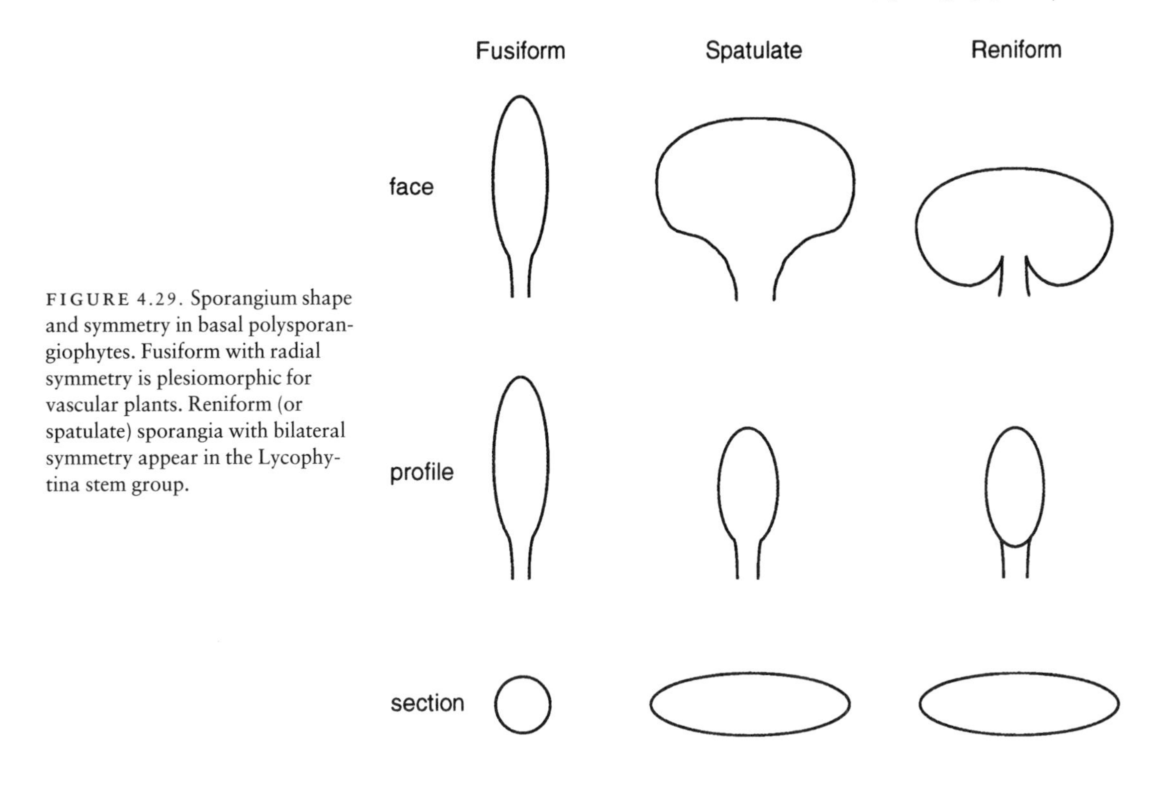

FIGURE 4.29. Sporangium shape and symmetry in basal polysporangiophytes. Fusiform with radial symmetry is plesiomorphic for vascular plants. Reniform (or spatulate) sporangia with bilateral symmetry appear in the Lycophytina stem group.

Sporangium abscission is treated as a binary character: 1 = sporangium abscission absent; 1 = sporangium abscission present. There were no inapplicable cases; one was missing data (see Appendix 3).

4.27 • SPORANGIUM SHAPE Sporangium shape varies considerably among the early polysporangiophytes, and there is much intergradation of form; nevertheless, certain basic morphologies are readily distinguished from each other (Figure 4.29). Sporangium shape in many zosterophylls is basically reniform, ranging from very reniform in *Oricilla* and *Tarella,* to oval or somewhat reniform in *Discalis, Zosterophyllum deciduum, Barinophyton,* and *Nothia* (Figure 5.13).

In extant homosporous lycopsids sporangia characteristically are reniform at maturity (Bierhorst 1971; Øllgaard 1987). In the early lycopsid-like fossils *Asteroxylon* and *Drepanophycus* and possibly also in *Baragwanathia,* sporangia are also oval to reniform in outline.

In putative rhyniophytes, such as *Uskiella* (Figure 4.10), *Sartilmania,* and *Yunia,* sporangia are rounded to somewhat elongate, resulting in a spatulate appearance. In other rhyniophytes such as *Cooksonia* (Figure 4.9) and *Renalia* (Figure 4.11), sporangia can be very reniform. Sporangium shape varies considerably in the genus *Cooksonia,* which includes taxa with clearly reniform sporangia *(C. caledonica),* with "hemispherical" sporangia that are little more than expansions of the axis tip *(C. hemisphaerica),* with more or less circular sporangia *(C. cambrensis),* and with sporangia that are horizontally extended into a marked ellipse *(C. pertonii).* Sporangia in *Horneophyton* (Figure 3.26; Eggert 1974; El-Saadawy and Lacey 1979b; Kidston and Lang 1920a) and *Caia* (Figure 4.19; Fanning, Edwards, and Richardson 1990) are cylindrical and axislike in that they are difficult to distinguish from the structure to which they are attached.

Early members of the Euphyllophytina bear basically fusiform sporangia. Sporangia of this kind have been described from *Psilophyton* (Figure 4.13; e.g., Banks, Leclercq, and Hueber 1975;

Doran 1980), *Estinnophyton* (Fairon-Demaret 1979), *Tetraxylopteris* (Figure 4.14; Bonamo and Banks 1967), *Rhacophyton* (Figure 4.15; Andrews and Phillips 1968), and *Pseudosporochnus* (Figure 4.17; Leclercq and Banks 1962). The sporangia of *Ibyka* were described as obovoid or pyriform (Figure 4.16; Skog and Banks 1973). Other early polysporangiophytes with fusiform sporangia include *Aglaophyton* (Figure 4.7), *Rhynia* (Figure 4.8), *Eogaspesiea,* and *Salopella.* More or less fusiform sporangia occur in such mosses as *Takakia* (Smith and Davison 1993), *Andreaea* (Figure 3.20), *Andreaeobryum* (Murray 1988), and the fossil mosslike sporophyte *Sporogonites* (Figure 4.18; Halle 1916a, 1936a). Other bryophytes included in this analysis have elongate and more or less cylindrical sporangia (e.g., *Haplomitrium:* Figure 3.17; *Anthoceros:* Figure 3.19), or more or less spherical sporangia (e.g., *Sphagnum; Sphaerocarpos:* Figure 3.18).

Sporangium shape is treated as an unordered multistate character with six states (Figure 4.29): 0 = sporangia more or less fusiform; 1 = sporangia more or less reniform; 2 = sporangia more or less spatulate; 3 = sporangia more or less spherical; 4 = sporangia axislike; 5 = sporangia elliptical. There were no inapplicable cases; none were missing data (see Appendix 3). See characters 3.40 and 5.13 for comparison.

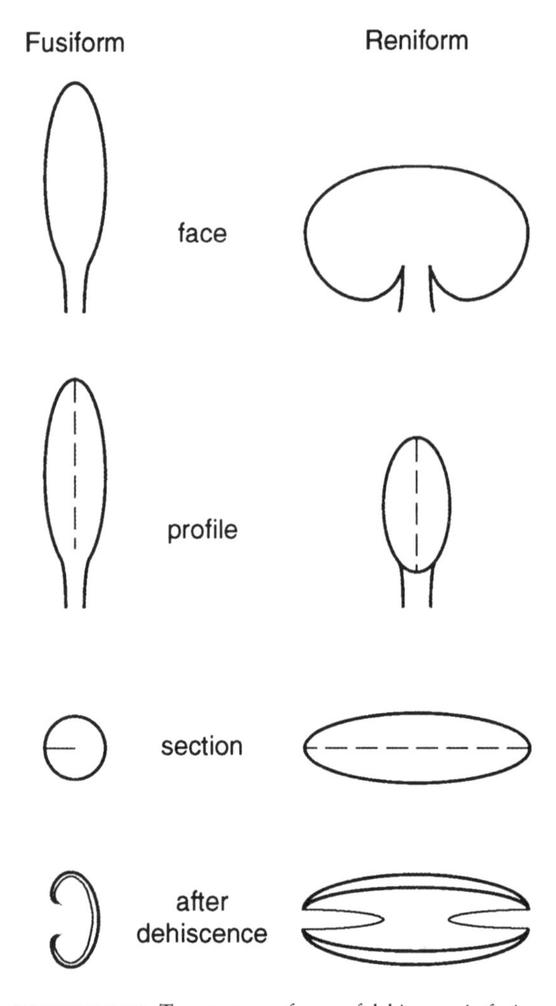

FIGURE 4.30. Two common forms of dehiscence in fusiform and reniform sporangia of early fossils. The fusiform sporangia of basal Euphyllophytina have a single dehiscence line on one side of the sporangium. The reniform sporangia of Lycophytina dehisce into two equal valves along the distal margin.

4.28 • SPORANGIUM SYMMETRY Sporangia that are radially symmetrical in cross section are typical of bryophytes and most polysporangiophytes. Among fossils, sporangium symmetry is best observed in permineralized material such as that of *Psilophyton dawsonii* (Banks, Leclercq, and Hueber 1975), *Rhynia gwynne-vaughanii* (D. S. Edwards 1980; Kidston and Lang 1920a), and *Aglaophyton major* (Figure 4.30; D. S. Edwards 1986). However, indications of sporangium symmetry can be obtained from compression material by observing the variations in sporangium outline among sporangia on the same specimen or on different specimens of the same species.

Many zosterophylls, lycopsids, and some rhyniophytes have sporangia with marked bilateral symmetry in cross section. In many instances it is clear that this is not a feature of compression during preservation because the bilateral symmetry can be observed in sporangia lying in the sediment at different angles. In addition, marked bilateral symmetry is evident in permineralized fossil material, in cuticles, and in such basal extant lycopsids as *Huperzia selago* (Øllgaard 1987). Permineralized material showing marked bilateral symmetry includes *Asteroxylon, Trichopherophyton, Zosterophyllum fertile, Z. llanoveranum, Nothia,* and *Uskiella.* Bilateral symmetry is also seen clearly in the well-preserved sporangium cuticles of such plants as *Renalia hueberi* (Gensel 1976).

Sporangium symmetry is treated as a binary character (Figure 4.30): 0 = sporangia radially symmetrical; 1 = sporangia bilaterally symmetrical. There were no inapplicable cases; two were missing data (see Appendix 3). See character 5.14 for comparison.

4.29 • SPORANGIUM DEHISCENCE In most zosterophylls, the sporangium splits along the distal margin into two valves of approximately equal size (isovalvate dehiscence; Figure 4.30). Anisovalvate dehiscence, which results in a small adaxial and a large abaxial valve, has been observed in *Crenaticaulis* and *Trichopherophyton.* In the early lycopsid-like fossils *Asteroxylon* and *Drepanophycus* and possibly also in *Baragwanathia,* dehiscence is isovalvate. Dehiscence in extant homosporous lycopsids is also isovalvate except in the genus *Lycopodiella,* in which it is anisovalvate (Øllgaard 1987).

Several dehiscence mechanisms have been described for the basal polysporangiophytes (Shute and Edwards 1989). In such plants as *Cooksonia pertonii* and *C. hemisphaerica,* there are no obvious dehiscence features. However, *C. caledonica* (D. Edwards 1970a) has a thickened rim around the distal edge of the sporangium, which is suggestive of an isovalvate mechanism similar to that in the zosterophylls. *Sartilmania* has spatulate sporangia and appears also to have isovalvate dehiscence (Fairon-Demaret 1986a). In *Uskiella* (Figure 4.10; formerly *Cooksonia* sp.: Croft and Lang 1942), dehiscence appears to be isovalvate, but unlike the condition in zosterophylls, the dehiscence line traces a helical path up the sporangium wall. In *Aglaophyton* and *Rhynia* (D. S. Edwards 1980, 1986; Hass and Remy 1991; Remy 1978), dehiscence may involve several vertical, or possibly helical, slits along the sporangium wall. In *Huvenia,* and possibly also in *Stockmansella,* dehiscence involves many helical slits, but the split segments of the sporangium wall all remain attached at the apex (Figures 4.27 and 4.28). A similar type of sporangium dehiscence involving one *(Takakia)* or many slits, producing valves that remain attached at the apex, is common in extant hornworts such as *Anthoceros* (Figure 3.19) and in valvate mosses such as *Andreaea* (Figure 3.20) and may also occur in the putative fossil bryophyte *Sporogonites* (Figure 4.18). The distinctive operculate dehiscence common in many of the bryopsid mosses (e.g., *Polytrichum:* Figure 3.22) has not been observed in the early fossil record, although one very poorly preserved fossil (*Steganotheca:* Figure 4.9) has indications of what might be an operculum.

In the euphyllophytes, such basal taxa as *Psilophyton dawsonii* (Banks, Leclercq, and Hueber 1975) and *P. crenulatum* (Doran 1980) dehisce via a single longitudinal split from the apex to the base along *one* side of the sporangium (Figure 4.30). In these plants the dehiscence line occurs on the facing sides of paired sporangia. Dehiscence is probably similar in other trimerophytes such as *Pertica* (Granoff, Gensel, and Andrews 1976) and in *Psilophyton charientos* and *P. forbesii* (Gensel 1979). A similar dehiscence mechanism has been noted in members of the Aneurophytales, such as *Tetraxylopteris* (Bonamo and Banks 1967), and possibly in *Rhacophyton* (Andrews and Phillips 1968) and *Pseudosporochnus* (Leclercq and Banks 1962).

In *Estinnophyton* (Fairon-Demaret 1979), a plant of uncertain phylogenetic position but which may be related to sphenophylls or lycopsids (Fairon-Demaret 1978, 1979; Gensel and Andrews 1984; Stewart 1983), the sporangia appear to dehisce through a distal slit (Fairon-Demaret 1979, 154). This slit is confined to the distal end of the sporangium and appears not to extend down the side to the base. In *Eviostachya høegii,* Leclercq (1957) described sporangia that dehisce apically through a wide pore. The distinction between an apical "pore" and an apical "slit" is difficult to make in this material and may represent aspects of a similar dehiscence mechanism.

Sporangium dehiscence is treated as an unordered multistate character with five states: 0 = dehiscence along four or more slits (multislit); 1 = dehiscence along one line on one side (single, one side); 2 = dehiscence by a single apical slit; 3 = dehiscence by operculum (operculate); 4 = no slits (indehiscent). There were no inapplicable cases; 10 were missing data (see Appendix 3). See character 3.9 for comparison.

4.30 • THICKENED SPORANGIUM VALVE RIM Marked thickening of the cell walls, an increase in the number of cell layers bordering the dehiscence line, or both occur in some early polysporangiophytes. In compression fossils, the dehiscence line is often a thickened, coalified, double rim where the two valves have separated and twisted slightly such that the underlying valve becomes visible (e.g., Gensel 1982b). Rarely, the cellular details of the dehiscence mechanism are observable in permineralized sporangia (Banks, Leclercq, and Hueber 1975; D. Edwards 1969a; El-Saadawy and Lacey 1979a; Lyon 1964; Lyon and Edwards 1991) and in well-preserved cuticles (Gensel 1976). The cells bordering the dehiscence line often have thickened walls *(Renalia, Nothia)*. The layer bordering the dehiscence line may be only a few cells thick *(Renalia, Nothia, Trichopherophyton)* or may develop into a structure up to nine cells thick *(Zosterophyllum llanoveranum)*. In many compression fossils that exhibit a prominent thickened rim, the dehiscence line is marked by a distinct groove in the sporangium wall that has fewer and smaller cells (e.g., *Uskiella, Z. llanoveranum, Trichopherophyton*). Although some authors have viewed the difference in cellular structure of these dehiscence lines as evidence of homoplasy (Shute and Edwards 1989, 133), it is equally possible that the differences represent simple character-state transitions involving an increasing number of cell tiers at the dehiscence line. Given the conspicuous nature and *general* similarity of the thickened dehiscence border in many of these taxa, we feel justified in proposing a hypothesis of homology.

In most early polysporangiophyte taxa and basal euphyllophytes, dehiscence lines are not bordered by conspicuously thickened cells or by cell layers (e.g., *Psilophyton dawsonii, Aglaophyton, Rhynia*).

The presence or absence of a thickened sporangium valve rim is treated as a binary character: 0 = thickened rim absent; 1 = thickened rim present. There were no inapplicable cases; one was missing data (see Appendix 3).

4.31 • SPORANGIUM BRANCHING Branched sporangia are a unique feature of several early fossil land plants. Of the taxa included in our analysis only *Horneophyton* (Figure 3.26; Eggert 1974; El-Saadawy and Lacey 1979b), *Caia* (Figure 4.19), and *Tortilicaulis* have branched sporangia. Limited sporangium branching has also been described recently in fragmentary coalified fossils assigned to *Salopella* in which anatomical details, including spores, are well preserved (Edwards, Fanning, and Richardson 1994). In these fossils some sporangia have lobed outlines.

The presence or absence of sporangium branching is treated as a binary character: 0 = sporangium branching absent; 1 = sporangium branching present. There were no inapplicable cases; none were missing data (see Appendix 3).

4.32 • COLUMELLA The presence or absence of a columella is treated as a binary character: columella 0 = absent; 1 = columella present. There were no inapplicable cases; seven were missing data (see Appendix 3). See character 3.10 for detailed comparisons and discussion.

4.33 • ELATERS AND PSEUDOELATERS The presence or absence of elaters and pseudoelaters is treated as a binary character: 0 = elaters or pseudoelaters absent; 1 = elaters or pseudoelaters present. There were no inapplicable cases; eight were missing data (see Appendix 3). See character 3.11 for detailed comparisons and discussion.

Other Potentially Relevant Characters

4.34 • PHLOEM Parenti (1980) and Sluiman (1985) listed phloem as a synapomorphy of tracheophytes, but Mishler and Churchill (1984) concluded that phloem was a more inclusive character defining the moss-tracheophyte clade. This interpretation was based on the similarities between the leptoids of polytrichaceous mosses and the sieve elements in basal tracheophytes that were noted by

Hébant (1977) and Scheirer (1980, 1990). More recent studies confirm the general similarities between the sieve elements of mosses and those of vascular plants, including (1) size, appearance, and position in the stem, (2) similarities in modification of the protoplast, (3) presence of a degenerated nucleus at maturity, (4) refractive spherules (electron-dense, membrane-bound proteinaceous bodies that refract light) associated with endoplasmic reticulum, (5) oblique end walls containing numerous small pores, and (6) thickenings of the lateral walls (Scheirer 1990). Furthermore, detailed comparative studies based on TEM observations document many potentially informative ultrastructural features in mosses and basal vascular plants (Evert 1990; Scheirer 1990). Based on our experience with tracheids, it is likely that some of these phloem characters are homologous at the moss-tracheophyte level, whereas others are restricted to tracheophytes and even tracheophyte subgroups. In angiosperms, for example, the nucleus is lost at maturity, and the sieve cells have characteristic sieve areas and sieve plates in the cell wall (Gifford and Foster 1989). At a more general level, certain plastid types are thought to characterize the sieve elements of seed plants (Behnke and Sjolund 1990). Other subcellular features, such as the presence of refractive spherules, correlate well with the Euphyllophytina as defined here. Refractive spherules appear to be characteristic of leptosporangiate ferns, the Ophioglossales, the Psilotaceae, and the Equisetaceae but are absent in lycopsids (Evert 1990) and mosses (Scheirer 1990).

We excluded phloem as a character from our analysis because of the relative paucity of detailed comparative studies. In addition, sieve cell structure is too poorly understood in most of the early fossils relevant to this study (D. Edwards 1993). See character 3.48 for comparison.

4.35 • XYLEM The character "xylem" of some authors (e.g., Parenti 1980; Sluiman 1985) was broken down by Mishler and Churchill (1984) into a more general component (elongate water-conducting cells that lack protoplasts at maturity), which occurs in all vascular plants and some mosses, and a more restrictive component (ornamentation on xylem cells), which characterizes tracheophytes. Mishler and Churchill (1984) noted that the only unequivocal defining feature of tracheids was the presence of an ornamented secondary wall. This conclusion was based on earlier observations by Hébant (1977) and Scheirer (1980) that showed other tracheid features, such as the elongate shape, lack of protoplast at maturity, and possibly even the presence of polyphenolics in the cell wall, to be characters of a more inclusive group that included mosses. This idea has received further support from similarities between the water-conducting cells of early fossil polysporangiophytes and of mosses (D. S. Edwards 1986; Hébant 1977). These observations suggest that the polysporangiate condition is a more general feature in land plants than tracheids and raise the possibility that the secondary wall thickenings may not be homologous in lycopsids and other vascular plants (Crane 1990). Combining data on extant groups with new information on early fossils, Kenrick and Crane (1991) further analyzed the "ornamentation" character using a numerical approach and showed that one specific type of "ornament" or "thickening" (i.e., helical or annular) does appear to define tracheophytes. These results are supported by the distribution of other features and are clarified further by this analysis (below) in which the tracheophyte crown group (eutracheophytes—see below) is supported by an additional structural feature of the tracheid cell wall (conspicuous pitlets between secondary thickenings). See character 3.48 for comparison.

4.36 • LIGNIN A further potential defining feature of tracheophytes is the presence of "true" lignin (Mishler and Churchill 1984), but a more precise definition is currently impossible to obtain because details of the chemical structure are poorly understood for extant plants and are unknown in most fossil material (Logan and Thomas 1985; Thomas 1986). The occurrence of lignin in fossil vascular plants can only be inferred from the presence of a homogenous decay-resistant layer within the cell wall of certain tissue systems (e.g., wood). In many fossil taxa, the position of this layer within the cell wall is similar to that in extant groups (Kenrick and Crane 1991), and in our analysis, a thick, decay-resistant layer of this type is a synapomorphy supporting eutracheophytes (tracheophyte crown group). Furthermore, our analysis suggests that lignin deposition on the inner surface of the tracheid may be a synapomorphy at the more inclusive level of tracheophytes (including, for example, *Rhynia gwynne-vaughanii*). However, the exact level at which this character should apply remains equivocal because the presence or absence of a thin lignified layer in the critical nonvascular polysporangiophyte taxa *Aglaophyton* and *Horneophyton* is uncertain.

4.37 • TERPENES A potential defining character of tracheophytes recognized by Mishler and Churchill (1984) is the presence of monoterpenes and sesquiterpenes (which is convergent with liverworts). We did not include this character because of uncertainties regarding its occurrence in several extant taxa, as well as obvious difficulties with fossils. The distribution of this feature predicted by Mishler and Churchill (1984) does not conflict with our results.

4.38 • APICAL MERISTEM The histology of apical meristems is highly variable in vascular plants (White and Turner 1995). A single, well-defined apical cell is clearly present in some groups (e.g., *Equisetum,* leptosporangiate ferns). However, in many other taxa (e.g., Lycopodiaceae, Isoetaceae, Ophioglossaceae, Marattiales, and many seed plants), similarities in cell size make the presence of an apical cell difficult to determine, and several superficial apical initials are thought to occur. There have been several attempts at classifying the variability in the apical structure of vascular plants. Newman (1961, 1965) recognized three types based on the number of initials and the plane of cell division. The *monoplex apex* is characterized by a single cell that divides anticlinally to produce superficial daughter cells. The daughter cells divide periclinally to produce epidermal and stem tissue system initials. The *simplex apex* has one or more apical cells that divide both anticlinally and periclinally. The *duplex apex* resembles the simplex apex but contains two cell layers: the inner layer behaves like the simplex apex, but the outer layer cells divide only anticlinally. The monoplex and duplex have been classified together as the "seed plant type" by Philipson (1990, 1991). Apical meristem structure was not used as a character in this analysis because many details remain to be clarified in extant taxa, and the form of the apex is unknown for most fossils.

ANALYSIS

Our primary analysis (analysis 4.1) is based on a total of 34 extant and fossil taxa, including the extant liverworts *Haplomitrium* and *Sphaerocarpos* designated as outgroups (Table 4.4), and 33 characters. We used a subset of 18 of the best-known ingroup taxa as a "core" data set for further experimentation and stability tests (analysis 4.2). The remaining 16 ingroup taxa were added to the "core" group in secondary experimental analyses in order to test the robustness of the less-inclusive cladogram (Table 4.4). Eight of these additional 16 taxa have significant amounts of missing information. In all analyses, trees were generated using the Branch-and-Bound routine of PAUP (version 3.1.1; Swofford 1990), which finds all the most parsimonious solutions for a given data set. Branch-and-Bound analyses of all 34 taxa were completed in approximately 10 hours running on a PowerBook® 150. The Heuristic search routine of PAUP found over 99% of the trees in approximately 5 hours and gave the same topology. Trees were rooted using outgroup comparison with two outgroups. Consistency indexes given throughout were calculated excluding uninformative characters. Tree stability was tested using Bremer Support (Donoghue, Olmstead, et al. 1992; Källersjö, Farris, et al. 1992), which measures the relative amount of support for each clade. Character-state changes are based on ACCTRAN optimization which favors reversals over parallelisms.

RESULTS

Cladistic analysis based on the full data set of 34 taxa and 33 characters (with *Haplomitrium* and *Sphaerocarpos* designated as outgroups) yielded 3,294 equally parsimonious trees of 72 steps with consistency indexes of 0.75. The strict consensus tree is shown in Figure 4.31, and character changes (Table 4.6) are mapped onto a representative most-parsimonious tree (Figure 4.32). These trees support the monophyly of vascular plants (tracheophytes *sensu lato*) and polysporangiophytes, as well as the paraphyly of bryophytes (Chapter 3; Mishler and Churchill 1984, 1985b). The results also agree in general terms with previous phylogenetic interpretations of the early fossil record

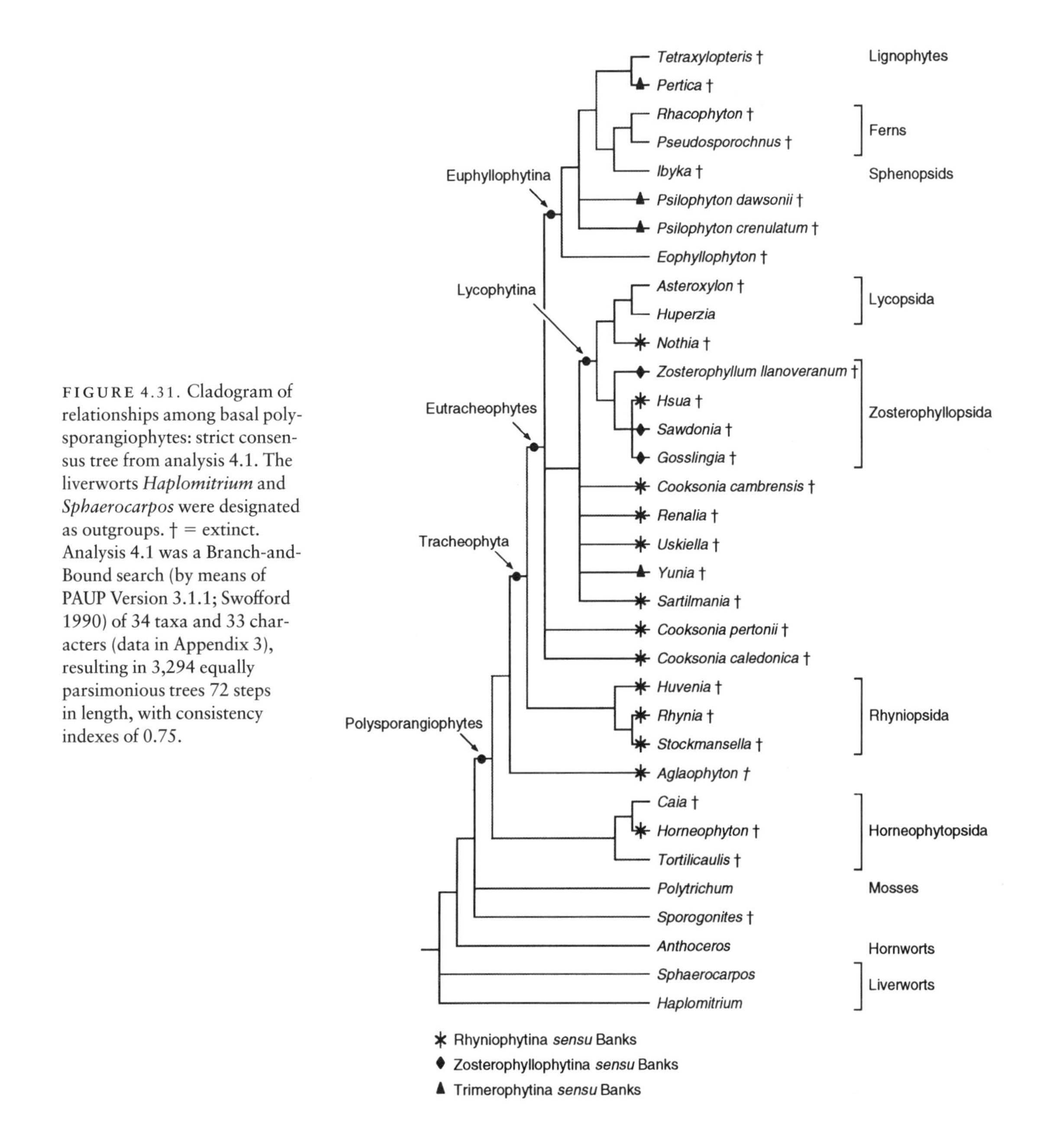

FIGURE 4.31. Cladogram of relationships among basal polysporangiophytes: strict consensus tree from analysis 4.1. The liverworts *Haplomitrium* and *Sphaerocarpos* were designated as outgroups. † = extinct. Analysis 4.1 was a Branch-and-Bound search (by means of PAUP Version 3.1.1; Swofford 1990) of 34 taxa and 33 characters (data in Appendix 3), resulting in 3,294 equally parsimonious trees 72 steps in length, with consistency indexes of 0.75.

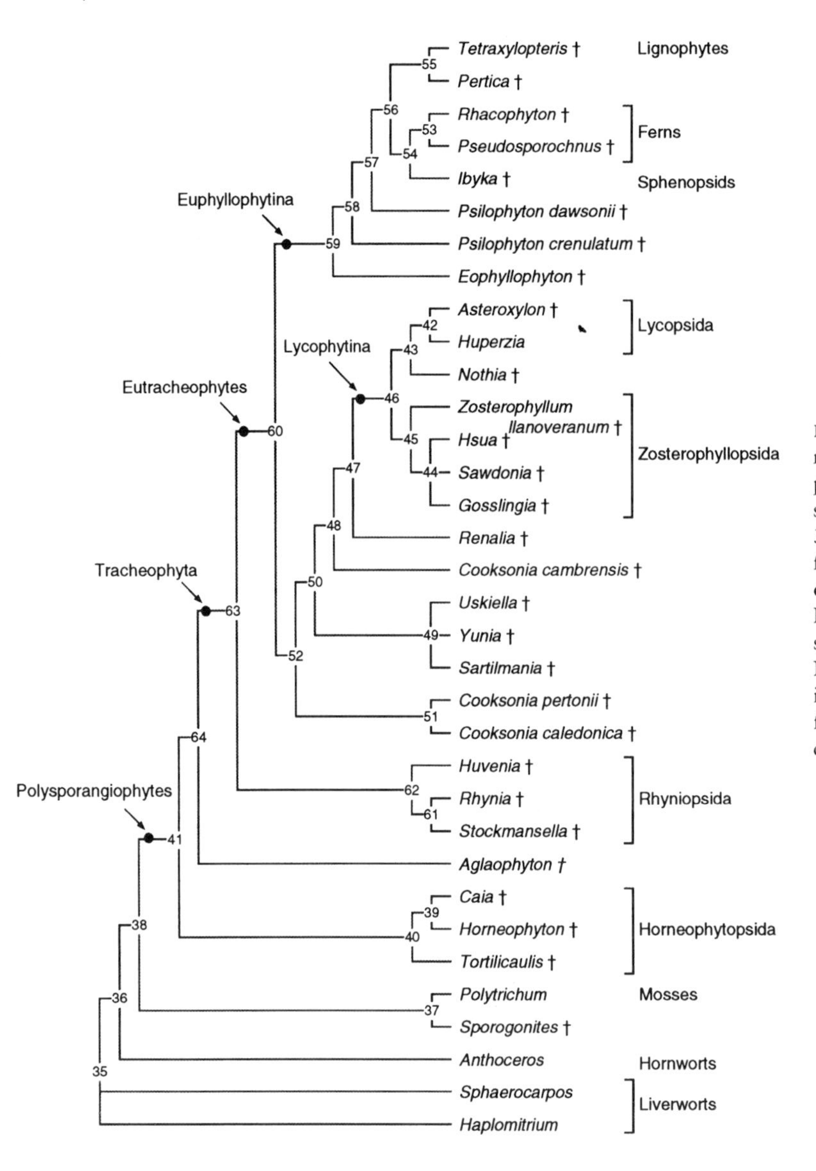

FIGURE 4.32. Cladogram of relationships among basal polysporangiophytes: character state changes on one of the 3,294 most parsimonious trees from analysis 4.1 (ACCTRAN optimization). See caption to Figure 4.31 for explanation of symbols and details of analysis. Numbers in the figure represent internal nodes. See Table 4.6 for character state changes corresponding to nodes.

by Banks (1970) and others (e.g., Chaloner and Sheerin 1979), which suggest an early divergence of lycopsids and zosterophylls from other vascular plants (Figures 4.4 and 4.5). The results support earlier suggestions of a protracheophyte grade of organization in the tracheophyte stem group (Kenrick and Crane 1991). Within vascular plants, a small extinct clade containing *Rhynia gwynne-vaughanii* and *Huvenia kleui* is a sister group to a large and diverse clade containing all other extinct and extant vascular plants (eutracheophytes) and so supports the results of previous analyses based on fewer taxa (Kenrick and Crane 1991).

Within the main tracheophyte clade (eutracheophytes), there are two distinct major groups, both with extant members. The Lycophytina comprise the Lycopsida (lycopsids *sensu stricto*) and their extinct sister group, the Zosterophyllopsida. This clade is extremely diverse in the Devonian and Carboniferous Periods, but comprises only three extant families (Lycopodiaceae, Selaginellaceae, Isoetaceae). Because of the importance of the Lycophytina in the early fossil record, detailed systematic treatments of the Lycopsida and Zosterophyllopsida are given in separate chapters (Chapters 5 and 6). The Euphyllophytina comprise all

TABLE 4.6

Character Changes on Nodes of Tree in Figure 4.32

Beginning node → end node		Character	Change in state	CI
Node 35 → node 36	4.12	Stomates	0 → 1	1.000
	4.32	Columella	0 → 1	0.500
Node 35 → *Sphaerocarpos*	4.27	Sporangium shape	0 → 3	0.833
	4.29	Sporangium dehiscence	0 → 4	1.000
Node 36 → node 38	4.2	Well-developed sporangiophore	0 → 1	1.000
	4.33	Elaters and pseudoelaters	1 → 0	1.000
Node 37 → *Polytrichum*	4.29	Sporangium dehiscence	0 → 3	1.000
Node 38 → node 41	4.1	Sporophyte independence or dependence	0 → 1	1.000
	4.3	Sporophyte branching	0 → 1	1.000
	4.15	Protoxylem	0 → 1	0.750
Node 40 → node 39	4.11	Conical sporangium emergences	0 → 1	1.000
	4.27	Sporangium shape	0 → 4	0.833
Node 41 → node 40	4.31	Sporangium branching	0 → 1	1.000
Node 41 → node 64	4.32	Columella	1 → 0	0.500
Node 42 → *Huperzia*	4.18	Metaxylem tracheid pitting	0 → 1	0.500
	4.19	Simple pitlets in tracheid wall	1 → 0	0.500
Node 43 → node 42	4.9	Multicellular appendages	0 → 2	0.333
	4.14	Xylem strand shape	0 → 2	1.000
	4.22	Sporangiotaxis	0 → 2	1.000
Node 43 → *Nothia*	4.17	Tracheid thickenings	1 → 0	0.500
Node 44 → *Sawdonia*	4.9	Multicellular appendages	0 → 1	0.333
Node 44 → *Hsua*	4.23	Sporangium attachment	1 → 0	0.500
Node 45 → node 44	4.4	Branching type	0 → 1	0.333
	4.5	Branching pattern	0 → 2	1.000
	4.6	Subordinate branching	0 → 1	1.000
Node 46 → node 45	4.8	Circinate vernation	0 → 1	0.667
	4.14	Xylem strand shape	0 → 1	1.000
	4.22	Sporangiotaxis	0 → 1	1.000
Node 46 → node 43	4.13	Sterome	1 → 0	0.667
Node 47 → node 46	4.23	Sporangium attachment	0 → 1	0.500
Node 47 → *Renalia*	4.4	Branching type	0 → 1	0.333
Node 49 → *Yunia*	4.9	Multicellular appendages	0 → 1	0.333
Node 50 → node 48	4.15	Protoxylem	1 → 2	0.750
Node 50 → node 49	4.27	Sporangium shape	1 → 2	0.833
Node 52 → node 50	4.27	Sporangium shape	5 → 1	0.833
	4.30	Thickened sporangium valve rim	0 → 1	1.000
Node 53 → *Rhacophyton*	4.5	Branching pattern	1 → 4	1.000
	4.21	Xylem rays	0 → 1	0.500
Node 53 → *Pseudosporochnus*	4.8	Circinate vernation	2 → 0	0.667
Node 54 → node 53	4.14	Xylem strand shape	3 → 4	1.000
Node 56 → node 55	4.5	Branching pattern	1 → 3	1.000
	4.21	Xylem rays	0 → 1	0.500
Node 56 → node 54	4.15	Protoxylem	1 → 3	0.750
Node 57 → *Psilophyton dawsonii*	4.6	Subordinate branching	0 → 2	1.000
Node 57 → node 56	4.13	Sterome	1 → 2	0.667
	4.14	Xylem strand shape	0 → 3	1.000
Node 58 → node 57	4.9	Multicellular appendages	1 → 0	0.333

Continues on next page

TABLE 4.6
Continued

Beginning node → end node		Character	Change in state	CI
Node 59 → *Eophyllophyton*	4.23	Sporangium attachment	0 → 1	0.500
Node 59 → node 58	4.18	Metaxylem tracheid pitting	0 → 1	0.500
	4.20	Aligned xylem	0 → 1	1.000
	4.27	Sporangium shape	5 → 0	0.833
	4.28	Sporangium symmetry	1 → 0	0.500
Node 60 → node 52	4.29	Sporangium dehiscence	1 → 2	1.000
Node 60 → node 59	4.4	Branching type	0 → 1	0.333
	4.5	Branching pattern	0 → 1	1.000
	4.8	Circinate vernation	0 → 2	0.667
	4.9	Multicellular appendages	0 → 1	0.333
	4.10	Dichotomous pinnulelike appendages	0 → 1	1.000
	4.24	Specialized fertile zone	0 → 1	1.000
Node 61 → *Stockmansella*	4.15	Protoxylem	1 → 0	0.750
Node 62 → *Huvenia*	4.9	Multicellular appendages	0 → 1	0.333
Node 62 → node 61	4.23	Sporangium attachment	0 → 2	0.500
Node 63 → node 60	4.13	Sterome	0 → 1	0.667
	4.16	Decay-resistance of tracheid cell wall	0 → 1	1.000
	4.19	Simple pitlets in tracheid wall	0 → 1	0.500
	4.27	Sporangium shape	0 → 5	0.833
	4.28	Sporangium symmetry	0 → 1	0.500
	4.29	Sporangium dehiscence	0 → 1	1.000
Node 63 → node 62	4.7	*Rhynia*-type adventitious branching	0 → 1	1.000
	4.25	Specialized sporangium-axis junction	0 → 1	1.000
	4.26	Sporangium abscission	0 → 1	1.000
Node 64 → node 63	4.17	Tracheid thickenings	0 → 1	0.500

Note: ACCTRAN optimization. CI = consistency index of character. Tree no. 318.

other vascular plants and is represented in our analysis by a selection of key early fossils. A detailed systematic treatment of this clade is beyond the scope of this book, but some higher level relationships have been considered with the inclusion of additional taxa on an experimental basis (see below).

Tree Stability

Tree stability was tested using Bremer Support (Decay Index = DI) measurements (Källersjö, Farris, et al. 1992). Because of the large number of taxa (34), tree stability measurements were made on 18 "core" taxa (with *Haplomitrium* and *Sphaerocarpos* designated as outgroups). This analysis (analysis 4.2) yielded three most parsimonious trees of 63 steps with consistency indexes of 0.83 (Figure 4.33). Tree topology for the taxa included was identical to that of the complete analysis. Support was strongest for the *Sawdonia-Gosslingia* clade within the zosterophylls (DI = 3). Intermediate levels of support (DI = 2) were found for the polysporangiophytes, eutracheophytes, Rhyniopsida, Lycophytina, and Euphyllophytina. All other branches were relatively weakly supported (DI = 1). Overall, however, the internal consistency of the data is high and the levels of homoplasy are relatively low. Much of the uncertainty in the data matrix arises because of inapplicable characters in the outgroups and in the ingroup taxon *Anthoceros*. Many characters relating to branching and

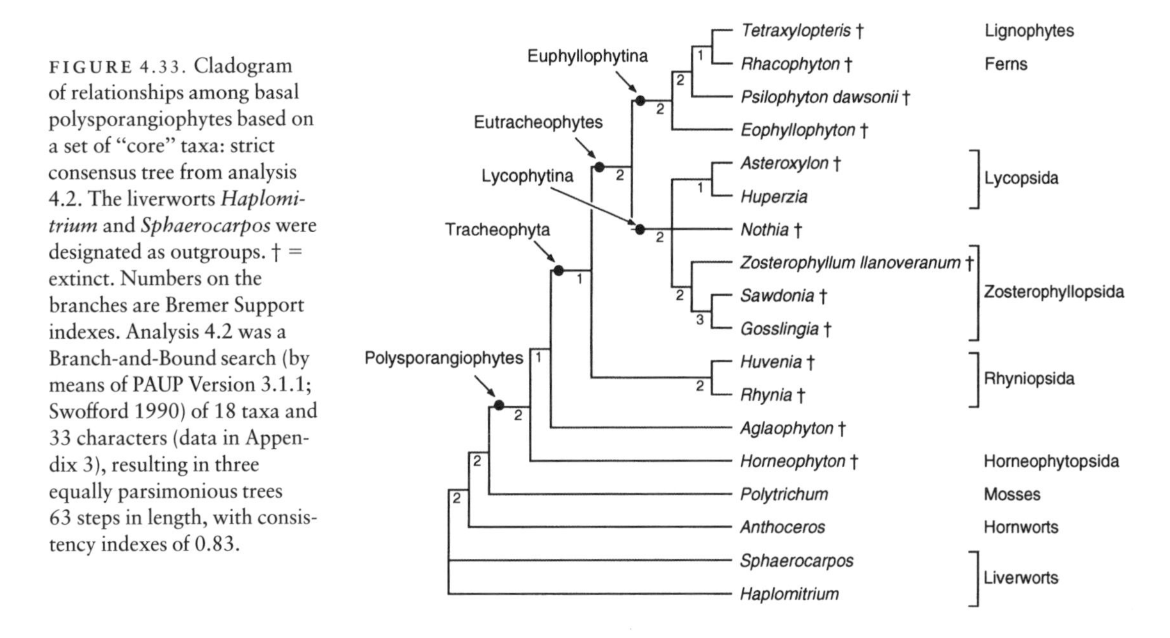

FIGURE 4.33. Cladogram of relationships among basal polysporangiophytes based on a set of "core" taxa: strict consensus tree from analysis 4.2. The liverworts *Haplomitrium* and *Sphaerocarpos* were designated as outgroups. † = extinct. Numbers on the branches are Bremer Support indexes. Analysis 4.2 was a Branch-and-Bound search (by means of PAUP Version 3.1.1; Swofford 1990) of 18 taxa and 33 characters (data in Appendix 3), resulting in three equally parsimonious trees 63 steps in length, with consistency indexes of 0.83.

sporangiophore anatomy were not strictly scorable in the outgroups and were represented with a "?" in our data matrix (see the handling of inapplicable characters in Chapter 3). The details of inapplicable and of unknown characters are given in Appendix 3. The addition of further taxa had little effect on tree topology but weakened the support for some nodes (see below).

Polysporangiophytes

DEFINITION Polysporangiophytes contain tracheophytes and such nonvascular taxa as *Aglaophyton,* Horneophytopsida, and possibly some other more poorly understood rhyniophytes (Figures 4.31 and 4.33). The defining features of polysporangiophytes (DI = 2) are the presence of multiple sporangia (a feature that is probably linked to apical dichotomy, which is not scored separately in our analysis), the independence of the sporophyte from the gametophyte, and more or less centrarch xylem maturation (Figure 4.32, Table 4.6: node 38 → node 41). Additional potential synapomorphies include isotomous and nonplanar branching, but these features are not observable in the unbranched outgroups. Gametophyte characters were not included in this analysis because they are unknown for most fossil taxa, but gametophyte morphology was considered in more detail in Chapter 3, which focused on basal embryophytes. The early, and rather poorly understood, monosporangiate fossil *Sporogonites* was placed outside the polysporangiophytes in an unresolved relationship with mosses and polysporangiophytes (Figure 4.31).

Both *Horneophyton* and *Aglaophyton* were described originally as vascular plants, but these assignments are problematic because the water-conducting cells do not have the distinctive wall thickenings that characterize tracheids. Kidston and Lang (1920a, 614) attributed the absence of thickenings to decay, but we consider this to be unlikely given that the cell walls are generally intact in other less robust tissue systems within these plants. The water-conducting cells of *Aglaophyton* have since been shown to lack thickenings (D. S. Edwards 1986), and although "tracheids" were illustrated for the basal regions of *Horneophyton,* the putative wall thickenings are highly irregular and weakly developed. In our view, there is no good evidence for such critical features as well-developed helical or annular thickenings and decay-resistant walls, as is found in the G-type tracheid typical of basal embryophytes (Figure 4.26). Furthermore, neither *Aglaophyton* nor *Horneo-*

phyton can be linked specifically with the Rhyniopsida, Euphyllophytina, and Lycophytina on any other characters. However, in our analysis, *Aglaophyton* is more closely linked to tracheophytes than to *Horneophyton,* based on the loss of the columella (Figure 4.32, Table 4.6: node 41 → node 64).

HORNEOPHYTOPSIDA The Horneophytopsida are the most basal polysporangiophytes (Figure 4.31). The group is based on the well-known Rhynie Chert plant *Horneophyton.* Defining features include branched sporangia (Figure 4.32, Table 4.6: node 41 → node 40), distinctive, small, multicellular protuberances from the sporangium surface; and a sporangium morphology that strongly resembles vegetative axes (Figure 4.32, Table 4.6: node 40 → node 39). We have been able to identify two other potential members of this clade. The early and rather poorly understood fossils *Caia langii* and *Tortilicaulis offaeus* were always grouped with *Horneophyton* on the basis of branched sporangia. A further similarity between *Caia* and *Horneophyton* is the presence of conical emergences on the sporangium surface. Other fragmentary early fossils with branched sporangia recently described by Fanning, Edwards, and Richardson (1992) may also be related to this clade.

Tracheophytes

DEFINITION Tracheophytes (Tracheophyta) are the vascular plant stem-based group and contain at least two clearly identifiable and well-defined clades that we term the Eutracheophytes and the Rhyniopsida (Figures 4.31 and 4.33, Chapter 7). The group is defined by annular or helical thickenings in the water-conducting cells (DI = 1) (Figure 4.32, Table 4.6: node 64 → node 63). Another possible synapomorphy at this level is lignin deposition on the inner surface of the tracheid cell wall, but we did not score this character in our analysis (see potential character 4.36). We apply the term *tracheophytes* to the group defined by the tracheid (water-conducting cell with *helical or annular* thickenings) because our analysis suggests that this cell type is homologous in all taxa. This is in contrast to previous suggestions (e.g., Crane 1990) that questioned the homology of the tracheid and have adopted a more inclusive definition of tracheophytes that includes all polysporangiophytes, whether or not they possess water-conducting cells with helical or annular thickenings.

RHYNIOPSIDA The Rhyniopsida are a small, poorly understood, yet distinctive, group of vascular plants (DI = 2; Figures 4.31 and 4.33) that is a sister group to the less inclusive eutracheophyte clade. The group contains the type species of Rhyniophytina *sensu* Banks *(Rhynia gwynne-vaughanii),* as well as other plants formerly placed in the genus *Taeniocrada,* of which *Stockmansella langii* and *Huvenia kleui* are the best understood. Rhyniopsida is supported as a group by three unequivocal synapomorphies: a distinctive type of adventitious branching (*Rhynia*-type), an abscission or isolation layer at the base of the sporangium, and sporangia attached to a "pad" of tissue (Figure 4.32, Table 4.6: node 63 → node 62). In addition, these three taxa possess the distinctive and unique S-type tracheid (Figures 4.25 and 4.26; Kenrick and Crane 1991).

Eutracheophytes

DEFINITION The taxon Eutracheophytes is the tracheophyte crown group and contains all extant vascular plants and most vascular plant fossils (Figures 4.31 and 4.33). The Eutracheophytes comprise two major clades: the Lycophytina and the Euphyllophytina. The eutracheophyte clade (DI = 2) is defined by two structural features of the tracheid cell wall: a thick, decay-resistant wall, and pitlets between thickenings or within pits (Figure 4.26). The thickness of the decay-resistant tracheid wall is characteristic of all of the eutracheophytes and distinguishes them from other land plants in which decay resistance is either absent from the water-conducting cells or is present only as a very thin inner layer in the distinctive S-type tracheids of the Rhyniopsida (Kenrick and Crane 1991). Other defining features of eutracheophytes include the reduction of sporangium dehiscence to a single well-defined slit and the presence of a sterome (Figure 4.32, Table 4.6: node 63 → node 60). Character optimization on our preferred most-parsimonious tree (Figure 4.32) also indicates a change in sporangium shape at this node from fusiform and radially symmetrical to elliptical with bilateral symmetry. This change is reversed in the Euphyllophytina from node 59 to node 58.

EUPHYLLOPHYTINA The Euphyllophytina comprise over 99% of extant vascular plant species and include *Equisetum,* seed plants, and many fossil taxa related to extant ferns, as well as many extinct "intermediate" groups. The Euphyllophytina (DI = 2) are supported as a group by six synapomorphies: (1) pseudomonopodial or monopodial branching, (2) a basically helical arrangement of branches, (3) small, "pinnulelike" vegetative branches (nonplanate in basal taxa), (4) recurvation of branch apexes, (5) sporangia in pairs grouped into terminal trusses, and (6) multicellular appendages (spines) (Figure 4.32, Table 4.6: node 60 → node 59). The Euphyllophytina stem group comprises a grade of trimerophytes, including *Eophyllophyton bellum, Psilophyton dawsonii,* and *Psilophyton crenulatum* (Figure 4.32). Other synapomorphies appearing in this stem group include (1) metaxylem pitting, (2) aligned xylem, (3) radial sporangium symmetry (reversal), (4) fusiform sporangia (reversal) (Figure 4.32, Table 4.6: node 59 → node 58), (5) the loss of multicellular spines (reversal) (Figure 4.32, Table 4.6: node 58 → node 57), (6) a lobed xylem strand, and (7) a change from a continuous sterome to discrete bundles of fibers in the stem cortex (Figure 4.32, Table 4.6: node 57 → node 56). This result supports the independent origins of pseudomonopodial branching, metaxylem pitting, and lobed xylem strands in the Lycophytina and the Euphyllophytina.

Relationships among the major clades of the Euphyllophytina are not dealt with in detail in this book. This is an area of great interest for future work in which the incorporation of the fossil record into cladistic treatments will be critical to a better understanding of relationships and homologies (see Chapter 7).

In the complete (34 taxa) analysis (Figures 4.31 and 4.32, analysis 4.1), two species of *Psilophyton* (Trimerophytina) and *Eophyllophyton* were placed consistently in a paraphyletic relationship to the Euphyllophytina crown group. No support was found for monophyly of the genus *Psilophyton.* Another trimerophyte, *Pertica,* was placed as a sister taxon to the progymnosperm *Tetraxylopteris,* supporting earlier suggestions that Trimerophytina *sensu* Banks is paraphyletic (Crane 1990). In our analysis, this relationship is based only on tetrastichous branch arrangement. A close relationship between these taxa would receive further support if earlier suggestions of a lobed stelar arrangement in *Pertica* are confirmed (Gensel 1984). The putative early ferns *Rhacophyton* and *Pseudosporochnus,* grouped as sister taxa based on the distinctive clepsydroid xylem shape, and the putative early sphenopsid *Ibyka* was a sister group to this clade as supported by the mesarch protoxylem position. Character distributions in this preliminary analysis support an independent origin of the xylem rays in progymnosperms and in *Rhacophyton.* Furthermore, our analysis indicates that, contrary to previous suggestions (Banks 1992; Hao and Beck 1991b), the problematic *Yunia dichotoma* is probably more closely related to the Lycophytina than to the Trimerophytina. The terete xylem strand of *Yunia* is plesiomorphic at this level, and the unique diarch protoxylem found in parts of the stem is autapomorphic.

LYCOPHYTINA The Lycophytina (DI = 2) are an important and diverse group in the early fossil record but represent less than 1% of extant land plant diversity. The group contains two major clades that we term Lycopsida and Zosterophyllopsida (extinct) (Figure 4.31). Relationships within the Lycophytina are dealt with in detail in Chapter 5 (Zosterophyllopsida: Zosterophyllophytina *pro parte*) and Chapter 6 (Lycopsida), but representative taxa were included in this analysis. The Lycophytina are defined by a change from terminal to lateral sporangia (reversed in *Hsua*) (Figure 4.32, Table 4.6:

node 47 → node 46). Other synapomorphies appearing in the Lycophytina stem group include (1) isovalvate dehiscence along the distal sporangium rim (Figure 4.32, Table 4.6: node 60 → node 52), (2) conspicuous cellular thickening of the dehiscence line, (3) reniform sporangia (Figure 4.32, Table 4.6: node 52 → node 50), and (4) exarch xylem differentiation (Figure 4.32, Table 4.6: node 50 → node 48).

One interesting aspect of this analysis is that many "intermediate rhyniophyte" fossils appear to be more closely related to the Lycophytina than to either *Rhynia gwynne-vaughanii* or the Euphyllophytina. *Renalia* and some other *Cooksonia*-like taxa *(C. caledonica, C. pertonii, C. cambrensis, Sartilmania, Uskiella)* were placed in the Lycophytina stem group mainly on the basis of sporangium morphology and dehiscence (Figure 4.31). Other species of *Cooksonia* (e.g., *C. hemisphaerica*) may have different relationships, based on their sporangium morphology, but we were unable to test this idea further because these taxa are very poorly understood. *Hsua* was placed within zosterophylls, and this placement implies that terminal sporangium arrangements in this taxon are a reversal from the lateral arrangements typical of zosterophylls (Figure 4.32, Table 4.6: node 44 → *Hsua)*. The problematic phylogenetic position of *Hsua* is considered in more detail in Chapter 5.

Contrary to our previous result that placed *Nothia* at the "protracheophyte grade" (Kenrick and Crane 1991), the genus is now placed in the Lycophytina (Figure 4.31). The affinities of *Nothia* have previously been considered problematic because although it shares many features with the zosterophylls (Meyen 1987), the absence of thickenings within the xylem cells and the more or less terete xylem strand have been used to suggest a more basal position in land plants (Banks 1975b, 1992; Crane 1990; Gensel and Andrews 1984; Kenrick and Crane 1991). Our analysis shows that the terete xylem shape is plesiomorphic at this level, and character optimization supports the *loss* of wall-thickenings in the xylem cells of *Nothia* (Figure 4.32, Table 4.6: node 43 → *Nothia)*. The implications of this reversal are discussed in greater detail in Chapter 7 (tracheid morphology).

DISCUSSION

Previous Hypotheses of Relationship

TRACHEOPHYTE MONOPHYLY In one of the earliest phylogenetic schemes for land plants, Haeckel (1868) suggested that tracheophytes are monophyletic, and many other early schemes gave cautious, and often less explicit, support to this view (e.g., Bower 1894; Campbell 1895; Fritsch 1916; Halle 1916b; Hallier 1902; Jeffrey 1902; Lignier 1908; Tansley 1907). While there have been several widely discussed polyphyletic hypotheses (e.g., Arber 1921; Church 1919), these are no longer considered plausible because of the weight of recent evidence supporting a green algal (charophycean) origin of land plants (Chapters 2 and 3).

As knowledge of Devonian floras increased from the 1920s to the 1960s, many botanists continued to support vascular plant monophyly as a parsimonious explanation of the new morphological data (Bower 1935; Chaloner and Sheerin 1979; Cronquist, Takhtajan, and Zimmermann 1966; Darrah 1939, 1960; Eames 1936; Emberger 1944, 1968; Høeg 1967; Kidston and Lang 1921; Kräusel 1950; Lam 1948, 1950; Mägdefrau 1942; Meyen 1987; Pichi-Sermolli 1959; Smith 1938, 1955; Stewart 1983; Takhtajan 1953; Zimmermann 1930, 1938, 1952, 1965). Others, however, viewed the near simultaneous appearance of different groups in the fossil record as evidence of tracheophyte polyphyly (Andrews 1947, 1959; Andrews and Alt 1956; Axelrod 1952, 1959; Banks 1970; Leclercq 1954, 1956). These two alternative interpretations were possible because proponents of polyphyletic hypotheses emphasized data on stratigraphic appearance over evidence from comparative morphology. The use of stratigraphic evidence for determining relationships has been extensively criticized (e.g., Hill and Crane 1982; Nelson and Platnick 1981; Smith 1994) for several reasons. First, establishing the relationships of fossils, at least at some minimal level, is necessary *before* stratigraphic data can be invoked. Second, the time of appearance of a group in the fossil record is not equal to its actual time of origin, and the fossil record provides only a *minimum* estimate of the age of a group (see Chapter 7). Third, arguments of polyphyly based on the more or less simultaneous appearance of different clades carry implicit assumptions about rates of morphological evolution for which direct evidence is lacking. Under these circumstances, we see no conflict between the long-standing hypothesis of tracheophyte monophyly supported by our results and the stratigraphic record.

We tested the notion of tracheophyte diphyly in the form suggested by Banks (1970: two vascular plant lineages independently derived and a close relationship between rhyniophytes and trimerophytes) by enforcing topological constraints during the analysis. Based on the results of our complete analysis (analysis 4.1; Figures 4.31 and 4.32), the most parsimonious interpretation of tracheophyte diphyly *sensu* Banks would place the Lycophytina clade at the base of polysporangiophytes on the node between the mosses and the Horneophytopsida. This topology retains the monophyly of embryophytes and polysporangiophytes and the paraphylly of the bryophytes. Trees with this topology are eight steps longer than our most parsimonious tree and require independent origins of such features as a sterome, tracheids (i.e., decay-resistant cell walls, helical thickenings, and simple pits), and similarities in sporangium morphology between the Lycophytina and *Renalia*-like rhyniophytes (i.e., reniform shape, bilateral symmetry, conspicuous cellular thickening of dehiscence line). Forcing an independent origin of lycophytes from within bryophytes introduces further homoplasy. This hypothesis was tested by specifying a sister group relationship between the Lycophytina and the liverwort *Sphaerocarpos* and forcing hornworts and mosses into the stem group of a rhyniophyte-trimerophyte clade. The resulting trees were 12 steps longer than our most parsimonious tree and required independent origins of such features as sporophyte independence, well-developed sporangiophores, the polysporangiate condition, a sterome, tracheids (i.e., decay-resistant cell walls, helical thickenings, and simple pits), and similarities in sporangium morphology between the Lycophytina and the *Renalia*-like rhyniophytes (i.e., reniform shape, bilateral symmetry, conspicuous cellular thickening of the dehiscence line). Forcing independent origins of two tracheophyte lineages from within green algae would require "reinvention" of the sporophyte generation, and thus would be even less parsimonious.

A few recent phylogenetic schemes based on comparative morphology challenge tracheophyte monophyly. Philipson (1990, 1991) argued for at least four independent origins of the sporophyte generation in tracheophytes from within a group of primitive land plants with gametophyte-dominated life cycles. Furthermore, he regarded lycopsids as diphyletic and suggested that the sporophyte generations of *Selaginella* and *Isoetes* also evolved independently. This hypothesis is based mainly on the analysis of differences in apical meristem structure. In the context of a general, parsimonious interpretation of land plant evolution, differences in meristem structure among groups are much more likely to reflect transformations of one "meristem type" into another, than to reflect independent origins of the sporophyte generations in different major tracheophyte groups (White and Turner 1995).

Few molecular cladistic studies have stringently tested vascular plant monophyly. Tracheophyte monophyly is consistent with the 18S and 26S rRNA study by Waters, Buchheim, et al. (1992; Figure 3.13) and the more recent 18S rRNA analyses by Kranz, Miks, et al. (1995) and Kranz and Huss (1996) that included lycopsids. Early results based on *rbcL* DNA sequences contradict all previous hypotheses as well as cladistic studies based on morphological data (Figure 3.15; Manhart 1994), and Manhart (1994) and others (Albert, Backlund, Bremer, Chase, et al. 1994; Mishler, Lewis, et al. 1994, Chapter 3) have questioned the utility of *rbcL* sequences for resolving relationships among major land plant clades.

MAJOR CLADES OF TRACHEOPHYTES Relationships within the tracheophytes have been widely discussed. The ferns, or megaphyllous pteridophytes, were initially considered to be the group from which seed plants and microphyllous pteridophytes *(Equisetum* and lycopsids including *Psilotum)* evolved (e.g., Haeckel 1868, 1894; Hallier 1902; Potonié 1899; Tansley 1907). This idea was successfully challenged around the turn of the century (e.g., Bower 1894; Jeffrey 1902; Lignier 1908), and the most influential arguments for a closer relationship of ferns to seed plants were advanced in a series of articles by Jeffrey (see Beck, Schmid, and Rothwell 1982; Schmid 1982). Jeffrey argued that the possession of leaf gaps, megaphyllous leaves, and abaxial sporangia suggest that ferns are more closely related to seed plants than to either lycopsids or *Equisetum* (Beck, Schmid, and Rothwell 1982). This idea was elaborated by Lignier (1908), who hypothesized that the microphylls of *Equisetum* had evolved by reduction from the larger leaves of earlier fossil forms and were therefore not homologous with those of lycopsids. *Equisetum* was therefore seen as probably closely related to ferns and seed plants (based on its megaphyllous but reduced leaves), and lycopsids were thus viewed as the earliest group to diverge from other tracheophytes. These ideas, which were based on comparative morphology, were reinforced by the recognition of the long and excellent fossil record of lycopsids (e.g., Scott 1900, 1924), as well as by the appearance of lycopsids shortly after the earliest fossil evidence of tracheophytes (Banks 1968, 1975a,b).

The main features of this phylogenetic scenario have been widely accepted (e.g., Banks 1968, 1970; Chaloner and Sheerin 1979; Darrah 1939, 1960; Emberger 1944, 1968; Kräusel 1950; Lam 1948, 1950; Leclercq 1954; Lignier 1908; Meyen 1987; Pichi-Sermolli 1959; Smith 1938; Thomas 1958; Zimmermann 1938, 1952). Furthermore, the characterization of three new fossil tracheophyte groups (Rhyniophytina, Zosterophyllophytina, Trimerophytina) by Banks (1968, 1975a,b) was broadly consistent with Lignier's ideas (1908), and it quickly became widely accepted that the Zosterophyllophytina are closely related to lycopsids, whereas the Trimerophytina are closely related to the *Equisetum*–fern–seed plant lineage.

Cladistic studies support the sister group relationship between lycopsids and other vascular plants (e.g., Bremer 1985; Crane 1989, 1990; Doyle and Donoghue 1986; Gensel 1992; Kenrick and Crane 1991; Parenti 1980), and studies that include early fossils support a sister group relationship between zosterophylls and lycopsids (Crane 1989, 1990; Gensel 1992; Kenrick and Crane 1991). Several treatments have also placed the small extant group Psilotaceae in a basal position within vascular plants (Bremer 1985; Bremer, Humphries, et al. 1987; Parenti 1980), but this interpretation is being increasingly challenged (see below and Chapter 7). Few molecular studies have addressed the relationships among pteridophytes and seed plants, but a restriction-site study analyzing major chromosome inversions (Raubeson and Jansen 1992) and 18S rRNA sequences (Kranz and Huss 1996) support an early divergence of lycopsids from other vascular plants. Interestingly, these studies also ally the Psilotaceae with non-lycopsid vascular plants.

The results of our analysis of relationships among the basal vascular plants are more highly resolved than those of previous treatments, but again there is strong support for an early divergence of lycopsids (Lycophytina) from other vascular plants (Euphyllophytina). The Lycophytina comprise lycopsids and their extinct sister group zosterophylls, and the relationships in these groups are discussed in greater detail in Chapters 5 and 6. Although our analysis of relationships within the Euphyllophytina is preliminary and based mainly on early fossil material, monophyly of the group is strongly supported. The extinct taxa placed in the Euphyllophytina have all been related to extant ferns, seed plants, or *Equisetum*.

Because extant ferns, seed plants, and *Equisetum* have highly divergent morphologies, some of the homology assignments employed in previous cladistic treatments are problematic and need to be reconsidered. Bremer, Humphries, et al. (1987) recognize the Euphyllophytina (which is equivalent to their *Equisetum*–fern–seed plant clade) on the basis of "sporangia on leaves" (character 69). This character is problematic because early fossils in these lineages are leafless. A more appropriate homology would correspond to our character 24 (specialized fertile zone), which recognizes a homology between the densely branched axes bearing paired sporangia in the sporangia-bearing portions of stem group taxa in the *Equisetum*–fern–seed plant clades (Figure 4.13). These structures appear to have been modified subsequently into peltate sporangiophores in *Equisetum,* into sporangium-bearing leaves in ferns, and into pollen organs and cupules containing seeds in seed plants. Bremer, Humphries, et al. (1987) recognize a fern–seed plant clade, but our most parsimonious solution groups early fossils related to ferns and *Equisetum (Pseudosporochnus, Rhacophyton, Ibyka)* together as a sister group to the progymnosperm–seed plant clade. Rigorous testing of relationships at this level requires a more detailed cladistic treatment that includes a greater sample of early fossil material than is possible here. Nevertheless, our results do not support the homology of megaphyllous leaves in ferns and seed plants (e.g., Bremer, Humphries, et al. 1987; Mishler, Lewis, et al. 1994) unless this homology is extended to incorporate *Equisetum* and redefined as our character 4.10 (dichotomous, pinnulelike appendages). The isotomously branching, small, sterile, ultimate appendages in many taxa of the Euphyllophytina are leaflike in several respects but are nonplanar and unwebbed in many early taxa (Figure 4.13). Our analysis supports the hypothesis of homology of these "megaphyll precursors" in the Euphyllophytina. Because such nonplanar and unwebbed structures occur in aneurophytalean progymnosperms (seed-plant stem group), it is likely that certain features of the "megaphylls" of seed plants and ferns (i.e., planation and webbing) evolved independently from these megaphyll precursors as suggested by Doyle and Donoghue (1986) and Crane (1990). A second character used to support the fern–seed plant clade, trichomes (Bremer, Humphries, et al. 1987), also appears to have arisen independently in these groups given that it appears to be absent in basal fossils.

Reevaluation of Rhyniophytina and Trimerophytina

RHYNIOPHYTINA Our analysis does not support the monophyly of the Rhyniophytina either in its original form *sensu* Banks (1975b), or in its subsequent redefinition (Table 4.2; Banks 1992; Edwards and Edwards 1986; Meyen 1987; T. N. Taylor 1988a). The Rhyniophytina of these authors is either strongly paraphyletic, or polyphyletic with elements that are related to several major polysporangiophyte clades (Figure 4.31). Furthermore, the group is paraphyletic by definition because current circumscriptions are based on features that are plesiomorphic at this level of plant evolution.

The original concept of Rhyniophytina was based mainly on three relatively well understood taxa from the Rhynie Chert: *Rhynia gwynne-vaughanii, Aglaophyton (Rhynia) major,* and *Horneophyton lignieri* (Banks 1975b). Of these taxa, only *R. gwynne-vaughanii* is currently placed unequivocally in a group with *Stockmansella langii* and *Huvenia kleui* (Rhyniopsida; Figure 4.31). Recent studies have shown that both *Aglaophyton* and *Horneophyton* lack the typical ornamented tracheids of vascular plants and that *Horneophyton* has a unique form of branched sporangium. Our analysis excludes these two taxa and places them in the tracheophyte stem group as part of a protracheophyte grade because they lack ornamented tracheids and there are no other features that unequivocally link them to other groups within tracheophytes (Figure 4.31). Other taxa usually associated with the Rhyniophytina include a range of often rather poorly understood *Cooksonia*-like plants (e.g., *Cooksonia, Renalia, Hsua, Uskiella, Dutoitea, Nothia*). Our analysis shows that these taxa form a grade of organization that are more closely related to the Lycophytina (lycopsids and zosterophylls) than to either *R. gwynne-vaughanii* or the Euphyllophytina (Figure 4.31).

The problematic *Nothia aphylla* is resolved as a plesiomorphic member of the tracheophyte subgroup Lycophytina. The absence of ornamentation in the tracheids of *Nothia* is best interpreted as a loss because of the weight of other features (i.e., sporangium morphology, dehiscence, and arrangement) that support relationships within lycopsids and zosterophylls. In this context, it is worth noting that the tracheids of *Nothia* otherwise have thick, decay-resistant cell walls that are more typical of those in lycopsids and zosterophylls than in those of water-conducting cells in the protracheophyte grade or mosses.

Based on these results, we reject the concept of the Rhyniophytina *sensu* Banks (1975b) in favor of several more inclusive groups within land plants (i.e., polysporangiophytes, tracheophytes, eutracheophytes). We have assigned former rhyniophyte taxa to higher level groups within polysporangiophytes, and where clear multitaxon clades occur, we have defined new groups on the basis of apomorphic features (Chapter 7).

TRIMEROPHYTINA Our analysis does not support monophyly of the Trimerophytina *sensu* Banks (1975b), which is resolved as paraphyletic with respect to other taxa related to seed plants, ferns, and sphenopsids (e.g., *Tetraxylopteris, Rhacophyton, Pseudosporochnus, Ibyka;* Figure 4.31). Furthermore, the definition of the Trimerophytina as a group is problematic because many of its features appear to be more general in polysporangiophytes and vascular plants. Similarly, the more restrictive aspects of the definition are clearly present in supposedly more advanced groups such as progymnosperms, suggesting that they, too, define more inclusive clades.

The concept of the Trimerophytina *sensu* Banks (1975b) is based mainly on three taxa *(Psilophyton, Pertica, Trimerophyton),* of which *Psilophyton* is the best known. Our analysis indicates that *Pertica* is more closely related to aneurophytalean progymnosperms (e.g., *Tetraxylopteris*) than to *Psilophyton,* although this conclusion is tentative because many aspects of the morphology of *Pertica* are still poorly understood. A close relationship with *Tetraxylopteris* is based on the tetrastichous branch arrangement, but more supporting evidence on stem anatomy (predicted to be lobed) is desirable. Our results confirm a close relationship between the recently described *Eophyllophyton bellum* and the trimerophyte grade plants (Hao and Beck 1993). The basal position of *E. bellum* in the Euphyllophytina provides the first clear evidence for plants with G-type tracheids in this clade. Cladistic analysis suggests that the putative trimerophyte *Yunia dichotoma* (Hao and Beck 1991b) is more closely related to the Lycophytina than to the Euphyllophytina (Figure 4.31).

Based on these results, we reject the concept of the Trimerophytina *sensu* Banks (1975b) in favor of several more inclusive groups. Former trimerophyte taxa are assigned to high level groups within the Euphyllophytina (Chapter 7).

5 Zosterophyllopsida and Basal Lycophytes

The zosterophylls are a distinctive group of leafless, herbaceous plants characterized by lateral, reniform sporangia and exarch protosteles (Banks 1975a,b). The zosterophylls were an important component of early terrestrial floras (Banks 1980; D. Edwards 1990; Edwards and Davies 1990; Knoll, Niklas, et al. 1984; Niklas, Tiffney, and Knoll 1985), and their diversity is greatest in the lower Devonian, where they are some of the largest and ecologically most prominent elements of early

TABLE 5.1

Species Examined in the Cladistic Analysis of Zosterophylls

Taxon	Exemplars
Zosterophylls	*Anisophyton gothanii* † [1,3]
	Bathurstia denticulata † [5]
	Crenaticaulis verruculosus † [1,3]
	Deheubarthia splendens † [1,2,3,4]
	Discalis longistipa † [1,2]
	Gosslingia breconensis † [1,2,3,4]
	Gumuia zyzzata † [1,2]
	Konioria andrychoviensis † [1,3]
	Oricilla bilinearis † [1,3]
	Rebuchia ovata † [1]
	Sawdonia ornata † [1,4]
	Serrulacaulis furcatus † [1]
	Tarella trowenii † [1]
	Thrinkophyton formosum † [1,3,4]
	Trichopherophyton teuchansii † [5]
	Zosterophyllum deciduum † [1,2]
	Zosterophyllum divaricatum † [1,2,3,4]
	Zosterophyllum fertile † [1,2]
	Zosterophyllum llanoveranum † [1,2]
	Zosterophyllum myretonianum † [1,2,3,4]
Putative zosterophylls	*Adoketophyton subverticillatum* † [1]
	Hicklingia edwardii † [1,2]
	Hsua robusta † [1,3,4]
	Huia recurvata † [1,2]
	Krithodeophyton croftii † [5]
	Nothia aphylla † [1,2]
Barinophytaceae	*Barinophyton citrulliforme* † [1,3]
	Barinophyton obscurum † [1,4]
	Protobarinophyton pennsylvanicum † [1,3,4]
Lycopsida	*Huperzia selago* [1,2,3,4]
	Asteroxylon mackiei † [1,2,3,4]
	Baragwanathia longifolia † [1]
	Drepanophycus spinaeformis † [1]
Rhyniopsida	*Rhynia gwynne-vaughanii* †* [1,2,3,4]
Euphyllophytina	*Psilophyton dawsonii* †* [1,2,3,4]

Notes: See Appendix 1 for taxon descriptions and Appendix 4 for data matrix. Superscript numbers indicate taxa included in various analyses. * = outgroup. † = extinct.

[1] Analysis 5.1, Figures 5.25 and 5.26, Heuristic search, 32 taxa.

[2] Analysis 5.2, Figure 5.27, Branch-and-Bound search, 16 taxa.

[3] Analysis 5.3, Figure 5.28, Branch-and-Bound search, 16 taxa.

[4] Analysis 5.4, Figure 5.29, Bremer Support measurements, Branch-and-Bound search, 13 taxa.

[5] Taxa included on an experimental basis.

terrestrial vegetation. As currently defined, zosterophylls are a predominantly Devonian group, and during this interval, they had a cosmopolitan distribution that included North America, northern Europe, eastern Europe, Siberia, China, and Australia. About 16 genera can be assigned unequivocally to the group, but there are other potentially closely related fossil taxa. These taxa include certain rhyniophytes and other enigmatic zosterophyll-like plants such as *Nothia, Gumuia,* and *Hsua*—the relationships of which are not well understood—as well as the heterosporous group Barinophytaceae (Table 5.1).

The zosterophylls are generally considered to be more closely related to lycopsids than to other extant vascular plants as determined by similarities in sporangium form and position, as well as by certain anatomical details (Chapter 4; Banks 1968, 1970, 1975a,b, 1992; Bateman 1996c; Chaloner and Sheerin 1979; Crane 1990; Gensel 1992; Meyen 1987; Niklas and Banks 1990; Selden and Edwards 1989; T. N. Taylor and E. L. Taylor 1993). However, zosterophylls also differ from lycopsids in several significant features: all are leafless, all have a simple unlobed protostele, and many taxa have planar branching patterns, circinate vernation, and distinctive linear arrangements of sporangia (Figures 5.1 and 5.2). Recent studies suggest that the zosterophylls are either strongly paraphyletic to lycopsids (e.g., Crane 1990) or that they are the lycopsid sister group (e.g., Gensel 1992; Kenrick and Crane 1992). It has also been suggested that the group should be extended to include the formerly enigmatic, heterosporous Barinophytaceae (Bateman and DiMichele 1994a; Kenrick and Crane 1992). In this chapter we evaluate relationships among the zosterophylls and potentially closely related groups, focusing mainly on early fossil taxa.

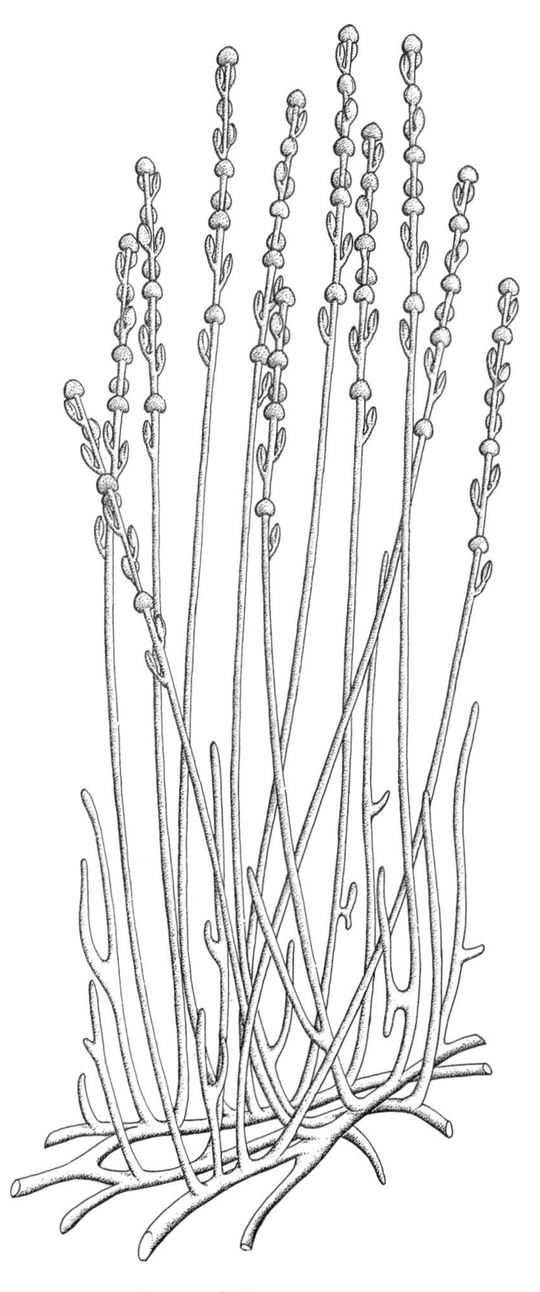

FIGURE 5.1. *Zosterophyllum myretonianum.* Reconstruction showing prostrate rhizome bearing upright axes with lateral reniform sporangia in a helical arrangement (×1). Modified after Walton 1964, Figure 1.

SYSTEMATICS: THE ORIGIN OF THE "ZOSTEROPHYLL" CONCEPT

The fossil genus *Zosterophyllum* was first described by Penhallow (1892) based on a specimen from the Old Red Sandstone (Figure 5.1). The original description provided few details (Arber 1921), and *Zosterophyllum* was regarded as highly enigmatic until the morphological investigations of Lang (1927) and Lang and Cookson (1930). These studies established that *Zosterophyllum* was a vas-

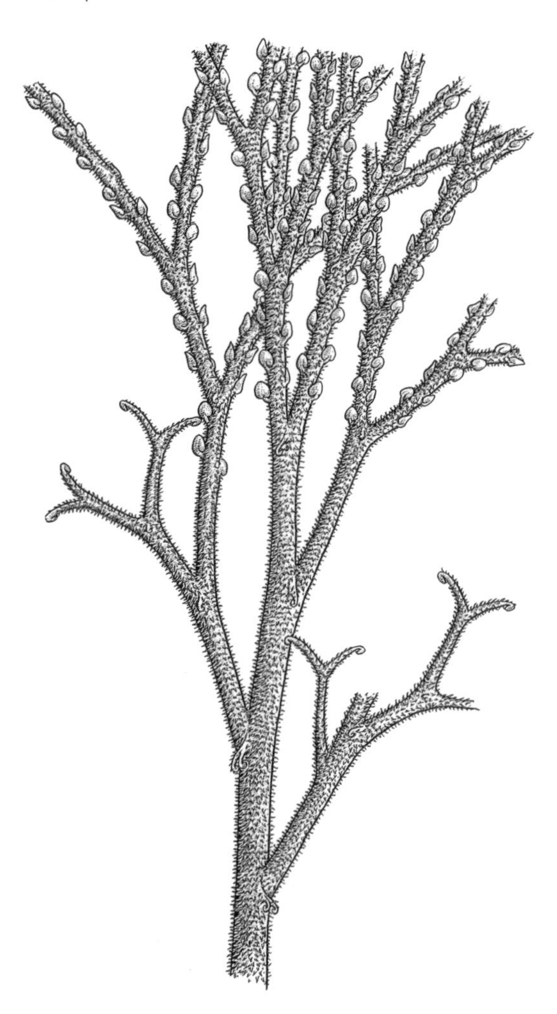

FIGURE 5.2. *Deheubarthia splendens.* Reconstruction of distal part of plant showing planar, pseudomonopodial axes bearing small spines and lateral reniform sporangia in two-rowed arrangement (×0.4). Note the presence of small, circinate, "axillary" branches and circinate tips of main branching system. Modified after Kenrick and Edwards 1988a, Figure 6.

cular plant and clarified many other aspects of its morphology, including critical information on the sporangium region. A close relationship with the Psilophytales was emphasized (Chapter 4), and this new information led to the recognition of a distinctive group of plants (Zosterophyllaceae) characterized by lateral, short-stalked, flattened sporangia with distinctive marginal dehiscence (Ananiev 1960; Andrews 1961; Arnold 1947; Høeg 1967; Kräusel 1938; Leclercq 1954; see Chapter 4 and Tables 4.1 and 5.2). Other fossils included originally within the Zosterophyllaceae were the *Zosterophyllum*-like *Ŗebuchia (Bucheria),* and later, the distinctive *Gosslingia breconensis* (Heard 1927). Also included by Kräusel (1938) and Ananiev (1960) were two rather poorly understood taxa: *Barinophyton* (Smith and White 1905), originally described as a lycopsid strobilus by Dawson (1861, 1862), and *Pectinophyton* (Høeg 1935). Both taxa had *Zosterophyllum*-like sporangium arrangements on more complex branches, but many other aspects of their morphology were unclear (Figures 5.3 and 5.4). These plants were later recognized as a distinct family-level group, the Barinophytaceae (Kräusel and Weyland 1961), and although similarities with zosterophylls continued to be recognized by some authors, a close relationship was considered unlikely because of differences in sporangium dehiscence and spore size (e.g., Arnold 1939, 1947; Høeg 1967; Kräusel and Weyland 1961).

As knowledge of Devonian plants increased, certain patterns of relationship within the Psilophytales began to emerge. Leclercq (1954, 1956) and Hueber (1964) drew attention to similarities between the Zosterophyllaceae and early lycopsid fossils and suggested that these groups formed a distinct lineage of plants within the Psilophytales. This phylogenetic hypothesis was subsequently developed by Banks (1968, 1970, 1975b) using morphologic and stratigraphic data. Banks argued that the evolution of lycopsids from early land plants at the zosterophyll level of organization was consistent with the sporangium morphology, the protoxylem maturation, and the early occurrence of both groups in the fossil record. Furthermore, this scenario was compatible with a widely held phylogenetic hypothesis that considered lycopsids as the most divergent extant tracheophyte group (Bower 1894, 1908; Jeffrey 1902; Lignier 1903, 1908). Emphasizing the morphological distinctness and early stratigraphic appearance of both lycopsids and zosterophylls, Banks (1970; see also Figure 4.4) suggested that this tracheophyte lineage may even have evolved independently from nonvascular ancestors (see Chapter 4).

In his influential reclassification of the Psilophytales, Banks (1968, 1975b; see Chapter 4 and Tables 4.1 and 5.2) elevated the Zosterophyllaceae of earlier authors to a subdivision (Zosterophyllophytina) of the Tracheophyta. By this time, six genera of zosterophylls were recognized, but the relationships of the Barinophytaceae were questioned, and the group was excluded from the Zosterophyllophytina and classified *incertae sedis* (Table 5.2). In defining the Zosterophyllophytina, Banks used much new morphological information and showed that in addition to being leafless and having lateral reniform-globose sporangia and distal dehiscence, zosterophylls are also characterized by a large, solid strand of tracheids that is elliptic or terete in transverse section. The smallest tracheids (in transverse section) are positioned centripetally, and therefore, exarch maturation of the xylem tissue was inferred (Banks 1975a; Banks and Davis 1969). The scope and definition of the Zosterophyllophytina *sensu* Banks (1975b) and the idea of a close relationship to lycopsids are currently widely accepted by both specialists and authors of general texts (Banks 1992; Bierhorst 1971; Gensel and Andrews 1984; Gifford and Foster 1989; Meyen 1987; Raven, Evert, and Eichhorn 1992; Stewart 1983; Stewart and Rothwell 1993; T. N. Taylor 1981; T. N. Taylor and E. L. Taylor 1993).

Throughout the last two decades, knowledge of the zosterophylls has increased dramatically. A recent review enumerated about 15 genera, most containing fewer than 3 species (Niklas and Banks 1990). Even accepting the possibility of unrecognized synonymy, *Zosterophyllum* is the largest genus and now contains over 18 species arranged in two subgenera. There are no widely recognized supergeneric groups. Banks (1968) initially recognized two families, the Zosterophyllaceae and the Gosslingiaceae, which were based on differences in branching and sporangium position. More recently, Niklas and Banks (1990) and Banks (1992) established two informal groupings based on the shoot apex: taxa with fertile shoots that terminate in a sporangium (e.g., *Zosterophyllum, Rebuchia, Gumuia*) and taxa with axes that do not terminate in sporangia (e.g., *Gosslingia, Sawdonia*).

PHYLOGENETIC QUESTIONS AND AIMS OF ANALYSIS

Much new information has recently emerged concerning early lycopsid-like plants, zosterophylls, and other potentially relevant fossil taxa. Many new species and genera have been described, but there has been no broad-based attempt to evaluate their relationships or to reexamine the circumscription of the Zosterophyllophytina *sensu* Banks. The aim of this chapter is to analyze the relationships among a broad range of zosterophylls and potentially related plants via numerical cladistic techniques. The specific phylogenetic questions to be addressed are outlined below.

Although a close phylogenetic relationship for zosterophylls and lycopsids is generally accepted, the exact nature of this relationship has never been defined clearly. The most explicit recent discussions imply that lycopsids are nested within zosterophylls, having been derived from spiny plants similar to *Sawdonia* (Figure 5.5; Gensel, Andrews, and Forbes 1975; Niklas and Banks 1990; Schweitzer 1983b). Central to this view is the hypothesis that the lycopsid microphyll is a transformational homology of the multicellular spines present in some zosterophylls: an idea consistent with the enation theory of microphyll evolution developed by Bower (1935, 552; Figure 7.21). In cladistic terms, this interpretation indicates that zosterophylls are paraphyletic with respect to lycopsids (Figure 4.6; Crane 1989, 1990) and implies that the sister group to lycopsids is to be found among spiny zosterophylls of the *Sawdonia*-type (e.g., *Deheubarthia:* Figure 5.2).

An alternative scenario favors a basically monophyletic interpretation of zosterophylls and a very early divergence of lycopsids from simple *Zosterophyllum*-type precursors as suggested by Banks (1970, 77; Figure 4.4), Chaloner and Sheerin

TABLE 5.2
Some Important Treatments of the Zosterophyllaceae and Barinophytaceae

Kräusel 1938	Banks 1975b[1]	Gensel and Andrews 1984[2]	Meyen 1987[3]
Psilophytales	Zosterophyllophytina	Zosterophyllophytina	Protopteridophyta
Zosterophyllaceae	Zosterophyllales	Zosterophyllales	Zosterophyllopsida
Zosterophyllum	Zosterophyllaceae	Zosterophyllaceae	*Zosterophyllum*
Barinophyton	*Zosterophyllum*	*Zosterophyllum*	*Sawdonia*
Pectinophyton	*Rebuchia* (=*Bucheria*)	*Rebuchia* (=*Bucheria*)	*Gosslingia*
Rebuchia (=*Bucheria*)	*Sawdonia*	*Sawdonia*	*Crenaticaulis*
	Gosslingia	*Gosslingia*	*Margophyton*
	Crenaticaulis	*Crenaticaulis*	*Oricilla*
	Bathurstia	*Bathurstia*	*Nothia*
	incertae sedis	*Serrulacaulis*	*Renalia*
	Barinophytales	*Oricilla*	*Hsua*
	Barinophytaceae	*Konioria*	*Konioria*
	Barinophyton	*Hicklingia*	*Hicklingia*
	Pectinophyton	*incertae sedis*	Pteridophyta
	Protobarinophyton	Barinophytales	Barinophytopsida
	Barinostrobus	Barinophytaceae	*Krithodeophyton*
		Barinophyton	*Barinophyton*
		Pectinophyton	*Protobarinophyton*
		Protobarinophyton	
		Barinostrobus	
		incertae sedis	
		Hsua	

[1] Some newly described taxa added; *Barinophyton*, etc., removed.

[2] Some new taxa, but very similar to Banks 1975b.

[3] Marked departure from other treatments through inclusion of "rhyniophytes" such as *Nothia* and the larger and more complex *Cooksonia*-like taxa *Renalia* and *Hsua*.

(1979; Figure 4.5), Meyen (1987; Figure 5.6), Selden and Edwards (1989; Figure 5.7), (Gensel 1992; Figure 5.8), and (Hueber 1992). According to this interpretation, lycopsids would be the sister group to a large and diverse zosterophyll clade, a scenario consistent with the virtually simultaneous appearance of simple zosterophylls (such as *Zosterophyllum*) and lycopsids in the fossil record.

Recent preliminary attempts to examine this problem using a numerical cladistic approach

Taylor 1988[4]	Banks 1992[5]
Zosterophyllophyta	Zosterophyllophytina
Zosterophyllum	Zosterophyllaceae
Sawdonia	"Terminate"
Deheubarthia	*Zosterophyllum*
Oricilla	*Rebuchia* (=*Bucheria*)
Discalis	*Gumuia*
Hicklingia	"Nonterminate, bilateral"
Serrulacaulis	*Sawdonia*
Gosslingia	*Gosslingia*
Tarella	*Crenaticaulis*
Thrinkophyton	*Bathurstia*
Konioria	*Serrulacaulis*
Crenaticaulis	*Oricilla*
Rebuchia (=*Bucheria*)	*Konioria*
Kaulangiophyton	*Margophyton*
incertae sedis	*Tarella*
Barinophyton	*Anisophyton*
Protobarinophyton	*Thrinkophyton*
	Discalis
	Deheubarthia
	incertae sedis
	Barinophytales
	Barinophytaceae
	Barinophyton
	Pectinophyton
	Protobarinophyton
	Barinostrobus
	incertae sedis
	Hsua
	Nothia
	Hicklingia
	Huia

[4] Similar to Banks 1975b but with newly described taxa added.

[5] Similar to Banks 1975b but with some newly described taxa added. Two informal groupings recognized within Zosterophyllophytina.

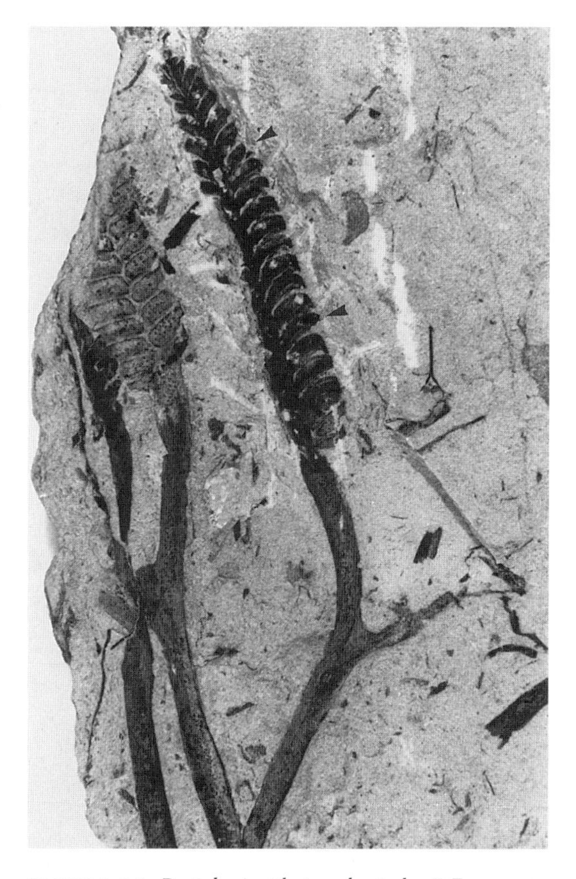

FIGURE 5.3. *Protobarinophyton obrutschevii.* Fragment of distal part of plant showing small strobilus *(arrows)* bearing two rows of lateral sporangia (one row obscured by sediment in specimen on right) on naked dichotomous axis (×1.2). Specimen from Natural History Museum, London, V51567. Photograph by P.K.

(Gensel 1992) favored an early divergence of the lycopsids from the zosterophylls and a sister group relationship for the two clades, although the position of *Zosterophyllum* itself was unresolved (Figure 5.8). This conclusion was based on a data set of 24 taxa (including 7 putative zosterophylls and 7 lycopsids) and 12 characters. One aim of our analysis in this chapter is to expand that previous cladistic study on the basis of a wider range of fossils and a larger number of characters in order to provide an explicit test of the two contrasting hypotheses of zosterophyll relationship. We have also adopted a different approach to characters and character coding. Of the 12 characters used by Gensel, 5 are ordered multistates, a condition that significantly constrains the analysis to preconceptions of character transformation and relationship. We have avoided such ordering because it effectively endorses a specific process-level explanation for the evolution of a particular feature and thus

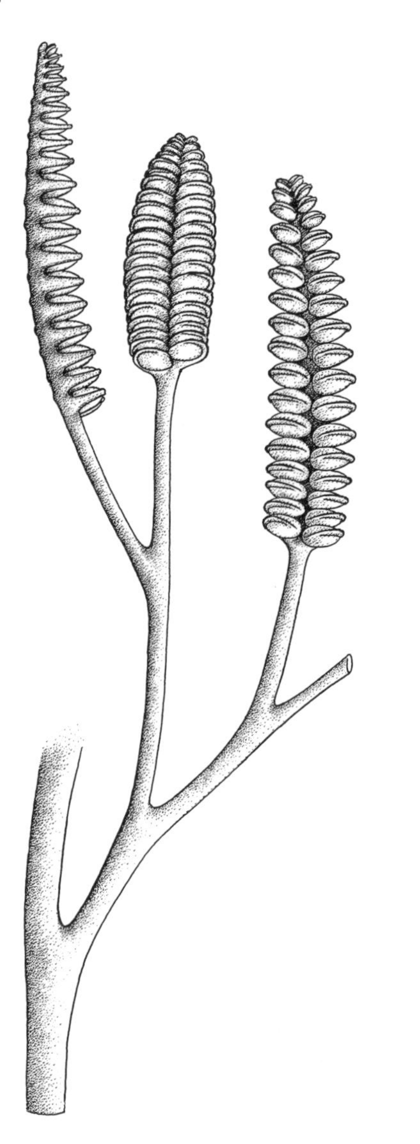

FIGURE 5.4. *Barinophyton citrulliforme.* Reconstruction of distal part of fertile lateral branch showing small strobili bearing two rows of lateral sporangia on a naked dichotomous axis (×1). Note also the sporangium orientation and attachment to stem. Modified after Brauer 1980, Figures 35–37.

precludes a posteriori testing of such ideas as part of the analysis.

A second significant phylogenetic question that our analysis addresses concerns the relationships of the Barinophytaceae. This group has been regarded as problematic because of difficulties in interpreting the structure of the fertile region, the peculiar form of sporangium heterospory, and the apparently unusual tracheid wall structure (Brauer 1980, 1981). The Barinophytaceae have been classified *incertae sedis* by many authors (Banks 1975b; Gensel and Andrews 1984; T. N. Taylor 1981), whereas others (e.g., Chaloner and Sheerin 1979; Meyen 1987; Figures 4.5 and 5.6) have acknowledged the barinophyte-zosterophyll similarities noted by Kräusel (1938), Ananiev (1960), and Høeg (1967).

A third general question concerns the position of zosterophylls and lycopsids with respect to potentially related Rhyniophytina (*sensu* Edwards and Edwards 1986). Our analysis in Chapter 4 places *Renalia, Yunia,* some species of *Cooksonia,* and other simple *Cooksonia*-like plants (e.g., *Uskiella*) in the Lycophytina stem group. Other rhyniophytes such as *Nothia* and the more complex *Hsua* are grouped within the Lycophytina clade. The relationships of these two species within zosterophylls are considered in more detail in this chapter. The recently described enigmatic zosterophyll-like plants *Huia, Gumuia,* and *Adoketophyton* are also considered.

CHOICE OF TAXA

As in the preceding analyses of embryophytes and polysporangiophytes (Chapters 3 and 4), our choice of zosterophyll taxa is a compromise dictated by the phylogenetic questions that we are seeking to address, by the quality of existing data, and by the need to include those fossils that appear to be of maximum interest with respect to phylogenetic issues at this level. The analysis is further complicated by the relatively large number of potential ingroup taxa (33). Similar problems in the previous chapters were addressed by using exemplars for well-supported monophyletic groups, but this approach was more problematic with zosterophylls because few unequivocal groups have been recognized previously. All of the taxa considered in our analysis of zosterophyll relationships are listed in Table 5.1. Brief descriptions and reviews of these taxa, along with citations of the relevant literature, are provided in Appendix 1.

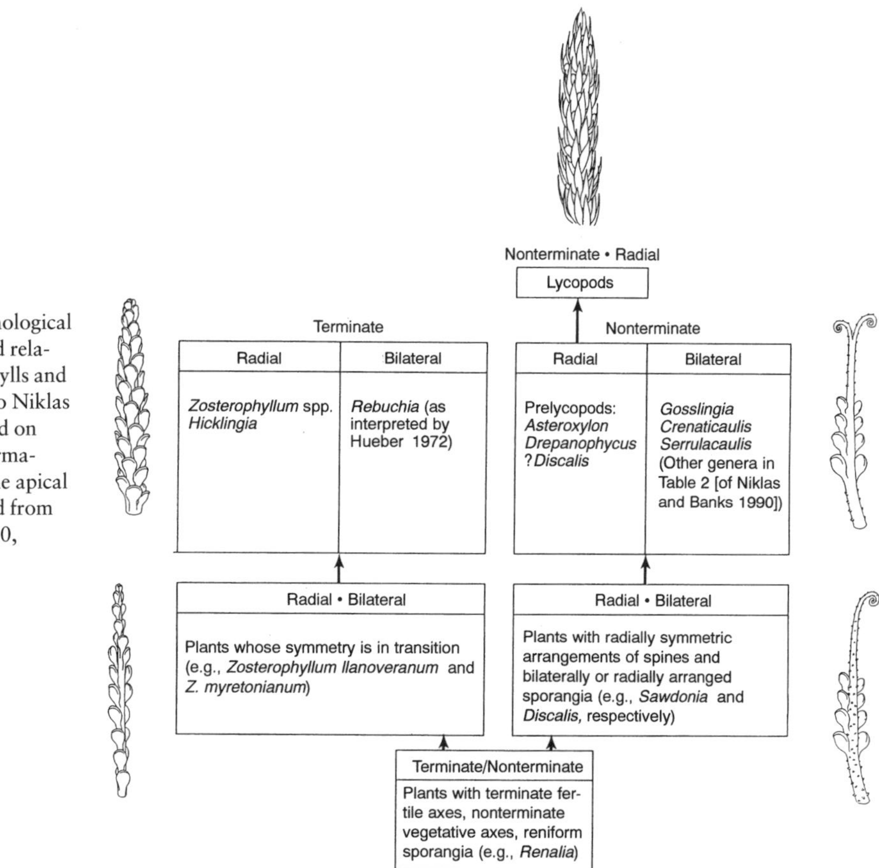

FIGURE 5.5. Morphological patterns and suggested relationships in zosterophylls and lycopsids, according to Niklas and Banks 1990, based on hypotheses of transformation in symmetry of the apical meristem. Reproduced from Niklas and Banks 1990, Figure 1.

As outgroups, we used *Rhynia gwynne-vaughanii* (Figure 4.8), a member of the Rhyniopsida (Figure 4.31), and *Psilophyton dawsonii* (Figure 4.13), a basal member of the Euphyllophytina clade (Figure 4.31). These plants were chosen because they are clearly not ingroup taxa (Chapter 4), but they are nevertheless sufficiently close and well understood to allow character polarization. As exemplars for the lycopsids, we included extant *Huperzia* (Figure 3.24), *Asteroxylon mackiei* from the early Devonian Rhynie Chert, *Baragwanathia longifolia* from the late Silurian and early Devonian of Australia, and the wide-ranging Devonian plant *Drepanophycus spinaeformis*. Our analyses in Chapter 6 indicate that these taxa are closely related and are relatively basal within the lycopsid clade.

Most zosterophyll genera are monotypic (Appendix 1), but where more than one species has been described, the type species is usually significantly more complete and better understood. One exception is the genus *Zosterophyllum*, which was created by Penhallow (1892) based on *Z. myretonianum* (Figure 5.1). At least 18 species have been assigned to the genus, but few are well known and some may prove to be synonymous. The known geographic range of *Zosterophyllum* is much greater than that of any contemporaneous genus, except perhaps that of *Cooksonia* (Chapter 4).

Lang (1927) made a useful early review of *Zosterophyllum* in which he recognized two subgenera: the subgenus *Zosterophyllum*, for plants with sporangia borne in a helical arrangement forming a terminal spike, and the subgenus *Platyzosterophyllum*, for plants with sporangia in two rows forming a dorsiventral spike (Croft and Lang 1942; Hueber 1972). More recent summaries of the literature on *Zosterophyllum* are given by

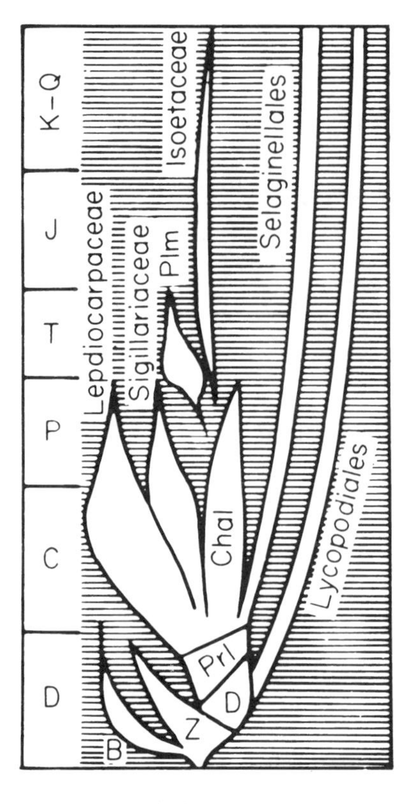

FIGURE 5.6. Proposed relationship among zosterophylls, lycopsids, and related groups, according to Meyen 1987. B = barinophytes; Z = zosterophylls; D = *Drepanophycus* and related plants; PrL = Protolepidodendrales; Chal = *Chaloneria* and related plants; Plm = *Pleuromeia* and related plants. Reproduced from Meyen 1987, Figure 16, with permission.

Høeg (1967), Hueber (1972), D. Edwards (1975), Li and Cai (1978), Gensel (1982a), Cai and Schweitzer (1983), Gensel and Andrews (1984), Gerrienne (1988), Hao (1989b), and Niklas and Banks (1990). One aim of our analysis was to test the monophyly of *Zosterophyllum,* and we included five of the better known species covering a range of morphologies: two from the subgenus *Zosterophyllum* (*Z. myretonianum:* Figure 5.1; *Z. deciduum:* Figure 5.9) and three from the subgenus *Platyzosterophyllum (Z. divaricatum:* Figure 5.10; *Z. fertile; Z. llanoveranum),* plus a range of similar fossils that appear to differ from *Zosterophyllum* in relatively minor features (e.g., *Discalis longistipa, Gumuia zyzzata, Rebuchia ovata, Adoketophyton subverticillatum, Hicklingia edwardii, Huia recurvata;* see Table 5.1).

We have also included in our analysis other relatively well understood taxa that were unequivocally placed in zosterophylls by other authors: *Anisophyton gothanii, Crenaticaulis verruculosus, Deheubarthia splendens, Gosslingia breconensis, Konioria andrychoviensis, Oricilla bilinearis, Sawdonia ornata, Serrulacaulis furcatus, Tarella trowenii,* and *Thrinkophyton formosum* (Table 5.1). Some of the more problematic taxa that we consider include the rhyniophyte-like *Nothia aphylla* and *Hsua robusta,* both of which we placed unequivocally within the Lycophytina in our analysis of polysporangiophytes (Chapter 4). We also consider the enigmatic Barinophytaceae, a group of heterosporous, predominantly upper Devonian to lower Carboniferous zosterophyll-like plants, which are represented in our analysis by *Barinophyton obscurum, B. citrulliforme,* and *Protobarinophyton pennsylvanicum.*

Three interesting taxa were found to be problematic because of incompleteness or uncertainty regarding certain critical aspects of morphology. *Trichopherophyton teuchansii,* from the Rhynie Chert, is highly informative with respect to apical stem and sporangium anatomy, but there is little evidence on sporangium arrangements and branching. Anatomical information for *Krithodeophyton croftii* is based on association evidence, and the morphology and arrangement of sporangia—in particular the presence or absence of a sporangium bract—are unclear (Appendix 1). *Bathurstia denticulata* is based on two poorly preserved axis fragments, and there is much missing data. These three taxa were included on an experimental basis.

CHARACTER DESCRIPTIONS AND CODING

Characters Scored for Almost All Taxa

5.1 • BRANCHING TYPE Pseudomonopodial branching results from unequal dichotomies that produce a wider, more or less straight or slightly

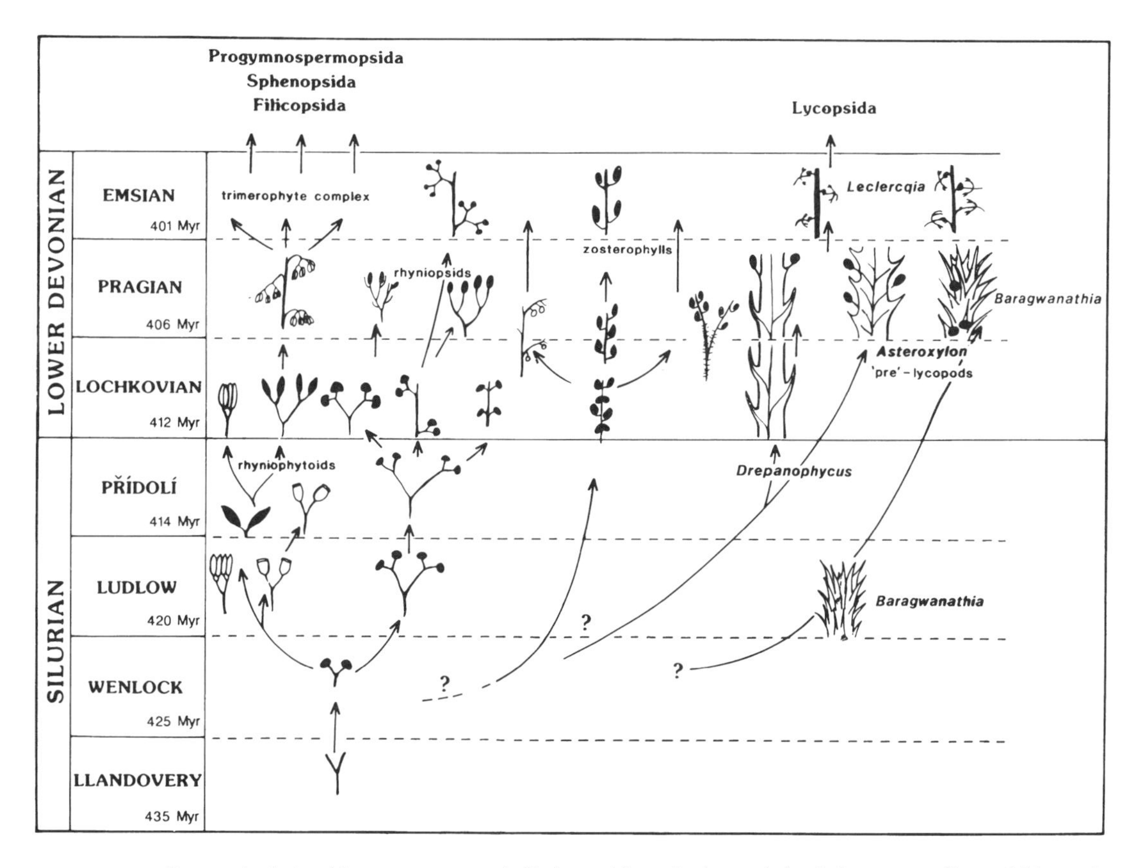

FIGURE 5.7. Proposed relationship among zosterophylls, lycopsids, and other early land plants, according to Selden and Edwards 1989. Reproduced from Selden and Edwards 1989, Figure 6.2.

zigzagged main axis from which narrower laterals depart alternately (Figure 5.2). This feature is easily observed in compression fossils, unless they are very fragmentary. In many zosterophylls, such as *Gosslingia, Anisophyton,* and *Thrinkophyton* (Figure 5.11), pseudomonopodial branching produces a more or less straight main axis, although strictly isotomous branching occurs in the distal regions of all these taxa. In such plants as *Zosterophyllum myretonianum* (Figure 5.1), *Discalis* (Figure 5.12), *Nothia* (Figure 5.13), and *Gumuia* (Figure 5.14), pseudomonopodial branching has not been recognized.

In the homosporous fossil lycopsids *Asteroxylon, Drepanophycus, Baragwanathia,* and extant *Huperzia* (Øllgaard 1987), branching of the main axes is basically isotomous. Bulbil production in *Huperzia selago* and lateral branching in *Asteroxylon mackiei* and *Baragwanathia longifolia* appear to have much in common (see Chapter 6) and may give the misleading appearance of a pseudomonopodial system. In both *Huperzia* and *Asteroxylon,* these bulbil branch traces arise in a similar manner to leaf traces (Kidston and Lang 1920b; Stevenson 1976) and are thus quite distinct from branch trace formation in a normal pseudomonopodial system. We treat the anisotomous branching in extant *Lycopodium* and *Lycopodiella* (Øllgaard 1987) as independently derived because these genera are clearly nested well within the lycopsid clade (Wagner and Beitel 1992; see also Chapter 6). Branching described as pseudomonopodial has also been attributed to certain members of the Protolepidodendrales, such as *Leclercqia* and *Minarodendron,* but this character requires more detailed study and is likely in any case to be a feature

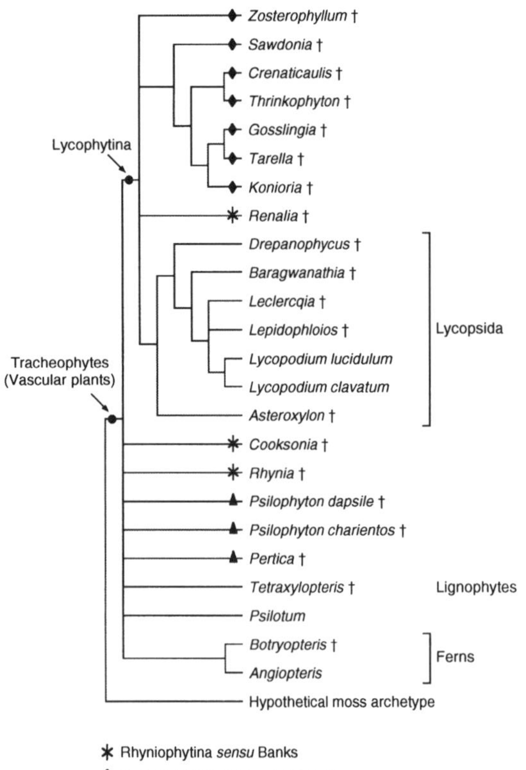

FIGURE 5.8. Cladistic relationships in early land plants, according to Gensel 1992: combinable component consensus tree based on 350 most parsimonious trees. Adapted from Gensel 1992, Figure 13.

derived within the lycopsid clade. Outgroup comparison with such rhyniophytes as *Aglaophyton, Rhynia, Uskiella,* and *Stockmansella* suggests that isotomous dichotomy is the primitive condition in polysporangiophytes.

Branching type is treated as a binary character: 0 = isotomous branching; 1 = pseudomonopodial branching. There were no inapplicable cases; six were missing data (see Appendix 4). See character 4.4 for comparison.

5.2 • BRANCHING PATTERN In many zosterophylls, branching in the main aerial axes occurs in one plane, giving a spreading, planated, branching system (Figures 5.2 and 5.11). This type of branching is typical of such plants as *Gosslingia, Anisophyton, Thrinkophyton, Sawdonia,* and *Deheubarthia.* In other taxa, such as *Zosterophyllum myretonianum* (Figure 5.1), *Discalis* (Figure 5.12), and *Gumuia* (Figure 5.14), spreading, planated branching systems are not present. In such plants as *Barinophyton, Bathurstia, Serrulacaulis,* and *Trichopherophyton,* the presence of this feature is equivocal. Planated branching systems of this type have not been recorded in primitive extant or fossil lycopsids. Strongly planated systems in Selaginellaceae appear to be a derived feature of the subgenus *Stachygynandrum* (Chapter 6). Outgroup comparison with rhyniophytes such as *Aglaophyton* and *Uskiella* suggests that nonplanar branching is the general condition in polysporangiophytes.

Branching pattern is treated as a binary character: 0 = nonplanar branching; 1 = planated aerial axes. There were no inapplicable cases; five had missing data (see Appendix 4). See character 4.5 for comparison.

5.3 • SUBORDINATE BRANCHING In addition to the main branching system, distinctive subordinate axes are common in many zosterophylls. One common type of branching pattern involves the production of a single, usually small, undeveloped, circinate axis slightly below each dichotomy of the main branching system (Figures 5.2 and 5.15). These small axes are oriented at right angles to the plane of the main branching system. Because of their orientation and distribution, they are usually observable in compression fossils as small conical bumps that indicate their point of attachment to the main axis (Figure 5.16). In compression fossils, features of this type have been termed *axillary tubercles* (D. Edwards 1970b) or *angular organs* (Remy, Hass, and Schultka 1986).

The pattern of vasculature associated with subordinate branch formation has been observed in *Gosslingia* and *Deheubarthia* and is similar in both plants (Figure 5.17). The vascular trace to the subordinate branch forms at the same time or slightly *after* the formation of the lateral branch trace and immediately bends through 90°. A further distinctive feature is that the xylem is terete at the base of the subordinate branch compared to the elliptical

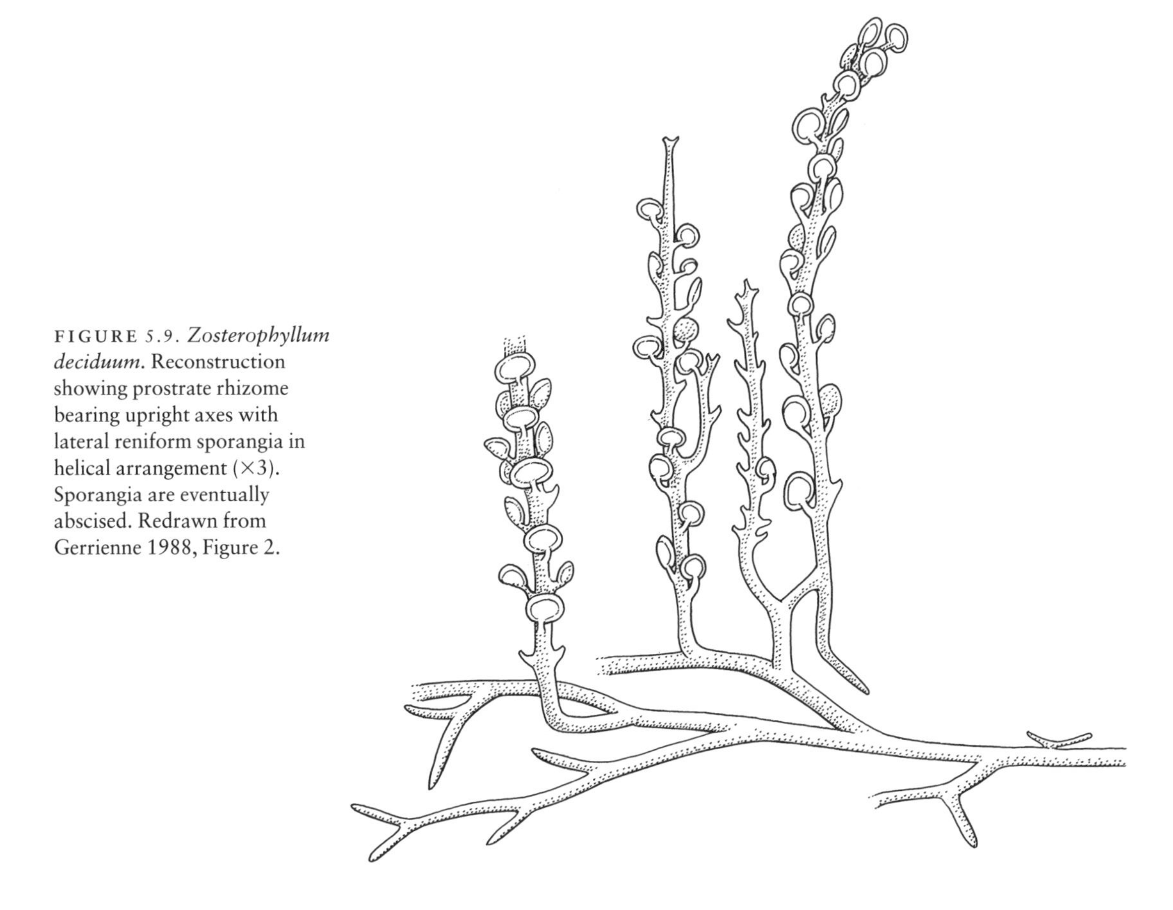

FIGURE 5.9. *Zosterophyllum deciduum*. Reconstruction showing prostrate rhizome bearing upright axes with lateral reniform sporangia in helical arrangement (×3). Sporangia are eventually abscised. Redrawn from Gerrienne 1988, Figure 2.

to strap-shaped xylem strands in the main axis and lateral branch.

Subordinate branches, which are usually small, circinate, and undeveloped, have been illustrated clearly in *Deheubarthia* (Figures 5.2 and 5.15) and *Tarella*. In *Tarella* and *Anisophyton potoniei*, these small circinate branches develop into a dichotomous branching system that is smaller than the main axial system (Remy, Hass, and Schultka 1986). Similar subordinate branching systems have been illustrated for *Sawdonia, Hsua,* and *Anisophyton gothanii* (Remy, Schultka, and Hass 1986).

In many zosterophylls, such as *Gosslingia, Deheubarthia, Anisophyton, Crenaticaulis,* and *Thrinkophyton,* the subordinate branch is positioned at, or slightly below, each dichotomy of the main axial system. We treat subordinate axes in this position as provisionally homologous with those that occur slightly above the dichotomy of the main axial system: on the main axes in the case of *Hsua* and on the lateral branches in the case of *Sawdonia.*

In addition to the conditions described above, in which subordinate axes occur singly and slightly above or below dichotomies of the main axial system, other arrangements and forms of subordinate branching have been observed in zosterophylls. In *Tarella,* subordinate circinate branches are numerous and scattered over the main axes. Similar features that require further investigation include short, noncircinate branches observed in the prostrate rhizomatous axes of *Zosterophyllum myretonianum* (Lele and Walton 1961) and *Z. deciduum* (Gerrienne 1988) and in the small circinate axes of *Z. divaricatum* (Gensel 1982a). Since rhizomatous axes are unknown for most zosterophylls, the systematic significance of these short axes is currently difficult to evaluate.

Subordinate axes comparable to those outlined above are absent in homosporous lycopsids. The

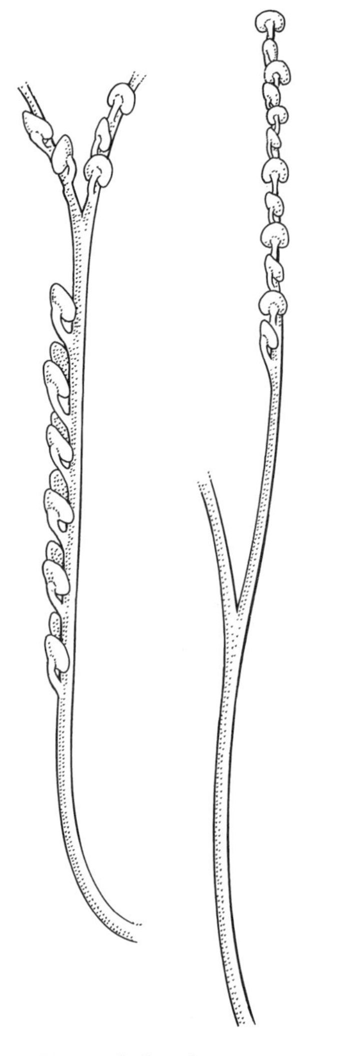

FIGURE 5.10. *Zosterophyllum divaricatum.* Reconstruction showing distal fragments of axis bearing lateral reniform sporangia in two-rowed arrangement (×1.7). Redrawn from Gensel 1982a, Figure 10.

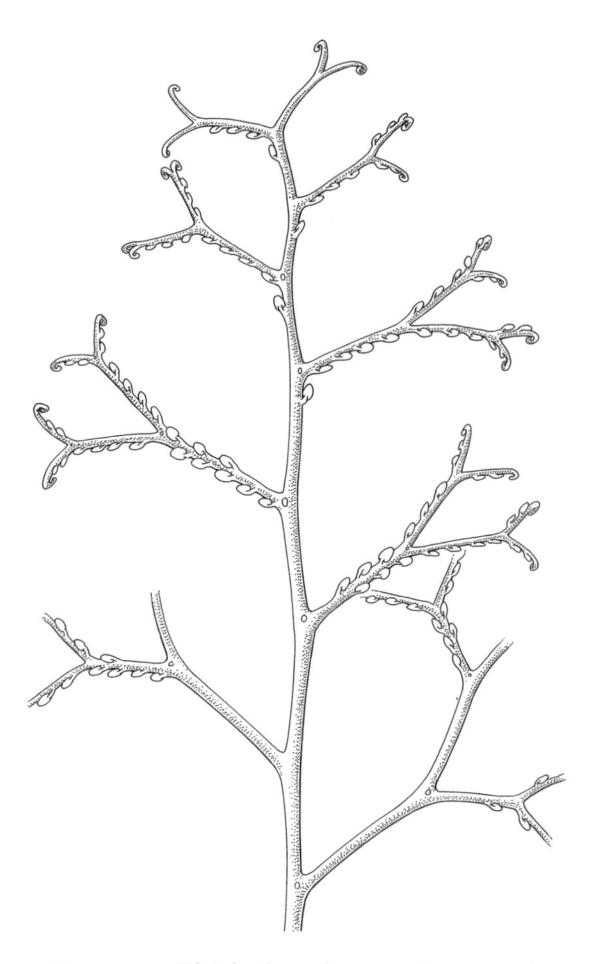

FIGURE 5.11. *Thrinkophyton formosum.* Reconstruction of distal part of plant showing planar, pseudomonopodial axes bearing lateral reniform sporangia in one- or two-rowed arrangements (×1.3). Note the circinate tips of the main branching system. Redrawn from Kenrick and Edwards 1988b, Figure 26.

noncircinate, subordinate branching systems described by Schweitzer (1980c) and Rayner (1984) in *Drepanophycus spinaeformis* were interpreted as roots on the basis of their orientation within the sediment and their attachment to leafless rhizomatous axes. Lateral branch formation in the fossil *Asteroxylon* and bulbil formation in extant *Huperzia selago* (Stevenson 1976) does not involve true dichotomy, as is the case with subordinate branching, but is leaflike in origin. Banks and Davis (1969) drew attention to the similarity between the position of the subordinate branches of *Crenaticaulis* and *Gosslingia* and that of the rhizophore of some Selaginellaceae. More recent investigations of the subordinate branch have shown it to be stemlike in all of its features, rather than rootlike, as in the rhizophore. In *Deheubarthia* and *Anisophyton,* the subordinate branch is clearly circinate and spiny, as in the main aerial axes. We consider that these differences are good grounds for provisionally interpreting *Selaginella* rhizophores and zosterophyll subordinate axes as nonhomologous.

In *Psilophyton dawsonii,* one of the branches produced by two successive, closely spaced di-

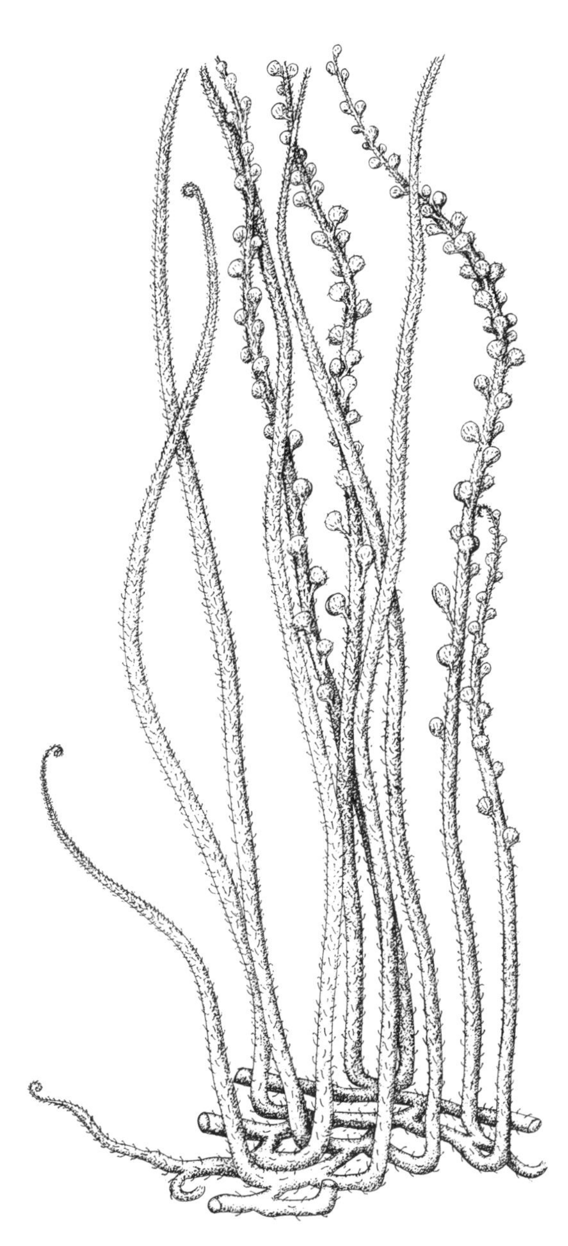

FIGURE 5.12. *Discalis longistipa.* Reconstruction showing prostrate spiny rhizome bearing upright spiny axes with lateral reniform sporangia in a helical arrangement (×0.6). Note the circinate vernation at the apex of branches. Reproduced, with permission, from Hao 1989b, Figure 6 (Amsterdam: Elsevier).

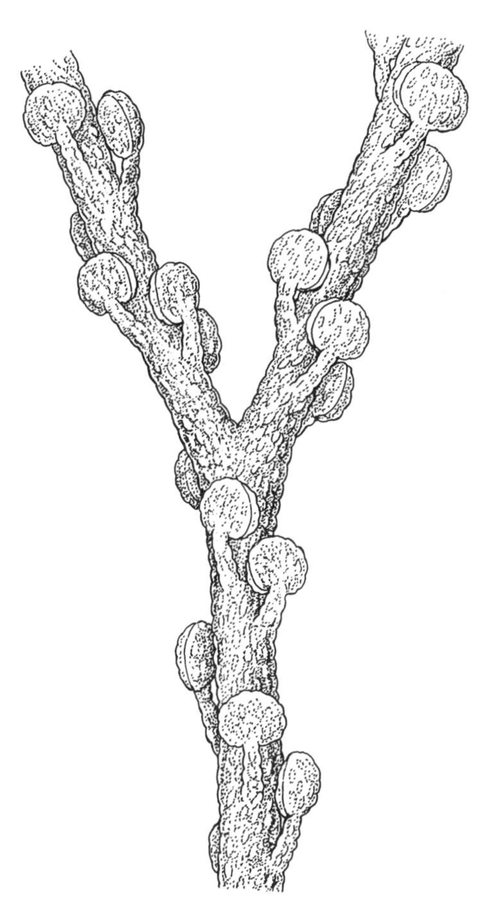

FIGURE 5.13. *Nothia aphylla.* Partial reconstruction of fragment from middle region of fertile branch showing lateral reniform sporangia positioned around stem (×3). Note that the stem and sporangium surfaces are covered by numerous small conical protrusions. Modified after El-Saadawy and Lacey 1979a, Figure 2.

chotomies (termed *double dichotomies* by Banks, Leclercq, and Hueber 1975) often aborts. The result is superficially similar to subordinate branching in zosterophylls; however, important differences are that in *P. dawsonii* any one of the three branches produced may abort and that this type of branching is confined to the lower part of the fertile region. Also, the aborted branch is neither basally oriented nor circinate as it is in zosterophylls. Outgroup comparison with such rhyniophytes as *Uskiella, Cooksonia* (Edwards and Edwards 1986), and *Steganotheca* suggests that the presence of any type of subordinate branch system is a derived feature.

Subordinate branching is treated as an unordered multistate character with three states: 0 = no subordinate branching (absent); 1 = more or less axillary branches; 2 = subordinate branches scattered over the main axes (unconfined) with no

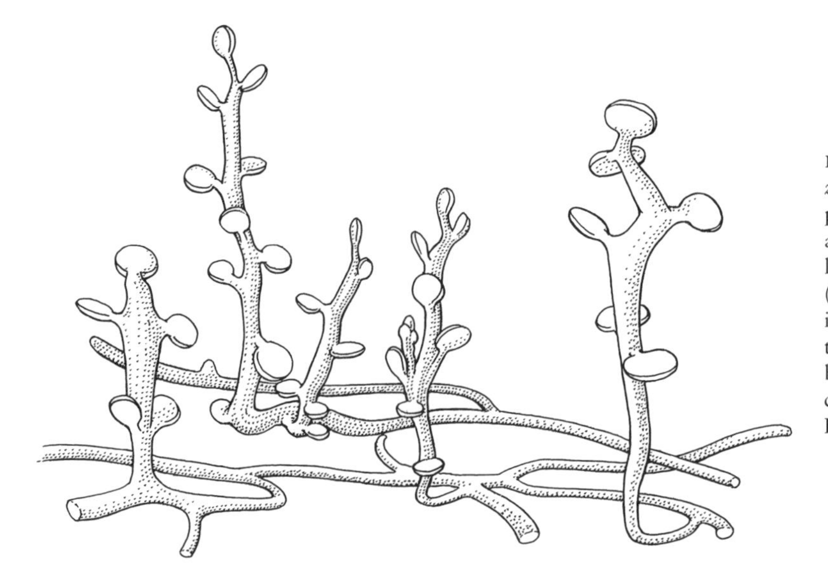

FIGURE 5.14. *Gumuia zyzzata.* Reconstruction showing prostrate axes with upright aerial axes bearing lateral spatulate sporangia on short stalks (×1.3). Note that in some instances sporangia appear to terminate short dichotomous branches, rather than being clearly lateral. Redrawn from Hao 1989a, Figure 2.

apparent order. This character is scored as inapplicable in the zosterophyll *Konioria* because the usual position of the subordinate branch is occupied by a sporangium. There was one inapplicable case; six were missing data (see Appendix 4). See character 4.6 for comparison.

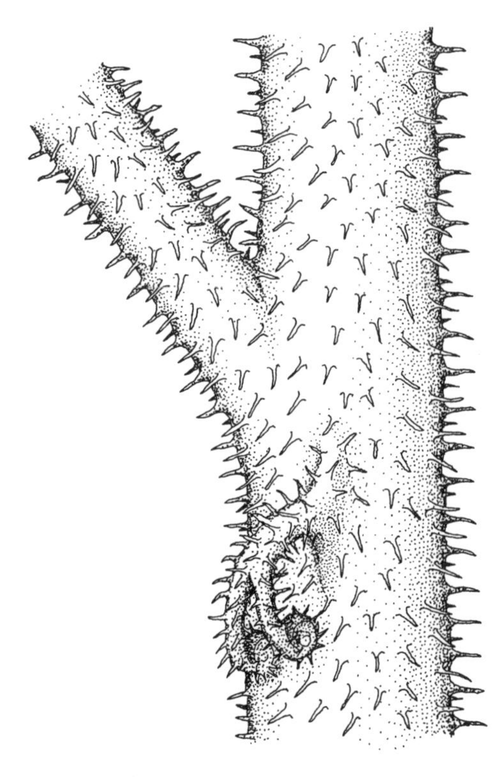

FIGURE 5.15. Axillary tubercle branch typical of some zosterophylls (e.g., *Deheubarthia splendens*) (×3).

5.4 • CIRCINATE VERNATION Many zosterophylls develop by expansion from a coiled apical region in a manner analogous to that of extant ferns. This type of development can be observed as circinate tips (Figure 5.18), both in the distal regions of the plant and in the subordinate axillary branches (Figure 5.15).

Circinate vernation has been observed in *Gosslingia, Sawdonia, Crenaticaulis, Discalis, Konioria, Oricilla, Serrulacaulis, Thrinkophyton, Tarella, Hsua, Anisophyton,* and *Bathurstia* and in two species of *Zosterophyllum* subgenus *Platyzosterophyllum (Z. llanoveranum, Z. divaricatum).* Circinate vernation has not been recorded in *Renalia, Hicklingia, Rebuchia,* and *Gumuia,* and it has not been observed in *Z. myretonianum* or *Z. deciduum.* In view of the quality and amount of material examined for these latter taxa, we infer that circinate growth was not part of their development. The mode of growth in *Barinophyton, Krithodeophyton, Huia,* and *Nothia* has not been established clearly.

Circinate vernation does not occur in any extant or fossil lycopsid. It is absent in such early lycopsid-like fossils as *Asteroxylon, Baragwanathia,* and *Drepanophycus.* Also, there is no evi-

FIGURE 5.16. *Protobarinophyton obrutschevii.* Part and counterpart showing dichotomous axis with conical projection (**a**, *arrow*) and depression (**b**, *arrow*) marking site of attachment of "axillary" branches characteristic of most zosterophylls (×3.7). Specimens from the Natural History Museum, London, V51567 and V44777. Photographs by P.K.

dence for circinate growth in any species of *Psilophyton* or in other members of Trimerophytina (Banks, Leclercq, and Hueber 1975; Doran 1980; Gensel 1979; Hueber and Banks 1967; Kasper, Andrews, and Forbes 1974; Kräusel and Weyland 1961). The presence of circinate vernation in Dawson's well-known original reconstruction of *Psilophyton princeps* (Dawson 1859, 1870; Figure 4.1)

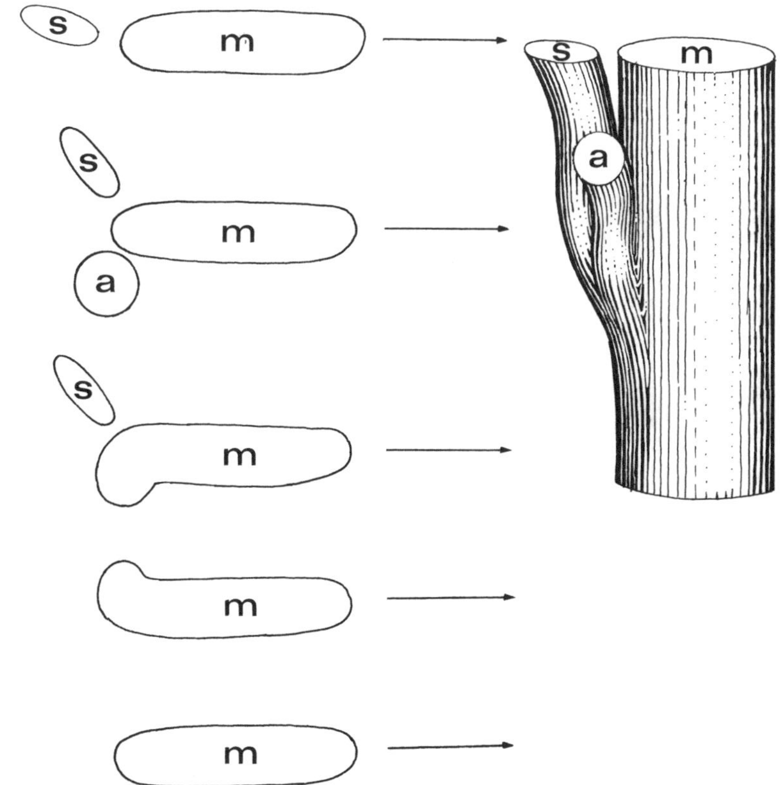

FIGURE 5.17. Diagrammatic representation of stelar morphology at a branching point in *Deheubarthia splendens.* m = main axis; s = lateral (second order) branch; a = subordinate "axillary" branch.

FIGURE 5.18. *Thrinkophyton formosum.* Compression fossils showing distal tips with circinate vernation *(arrows)* typical of many zosterophylls (×9). From Kenrick and Edwards 1988b, Figure 2.

is due to the confusion of two separate plants—one of which was later redescribed as *Sawdonia ornata,* a zosterophyll with circinate growth (Banks, Leclercq, and Hueber 1975; Hueber 1967). The vegetative apexes and terminal sporangia of species of *Psilophyton,* in common with other trimerophytes *(sensu* Banks; e.g., *Pertica),* are known to be slightly bent backwards (recurved), but this does not appear to result from a circinate pattern of development. The circinate growth of extant ferns and early fossil fernlike plants *(Rhacophyton)* is convergent based on our analysis of polysporangiophytes (Chapter 4). Outgroup comparison with rhyniophytes such as *Aglaophyton, Uskiella,* and *Cooksonia* (Edwards and Edwards 1986) suggests that circinate vernation is a derived condition among polysporangiophytes.

The presence or absence of circinate vernation is treated as a binary character: 0 = circinate vernation absent; 1 = circinate vernation present. There were no inapplicable cases; eight were missing data (see Appendix 4). See character 4.8 for comparison.

5.5 • MULTICELLULAR APPENDAGES Many zosterophylls are either naked or possess only unicellular papillae or hairs *(Gosslingia, Barinophyton, Oricilla, Thrinkophyton, Tarella, Krithodeophyton,* all species of *Zosterophyllum, Huia, Hsua, Renalia, Gumuia, Hicklingia, Rebuchia).* However, multicellular spines or teeth are a distinctive feature of several taxa (Figures 5.2, 5.19, 5.20). Filiform spines, 1–5 mm long, are typical of *Sawdonia, Deheubarthia,* and *Konioria* (Figure 5.19). Spines of a similar type, but possessing an expanded tip, occur in *Discalis* and *Anisophyton* (Figure 5.19). Deltoid spines are typical of *Crenaticaulis, Serrulacaulis,* and possibly also *Bathurstia* (Figure 5.19). *Nothia* has small emergences bearing a stomate at the tip and is unique in this respect. Some taxa, such as *Sawdonia acanthotheca,* bear forked spines (Gensel 1991; Gensel, Andrews, and Forbes 1975). None of these prominent multicellular enations are vascular, nor have they been shown to influence the differentiation of vascular tissue in the underlying axis.

The microphylls of the primitive homosporous lycopsids are larger and generally much less numerous than the spines discussed above and are either fully vascularized or at least influence the differentiation of vascular tissue in the axis (e.g., *Asteroxylon*). The presence of a single vascular strand within the microphyll is characteristic of lycopsids *sensu stricto* and is also a feature of early fossils such as *Baragwanathia* and *Drepanophycus.* Leaf traces are present in the axis of *Asteroxylon* but do not differentiate within the leaf (Kidston and Lang 1920b) despite the presence of elongated cells resembling a procambium (Hueber 1992, Figure 6B). *Adoketophyton* has a unique sporangium-shaped sporophyll but no vegetative leaves. These sporophylls share strong positional similarities with lycopsid sporophylls, and we suggest that they may be homologous structures de-

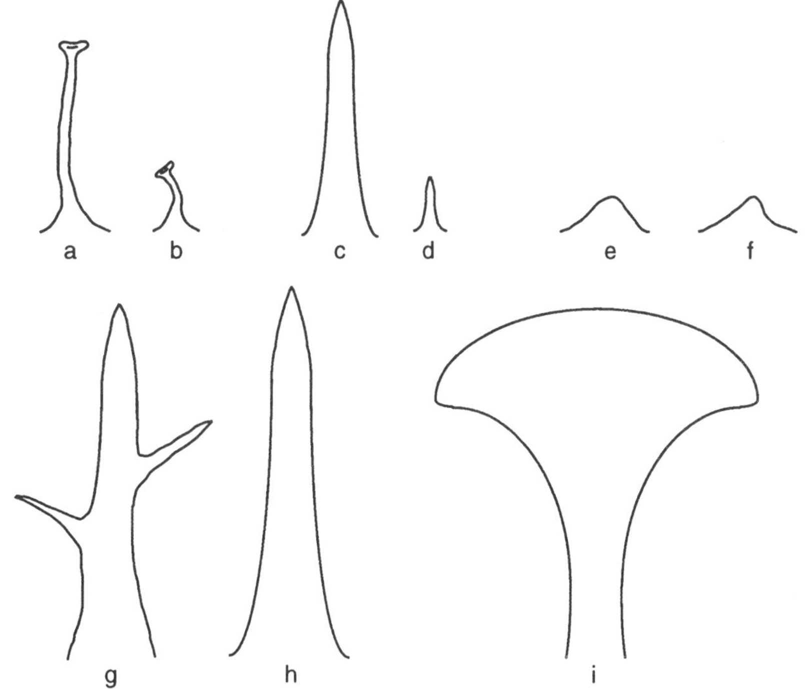

FIGURE 5.19. Diagrammatic representation of lateral structures (nonvascular enations) in zosterophylls and related taxa (×8). (**a**) *Anisophyton* (Remy, Schultka, and Hass 1986). (**b**) *Discalis* (Hao 1989b). (**c**) *Sawdonia* (Gensel 1991). (**d**) *Deheubarthia* (Edwards, Kenrick, and Carluccio 1989; Kenrick 1988). (**e**) *Crenaticaulis* (Banks and Davis 1969). (**f**) *Serrulacaulis* (Hueber and Banks 1979). (**g, h**) *Psilophyton crenulatum* (Doran 1980). (**i**) Sporophyll of *Adoketophyton* (probably vascularized) (Li and Edwards 1992).

spite the difference in shape. To avoid weighting the analysis in favor of this hypothesis, we scored this feature as an independent character state—"bracts."

In our analysis of polysporangiophytes, multicellular spines introduced significant homoplasy, and thus spinelike outgrowths appear to have evolved several times in different lineages of land plants. Outgroup comparison with rhyniophytes suggests that the presence of prominent multicellular spines is a derived feature. *Huvenia kleui* (Hass and Remy 1991) bears prominent enations

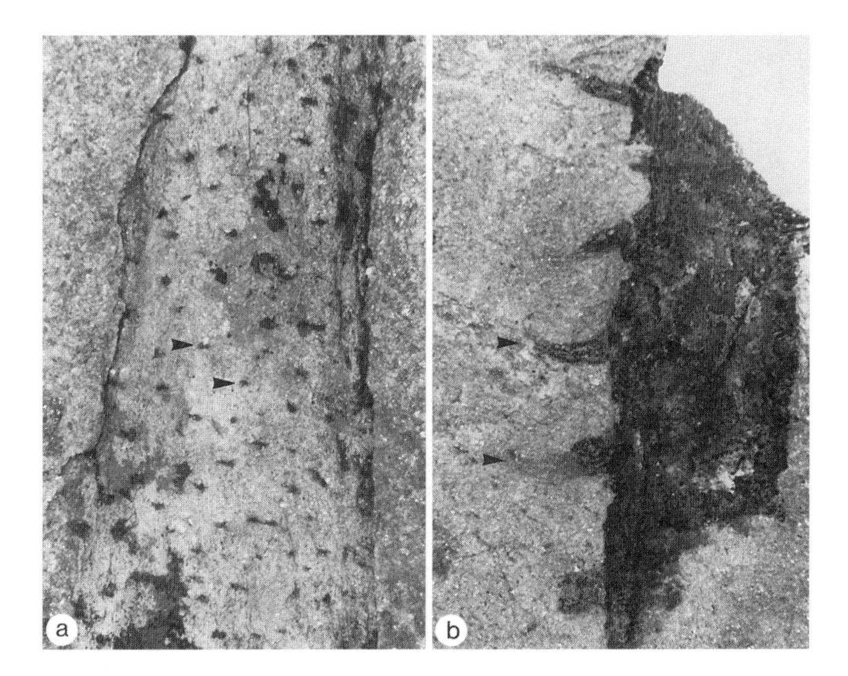

FIGURE 5.20. Multicellular spines and spine bases in *Deheubarthia splendens*. (**a**) Spine bases visible as small black specks on the surface of a stem impression *(arrows)* (×7). (**b**) Spines in profile *(arrows)* (×7).

in certain areas, as does the poorly understood genus *Dutoitea pulchra* (Høeg 1967; Rayner 1988). Prominent filiform enations with pointed or expanded tips are characteristic of several species of *Psilophyton.*

Multicellular appendages are treated as an unordered multistate character with four states: 0 = no appendages (absent); 1 = microphylls (distinguished from "spines" primarily by their ability to influence the differentiation of vascular tissue); 2 = multicellular filiform spines; 3 = "bracts" (as in *Adoketophyton*). There were no inapplicable cases; none were missing data (see Appendix 4). See character 4.9 for comparison.

5.6 • APPENDAGE PHYLLOTAXY The multicellular enations of zosterophylls are not regularly arranged, except in *Crenaticaulis* and *Serrulacaulis,* where they are in one or two vertical rows. There is also no detectable arrangement of these structures in enation-bearing trimerophytes and rhyniophytes. In general, the microphylls of lycopsids and early fossil lycopsid-like plants are helically arranged, although accurate measures of divergence angles have so far failed to establish a classic Fibonacci series (W. E. Stein, pers. comm. to P.R.C., 1995).

Appendage phyllotaxy is treated as an unordered multistate character with four states: 0 = no detectable pattern (absent); 1 = helical arrangement; 2 = two-rowed; 3 = four-rowed. This character is inapplicable in taxa without spines or microphylls. There were 22 inapplicable cases; none were missing data (see Appendix 4).

5.7 • XYLEM STRAND SHAPE Zosterophylls are protostelic, and in most taxa, the xylem strand of the main axes is solid and elliptical to strap-shaped in transverse section (Figure 5.21). An elliptical stele has been illustrated in *Gosslingia, Sawdonia, Barinophyton, Crenaticaulis, Deheubarthia, Konioria, Thrinkophyton, Zosterophyllum llanoveranum, Huia,* and possibly *Krithodeophyton.* Ontogenetic variation in xylem shape is poorly understood, but in many taxa the strand is distinctly strap-shaped in larger axes but more elliptical or possibly even terete in outline in smaller distal axes. Strands that are distinctly strap-shaped have been noted in *Sawdonia, Gosslingia* (Kenrick, pers. obs.; Figure 5.21), *Barinophyton,* and *Deheubarthia.* Also, we interpret the shape of the xylem strand in *Hsua* as elliptical to strap-shaped (Li 1982). In both *Barinophyton* and *Deheubarthia,* a transition from a strongly elliptical or strap-shaped main axis strand to a more weakly elliptical lateral branch strand has been clearly demonstrated.

The shape of the xylem strand at the base of subordinate lateral branches (see earlier) is clearly terete but has been described only in *Gosslingia* and *Deheubarthia.* In some of the smaller taxa, the xylem strand is only weakly elliptical or terete. In *Zosterophyllum llanoveranum* the strand is elliptical within (or close to) the fertile spike but terete in the region below (D. Edwards 1969a). In *Z. fertile* the strand appears also to be terete, at least in part of the plant (D. Edwards 1969b). In *Nothia* the strand is also weakly elliptical, and in *Huia recurvata,* and possibly *Krithodeophyton,* it is elliptical to terete. In *Trichopherophyton* the strand is more or less terete, whereas in *Yunia dichotoma,* the xylem strand is clearly terete (Hao and Beck 1991b). In other taxa the shape of the xylem strand is unknown.

Our analysis of lycopsids (Chapter 6) suggests that the solid stellate protostele of extant *Huperzia* (Wardlaw 1924) and of the early fossil lycopsids *Asteroxylon, Baragwanathia,* and *Drepanophycus* is basic within the group and thus distinct from the situation in zosterophylls (see the discussion of lycopsid characters in Chapter 6). Strap-shaped or U-shaped xylem strands are common in many species of *Selaginella* having dorsiventrally planated shoots, and these may be very similar in appearance to the xylem strands of some zosterophylls (Bierhorst 1971; Bower 1935, 225). In view of the derived position of *Selaginella* among the lycopsids, we interpret this as an interesting, and perhaps functionally related, parallelism that deserves further developmental investigation.

Outgroup comparison with trimerophytes, such as *Psilophyton dawsonii,* and rhyniophytes, such

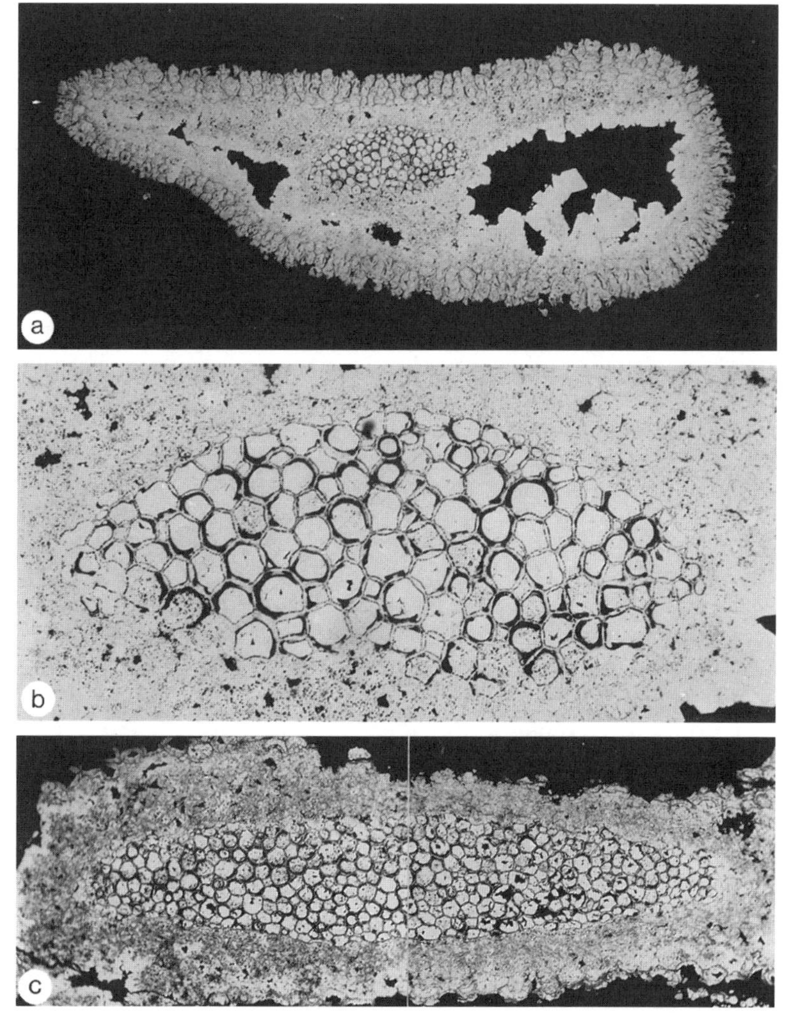

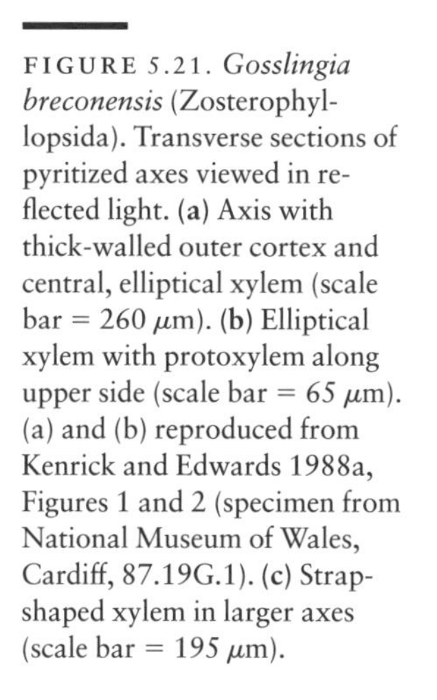

FIGURE 5.21. *Gosslingia breconensis* (Zosterophyllopsida). Transverse sections of pyritized axes viewed in reflected light. (**a**) Axis with thick-walled outer cortex and central, elliptical xylem (scale bar = 260 μm). (**b**) Elliptical xylem with protoxylem along upper side (scale bar = 65 μm). (a) and (b) reproduced from Kenrick and Edwards 1988a, Figures 1 and 2 (specimen from National Museum of Wales, Cardiff, 87.19G.1). (**c**) Strap-shaped xylem in larger axes (scale bar = 195 μm).

as *Rhynia gwynne-vaughanii,* shows that a solid terete protostele is general among basal tracheophytes. This interpretation is also supported by the terete stele in subordinate branches (axillary tubercles), in microphylls, and in early ontogenetic stages (e.g., extant lycopsids).

Xylem strand shape is treated as an unordered multistate character with three states: 0 = more or less terete; 1 = elliptical to strap-shaped; 2 = stellate. There were no inapplicable cases; 16 were missing data (see Appendix 4). See character 4.14 for comparison.

5.8 • PROTOXYLEM A transverse section through the primary xylem of extant vascular plants shows cells with a range of diameters. One source of variation in cell diameter is caused by the tapering end-regions of tracheids, but further variation is introduced by size differences related to the order in which the xylem cells mature. Protoxylem cells are usually narrower than metaxylem and are generally confined to small, relatively easily identified zones within the primary xylem. Similar marked cell-size zonations have been noted in the xylem of zosterophylls, and these have been interpreted as indicating the presence of protoxylem.

In most zosterophylls, protoxylem maturation is described as exarch, but the exact distribution of protoxylem around the edge of the xylem strand has not been investigated in detail and may provide further useful characters (see character 5.22). An exarch xylem strand has been described in *Gosslingia* (Figure 5.21), *Sawdonia, Deheubarthia, Barinophyton, Crenaticaulis, Konioria, Thrinkophyton, Trichopherophyton,* and possibly also *Zosterophyllum llanoveranum* and *Z. fertile.*

Three taxa—*Huia, Hsua,* and *Nothia*—have been described as centrarch, but because the accompanying illustrations are unclear, these interpretations should be treated with caution. Xylem maturation in *Yunia* is unusual and may be described as diarch—the protoxylem is clearly embedded well within metaxylem for most of the axis. However, immediately after a dichotomy, each daughter axis has a single centrarch strand.

The lycopsids are characterized by more or less exarch protoxylem, but the more general centrarch condition is widespread in early ontogenetic stages (Bower 1908, 333). In extant Lycopodiaceae, xylem maturation in the stem is almost uniformly exarch with protoxylem situated at the outer edge of each xylem arm. Only rarely in large stems is weak mesarchy observed (Bierhorst 1971). Xylem maturation in the early fossil lycopsid *Drepanophycus* appears to be exarch (Rayner 1984) or possibly mesarch (Fairon-Demaret 1971). In *Asteroxylon* both mesarch and exarch conditions seem to occur (Kidston and Lang 1920b).

In outgroups such as the trimerophytes, xylem maturation is clearly centrarch (Banks, Leclercq, and Hueber 1975), but in rhyniophytes the situation is unclear. In *Aglaophyton* and *Sennicaulis hippocrepiformis* (Kenrick, Edwards, and Dales 1991), there is a clear inner zone of smaller cells but this is much more extensive than typical protoxylem, which we therefore consider to be absent. In *Rhynia* the xylem strand is very small and we tentatively interpret the protoxylem position as centrarch. In bryophytes with water conducting tissues, there does not appear to be a functional or positional equivalent of protoxylem (Hébant 1977).

The presence or absence and type of protoxylem is treated as an unordered multistate character with four states: 0 = centrarch; 1 = exarch; 2 = diarch-centrarch; 3 = no differentiation into protoxylem (absent). There were no inapplicable cases; 20 were missing data (see Appendix 4). See character 4.15 for comparison.

5.9 • TRACHEID THICKENINGS Nearly all vascular plants possess tracheids with helical-annular type thickenings. In many early fossil taxa, this is the only type of tracheid that occurs in the xylem (Figures 4.25, 4.26, 5.22, 5.23; Kenrick and Crane 1991). In contrast, in extant groups various thick-

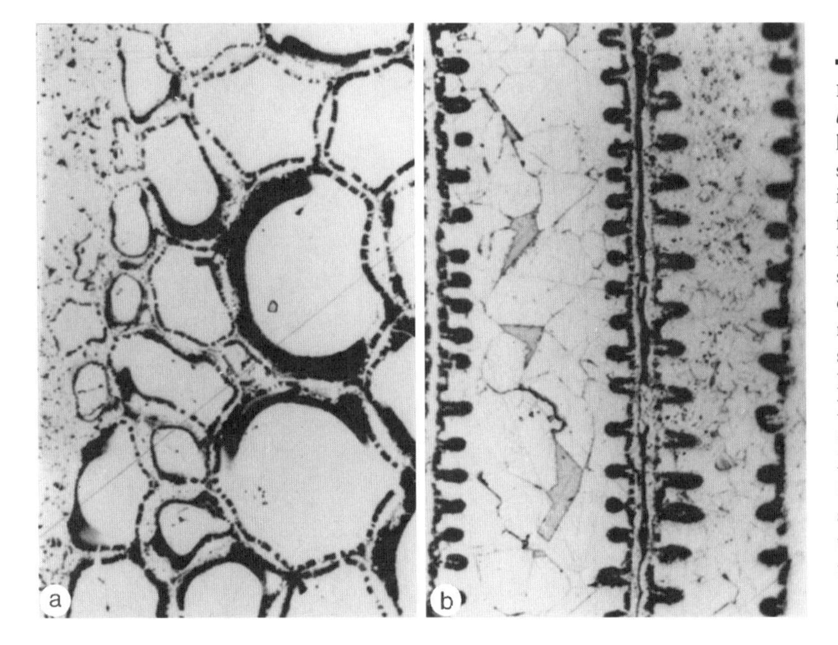

FIGURE 5.22. *Gosslingia breconensis* (Zosterophyllopsida). Details of tracheid structure. Cells are preserved in pyrite and viewed with reflected light. Scale bar = 30 μm for both views. (**a**) Transverse section of tracheids at edge of xylem showing large metaxylem cells (right) and smaller protoxylem (left). Reproduced from Kenrick and Edwards 1988a, Figure 5 (specimen from National Museum of Wales, Cardiff, 87.19G.1). (**b**) Longitudinal section of parts of two tracheids showing wall thickenings.

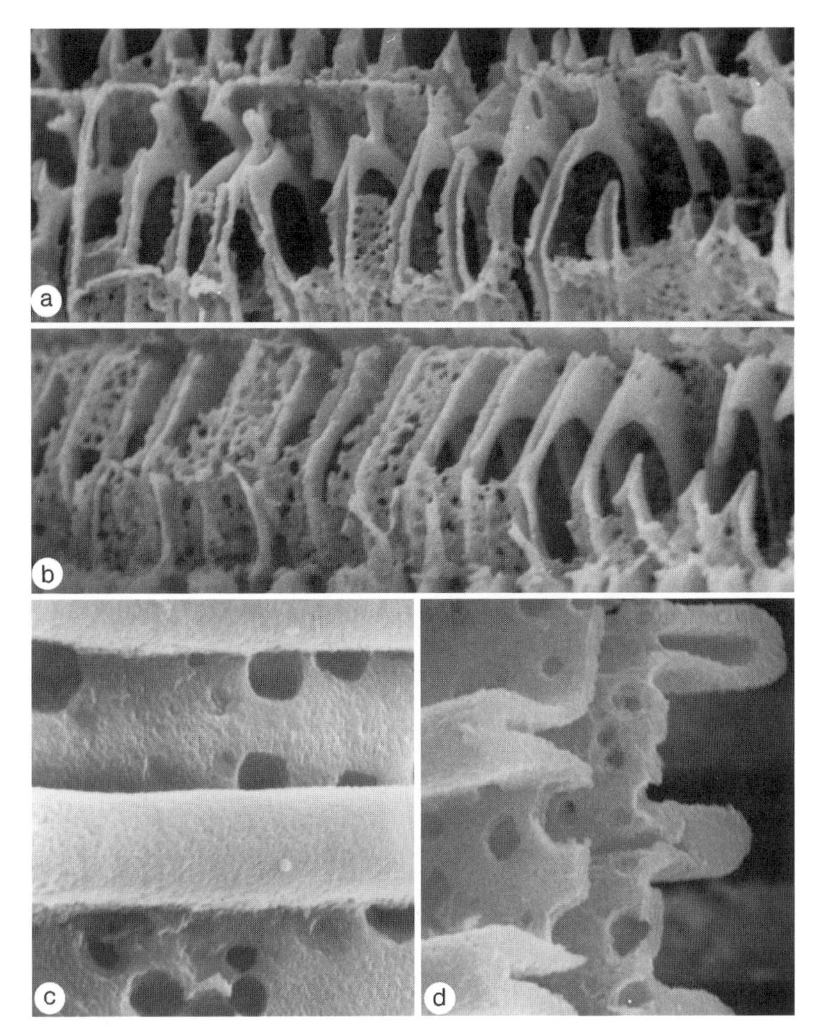

FIGURE 5.23. *Gosslingia breconensis* (Zosterophyllopsida). Details of tracheid structure. Specimen from National Museum of Wales, Cardiff, 87.19G.5. Photographs by P.K. Pyrite was removed with nitric acid and coalified remains of cell wall are viewed with SEM. (**a, b**) Exterior view of cells and thickenings (scale bar = 10 μm). Note that the wall layer between thickenings has fallen off from parts of the cells, allowing views of cell interior. (**c**) Detailed view of thickening and wall between thickenings from cell interior (scale bar = 2 μm). (**d**) Details of wall of two adjacent cells in section view (scale bar = 3 μm).

ening types occur. These are usually located in specific areas and are often associated with different stages in xylem ontogeny. Early-formed xylem elements are always of the helical-annular type, but this type of thickening is usually absent from the walls of later formed elements, which contain complex pits (Bierhorst 1960). Distinctive thickenings are absent from the water-conducting cells of such stem group vascular plants such as *Aglaophyton* and *Horneophyton* (Chapter 4). Thickenings are also absent in the enigmatic *Nothia aphylla,* a plant thought to be closely related to zosterophylls. Our analysis of relationships in polysporangiophytes (Chapter 4) implies a loss of wall thickenings in the tracheids of *Nothia.* The type of pitting in tracheids is not scored in this part of the analysis because at this level it is an autapomorphy of *Huperzia.*

The presence or absence of tracheid thickenings is treated as a binary character: 0 = tracheid thickenings absent; 1 = annular or helical thickenings present in metaxylem or protoxylem or all xylem elements. There were no inapplicable cases; 18 were missing data (see Appendix 4). See characters 4.17 and 4.18 for comparison.

5.10 • SPORANGIOTAXIS Among those taxa with lateral sporangia, there are several easily recognizable patterns of sporangium arrangement *(sporangiotaxis).* Most commonly there are two vertical rows of sporangia that may become one row in the basal or distal regions (Figure 5.24). Where two vertical rows are present, they are usually on opposite sides of the axis and in the same plane as the major plane of branching. Such an arrangement

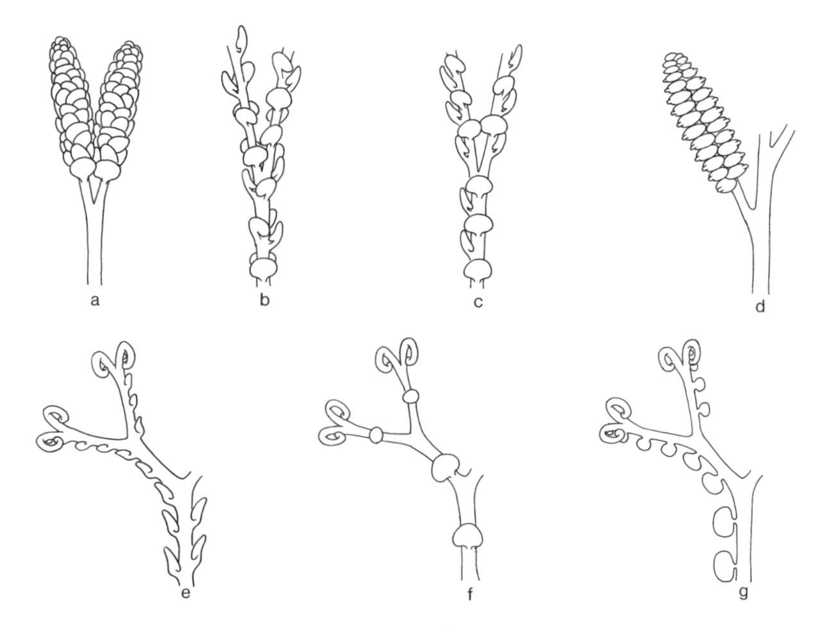

FIGURE 5.24. Diagrammatic representation of patterns of sporangiotaxis and sporangium orientation in Zosterophyllopsida and basal Lycopsida. (**a**) "Upright" orientation, helical sporangiotaxis in compact unbranched spike (e.g., *Zosterophyllum fertile*). (**b**) "Upright" orientation, helical sporangiotaxis over two or more orders of branching (e.g., *Zosterophyllum deciduum*). (**c**) "Upright" orientation, two-rowed sporangiotaxis over two or more orders of branching (e.g., *Zosterophyllum divaricatum*). (**d**) "Clasping" orientation, two-rowed sporangiotaxis) in compact unbranched spike (e.g., *Barinophyton citrulliforme*). (**e**) "Upright" orientation, one- and two-rowed sporangiotaxis over two or more orders of branching (e.g., *Thrinkophyton formosum*). (**f**) "Upright" orientation, "dorsal," one-rowed sporangiotaxis over two or more orders of branching (e.g., *Anisophyton gothanii*). (**g**) "Auricular" orientation, one- and two-rowed sporangiotaxis over two or more orders of branching (e.g., *Gosslingia breconensis*).

occurs in *Zosterophyllum* subgenus *Platyzosterophyllum* (Gensel 1982a; Gerrienne 1988), *Crenaticaulis, Gosslingia, Oricilla, Tarella, Sawdonia, Deheubarthia, Serrulacaulis, Rebuchia, Krithodeophyton, Bathurstia,* and *Thrinkophyton.* In *Barinophyton* and *Protobarinophyton,* sporangia are inserted in two rows on opposite sides of the axis but curvature of the sporangium stalks results in a single row of sporangia along only one side of the axis (Figures 5.3, 5.4, 5.24).

In the second common arrangement the sporangia are pedicellate and arrayed helically around the axis. This disposition is characteristic of *Zosterophyllum*-like plants such as *Huia,* as well as of species of *Zosterophyllum* subgenus *Zosterophyllum* (e.g., *Z. myretonianum, Z. deciduum*). In some zosterophylls, the sporangia are known to be arranged around the axis but the exact sporangiotaxis has not yet been clarified. In these plants, for example *Discalis, Nothia, Gumuia,* and *Hicklingia,* it is probable that the sporangiotaxy is helical, and provisionally we treat it as such. Helical sporangiotaxy is also characteristic of basal members of extant Lycopodiaceae (Bierhorst 1971; Øllgaard 1987). Among early fossil lycopsids, the sporangia appear to be helical in *Baragwanathia,* and in *Drepanophycus* and *Asteroxylon,* the sporangia are probably arranged around the axis, but the sporangiotaxy is poorly understood. The arrangement of fertile lateral branches with respect to the main axis in *Renalia* and *Hsua* is unknown.

The sporangium position in zosterophyll outgroups is usually terminal, either on the main axes or on smaller lateral branches, and hence these taxa cannot be scored for sporangiotaxis. Single sporangia terminal on main axes are characteristic of rhyniophytes, such as *Cooksonia pertonii* and *C. hemisphaerica, Uskiella,* and *Sporogonites.* Pairs of terminal sporangia, often clustered in groups, are characteristic of trimerophytes. In two recently described rhyniophytes, *Stockmansella* and *Huvenia,* the sporangia were shown to be terminal on short lateral branches or virtually sessile on main axes. In both cases, no definite sporangiotaxis was detectable.

Sporangiotaxis is treated as an unordered multistate character with three states: 0 = helical sporangia; 1 = sporangia in one to two rows; 2 = sporangia in four-rowed arrangements (e.g., *Adoketophyton*). This character is inapplicable in taxa with terminal sporangia. There were three inapplicable cases; three were missing data (see Appendix 4). See character 4.22 for comparison.

5.11 • SPORANGIUM ROWS This character is applicable only to those zosterophylls having sporangia organized in vertical rows and distinguishes the different positions of such rows with respect to the plane of branching. In many zosterophylls, there are two vertical rows of sporangia, which are usually arranged on opposite sides of the axis and in the same plane as the major plane of branching (Figure 5.24). However, another arrangement has been observed in *Konioria* and *Anisophyton,* in which there is one dorsiventral row of sporangia (Figure 5.24). A third arrangement occurs in *Serrulacaulis,* in which two rows occur on the same (presumed ventral) surface (Figure 5.24).

The location of sporangium rows is treated as an unordered multistate character with three states: 0 = ventral; 1 = dorsiventral; 2 = lateral. This character is inapplicable in taxa in which sporangia are not in rows. There were 15 inapplicable cases; one was missing data (see Appendix 4).

5.12 • SPORANGIUM DISTRIBUTION The distribution of sporangia is difficult to circumscribe in many zosterophylls. Frequently, there is an extended fertile region over two to several distal orders of branching (Figures 5.9 and 5.24). This is characteristic of taxa such as *Gosslingia, Deheubarthia* (Figure 5.2), *Oricilla, Thrinkophyton* (Figure 5.18), *Tarella, Zosterophyllum divaricatum* (Figure 5.10), *Nothia* (Figure 5.13), and *Anisophyton.* In other zosterophylls, sporangium distribution is mainly confined to the terminal branch (Figures 5.1 and 5.2) but may extend downward a short distance below the ultimate dichotomy (Figure 5.11). This is characteristic of *Zosterophyllum myretonianum, Z. llanoveranum,* and *Discalis.* In *Barinophyton* and *Protobarinophyton,* sporangia occur in very compact, unbranched spikes.

In the extant homosporous lycopsids *Phylloglossum, Lycopodium,* and *Lycopodiella,* the sporangia occur in compact, unbranched *strobili* (aggregations of micro- and megasporangiophores, typically around an unbranched determinate axis) (Øllgaard 1987). In *Huperzia,* the sporangia are either produced seasonally in all leaf axils of more or less distinct zones along the stem, or are produced continuously along the terminal stem divisions (Øllgaard 1987). Sporangium distribution in relation to branching in early fossil lycopsids is unknown. In most basal polysporangiophytes, sporangia are terminal and solitary, and a similar condition occurs in some potential ingroup taxa such as *Renalia* and *Hsua.*

Sporangium distribution is treated as an unordered multistate character with three states: 0 = fertile zone more or less unbranched; 1 = loose fertile zone covering two to several orders of branching (+2 branches); 2 = fertile zone compact (as seen in *Barinophyton* and *Protobarinophyton*). This character was scored as inapplicable for taxa with terminal sporangia. There were three inapplicable cases; seven were missing data (see Appendix 4).

5.13 • SPORANGIUM SHAPE The sporangia of most zosterophylls are oval to strongly reniform in outline (Figures 4.29, 4.30, 5.24). Sporangium shape can be difficult to determine in compact spikes, such as those of *Zosterophyllum, Krithodeophyton,* and *Barinophyton,* because of overlap as well as because of folding or compression of sporangia that are oriented at different angles in the sediment. Sporangium shape for most taxa ranges from flattened and strongly reniform in

Renalia, Oricilla, and *Tarella* to flattened, oval, or weakly reniform in *Discalis, Zosterophyllum deciduum, Barinophyton,* and *Nothia.* In many taxa, immature sporangia are less distinctly reniform in outline (D. Edwards 1970b). In *Sawdonia acanthotheca* (Gensel, Andrews, and Forbes 1975) and *Gumuia,* the sporangia appear to be more rounded in shape, whereas in *Hicklingia,* they are spatulate (Hao 1989a).

The sporangia of extant homosporous lycopsids characteristically are reniform at maturity (Bierhorst 1971; Øllgaard 1987). In the primitive lycopsid-like fossils *Asteroxylon, Drepanophycus,* and possibly *Baragwanathia,* the sporangia are also oval to reniform in outline.

Outgroup comparison with such rhyniophytes as *Aglaophyton, Rhynia, Stockmansella, Huvenia,* and *Salopella allenii* (Edwards and Richardson 1974), as well as with such trimerophytes as *Psilophyton dawsonii* (Banks, Leclercq, and Hueber 1975), suggests that elongated fusiform sporangia are basic in polysporangiophytes.

Sporangium shape is treated as an unordered multistate character with three states: 0 = sporangia more or less fusiform; 1 = sporangia more or less reniform (including oval to strongly reniform sporangia); 2 = sporangia more or less spatulate. There were no inapplicable cases; none were missing data (see Appendix 4). See character 4.27 for comparison.

5.14 • SPORANGIUM SYMMETRY One feature that has received little attention in lycopsid-zosterophyll systematics is sporangium symmetry. We recognize two conditions based on the presence or absence of marked dorsiventral flattening. The sporangia of most zosterophylls and homosporous lycopsids (e.g., *Huperzia selago*) are reniform and strongly flattened (Figures 4.29 and 4.30). However, these features are not absolutely correlated because in some heterosporous lycopsids, such as *Isoetes,* there is a marked radial elongation of the basically reniform sporangium (Figure 6.11). In compression fossils, sporangium flattening is manifested as variation in the width of sporangia attached around an axis (Figure 5.24). This marked flattening has also been observed in rare examples of permineralized sporangia in *Zosterophyllum,* as well as in *Nothia, Trichopherophyton,* and *Asteroxylon* from the Rhynie Chert. It has also been observed in well-preserved sporangium cuticles such as those of *Renalia.*

Sporangium symmetry in the outgroups, as observed in permineralized sporangia of *Aglaophyton, Rhynia,* and *Psilophyton dawsonii* (Banks, Leclercq, and Hueber 1975), is clearly radial.

Sporangium symmetry is treated as a binary character: 0 = sporangia radially symmetrical; 1 = sporangia markedly bilaterally symmetrical. There were no inapplicable cases; none were missing data (see Appendix 4). See character 4.28 for comparison.

5.15 • RELATIVE VALVE SIZE In most zosterophylls, sporangium dehiscence is isovalvate: the sporangium splits along the distal margin into two valves of approximately equal size. In compression fossils, the dehiscence line is observed often as a thickened, coalified, double rim, and the two valves sometimes separate and twist slightly such that the underlying valve becomes visible (e.g., see Gensel 1982b). Permineralized sporangia (D. Edwards 1969a; El-Saadawy and Lacey 1979a; Lyon 1964; Lyon and Edwards 1991) and well-preserved cuticles (Gensel 1976) show cellular details of the sporangium valves, and the structural features of the dehiscence mechanism can be observed. Anisovalvate dehiscence, involving a small adaxial and a large abaxial valve, occurs in *Crenaticaulis* and *Trichopherophyton.*

Dehiscence in extant homosporous lycopsids is isovalvate except for the genus *Lycopodiella,* in which dehiscence is anisovalvate (Øllgaard 1987). In the early lycopsid-like fossil *Asteroxylon,* dehiscence is also isovalvate, but in *Drepanophycus* and *Baragwanathia,* the details of dehiscence are unclear.

Relative valve size is treated as a binary character: 1 = sporangia isovalvate; 2 = sporangia anisovalvate. Taxa with univalvate or indehiscent sporangia were scored as inapplicable. There were two inapplicable cases; eight were missing data (see Appendix 4).

5.16 • SPORANGIUM DEHISCENCE In most zosterophylls, Lycopodiaceae (e.g., Øllgaard 1987), Selaginellaceae (e.g., Koller and Scheckler 1986), Protolepidodendrales (e.g., Banks, Bonamo, and Grierson 1972), and arborescent lycopsids (Bateman, DiMichele, and Willard 1992), dehiscence splits the sporangia into two valves. In *Barinophyton,* some arborescent lycopsids (Bateman, DiMichele, and Willard 1992), and *Isoetes* (Bierhorst 1971), there is no marked dehiscence feature and spore release occurs by fragmentation of the sporangium wall.

Valve number is difficult to score for potential outgroups among basal polysporangiophytes because the dehiscence mechanism for many taxa is unknown.

Sporangium dehiscence is treated as an unordered multistate character with three states: 0 = sporangia with one valve (as in *Psilophyton dawsonii*); 1 = sporangia with two valves; 2 = sporangia with no marked dehiscence feature (none). There were no inapplicable cases; seven were missing data (see Appendix 4). See character 4.29 for comparison.

5.17 • THICKENED SPORANGIUM VALVE RIM A marked thickening of cell walls, an increase in the number of cell layers bordering the dehiscence line, or both occurs in some early polysporangiophytes. This feature is discussed in detail under character 4.30.

The presence or absence of a thickened sporangium valve rim is treated as a binary character: 0 = thickened rim absent; 1 = thickened rim present. There were no inapplicable cases; eight were missing data (see Appendix 4). See character 4.30 for comparison.

5.18 • SPORANGIUM ATTACHMENT The sporangia of most zosterophylls are attached to the main axes by short, narrow unbranched stalks (Figure 5.1). In *Barinophyton* the stalks are curved and 10–12 mm in length ("sporangiferous appendages": Brauer 1980, 1187), in *Discalis* they are straight and 2–5 mm long but oriented distally, and in *Oricilla* they are straight and 0.2–0.4 mm long but oriented perpendicular to the axes. In *Krithodeophyton* and *Trichopherophyton* the sporangia are described as sessile. A similar form of sporangium attachment occurs in extant homosporous lycopsids (Bierhorst 1971; Gifford and Foster 1989), as well as in such early fossil lycopsid-like plants as *Asteroxylon, Drepanophycus,* and possibly also *Baragwanathia.*

In most other early land plants, the sporangia are terminal. In *Renalia* (Figure 4.11), the sporangia are terminal on branched laterals that dichotomize once or twice to produce a system 3–10 mm long and bearing one to four sporangia. In *Hsua,* lateral branches dichotomize up to four times to produce a branch system up to 11 cm long and bearing as many as 16 sporangia (Figure 4.12). In most outgroup taxa, the sporangia are terminal on main axes, but in *Rhynia* and related taxa, the sporangia are attached to special "pads" of tissue that are sessile on the main axes or terminal on short, branched laterals.

Sporangium attachment is treated as an unordered multistate character with four states: 0 = sporangia terminal; 1 = sporangia lateral on short, unbranched stalks; 2 = sporangia lateral sessile; 3 = sporangia attached to special "pads" of tissue (i.e., *Rhynia*). There were no inapplicable cases; three were missing data (see Appendix 4). See character 4.23 for comparison.

5.19 • SPORANGIUM ORIENTATION Three different types of sporangium orientation (with respect to the main axis) can be recognized in zosterophylls. In most zosterophylls, in all extant Lycopodiaceae (Bierhorst 1971; Øllgaard 1987), and in such early fossil lycopsid-like plants as *Asteroxylon, Drepanophycus,* and *Baragwanathia,* the sporangia have an "upright" orientation (Figure 5.24a–c). In *Gosslingia, Tarella,* and *Oricilla,* the sporangia have an "auricular" orientation (Figure 5.24g), and in *Barinophyton* and *Protobarinophy-*

ton, the sporangia have a "clasping" orientation (Figure 5.24d).

Outgroup comparison with such rhyniophytes as *Uskiella, Sartilmania, Cooksonia pertonii,* and *C. hemisphaerica* suggests that the "upright" orientation is the general condition in polysporangiophytes.

Sporangium orientation is treated as an unordered multistate character with three states: 0 = sporangia upright; 1 = sporangia auricular; 2 = sporangia clasping. There were no inapplicable cases; one was missing data (see Appendix 4).

5.20 • SPORE SIZE Most zosterophylls are thought to be homosporous. Megaspores are known from only two species of *Barinophyton* (Arnold 1939; Brauer 1980; Pettitt 1965; Taylor and Brauer 1983) and from one species of *Protobarinophyton* (Brauer 1981). Descriptions of in situ microspores often suffer from problems of small sample size and poor preservation, and this situation results in incomplete descriptions and inconclusive comparisons with the better-described, dispersed spores (Allen 1980; Gensel 1980). Occasionally, material is more abundant and more complete descriptions are possible (e.g., Cichan, Taylor, and Brauer 1984; Gensel and White 1983; Gerrienne 1988; T. N. Taylor 1990; Taylor and Brauer 1983; W. A. Taylor and T. N. Taylor 1990).

The presumed homosporous status of most taxa is based on the identification of microspores in one or, sometimes, several sporangia. Microspores range in size from a minimum of 18 μm in *Hsua,* or 25 μm in *Zosterophyllum myretonianum* to a maximum of 92 μm in *Oricilla* (Gensel 1982b), having an average size of about 52 μm. Megaspores in *Barinophyton* range from 650 to 900 μm. All extant Lycopodiaceae are homosporous (Bierhorst 1971; Gifford and Foster 1989; Wilce 1972), and microspores averaging about 50 μm in diameter have been recorded from two early fossil lycopsid-like plants: *Asteroxylon* and *Baragwanathia.* Outgroup comparison with rhyniophytes and trimerophytes shows that the homosporous condition is general for polysporangiophytes (Allen 1980; Gensel 1980).

Spore size is treated as a binary character: 0 = spores all of one size (homosporous); 1 = spores of two sizes in the same sporangium (unusual form of heterospory seen in *Barinophyton* and *Protobarinophyton*) (heterosporous). There were no inapplicable cases; nine were missing data (see Appendix 4).

Other Potentially Relevant Characters

Although we have attempted to make our compilation of phylogenetically informative characters in zosterophylls and related plants as complete as possible, there are many other potentially informative features that we have been unable to include, either because their taxonomic distribution is not adequately understood or because they have not yet been studied in sufficient detail in any taxon. Several of these are discussed briefly below to emphasize the need for further study.

5.21 • RHIZOMES *Rhizomes,* or prostrate axes of rootlike appearance that give rise to aerial axes or leaves, have been described in only a few zosterophylls. *Serrulacaulis* (Hueber and Banks 1979) and *Discalis* (Figure 5.12; Hao 1989b) have spiny rhizomes that are similar to the fertile aerial axes. Unicellular rhizoids that develop from epidermal cells were observed in the rhizome of *Serrulacaulis* (Hueber and Banks 1979), and a questionable rhizome with rhizoids was also illustrated in the original description of *Gosslingia* (Heard 1927). A second type of rhizome, characterized by prolific, closely spaced dichotomies, has been observed in *Zosterophyllum myretonianum* (Figure 5.1; Lang 1927; Lele and Walton 1961), in *Z. deciduum* (Gerrienne 1988), and in the *Zosterophyllum*-like plant *Gumuia* (Hao 1989a). Outgroup comparison with rhyniophyte taxa in which rhizomes have been described, such as *Aglaophyton* (D. S. Edwards 1986; Kidston and Lang 1917) and *Rhynia gwynne-vaughanii* (D. S. Edwards 1980, 1986; Kidston and Lang 1920a), suggests that overall similarity of the rhizome to aerial axes and the presence of unicellular rhizoids are plesiomorphic features within zosterophylls. Given this polarity assessment, prolific dichotomous branching of the rhizome may be of both systematic and ecological significance in some *Zosterophyllum*-like taxa. The putative rhizome of the early lycopsid *Asteroxylon* is leafless and, despite excellent preservation, rhizoids are absent (Kidston and Lang 1920b). Rhizome characters were not included in this analysis because they are unknown for a high proportion of the ingroup taxa.

5.22 • PROTOXYLEM DISTRIBUTION Xylem differentiation in most zosterophylls is described as exarch, but the exact distribution of protoxylem around the edge of the strand has not been investigated in detail and may provide additional characters on further study. One potentially useful feature of *Gosslingia* (Heard 1927; Kenrick and Edwards 1988a), and possibly also of *Deheubarthia* and *Konioria,* is that the protoxylem appears to be restricted to one side of the xylem only (Figure 5.21). This character was not included in the analysis because it is unknown for most taxa considered.

5.23 • AXIS EPIDERMAL CELL PATTERNS A distinctive epidermal cell pattern that consists typically of six to eight elongate cells radiating from a central isodiametric cell has been noted in several zosterophylls including *Sawdonia* (Rayner 1983; Zdebska 1972), *Crenaticaulis* (Banks and Davis 1969), *Deheubarthia* (Edwards, Kenrick, and Carluccio 1989), *Oricilla* (Gensel 1982b), and *Serrulacaulis* (Hueber and Banks 1979). It has been suggested that the central isodiametric cell might represent the base of a unicellular hair (Lang 1932), and similar cellular patterns are associated with hairs in *Trichopherophyton* (Lyon and Edwards 1991). This epidermal cell pattern was not noted in *Zosterophyllum myretonianum* (Lele and Walton 1961), despite well-preserved cuticles, and has not been recorded in any relevant outgroup taxa. Characters of epidermal cell patterns were not included in this analysis because they are unknown in many taxa.

5.24 • SPORANGIUM EPIDERMIS Significant differences in the morphology of epidermal cells in the axis and sporangium have been noted in several zosterophylls. In most taxa for which these details are known, the sporangium epidermal cells are more or less isodiametric whereas those of the main axes are predominantly elongate. In *Crenaticaulis,* some sporangium cells are also papillate (Banks and Davis 1969). In *Oricilla* (Gensel 1982b) and *Zosterophyllum deciduum* (Gerrienne 1988), the sporangium cell walls have been described as thicker than those of the axis. In *Serrulacaulis* (Hueber and Banks 1979) and *Hsua* (Li 1982), the distal half of the sporangium has isodiametric cells, whereas proximally the epidermal cells are more similar to those of the axis. In the early lycopsids *Baragwanathia, Drepanophycus,* and *Asteroxylon,* sporangium epidermal cell morphology is unknown, but in the extant Lycopodiaceae, these cells are more or less sinuate. Characters of the sporangium epidermis were omitted from this analysis because they are unknown in many taxa.

5.25 • TERMINATE OR NONTERMINATE FERTILE AXES The presence (terminate) or absence (nonterminate) of a sporangium terminating the fertile axis in zosterophylls and lycopsids was scored for a variety of taxa by Niklas and Banks (1990). On the basis of implicit outgroup comparison, it was concluded that the nonterminate condition is a derived feature shared by lycopsids and most zosterophylls. Zosterophylls possessing terminate axes were exclusively of the *Zosterophyllum* type. In our view this character is problematic because it is very difficult to evaluate in the critical *Zosterophyllum*-type zosterophylls in which the apical meristem is obscured by overlapping sporangia. Furthermore, substantial similarities in sporangium ontogeny within lycopsids imply that the identification of "truly terminal" sporangia is highly problematic in fossil material. In both lycopsids and zosterophylls, sporangia develop rapidly close to the apex and may appear to be terminal.

We have omitted this character from our analysis because we were unable to assign character states confidently, especially in the critical *Zosterophyllum*-type zosterophylls.

5.26 • SPORANGIUM VASCULATURE The presence or absence of a vascular supply to the sporangium is a potentially useful feature but has a poorly understood systematic distribution. Vascularized sporangia have been observed in *Zosterophyllum myretonianum* (Lang 1927), *Nothia* (El-Saadawy and Lacey 1979a), *Gumuia* (Hao 1989a), and *Trichopherophyton* (Lyon and Edwards 1991), but the condition in other zosterophylls is unknown. Vascularization of the sporangium also occurs in the early lycopsid fossil *Asteroxylon* (Lyon 1964). In extant lycopsids, the sporophyll is vascularized but the sporangium is not directly supplied with vascular tissue. Outgroup comparison with rhyniophyte taxa suggests that sporangium vascularization is a plesiomorphic feature in zosterophylls and lycopsids, and that the absence of a vascular supply to the sporangia in extant lycopsids is best interpreted as a loss, possibly incurred with the development of a close association between a sporangium and its subtending sporophyll. This character was omitted from our analysis because it is unknown for most of the taxa we consider.

5.27 • SPORANGIUM SPINES Multicellular spine-like outgrowths are a distinctive feature of many zosterophylls, and the details of spine distribution may provide additional systematic characters. In some taxa, for example, multicellular spines occur on the sporangia in addition to the main axes. Spines on sporangia have been recorded in *Discalis* (Hao 1989b), *Sawdonia acanthotheca* (Gensel 1991; Gensel, Andrews, and Forbes 1975), and *Konioria* (Zdebska 1982). In both *S. acanthotheca* and *Konioria,* the spines on the sporangia are significantly smaller than those on the main axes. In *Crenaticaulis* (Banks and Davis 1969) and *Serrulacaulis* (Hueber and Banks 1979), spines do not occur on the sporangia, but in *Crenaticaulis* they are found on the sporangium stalks. Similarly, in *Yunia,* the axes are spiny but sporangia—if correctly attributed to the spiny stems—are naked. The presence or absence of spines on sporangia is unknown or equivocal for other spiny taxa such as *Deheubarthia* (Edwards, Kenrick, and Carluccio 1989), *Anisophyton* (Remy, Schultka, and Hass 1986), and *Bathurstia* (Hueber 1971a). This potential character was omitted from our analysis because it is unknown or equivocal in many taxa.

5.28 • SPORE WALL STRUCTURE In their reviews of Devonian in situ spores, Allen (1980) and Gensel (1980) recognized several general features of zosterophyll microspores. Allen noted that all were azonate (exine equally thick throughout) and lacked any thickening of the equatorial exine (crassitude). Gensel recognized two spore types: smooth forms referable to the dispersed spore genus *Calamospora* and ornamented forms probably referable to *Apiculiretusispora.* She also noted that some of the features interpreted as wall sculpture in certain zosterophylls could potentially be tapetal residue. Other taxa not included in zosterophylls by Allen (1980) and Gensel (1980), but from which spores are known, include the Barinophytaceae and *Renalia.*

Since 1980, in situ spores have been described from several new zosterophylls or related taxa; these include *Discalis* (Hao 1989b), *Oricilla* (Gensel 1982b), *Serrulacaulis* (Hueber and Banks 1979), *Tarella* (Edwards and Kenrick 1986), *Zosterophyllum deciduum* (Gerrienne 1988), *Z. divaricatum* (Gensel 1982a), *Huia* (Geng 1985), *Hsua* (Li 1982), *Gumuia* (Hao 1989a), *Trichopherophyton* (Lyon and Edwards 1991), and *Sawdonia ornata* (Rayner 1983).

Microspores (in situ) have been recorded in about three quarters of the described zosterophyll species, range from 18 to 92 μm in diameter (mean = 52 μm), and are circular to subtriangular. Trilete marks have been recorded in most taxa but have not been observed in *Gosslingia* (D. Edwards 1970b), *Tarella* (Edwards and Kenrick 1986), *Zosterophyllum myretonianum* (Lang and Cookson 1930), and *Rebuchia ovata* (Hueber 1972) (probably due to a small sample size and poor preservation). *Laesurae* (contact scars) typically extend for 30–80% of the spore radius. All of these spores conform to the generalizations made by Allen (1980) and Gensel (1980) and are azonate and more or less smooth-walled. Exine ornamentation with units of less than 1 μm in diameter, which may represent tapetal residue, has been noted in most taxa.

Outgroup comparison with rhyniophytes suggests that all of the features recognized in zosterophyll spores are probably plesiomorphic at this hierarchical level. The presence of an equatorial crassitude (thickening of exine) in some outgroup taxa (Fanning, Richardson, and Edwards 1991) also means that polarization of the azonate-zonate character is equivocal.

In the early lycopsid fossils *Asteroxylon* and *Baragwanathia,* spores are poorly preserved but appear to be very simple and similar to those in zosterophylls. More distinctive spore wall morphologies are present in other lycopsid groups. In the Lycopodiaceae a foveolate-fossulate ornament predominates (Wilce 1972), whereas in *Leclercqia* (Streel 1972), *Selaginella* (Tryon and Lugardon 1991), and many arborescent taxa (e.g., *Oxroadia:* Bateman 1992), microspores have a prominent echinate ornamentation. A TEM study of the microspore wall of *Barinophyton citrulliforme* (Taylor and Brauer 1983) concluded that the sporoderm (0.8–1.5 μm thick) is three-layered, comprising a thin (50 nm) outer layer (which is absent in some spores), an amorphous middle layer (1.4 μm) with lamellae along the outer edge, and an inner, dark layer (30–50 nm). Similar sporoderm structure has been observed in poorly preserved material of *Protobarinophyton pennsylvanicum* (Cichan, Taylor, and Brauer 1984), but the wall layers are not as distinct. The delicate inner layer was not observed and may not have been preserved. The thin outer layer appears to be part of the exine and not a tapetal residue as is inferred in *B. citrulliforme.*

Among zosterophylls, megaspores are known only in the Barinophytaceae. Megaspores of *Barinophyton* are circular to oval, smooth, trilete, and 650–900 μm in size, having a wall 6.5–10 μm thick (Brauer 1980). The sporoderm contains numerous lacunae (seen with TEM) and is comparable with megaspore walls in heterosporous lycopsids (Taylor and Brauer 1983). Megaspores of *Protobarinophyton* are smaller (410–560 μm) and possibly alete but resemble those of *Barinophyton* in other aspects of wall structure (Cichan, Taylor, and Brauer 1984).

Spore wall features were omitted from our analysis of zosterophyll relationships because of the difficulty of recognizing unambiguous characters. Sporoderm ultrastructure is known only for *Barinophyton* and *Protobarinophyton.*

ANALYSES

A single numerical analysis of our entire data set using the Branch-and-Bound algorithm of PAUP (version 3.1.1; Swofford 1990) was impractical because the number of taxa involved precludes finding all of the most parsimonious trees (Table 5.1). Furthermore, our data matrix has fewer character states than taxa, implying that a fully resolved solution is unlikely and that many equally parsimonious relationships exist in parts of the tree. Our approach therefore has been to develop a robust cladogram for 32 taxa (Table 5.1, analysis 5.1) based on the Heuristic algorithm and to compare this result with Branch-and-Bound analyses (analyses 5.2–5.4) of subsets of species chosen for their likely systematic interest and relative completeness of available data. The results of the Branch-and-Bound analyses are summarized as strict consensus trees and show similar patterns of relationships to the more comprehensive but analytically incomplete Heuristic search. The results of the Heuristic analysis (analysis 5.1) were used as a framework for examining character evolution and the relationships of the more poorly understood taxa on an experimental basis.

Characters were polarized by outgroup comparison using two well-understood fossils—*Rhynia gwynne-vaughanii* and *Psilophyton dawsonii*—for the reasons outlined earlier. Consistency indexes given throughout are calculated excluding uninformative characters. Bremer Support (Decay Indexes = DI) was calculated for a representative subset of taxa (analysis 5.4).

RESULTS

A Heuristic analysis based on 32 taxa and 20 characters yielded trees of 45 steps with consistency indexes of 0.74 (Figures 5.25 and 5.26, analysis 5.1). No shorter trees were found in repeated sampling

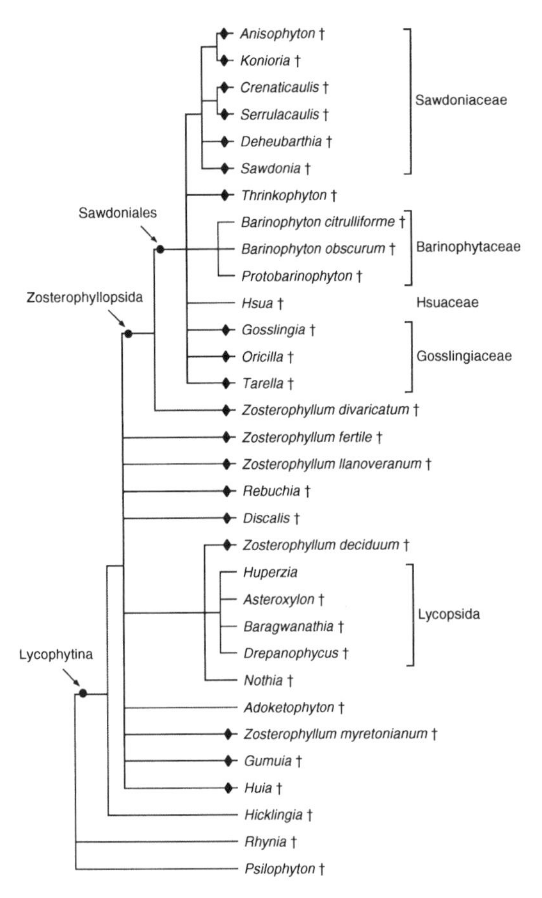

◆ Zosterophyllophytina *sensu* Banks

FIGURE 5.25. Cladogram of relationships among zosterophylls and basal lycopsids: strict consensus tree from analysis 5.1. *Psilophyton* and *Rhynia* were designated as outgroups. † = extinct. Analysis 5.1 was a Heuristic search (by means of PAUP Version 3.1.1; Swofford 1990) of 32 taxa and 20 characters (data in Appendix 4). The search, which was terminated before completion because of time and hardware limitations, resulted in 18,500 equally parsimonious trees 45 steps in length, with consistency indexes of 0.74.

(100 searches) of the data matrix using random addition starting trees (TBR branch swapping, MULPARS selected). The analysis was terminated at 18,500 trees because of hardware (memory) and time limitations. The results are summarized as a strict consensus tree (Figure 5.25), and the character state changes are plotted on a representative most parsimonious tree (Figure 5.26, Table 5.3). These trees support the monophyly of the Lycophytina (zosterophylls plus lycopsids) and lycop-

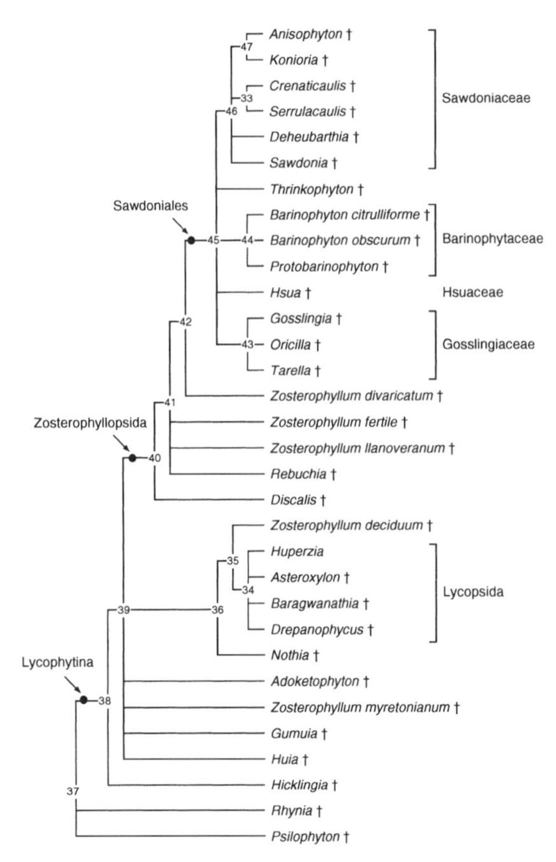

FIGURE 5.26. Cladogram of relationships among zosterophylls and basal lycopsids: representative most parsimonious tree from analysis 5.1. See caption to Figure 5.25 (the strict consensus tree) for explanation of symbols and details of analysis. Numbers on nodes refer to character changes listed in Table 5.3 (ACCTRAN optimization).

sids (Lycopsida). There is support for a zosterophyll clade (Zosterophyllopsida) containing most, but not all, Zosterophyllophytina *sensu* Banks (1975b, 1992) and several additional groups. Within the Zosterophyllopsida, there is a large and diverse clade characterized by planar, pseudomonopodial branching and subordinate (axillary) branches (Sawdoniales: Figure 5.26, Table 5.3: node 42 → node 45). Within the Sawdoniales are several clades, or possible clades, with varying degrees of support. We provisionally recognize a heterosporous clade (Barinophytaceae: Figure 5.26, Table 5.3: node 45 → node 44), a group with distinctive auricular sporangium arrangements (Gosslingiaceae: Figure 5.26, Table 5.3: node

TABLE 5.3

Character Changes on Nodes of Tree in Figure 5.26

Beginning node → end node		Character	Change in state	CI
Node 33 → *Crenaticaulis*	5.15	Relative valve size	1 → 2	1.000
Node 33 → *Serrulacaulis*	5.11	Sporangium rows	2 → 0	1.000
Node 35 → node 34	5.5	Multicellular appendages	0 → 1	0.750
Node 36 → *Nothia*	5.9	Tracheid thickenings	1 → 0	1.000
Node 36 → node 35	5.7	Xylem strand shape	1 → 2	1.000
Node 37 → node 38	5.7	Xylem strand shape	0 → 1	1.000
	5.8	Protoxylem	0 → 1	1.000
	5.13	Sporangium shape	0 → 1	1.000
	5.14	Sporangium symmetry	0 → 1	1.000
	5.16	Sporangium dehiscence	0 → 1	0.667
	5.18	Sporangium attachment	0 → 1	0.750
Node 37 → *Psilophyton*	5.1	Branching type	0 → 1	0.333
Node 37 → *Rhynia*	5.16	Sporangium dehiscence	0 → 2	0.667
	5.18	Sporangium attachment	0 → 3	0.750
Node 38 → node 39	5.17	Thickened sporangium valve rim	0 → 1	0.500
Node 38 → *Hicklingia*	5.13	Sporangium shape	1 → 2	1.000
Node 39 → *Adoketophyton*	5.5	Multicellular appendages	0 → 3	0.750
	5.6	Appendage phyllotaxy	0 → 3	1.000
	5.10	Sporangiotaxis	0 → 2	1.000
Node 39 → *Huia*	5.1	Branching type	0 → 1	0.333
	5.2	Branching pattern	0 → 1	0.333
	5.18	Sporangium attachment	1 → 2	0.750
Node 39 → node 36	5.6	Appendage phyllotaxy	0 → 1	1.000
	5.12	Sporangium distribution	0 → 1	0.667
Node 39 → node 40	5.4	Circinate vernation	0 → 1	0.500
Node 40 → node 41	5.10	Sporangiotaxis	0 → 1	1.000
Node 40 → *Discalis*	5.5	Multicellular appendages	0 → 2	0.750
Node 41 → node 42	5.12	Sporangium distribution	0 → 1	0.667
Node 41 → *Rebuchia*	5.4	Circinate vernation	1 → 0	0.500
Node 42 → node 45	5.1	Branching type	0 → 1	0.333
	5.2	Branching pattern	0 → 1	0.333
	5.3	Subordinate branching	0 → 1	0.667
Node 43 → *Oricilla*	5.3	Subordinate branching	1 → 0	0.667
Node 43 → *Tarella*	5.3	Subordinate branching	1 → 2	0.667
Node 45 → node 43	5.19	Sporangium orientation	0 → 1	1.000
Node 45 → node 44	5.12	Sporangium distribution	1 → 2	0.667
	5.16	Sporangium dehiscence	1 → 2	0.667
	5.17	Thickened sporangium valve rim	1 → 0	0.500
	5.19	Sporangium orientation	0 → 2	1.000
	5.20	Spore size	0 → 1	1.000
Node 45 → node 46	5.5	Multicellular appendages	0 → 2	0.750
Node 45 → *Hsua*	5.18	Sporangium attachment	1 → 0	0.750
Node 46 → node 33	5.6	Appendage phyllotaxy	0 → 2	1.000
Node 46 → node 47	5.11	Sporangium rows	2 → 1	1.000
Node 47 → *Konioria*	5.2	Branching pattern	1 → 0	0.333

Notes: ACCTRAN optimization. CI = consistency index of character.

45 → node 43), and a clade distinguished by large multicellular spines (Sawdoniaceae: Figure 5.26, Table 5.3: node 45 → node 46). Our analysis indicates that the genus *Zosterophyllum* is paraphyletic to the Sawdoniales and possibly also to the Lycopsida. Species having marked linear sporangium arrangements and circinate vernation (e.g., *Z. divaricatum*) are placed within the Zosterophyllopsida in the Sawdoniales stem group. The position of other taxa with helical sporangium arrangements and no circinate growth (e.g., the type species, *Z. myretonianum*) is ambiguous with respect to the Zosterophyllopsida and the Lycopsida sister group (Figure 5.25). The poor resolution among basal *Zosterophyllum*-like plants in this analysis is attributable to the relatively small number of characters (20) and larger number of taxa (32) combined with the effects of missing data, rather than to high levels of homoplasy.

Two additional tests of this general pattern of relationship were performed on subsets of taxa using the Branch-and-Bound algorithm of PAUP, which finds all the most parsimonious trees. The taxon choice for these analyses was based on the tree topology generated in the Heuristic search (Figure 5.25, analysis 5.1). Analysis 5.2 (Figure 5.27) focused on the basal relationships in the Lycophytina using exemplars to represent the Sawdoniales *(Deheubarthia, Gosslingia, Protobarinophyton)*. This analysis was based on 16 taxa and 20 characters and yielded 342 equally parsimonious trees of 28 steps with consistency indexes of 0.76. Analysis 5.3 (Figure 5.28) focused on the Sawdoniales using exemplars to represent basal the Lycophytina *(Asteroxylon, Huperzia, Zosterophyllum myretonianum*, and *Z. divaricatum)*. This analysis was based on 16 taxa and 20 characters and yielded 1,176 equally parsimonious trees of 32 steps with consistency indexes of 0.81. Both analyses found similar tree topologies to the more comprehensive Heuristic search.

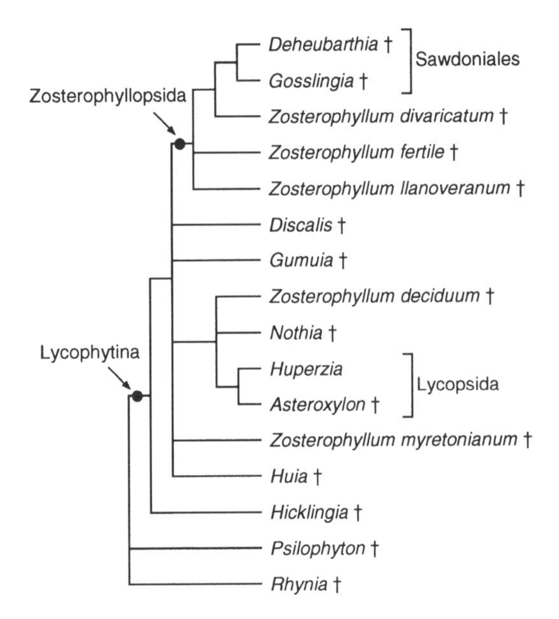

FIGURE 5.27. Cladogram of relationships among zosterophylls and basal lycopsids: strict consensus tree from analysis 5.2, which focused on putatively plesiomorphic taxa. *Psilophyton* and *Rhynia* were designated as outgroups. † = extinct. Analysis 5.2 was a Branch-and-Bound search (by means of PAUP Version 3.1.1; Swofford 1990) of 16 taxa and 20 characters (data in Appendix 4). All most parsimonious trees were found (342 equally parsimonious trees 28 steps in length, with consistency indexes of 0.76).

Tree Stability

Tree stability was tested using Bremer Support (Decay Index = DI) measurements (Källersjö, Farris, et al. 1992). Bremer Support measurements were impractical with the larger data sets (analyses 5.1–5.3) because of memory limitations caused by the high number of trees saved and excessive time taken for the analysis. We performed Bremer Support measurements on a representative subset of taxa (Figure 5.29, analysis 5.4) to obtain estimates of the relative support for clades. Support was strongest for the Lycophytina (DI = 3), the Barinophytaceae (DI = 2), and the Sawdoniales (DI = 2) (Figure 5.25). All other branches were relatively weakly supported (DI = 1). Much of the uncertainty in the data matrix is attributable to the relatively small number of characters (20) and larger number of taxa (33) combined with the effects of missing data, rather than to high levels of homoplasy (Appendix 4). The addition of poorly understood or problematic taxa had little effect on the general tree topology, but experimentation showed

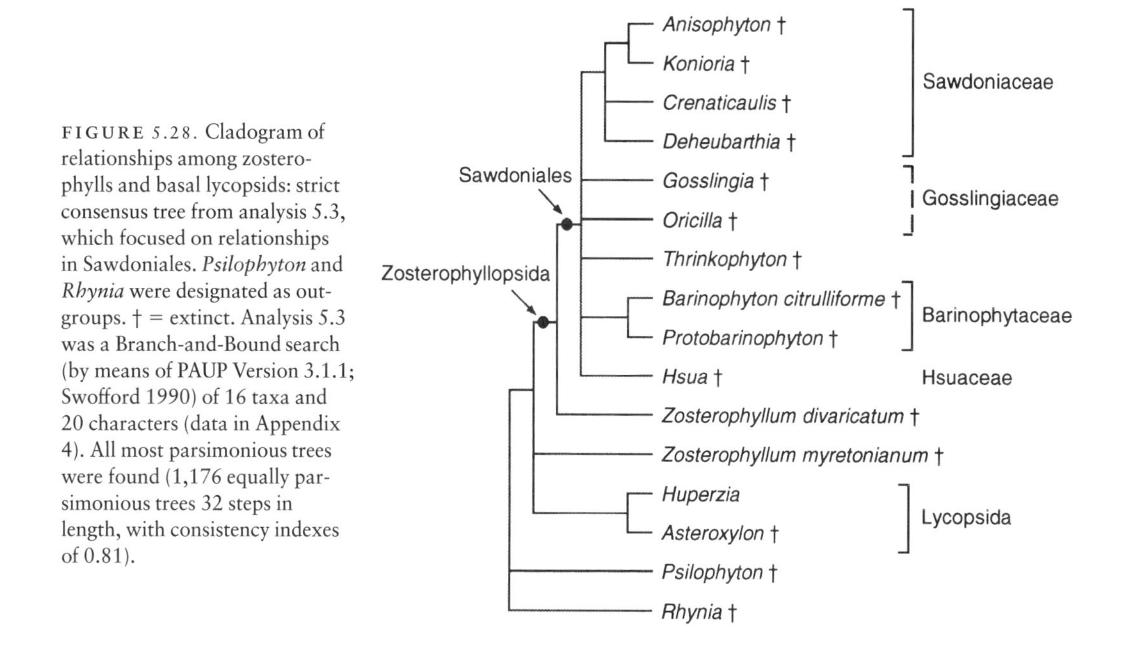

FIGURE 5.28. Cladogram of relationships among zosterophylls and basal lycopsids: strict consensus tree from analysis 5.3, which focused on relationships in Sawdoniales. *Psilophyton* and *Rhynia* were designated as outgroups. † = extinct. Analysis 5.3 was a Branch-and-Bound search (by means of PAUP Version 3.1.1; Swofford 1990) of 16 taxa and 20 characters (data in Appendix 4). All most parsimonious trees were found (1,176 equally parsimonious trees 32 steps in length, with consistency indexes of 0.81).

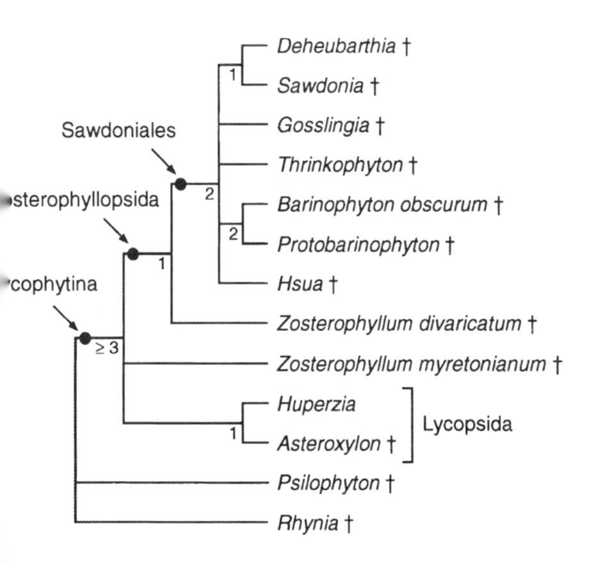

FIGURE 5.29. Cladogram of relationships among zosterophylls and basal lycopsids: strict consensus tree from analysis 5.4, which focused on a subset of taxa chosen for Bremer Support measurements (represented by the numbers on the branches). *Psilophyton* and *Rhynia* were designated as outgroups. † = extinct. Analysis 5.4 was a Branch-and-Bound search (by means of PAUP Version 3.1.1; Swofford 1990) of 13 taxa and 20 characters (data in Appendix 4). All the most parsimonious trees were found (270 equally parsimonious trees 25 steps in length, with consistency indexes of 0.91).

that some combinations of taxa did cause branches to collapse in the consensus trees (see below).

Lycophytina

The Lycophytina are a well-supported group (DI = 3) containing two major clades which we term Lycopsida (lycopsids *sensu lato*) and Zosterophyllopsida (extinct; Zosterophyllophytina *sensu* Banks in part) (Figure 5.25). The relationship of the Lycophytina to other land plants was dealt with in greater detail in Chapter 4, and based on the analysis there, the Lycophytina are defined by a change from terminal to lateral sporangia (reversed in *Hsua;* Figure 4.33, Table 4.6: node 47 → node 46). Other synapomorphies appearing in the Lycophytina stem group include (1) isovalvate dehiscence along the distal sporangium rim (node 60 → node 52), (2) conspicuous cellular thickening of the dehiscence line, (3) reniform sporangia (node 52 → node 50), and (4) exarch xylem differentiation (node 50 → node 48). Another possible defining feature is an elliptical-shaped xylem strand. The appropriate hierarchical level for characters based on xylem anatomy is equivocal because their distribution is poorly understood in stem group taxa as well as in the plesiomorphic species of *Zosterophyllum.* The Lycophytina stem group contains a range of extinct rhyniophytes (e.g., *Renalia, Uskiella,* some *Cooksonia* species) that resemble lycopsids and zosterophylls in sporangium shape and dehiscence but retain the plesiomorphic condition of terminal sporangia (Chapter 4). The least inclusive defining feature of the Lycophytina is sporangium insertion on short lateral stalks (character 5.18).

The relationships of many basal taxa within the Lycophytina are ambiguous (Figures 5.25 and 5.27, analyses 5.1 and 5.2). Consistently obtained patterns included sister group relationships between *Hicklingia* and the Lycophytina, between *Z. divaricatum* and the Sawdoniales, and among *Z. deciduum, Nothia,* and the Lycopsida. *Z. fertile* and *Z. llanoveranum* are often grouped with *Z. divaricatum* in the Sawdoniales stem group (Figure 5.27). The relationships of other *Zosterophyllum* species, as well as of *Huia, Discalis,* and *Rebuchia,* were unresolved in the consensus tree (Figures 5.25 and 5.27). Analysis of alternative tree topologies showed that some species of *Zosterophyllum* (e.g., *Z. llanoveranum, Z. divaricatum, Z. fertile*) and some *Zosterophyllum*-like taxa (e.g., *Rebuchia*) have clear similarities with the Zosterophyllopsida in such features as rowed sporangium arrangements and, in some taxa, circinate vernation. Other species of *Zosterophyllum* (e.g., *Z. myretonianum, Z. deciduum*) and the *Zosterophyllum*-like *Nothia,* all of which have helical sporangium arrangements, appear to be more closely related to the Lycopsida based on sporangium distributions that extend over several orders of branching (an apparent parallelism with some members of the Sawdoniales). Our analysis suggests that *Zosterophyllum* is paraphyletic to both the Lycopsida and the Zosterophyllopsida and perhaps also to the Lycophytina. Other *Zosterophyllum*-like taxa (e.g., *Huia, Discalis, Nothia, Adoketophyton*) have similar ambiguous relationships within the basal Lycophytina. Our analysis indicates that many of these plesiomorphic taxa are currently too incomplete and poorly understood to allow more detailed and unambiguous groupings.

Zosterophyllopsida

The Zosterophyllopsida are a diverse, extinct group of predominantly Devonian plants (DI = 1; Figure 5.25). The group is defined by (1) circinate growth (a parallelism with many extant ferns) and (2) a unique two-rowed form of sporangium arrangement (Figures 5.25 and 5.26, Table 5.3: node 39 → node 40 and node 40 → node 41). A third possible defining feature is an elliptical xylem strand, but this character may be more general in the Lycophytina because anatomical information is lacking for many plesiomorphic *Zosterophyllum*-like plants. The Zosterophyllopsida comprise the diverse Sawdoniales clade, as well as *Zosterophyllum*-like plants, such as *Z. divaricatum* and possibly *Z. llanoveranum, Rebuchia,* and *Discalis,* which fall into the Sawdoniales stem group.

Sawdoniales

The Sawdoniales (DI = 2; Figures 5.25, 5.26, 5.28, analyses 5.1 and 5.3) are defined by (1) pseudomonopodial branching (parallelism with the Euphyllophytina; see Chapter 4), (2) planated branching system (lost in some taxa, e.g., *Konioria*), and (3) a unique form of subordinate branching (character 5.3) (Figure 5.26; Table 5.3: node 42 → node 45). Another possible defining feature is sporangium distributions that extend over several orders of branches (Figure 5.26, Table 5.3: node 41 → node 42). The Sawdoniales comprise several distinctive subgroups: the Sawdoniaceae, the Barinophytaceae, and the Gosslingiaceae, as well as the problematic *Hsua* (see below) and plesiomorphic taxa that are difficult to position unambiguously (e.g., *Thrinkophyton*). Two of these families, the Sawdoniaceae and the Gosslingiaceae, are supported by only one distinctive, unambiguous character each, but despite this low level of support, we feel that the recognition of these two putative groups within Sawdoniales is important to highlight.

Sawdoniaceae

The Sawdoniaceae (DI = 1) are a large clade defined on the presence of conspicuous multicellular spines (Figure 5.26, Table 5.3: node 45 → node 46), and the family contains such taxa as *Sawdonia, Anisophyton, Konioria, Deheubarthia, Crenaticaulis,* and *Serrulacaulis* (Figure 5.29). Two possible subfamily groupings include a clade in which sporangium attachment has been modified to the ventral side of the main axis (*Anisophyton, Konioria;* character 5.11) and a clade in which spines are ordered in distinctive rows (*Crenaticaulis* and *Serrulacaulis;* character 5.6).

Barinophytaceae

The Barinophytaceae (DI = 2) are defined by (1) a unique form of heterospory (megaspores and microspores in the same sporangium), (2) a very compact, unbranched strobilus, (3) a unique form of "clasping" sporangium orientation (character 5.19), (4) the loss of a thickened dehiscence line, and (5) the loss of valvate dehiscence (Figure 5.26, Table 5.3: node 45 → node 44). In the context of our cladogram the absence of valvate dehiscence and the absence of cellular thickening of the sporangium wall near the dehiscence line are interpreted as losses and are probably developmentally correlated. Details of sporangium dehiscence are poorly understood in *Barinophyton obscurum* and *Protobarinophyton pennsylvanicum* and were treated as missing data in our analysis. As currently defined, the Barinophytaceae comprise two genera: *Barinophyton* and *Protobarinophyton.*

Gosslingiaceae

The Gosslingiaceae are a clade defined by a unique form of "auricular" sporangium orientation (character 5.19; Figure 5.26, Table 5.3: node 45 → node 43), and the family comprises *Gosslingia, Tarella,* and *Oricilla.* The relationships among these three taxa and other clades of the Sawdoniales were unresolved (DI = 0) in our consensus trees (Figures 5.25 and 5.28), although the group formed by these three genera does not appear to be contradicted by the distribution of other characters.

Hsuaceae

The Hsuaceae are a clade of the Sawdoniales defined by lateral branches bearing terminal sporangia (characters 4.23 and 5.18; Figure 5.26, Table 5.3: node 45 → *Hsua*). The phylogenetic position of the Hsuaceae is discussed below. Only one taxon has been described: *Hsua robusta.*

Lycopsida

The Lycopsida are a large monophyletic group (DI = 1) that was represented in this analysis by extant *Huperzia* and the early fossils *Asteroxylon, Baragwanathia,* and *Drepanophycus.* Relationships within the Lycopsida are dealt with in more detail in Chapter 6. The Lycopsida are defined on the basis of the presence of microphylls (Figure 5.26, Table 5.3: node 35 → node 34). Other potential synapomorphies appearing in the Lycopsida stem group include (1) helical phyllotaxis, (2) stellate xylem anatomy, and (3) sporangium distributions extending over several orders of branches (parallelism in Sawdoniales) (Figure 5.26, Table 5.3: node 39 → node 36, node 36 → node 35). Helical phyllotaxis is an ambiguous character at this level because it cannot be scored in many of the leafless outgroups, and it could be regarded as more general if microphylls are interpreted as transformational homologues of sporangia (see Chapter 7). Likewise, the form of the xylem strand is unknown in many potential Lycopsida stem group species. Our results clearly indicate that the Lycopsida emerge from a basal grade of *Zosterophyllum*-like Lycophytina. Furthermore, the putative lycopsids *Baragwanathia, Drepanophycus,* and *Asteroxylon* are always grouped with

Huperzia. Possible sister groups of the Lycopsida include the Zosterophyllopsida and some of the plesiomorphic *Zosterophyllum*-like taxa such as *Z. myretonianum, Z. deciduum, Adoketophyton,* and *Nothia aphylla. Adoketophyton* is problematic because it possess unique sporangium-shaped ("fan-shaped") sporophylls but no vegetative leaves as do the lycopsids. The genus has been delineated on the basis of fragments of fertile axes, and there is no information on stem anatomy. In recognition of the problematic nature of the sporophylls, we scored them as a unique character state (autapomorphy) of the genus. Our analysis places *Adoketophyton* in a basal position within the Lycophytina, and many trees suggest a sister group relationship with *Zosterophyllum myretonianum* and the Lycopsida. This conclusion raises the interesting possibility of homology between the microphylls of lycopsids and the sporangium-shaped sporophylls of *Adoketophyton* (see Chapter 7).

Problematic Taxa

The problematic taxa in our analysis are of two general kinds: those with relatively large amounts of missing data and those that introduce homoplasy. The only taxon that introduced striking morphological homoplasy was *Hsua,* and despite the presence of "terminal" sporangia, it was placed in the Sawdoniales in this analysis (Figures 5.25 and 5.26) and in the more general analysis in Chapter 4. *Hsua* possesses the circinate vernation and elliptical xylem of the Zosterophyllopsida and the planar, pseudomonopodial, and subordinate branching of the Sawdoniales. However, instead of the usual lateral sporangia typical of the Lycophytina, *Hsua* has branched fertile lateral axes resembling the plesiomorphic condition for polysporangiophytes observed in the Lycophytina stem group (e.g., *Renalia*) (character 5.18). This combination of characters implies either that the five characters linking *Hsua* to the Sawdoniales evolved independently at least twice or that the "terminal" sporangium arrangements in *Hsua* represent a striking morphological reversal. Since the inclusion of *Hsua* in the Sawdoniales was relatively strongly supported in two analyses (Chapters 4 and 5), we provisionally accept the latter interpretation.

Missing data was a problem with many taxa and was particularly acute for anatomical characters. Furthermore, several interesting taxa are known only from small terminal fragments of fertile axes (e.g., *Trichopherophyton, Bathurstia*), whereas with others, disarticulation of the original material raises questions over the validity of the whole-plant concept (e.g., *Krithodeophyton*). Separate analyses were run using a subset of taxa representing the major elements of diversity in the Lycophytina to analyze the effect of these rather problematic plants. *Bathurstia* was the most incomplete taxon in our analysis, scorable for only 7 out of 20 (35%) of characters, but was always placed unequivocally within the Sawdoniaceae.

Trichopherophyton is based on only a fragment of fertile axis, but much anatomical information is known because it occurs in the Rhynie Chert lagerstätten. *Krithodeophyton* is problematic because it has been defined on the basis of highly disarticulated fragments and because there are some interpretational difficulties with the structure of the strobilus (Appendix 1). This taxon was originally associated with the Barinophytaceae. Our analyses placed both plants in the Lycophytina in an unresolved basal relationship with *Zosterophyllum*-like plants.

Experiments

The taxa examined in analysis 5.3 (Figure 5.28) were used to conduct several specific tests of relationship and homology. Alternative hypotheses of relationship were tested through the implementation of constrained tree topologies during Branch-and-Bound analyses. The relative parsimony was then evaluated against our initial unconstrained analysis based on the same data. Specific homology tests were constructed by rescoring certain character states to favor a particular hypothesis and observing the effect on subsequent analyses.

One previous hypothesis of relationship in the Lycophytina proposes that lycopsids are closely related to spiny zosterophylls. This hypothesis has been favored because the vascular microphyll of lycopsids might be explained as an elaboration of the nonvascular, spinelike appendages in some zosterophylls. It also implies that zosterophylls are strongly paraphyletic (see Crane 1990). Our analysis supports the alternative hypothesis—that most zosterophylls form a monophyletic group that is a sister group to lycopsids. Forcing lycopsids into a sister group relationship with spiny zosterophylls in the Sawdoniaceae added five steps to our most parsimonious tree. Additional character-state changes required by this scenario involve five reversals in basal lycopsids of characters acquired in the Zosterophyllopsida: (1) reversal of strongly planar branching to nonplanar branching, (2) reversal of pseudomonopodial to isotomous branching, (3) loss of the distinctive axillary branch of Sawdoniales, (4) loss of circinate vernation, and (5) reversal of sporangiotaxis from two-rowed to helical.

In a variation of the above hypothesis, Schweitzer (1983b) suggested that the lycopsid-like *Asteroxylon* is related to spiny zosterophylls (Sawdoniaceae) whereas true lycopsids (e.g., *Huperzia*) are not. This hypothesis adds at least six steps to our most parsimonious tree. Character-state changes involved the loss of the five characters defining the Zosterophyllopsida and the Sawdoniales (above), as well as independent origins of the microphyll and stellate xylem strands in *Asteroxylon* and *Huperzia*.

In our original analysis, we took a neutral stance on the origin of the lycopsid microphyll by scoring the spinelike appendages in the Sawdoniaceae and the vascular microphylls of lycopsids as two independent states of the same character (multicellular appendages, character 5.5). This treatment was justified because the positional and structural similarities between these two organs were thought to be weak. To further test the idea that these features may be homologous, we assigned them the same state so as weight in favor of homology. Rerunning analysis 5.3 based on our original 16 taxa (Figure 5.28) yielded the same tree topology with a reduced consistency index (0.77). Examination of the possible character-state optimizations for character 5.5 (microphyll = spine) showed independent gains in both the Sawdoniaceae and the Lycopsida. We therefore reject the notion of homology between lycopsid microphylls and spinelike enations in the Zosterophyllopsida (see Chapter 7).

DISCUSSION

Previous attempts at evaluating relationships among zosterophylls and lycopsids (e.g., Banks 1968, 1970, 1975b, 1992; Chaloner and Sheerin 1979; Meyen 1987; Niklas and Banks 1990; Selden and Edwards 1989) using a variety of approaches and methods have generated useful perspectives, but the lack of explicit details makes these hypotheses difficult to evaluate and compare (Figures 4.4, 4.5, 5.5–5.7). The hypothesis of a close relationship between zosterophylls and lycopsids was first developed in detail by Banks (1968, 1975b) based mainly on similarities in sporangium morphology and on the early occurrence of both groups in the fossil record. Our cladistic analysis provides clear support for this hypothesis (Chapter 4). The close relationship between zosterophylls and lycopsids has been challenged only indirectly by authors who have questioned lycopsid monophyly (e.g., Stewart 1983; Stewart and Rothwell 1993). Under these circumstances, one or more groups of heterosporous lycopsids are thought to be more closely related to taxa that we would place in the basal Euphyllophytina. Our analysis indicates that polyphyly of lycopsids is highly unparsimonious (see Chapter 6).

Although polyphyly can be ruled out, critical examination of the Zosterophyllophytina *sensu* Banks (1968, 1975b) discloses that the group is not strictly monophyletic, although it contains a large monophyletic component (our Zosterophyllopsida). As originally defined by Banks, the Zosterophyllophytina are broadly equivalent to our Lycophytina, and this similarity is consistent with the phylogenetic diagrams in Banks (1970; see Figure 4.4), Meyen (1987; see Figure 5.6) and Niklas and Banks (1990; see Figure 5.5), which show lycopsids originating within basal zosterophylls. In other words, as originally defined, the group is paraphyletic unless it also includes lycopsids *(sensu stricto)*. We therefore raise the possibility of redefining the Zosterophyllophytina (Zosterophyllopsida) to include only those taxa with one- or two-rowed sporangium arrangements. Imposing this condition allows most of the taxa originally classified in the Zosterophyllophytina to be retained within the Zosterophyllopsida as recognized here. The relationships of the remaining plesiomorphic members of the original zosterophylls (i.e., those with helical sporangium arrangements) are ambiguous with respect to both the Lycopsida and the Zosterophyllopsida. Our analysis implies that *Zosterophyllum* is paraphyletic: species possessing rowed sporangium arrangements and circinate vernation are basal within the Zosterophyllopsida, whereas those having helical arrangements and lacking circinate vernation may be more closely related to the Lycopsida.

Two recent noncladistic analyses of relationships among zosterophylls and lycopsids by Niklas and Banks (1990; see Figure 5.5) and Hueber (1992) adopted a developmental perspective, focusing on sporangium position and ontogeny. Niklas and Banks recognized two types of plants characterized by determinate and indeterminate growth. These two categories were further subdivided into radial and bilateral groups, yielding a total of four categories. We regard this approach, and the resulting subdivisions, as problematic for three reasons. First, we have found no evidence for multiple origins of bilateral and radial symmetry among zosterophylls, and both features are explicable by one character-state change in our analysis. Second, we are not convinced that it is possible to determine whether the fertile shoot is actually terminated by a sporangium in any of the "terminate" taxa (e.g., *Zosterophyllum* and *Zosterophyllum*-like plants). Third, our interpretation of the groups recognized by Niklas and Banks is that they are based on both plesiomorphic and potential synapomorphic features that provide an unparsimonious explanation of character distributions.

Hueber (1992) recognized the monophyly of zosterophylls and lycopsids and suggested separate origins of these groups from rhyniophytes with reniform sporangia. According to this hypothesis, the lateral sporangia of both groups evolved independently from a *Renalia*-like ancestor (Figure 4.11). This interpretation is based on the view that there are fundamental developmental differences between lycopsid and zosterophyll sporangia—the sporangia of zosterophylls are cauline and vascularized, whereas those in extant lycopsids are associated with a sporophyll and do not possess their own vascular supply (although the sporophyll does). Our analysis is at odds with this interpretation and supports a single origin of lateral sporangia in the Lycophytina. Morphological differences in lycopsids represent modifications to a basic *Zosterophyllum*-like sporangium morphology, including the loss of sporangium vascular tissue and the evolution of the sporophyll (see Chapter 7). Our interpretation is supported by evidence from early fossil lycopsids that show greater similarities to zosterophylls than to extant lycopsids. Vascularized strands have been documented in the sporangium stalks of the early fossil lycopsids *Asteroxylon* (Lyon 1964) and *Drepanophycus* (Li and Edwards 1995), and the sporangia are probably cauline in both taxa.

Our inclusion of the Barinophytaceae and of *Hsua* within the Zosterophyllopsida is a novel departure from most previous treatments that have viewed the systematic relationships of both groups as problematic. Meyen (1987), however, treated the Barinophytaceae as an early offshoot from zosterophylls. Kräusel (1938) and Ananiev (1960) had originally included the Barinophytaceae in the Zosterophyllaceae, but in the original circumscription of the Zosterophyllophytina, Banks (1968) removed this family, which he placed *incertae sedis*. At the time, this treatment was justifiable because many aspects of the morphology of *Barinophyton* were problematic. However, detailed studies by Brauer (1980) have shown subsequently that *Barinophyton* resembles zosterophylls in many morphological details, but Brauer did not specify a close relationship between the two groups because of the seemingly unusual tracheid cell wall, the unique form of heterospory, and the small differences in sporangium orientation in *Barinophyton*. In cladistic terms the unusual heterospory and distinctive features of the strobilus are autapomorphic with respect to zosterophylls and lycopsids. The distinctive tracheid type in *Barinophyton* (G-type) has now been shown to be widespread in zosterophylls and early lycopsids (Kenrick and Crane 1991; Kenrick and Edwards 1988a). The consequences of including the Barinophytaceae within zosterophylls are that heterospory can now be recognized as having evolved within the group (see Bateman 1996b; Bateman and DiMichele 1994a), that the stratigraphic range of zosterophylls is extended from the upper Devonian (Frasnian) into the lower Carboniferous (Tournaisian), and that the diversity of the group is expanded considerably in the upper Devonian.

The only published numerical analysis of relationships in zosterophylls and lycopsids (Gensel 1992) prior to that presented here resolved zosterophylls as basically monophyletic (Figure 5.8). This result is consistent with our findings. Furthermore, both analyses support the monophyly of zosterophylls that have marked bilateral symmetry, the monophyly of lycopsids, and a sister group relationship for these clades. Because only a single species of *Zosterophyllum* was included in Gensel's analysis, paraphyly of this genus went undetected, but both analyses nevertheless found that the relationships of some species of *Zosterophyllum* are ambiguous with respect to the main zosterophyll and lycopsid clades. The main difference between Gensel's analysis and that outlined here is that, following traditional ideas on group membership, the possibility of a relationship with the Barinophytaceae and such taxa as *Hsua* was not considered. A further difference concerns the position of nonspiny "higher" zosterophylls such as *Gosslingia* and *Tarella* (our Gosslingiaceae). Gensel places these taxa within a clade characterized by multicellular, spinelike enations. In contrast, in our analysis, the multicellular spine character defines a clade within "higher" zosterophylls (Sawdoniaceae): taxa with naked axes are placed in the stem group. In both analyses this character was a source of homoplasy: in Gensel's analysis, spines are lost in *Gosslingia* and other taxa, whereas in our analysis they are gained independently—once in higher zosterophylls and also in the "lower" *Zosterophyllum*-like taxon *Discalis*. Since a greater number of taxa are considered in our analysis and there is firm character support for relationships within our main zosterophyll clade, we strongly favor the parallel acquisition of spines within the Zosterophyllopsida.

6 Lycopsida

The Lycopsida (lycopsids) are a diverse and ancient group of free-sporing vascular plants; extant taxa are characterized by lateral reniform sporangia, exarch protosteles, and microphyllous leaves (see Chapter 5). The Lycopsida comprise only three extant families—the Lycopodiaceae, the Selaginellaceae, and the Isoetaceae—but the group has a long and diverse fossil record (DiMichele and Skog 1992). They are first known in the early Devonian (or possibly late Silurian) fossil record, and by the late Carboniferous (Pennsylvanian), they dominated the peat-forming vegetation of tropical swamps. The Lycopsida comprise almost half of all known fossil plants of this age (Niklas, Tiffney, and Knoll 1980, 1983, 1985; Phillips and DiMichele 1992). The remains of arborescent lycopsids are largely responsible for the extensive, economically important, Pennsylvanian coals of Euramerica. Through the late Paleozoic, the lycopsids included great trees over 40 meters in height as well as small herbs, and this variety in growth form is mirrored by a diversity of reproductive strategies spanning homospory, heterospory, and an extreme derivative of heterospory that approached the seed habit. Compared to their extinct relatives, the three extant families of lycopsids encompass a far more limited range of body plans, show less diversity in reproduction, and contain a relatively small number of species (c. 700–1,000: Hickey 1986; Jermy 1990a,b; Øllgaard 1987, 1990, 1992).

Lycopsids have long been recognized as a distinct group of vascular plants that are only distantly related to other extant pteridophytes and seed plants (Bower 1908; Lignier 1903, 1908). A basal relationship within vascular plants is supported by comparative morphology and is consistent with the early appearance of the group in the fossil record. A close relationship between lycopsids and the extinct early land plant group zosterophylls was recognized by Banks (1968, 1975b; Figure 4.4), and this idea has received widespread acceptance (Chapter 5; Bateman 1994, 1996c; Chaloner and Sheerin 1979; DiMichele and Skog 1992; Gensel and Andrews 1984; Meyen 1987; T. N. Taylor and E. L. Taylor 1993). Lycopsids can be distinguished unequivocally from zosterophylls by the presence of microphyllous leaves, and many zosterophylls also differ in having rowed sporangium arrangements and circinate vernation (Chapter 5; Niklas and Banks 1990). As currently circumscribed, lycopsids comprise several major groups including the three extant families as well as the extinct Protolepidodendrales and tree lycopsids (rhizomorphic arborescent taxa *sensu* Bateman, DiMichele, and Willard [1992]). In addition, there are several other early taxa that have been considered intermediate between zosterophylls and lycopsids, including *Asteroxylon, Drepanophycus,* and *Baragwanathia.*

Most cladistic studies support the monophyly of lycopsids and a close relationship with zosterophylls (Bateman 1992, 1996c; Bateman, DiMichele, and Willard 1992; Crane 1990; Gensel 1992; Kenrick and Crane 1991, 1992; Chapter 5; Figures 6.1 and 6.2). Within lycopsids, there is support for a ligulate clade and a heterosporous clade, with homosporous taxa such as the Lycopodiaceae and *Asteroxylon* placed in the ligulate stem group. Character support for the monophyly of the Lycopodiaceae and the Selaginellaceae is weak to nonexistent (Bateman 1992; Crane 1990; Gensel 1992; Wagner and Beitel 1992). Possible alternative topologies include the Lycopodiaceae as paraphyletic to the ligulate clade, and the Selaginellaceae as paraphyletic to the arborescent clade (Bateman 1992). This chapter evaluates the relationships among major groups of lycopsids, focusing in particular on basal relationships involving zosterophylls, putative early fossil lycopsids such as *Asteroxylon,* Protolepidodendrales, Lycopodiaceae, Selaginellaceae, Isoetaceae, and extinct arborescent taxa.

SYSTEMATICS

In 1841 Endlicher published an influential and comprehensive classification of plants that took the novel approach of including both living and fossil groups. This classification was the first to fully recognize a close relationship among *Isoetes,* the ex-

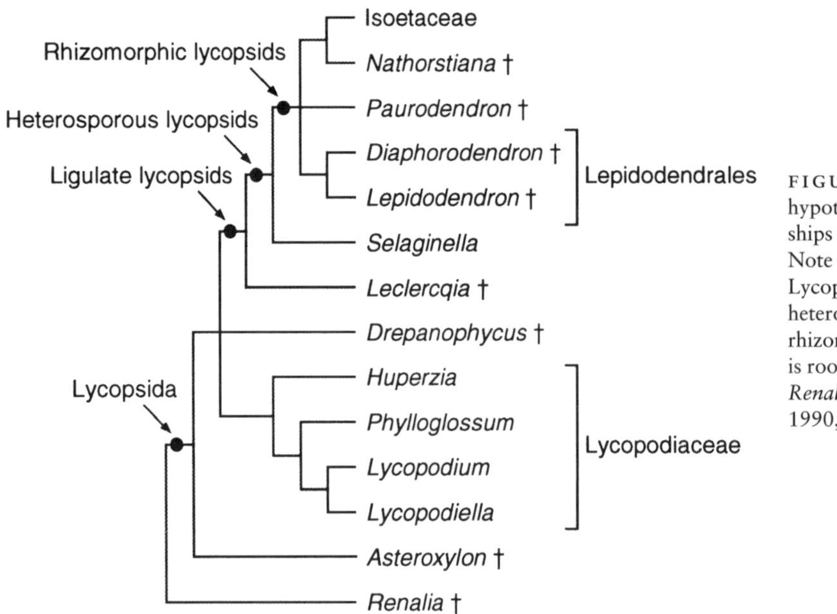

FIGURE 6.1. Preliminary hypothesis of cladistic relationships in lycopsids. † = extinct. Note the monophyly of the Lycopsida, ligulate lycopsids, heterosporous lycopsids, and rhizomorphic lycopsids. Tree is rooted on the rhyniophyte *Renalia*. Redrawn from Crane 1990, Figure 1.

tant herbaceous lycopsid families (Lycopodiaceae, Selaginellaceae), and arborescent fossil forms (e.g., *Lepidodendron*) by placing them all in the same class ("Classis IX. Selagines"). *Psilotum* (Swartz 1801) and *Tmesipteris* (Bernhardi 1801) were also classified in the Lycopodiaceae at this time. The modern concept of lycopsids only emerged half a century later when Sykes (1908) removed these two genera from the group.

Most phylogenetic hypotheses depict lycopsids as monophyletic, and all except a few older interpretations agree that the Isoetaceae and extinct arborescent forms comprise a closely related (implicitly monophyletic) group (e.g., Axelrod 1959; Banks 1968; Chaloner 1967a; Darrah 1939, 1960; Emberger 1944, 1960, 1968; Haeckel 1868; Kräusel 1950; Lam 1948; Lignier 1908; Meyen 1987; Smith 1938; Zimmermann 1930, 1938, 1952,

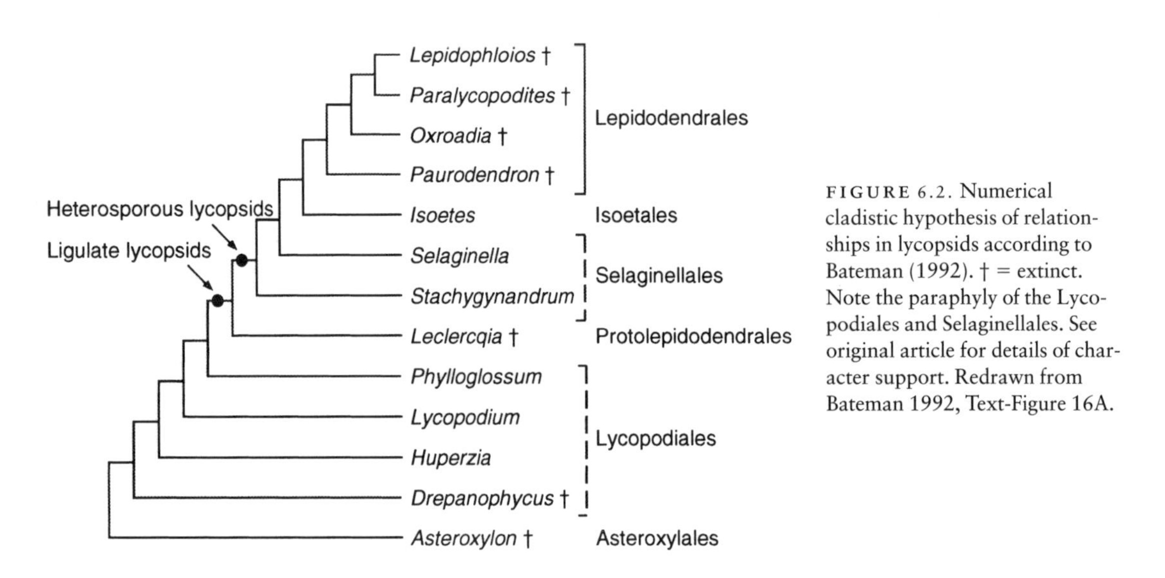

FIGURE 6.2. Numerical cladistic hypothesis of relationships in lycopsids according to Bateman (1992). † = extinct. Note the paraphyly of the Lycopodiales and Selaginellales. See original article for details of character support. Redrawn from Bateman 1992, Text-Figure 16A.

1965). Several early treatments recognized a major dichotomy corresponding to eligulate homosporous and ligulate heterosporous forms (Bower 1908, 1935; Engler and Prantl 1902; Fries 1897; Van Tieghem 1891), and this division has been upheld by many subsequent workers (e.g., Bold 1957; Emberger 1960). These ideas on relationship have received strong support from recent morphology-based cladistic studies (Bateman 1992, 1996c; Bateman, DiMichele, and Willard 1992; Crane 1990; DiMichele and Skog 1992; Gensel 1992) and molecular sequence data (Kranz and Huss 1996; Manhart 1994). In particular, the recent cladistic treatments by Bateman (1992) and Bateman, DiMichele, and Willard (1992) have clarified relationships among the important and diverse rhizomorphic clade forming the basis for a more stable classification of these groups (DiMichele and Bateman, in press).

In many recent treatments of lycopsids, there has been reluctance to circumscribe the higher-level taxa and a general unwillingness to express any phylogenetic interpretations in a supraordinal classification. Most classifications simply recognize from five to seven orders (Bateman 1992; Bateman, DiMichele, and Willard 1992; Bell and Woodcock 1971; Bierhorst 1971; Bold 1957; Bower 1935; Chaloner 1967a; Darrah 1939; Eames 1936; Emberger 1944, 1968; Gifford and Foster 1989; Pichi-Sermolli 1959; Smith 1938; Sporne 1970; Stewart 1983; Stewart and Rothwell 1993; T. N. Taylor 1981; T. N. Taylor and E. L. Taylor 1993; Thomas and Brack-Hanes 1984; Zimmermann 1930; see Table 6.1). The classifications of Meyen (1987) and DiMichele and Bateman (in press) formally acknowledge the widely accepted relationship between the extinct arborescent forms and extant Isoetaceae by placing them together in the order Isoetales.

Despite widespread acceptance of their monophyly, lycopsids are occasionally interpreted as either polyphyletic or paraphyletic to other groups. Early ideas focused on a relationship with conifers that was based on perceived similarities to extant Araucariaceae, including strobilus morphology, phyllotaxy, leaf shape, and in some taxa, stomatal distribution (Greguss 1955; Mägdefrau 1932; Seward and Ford 1906). A relationship between lycopsids and conifers is not explicitly tested here but is clearly highly unparsimonious based on character-state distributions in basal polysporangiophytes (Chapter 4) and on the results of other recent cladistic treatments of seed plants (Crane 1985; Doyle and Donoghue 1986, 1992; Nixon, Crepet, et al. 1994; Rothwell and Serbet 1994). More recently, Stewart (1983) and Stewart and Rothwell (1993) suggested an independent origin of the Lycopodiaceae and ligulate lycopsids from within vascular plants (Figure 6.3). Others have argued that the Selaginellaceae have an independent origin from plants at the charophycean level of organization, based on apical meristem morphology (Figure 6.4; Philipson 1990, 1991; see Chapter 3), and Garbary, Renzaglia, and Duckett (1993) resolved lycopsids as paraphyletic to both bryophytes and vascular plants in a cladistic analysis based on male gamete structure and gametogenesis (Chapter 3, Figure 3.10). These interpretations are evaluated against the results of our analysis in the discussion concluding this chapter.

Among the more important current areas of uncertainty in lycopsid phylogeny are relationships among the predominantly extinct arborescent clade (including *Isoetes*), extant Selaginellaceae, the Lycopodiaceae, early fossil taxa such as *Asteroxylon, Drepanophycus* and *Baragwanathia,* and extinct Protolepidodendrales. In many phylogenies, arborescent lycopsids and the Selaginellaceae are grouped as sister taxa based on heterospory and the presence of a ligule (Figures 6.1 and 6.2; Bateman 1992; Bateman, DiMichele, and Willard 1992; Crane 1990; DiMichele and Skog 1992; Lam 1948; Meyen 1987; Smith 1938; Stewart 1983; Zimmermann 1930), whereas others have grouped the Lycopodiaceae and the Selaginellaceae together based on their herbaceous habit (Axelrod 1959; Banks 1968; Chaloner 1967a; Kräusel 1950; Lam 1950). Similarly, whereas the early fossil *Asteroxylon* is sometimes included at the base of the lycopsid tree (Banks 1968; Bateman,

TABLE 6.1
Contrasting Systematic Treatments of Lycopsida

Bower 1935	Bold 1967	Chaloner 1967	Bierhorst 1971
Lycopodiales	Microphyllophyta	Lycophyta	Tracheophyta
Eligulatae	Aglossopsida	Lycopsida	Lycopsida
Lycopodium	Asteroxylates†	Drepanophycales†	Asteroxylales†
Phylloglossum	Lycopodiales	Lycopodiales	Lycopodiales
Lycopodites†	Protolepidodendrales†	Protolepidodendrales†	Protolepidodendrales†
Ligulatae	Glossopsida	Selaginellales	Selaginellales
Selaginella	Selaginellales	Lepidodendrales†	Lepidodendrales†
Lepidodendraceae†	Lepidodendrales†	Isoetales	Isoetales
Sigillariaceae†	Isoetales	Miadesmiales†	
Bothrodendraceae†	Pleuromeiales†		
Isoetes			
Pleuromeiaceae†			

Notes: Lycopsida of Stewart (1983) is acknowledged as polypheletic. † = extinct.

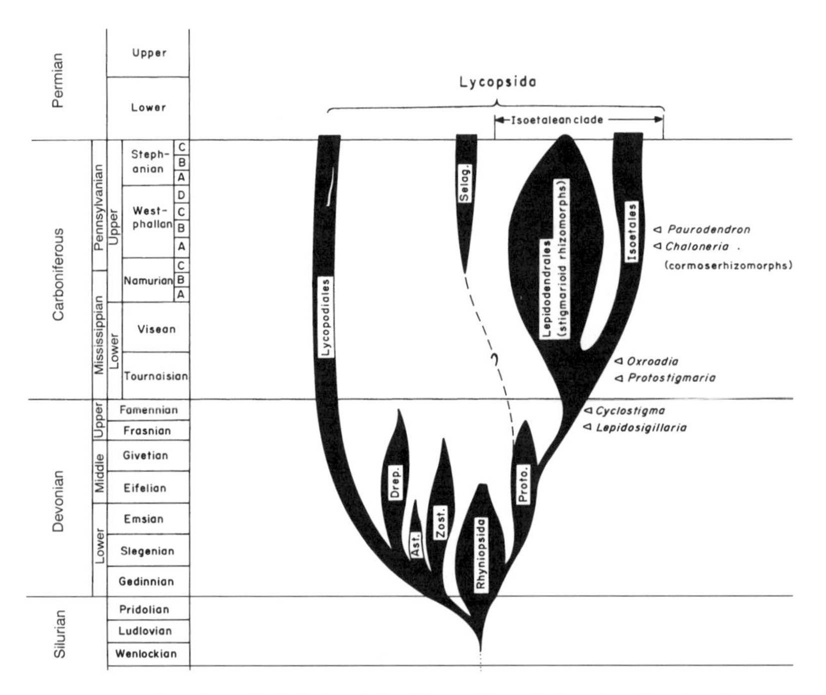

FIGURE 6.3. Putative polyphyletic origin of lycopsids and changing diversity of groups through time, according to Stewart and Rothwell (1993). The Selaginellaceae and rhizomorphic lycopsids (isoetalean clade) evolve from Protolepidodendrales and ultimately rhyniophyte ancestors. The Lycopodiaceae and early fossils such as *Asteroxylon* and *Drepanophycus* originate from ancestors similar to basal zosterophylls. Drep. = *Drepanophycus;* Ast. = *Asteroxylon;* Zost. = zosterophylls; Proto. = Protolepidodendrales; Selag. = Selaginellaceae. Reproduced from Stewart and Rothwell 1993, Chart 11.1, with the permission of Cambridge University Press.

Taylor 1981	Stewart 1983	Thomas and Brack-Hanes 1984	Meyen 1987
Lycophyta	Tracheophyta	Lycophyta	Pteridophyta
Lycopodiales	Lycopsida	Lycopsida	Lycopodiopsida
Protolepidodendrales†	Drepanophycales†	Drepanophycales†	Drepanophycales†
Selaginellales	Lycopodiales	Lycopodiales	Lycopodiales
Lepidodendrales†	Protolepidodendrales†	Protolepidodendrales†	Protolepidodendrales†
Isoetales	Selaginellales	Selaginellales	Selaginellales
Pleuromeiales†	Lepidodendrales†	Lepidocarpales†	Isoetales
	Isoetales	Isoetales	
		Miadesmiales†	

DiMichele, and Willard 1992; Crane 1990; Gensel 1992), it is often excluded from lycopsids *sensu stricto* because the leaf traces are incomplete (Banks 1968; Chaloner 1967a; Høeg 1967). There is comparable disagreement concerning the position of the Protolepidodendrales. Some authors interpret these taxa as the sister group, or part of the stem group, to heterosporous lycopsids (Crane 1990; Lam 1948; Meyen 1987; Stewart 1983; Zimmermann 1930); others interpret them as "di-

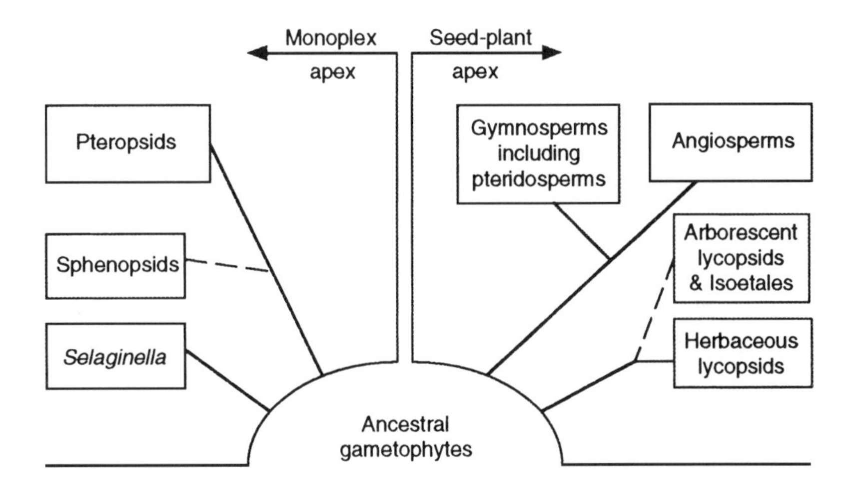

FIGURE 6.4. Philipson's schematic representation of hypothesized polyphyletic origin of the sporophyte generation in land plants. According to this hypothesis, the sporophyte generation of the Selaginellaceae, of sphenopsids and ferns (pteropsids), of the Lycopodiaceae (herbaceous lycopsids) and rhizomorphic lycopsids (arborescent lycopsids and Isoetales), and of seed plants evolved independently from green algal ancestors at the charophycean level of organization. Putative fundamental differences in the morphology of the apical meristem are cited in support of this hypothesis. Redrawn from Philipson 1990, Figure 1.

rectly ancestral" to arborescent lycopsids (Axelrod 1959; Kräusel 1950); and still others interpret the Protolepidodendrales and the Lycopodiaceae as sister taxa (Smith 1938). Most authors have been unwilling to specify a precise relationship between the Protolepidodendrales and any "higher" lycopsid group (e.g., Banks 1968; Bonamo, Banks, and Grierson 1988; Chaloner 1967a).

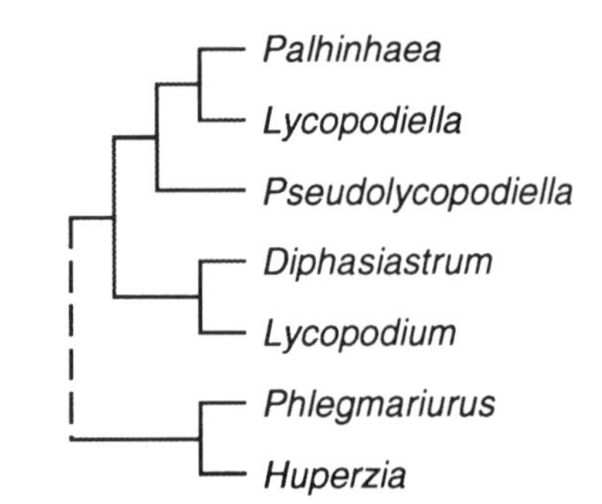

FIGURE 6.5. Numerical cladistic hypothesis of relationships among North American Lycopodiaceae, according to Wagner and Beitel 1992. The broken line indicates that the assumed monophyly of the Lycopodiaceae lacks character support. See original article for details of character support. Redrawn from Wagner and Beitel 1992, Figure 1.

PHYLOGENETIC QUESTIONS AND AIMS OF ANALYSIS

Based on the current literature, most disagreement concerning the higher-level phylogeny of lycopsids can be attributed to the approaches used to determine relationships rather than to major character conflicts. Cladistic analyses converge on similar solutions and highlight similar problematic areas. Using a relatively small set of characters (Figure 6.1), Crane (1990) made a preliminary cladistic assessment that favored the monophyly of the following groups: lycopsids, ligulate lycopsids, heterosporous lycopsids, and rhizomorphic lycopsids (arborescent lycopsids plus *Isoetes*). This is consistent with many of the ideas about the classification and phylogeny of the group that have been published over the last 120 years, including the recent brief review by DiMichele and Skog (1992). The main area of conflict between Crane's hypothesis and several previous classifications is the rejection of a herbaceous clade containing the Selaginellaceae and the Lycopodiaceae (Axelrod 1959; Chaloner 1967a; Kräusel 1950; Lam 1950).

A numerical cladistic analysis by Bateman (1992; see Figure 6.2; see also Bateman, DiMichele, and Willard 1992) found similar general relationships but questioned the monophyly of the Lycopodiaceae and Selaginellaceae. The Lycopodiaceae were resolved as paraphyletic to the ligulate clade, and at a less inclusive level, the Selaginellaceae were paraphyletic to an arborescent (woody) clade (rhizomorphic lycopsids). Furthermore, the early fossil *Baragwanathia* was found to be indistinguishable from modern *Huperzia* based on the characters coded. These results are problematic for basal relationships among lycopsids because trees were rooted on the early fossil *Asteroxylon,* which may be nested among extant herbaceous species. An alternative rooting of the Bateman tree might support the monophyly of the Lycopodiaceae but only with the inclusion of *Drepanophycus, Asteroxylon,* and *Baragwanathia.* The monophyly of the Lycopodiaceae was also shown to be questionable in a recent analysis (Wagner and Beitel 1992) based largely on extant taxa that was also rooted with reference to the early fossil *Asteroxylon.* Wagner and Beitel were also unable to find a family-level synapomorphy (Figure 6.5).

The aim of the analysis we present in this chapter is to provide an improved cladistic treatment of lycopsids by including a more acceptable configuration of fossil outgroups based on our previous analysis of zosterophylls (Chapter 5). We focus particularly on relationships among basal taxa because we are concerned primarily with clarifying the position of lycopsids with respect to other major clades of land plants, establishing major ingroup patterns, and examining the position of early fossils with respect to extant taxa. Specific aims include testing the monophyly of the Lycopodiaceae, clarifying the relationships of the Devonian taxa *Asteroxylon, Baragwanathia,* and *Drepanophycus,* and evaluating the position of the Protolepidodendrales.

TABLE 6.2
Genera and Species Examined in the Cladistic Analysis of Lycopsida

Taxon	Exemplars
Lycopodiaceae	*Huperzia selago*
	Huperzia squarrosa
	Phylloglossum drummondii
	Lycopodium sp.
	Lycopodiella sp.
Protolepidodendrales	*Minarodendron cathaysiense*†
	Leclercqia complexa†
Selaginellaceae	*Selaginella selaginoides*
	Selaginella subgenus *Tetragonostachys*
	Selaginella subgenus *Stachygynandrum*
Lepidodendrales	*Paralycopodites*†
Isoetaceae	*Isoetes* sp.
Zosterophylls	*Gosslingia breconensis*†
	Sawdonia ornata†
	Zosterophyllum llanoveranum†
	Zosterophyllum myretonianum†
Putative lycopsids	*Asteroxylon mackiei*†
	Baragwanathia longifolia†
	Drepanophycus qujingensis†
Rhyniopsida	*Rhynia gwynne-vaughanii*† *

Notes: See Appendix 1 for taxon descriptions and Appendix 5 for data matrix. † = extinct. * = outgroup.

CHOICE OF TAXA

As in the preceding analyses of embryophytes, polysporangiophytes, and zosterophylls (Chapters 3–5), our choice of lycopsid taxa is a compromise dictated by the phylogenetic questions that we are seeking to address, the quality of the existing data, and the need to include those fossils that appear to be of maximum interest with respect to the phylogenetic issues at this level. All of the taxa considered in the analysis of lycopsid relationships are listed in Table 6.2. Brief descriptions and reviews of these taxa, along with citations of the relevant literature, are provided in Appendix 1.

As an outgroup we used *Rhynia gwynne-vaughanii* (Rhyniopsida; Chapter 4, Figure 4.8) because it is clearly not an ingroup taxon, but is nevertheless sufficiently close to allow polarization of many characters (Chapter 4). We also included a range of zosterophylls in the ingroup to critically assess lycopsid monophyly and to test putative homologies with various lycopsid subgroups (Table 6.2). Silurian-Devonian fossil lycopsids were represented by *Asteroxylon, Drepanophycus* (representing the Drepanophycales), and *Baragwanathia.* Additional taxonomic diversity among lycopsids was sampled from the Selaginellaceae *(Selaginella kraussiana, S. martensii:* Figure 6.6; *S. selaginoides:* Figure 6.7), Lycopodiaceae (*Huperzia selago:* Figure 3.24; *Lycopodium clavatum:* Figure 6.8; *Phylloglossum drummondii:* Figure 6.9), Protolepidodendrales (*Leclercqia complexa:* Figure 6.10), Lepidodendrales, and Isoetaceae (*Isoetes bolanderi, I. nuttallii:* Figure 6.11).

Zosterophylls

Most zosterophylls fall into a well-defined monophyletic group comprising a large, diverse clade with planated pseudomonopodial branching and

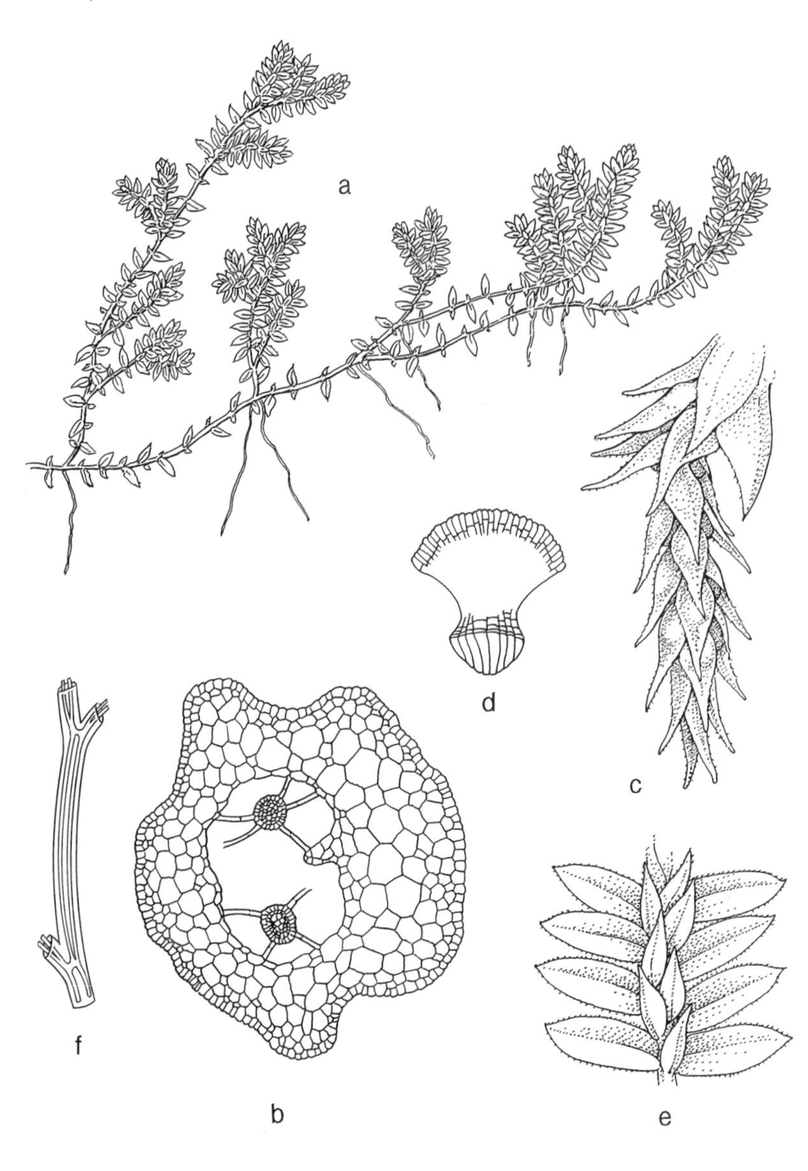

FIGURE 6.6. Rhizophoric Selaginellaceae. (**a**) *Selaginella kraussiana:* shoot system with leaves and rhizophores (×1.5). (**b**) *S. kraussiana:* details of stem anatomy showing two steles suspended in stem cavity by endodermal cells (×75). Redrawn from Smith 1938, Figure 102. (**c**) *S. kraussiana:* strobilus with larger basal vegetative leaves (×7). (**d**) *Selaginella martensii:* details of ligule morphology (×35). Redrawn from Engler and Prantl 1902, Figure 398A. (**e**) *S. kraussiana:* dimorphic leaves (×7). (**f**) *S. kraussiana:* details of vascular strand in stem (×8).

elliptical to strap-shaped steles (Sawdoniales) together with a range of *Zosterophyllum*-like stem group taxa (Chapter 5). Three of the better known taxa were chosen to represent this range of diversity: *Gosslingia breconensis* and *Sawdonia ornata* for the Sawdoniales and *Zosterophyllum llanoveranum* for the Sawdoniales stem group (Table 6.2).

Putative Lycopsids

Putative early lycopsids were represented by three of the better known early fossils—*Asteroxylon mackiei, Baragwanathia longifolia,* and *Drepanophycus qujingensis.* All three taxa appear to possess some but not all of the defining features of extant lycopsids.

Lycopodiaceae

The Lycopodiaceae are a diverse, cosmopolitan family comprising slender, dichotomously branched homosporous plants. Most species are pendant epiphytes that grow in wet tropical forests, often in mountainous areas, whereas extratropical species are almost exclusively terrestrial or rupestral and erect to prostrate in habit. Most

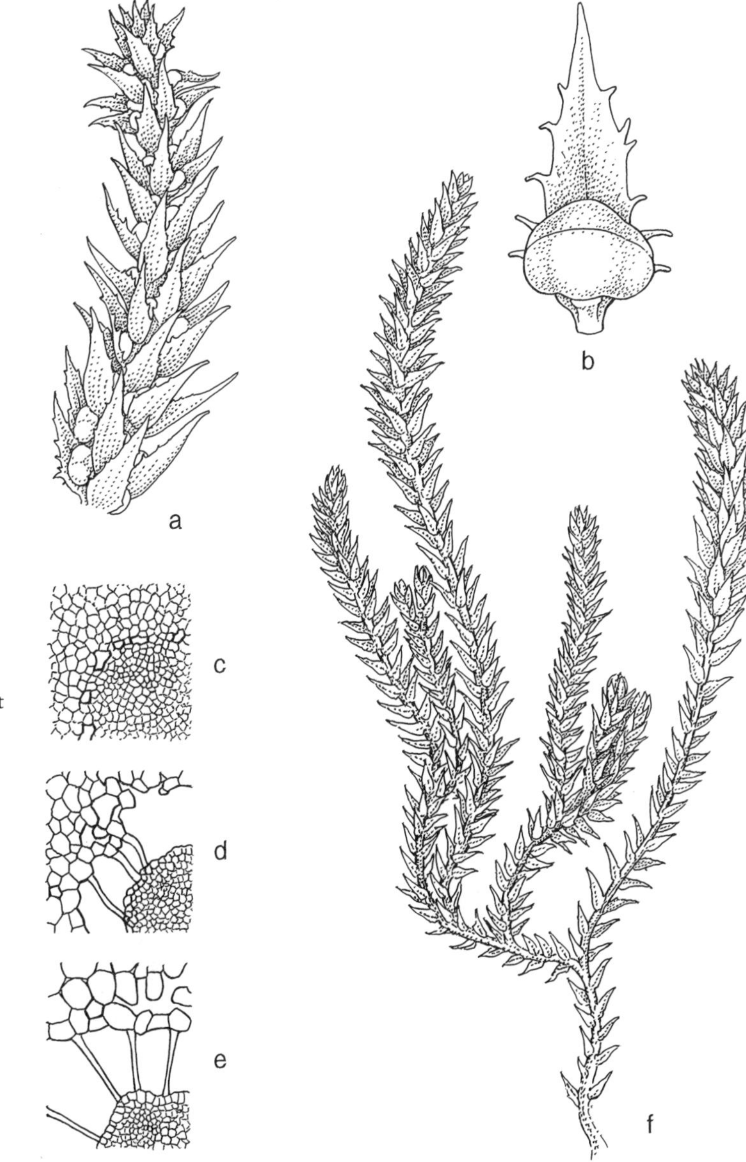

FIGURE 6.7. *Selaginella selaginoides.* (**a**) Distal strobilate zone (×2.5). (**b**) Sporophyll with megasporangium (×10). (**c, d, e**) Ontogeny of stelar cavity and suspended stele (×150). Redrawn from Wardlaw 1925, Figure 14. (**c**) Distal end of axis showing first appearance of cavity. (**d, e**) Development of stem cavity and suspended stele. (**f**) Habit sketch showing long, determinate, upright fertile axis and smaller, indeterminate basal vegetative axes (×1.3).

botanical texts recognize only two genera of homosporous lycopsids: *Lycopodium* (200–500 species) and *Phylloglossum* (1 species), but more critical treatments subdivide *Lycopodium* into several genera. Øllgaard (1987, 1990) recognized four genera *(Huperzia, Lycopodium, Lycopodiella, Phylloglossum)* and his classification is followed here. We provisionally treat *Lycopodium* and *Lycopodiella* as monophyletic groups because they appear to be relatively well defined (Øllgaard 1987); *Huperzia* is also treated as monophyletic, although it is less well circumscribed. An alternative to Øllgaard's treatment of the Lycopodiaceae was presented by Wagner and Beitel (1992), who recognized seven genera (in three subfamilies; Figure 6.5) among the North American species of the family. The relationships defined by their cladistic analysis are consistent with the monophyly of Øllgaard's four genera.

Our treatment of the genus *Huperzia* follows closely that of Øllgaard (1987). *Huperzia* is by far the most speciose genus of extant homosporous ly-

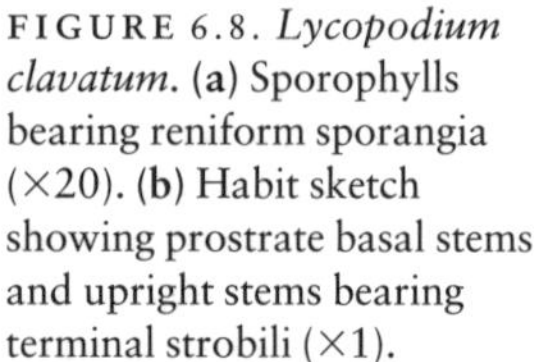

FIGURE 6.8. *Lycopodium clavatum.* (**a**) Sporophylls bearing reniform sporangia (×20). (**b**) Habit sketch showing prostrate basal stems and upright stems bearing terminal strobili (×1).

copsids, and diversity estimates range from 200 to more than 400 species. The genus is virtually cosmopolitan in distribution, but the greatest diversity is in tropical evergreen montane forests, and in the Andean paramos and subparamos of South America (Øllgaard 1987). A preliminary cladistic analysis of *Huperzia* species groups, based mainly on the Øllgaard's data (1987), is presented in Figure 6.12 as a working hypothesis for the examination of relationships among lycopsids as a whole.

We recognize a large clade of pantropical epiphytes with predominantly pendulous growth habit, which includes most *Huperzia* species. The partially resolved sister group to the epiphyte clade consists of four species groups containing erect, terrestrial species that occur predominantly in central and southern America and montane Africa. This unresolved grade includes the *H. brongniartii, H. reflexa, H. saururus,* and *H. brevifolia* species groups of Øllgaard (1987). The *H. selago* species group is basal within the genus and consists of erect, terrestrial species with a predominantly Northern

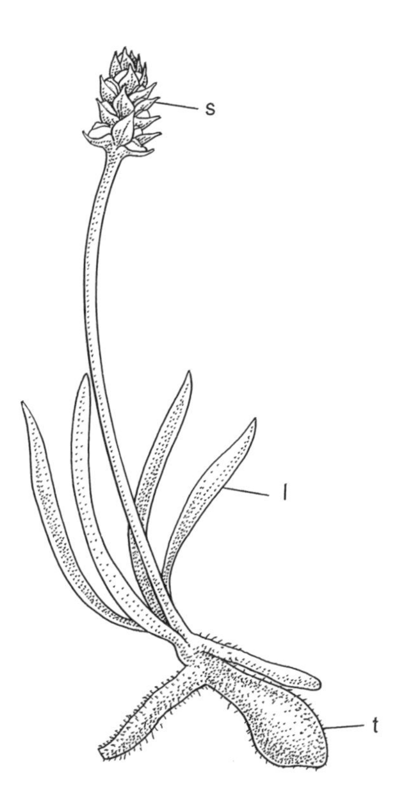

FIGURE 6.9. *Phylloglossum drummondii.* Whole plant showing subterranean tuber *(t)*, basal vegetative leaves *(l)*—strongly resembling juvenile leaves in other Lycopodiaceae—and strobilus *(s)* on a leafless stalk (×3).

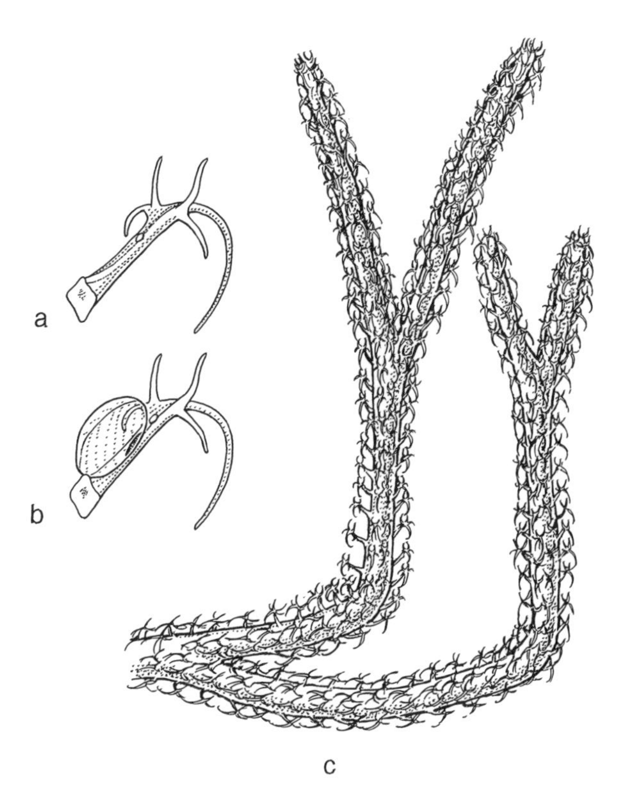

FIGURE 6.10. *Leclercqia complexa.* Drawings based on Bonamo, Banks, and Grierson 1988, Figures 57–59. (**a**) Vegetative leaf with ligule pit (×5). (**b**) Sporophyll with sporangium (×5). (**c**) Habit sketch showing distal leafy axes (×1).

Hemispheric (temperate, arctic, alpine) and Paleotropical (montane, alpine) distribution (Øllgaard 1987). We tentatively treat the genus *Huperzia* as monophyletic because of the presence of pluricellular, uniseriate hairs among the gametangia and of nonlignified cortical "sclerenchyma." In addition, the genus is defined by the possession of cortical roots that are initiated near the stem apex but grow down through the cortex to emerge proximally (Øllgaard 1987). For the purpose of our analysis, *Huperzia* was represented by two species groups: the "*Huperzia selago* group" and the "*Huperzia squarrosa* group." Our *Huperzia selago* group corresponds to *Huperzia sensu* Wagner and Beitel (1992), while the *Huperzia squarrosa* group is represented in North America by the single species *Phlegmarius phlegmarius sensu* Wagner and Beitel (Figure 6.5).

Our treatment of *Lycopodium* and *Lycopodiella* closely follows Øllgaard's (1987). *Lycopodium* consists of about 40 species, most of which have a virtually cosmopolitan distribution. We provisionally treat *Lycopodium sensu* Øllgaard as monophyletic, based on the presence of parallel bands of xylem in major branches and main roots and of microspores with a reticulate sculpture (Øllgaard 1987; Wilce 1972). Within the genus, Wagner and Beitel (1992) recognize a less inclusive *Lycopodium* as a sister group to *Diphasiastrum* (Figure 6.5). *Lycopodiella sensu* Øllgaard (1987) consists of about 40 species, most of which occur in the Americas, although some are found in almost all moist temperate and tropical regions of the world. We provisionally treat *Lycopodiella* as monophyletic because of the presence of anisovalvate sporangia (except for section *Caroliniana,* treated provisionally as a reversal) and of microspores with

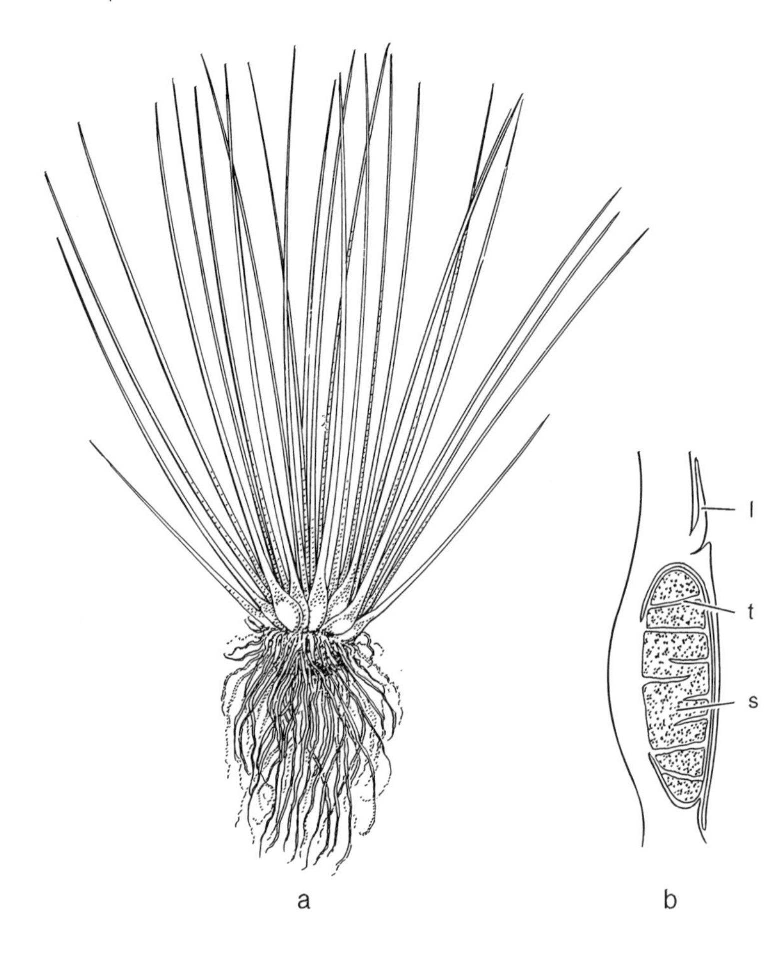

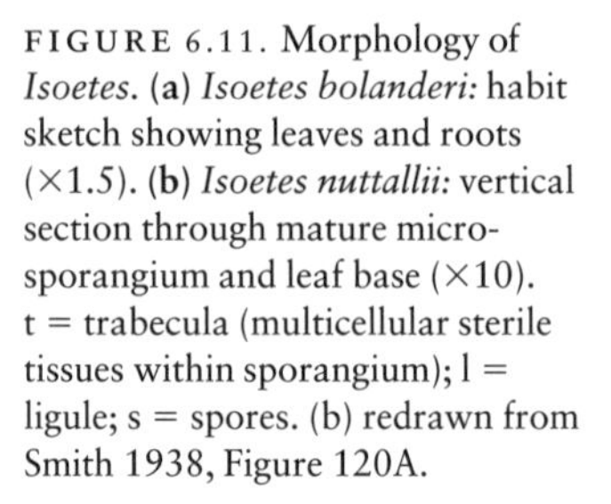

FIGURE 6.11. Morphology of *Isoetes*. (**a**) *Isoetes bolanderi:* habit sketch showing leaves and roots (×1.5). (**b**) *Isoetes nuttallii:* vertical section through mature micro-sporangium and leaf base (×10). t = trabecula (multicellular sterile tissues within sporangium); l = ligule; s = spores. (b) redrawn from Smith 1938, Figure 120A.

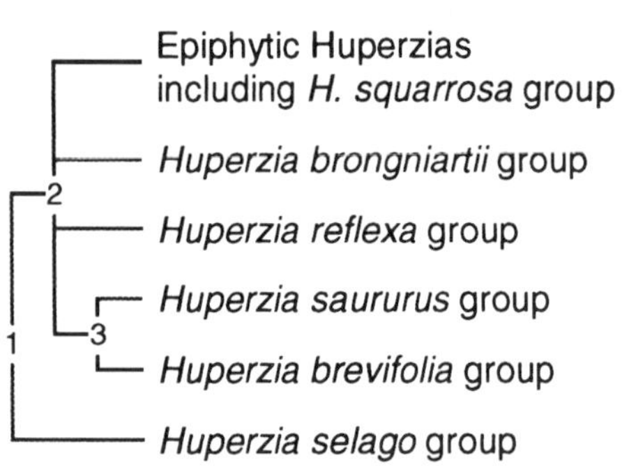

FIGURE 6.12. Preliminary, nonnumeric hypothesis of basal relationships in *Huperzia* (Lycopodiaceae), based on the species groups and biogeographic data recognized by Øllgaard (1987). Synapomorphies on nodes: node 1 → *H. selago,* (1) bulbils and (2) spore shape (triangular with truncate corners and concave sides); node 1 → node 2, lanceolate leaves; node 2 → node 3, air sac in leaf; node 2 → epiphytic *Huperzia* species, epiphytic-rupestral ecology. This hypothesis of relationship needs to be tested by more detailed analysis of both molecular and morphological data.

rugate sculpture (Øllgaard 1987; Wilce 1972). Within this concept of the genus, Wagner and Beitel (1992) recognize a less inclusive *Lycopodiella* as a sister group to *Palhinhaea*. *Pseudolycopodiella* is the sister group to the *Lycopodiella-Palhinhaea* clade (Figure 6.5).

Selaginellaceae

The Selaginellaceae are a cosmopolitan family of slender, prostrate-creeping to erect, mainly terrestrial or rupestral (rarely pendant epiphytic), heterosporous plants. Most species grow in mesic, shaded forests, but the genus also occurs in other environments, including wet primary and secondary rain forest, arctic-alpine turf, rocky deserts, and xeric scrub land (Tryon and Tryon 1982). As currently circumscribed, the family consists of one extant genus, *Selaginella* (c. 415 species), but sev-

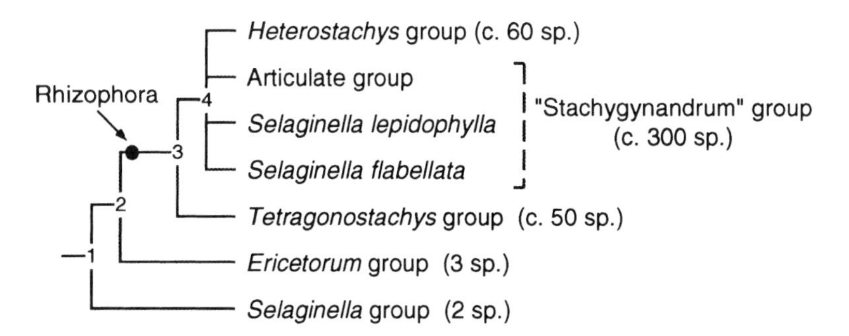

FIGURE 6.13. Preliminary, nonnumeric hypothesis of cladistic relationships in Selaginellaceae, based on taxa recognized by Jermy (1990b) (approximate species numbers in parentheses). Quotation marks = paraphyletic group. Note that 99% of extant species diversity is in the Rhizophora clade, which comprises the subgenera *Tetragonostachys, "Stachygynandrum,"* and *Heterostachys.* Possible synapomorphies of Selaginellaceae include (1) suspension of stele in stem cavity by trabeculate endodermis and (2) four megaspores per megasporangium. Synapomorphies on nodes: node 1 → subgenus *Selaginella,* (1) roots emerging from hypocotylar node, (2) new primary shoots developing from base of plant, and (3) possibly vessel elements in xylem; node 1 → node 2, sporophylls in four rows; node 2 → subgenus *Ericetorum,* decussate phyllotaxy; node 2 → node 3, (1) strap-shaped stele with lateral protoxylem, (2) prostrate habit, and (3) branched rhizophores; node 3 → node 4, (1) dimorphic leaves, (2) planar branching, and (3) possibly polystelic anatomy; node 4 → Articulate group, (1) articulations (swellings) just below stem dichotomies, (2) strobilus with single basal megasporangium, (3) possibly megaspores with opalescent iridescence (caused by pattern of sporopollenin units in megaspore wall), and (4) specialized ejector mechanism for spore dispersal; node 4 → subgenus *Heterostachys,* (1) dimorphic sporophylls and (2) pteryx. This hypothesis of relationship needs to be tested by more detailed analysis of both molecular and morphological data.

eral subgeneric schemes have been proposed recently (Jermy 1990b; Thomas and Quansah 1991; Valdespino 1992). We follow Jermy (1990b) and recognize five groups: *Selaginella, Ericetorum, Tetragonostachys, Stachygynandrum, and Heterostachys.* A preliminary working cladistic hypothesis for the group (nonnumeric) is presented in Figure 6.13.

The *Stachygynandrum* group (c. 300 species) and the *Heterostachys* group (c. 60 species) have a pantropical distribution and occur predominantly in lowland to midmontane primary rain forest (Jermy 1990b). A few species are adapted for seasonal drought and live in cooler temperate regions (e.g., *S. denticulata, S. lepidophylla*). These taxa are probably nested within tropical clades (Figure 6.13). The *Stachygynandrum* group is characterized by elaborate, planated, compound branching systems with clearly dimorphic vegetative leaves in four rows. The *Heterostachys* group is similar to the *Stachygynandrum* group but has the additional features of dimorphic sporophylls, complanate strobili, and a characteristic folding of the sporophyll lamina that partially encloses the sporangium (*pteryx* of Quansah and Thomas 1985). The *Tetragonostachys* group (c. 50 species) is widely distributed in seasonally dry tropical areas, but a few species occur in temperate mesophytic woodlands of the Northern Hemisphere. Many taxa are characterized by root formation throughout the stems, which is associated with a mat-forming habit, and thick, cutinized leaves with fine hair points, which protect the shoot apex. We provisionally treat *Tetragonostachys* as monophyletic, but *Stachygynandrum* appears to be paraphyletic to *Heterostachys* (Figure 6.13).

The *Ericetorum* group (3 species) comprises two Australian and one southern African species that occur in proteaceous heathlands or equivalent habitats. The group is characterized by a creeping stem and decussately arranged lower leaves. The *Selaginella* group (2 species) comprises one widely distributed species having a circumboreal distribution in the Northern Hemisphere *(S. selaginoides)* and a maximum southern range of approximately 30°N and one tropical species *(S. deflexa)* confined to the Hawaiian archipelago. This group is characterized by primary shoots that originate at the base of the plant and by roots attached to a small swollen region at the base of the hypocotyl (Karrfalt 1981; Rothwell and Erwin 1985). The roots originate endogenously within the cortex, and typically not more than eight are produced. Our preliminary analysis suggests that the *Selaginella* group is a sister group to all other Selaginellaceae (Figure 6.13).

For our numerical analysis we selected three subgenera to represent diversity within *Selaginella.* The *Stachygynandrum* subgenus is used as an exemplar for a putative *Stachygynandrum-Heterostachys* clade. Our preliminary assessment of character distributions within the Selaginellaceae suggests that the *Stachygynandrum* group is paraphyletic to *Heterostachys* and that together these taxa form a large, well-defined clade (Figure 6.13). Also included in our analysis is the subgenus *Tetragonostachys,* which we provisionally treat as monophyletic. We also include the *Selaginella* group, represented by *S. selaginoides,* because of its probable basal position within the family and its divergent morphology.

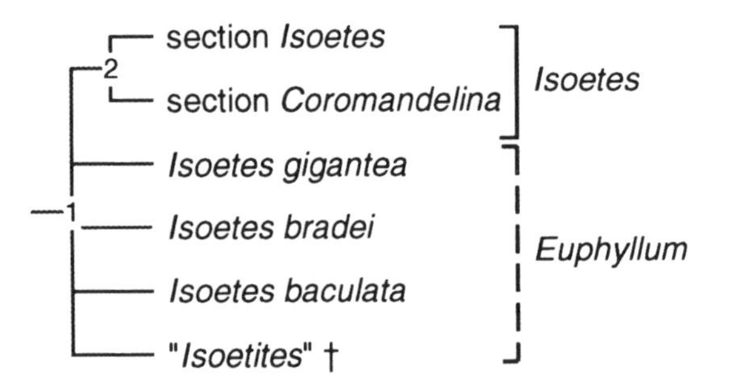

FIGURE 6.14. Preliminary, nonnumeric hypothesis of cladistic relationships in the Isoetaceae. Compiled from data in Hickey 1986, 1990; Jermy 1990a; and Taylor and Hickey 1992. † = extinct; quotation marks = paraphyletic group. Possible synapomorphies of Isoetaceae include (1) lacunate leaves and (2) reduced apical growth. Other potential synapomorphies of the Isoetaceae that may be more general in rhizomorphic lycopsids include (1) a velum, (2) indehiscent sporangia, (3) multiflagellate sperm, (4) trabeculae (sterile strands of tissue within the sporangium), and (5) the sporangium embedded in the leaf. Synapomorphies on nodes: node 1 → node 2, (1) restriction of leaf lamina (ala) at distal end and (2) fibrous bundles in leaf; node 2 → section *Coromandelina,* protective papyraceous leaf bases; node 2 → section *Isoetes,* (1) scales (arrested leaf primordia, lost in some species), (2) sporangium pigmentation, and (3) phyllopodia (sclerified remains of leaf bases, lost in some species). This hypothesis of relationship needs to be tested by more detailed analysis of both molecular and morphological data.

Isoetaceae

The Isoetaceae are a small (c. 150 extant species), distinctive, and divergent family of vascular plants that have a cosmopolitan distribution (Bierhorst 1971; Hickey 1986, 1990; Jermy 1990a; Taylor and Hickey 1992) but are confined to habitats where the soil is saturated with water for at least part of the year. A provisional outline of the phylogenetic relationships within the family is given in Figure 6.14. The monophyly of the Isoetaceae is supported by the following characters: (1) sporangia embedded in the leaf base, (2) a *velum*—a thin membrane covering the sporangium, (3) indehiscent sporangia, (4) trabeculae—sterile strands of tissue transversing the sporangia, (5) multiflagellate sperm, a trait convergent with that of other non-lycopsid vascular plants, (6) lacunate leaves, and (7) reduced axial growth (Taylor and Hickey 1992). Based on the synapomorphy of lacunate leaves, Hickey (1986) interpreted the family as primitively aquatic. We follow Taylor and Hickey (1992) in recognizing one extant genus—*Isoetes*—and two subgenera. The subgenus *Euphyllum* comprises plesiomorphic Isoetaceae with fully laminate (alate) leaves (Hickey 1990). Taxa in *Euphyllum* include fossils in the genus *Isoetites* plus extant *Isoetes baculata, I. bradei,* and *I. gigantea.* The remaining taxa comprise a much larger clade, the subgenus *Isoetes,* which is characterized by the restriction of the leaf lamina to the proximal portion of the leaf and by leaves with longitudinal fibrous bundles (Taylor and Hickey 1992). The subgenus *Isoetes* is further divided into two sections: *Isoetes* and *Coromandelina. Isoetes* is characterized by the ability to produce sporangium pigmentation, the presence of scales (arrested leaf primordia), and the presence of *phyllopodia* (sclerified leaf bases that remain attached to the base of the plant after the distal portion of the leaf has eroded). *Isoetes* contains most extant species and has a cosmopolitan distribution. The predominantly Indian subgenus *Coromandelina* comprises nonsclerotic taxa and is characterized by protective, light brown, papyraceous leaf bases. As currently defined, the subgenus *Euphyllum* is problematic because it is probably paraphyletic to the subgenus *Isoetes.* Hickey (1986) and Taylor and Hickey (1992) do not recognize the genus *Stylites* (Amstutz 1957; Rauh and Falk 1959) because it is considered to represent a single species within the subgenus *Isoetes,* section *Isoetes.*

The fossil record of the Isoetaceae comprises approximately 16 megafossil species and over 30 occurrences ranging from the Triassic Period to the upper Tertiary (Pigg 1992). The fossil record of megaspores is also extensive, but unequivocal assignment to the Isoetaceae on the basis of spore morphology is not possible because of general similarities to those of the arborescent lycopsids and the Selaginellaceae (Collinson 1991; Kovach 1989; Kovach and Batten 1989, 1993). Tuberculate megaspores are considered to be plesiomorphic (Taylor and Hickey 1992). Unlike all other extant lycopsids, the Isoetaceae produce secondary xylem and are generally regarded as the closest living relatives of the extinct arborescent groups of the late Paleozoic. Cladistic analysis strongly supports a close relationship between the Isoetaceae and the extinct arborescent lycopsids (Bateman, DiMichele, and Willard 1992; Crane 1990).

We follow Meyen (1987) and treat the Isoetaceae within a monophyletic arborescent group (Isoetales) that includes the Sigillariaceae, the Diaphorodendraceae, and the Lepidodendraceae (*sensu* Bateman [1992]; Bateman, DiMichele, and Willard 1992). Arborescent fossil taxa are represented by the plesiomorphic *Paralycopodites* (Bateman 1992; Bateman, DiMichele, and Willard 1992; DiMichele and Bateman, in press; Pearson 1986; Phillips and DiMichele 1992). The Protolepidodendrales is an exclusively fossil group recognized by previous authors and generally placed close to the arborescent clade (Bateman 1992) but includes very few well-known taxa. We adopt a conservative approach that does not accept the Protolepidodendrales as an a priori monophyletic group but represents them by the relatively well known taxa *Leclercqia complexa* and *Minarodendron cathaysiense* (Appendix 1).

CHARACTER DESCRIPTIONS AND CODING

Characters Scored for Almost All Taxa

6.1 • CIRCINATE VERNATION Many zosterophylls develop by expansion from a coiled apical region in a manner analogous to extant ferns. This type of development can be observed as circinate vernation (Figure 5.18) in the distal regions of the plant. Circinate vernation does not occur in lycopsids.

The presence or absence of circinate vernation is treated as a binary character: 0 = circinate vernation absent; 1 = circinate vernation present. There were no inapplicable cases; one was missing data (see Appendix 5). See characters 4.8 and 5.4 for comparison.

6.2 • BRANCHING TYPE Polysporangiophytes often exhibit several orders of branching (Chapter 4). A dichotomy resulting in daughter axes of approximately equal diameter (isotomous) is the general condition for the group. Within the category of isotomous branching, we also include provisionally anisotomous systems in which the wider of the two daughter axes is not structurally dominant, as in pseudomonopodial systems (e.g., *Lycopodium, Lycopodiella;* see below). In many taxa, the dichotomy of the larger axes is strongly anisotomous, and the wider axis is almost straight and overtops narrower laterals (pseudomonopodial). Pseudomonopodial branching is typical of the Sawdoniales (Zosterophyllopsida: Chapter 5) and of some lycopsids, including the fossils *Minaro-*

dendron and *Drepanophycus* and many Selaginellaceae (e.g., *Stachygynandrum*).

Branching type is treated as a binary character: 0 = more or less isotomous branching predominating; 1 = pseudomonopodial branching. There were no inapplicable cases; none were missing data (see Appendix 5). See characters 3.4, 4.4, 4.5, and 5.1 for comparison.

6.3 • BRANCHING PATTERN Significant planation of aerial branch systems is characteristic of many Zosterophyllopsida and of some groups within the Selaginellaceae, such as the subgenus *Stachygynandrum* (but not the subgenus *Selaginella*). In the Selaginellaceae, this growth form is often associated with recumbent taxa (Jermy 1990b), but in the Zosterophyllopsida, the details of growth habit are poorly understood. The relationship between planar branching in the Selaginellaceae and the Zosterophyllopsida is tested in this analysis by scoring these two characters as homologous.

Branch orientation is treated as a binary character: 0 = nonplanar branching; 1 = planated aerial axes. There were no inapplicable cases; none were missing data (see Appendix 5). See characters 4.5 and 5.2 for comparison.

6.4 • SUBORDINATE BRANCHING Subordinate branching in the Zosterophyllopsida occurs with the production of smaller, usually less differentiated axes in association with major dichotomies of the main branching system (see character 5.3). These axes have stemlike features, such as circinate vernation and spines in the spiny taxa. Subordinate branches in the Zosterophyllopsida generally remain in the circinate, unexpanded state, but they are known to develop into typical aerial branches in some taxa.

Subordinate branching in the Zosterophyllopsida shares positional similarities with the rhizophore typical of many Selaginellaceae. In addition to primary root formation at the base of the plant during early ontogeny, most Selaginellaceae produce an additional rooting structure *(rhizophore)* that develops *exogenously* from points along the aerial stem (Figure 6.6). Rhizophores often arise in stem dichotomies (ventrally, dorsally, or both), but may form only at the base of the plant or throughout the stem (Bierhorst 1971; Bower 1935; Hieronymus 1901; Jermy 1990b; Jernstedt, Cutter, et al. 1992). The rhizophore may remain unbranched or branch dichotomously once or twice until contact with the soil induces *endogenous* root formation. Anatomically, the rhizophores in some taxa (e.g., *S. kraussiana*) possess the distinctive crescent-shaped, bilaterally symmetrical stele typical of roots of the Lycopodiaceae and the Selaginellaceae (see characters 6.30–6.32). In other taxa the stele is terete, centrarch, and radially symmetrical (Gifford and Foster 1989, 130, Figure 9-22B). Within the Selaginellaceae, rhizophores occur in such taxa as the *Tetragonostachys* and *Stachygynandrum* groups (Jermy 1990b) but are absent in the *Selaginella* group (Karrfalt 1981). The exogenous origin from the aerial stem, the initial absence of a root cap, and the occasional differentiation of rhizophores into leafy shoots have led to much discussion over the homologies of this structure in the context of the traditional morphological distinction between "root" and "stem" (e.g., Bierhorst 1971; Gifford and Foster 1989; Imaichi and Kato 1991; Webster 1992). The relationship between subordinate branching in the Zosterophyllopsida and rhizophore formation in the Selaginellaceae is tested in this analysis by scoring both structures as homologous.

The presence or absence of subordinate branching is treated as a binary character: 0 = subordinate branching absent; 1 = subordinate or rhizophoric branching present. There were no inapplicable cases; none were missing data (see Appendix 5). See characters 4.6 and 5.3 for comparison.

6.5 • BULBILS Small leafy branches *(bulbils* or *gemmae)* are characteristic of the *H. selago* species group within extant Lycopodiaceae (Figure 3.24; Øllgaard 1987). Bulbil formation strongly resembles leaf formation in position and in the development of the xylem strand (Stevenson 1976). Similar structures have also been observed in the early fossil lycopsids *Asteroxylon* from the Rhynie Chert (Kidston and Lang 1920b, 657), *Baragwanathia* ("lateral buds": Hueber 1983, 69), and *Drepanophycus* (Li and Edwards 1995). We score this character as provisionally homologous in *Huperzia* and in the early fossil lycopsids, based on positional, morphological, and developmental similarities.

The presence or absence of bulbils is treated as a binary character: 0 = bulbils absent; 1 = bulbils present. There were no inapplicable cases; none were missing data (see Appendix 5).

6.6 • MICROPHYLLS We define the microphyll as a nonfertile stem outgrowth that influences the differentiation of vascular tissue. This character can be observed in well-preserved compression fossils and permineralizations. Typically, microphylls in homosporous lycopsids have a simple vascular strand within the leaf, but in the fossil *Asteroxylon,* leaf-trace differentiation stops at the leaf base (Hueber 1992; Kidston and Lang 1920b). A putative homology between the nonvascular, multicellular spines of some Zosterophyllopsida (Sawdoniales) and the vascular microphylls of lycopsids was rejected in an earlier analysis (Chapter 5, "Experiments").

The presence or absence of microphylls is treated as a binary character: 0 = microphylls absent; 1 = microphylls present. There were no inapplicable cases; none were missing data (see Appendix 5). See characters 4.9, 5.5, and 6.7 for comparison.

6.7 • LEAF VASCULATURE The presence of a single vascular strand within the leaf is characteristic of lycopsid microphylls (Bierhorst 1971; Gifford and Foster 1989). This feature can often be observed in well-preserved compression fossils and permineralizations. Exceptions to the single, unbranched leaf trace are found in the fossil *Asteroxylon,* in which full differentiation of vascular tissue stops at the leaf base (Hueber 1992; Kidston and Lang 1920b) and rarely in certain Selaginellaceae, in which branched venation can occur (Wagner, Beitel, and Wagner 1982). The occurrence of branched venation within some Selaginellaceae is viewed by Wagner, Beitel, and Wagner (1982) as independently derived within this clade.

Leaf vasculature is treated as a simple, binary character: 0 = partial; 1 = full. This character is inapplicable in taxa without microphylls. There were five inapplicable cases; none were missing data (see Appendix 5). See characters 4.9, 5.5, and 6.6 for comparison.

6.8 • PHYLLOTAXY Helical phyllotaxy predominates in lycopsids (Bierhorst 1971; Gifford and Foster 1989), although Fibonacci series are absent (W. E. Stein, pers. comm. to P.R.C., 1995) and the helixes are prone to perturbation that are developmentally related to branching (Bateman 1992). This character is relatively easily observed in fossil material. Departures from a helical arrangement are found among extant homosporous lycopsids in certain sections of *Lycopodium* and *Lycopodiella* (Øllgaard 1987), in some sections of Selaginellaceae, and in some arborescent lycopsids (Phillips 1979). In these cases the whorled condition is common, but ranked or decussate phyllotaxies occur also in some Lycopodiaceae (Øllgaard 1987) and Selaginellaceae (Jermy 1990b).

Phyllotaxy is treated as a binary character: 0 = helical; 1 = 4-rowed. This character is inapplicable in taxa without microphylls. There were five inapplicable cases; none were missing data (see Appendix 5). See character 5.6 for comparison.

6.9 • LEAF SHAPE Lycopsid microphylls characteristically are simple, unbranched, and entire (Bierhorst 1971; Gifford and Foster 1989; Øllgaard 1987), and their size varies enormously, from small and scalelike in extant *Lycopodium densum* (Bierhorst 1971) to over a meter in length in some extinct arboreous species (T. N. Taylor 1981). Certain extinct herbaceous taxa are characterized by leaves that are sagittate (e.g., *Haskinsia*) or once or twice forked (e.g., *Leclercqia, Colpodexylon*). Careful preparation of fossil material is required in order to obtain an accurate reconstruction of leaf shape (Bonamo, Banks, and Grierson 1988; Edwards and Benedetto 1985; Hueber 1992; Li and Edwards 1995).

Leaf shape is treated as a binary character: 0 = simple; 1 = forked. This character is inapplicable in taxa without microphylls. There were five inapplicable cases; none were missing data (see Appendix 5).

6.10 • ANISOPHYLLY Isophylly is typical of most lycopsids, but distinctive dimorphic leaves are characteristic of most species of Selaginellaceae (Figure 6.6). *Anisophylly* is the occurrence of vegetative leaves of different sizes at the same point on the stem. Usually this is characteristic of plagiotropic branches, which branch in a single plane and on which leaves on the upper side are smaller than those on the lower side. Anisophylly is typical (over at least part of the aerial system) of the *Tetragonostachys* and *Stachygynandrum* groups in *Selaginella,* but isophylly is characteristic of the *Selaginella* group (Jermy 1990b). Isophylly is typical of most species of Lycopodiaceae, but anisophylly occurs in four sections of *Lycopodium* (*Diphasiastrum sensu* Wagner and Beitel [1992]) and in one section of *Lycopodiella* (*Pseudolycopodiella sensu* Wagner and Beitel [1992]) (Øllgaard 1987). In both cases, parsimony analysis suggests that anisophylly in these taxa evolved independently (Wagner and Beitel 1992). We score both *Lycopodium* and *Lycopodiella sensu* Øllgaard (1987) as plesiomorphically isophyllous.

The presence or absence of anisophylly is treated as a binary character: 0 = anisophylly absent; 1 = anisophylly present. This character is inapplicable in taxa without microphylls. There were five inapplicable cases; none were missing data (see Appendix 5).

6.11 • LIGULE The *ligule* is a small, leaflike structure situated on the adaxial side of vegetative leaves and sporophylls (Figure 6.6; Bierhorst 1971; Gifford and Foster 1989). Once thought characteristic only of heterosporous lycopsids, a ligule has recently been documented in the homosporous fossil lycopsid *Leclercqia complexa* (Bonamo, Banks, and Grierson 1988; Grierson and Bonamo 1979). The ligule is difficult to observe in compression fossils, but in arboreous forms that shed their leaves, a ligule pit is sometimes observable on stem compressions just above the leaf scar. The ligule in the herbaceous *L. complexa* differs from that in other lycopsids in being further out on the adaxial surface of the leaf (i.e., more distant from the stem; Bonamo, Banks, and Grierson 1988). In this position the ligule is difficult to observe in most compression material, and therefore the unequivocal absence of a ligule is often not possible to establish. We score the ligule as unknown in species described from compression material. For well-preserved permineralized fossils such as *Asteroxylon* (Kidston and Lang 1920b) and *Baragwanathia abitibiensis* (Hueber 1983), we are more confident that the ligule is truly absent.

The presence or absence of a ligule is treated as a binary character: 0 = ligule absent; 1 = ligule present. There were no inapplicable cases; two were missing data (see Appendix 5).

6.12 • MUCILAGE CANALS Cavities containing mucilage are common in the leaves and sporophylls of certain extant homosporous lycopsids (Bruce 1976a; Øllgaard 1987). These cavities are either elongate and situated on the underside of the vein in leaves and sporophylls, or they form an extensive cylinder in the outer cortex of sporophylls. In extant lycopsids, such cavities appear to be confined to the homosporous genera *Lycopodium* and *Lycopodiella* (Øllgaard 1987). No mucilage canals have been observed in anatomically well-preserved homosporous fossil lycopsids such as *Baragwanathia abitibiensis* (Hueber 1983) and *Asteroxylon mackiei* (Kidston and Lang 1920b).

The presence or absence of mucilage canals is treated as a binary character: 0 = mucilage canals absent; 1 = mucilage canals present. This character is inapplicable in taxa without microphylls. There were five inapplicable cases; two were missing data (see Appendix 5).

6.13 • SPORANGIUM SHAPE Sporangia in the Lycopsida are characteristically reniform. Even in taxa with radially elongate sporangia, such as *Isoetes,* the shape is reniform in tangential section. In arborescent taxa, the shape of the megasporangium in transverse section ranges from approximately circular to strongly flattened in either horizontal or vertical planes (Bateman, DiMichele, and

Willard 1992). In the megasporangia of heterosporous groups, such as the Selaginellaceae, the growth of the megaspores distorts sporangium shape during ontogeny. A distinctive fusiform sporangium shape is the general condition in polysporangiophytes (Chapter 4).

Sporangium shape is treated as a binary character: 0 = fusiform; 1 = reniform. There were no inapplicable cases; none were missing data (see Appendix 5). See characters 4.27 and 5.13 for comparison.

6.14 • SPORANGIUM ATTACHMENT In zosterophylls and lycopsids the sporangia are attached to strobilate branches laterally by short stalks. Terminal sporangium attachment on longer axes is the general condition in polysporangiophytes (Chapter 4).

Sporangium attachment is treated as a binary character: 0 = terminal; 1 = lateral-stalked. There were no inapplicable cases; none were missing data (see Appendix 5). See characters 4.23 and 5.18 for comparison.

6.15 • SPOROPHYLL The association of a single sporangium with a sporophyll is a unique feature of lycopsids (Gifford and Foster 1989). Usually, the sporangium is axillary to the sporophyll or positioned on the basal part of its adaxial surface, but in some taxa it is well out on the leaf (e.g., *Leclercqia complexa*). In extant lycopsids, sporangium ontogeny usually begins with surface initials in the axil of the leaf (Bierhorst 1971; Bower 1894), but in some Selaginellaceae, the sporangium develops from a group of cells on the stem immediately above those that give rise to the leaf itself (Bower 1894). In certain early fossils such as *Asteroxylon* (Lyon 1964) and *Drepanophycus* (Li and Edwards 1995; Schweitzer 1980c), sporangia clearly are cauline and do not appear to be associated with specific "sporophylls," but in others, such as *Baragwanathia,* the condition is unclear (Lang and Cookson 1935). It is uncertain whether the cauline sporangia of such plants as *Asteroxylon* are loosely associated with a leaf or completely disassociated from leaf phyllotaxy (Hueber 1992; Li and Edwards 1995). Cauline sporangium attachment is the general condition within the Zosterophyllopsida (Chapter 5).

The presence or absence of a sporophyll is treated as a binary character: 0 = sporophyll absent; 1 = sporophyll present. There were no inapplicable cases; one was missing data (see Appendix 5).

6.16 • SPOROPHYLL SHAPE Some lycopsids show modifications of sporophyll morphology into peltate or subpeltate structures. This type of modification is most marked in certain groups within the Lycopodiaceae (Øllgaard 1987, 1990; Wagner and Beitel 1992) and especially in arboreous lycopsids (Figure 6.8; Bateman, DiMichele, and Willard 1992; Phillips and DiMichele 1992). In the Lycopodiaceae, all taxa have subpeltate or peltate sporophylls except *Huperzia* and two sections within *Lycopodium* (*Pseudodiphasium* and *Pseudolycopodium;* Øllgaard 1987). Sporophylls or microphylls associated with sporangia strongly resemble vegetative leaves in other homosporous fossils, such as *Asteroxylon* and Protolepidodendrales, and in the Isoetaceae. In the Selaginellaceae, the sporophylls strongly resemble vegetative leaves except for occasional small size differences (Jermy 1990b). In *Heterostachys,* folding of the microphyll lamina in some of the larger sporophylls partially encloses the sporangium, forming the so-called pteryx (Quansah and Thomas 1985), but otherwise, the sporophylls resemble vegetative leaves. The most highly modified sporophylls occur in the compact cones of arborescent lycopsids and these are sometimes strongly peltate (Phillips and DiMichele 1992).

Sporophyll shape is treated as a binary character: 0 = more or less unmodified; 1 = subpeltate to peltate. This character is inapplicable in taxa without sporophylls. There were five inapplicable cases; none were missing data (see Appendix 5).

6.17 • SPORANGIOTAXIS There are several readily recognizable patterns of sporangium arrangement among the zosterophylls and lycopsids that have lateral sporangium attachment. Two-rowed arrangements characterize many of the Zosterophyllopsida, whereas helical arrangements are the general condition within lycopsids and many *Zosterophyllum*-like plants (Chapter 5). Sporangiotaxis may be linked to phyllotaxis in many lycopsids but this may not be the case in such early fossils as *Asteroxylon* and *Drepanophycus* (see Hueber 1992).

Sporangiotaxis is treated as an unordered multistate character with three states: 0 = helical; 1 = 2-rowed; 2 = 4-rowed. This character is inapplicable in taxa without lateral sporangia. There was one inapplicable case; two were missing data (see Appendix 5). See characters 4.22 and 5.10 for comparison.

6.18 • SPORANGIUM VASCULATURE The presence of a vascular strand to the sporangium is a potentially useful feature with a poorly understood systematic distribution. Vascularized sporangia have been observed in *Zosterophyllum myretonianum* (Lang 1927), *Nothia* (El-Saadawy and Lacey 1979a), *Gumuia* (Hao 1989a), and *Trichopherophyton* (Lyon and Edwards 1991), but the condition in other zosterophylls is unknown. Vascularization of the sporangium also occurs in the early lycopsid fossils *Asteroxylon* (Lyon 1964) and *Drepanophycus* (Li and Edwards 1995). In extant lycopsids the sporophyll is vascularized, but the sporangium is not directly supplied with vascular tissue.

The presence or absence of sporangium vasculature is treated as a binary character: 0 = sporangium vasculature absent; 1 = sporangium vasculature present. There were no inapplicable cases; five were missing data (see Appendix 5). See character 5.26 for comparison.

6.19 • SPOROPHYLL STRUCTURE Sporophylls are more or less identical to vegetative leaves in many lycopsids, including in such early fossils as *Leclercqia* (Bonamo, Banks, and Grierson 1988), *Colpodexylon,* (Banks 1944) and *Minarodendron* (Li 1990). Similarly, in *Selaginella* and *Isoetes,* sporophylls resemble vegetative leaves (Jermy 1990a,b). Marked departures from the prevalent unmodified condition occur in some extant homosporous forms and in some extinct heterosporous arborescent taxa. In *Phylloglossum, Lycopodium,* and *Lycopodiella,* sporophylls are highly differentiated ephemeral organs that are peltate, subpeltate, or paleate. This condition is treated as "ephemeral" (Øllgaard 1987). In the heterosporous arborescent lycopsids of the *Lepidostrobus, Mazocarpon, Achlamydocarpon,* and *Lepidocarpon* types, the sporophyll is typically modified into an abaxial projecting heel and an adaxially extended, distal lamina (Phillips 1979). This condition is treated here as "complex." In *Heterostachys* (Selaginellaceae), sporophylls are folded, forming a pteryx (Quansah and Thomas 1985).

Sporophyll structure is treated provisionally as an unordered multistate character with three states: 0 = simple; 1 = ephemeral; 2 = complex. This character is inapplicable in taxa without sporophylls. There were five inapplicable cases; none were missing data (see Appendix 5).

6.20 • SPORANGIUM SYMMETRY Marked dorsiventral flattening of the sporangium is the general condition within zosterophylls and lycopsids, whereas strong radial symmetry is more general in polysporangiophytes (Chapter 4). The sporangium of extant homosporous lycopsids characteristically is radially flattened at maturity (Bierhorst 1971), and the shape of the sporangia is similar in such early homosporous fossils as *Asteroxylon* (Lyon 1964), *Drepanophycus* (Kräusel and Weyland 1935), and the Zosterophyllopsida (Chapter 5). The microsporangium in the Selaginellaceae is more spherical, having a tendency toward radial symmetry, and the shape of the megasporangium is influenced by the enlarging megaspores (Bierhorst 1971). In *Isoetes* (Bower 1894), arborescent lycopsids (Bower 1894; Chaloner 1967a; Phillips 1979), and Protolepidodendrales such as *Colpodexylon* (Banks 1944), *Minarodendron* (Li 1990), and *Leclercqia* (Banks, Bonamo, and Grierson 1972; Bonamo, Banks, and Grierson 1988), the sporangium exhibits marked radial extension (Figures 6.10 and 6.11).

Sporangium symmetry is treated as an unordered multistate character with three states: 0 = radial; 1 = flattened; 2 = elongated. There were no inapplicable cases; none were missing data (see Appendix 5). See characters 4.28 and 5.14 for comparison.

6.21 • SPORANGIUM DEHISCENCE Dehiscence of the sporangium into two valves is a common feature of lycopsids. In homosporous lycopsids, such as the Lycopodiaceae (Bierhorst 1971), *Asteroxylon* (Lyon 1964), *Baragwanathia* (Lang and Cookson 1935), and *Drepanophycus* (Li and Edwards 1995; Schweitzer 1987), sporangia dehisce into two valves along a line that is tangential to the stem. Typically the two valves are approximately equal in size, but in some extant taxa, such as *Lycopodiella* (except section *Caroliniana*) *sensu* Øllgaard (1987), they are anisovalvate. Similarly, in heterosporous species of *Selaginella,* dehiscence of the microsporangia is tangential with respect to the stem (Bierhorst 1971; Koller and Scheckler 1986). In protolepidodendralean fossils, such as *Leclercqia* (Banks, Bonamo, and Grierson 1972; Bonamo, Banks, and Grierson 1988), *Minarodendron* (Li 1990), and possibly *Colpodexylon* (Banks 1944), dehiscence into two equal valves occurs along a line that is radial with respect to the stem. A third type of dehiscence is characterized by simple decay without a preformed dehiscence line (e.g., *Isoetes:* Bierhorst 1971). Several dehiscence mechanisms have been identified in the sporangia of arborescent lycopsids. Bateman, DiMichele, and Willard (1992) recognize four mechanisms, of which longitudinal dehiscence is interpreted as plesiomorphic. In *Paralycopodites,* dehiscence is longitudinal (radial), thus resembling the protolepidodendralean taxon *Leclercqia.*

Sporangium dehiscence is treated as an unordered multistate character with three states: 0 = tangential; 1 = radial; 2 = absent. There were no inapplicable cases; one was missing data (see Appendix 5). See characters 4.29 and 5.16 for comparison.

6.22 • SPORANGIUM DISTRIBUTION In extant *Huperzia* (Øllgaard 1987) and in many fossils such as *Minarodendron* (Li 1990) and probably also *Leclercqia* (Banks, Bonamo, and Grierson 1972), sporangia occur over a relatively extended fertile area that may alternate with sterile regions. In other early fossils, such as *Asteroxylon* (Lyon 1964), *Drepanophycus* (Kräusel and Weyland 1935; Li and Edwards 1995; Schweitzer 1980c), and *Baragwanathia* (Lang and Cookson 1935), the distribution of the fertile zone is poorly understood. In the extant homosporous genera *Phylloglossum* (Figure 6.9), *Lycopodium,* and *Lycopodiella* (Øllgaard 1987), as well as in the arborescent lycopsids (Bateman, DiMichele, and Willard 1992; DiMichele and Phillips 1985; Phillips 1979; T. N. Taylor 1981), *Selaginella* (Gifford and Foster 1989), and *Isoetes* (Bierhorst 1971), sporangia occur in a more or less compact strobilus.

Sporangium distribution is treated as a binary character: 0 = extended; 1 = compact. There were no inapplicable cases; three were missing data (see Appendix 5). See character 5.12 for comparison.

6.23 • XYLEM STRAND SHAPE Xylem shape in transverse section has a variety of forms in zosterophylls and lycopsids (Figure 6.15). In extant homosporous lycopsids, xylem shape typically is stellate

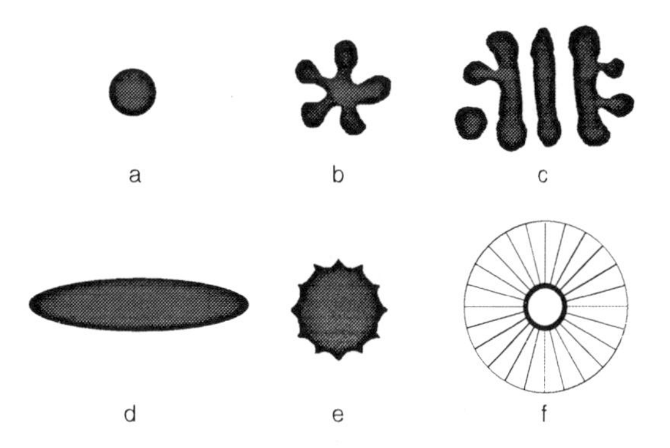

FIGURE 6.15. Comparison of xylem shape in transverse sections of selected Lycopsida (not drawn to scale). See character 6.23. (**a**) Terete xylem in early ontogenetic stages of most lycopsids and in mature stages of some taxa (e.g., *Selaginella selaginoides*). (**b**) Stellate xylem present at some point during the ontogeny of most Lycopodiaceae and early stem group Lycopsida such as the fossils *Asteroxylon, Drepanophycus,* and *Baragwanathia.* (**c**) Inclusion of parenchyma in the xylem in larger axes of some species of Lycopodiaceae. (**d**) Elliptical to strap-shaped xylem typical of many species of Selaginellaceae and Zosterophyllopsida. (**e**) Ribbed xylem of Protolepidodendrales and distal branches of some rhizomorphic lycopsids. (**f**) Siphonostele with secondary xylem in some ontogenetic stages of arborescent lycopsids. Thick black line = primary xylem; thin, radiating lines = secondary xylem.

in the axes of smaller species, in the distal branches of larger species, and in sporelings (Wardlaw 1924). In larger species, the stellate shape is retained in the larger axes by an increase in the number of lobes up to a maximum of about nine. Further increases in the size of the axis are followed by separation and division of the lobes, resulting in a change from the stellate condition to a stele that has a spongy appearance. Wardlaw (1924) showed that these changes in shape were related to the size of the stem, thus supporting the functional interpretation offered by Bower (1923b) that such changes were necessary to maintain a relatively high proportion of surface area to volume in the stele. Wardlaw (1924) also used ontogenetic evidence to support the idea that the stellate condition is primitive in extant homosporous lycopsids. In contrast to the genera *Huperzia, Lycopodium,* and *Lycopodiella,* the stele of *Phylloglossum* is highly distinctive. According to Hackney (1950), the xylem of *Phylloglossum* does not pass through a stellate phase at any time in its development.

The stellate xylem strand of extant homosporous lycopsids is strikingly similar in shape in transverse section to that of the fossils *Asteroxylon* (Figure 6.16; Kidston and Lang 1920b), *Drepanophycus* (Li and Edwards 1995; Rayner 1984), and *Baragwanathia* (Lang and Cookson 1935). Similarities include a stellate shape with deep bays between adjacent lobes and the presence of protoxylem in the enlarged ends of the xylem lobes. In the fossil taxa *Archaeosigillaria vanuxemii* (Grierson and Banks 1963), *Haskinsia colophylla* (Grierson and Banks 1983), *Colpodexylon deatsii* (Banks 1944), *Minarodendron cathaysiense* (Li 1990), and *Leclercqia complexa* (Banks, Bonamo, and Grierson 1972), the xylem is circular with many (8–18) small, distinctive ridges of protoxylem. In *Colpodexylon,* some lobing of the strand occurs, but unlike the stellate strands of extant homosporous lycopsids and of fossils such as *Asteroxylon,* up to four distinct protoxylem ridges have been observed on the outer edge of the lobes (Banks 1944). The protosteles of arborescent lycopsids are similarly circular with many small, distinctive ridges of protoxylem (Bateman, DiMichele, and Willard 1992).

In the Selaginellaceae, plants with radial symmetry (e.g., *Selaginella*) have a simple circular protostele within the stem, whereas many strongly dorsiventral species (e.g., *Tetragonostachys* and *Stachygynandrum* groups) have strap-shaped steles (Harvey-Gibson 1894, 1902; Jermy 1990b; Mitchell 1910; Wardlaw 1925). The polystelic condition is also a feature of some *Selaginella* species. The circular type generally is considered ple-

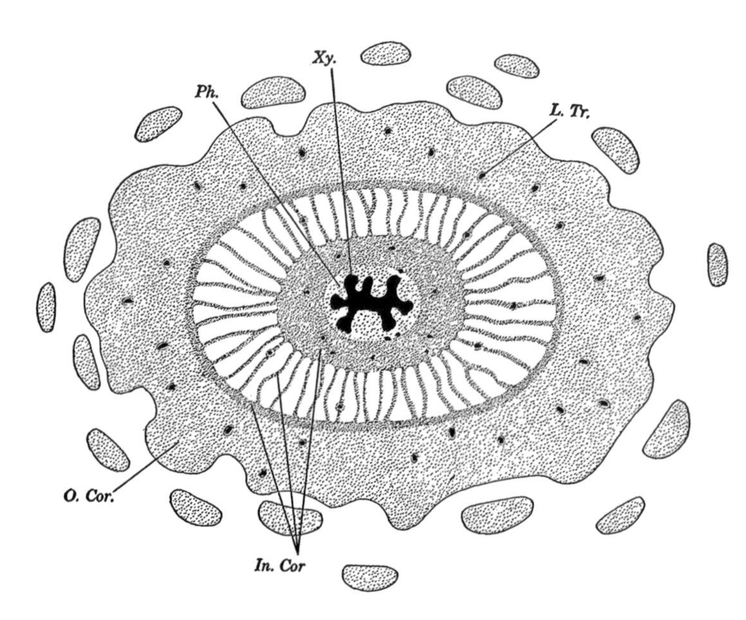

FIGURE 6.16. Transverse section of leafy shoot of the fossil plant *Asteroxylon mackiei* from the Rhynie Chert, lower Devonian (×6). *In. Cor.* = inner cortex; *O. Cor.* = outer cortex; *L. Tr.* = leaf trace; *Ph.* = phloem; *Xy.* = xylem. Reproduced from Smith 1938, Figure 83.

siomorphic and contains variants with slightly projecting flanges of protoxylem comparable to those in the Protolepidodendrales (Bower 1935, 225). Bateman (1992) scored the larger strap-shaped steles in some species of the *Stachygynandrum* group as homologous to the flattened steles in some of the larger Lycopodiaceae. We rejected this notion of homology because of the distinct differences in xylem ontogeny in the Lycopodiaceae and Selaginellaceae (Wardlaw 1924, 1925, 1928). In the Lycopodiaceae, a pseudopolystelic condition arises through dissection and elaboration of a stellate stele into stelar fragments, some of which may be strap-shaped. In the Selaginellaceae, a strap-shaped stele appears to develop in response to planar branching. Furthermore, the polystelic condition of some taxa does not develop through dissection of a stellate stele. The xylem strand of *Isoetes* is a distinctive, and unique, "anchor-shape" in transverse section (Bierhorst 1971).

Xylem strand shape is treated as an unordered multistate character with five states: 0 = more or less circular (terete); 1 = stellate; 2 = elliptical 3 = *Phylloglossum*-type (P-type); 4 = *Isoetes*-type (I-type). There were no inapplicable cases; one was missing data (see Appendix 5). See characters 4.14 and 5.7 for comparison.

6.24 • PROTOXYLEM Protoxylem is characteristically exarch in zosterophylls and lycopsids, and the general condition for polysporangiophytes is centrarch (see Chapters 4 and 5). Protoxylem position was scored as unknown in *Isoetes* because no clear-cut developmental sequence is visible in the primary tissues (Bower 1935).

The type of protoxylem is treated as a binary character: 0 = centrarch; 1 = exarch. There were no inapplicable cases; four were missing data (see Appendix 5). See characters 4.15, 5.8, and 5.22 for comparison.

6.25 • STELAR SUSPENSION A characteristic anatomical feature of the stems of extant Selaginellaceae is the presence of a conspicuous cavity surrounding the stele. The stele is suspended within the cavity by a few endodermal cells with casparian strips that are attached to the cavity walls (Bierhorst 1971; Gifford and Foster 1989; Harvey-Gibson 1894, 1902). Recognition of this character is difficult in the fossil record because of problems distinguishing taphonomic changes (e.g., decay) from real anatomical features. Poor cellular preservation in the tissue systems immediately surrounding the xylem often results in a mineralized cavity in the fossil, and the recognition of a few delicate endodermal cells attaching the stele to the wall of the cavity is likely to be problematic. The interpretation of fossils is further complicated by evidence from exceptionally well preserved material in the Rhynie Chert that indicates that cavities surrounding the stele may be more general in lycopsids. The early lycopsid *Asteroxylon* has large intercellular spaces surrounding the stele that are transversed by plates of cells (Kidston and Lang 1920b). The cellular matrix in the cortex of *Asteroxylon* is much more extensive than that in equivalent areas of the Selaginellaceae. Because of these taphonomic and interpretational problems, we were forced to score many fossil taxa as unknown.

Suspension of the stele is treated as an unordered multistate character with three states: 0 = absent; 1 = stelar cavity; 2 = partial cavity *(Asteroxylon)*. There were no inapplicable cases; nine were missing data (see Appendix 5).

6.26 • CORTICAL SCLERENCHYMA A well-marked band, either continuous or discontinuous, of cortical sclerenchyma is a common feature of many zosterophylls and lycopsids. In extant Lycopodiaceae the cortical sclerenchyma is lignified in *Lycopodium* and *Lycopodiella,* but is unlignified in *Huperzia* (Øllgaard 1987) and probably absent in *Phylloglossum* (Hackney 1950). Lignin cannot be unequivocally identified in early fossils, but an approximation can be obtained by comparing decay resistance and the optical qualities of cell walls in various tissue systems against unequivocally lignified cells (e.g., wood) within the same axis. This comparison provides a crude standardization of the relative decay resistance. Judged by these criteria, a "lignified" sclerenchymatous band similar to that in extant Lycopodiaceae is present in all Zosterophyllopsida (character 4.13: sterome). Sclerenchymatous tissues have not been observed in the rather incomplete anatomical specimens of *Baragwanathia* and *Drepanophycus,* but they are probably absent in *Asteroxylon* (Kidston and Lang 1920b). Cortical sclerenchyma has been observed in some Protolepidodendrales (e.g., *Leclercqia* and *Haskinsia*) and *Isoetes* and is common in the Selaginellaceae (Gifford and Foster 1989) and arborescent lycopsids (Bateman, DiMichele, and Willard 1992).

The absence or presence and type of cortical sclerenchyma is treated as an unordered multistate character with three states: 0 = sclerenchyma absent; 1 = lignified sclerenchyma; 2 = nonlignified sclerenchyma. There were no inapplicable cases; six were missing data (see Appendix 5). See characters 4.13 and 6.27 for comparison.

6.27 • SCLERENCHYMA POSITION Cortical sclerenchyma is more or less peripheral in all of the Zosterophyllopsida and in most lycopsids. In some extant Lycopodiaceae (e.g., *Lycopodium, Lycopodiella*) the cortical sclerenchyma is subperipheral and situated next to the endodermis (Øllgaard 1987).

Sclerenchyma position is treated as a binary character: 0 = peripheral; 1 = subperipheral. This character is inapplicable in taxa lacking sclerenchyma. There was one inapplicable case; six were missing data (see Appendix 5). See characters 4.13 and 6.26 for comparison.

6.28 • SECONDARY GROWTH A cambium producing secondary xylem occurs in the aerial stems of arboreous lycopsids (Cichan 1985a; Eggert and Kanemoto 1977), as well as in such plants as *Paurodendron* (Rothwell and Erwin 1985) and *Isoetes* (Bierhorst 1971; Gifford and Foster 1989; Karrfalt 1982). Character analysis supports the widely held hypothesis that the cambium of arborescent lycopsids is independently derived from the cambium of seed plants and sphenopsids (Crane 1990). This conclusion is further reinforced by significant differences in cambial function and by ontogeny in these groups (Cichan 1985a,b, 1986a,b; Cichan and Taylor 1990; Rothwell and Pryor 1991). In seed plants, cambial activity produces both secondary xylem and phloem. The lycopsid cambium is unifacial (i.e., it produces significant quantities of secondary xylem only: Cichan 1985a; Eggert and Kanemoto 1977). A unifacial, orbifacial cambium also occurs in some other Carboniferous pteridophytes, including *Sphenophyllum* (E. L. Taylor 1990). A further unique feature of cambial activity in lycopsids that also occurs in *Sphenophyllum* is that the increase in trunk circumference is accompanied by an increase in the size of the cortical cells (supplemented by interdigitating growth) rather than an increase in the number of cambial initials through anticlinal divisions (Cichan 1985a).

Secondary xylem in stigmarian axes (the rooting organs of the Lepidodendraceae) exhibits little increase in thickness toward the base of the plant, in contrast to the aerial stems and the woody organs of seed plants, which show dramatic increases. A determinate cambium has been invoked to explain this phenomenon (Eggert 1961; Eggert and Kanemoto 1977), but more recent observations on wood ontogeny and meristem structure in *Stigmaria* and *Isoetes* suggest an intriguing alternative explanation. Much of the "secondary xylem" in the rooting structures of lepidodendralean lycopsids could be the product of a primary thickening meristem (Karrfalt 1982, 1984a,b; Phillips and DiMichele 1992; Rothwell and Pryor 1991). According to this hypothesis, all of the stelar tissues are produced in a rapid burst of meristematic activity near the apex, resulting in a dome-shaped apical zone and a trunk that increases little in girth basipetally, in a manner analogous to some extant monocotyledons. This hypothesis is supported by recent observations on the apex of *Stigmaria ficoides* (Rothwell and Pryor 1991).

A well-marked cambium with secondary growth is absent in the Selaginellaceae. Some cell alignment occurs in the rootstock of *Selaginella selaginoides,* but a cambium has not been identified (Karrfalt 1981).

The presence or absence of secondary growth is treated as a binary character: 0 = secondary xylem absent; 1 = secondary xylem present (cambium present). There were no inapplicable cases; one was missing data (see Appendix 5).

6.29 • METAXYLEM TRACHEID PITTING In extant homosporous lycopsids the metaxylem is composed of tracheids having circular and scalariform bordered pits, except for *Phylloglossum,* in which the thickenings are simple, annular rings that have occasional direct connections between them (Bierhorst 1971; Hackney 1950). In the protoxylem of all lycopsids, the thickenings are annular and connected by thin strands of additional wall material (Bierhorst 1960). In *Selaginella,* tracheids of the metaxylem have scalariform bordered pitting, and in some species vessel-like elements also occur. Protoxylem is composed of annular and helical elements with occasional reversals in the direction of the helical gyres (Bierhorst 1971). The primary xylem of *Isoetes* is composed entirely of annular and helical elements—pitted elements have not been described (Bierhorst 1960). In *Lepidophloios* and *Stigmaria,* metaxylem tracheids have simple (unbordered) scalariform pits with a perforate sheet of secondary wall material extending across the pit aperture between the horizontal wall thickenings. This structure has been termed a *pitlet sheet* (Li 1990) and is not found in extant lycopsids. It is formed from an extensive meshwork of secondary wall material (Cichan, Taylor, and Smoot 1981; Geng 1990). Tracheids of the secondary xylem also have scalariform pits, but in contrast to cells of the primary xylem, the pitlet sheet consists of many vertical strands *(Williamson striations)* having few horizontal connections (Cichan, Taylor, and Smoot 1981; Wesley and Kuyper 1951). Oval to circular bordered pitting predominates in the metaxylem of *Leclercqia complexa* (Grierson 1976) and *Haskinsia colophylla* (Grierson and Banks 1983), but there is no pitlet sheet. The early-formed protoxylem has helical and annular thickenings, and the transitional protoxylem is reticulate. In *Minarodendron cathaysiense,* the pitting in the metaxylem is mainly scalariform, and there is a pitlet sheet that is composed of one or two rows of pitlets (Li 1990). The thickenings of the early-formed protoxylem are described as helical.

In marked contrast to other lycopsids, the early fossils *Drepanophycus spinaeformis* (Hartman 1981; Kenrick and Edwards 1988a; Rayner 1984), *D. qujingensis* (Li and Edwards 1995), *Baragwanathia abitibiensis* (Hueber 1983), and *Asteroxylon mackiei* (Kenrick and Crane 1991; Kidston and Lang 1920b; Lemoigne and Zdebska 1980) do not have pitted tracheids. In these plants both the protoxylem and metaxylem tracheids have annular thickenings, with occasional forking of the annular bars, and are indistinguishable from the distinctive G-type tracheids of the Zosterophyllopsida (Kenrick and Crane 1991). A perforate sheet of secondary wall material extends between adjacent annular bars, and we consider this to be homologous with the pitlet sheet found in the stratigraphically younger lycopsids discussed above. Protoxylem in *Drepanophycus, Baragwanathia,* and *Asteroxylon* appears to be composed of cells similar to those of the metaxylem but with a smaller diameter. The structure of the pitting in the fossil taxa *Archaeosigillaria vanuxemii* (Grierson and Banks 1963) and *Colpodexylon deatsii* (Banks 1944) is not well understood.

Outgroup comparison indicates that annular or helical thickenings *and* pitlet sheets is the general condition in tracheophytes (Kenrick and Crane 1991).

Metaxylem tracheid pitting is treated as a binary character: 0 = tracheid pitting absent; 1 = tracheids pitting present. There were no inapplicable cases; one was missing data (see Appendix 5). See characters 4.16–4.19 and 5.9 for comparison.

6.30 • PSEUDOBIPOLAR GROWTH The rooting organs of the lycopsids *Paurodendron, Sigillaria, Nathorstiana,* and *Isoetes,* as well as those of *Chaloneria* and other Lepidodendrales, are interpreted as a modified shoot system that is not homologous with the primary root system of seed plants or progymnosperms (Rothwell and Erwin 1985). The rooting organ of the rhizomorphic lycopsids forms at the first dichotomy of the stem—one branch growing upward to produce the shoot and the other growing downward to produce the root *(pseudobipolar growth).* Rootlets in these plants are interpreted as modified leaves (Karrfalt 1981; Rothwell and Erwin 1985). The first root of *Isoetes* is interpreted as a rooting leaf developed from a rooting shoot that does not elongate axially (Rothwell and Erwin 1985; Stubblefield and Rothwell 1981). Such rooting organs are not homologous with the adventitious roots of rhizomatous lycopsids in the Lycopodiaceae and Selaginellaceae; those structures usually develop endogenously along the stem. The Lycopodiaceae and Selaginellaceae also have strictly unipolar (as opposed to bipolar) growth (Karrfalt 1981).

Rooting organs in the earliest herbaceous fossil lycopsids are poorly understood because the available material is fragmentary. Such structures are unknown in *Baragwanathia, Archaeosigillaria, Haskinsia, Colpodexylon, Minarodendron,* and *Leclercqia.* Kidston and Lang (1920b) described dichotomously branching, leafless rhizomes in *Asteroxylon,* but these do not produce endogenous roots as in the extant Lycopodiaceae, nor is *Asteroxylon* considered to have pseudobipolar growth (Rothwell and Erwin 1985). Naked, rhizomatous axes bearing narrow, much branched, rootlike structures have been described in *Drepanophycus spinaeformis* (Rayner 1984; Schweitzer 1980c) and *D. qujingensis* (Li and Edwards 1995). These axes bear a superficial resemblance to endogenous roots, but the critical anatomical information needed to pursue this comparison is not available.

Comparison of lycopsid rooting structures with outgroup taxa in the zosterophylls is difficult because there are few clear examples in which such structures are preserved. Prostrate, much-branched, dichotomous, basal axes of similar morphology to those of *Asteroxylon* have been described in *Discalis longistipa* (Hao 1989b), in *Zosterophyllum myretonianum* (Lang 1927; Lele and Walton 1961), and in *Z. deciduum* (Gerrienne 1988). Endogenous roots and rhizoids have not been described, but their apparent absence may result from poor preservation. These prostrate axes are most comparable to the naked, rhizomatous axes of *Asteroxylon* and *Drepanophycus.* A second type of rhizomatous axis found in the zosterophylls was described from *Serrulacaulis* (Hueber and Banks 1979). Here the rhizome is of similar morphology to the aerial axes except for the presence of nonseptate rhizoids and the absence of thickened epidermal cells.

Pseudobipolar growth is treated as a binary character: 0 = monopolar growth; 1 = pseudobipolar growth. There were no inapplicable cases; none were missing data (see Appendix 5).

6.31 • ROOTLET ANATOMY The roots of *Isoetes* and the rootlets of arboreous lycopsids have a unique structure and are strikingly similar. Each rootlet contains a single, monarch vascular bundle attached to one side of a large central cavity that is surrounded by cortex *(stigmarian rootlet anatomy).* Phloem forms on the inner side of the vascular bundle (Stewart 1947).

Rootlet anatomy is treated as a binary character: 0 = simple; 1 = rhizomorphic (stigmarian rootlet anatomy). There were no inapplicable cases; eight were missing data (see Appendix 5). See characters 5.21, 6.30, and 6.32 for comparison.

6.32 • ROOT STELE SYMMETRY The roots of most lycopsids are characterized by marked bilateral symmetry: phloem is situated on one side of the stele only (Figure 6.17; Bierhorst 1971). Bilateral symmetry in xylem and phloem distribution occurs in all of the Lycopodiaceae (except in the larger parts of the rooting system in some taxa within the subgenus *Lycopodium*), in the Selaginellaceae, in arboreous lycopsids, and in *Isoetes.* In the early fossil lycopsid *Asteroxylon,* the stele in the putative rooting system is radially symmetrical

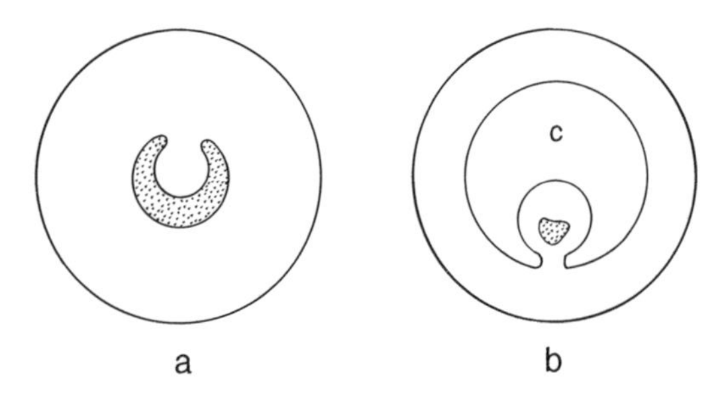

FIGURE 6.17. Comparison of rootlet anatomy in Lycopsida (*stippled areas,* xylem). (**a**) *Lycopodium serratum:* transverse section of root (×100). Redrawn from Smith 1938, Figure 96. (**b**) *Paurodendron (Selaginella) fraipontii:* transverse section of root showing distinctive monarch xylem strand suspended in root cavity *(c)* (×100). Redrawn from Stewart and Rothwell 1993, Figure 11.29.

(Kidston and Lang 1920b, 647). Radial symmetry also occurs in the rhizophores of some Selaginellaceae (Gifford and Foster 1989, 130, Figure 9-22B), whereas marked bilateral symmetry occurs in others (e.g., *S. kraussiana,* Bierhorst 1971). The anatomy of rooting systems is unknown in nearly all of the early fossil lycopsids and zosterophylls. Rhizomes in such outgroup taxa as *Rhynia* are probably modified stems and as such are not strictly comparable to roots.

Root stele symmetry is treated as a binary character: 0 = radial; 1 = bilateral. This character is inapplicable in taxa lacking roots. There was one inapplicable case; eight were missing data (see Appendix 5). See characters 5.21, 6.30, and 6.31 for comparison.

6.33 • ROOT XYLEM SHAPE The roots of many lycopsids are characterized by a xylem strand that is crescent-shaped in transverse section (Figure 6.17; Bierhorst 1971). Such xylem is typical of the Lycopodiaceae (except for the larger roots in the subgenus *Lycopodium* and in some sections of *Lycopodiella,* Øllgaard 1987) and Selaginellaceae (Bierhorst 1971). In *Isoetes* and in arboreous lycopsids, the root xylem is more or less circular (Bierhorst 1971). There is little anatomical information on the rooting organs of early fossils. In *Asteroxylon* the xylem of putative rooting axes was described as cylindrical (Kidston and Lang 1920b).

Root xylem shape is treated as a binary character: 0 = more or less circular; 1 = crescent-shaped. This character is inapplicable in taxa lacking roots. There was one inapplicable case; eight were missing data (see Appendix 5).

6.34 • CORTICAL ROOTS In lycopsids, roots originate endogenously from the pericycle of other roots or from comparable regions within stems. The unique cortical roots of *Huperzia* also develop endogenously but grow vertically downward through the stem cortex to emerge lower on the plant. Because cortical roots are visible within stem sections, this character can be evaluated for many fossil taxa; however, cortical roots have not been observed in anatomically preserved specimens of the early fossil lycopsids *Asteroxylon, Drepanophycus,* and *Baragwanathia.*

The presence or absence of cortical roots is treated as a binary character: 0 = cortical roots absent; 1 = cortical roots present. There were no inapplicable cases; one was missing data (see Appendix 5).

6.35 • SPORE SIZE *Isoetes, Selaginella,* and probably all of the extinct arborescent lycopsids are heterosporous (Bateman and DiMichele 1994a; Bateman, DiMichele, and Willard 1992; Bierhorst 1971; DiMichele and Phillips 1985; Gifford and Foster 1989; Phillips 1979), producing microspores and megaspores in separate sporangia. In herbaceous fossils, the detection of heterospory is often difficult (Bateman and DiMichele 1994a), and for many taxa, spore morphology is unknown (e.g., *Drepanophycus, Archaeosigillaria, Haskinsia, Colpodexylon, Minarodendron*). Extant Lycopodiaceae are homosporous (Øllgaard 1987). The inference of homospory in *Baragwanathia* (Lang and Cookson 1935) and *Asteroxylon* (Lyon 1964) is based on microspores isolated from only a few sporangia, but the homosporous status of *Leclercqia* is more secure because hundreds of sporangia have yielded only microspores (Grierson and Bonamo 1979).

Spore size is treated as a binary character: 0 = homosporous; 1 = heterosporous. There were no inapplicable cases; two were missing data (see Appendix 5). See character 5.20 for comparison.

6.36 • MICROSPORE ORNAMENTATION Microspore morphology in many early polysporangiophytes, in the Zosterophyllopsida, and in basal Lycophytina offers few unequivocal characters (Chapter 5; Fanning, Edwards, and Richardson 1990, 1991; T. N. Taylor 1990). In the early lycopsid fossils *Asteroxylon* and *Baragwanathia,* few details of spore morphology are known except that conspicuous ornamentation is absent. More distinctive spore wall morphologies are present in other lycopsid groups. A basic foveolate-fossulate ornamentation predominates in the Lycopodiaceae, and the family can be further subdivided on the basis of spore wall morphology (Wilce 1972).

The protolepidodendralean fossil *Leclercqia* has echinate spore ornamentation (Streel 1972), which also predominates in many Selaginellaceae and Isoetaceae, although other forms of wall ornament have also been described in these groups (W. A. Taylor and T. N. Taylor 1990; Tryon and Lugardon 1991). Within the Selaginellaceae, echinate ornamentation occurs in the *Selaginella* and *Stachygynandrum* groups, but in the *Tetragonostachys* group, spore ornamentation is granulate to papillate. We provisionally treat echinate spore ornamentation as plesiomorphic in *Isoetes*, but additional careful analysis of subfamily relationships is required to substantiate this hypothesis. Bateman, DiMichele, and Willard (1992) recognized 14 microspore wall characters among rhizomorphic lycopsids. We were unable to use most of these features at this level of analysis either because of their absence or because of inadequate preservation. Microspores with echinate ornamentation occur in the plesiomorphic arborescent lycopsids *Paurodendron* and *Oxroadia*, but in *Paralycopodites*, microspores are densely granulate on the distal surface only (Bateman, DiMichele, and Willard 1992).

Microspore ornamentation is treated as an unordered multistate character with five states: 0 = absent; 1 = foveolate-fossulate; 2 = echinate; 3 = densely granulate; 4 = papillate. There were no inapplicable cases; two were missing data (see Appendix 5). See character 5.28 for comparison.

6.37 • MICROGAMETOPHYTE DEVELOPMENT In *Selaginella* and *Isoetes*, microgametophytes are entirely endosporic (Bierhorst 1971), and those of the extinct arborescent lycopsids are also thought to be at least partly endosporic (Brack-Hanes 1978; DiMichele and Phillips 1985). Brack-Hanes and Vaughn (1978) observed that the initial two divisions of the male gametophyte of *Lepidostrobus schopfii* were endosporic; however, the development of the antheridia and gametes has not been observed. Megagametophytes of *Selaginella, Isoetes,* and the arborescent lycopsids also undergo most of their development within the megaspore wall (Bierhorst 1971; Brack-Hanes 1978). In extant homosporous lycopsids, initial cell divisions of the gametophyte may occur within the spore wall but these are followed by the germination and development of a free-living plant (Bierhorst 1971; Øllgaard 1987). Gametophytes of the early herbaceous fossil taxa are unknown.

Microgametophyte development is treated as a binary character: 0 = exosporic; 1 = endosporic. There were no inapplicable cases; 10 were missing data (see Appendix 5).

6.38 • MEGAGAMETOPHYTE DEVELOPMENT In the early development of the megagametophytes of *Selaginella* and *Isoetes*, there is a period of free nuclear division—the nucleus of the megaspore divides, and its derivatives divide without associated cytokinesis (Bierhorst 1971). In the gametophytes of extant homosporous lycopsids, free nuclear division has not been detected, and gametophyte development in these taxa is scored as cellular. Free nuclear division has also not been recorded in the spores of relevant fossil plants. Convincing evidence of fossil nuclei is rare, but structures interpreted as chromosomes were described by Brack-Hanes and Vaughn (1978) in the microspores of *Lepidostrobus schopfii*.

Megagametophyte development is treated as a binary character: 0 = entirely cellular; 1 = free nuclear division. There were no inapplicable cases; 11 were missing data (see Appendix 5).

6.39 • GAMETOPHYTE HABIT A unique form of subterranean, holosaprophytic (exclusively saprophytic and non-photosynthetic) gametophyte occurs in some groups within the Lycopodiaceae. Such gametophytes are typical of *Huperzia* and *Lycopodium*, whereas the gametophytes of *Lycopodiella* and *Phylloglossum* are green and grow on the surface (Øllgaard 1987). We score the predominantly endoscopic gametophytes of the heterosporous groups as superficial also. Gametophytes are unknown for most early homosporous taxa, but recent discoveries in the Rhynie Chert indicate that the gametophyte generation in early polysporangiophytes was superficial and had a highly complex morphology (see reviews in Kenrick 1994; Remy, Gensel, and Hass 1993).

Gametophyte habit is treated as a binary character: 0 = superficial; 1 = subterranean. There were no inapplicable cases; 10 were missing data (see Appendix 5).

Other Potentially Relevant Characters

6.40 • MEGASPORE WALL STRUCTURE Megaspore wall structure provides a range of potentially useful systematic characters from surface ornamentation to spore wall ultrastructure (Bateman 1992; Bateman, DiMichele, and Willard 1992; Collinson 1991; Collinson, Batten, et al. 1985; Hemsley 1993; Hemsley, Collinson, and Brain 1992; Hemsley and Scott 1991; Kovach 1989; Kovach and Batten 1989; Kovach and Batten 1993; Minaki 1984; Morbelli and Rowley 1993; Stafford 1991; T. N. Taylor 1990; Taylor and Brauer 1983; W. A. Taylor 1989, 1990, 1992; W. A. Taylor and T. N. Taylor 1990; Tryon and Lugardon 1991). However, we have not included megaspore wall structure in our analysis because most of the features appear to be more relevant to understanding relationships at the restricted subgeneric level rather than at the more general level considered here. One general feature of the megaspores of heterosporous lycopsids is the deposition of relatively large quantities of silica within the cell wall.

There are few unequivocal megaspore synapomorphies of such putative groups as the Selaginellaceae and the Isoetaceae. Seemingly major differences in exospore ultrastructure are mainly quantitative when examined in more detail. In general, in the Isoetaceae there is a three dimensional reticulate network of threads and air spaces oriented more or less parallel to the surface, whereas in the Selaginellaceae, there is a mesh of tightly packed threads, robust rods, or both; less airspace; and elements oriented at various angles to the spore surface (Collinson, Batten, et al. 1985; Kovach and Batten 1989; Skog and Hill 1992). Comparison with Paleozoic arborescent lycopsids makes these distinctions more ambiguous (W. A. Taylor 1990). The occurrence of associated adherent microspores can be used to identify dispersed megaspores of the Isoetaceae crown group (Collinson 1991). Microspores in the Isoetaceae are monolete or mixed monolete-trilete, as opposed to the general trilete condition in lycopsids. Occurrence of adherent trilete microspores on the other hand is unlikely to characterize the Selaginellaceae because the trilete condition is plesiomorphic and microspore adherence is of more general occurrence in heterosporous lycopsids.

Despite the absence of unequivocal synapomorphies, certain subgroups within the Selaginellaceae and Isoetaceae have highly diagnostic megaspores. Opalescent megaspores with characteristic wall ultrastructure are diagnostic of certain Selaginellaceae, and these distinctive spores are first found in the fossil record of the Jurassic (Collinson 1991; Hemsley, Collinson, and Brain 1992; Kovach and Batten 1993; Minaki 1984). This spore type has a highly restricted systematic distribution in extant Selaginellaceae, but the subfamily distribution pattern of this important character has not yet been evaluated in a cladistic context. In extant Selaginellaceae, spores with an opalescent structure appear to be restricted to certain members of *Stachygynandrum,* a group confined mainly to lowland and midmontane primary tropical rain forest. Cladistic analysis of the distribution of this feature within the family may lead to interesting possibilities for assessing the persistence and historical development of tropical rain forests.

6.41 • MEGASPORE NUMBER The number of megaspores is characteristically four per megasporangium in all of the major groups within the Selaginellaceae (e.g., *Selaginella, Tetragonostachys, Stachygynandrum;* Duerden 1929; Jermy 1990b). We regard the higher number occasionally found in some species as probably a derived condition within this clade. Low megaspore numbers are also common in early plesiomorphic woody lycopsids such as *Paurodendron* (four per sporangium: Schlanker and Leisman 1969), *Oxroadia* (4 in *O. gracilis,* c. 16 in *O. conferta:* Bateman 1992), but reduction to a single functional megaspore is a derived feature within arborescent lycopsids (Bateman, DiMichele, and Willard 1992). In extant *Isoetes,* much greater numbers of megaspores per sporangium are produced (100–300) (Bierhorst 1971; Gifford and Foster 1989).

6.42 • STELE NUMBER Stele number in the Selaginellaceae varies from monostelic (e.g., *Selaginella* group) to bistelic and polystelic in the larger axes of some taxa (Jermy 1990b). The polystelic condition in some Selaginellaceae is unique within the zosterophyll-lycopsid clade.

6.43 • DETERMINATE GROWTH One aspect of growth architecture scored by Bateman (1992) was the presence of determinate or indeterminate growth. Determinate growth was scored as present in rhizomorphic lycopsids and in *Selaginella* subgenus *Selaginella,* and this was one of two characters that supported the paraphyly of the Selaginellaceae (see Discussion). We disagree with this interpretation of growth architecture in the subgenus *Selaginella,* which we view as essentially indeterminate, as in other Selaginellaceae and Lycopodiaceae, although there may be a limit to the number of shoots and roots that a plant can produce. In both families, a much-branched, prostrate, indeterminate system bears upright axes that may be determinate. Determinate growth is usually associated with strobilus formation in the Selaginellaceae and Lycopodiaceae (except in *Huperzia*). A similar growth habit, in which a much-branched, horizontal system bears shorter, often determinate, fertile axes occurs in zosterophylls (see character 5.21) and may be more general in polysporangiophytes. The subgenus *Selaginella* differs from other Selaginellaceae in that the indeterminate axes in the horizontal system are smaller than the determinate, upright, strobilate axes. We agree with Bateman (1992) that growth architecture is essentially determinate in rhizomorphic lycopsids and fundamentally different from that in the Lycopodiaceae and Selaginellaceae in this respect, but this difference appears to be the result of a major modification of growth form. In our view, the determinate form of rhizomorphic lycopsids is linked to pseudobipolar growth (see character 6.30 and Bateman 1994).

6.44 • ROOTSTOCK In an earlier analysis of relationships in lycopsids, Bateman (1992) scored the "rootstock" of *Selaginella selaginoides, Phylloglossum drummondii,* and arborescent lycopsids as homologous. Subsequent numerical analysis rejected the notion of homology between the rootstock of *Phylloglossum* and that of the other taxa but was consistent with the homology of rootstocks of *S. selaginoides* and of arborescent taxa. However, we have not scored the rootstocks in these two groups as homologous because of fundamental structural and positional differences.

S. selaginoides is unique in the Selaginellaceae in that all of its roots are attached to a small swollen region at the base of the hypocotyl (Karrfalt 1981). They originate endogenously within the cortex, and typically not more than eight are produced. Meristematic activity gives rise to the roots, but there is no clear evidence of a cambium and no secondary xylem (Karrfalt 1981). Confusion over the presence or absence of a cambium led to early comparisons with arborescent lycopsids by Bower (1908, 1935) and others, and this idea was further reinforced through the morphological comparisons of the fossil *Paurodendron (Selaginella) fraipontii* and extant *S. selaginoides* (Phillips and Leisman 1966; see critical discussions in Karrfalt 1981; Rothwell and Erwin 1985). We follow Karrfalt (1981) and Rothwell and Erwin (1985) and reject any notion of homology between the swollen hypocotyl of *S. selaginoides* and the rhizomorphic root systems of arborescent lycopsids on the grounds that there are no real similarities between these structures. Some of the major structural and positional similarities typical of arborescent lycopsid rhizomorphs that are absent in the hypocotyl of *S. selaginoides* are radial symmetry, apical growth, and secondary xylem and periderm (Karrfalt 1981). Furthermore, root formation in *S. selaginoides* has more in common with root formation in early ontogenetic (embryonic) stages of other Selaginellaceae. These observations strongly suggest that *S. selaginoides* has retained a juvenile form of root production in the adult plant (Karrfalt 1981).

The underground tuber produced seasonally by *Phylloglossum* is a leafless geotropic branch that has evolved a special storage and perennating function that allows the plant to tolerate extreme seasonal variations in water supply (Hackney 1950). The axial origin of the tuber suggests comparisons with the rhizomorphic root systems of arborescent lycopsids, also of stem origin, but there are no other similarities. The tuber of *Phylloglossum* is not a rooting but a storage organ. Roots in *Phylloglossum* are usually produced on the lower sides of aerial stems. The possibility of homology between these structures in *Phylloglossum* and those in the rhizomorphs of arborescent lycopsids was rejected on the basis of parsimony by Bateman (1992).

ANALYSIS

Our analysis of phylogenetic relationships among major basal groups of lycopsids is based on 20 taxa (11 extinct, 9 extant; Table 6.2) and 39 characters. In all of the analyses, trees were generated using the Branch-and-Bound search routine of PAUP (version 3.1.1; Swofford 1990) which finds all the most parsimonious solutions. Trees were rooted by outgroup comparison using the early fossil vascular plant *Rhynia gwynne-vaughanii*. Consistency indexes given throughout are calculated excluding uninformative characters. Tree stability was tested using Bremer Support (Donoghue, Olmstead, et al. 1992; Källersjö, Farris, et al. 1992), which provides a measure of the relative amount of support for each clade. Character-state changes are based on ACCTRAN optimization which favors reversals over parallelisms.

RESULTS

An analysis based on all 20 taxa (Table 6.2) with *Rhynia gwynne-vaughanii* designated as the outgroup yielded 78 trees of 66 steps with consistency

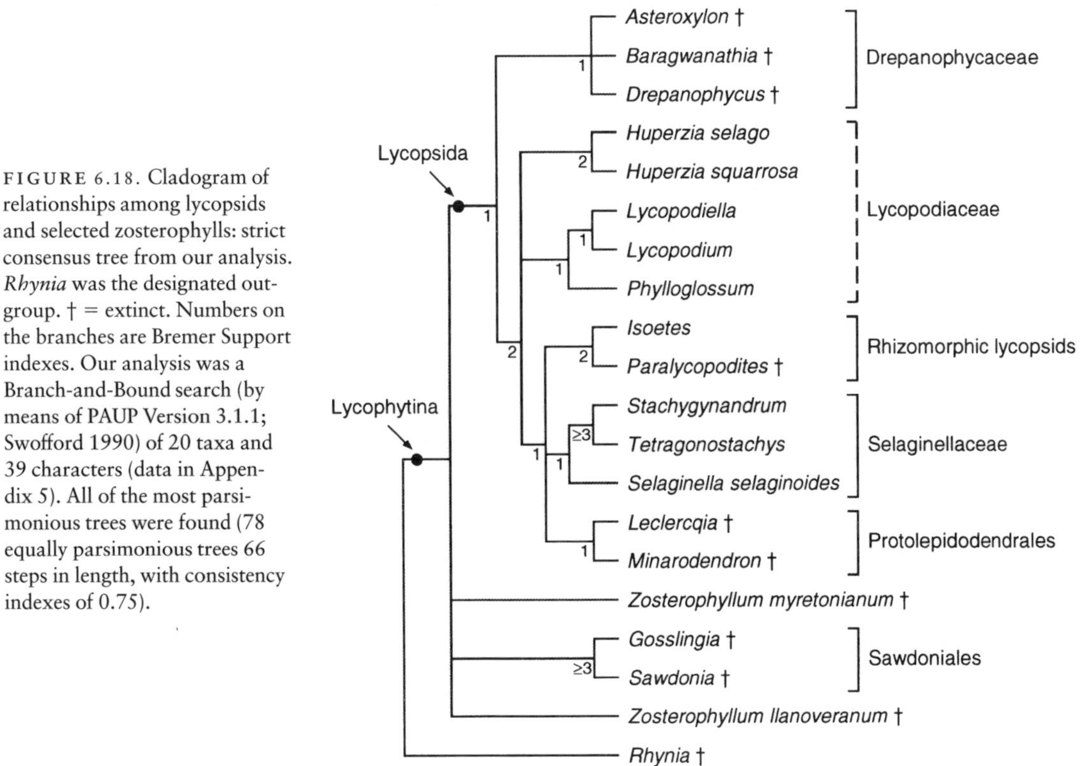

FIGURE 6.18. Cladogram of relationships among lycopsids and selected zosterophylls: strict consensus tree from our analysis. *Rhynia* was the designated outgroup. † = extinct. Numbers on the branches are Bremer Support indexes. Our analysis was a Branch-and-Bound search (by means of PAUP Version 3.1.1; Swofford 1990) of 20 taxa and 39 characters (data in Appendix 5). All of the most parsimonious trees were found (78 equally parsimonious trees 66 steps in length, with consistency indexes of 0.75).

TABLE 6.3

Character Changes on Nodes of Tree in Figure 6.19

Beginning node → end node		Character	Change in state	CI
Node 22 → node 21	6.2	Branching type	0 → 1	0.333
	6.3	Branching pattern	0 → 1	0.500
	6.4	Subordinate branching	0 → 1	0.500
	6.23	Xylem strand shape	0 → 1	0.667
Node 23 → node 22	6.1	Circinate vernation	0 → 1	1.000
	6.17	Sporangiotaxis	0 → 1	1.000
Node 23 → node 24	6.22	Sporangium distribution	0 → 1	0.333
	6.23	Xylem strand shape	0 → 2	0.667
Node 23 → *Rhynia gwynne-vaughanii*	6.13	Sporangium shape	1 → 0	1.000
	6.14	Sporangium attachment	1 → 0	1.000
	6.20	Sporangium symmetry	1 → 0	0.667
	6.24	Protoxylem	1 → 0	1.000
	6.26	Cortical sclerenchyma	1 → 0	1.000
Node 24 → node 36	6.6	Microphylls	0 → 1	1.000
Node 25 → *Huperzia selago*	6.5	Bulbils	0 → 1	0.500
Node 26 → *Lycopodium*	6.39	Gametophyte habit	0 → 1	0.500
Node 27 → node 26	6.12	Mucilage canals	0 → 1	1.000
Node 27 → *Phylloglossum*	6.23	Xylem strand shape	2 → 3	0.667
	6.29	Metaxylem tracheid pitting	1 → 0	0.500
Node 28 → node 25	6.22	Sporangium distribution	1 → 0	0.333
	6.26	Cortical sclerenchyma	1 → 2	1.000
	6.34	Cortical roots	0 → 1	1.000
	6.39	Gametophyte habit	0 → 1	0.500
Node 28 → node 27	6.16	Sporophyll shape	0 → 1	0.500
	6.19	Sporophyll structure	0 → 1	1.000
	6.27	Sclerenchyma position	0 → 1	1.000
Node 29 → *Isoetes*	6.21	Sporangium dehiscence	1 → 2	0.667
	6.23	Xylem strand shape	0 → 4	0.667
Node 29 → *Paralycopodites*	6.16	Sporophyll shape	0 → 1	0.500
	6.19	Sporophyll structure	0 → 2	1.000
	6.36	Microspore ornamentation	2 → 3	1.000
Node 30 → *Stachygynandrum*	6.8	Phyllotaxy	0 → 1	1.000
	6.17	Sporangiotaxis	0 → 2	1.000
Node 30 → *Tetragonostachys*	6.36	Microspore ornamentation	2 → 4	1.000
Node 31 → node 30	6.2	Branching type	0 → 1	0.333
	6.3	Branching pattern	0 → 1	0.500
	6.4	Subordinate branching	0 → 1	0.500
	6.10	Anisophylly	0 → 1	1.000
	6.23	Xylem strand shape	0 → 1	0.667
Node 32 → node 29	6.28	Secondary growth	0 → 1	1.000
	6.30	Pseudobipolar growth	0 → 1	1.000
	6.31	Rootlet anatomy	0 → 1	1.000
	6.33	Root xylem shape	1 → 0	0.500
Node 32 → node 31	6.20	Sporangium symmetry	2 → 0	0.667
	6.21	Sporangium dehiscence	1 → 0	0.667
	6.25	Stelar suspension	0 → 1	1.000
Node 33 → *Minarodendron*	6.2	Branching type	0 → 1	0.333

TABLE 6.3
Continued

Beginning node → end node		Character	Change in state	CI
Node 34 → node 32	6.35	Spore size	0 → 1	1.000
Node 34 → node 33	6.9	Leaf shape	0 → 1	1.000
	6.22	Sporangium distribution	1 → 0	0.333
Node 35 → node 34	6.11	Ligule	0 → 1	1.000
	6.20	Sporangium symmetry	1 → 2	0.667
	6.21	Sporangium dehiscence	0 → 1	0.667
	6.23	Xylem strand shape	2 → 0	0.667
	6.36	Microspore ornamentation	1 → 2	1.000
	6.37	Microgametophyte development	0 → 1	1.000
	6.38	Megagametophyte development	0 → 1	1.000
Node 36 → node 35	6.15	Sporophyll	0 → 1	1.000
	6.18	Sporangium vasculature	1 → 0	1.000
	6.29	Metaxylem tracheid pitting	0 → 1	0.500
	6.32	Root stele symmetry	0 → 1	1.000
	6.33	Root xylem shape	0 → 1	0.500
	6.36	Microspore ornamentation	0 → 1	1.000
Node 36 → node 37	6.5	Bulbils	0 → 1	0.500
	6.25	Stelar suspension	0 → 2	1.000
Node 37 → *Asteroxylon*	6.7	Leaf vasculature	1 → 0	1.000

Notes: ACCTRAN optimization. CI = consistency index of character.

indexes of 0.75. The strict consensus tree is shown in Figure 6.18, and character changes (Table 6.3) are mapped onto a representative most parsimonious tree in Figure 6.19. These trees support the monophyly of the Lycopsida, ligulate lycopsids, the Selaginellaceae, and a rhizomorphic (arborescent) clade. The early fossil lycopsids *Drepanophycus, Asteroxylon,* and *Baragwanathia* (Drepanophycales) form a clade that is a sister group to extant Lycopsida. Monophyly of the Lycopodiaceae and heterosporous lycopsids was not supported by the consensus tree. Examination of alternative tree topologies in the 78 most parsimonious trees revealed two alternative relationships for both groups. The Lycopodiaceae were resolved either as monophyletic or paraphyletic to ligulate lycopsids. Within the ligulate lycopsids, the Protolepidodendrales were resolved either as a sister group to a heterosporous clade comprising the Selaginellaceae and arborescent lycopsids or as a sister group to arborescent lycopsids only.

Tree Stability

Tree stability was tested using Bremer Support (Decay Index = DI) measurements (Källersjö, Farris, et al. 1992) on all 20 taxa (see Chapter 3). Support is strongest for a rhizophoric clade within the Selaginellaceae and for the monophyly of the Sawdoniales within zosterophylls (DI ≥ 3). Intermediate levels of support were found for the Lycopsida crown group, *Huperzia,* and rhizomorphic lycopsids (DI = 2). Other clades were weakly supported (DI = 1).

Lycopsida

Our analysis supports the monophyly of the Lycopsida (DI = 1) based on the presence of microphylls (Figure 6.19, Table 6.3: node 24 → node 36). The stellate xylem strand is an additional potential synapomorphy (Figure 6.19, Table 6.3: node 23 → node 24), the position of which is

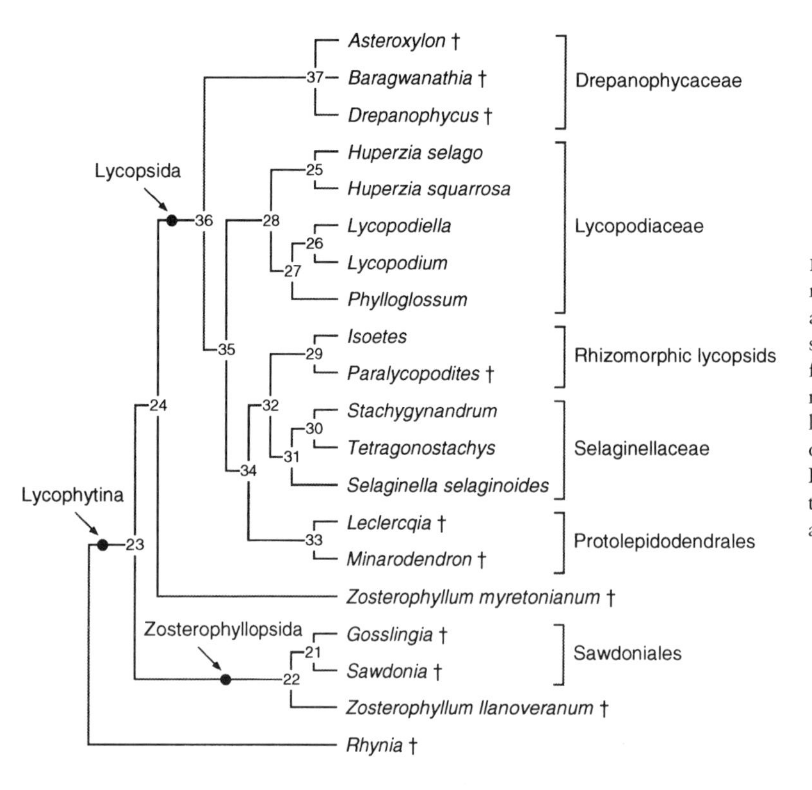

FIGURE 6.19. Cladogram of relationships among lycopsids and selected zosterophylls: representative most parsimonious tree from our analysis. Numbers on nodes refer to character changes listed in Table 6.3 (ACCTRAN optimization). See caption to Figure 6.18 (the strict consensus tree) for explanation of symbols and details of the analysis.

equivocal in Figure 6.19 because of the absence of anatomical information for *Zosterophyllum myretonianum*. We identified a stem-based group comprising the early fossils *Asteroxylon, Baragwanathia,* and *Drepanophycus* (Drepanophycales). Characters supporting the lycopsid crown group (DI = 2) include (1) the close developmental association of sporangium and microphyll (i.e., the presence of sporophylls), (2) pitted tracheids, (3) a foveolate-fossulate microspore exine, and (4) the loss of sporangium vasculature (Figure 6.19, Table 6.3: node 36 → node 35). Two other potential synapomorphies involve features of root anatomy—bilateral root stele symmetry and a crescent-shaped xylem strand—and are equivocal because their states are unknown in closely related zosterophylls.

Drepanophycales

Drepanophycales (DI = 1) comprise the three early fossil lycopsids *Asteroxylon, Drepanophycus,* and *Baragwanathia*. Drepanophycales is a sister group to a clade comprising all of the other lycopsids (Lycopsida crown group) (Figure 6.19, Table 6.3: node 36 → node 37). The monophyly of the Drepanophycales is supported by the presence of bulbils or small lateral buds (a parallelism with extant Lycopodiaceae in the *Huperzia selago* group). An additional potential synapomorphy is the development of a cavity surrounding the stele that is traversed by plates of cells (e.g., *Asteroxylon:* Figure 6.16) resembling those of extant Selaginellaceae (a parallelism). The state of this character is unknown in *Baragwanathia* and *Drepanophycus* because of inadequate anatomical preservation.

Lycopodiaceae

The Lycopodiaceae were not resolved as monophyletic in our consensus tree (Figure 6.18), although some trees support the monophyly of the Lycopodiaceae based on a foveolate-fossulate microspore sculpture. Other trees supported a paraphyletic relationship for the Lycopodiaceae with ligulate lycopsids on the basis of compact strobili that are present in the *Lycopodium-Lycopodiella-Phylloglossum* clade and in the ligulate taxa except in the Protolepidodendrales. The nonstrobilate *Huperzia* species groups are monophyletic based on the presence of cortical roots, the presence of nonlignified cortical sclerenchyma, and possibly also the presence of subterranean gametophytes and the loss of compact strobili (Figure 6.19, Table 6.3: node 28 → node 25). In the strobilate clade, *Phylloglossum* is resolved as a sister group to a *Lycopodium-Lycopodiella* group. The strobilate clade is supported by highly differentiated peltate sporophylls and a peripheral band of lignified sclerenchyma in the stem (Figure 6.19, Table 6.3: node 28 → node 27). *Lycopodium* plus *Lycopodiella* are monophyletic based on the presence of mucilage canals in the leaves (Figure 6.19, Table 6.3: node 27 → node 26).

Ligulate Lycopsids

A ligulate clade (DI = 1) comprising the Protolepidodendrales (represented by *Leclercqia*) and heterosporous lycopsids is defined by (1) the presence of a ligule, (2) radial extension of the sporangium, and (3) a more or less terete, ribbed stele. Other potential synapomorphies include (1) radial as opposed to tangential sporangium dehiscence (reversed in the Selaginellaceae), (2) echinate microspore ornament, and (3) two features of gametophyte development unknown in fossils (Figure 6.19, Table 6.3: node 35 → node 34).

Protolepidodendrales

Characters supporting the monophyly of the extinct Protolepidodendrales (DI = 1) include (1) the unique forked microphylls and (2) a reversal of the strobilate condition to a more extended fertile zone (Figure 6.19, Table 6.3: node 34 → node 33). Other putative synapomorphies include (1) anisotomous branching, (2) nonsinuate sporangium epidermal cells, and (3) the position of the ligule on the leaf (i.e., substantially further from the leaf axil than in other lycopsids).

Heterosporous Lycopsids

The monophyly of heterosporous lycopsids was not supported in our strict consensus tree (Figure 6.18), but examination of alternative trees revealed that it is one of the two most parsimonious topologies. Characters supporting a Protolepidodendrales-arborescent clade include (1) sporangium symmetry (sporangia are radially extended) and (2) elongate as opposed to tangential sporangium dehiscence. This topology implies that heterospory evolved twice in the lycopsids—once in the Selaginellaceae and once in the rhizomorphic clade—or that it evolved once and has been lost in the Protolepidodendrales. The alternative heterosporous Selaginellaceae-rhizomorphic clade is supported on the basis of spore size (Figure 6.19, Table 6.3: node 34 → node 32). Other potential synapomorphies of the heterosporous clade include gametophyte characters, which could not be scored for the extinct Protolepidodendrales (e.g., reduction of gametophyte, endosporic microgametophyte, and free nuclear divisions in the early stages of megagametophyte development).

Selaginellaceae

The monophyly of the Selaginellaceae is not strongly supported in our analysis, although most of the diversity within the group falls into a well-supported monophyletic "rhizophoric" clade. The most problematic taxon in the family is the widespread but divergent *Selaginella selaginoides*. The monophyly of the Selaginellaceae (DI = 1) is based on more or less spherical microsporangia. Other possible synapomorphies include (1) transverse sporangium dehiscence (optimized on our tree as a reversal) and (2) suspension of stele in a stem cavity by distinctive trabeculate endodermal cells (Figure 6.19, Table 6.3: node 32 → node 31). Stele suspension by trabeculate endodermal cells—a unique and distinctive feature of extant Selaginellaceae—

is equivocal because it cannot be scored in potential extinct sister groups. Tissues immediately surrounding the xylem (e.g., phloem) are often not present, even in well-preserved fossil material. Character support for the rhizophoric clade within the Selaginellaceae (DI ≥ 3) includes (1) pseudomonopodial branching, (2) planar branching (plagiotropic shoots), (3) a rhizophore (subordinate branch, a parallelism in the Sawdoniales), (4) anisophylly, and (5) elliptical to strap-shaped xylem strands (a parallelism in the Sawdoniales; Figure 6.19, Table 6.3: node 31 → node 30).

Isoetales–Rhizomorphic Lycopsids

The rhizomorphic (woody) lycopsids comprise extant *Isoetes* and closely related, extinct arborescent taxa (Isoetales *sensu* DiMichele and Bateman [in press]). Relationships among the major groups within rhizomorphic lycopsids have recently been considered in detail by Bateman (1992) and Bateman, DiMichele, and Willard (1992; see Figure 6.20). In agreement with these previous cladistic studies, our analysis supports the monophyly (DI = 2) of this large clade based on the presence of

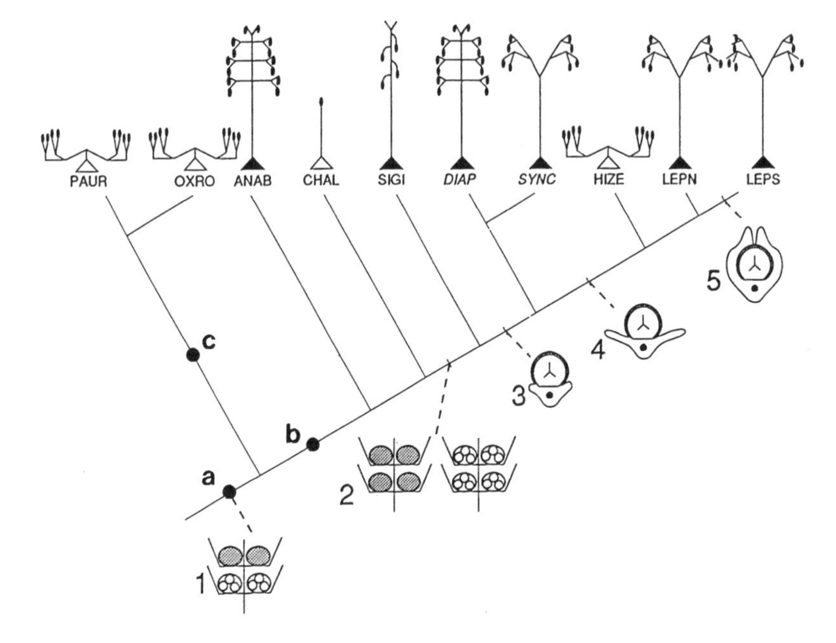

FIGURE 6.20. Preferred most parsimonious cladogram of rhizomorphic lycopsids. PAUR = *Paurodendron;* OXRO = *Oxroadia;* ANAB = *Anabathra (Paralycopodites);* CHAL = *Chaloneria;* SIGI = *Sigillaria;* DIAP = *Diaphorodendron;* STNC = *Synchysidendron;* HIZE = *Hizemodendron;* LEPN = *Lepidodendron;* LEPS = *Lepidophloios.* Trees are indicated by solid triangles, smaller-bodied genera by open triangles. Key reproductive innovations are shown on the major axis: *(1)* bisexual cone, *(2)* unisexual cones bearing either megasporophylls or microsporophylls, *(3)* reduction to single functional megaspore, *(4)* lateral expansion of megasporophylls, and *(5)* enclosure of megasporophyll producing seedlike structure. Character-state changes on labeled nodes: *(a)* woody stem-base and rhizomorph with stigmarian rootlets, three-zoned cortex with periderm, ligule, heterospory; *(b)* arboreous habit, tall upright stem with minimal crown development but extensive lateral branching, xylem bundle a medullated protostele with continuous sheath of protoxylem lacking discernible longitudinal strands and with superficial leaf trace emission, foliar parichnos in leaf base; *(c)* microspore with echinate proximal and distal hemispheres. Reproduced from Bateman 1992, Text-Figure 15, with the permission of E. Schweizerbart'sche Verlagsbuchhandlung (Nägele u. Obermiller), Johannesstrasse 3A, D-70176 Stuttgart.

(1) a unifacial cambium, (2) pseudobipolar growth involving a rhizomorphic root system, and (3) the distinctive monarch xylem strand of the roots. Rhizomorphic lycopsids are a diverse group comprising a range of small pseudoherbaceous taxa as well as large trees (Bateman 1992, 1994; Bateman, DiMichele, and Willard 1992). Cladistic analysis supports the traditional idea that a close relationship exists between extant *Isoetes* and medium- to small-sized extinct arborescent rhizomorphic lycopsids with bilaterally symmetrical rhizomorphs, such as *Chaloneria, Pleuromeia, Takhtajanodoxa,* and *Nathorstiana* (Bateman, DiMichele, and Willard 1992; Pigg 1992).

DISCUSSION

Lycopsid Monophyly

Current classifications implicitly treat lycopsids as a monophyletic group containing between five and nine orders (Bateman 1992; Bell and Woodcock 1971; Bierhorst 1977; Bold 1957; Bower 1935; Chaloner 1967a; Gifford and Foster 1989; Meyen 1987; Sporne 1970; T. N. Taylor 1981; T. N. Taylor and E. L. Taylor 1993; Thomas and Brack-Hanes 1984; see Table 6.1), and our analysis also strongly supports lycopsid monophyly. Shared derived features of the Lycopsida crown group include the close developmental association of the sporangium and the microphyll (i.e., the presence of sporophylls), the presence of pitted tracheids, and the loss of sporangium vasculature. Three other potential synapomorphies are equivocal because their states are unknown in closely related fossils. These characters include a foveolate-fossulate microspore exine, bilateral root stele symmetry, and a crescent-shaped xylem strand in the root. Shared derived features of the more inclusive Lycopsida stem-based group (the crown group plus the Drepanophycales) include the presence of microphylls and a stellate xylem strand in transverse section.

Stewart (1983) and Stewart and Rothwell (1993) suggested a diphyletic origin of lycopsids from within vascular plants (Figure 6.3). According to this hypothesis, the Lycopodiaceae and early homosporous forms such as *Drepanophycus* and *Baragwanathia* are thought to represent one distinct lineage of homosporous eligulate taxa that evolved from zosterophylls, whereas the Protolepidodendrales and heterosporous groups form an independent lineage that evolved from rhyniophytes or *Protohyenia*-type plants. Arguments supporting this hypothesis are based on different putative evolutionary pathways for the origin of the microphyll in ligulate and eligulate lycopsids as well as on when the groups appeared in the fossil record.

Based on the transformational hypotheses of the Telome Theory (Zimmermann 1952), Stewart (1983) interpreted the forked leaves of the Protolepidodendrales (an apomorphic feature in our analysis) as intermediate structures between a terminal fertile branch in rhyniophyte-like plants and a sporophyll in ligulate lycopsids. A process of reduction and simplification was then used to explain the origin of the sporangium-leaf association in terms of preexisting structures, and in this context, the unique forked leaf tips of the Protolepidodendrales can be viewed as evidence of vestigial branches. Evolution of microphylls in the Selaginellales, Lepidodendrales, and Isoetales would have involved further loss of the forked leaf tips to provide an unbranched leaf, a shift of sporangium position from the leaf to the leaf axil, and a change in the orientation of dehiscence from radial to transverse in the Selaginellales. Stewart suggested that possible intermediate conditions were present in such taxa as *Protohyenia* and *Estinnophyton.* Evolution of similar microphylls in the Lycopodiaceae and the Drepanophycales was viewed as convergent on the condition in heterosporous lycopsids and in the Protolepidodendrales. Invoking the enation theory (Bower 1908, 1935), Stewart (1983) suggested that microphylls in groups such as the Lycopodiaceae evolved independently from unbranched spines in zosterophylls.

We view Stewart's (1983) diphyletic hypothesis as problematic for two reasons. First, because it focuses on the evolutionary origin of the microphyll through transformations of preexisting structures, we are concerned that one aspect of lycopsid morphology has been overemphasized in interpreting lycopsid evolution. Second, in the formulation of the diphyletic hypothesis, putative evolutionary processes (e.g., reduction) are given greater weight than patterns of character distribution. Manipulation of our data to conform to the diphyletic hypothesis shows that it is at least seven steps longer than our most parsimonious tree because it requires the independent origin in the Lycopodiaceae and ligulate lycopsids of the following: microphylls, helical phyllotaxy, a sporangium-leaf association, reniform sporangia in tangential section, a bilateral distribution of phloem and xylem in the root stele, a crescent-shaped root xylem strand, and pitted tracheids. Furthermore, independent evolution of the microphyll in lycopsids by two radically different pathways is not supported by ontogenetic studies, which show that leaf and sporangium development are very similar in the Lycopodiaceae, Selaginellaceae, and Isoetaceae (Bower 1894). There are also difficulties with the inclusion of *Estinnophyton* as a potential intermediate between rhyniophytes and a Protolepidodendrales–heterosporous lycopsids group. *Estinnophyton* is very poorly understood but differs from lycopsids in leaf and sporangium morphology. It is probably more closely related to sphenopsids, but much basic morphological and anatomical information is required before the relationships of this plant can be evaluated critically.

An alternative view of lycopsid phylogeny to Stewart's (1983) was provided by Philipson (1990, 1991) as part of a general hypothesis on the origin of land plants from primitive "bryophytic plants" with little or no sporophyte development (Figure 6.4). According to this hypothesis, which is based on the organization of the apical meristem, the Selaginellaceae are viewed as independently derived from a Lycopodiaceae-Isoetaceae clade. Philipson's suggestions are equivalent to a hypothesis of embryophyte polyphyly from within charophycean algae, which our analysis of general relationships in embryophytes found to be highly unparsimonious (Chapter 3). The independent evolution of the Selaginellaceae and its sporophyte generation requires the separate origin of numerous sporophyte features (e.g., eusporangium, stomates, internally differentiated sporophyte axis, tracheids, dichotomous branching, lignified hypodermis, roots), as well as the independent acquisition of both general and derived lycopsid features, such as reniform sporangium shape, bivalvate dehiscence, lateral sporangium position, microphylls, helical phyllotaxy, sporangium-microphyll association, nonvascular sporangia, ligules, heterospory, strobili, gametophyte reduction, an endosporic microgametophyte, free nuclear cell divisions in early stages of megagametophyte development, bilateral distribution of phloem and xylem in the root stele, and a crescent-shaped root xylem strand. It is clearly more straightforward and more parsimonious to treat these numerous features as homologous because of their similarity and congruence and to consider the possibility that the architecture of the sporophyte apex has been modified in the Selaginellaceae.

In a cladistic analysis of relationships in embryophytes based on male gametogenesis, Garbary, Renzaglia, and Duckett (1993) suggested that lycopsids are paraphyletic to other land plants (see Chapter 3, Figure 3.10). The Selaginellaceae were found to be a sister group to bryophytes, whereas the Lycopodiaceae were resolved as a sister group to other vascular plants. This hypothesis requires either that bryophytes evolved by reduction from lycopsids, which involves homoplastic losses of lycopsid sporophyte features, or that the sporophyte generation of the Selaginellaceae evolved independently with many homoplastic gains of sporophyte features. The relative parsimony of this hypothesis was examined in the context of general relationships in embryophytes (Chapter 3) and was found to be more than six steps less parsimonious than our solution.

Relationships within Lycopsids

Existing classifications show substantial agreement concerning the scope of lycopsid orders; most of the variation among the different systems is in the relative ranking of such arborescent heterosporous taxa as the Lepidodendrales (Lepidocarpales), the Pleuromeiales, and extant Isoetales. In some classifications these are recognized as separate orders (e.g., Bold 1957; Gifford and Foster 1989; T. N. Taylor 1981); in others, the Pleuromeiales are subsumed in the Lepidodendrales (e.g., Chaloner 1967a), while in still others, the Pleuromeiales and Lepidodendrales are both placed in the Isoetales (DiMichele and Bateman, in press; Meyen 1987) or Rhizomorpha (Bateman 1992). The Miadesmiales of Chaloner (1967a) and Thomas and Brack-Hanes (1984) is a poorly understood group that may be closely related to the Selaginellales or Rhizomorpha. Most authors place nonligulate homosporous fossils in the Drepanophycales (Asteroxylales *sensu* Bold [1957]) and ligulate homosporous fossils in the Protolepidodendrales. Sporne (1970) and T. N. Taylor (1981) include the Drepanophycales within Protolepidodendrales, whereas Gifford and Foster (1989) place all early homosporous forms in the Lycopodiales.

Our cladistic analysis supports the separation of the Drepanophycales (Asteroxylales) and Protolepidodendrales from each other and also from the Lycopodiales (Bierhorst 1971; Bold 1957; Chaloner 1967a; Meyen 1987; Stewart 1983; Thomas and Brack-Hanes 1984). Our analysis also supports a close relationship for heterosporous arborescent lycopsids and the Isoetaceae and therefore is compatible with the classifications by Meyen (1987), Bateman (1992), and DiMichele and Bateman (in press). The phylogenetic hypothesis of Meyen (1987; see Figure 5.6), although less explicit and less well resolved, also recognizes the same major groups as in our analysis (Figure 6.18) and can be interpreted as showing a broadly similar phylogenetic branching pattern. Bateman's (1992) more detailed numerical cladistic treatment differs principally from our analysis by resolving the Selaginellaceae as paraphyletic to rhizomorphic lycopsids and the Drepanophycales as a basal grade in the Lycopsida (see below and Figure 6.2).

Our results correspond closely to those shown in Crane's manually generated cladogram (1990; see also Figure 6.1); however, one difference is that Crane resolved *Asteroxylon* as a sister group to all of the lycopsids (including *Drepanophycus*), whereas in our analysis, *Asteroxylon, Baragwanathia,* and *Drepanophycus* form a clade that is a sister group to all of the other lycopsids. In this context, one surprising aspect of our analysis is that the characteristic vascularization of the microphylls in *Asteroxylon,* which is frequently interpreted as an intermediate stage in the evolution of the true lycopsid microphyll, is actually uninformative with respect to relationships with lycopsids in general. In *Asteroxylon* the vascular strand to the leaf stops at the leaf base, whereas in all other lycopsids it extends to the tip of the leaf, and this feature has been used to exclude *Asteroxylon* from lycopsids *sensu stricto.* Our analysis shows that this is unjustified because, as currently understood, the feature is autapomorphic in *Asteroxylon.* In the absence of an a priori notion of the process of microphyll evolution (such as might be embedded in an ordered multistate character), it is equally parsimonious to interpret the absence of true microphyll vascularization as a loss.

Herbaceous Lycopsids

One distinctly different phylogenetic scheme of lycopsid classification from that presented here hypothesizes that the main division is between herbaceous and arborescent groups rather than between homosporous eligulate and heterosporous ligulate taxa (Chaloner 1967a, 783). Chaloner pragmatically de-emphasized the importance of the ligule and heterospory because these two characters are unknown for many fossils. Experiments with our data show that a herbaceous clade is only one step longer than our most parsimonious tree and requires independent origins in the Selaginellaceae (or losses in the Lycopodiaceae) of the ligule, heterospory, and various gametophyte characters such as a small, reduced gametophyte, an endosporic microgametophyte, and free nuclear divisions in early stages of megagametophyte development. Also, according to Chaloner's hypothesis, independent origins of stellate xylem strands in the Lycopodiaceae are likely.

In a cladistic context the herbaceous habit shared by the Lycopodiaceae and Selaginellaceae is symplesiomorphic (present in the Protolepidodendrales, Drepanophycales, and the outgroup, zosterophylls) and does not provide a useful defining feature (i.e., is not a synapomorphy unless it is a reversal in a fundamentally arboreous group). However, our analysis of lycopsid characters revealed two features that could potentially be interpreted as synapomorphies for a putative herbaceous clade: a crescent-shaped xylem strand in the root and sinuate outlines of sporangium epidermal cells. The distribution of these features is poorly understood in homosporous fossils, and in our analysis they are outweighed by the characters that place the Selaginellaceae in a sister group relationship with rhizomorphic lycopsids.

Bateman's numerical cladistic analysis (1992) placed the early fossil *Baragwanathia* in the lycopsid crown group because its features were scored as identical to modern *Huperzia* (Figure 6.2), based on Hueber's description (1992). Our analysis supports previous suggestions of a stem group relationship for *Baragwanathia* with other early taxa such as *Asteroxylon* and *Drepanophycus*. Five critical characters at this level are (1) the type of sporangia (cauline versus foliar), (2) the type of tracheids (pitted versus G-type), (3) the presence of bulbils, (4) the presence of vascularized sporangia, and (5) the type of branching (anisotomous or isotomous). Bateman (1992) scored foliar sporangia as present in *Drepanophycus* and *Baragwanathia,* supporting a crown group relationship, but the state of this character is unclear in *Baragwanathia,* and new data (Li and Edwards 1995) support earlier interpretations of cauline sporangia in *Drepanophycus* (Gensel and Andrews 1984). A stem group relationship for these early fossils is also supported by the presence of sporangium vascularization in *Drepanophycus* and *Asteroxylon* (which is lost in the Lycopsida crown group), as well as by the occurrence of G-type tracheids rather than true pitted cells. The presence of bulbil-like branches is one character supporting a relationship among *Asteroxylon, Baragwanathia,* and some Huperzias (*Huperzia selago* group). Our analysis resolves this character as a parallelism in the Drepanophycales and the *Huperzia selago* group. Other types of branching (anisotomous versus isotomous) are more problematic at this level because they are difficult to score for many early fossils and are among the most homoplastic characters in our analysis. In view of these problems, we regard the monophyly of the Drepanophycales as weakly supported. More detailed information, particularly of strobilate regions, is required from early fossil taxa.

Lycopodiaceae

Two alternative hypotheses of relationship for the Lycopodiaceae were supported in our analysis. One set of trees favored the monophyly of the Lycopodiaceae based on the morphology of the spore wall surface. The pattern of relationships in these trees (Figure 6.18) corresponds well to Crane's interpretation (1990) of relationships among Øllgaard's (1987) groups (Figure 6.1). This topology is also consistent with cladistic relationships among the nine genera of North American Lycopodiaceae recognized by Wagner and Beitel (1992; see Figure 6.5). In light of the weak character support for the monophyly of the Lycopodiaceae in our analysis, it is also interesting that Wagner and Beitel do not list any family synapomorphies. Some of our trees supported a paraphyletic relationship between the Lycopodiaceae and ligulate lycopsids. One set of trees supported a sister group relationship for a *Lycopodium-Lycopodiella-Phylloglossum* clade and ligulate taxa, based on the presence of a strobilus (secondarily lost in the Protolepidodendrales).

Our analysis supports a close relationship between *Phylloglossum* and the *Lycopodium-Lycopodiella* clade, suggesting that *Phylloglossum* is nested within *Lycopodium sensu lato* (Crane 1990). Bateman (1992) resolved *Phylloglossum* as a sister group to ligulate lycopsids based on its "not deeply lobed" stelar morphology. We regard this hypothesis as poorly supported. The stelar morphology in *Phylloglossum* is highly divergent (Hackney 1950) and shows few similarities to steles in ligulate lycopsids. We therefore scored the stelar morphology of *Phylloglossum* as an independent character state.

Selaginellaceae

We resolved the Selaginellaceae as a weakly supported monophyletic group containing a large, strongly supported rhizophoric clade (Jermy's [1990b] *Tetragonostachys-Stachygynandrum-Heterostachys* group) and a small number of divergent basal taxa (e.g., *S. selaginoides;* Figure 6.18). This solution contrasts with Bateman's (1992) preferred most parsimonious tree, which resolved the Selaginellaceae as paraphyletic to the large rhizomorphic lycopsid clade (Figure 6.2). He resolved the *Stachygynandrum* group as a sister group to a *Selaginella*-rhizomorphic clade supported by the presence of determinate growth and a rootstock (Bateman 1992). In our view, both of these putative synapomorphies are problematic (see characters 6.43 and 6.44); we follow Karrfalt (1981) and Rothwell and Erwin (1985) in rejecting the homology between the swollen hypocotyl of *S. selaginoides* and the rhizomorphic root systems of arborescent lycopsids on the grounds that there are no significant similarities between these structures (see character 6.44). We also disagree with the idea that growth in the subgenus *Selaginella* is determinate, as it is in rhizomorphic lycopsids, and is therefore fundamentally different from growth in other Selaginellaceae, Lycopodiaceae, and Zosterophyllopsida (see character 6.43). The horizontal shoot system of the subgenus *Selaginella* is indeterminate; only the upright strobilate axes are determinate.

Bateman (1992) resolved the problematic *Paurodendron (Selaginella) fraipontii* as basal in rhizomorphic lycopsids and as having a probable close relationship to the pseudoherbaceous *Oxroadia,* rather than to early Selaginellaceae as had been previously suggested (Schlanker and Leisman 1969; Stewart 1983; T. N. Taylor 1981; T. N. Taylor and E. L. Taylor 1993; Figures 6.2 and 6.20). Although we did not explicitly test Bateman's hypothesis in our analysis, we support it because of the clear similarities between *P. fraipontii* and other rhizomorphic taxa in such characters as the presence of rootstock, bipolar growth, the presence of a rhizomorphic root system with characteristic monarch xylem anatomy, the presence of a cambium, the anatomy of the cortex, and longitudinal sporangium dehiscence (Bateman 1992; Rothwell and Erwin 1985; Stewart and Rothwell 1993). Similarities to the Selaginellaceae, such as small, bisporangiate cones, appear to be symplesiomorphic features because they occur in other basal rhizomorphic lycopsids such as *Oxroadia* (Bateman 1992, 1994).

Ligulate Lycopsids

Implicit in the earlier classifications of Bower (1935) and Bold (1957) is the recognition of two divergent lineages comprising the homosporous eligulate (Eligulatae, Aglossopsida) and heterosporous ligulate forms (Ligulatae, Glossopsida). Bower (1935, 285) viewed the basic condition in eligulate lycopsids as being represented by *Huperzia (Lycopodium) selago,* and in ligulate lycopsids by *Selaginella selaginoides (spinosa).* In general terms, these interpretations correspond well with previous cladistic treatments (Bateman 1992; Crane 1990). The discovery of a ligule in the homosporous fossil lycopsid *Leclercqia* (Grierson and Bonamo 1979) implies that the ligule is a synapomorphy at a more inclusive level than is heterospory: in evolutionary terms, the ligule evolved before heterospory. Our scheme predicts that in other members of the Protolepidodendrales, the ligule will be found to be situated some distance from the leaf base as in *Leclercqia* (Bonamo, Banks, and Grierson 1988).

Protolepidodendrales

Uncertainties regarding the position of the Protolepidodendrales were not fully resolved in our analysis. Our preferred most parsimonious tree resolves heterosporous lycopsids as monophyletic, as in previous cladistic studies (Bateman 1992; Crane 1990). However, an alternative, equally parsimonious hypothesis united the Protolepidodendrales with the rhizomorphic clade on the basis of sporangium morphology and dehiscence. This alternative hypothesis implies a double origin for heterospory in lycopsids or a single origin in ligulate lycopsids and a loss in the Protolepidodendrales. A better understanding of the morphology of early stem group Selaginellaceae will be critical to resolving this problem. This result also raises the possibility of heterospory in *Leclercqia,* and even though the hundreds of sporangia examined by Grierson and Bonamo (1979) yielded only microspores, this possibility is still difficult to exclude completely.

Isoetales–Rhizomorphic Lycopsids

The close phylogenetic relationship between *Isoetes* and the woody, heterosporous, arborescent Lepidodendrales and Pleuromeiales, supported strongly by many previous authors (Bateman 1992; Bateman, DiMichele, and Willard 1992; DiMichele and Bateman, in press; Meyen 1987; Rothwell 1984; Rothwell and Erwin 1985; Stewart 1947, 1983) is also well supported in our analysis. Shared derived features of this group include secondary xylem (cambium), pseudobipolar growth, distinctive "stigmarian" rootlet anatomy, a more or less terete root xylem strand, and the reduction or elimination of valvate dehiscence in the megasporangium.

Cladistic relationships among the diverse rhizomorphic lycopsids have recently been considered in detail by Bateman (1992) and Bateman, DiMichele, and Willard (1992; see Figure 6.20). Genera with bisporangiate cones (i.e., *Paurodendron, Oxroadia, Paralycopodites, Chaloneria*) form a poorly resolved basal paraphyletic group within which there is a clade characterized by monosporangiate cones (Figure 6.20). The monosporangiate clade contains three well-defined families: the Sigillariaceae, the Lepidodendraceae, and the recently recognized Diaphorodendraceae (DiMichele and Bateman 1992). Progressive reduction of the megaspore number per sporangium within this clade results in seedlike reproduction in the Diaphorodendraceae and Lepidodendraceae.

Convergence in the Selaginellaceae (Lycopsida) and Sawdoniales (Zosterophyllopsida)

Our analysis of relationships in the basal Lycopsida highlights several striking similarities between certain extant Selaginellaceae in the well-supported "rhizophoric" clade, in particular the *Stachygynandrum* group, and certain extinct zosterophylls in the Sawdoniales. Examination of character-state distributions shows that much of the homoplasy in our analysis involved parallel gains of certain features in these two groups (e.g., planar, pseudomonopodial branching; elliptical to strap-shaped xylem strands; and "axillary branches or rhizophores"), which are interpreted a posteriori as convergences. Similarity between the rhizophore in *Stachygynandrum* and the "axillary tubercle branch" in the Sawdoniales was pointed out by Banks and Davis (Banks and Davis 1969) and discussed in detail by Edwards, Kenrick, and Carluccio (1989). The similarity extends to position (associated with the dichotomies of the main branching system and positioned on one side of the stem), exogenous development, and possibly also retarded or suppressed development. However, the axillary tubercle branch in zosterophylls differs in its basically shootlike morphology, which is expressed as circinate vernation and spinelike emergences in taxa with spiny stems. A stringent test of homology between these two organs was built into our analysis by scoring "rhizophore" and "axillary branching" as the same state (i.e., weighting in favor of homology). Our examination of character-state distribution shows two independent acquisitions of this feature and therefore clearly rejects the hypothesis of homology. The notion of homology can be tested further by changing the tree topology to favor homology in this organ and observing the effect on other characters. Placing the Sawdoniales within the lycopsids in a sister group relationship with the rhizophoric groups in *Selaginella* was five steps longer than our most parsimonious tree. Important homoplastic changes include loss of microphyll and sporophyll, loss of ligule, loss of cone and pitted tracheids, loss of heterospory, and loss of microspore ornamentation. Placing the Selaginellaceae outside the lycopsids in a sister group relationship with the Sawdoniales was even less parsimonious (nine additional steps).

7 Perspectives on the Early Evolution of Land Plants

Current perspectives on the early diversification of land plants have been influenced strongly by the wealth of comparative data that has accumulated over the last century—much of which has been reviewed and partially systematized in the preceding chapters. Using these data to develop meaningful phylogenetic and evolutionary interpretations requires an analytical approach, which in turn requires an appropriate theoretical framework. Although many modern studies of plant diversity have eschewed theoretical considerations, it is undeniable that various theories are deeply embedded in previous and current concepts of early land plant phylogeny. Hypotheses of life-cycle evolution (e.g., Bower 1908; Fritsch 1916, 1945) were used to bolster contrasting views of bryophyte phylogeny, while ideas of organ evolution (e.g., the Telome Theory: Zimmermann 1952, 1965) continue to have a substantial impact on the phylogenetic interpretations of lycopsids, sphenopsids, and many other groups of land plants (e.g., Stewart and Rothwell 1993). In general, however, the utility of such theories has been limited by the absence of an analytical component and hence an explicit method for discriminating among alternative phylogenetic possibilities. The development of cladistics—and associated advances in systematic theory—has introduced a discipline that overcomes some of these difficulties.

A further innovation of cladistics has been to shift the emphasis in phylogenetic studies away from theories that incorporate a priori notions of evolutionary process toward the more straightforward task of recognizing patterns of character distribution. As a result, there has also been substantial clarification of the way in which groups can be recognized, of the nature of the evidence required to substantiate relationships, and of the basis for evaluating competing phylogenetic hypotheses. Our focus in this book on character analysis reflects this change in approach and emphasizes our view that well-established systematic patterns are a prerequisite for well-founded interpretations of most evolutionary events and processes.

In this chapter we summarize the main systematic conclusions developed in Chapters 3–6, discuss some of the remaining problems, and based on these results, reiterate the classification of major embryophyte groups presented earlier (Tables 1.2 and 7.1). This systematic framework is then used to examine temporal and geographic patterns in the early diversification of land plants. Finally, we examine the implications of our phylogenetic hypotheses for interpreting the evolution of morphology, anatomy, and life history patterns in the initial phases of land plant evolution.

SYSTEMATICS AND SUMMARY OF RELATIONSHIPS

Embryobiotes

The Embryobiotes (land plants) are one of the most well-supported major groups in the Plant Kingdom (Figure 7.1). Our analysis confirms other cladistic studies that support the monophyly of land plants based on morphological characters (Bremer 1985; Bremer, Humphries, et al. 1987; Donoghue 1994; Garbary, Renzaglia, and Duckett 1993; Graham, Delwiche, and Mishler 1991; Mishler and Churchill 1984; 1985b; Mishler, Lewis, et al. 1994). Synapomorphies of the Embryobiotes include the presence of archegonia and embryos (possibly developmentally linked), antheridia, cuticles, sporangia (spore mass encapsulated within a sterile, cellular wall), and spore walls containing sporopollenin, as well as ultrastructural details of male gametogenesis (Table 7.2; Garbary, Renzaglia, and Duckett 1993; Graham, Delwiche, and Mishler 1991; Graham and Repavich 1989) and cell division (Brown and Lemmon 1990a,b, 1991a,b). Embryophyte monophyly is also consistent with recent studies of 18S and 26S rRNA (Chapman and Buchheim 1991; Kranz and Huss 1996; Kranz, Miks, et al. 1995; Mishler, Lewis, et al. 1994; Mishler, Thrall, et al. 1992; Waters, Buchheim, et al. 1992) and the *rbcL* chloroplast gene (Albert, Backlund, Bremer, Chase, et al. 1994; Manhart 1994).

Cladistic studies show that hypotheses of embryophyte polyphyly from within green algae (e.g., Crandall-Stotler 1984; Duckett and Renzaglia 1988a; Philipson 1990, 1991; Schuster 1984a; Stewart and Rothwell 1993) are highly

TABLE 7.1
Cladistic Classification of Green Plants

SUPERKINGDOM EUKARYOTA (DOMAIN EUCARYA)

KINGDOM CHLOROBIOTA (METAPHYTAE, PLANTAE)

Subkingdom "Micromonadobionta" *incertae sedis*

DIVISION "MICROMONADOPHYTA"
- Class "Micromonadophyceae" ("Micromonadophytopsida")

Subkingdom Ulvobionta

DIVISION ULVOPHYTA
- Class "Ulvophyceae" ("Ulvophytopsida")
- Class Pleurastrophyceae (Pleurastrophytopsida)
- Class Chlorophyceae (Chlorophytopsida)

Subkingdom Streptobionta

INFRAKINGDOM CHLOROKYBIOTES
DIVISION CHLOROKYBOPHYTA
- Class Chlorokybophyceae (Chlorokybophytopsida)

INFRAKINGDOM "KLEBSORMIDIOBIOTES"
DIVISION "KLEBSORMIDIOPHYTA"
- Class "Klebsormidiophyceae" ("Klebsormidiophytopsida")

INFRAKINGDOM ZYGNEMOBIOTES
DIVISION ZYGNEMOPHYTA
- Class Zygnemophyceae (Zygnemophytopsida)

INFRAKINGDOM CHAROBIOTES *INCERTAE SEDIS*
DIVISION CHAROPHYTA
- Class Charophyceae (Charophytopsida)

Superdivision Polysporangiomorpha
PLESION HORNEOPHYTOPSIDA
PLESION *AGLAOPHYTON MAJOR*
DIVISION TRACHEOPHYTA
Plesion Rhyniopsida
Subdivision Lycophytina
- Plesion *Zosterophyllum myretonianum incertae sedis*
- Class Lycopsida (Lycophyceae)
  - Plesion Drepanophycales
  - Order "Lycopodiales"
  - Plesion Protolepidodendrales
  - Order Selaginellales
  - Order Isoetales
- Plesion Zosterophyllopsida (Zosterophyllophyceae)
  - Plesion *Zosterophyllum divaricatum*
  - Plesion Sawdoniales (families *sedis mutabilis*)
    - Plesion Sawdoniaceae
    - Plesion Barinophytaceae
    - Plesion "Gosslingiaceae"
    - Plesion Hsuaceae

Subdivision Euphyllophytina
Plesion *Eophyllophyton bellum*
Plesion *Psilophyton dawsonii*
Infradivision Moniliformopses (classes *sedis mutabilis*)
- Plesion "Cladoxylopsida" ("Cladoxylophyceae") (subclasses *sedis mutabilis*)
  - PLESION "CLADOXYLIIDAE"
  - PLESION STAUROPTERIDAE
  - PLESION ZYGOPTERIDAE
- Class Equisetopsida (Equisetophyceae)
- Class Filicopsida (Filicophyceae) (subclasses *sedis mutabilis*)
  - SUBCLASS OPHIOGLOSSIDAE
  - SUBCLASS PSILOTIDAE
  - SUBCLASS MARATTIIDAE
  - SUBCLASS POLYPODIIDAE

INFRAKINGDOM CHAETOSPHAERIDIOBIOTES

INCERTAE SEDIS

DIVISION CHAETOSPHAERIDIOPHYTA

Class Chaetosphaeridiophyceae (Chaetosphaeridiophytopsida)

INFRAKINGDOM "COLEOCHAETOBIOTES"

INCERTAE SEDIS

DIVISION "COLEOCHAETOPHYTA"

Class "Coleochaetophyceae" ("Coleochaetophytopsida")

INFRAKINGDOM EMBRYOBIOTES

Superdivision Marchantiomorpha

DIVISION MARCHANTIOPHYTA

Class Marchantiopsida (Marchantiophyceae)

Order Sphaerocarpales
Order Monocleales *incertae sedis*
Order Marchantiales *incertae sedis*
Order Calobryales
Order "Metzgeriales"
Order Jungermanniales

Superdivision Anthoceromorpha

DIVISION ANTHOCEROPHYTA

Class Anthocerotopsida (Anthocerotophyceae)

Superdivision Bryomorpha

DIVISION BRYOPHYTA

Class Bryopsida (Bryophyceae)
SUBCLASS SPHAGNIDAE
SUBCLASS ANDREAEIDAE
Order Takakiales
Order Andreaeales
Order Andreaeobryales
SUBCLASS TETRAPHIDAE
SUBCLASS POLYTRICHIDAE
SUBCLASS BUXBAUMIIDAE
SUBCLASS BRYIDAE

Infradivision Radiatopses
PLESION *PERTICA VARIA*
SUPERCOHORT LIGNOPHYTIA (COHORTS *SEDIS MUTABILIS*)
Plesion "Aneurophytales"
Plesion "Archaeopteridales"
Plesion "Protopityales"
Cohort Spermatophytata
PLESION "CALAMOPITYACEAE" *INCERTAE SEDIS*
PLESION "HYDRASPERMACEAE"
PLESION "LYGINOPTERIDACEAE"
PLESION MEDULLOSACEAE
SUBCOHORT EUSPERMATOCLIDES (INFRACOHORTS *SEDIS MUTABILIS*)
Infracohort Cycadatae
Plesion Callistophytaceae
Infracohort Coniferophytatae
PLESION CORDAITIDRA
SUPERCLASS CONIFERIDRA (PINIDRA)
Plesion Glossopteridaceae
Plesion Czekanowskiaceae
Infracohort Ginkgoatae
Plesion "Peltaspermaceae"
Plesion "Corystospermaceae" ("Umkomastiaceae")
Plesion Caytoniaceae
Infracohort Anthophytatae (superclasses *sedis mutabilis*)
PLESION PENTOXYLALES
PLESION BENNETTITALES
SUPERCLASS GNETIDRA
SUPERCLASS MAGNOLIDRA
Class "Magnoliopsida" (Magnoliophyceae)
Class Liliopsida (Liliophyceae)
Class Hamamelidopsida (Hamamelidophyceae)
SUBCLASS RANUNCULIDAE
SUBCLASS HAMAMELIDIDAE (INFRACLASSES *SEDIS MUTABILIS*)
Infraclass Caryophyllidna
Infraclass "Rosidna"
Infraclass "Dilleniidna"
Infraclass Lamiidna
Infraclass Asteridna

See notes on next page.

unparsimonious. Generally, such polyphyletic hypotheses have been supported on the basis of the divergent morphology of the basal embryophyte groups or by perceived inconsistencies in the stratigraphic occurrence of fossils. From a cladistic perspective, the highly divergent morphology of such groups as hornworts (Anthocerotopsida) is best interpreted as autapomorphic—that is, a result of changes that occurred *after* splitting from other embryophyte lineages. Similarly, interpretations based on perceived stratigraphic inconsistencies contain assumptions about the rates of evolutionary change and about the completeness of the fossil record that have little empirical support. Based on the analyses presented here, hypotheses of embryophyte polyphyly are unparsimonious, are incongruent with each other, fail to specify alternative green algal sister groups, and frequently imply groupings of taxa that are based on plesiomorphic characters.

Recent systematic studies of green algae strongly support an origin of land plants from the grade of charophycean algae (Bremer 1985; Bremer, Humphries, et al. 1987; Chapman and Buchheim 1991; Graham 1993; Graham, Delwiche, and Mishler 1991; Kenrick 1994; Manhart 1994; Mattox and Stewart 1984; Mishler and Churchill 1985b; Kranz, Miks, et al. 1995; Mishler, Lewis, et al. 1994). Cladistic studies based on morphological and biochemical data, as well as molecular sequences, resolve charophycean algae as paraphyletic to land plants, but there is disagreement over the extant sister group. A sister group relationship between the Charales and a *Coleochaete*-embryophyte clade is supported by morphological characters (Figure 7.1; Graham 1993; Graham, Delwiche, and Mishler 1991; Mishler and Churchill 1985b), and according to this hypothesis, *Coleochaete* may be the sister group to embryophytes. More controversially, it has been suggested that the genus *Coleochaete* itself is paraphyletic to embryophytes (Mishler and Churchill 1985b). Such a relationship seems unlikely because it requires the loss of the distinctive hairs *(chaete)* that are present in all *Coleochaete* species, and also because it implies derivation of the large, diverse, and ancient embryophyte clade from within a small and well-defined extant genus. There is also reason for caution because molecular studies give conflicting results on the question of which green algae are the closest relatives of embryophytes. Different studies favor principally either the Charales or *Coleochaete* or both, as the sister group to embryophytes (Chapman and Buchheim 1991; Kranz, Miks, et al. 1995; Manhart 1994; Mishler, Lewis, et al. 1994; Wilcox, Fuerst, and Floyd 1993).

Relationships among basal groups of embryophytes are still poorly resolved. Conflicting results have been obtained from studies of comparative morphology (Mishler and Churchill 1984, 1985b), of the ultrastructural features of male gametes and gametogenesis (Garbary, Renzaglia, and Duckett 1993), and of *rbcL* sequences (Manhart 1994). More recently, an analysis incorporating data on morphology, molecular sequences, and male gametogenesis produced a completely resolved topology consistent with the bryophyte paraphyly hypothesis of Mishler and Churchill (1984, 1985b) and this study (Chapter 3).

Notes to Table 7.1: Outline classification of the plant kingdom based mainly on the work reviewed in this book and on previous cladistic classifications of Bremer (1985) and Bremer, Humphries, et al. (1987). The classification of seed plants is based on broad areas of agreement at the most basic level in recent cladistic studies, although many uncertainties remain (Crane 1985; Doyle and Donoghue 1986, 1992; Nixon, Crepet, et al. 1994; Rothwell and Serbet 1994). To accommodate additional hierarchical levels, while at the same time preserving common usage (especially at the level of angiosperms), the recommendations of the International Code of Botanical Nomenclature have been extended by (1) use of the prefix *infra-* to designate ranks immediately below the levels subkingdom, subclass, and subdivision; (2) use of the prefix *super-* in conjunction with division, cohort, and class; and (3) the introduction of the rank cohort between division and class. This usage differs from that in some zoological treatments in which cohort is used as broadly equivalent to class. At the level of class the two alternative standardized endings given in the ICBN for classes of algae *(-ophyceae)* and "higher" green plants *(-opsida)* are shown throughout to emphasize their application to the same level of the hierarchy. We deeply appreciate the assistance of Fred Barrie, Dan H. Nicolson, and Sandy Knapp with respect to the nomenclature of this classification. Any remaining errors are, however, our responsibility. Quotation marks = paraphyletic group.

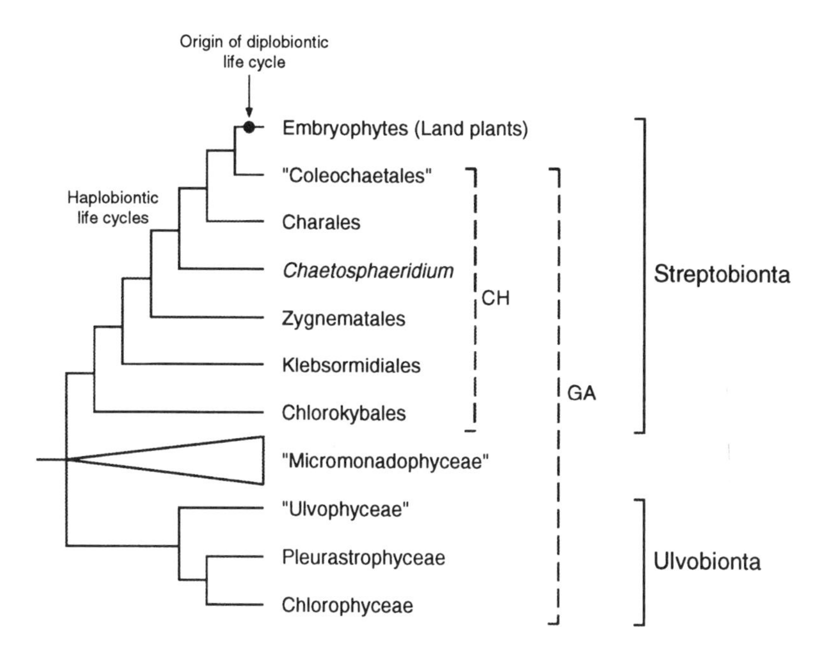

FIGURE 7.1. Cladistic relationships among Chlorobiota (green plants). Adapted from Bremer 1985; Graham, Delwiche, and Mishler 1991; Kenrick 1994; Mattox and Stewart 1984; Mishler and Churchill 1985b; and Mishler, Lewis, et al. 1994. CH = Charophyceae; GA = green algae; quotation marks = paraphyletic group. The Micromonadophyceae are not monophyletic and their relationships are currently poorly understood. Ulvophyceae are possibly not monophyletic. The cladogram implies that diplobiontic life cycles of land plants evolved from haploid, haplobiontic life cycles in charophycean algae. See Chapter 2 for further discussion and synapomorphies.

TABLE 7.2
Synapomorphy-Based Definitions of Monophyletic Higher Taxa

Clade	Synapomorphies
Embryobiotes	(1) Multicellular sporophytes; (2) cuticle; (3) archegonia; (4) antheridia; and (5) sporangium (Chapter 3; Mishler and Churchill 1985b). Other synapomorphies include (6) details of spermatozoid ultrastructure (Garbary, Renzaglia, and Duckett 1993; Graham and Repavich 1989; Mishler, Lewis, et al. 1994); (7) details of cell division (Brown and Lemmon 1990a); and (8) sporopollenin in the spore wall (Graham 1990). Molecular studies summarized by Manhart (1994, 1995) and Mishler, Lewis, et al. (1994)
Marchantiopsida	(1) Oil bodies; (2) spermatozoid ultrastructure; and possibly (3) presence of lunularic acid (Chapter 3; Garbary, Renzaglia, and Duckett 1993; Mishler, Lewis, et al. 1994). Molecular studies summarized by Manhart (1994, 1995) and Mishler, Lewis, et al. (1994)
Stomatophytes	(1) Stomates; possibly (2) columellate sporangium; and (3) D-methionine (Chapter 3; Mishler and Churchill 1984)
Anthocerotopsida	(1) Apical cell shape; (2) pyrenoid in chloroplast; (3) mucilage cells in thallus; (4) cavities in thallus; and (5) endogenous antheridia. Other putative synapomorphies include (6) sunken archegonium; (7) vertical division of zygote; (8) meristem at base of sporangium (Chapter 3; Hässel de Menéndez 1988; Mishler and Churchill 1985b); and (9) spermatozoid ultrastructure (Garbary, Renzaglia, and Duckett 1993; Mishler, Lewis, et al. 1994)

Continues on next page

Notes: Taxa listed in approximate taxonomic sequence. See also classifications in Tables 7.1 and 7.3.

TABLE 7.2
Continued

Clade	Synapomorphies
Bryopsida and Polysporangiomorpha	(1) Axial gametophyte; (2) terminal gametangia; (3) perine layer on spores; (4) persistent and internally differentiated sporophytes; and possibly (5) details of auxin metabolism (Sztein, Cohen, et al. 1995) (see also Chapter 3; Mishler, Lewis, et al. 1994)
Bryopsida	(1) Multicellular gametophytic rhizoids; (2) gametophytic leaves; and (3) spermatozoid ultrastructure (Chapter 3; Mishler, Lewis, et al. 1994)
Polysporangiomorpha	(1) Multiple sporangia (sporophyte branching); (2) independent alternation of generations; and possibly (3) sunken archegonia (Chapter 4; Kenrick and Crane 1991)
Horneophytopsida	(1) Branched sporangia; (2) small, multicellular protuberances from the sporangium surface; and possibly (3) dehiscence through an apical slit or pore (Chapter 4; Kenrick and Crane 1991)
Tracheophyta	(1) Annular-helical thickenings in tracheids and possibly (2) lignin deposition on the inner surface of the tracheid cell wall (Chapter 4; Kenrick and Crane 1991)
Rhyniopsida	(1) Distinctive adventitious branching (*Rhynia*-type); (2) abscission or isolation layer at base of sporangium; and (3) sporangia attached to a "pad" of tissue (Chapter 4; Kenrick and Crane 1991)
Eutracheophytes	(1) Thick, lignified wall layer in tracheid; (2) pitlets between thickenings or within pits in tracheid; and (3) sterome (Chapter 4; Kenrick and Crane 1991)
Lycophytina	(1) More or less reniform sporangia; (2) marked sporangium dorsiventrality; (3) isovalvate dehiscence; (4) conspicuous cellular thickening of the dehiscence line; (5) sporangia on short, laterally inserted stalks; and (6) exarch xylem differentiation (see also Chapters 4 and 6; DiMichele and Bateman, in press; Hueber 1992)
Lycopsida	(1) Microphylls; (2) stellate xylem strand; (3) close developmental association of sporangium and microphyll; (4) pitted tracheids; and (5) loss of sporangium vasculature (see also Chapter 6; DiMichele and Bateman, in press)
Drepanophycales	Bulbils or small lateral buds (a parallelism with extant Lycopodiaceae in the *Huperzia selago* group: Chapter 6)
Lycopodiales	"Foveolate-fossulate" microspore wall morphology (see also Chapter 6; Bateman 1992; Crane 1990; DiMichele and Bateman, in press; Wagner and Beitel 1992)
Ligulate lycopsids	(1) Ligule; (2) terete, ribbed stele; and possibly (3) radial extension of sporangium (reversed in Selaginellaceae) (see also Chapter 6; DiMichele and Bateman, in press)
Protolepidodendrales	(1) Forked microphylls and possibly (2) anisotomous branching; (3) nonsinuate sporangium epidermal cells; and (4) radial dehiscence of sporangium (Chapter 6)
Heterosporous lycopsids	(1) Heterospory; (2) strobili; and possibly (3) reduction of gametophyte; (4) endosporic microgametophyte; and (5) free nuclear cell divisions in early stages of megagametophyte (see also Chapter 6; DiMichele and Bateman, in press)
Selaginellales	(1) More or less spherical microsporangia; and possibly (2) distal dehiscence (reversal to plesiomorphic condition from radial dehiscence); (3) four megaspores per sporangium; (4) suspension of stele in cavity by trabeculate endodermal cells; and (5) echinate microspores (Chapter 6)
Isoetales	(1) Cambium; (2) pseudobipolar growth involving rhizomorphic root system; and (3) monarch xylem strand in root (Chapter 6; Bateman 1992; Bateman, DiMichele, and Willard 1992; Crane 1990; DiMichele and Bateman, in press)
Zosterophyllopsida	(1) Circinate growth; (2) two-rowed sporangial arrangement; and possibly (3) elliptical xylem strand (see also Chapters 4 and 5; Hueber 1992)

TABLE 7.2
Continued

Clade	Synapomorphies
Sawdoniales	(1) Pseudomonopodial branching; (2) planated branching system; and (3) a unique form of subordinate axillary branching (Chapter 5)
Sawdoniaceae	Multicellular spines (Chapter 5)
Barinophytaceae	(1) Unique form of heterospory (megaspores and microspores in same sporangium); (2) compact, unbranched strobilus; (3) a unique form of "clasping" sporangium orientation; and possibly (4) loss of well-defined sporangium dehiscence (Chapter 5; Brauer 1980, 1981; Cichan, Taylor, and Brauer 1984; Taylor and Brauer 1983)
Gosslingiaceae	Unique "auricular" sporangium orientation (Chapter 5)
Hsuaceae	Reversal to terminal sporangia on planar lateral branches (Chapter 5)
Euphyllophytina	(1) Pseudomonopodial or monopodial branching; (2) helical arrangement of branches; (3) small, pinnulelike vegetative branches (nonplanated in basal taxa); (4) "recurvation" of branch apexes; (5) tracheids with scalariform bordered pits; (6) sporangia in pairs grouped into terminal trusses; (7) sporangium dehiscence along one side through a single slit; (8) radially aligned xylem in larger axes; and possibly (9) multiflagellate spermatozoids (convergent in *Isoetes*) (Chapters 4 and 7; Banks 1968, 1970; Bremer 1985; Chaloner and Sheerin 1979; Crane 1990; Doyle and Donoghue 1986; Hueber 1992; Kenrick and Crane 1991; Raubeson and Jansen 1992; Stein, Wight, and Beck 1984)
Moniliformopses	Mesarch protoxylem confined to lobes of xylem strand (Chapter 7; Beck and Stein 1993; Stein 1993)
Equisetopsida	(1) Whorled appendages; (2) sporangiophore morphology; (3) stelar morphology; (4) regular alternation of appendages at successive nodes; (5) microphyllous "leaves"; possibly (6) cambium (lost in Equisetaceae); and (7) the presence of a perispore (possibly more general) (Stein, Wight, and Beck 1984); additional characters noted by Bateman (1991) include (1) a medullated stele; (2) operculate strobili; and (3) columnar wall thickenings on sporangium epidermis; characters supporting a close relationship between the early fossil *Ibyka* and sphenopsids include (1) whorled branching (Stein, Wight, and Beck 1984) and (2) protoxylem disintegration to form lacunae (Chapter 7; Skog and Banks 1973)
Filicopsida	Leptosporangiate ferns, Ophioglossales and Marattiales: (1) "fern megaphylls"; (2) circinate vernation (present, but less pronounced in Ophioglossales [Tryon and Tryon 1982]; lost in Psilotaceae); and (3) septate rhizoids on gametophyte (Bierhorst 1971; Camus 1990); leptosporangiate ferns (including Osmundaceae): (1) distinctive annulate dehiscence of sporangium; (2) superficial antheridia; (3) operculate cell in antheridium; (4) C-shaped leaf trace (Bierhorst 1971); and (5) possibly siphono-dictyostelic anatomy (see also Chapter 7; Pryer, Smith, and Skog 1996)
Radiatopses	(1) Tetrastichous branching and (2) a distinctive form of protoxylem ontogeny with multiple strands occurring along the midplanes of the primary xylem ribs (Chapter 7; Beck and Stein 1993; Stein 1993)
Lignophytia	Bifacial cambium producing secondary xylem, phloem, and wood rays (Crane 1985, 1990; Doyle and Donoghue 1986). Certain aspects of cambial activity convergent with Isoetales, sphenopsids, and some early Cladoxylopsida (Chapter 7)
Spermatophytata	(1) Single megaspore per megasporangium and (2) integument. The medullosan-platysperm clade is further defined on (1) loss of lagenostome; (2) presence of pollen chamber; and (3) possibly also bilaterally symmetrical pollen (Crane 1985, 1990; Doyle and Donoghue 1986; Rothwell and Serbet 1994)

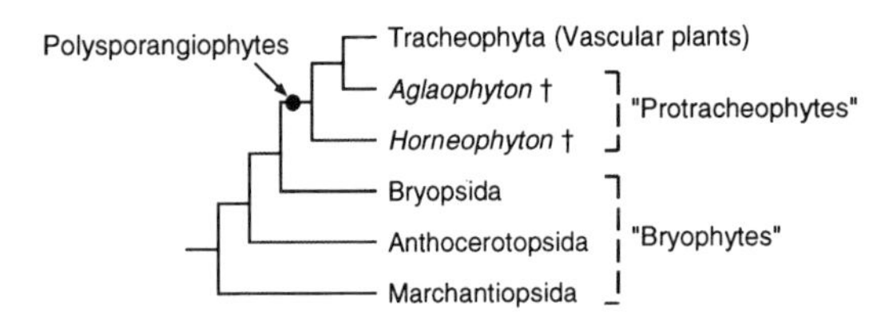

FIGURE 7.2. Cladistic relationships among basal land plants, showing the relationship of bryophytes to polysporangiophytes and protracheophytes. † = extinct; quotation marks = paraphyletic group. The sister group relationship between Bryopsida (mosses) and polysporangiophytes is currently weakly supported (see Chapter 3; and Mishler, Lewis et al. 1994).

Our study supports the monophyly of liverworts (Marchantiopsida), hornworts (Anthocerotopsida), mosses (Bryopsida), and polysporangiophytes and the paraphyly of bryophytes with respect to polysporangiophytes (Figures 7.1 and 7.2). We resolved mosses as a sister group to vascular plants but obtained conflicting results on relationships among liverworts, hornworts, and the moss-tracheophyte clade. The stomatophyte clade (hornworts-mosses-polysporangiophytes) of Mishler and Churchill (1984, 1985b) and Kenrick and Crane (1991) was recognized in a set of equally parsimonious trees that also included alternative topologies supporting a liverwort-hornwort clade or a liverwort-moss-polysporangiophyte clade. New data on auxin (indole-3-acetic acid, IAA) metabolism in land plants is consistent with a stomatophyte clade and a basal position for liverworts in land plants (Sztein, Cohen, et al. 1995). Characters supporting a moss-polysporangiophyte clade include axial gametophytes, terminal gametangia (both independently derived in such liverworts as the Calobryales and Jungermanniales), a perine layer on the spores, and persistent and internally differentiated sporophytes with "xylem" surrounded by "phloem" (sporangiophores: lost in mosses such as *Andreaea* and *Sphagnum*). Problems at this level include the relatively small amount of well-sampled comparative data, the difficulty of polarizing multicellular sporophyte characters because of unicellularity in the green algal outgroups, the divergent morphologies of the major embryophyte groups (probably exacerbated by extinction in the stem lineages), and the poor early macrofossil record of charophycean algae as well as of certain critical land plant groups (i.e., liverworts, hornworts, mosses; see below). Macrofossil evidence of the early members of these groups could have a significant impact on our understanding of higher-level relationships in embryophytes.

Marchantiopsida

Our analysis is consistent with the monophyly of the Marchantiopsida as argued by Mishler and Churchill (1985b), but character support for this group is relatively weak (Chapter 3; Table 7.2). Based on the taxonomy of Schuster (1984b), the cladistic study by Mishler and Churchill (1985b) supported the monophyly of the Sphaerocarpales, Marchantiales, Monocleales, and Jungermanniales but questioned the monophyly of the Metzgeriales and the Calobryales (Figure 7.3). The Metzgeriales are probably paraphyletic to the Jungermanniales. Similarly, the monophyly of the Calobryales (e.g., *Haplomitrium, Takakia*) was not strongly supported, and the relationship of *Takakia* (sporophyte unknown at the time) to the Marchantiopsida was questioned because of the possibility that this problematic taxon is more closely related to the Bryopsida (see Murray 1988). The Marchantiidae (Sphaerocarpales, Marchantiales,

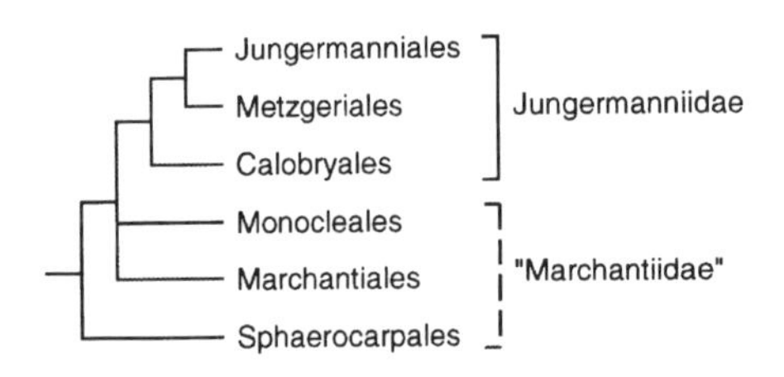

FIGURE 7.3. Cladistic relationships of the Marchantiopsida (liverworts), from Mishler and Churchill 1985b, Figure 3, based on the classification of Schuster 1984b. Quotation marks = paraphyletic group. Note that significant new information from spermatozoid ultrastructure and the recent discovery of the sporophyte generation indicate that *Takakia* (Calobryales: Schuster 1984b) is more closely related to the Bryopsida (Figure 7.5). See Chapter 3 for further discussion and Mishler and Churchill 1985b for synapomorphies.

Monocleales) were found to be paraphyletic to the clearly monophyletic Jungermanniidae (Calobryales, Metzgeriales, Jungermanniales: Figure 7.3; Mishler and Churchill 1985b). Relationships within the Marchantiidae were poorly resolved, but the Sphaerocarpales were placed as a sister group to all other liverworts.

More recent studies based on rRNA and *rbcL* sequences and data on spermatozoid ultrastructure give contradictory results. Early results from spermatozoid data (seven taxa) show similarities with the Mishler and Churchill cladogram (1985b), except that liverworts are rooted near the Calobryales and Metzgeriales (Garbary, Renzaglia, and Duckett 1993). Reanalysis of a modified spermatozoid data set did not resolve liverwort monophyly (Mishler, Lewis, et al. 1994). Spermatozoid data indicate that *Takakia* is more closely related to the Bryopsida than to the Marchantiopsida (Garbary, Renzaglia, and Duckett 1993; Mishler, Lewis, et al. 1994), a result that is supported by the recent description of antheridia and the discovery of the sporophyte generation (Davison, Smith, and McFarland 1989; Renzaglia, Smith, et al. 1991; Smith and Davison 1993). Initial analyses of rRNA and *rbcL* sequences provide a range of contradictory results, some of which support liverwort monophyly (Mishler, Thrall, et al. 1992), and some of which indicate other relationships (Manhart 1994; Waters, Buchheim, et al. 1992). Contradictions in the trees generated using molecular sequences, the ultrastructure of male gametes, and gametogenesis may be partly explained by the relatively small sample of liverwort diversity included in these early studies, as well as by the absence of critical land plant taxa (Chapter 3). The fossil record of the Marchantiopsida has not yet contributed significant information to cladistic studies of this group.

Anthocerotopsida

The monophyly of the Anthocerotopsida is well supported and uncontroversial (Table 7.2; Crandall-Stotler 1981, 1984; Garbary, Renzaglia, and

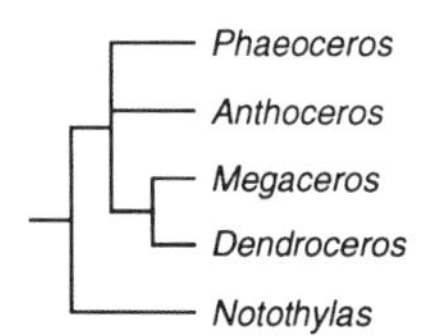

FIGURE 7.4. Cladistic relationships of genera in the Anthocerotopsida (hornworts), from Mishler and Churchill 1985b, Figure 4. See Chapter 3 for further discussion and Mishler and Churchill 1985b for synapomorphies.

Duckett 1993; Hasegawa 1988; Hässel de Menéndez 1988; Mishler and Churchill 1985b; Renzaglia 1978; Schuster 1984c). Mishler and Churchill (1985b) cited 11 synapomorphies for the group (Table 3.3), and Hässel de Menéndez (1988) identified a total of 17. Monophyly is also consistent with recent studies based on molecular sequences (Mishler, Lewis, et al. 1994; Waters, Buchheim, et al. 1992); however, relationships of genera within the Anthocerotopsida are currently poorly resolved. Such taxa as *Phaeoceros, Anthoceros, Megaceros,* and *Dendroceros* are poorly defined and may be paraphyletic (see Mishler and Churchill 1987; Whittemore 1987). In addition, several new generic entities have been recognized recently (e.g., Hasegawa 1988; Hässel de Menéndez 1988; Renzaglia 1978; Schuster 1984c). Mishler and Churchill (1985b) resolved *Notothylas* as a sister group to an *Anthoceros-Dendroceros-Megaceros-Phaeoceros* clade (Figure 7.4). An alternative, less parsimonious solution interprets the small, simple sporophytes of *Notothylas* as derived and taxa with larger sporophytes such as *Dendroceros* and *Megaceros* as primitive. The fossil record of the Anthocerotopsida is poor and has not yet contributed important information to cladistic studies of this group.

Bryopsida

Our analysis is consistent with the monophyly of the Bryopsida (Table 7.2). The cladistic study by Mishler and Churchill (1984) supported the monophyly of the peristomate mosses (Bryales, Buxbaumiales, Polytrichales) and a sister group relation-

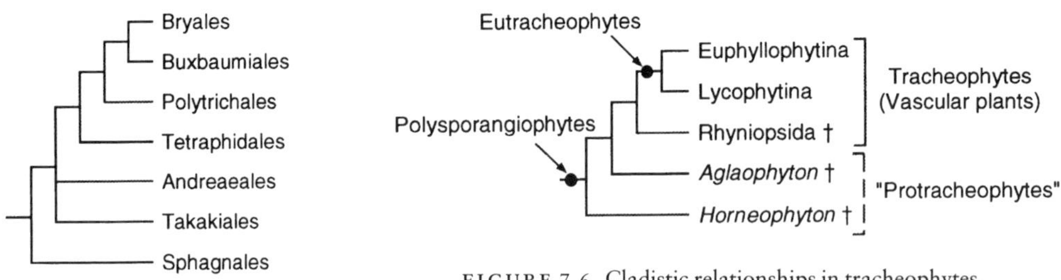

FIGURE 7.5. Cladistic relationships in the Bryopsida (mosses). Modified after Garbary, Renzaglia, and Duckett 1993, Figure 2, and Mishler, Lewis, et al. 1994, Figure 12. Note the position of the Takakiales (formerly the Calobryales, Marchantiopsida: Schuster 1984) in basal Bryopsida, based on new information on sporophyte morphology and male gametes and gametogenesis. See Chapter 3 for further discussion and Mishler, Lewis, et al. 1994 for synapomorphies.

FIGURE 7.6. Cladistic relationships in tracheophytes, based on comparative morphology of extant and extinct groups. † = extinct; quotation marks = paraphyletic group. The eutracheophytes are the vascular plant crown group. See Table 7.2 and Chapter 4 for synapomorphies.

ship with the valvate Andreaeales. The problematic and distinctive Sphagnales were resolved as a sister group to all other mosses. Our analysis supports other morphology-based cladistic studies that place *Takakia* among the basal Bryopsida (Garbary, Renzaglia, and Duckett 1993; Mishler, Lewis, et al. 1994), as Murray (1988) suggested (Figure 7.5). Prior to the discovery of the sporophyte generation, Schuster (1966, 1984b) argued strongly for a close relationship with *Haplomitrium* and placed *Takakia* in the Calobryales (Marchantiopsida). Molecular sequence data have not yet addressed in detail the relationships within mosses, and early studies give contradictory results. Analyses of rRNA sequences provide some support for the Mishler and Churchill (1984) tree (Bopp and Capesius 1996; Mishler, Lewis, et al. 1994; Mishler, Thrall, et al. 1992). Recent results from *rbcL* sequences yield a range of mutually contradictory results (Manhart 1994). The fossil record of the Bryopsida is poor and has not yet contributed important information to cladistic studies of this group.

Polysporangiomorpha

Inclusion of fossils in phylogenetic analyses of the basal embryophytes results in the recognition of several groups that cannot be discriminated on the basis of extant taxa. The Polysporangiomorpha (polysporangiophytes) are one of the most inclusive groups of land plants and are diagnosed by the presence of axial, branching sporophytes that bear multiple sporangia (Figures 7.2 and 7.6, Table 7.2; Kenrick and Crane 1991). The group comprises tracheophytes (vascular plants), as well as a grade of extinct, nonvascular stem group taxa that we term *protracheophytes* (Kenrick and Crane 1991; see Figure 7.2). The protracheophyte grade includes *Horneophyton, Caia, Aglaophyton,* and some species of *Cooksonia*. Further clarification of relationships at this level requires more detailed anatomical information from stem group fossils. Such data can potentially be obtained from exceptionally well preserved material, such as the silicified plants from the Rhynie Chert (Remy, Gensel, and Hass 1993), as well as from fusainized or pyritized plant fossils (e.g., Edwards, Fanning, and Richardson 1986, 1994; Fanning, Edwards, and Richardson 1992; Kenrick and Crane 1991).

Tracheophyta

The Tracheophyta contain at least two clearly identifiable and well-defined clades that we term Eutracheophytes and Rhyniopsida (Figure 7.6; see also Kenrick and Crane 1991). Tracheophytes are defined by the annular or helical thickenings in water-conducting cells and possibly also by lignin deposition on the inner surface of the tracheid cell wall (Table 7.2). The Eutracheophytes are the tracheophyte crown group (Tracheidatae *sensu* Bremer [1985]) and contain all extant vascular plants, together with most early fossils. The group is defined by two structural features of the tracheid cell wall—a thick, decay-resistant (presumably lignified) wall and pitlets between thickenings or within pits—as well as by the presence of a sterome (Table 7.2). These features of eutracheophytes distinguish them from other land plants in which decay-resistant material is either absent from the water-conducting cells or appears to be present only as a very thin inner layer (e.g., S-type tracheids of Rhyniopsida: Kenrick and Crane 1991). Other possible defining features of eutracheophytes include the specialization of sporangium dehiscence—from several slits (which are frequently somewhat irregular) to a single slit that is typically well defined—and the ability to synthesize a complex array of auxin conjugates (Sztein, Cohen, et al. 1995).

Our numerical cladistic treatment confirms previous studies that recognize an early divergence in eutracheophytes (tracheophyte crown group) between lycopsids and other vascular plants (Figure 7.6; Banks 1968, 1970, 1975b; Bremer 1985b; Chaloner and Sheerin 1979; Crane 1990; Doyle and Donoghue 1986; Hueber 1992; Kenrick and Crane 1991, 1992; Kranz and Huss 1996; Raubeson and Jansen 1992; Stein, Wight, and Beck 1984). The Lycophytina comprise lycopsids and their extinct sister group, zosterophylls (DiMichele and Skog 1992; Kenrick and Crane 1991). The Euphyllophytina comprise all other extant vascular plants, and early members of this lineage include the *Psilophyton*-like fossils assigned to the Trimerophytina of Banks. The Euphyllophytina correspond to the widely recognized but previously unnamed trimerophyte–sphenopsid–fern–progymnosperm–seed plant clade of Banks (1968, 1970), Chaloner and Sheerin (1979), Stein, Conant, et al. (1984), Bremer (1985), Bremer, Humphries, et al. (1987), Doyle and Donoghue (1986), Crane (1990), Kenrick and Crane (1991, 1992), Hueber (1992) and Raubeson and Jansen (1992) (Figure 7.6).

The results presented in this book and summarized above are identical to those recognized in our previous numerical cladistic study (Kenrick and Crane 1991), except for the position of the early zosterophyll-like fossil *Nothia aphylla*. In the earlier study, *Nothia* was resolved as a protracheophyte based on the absence of wall thickenings in the xylem elements. The more detailed evaluation in Chapter 4 shows that absence of wall thickenings in *Nothia* is most parsimoniously interpreted as a loss. *Nothia* is now resolved as a plesiomorphic member of the tracheophyte group Lycophytina based on its possession of such lycopsid-like features as lateral reniform sporangia.

Rhyniopsida

The Rhyniopsida are a small, poorly understood, yet distinctive group of vascular plants that is resolved as a sister group to the eutracheophyte clade (Figure 7.6). The Rhyniopsida are not equivalent to the Rhyniophytina *sensu* Banks (Banks 1975b; Edwards and Edwards 1986; Gensel and Andrews 1984; Taylor 1988a) but do contain *Rhynia gwynne-vaughanii*—one of the well-characterized early fossils from the Rhynie Chert. The Rhyniopsida are supported by three unequivocal synapomorphies: (1) a distinctive type of adventitious branching (*Rhynia*-type), (2) an abscission or isolation layer at the base of the sporangium, and (3) sporangia attached to a pad of tissue (Table 7.2). In addition, taxa in Rhyniopsida possess the distinctive and unique S-type tracheid (Kenrick and Crane 1991). Further characterization of this group, and the possibilities for incorporating other taxa, will require detailed study of the distinctive, but as yet rather poorly understood, *Taeniocrada*-like fossils that have been described from many Devonian localities.

Lycophytina

The Lycophytina comprise the Lycopsida (lycopsids *sensu stricto*) and their extinct sister group the Zosterophyllopsida (Figure 7.7). Plesiomorphic members of this clade are *Zosterophyllum*-like plants having helical arrangements of the sporangia (e.g., *Zosterophyllum myretonianum, Nothia aphylla*). The Lycophytina are defined by the presence of (1) more or less reniform sporangia, (2) marked sporangium dorsiventrality, (3) isovalvate dehiscence along the distal sporangium

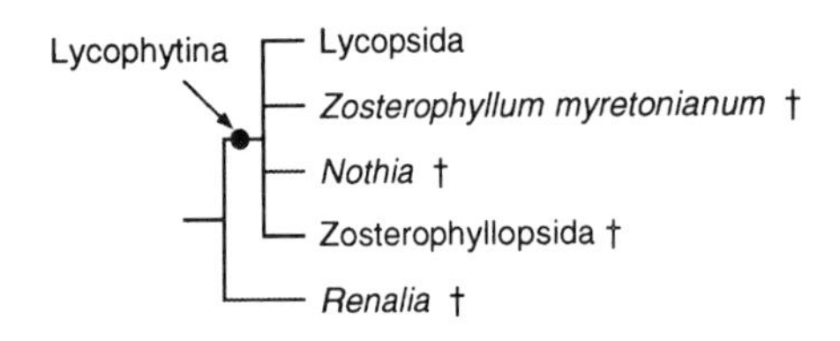

FIGURE 7.7. Summary of cladistic relationships in basal Lycophytina. † = extinct. See Table 7.2 and Chapters 4 and 5 for synapomorphies.

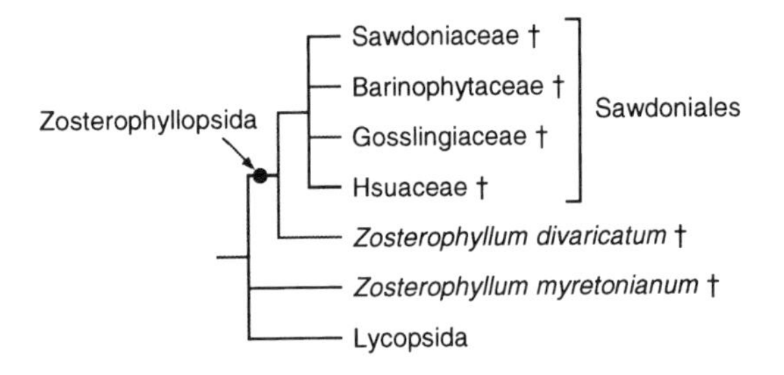

FIGURE 7.8. Summary of cladistic relationships in the Zosterophyllopsida. † = extinct. See Table 7.2 and Chapter 5 for synapomorphies.

rim, (4) conspicuous cellular thickening along the line of sporangium dehiscence, (5) sporangia on short, laterally inserted stalks, and (6) exarch xylem differentiation (Table 7.2). A strongly paraphyletic relationship between zosterophylls and lycopsids, which has been suggested previously (Crane 1990; Niklas and Banks 1990) and is founded on the widely perceived similarities between spines in the Sawdoniales (Zosterophyllopsida) and microphylls in lycopsids, is unparsimonious (Gensel 1992; Kenrick and Crane 1992). Experimental manipulation of our data also shows that the "spines" in *Sawdonia*-like zosterophylls and the "microphylls" in lycopsids are not homologous.

Many rather fragmentary and poorly understood fossils formerly assigned to Rhyniophytina *sensu* Banks, including some species of *Cooksonia*, were resolved as a grade of organization in the Lycophytina stem group. Further resolution of relationships in this stem lineage requires more detailed knowledge of stem anatomy and whole-plant morphology in such critical taxa as *Renalia*, *Uskiella*, *Yunia*, and *Cooksonia*.

The Zosterophyllopsida are an abundant and diverse group of Devonian plants containing many taxa formerly included in the Zosterophyllophytina (Banks) and Barinophytaceae (Figure 7.8; Kräusel and Weyland 1961). The group is characterized by (1) circinate growth and (2) a unique two-rowed arrangement of sporangia. A third potential defining feature is an elliptical xylem strand (Table 7.2). Within the Zosterophyllopsida, most taxa are contained within a large clade (Sawdoniales) defined by (1) pseudomonopodial branching, (2) planated branching, and (3) the presence of unique subordinate axes associated with dichotomies in the main axial system (Table 7.2). The Sawdoniales stem group contains *Zosterophyllum*-like species with marked bilateral symmetry (e.g., *Z. divaricatum*). The Sawdoniales comprise four distinctive subgroups: the Sawdoniaceae, which contain spiny taxa such as *Sawdonia;* the Barinophytaceae, which are a predominantly late-Devonian group of heterosporous zosterophylls; the Gosslingiaceae, which comprise taxa with characteristic auricular sporangia; and the Hsuaceae, which have lateral branches bearing terminal sporangia, a trait that reflects substantial modification of the two-rowed sporangium arrangements that are typical of the Sawdoniales. The Hsuaceae and Barinophytaceae are currently the two most poorly understood groups of Sawdoniales.

The Lycopsida have a large and important Paleozoic fossil record, but diversity in the three extant families (Lycopodiaceae, Selaginellaceae, Isoetaceae) is comparatively low. The Lycopsida is defined by the presence of (1) microphylls, (2) a stellate xylem strand, (3) leaf vasculature, (4) sporophylls (close association of sporangium and leaf), and (5) the loss of sporangium vasculature (Table 7.2). Early fossils such as *Asteroxylon* and *Baragwanathia* share some of the defining features of the Lycopsida and are placed in the stem group (Figure 7.9). Experimental manipulation of our data showed that lycopsid polyphyly, as proposed by Stewart (1983) and Stewart and Rothwell (1993), is highly unparsimonious (Chapter 6). According to Stewart's hypothesis, the Lycopodiaceae evolved from zosterophylls, while the ligulate ly-

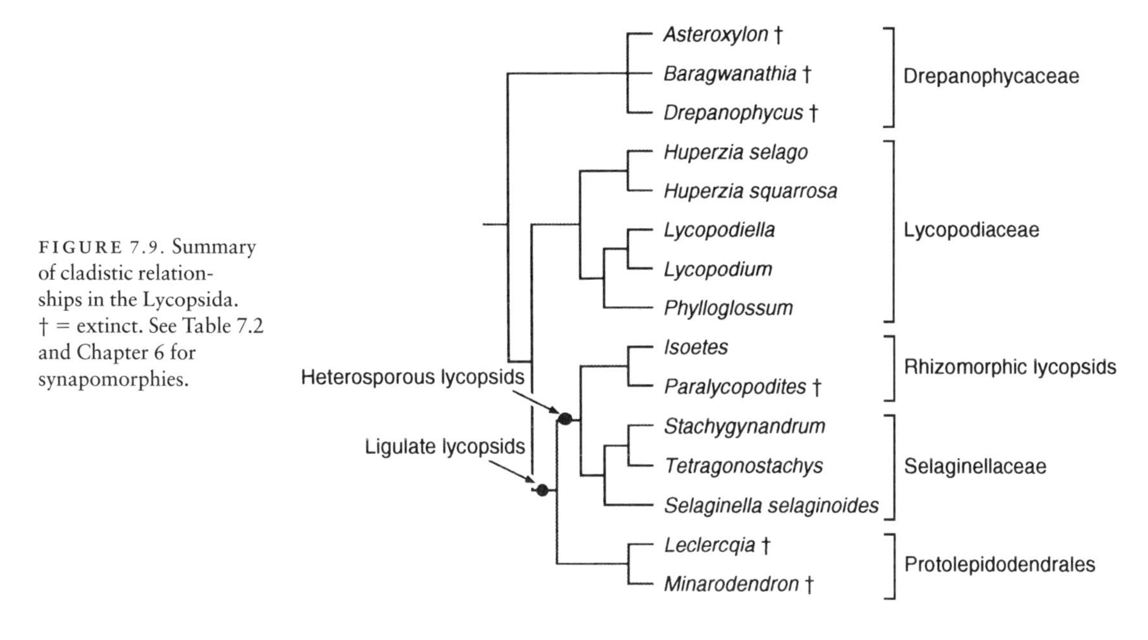

FIGURE 7.9. Summary of cladistic relationships in the Lycopsida. † = extinct. See Table 7.2 and Chapter 6 for synapomorphies.

copsids evolved independently from rhyniophyte ancestors. The fossil evidence cited in support of such polyphyly is also unconvincing because the morphology of the critical fossils *Protohyenia* and *Estinnophyton* is poorly understood. These taxa are probably more closely related to the Equisetopsida.

There is no comprehensive cladistic treatment of the Lycopodiaceae, but our analysis confirms other recent preliminary studies that show that the group is poorly defined (Table 7.2; Bateman 1992; Crane 1990; Wagner and Beitel 1992). The Lycopodiaceae are either a sister group to, or paraphyletic to, the ligulate lycopsid clade. Our preferred most parsimonious tree recognizes the monophyly of the Lycopodiaceae based on spore type (Figure 7.9). Our analysis confirms earlier suggestions (Crane 1990) that the monospecific Australian genus *Phylloglossum* is more closely related to a *Lycopodium-Lycopodiella* clade (*sensu* Øllgaard 1987) than to other Lycopodiaceae. No support was found for a sister group relationship between *Phylloglossum* and ligulate lycopsids (Bateman 1992). Further progress in clarifying relationships in the Lycopodiaceae requires a detailed cladistic treatment of extant taxa based on a thorough global sampling of the diversity within the group.

Ligulate lycopsids comprise the Protolepidodendrales, the Selaginellales, and the rhizomorphic clade *sensu* Bateman (1992) (Figure 7.9). The Protolepidodendrales are a poorly understood early fossil group represented by taxa such as *Leclercqia, Minarodendron, Haskinsia, Archaeosigillaria,* and *Colpodexylon.* The monophyly of the Protolepidodendrales is weakly supported, and is based only on their characteristic forked leaves (Table 7.2).

There is no cladistic treatment of the Selaginellaceae, but our preliminary study indicates that although the monophyly of the family is not strongly supported, most diversity falls into a well-supported rhizophoric clade. The most divergent and problematic taxon is *Selaginella selaginoides,* which is a sister group to the rhizophoric clade. The monophyly of the Selaginellaceae is based on more or less spherical microsporangia; other possible synapomorphies include (1) transverse distal dehiscence (a reversal to the plesiomorphic condition from protolepidodendralean and rhizomorphic lycopsid type), (2) four megaspores per sporangium, (3) suspension of the stele in a cavity by distinctive trabeculate endodermal cells, and (4) echinate microspores (Table 7.2). The spore characters are equivocal because one or more also occur in potential fossil sister taxa such as *Paurodendron, Leclercqia,* and *Oxroadia.* The well-supported rhizophoric clade is based on (1) anisotomous branching, (2) planar branching (plagiotropic shoots), (3) the presence of a rhizophore, (4) vegetative leaves in basically four-rowed arrangements, (5) anisophylly, and possibly (6) elliptical to strap-shaped xylem strands and (7) the polystelic condition.

Our analysis supports the monophyly of the rhizomorphic clade within the ligulate lycopsids and its close relationship to the Isoetaceae, as have been recognized in other cladistic treatments (Bateman 1992; Bateman, DiMichele, and Willard 1992; Crane 1990). Cladistic relationships among rhizomorphic lycopsids (excluding *Isoetes*) have been analyzed in detail by Bateman (1992) and Bateman, DiMichele, and Willard (1992). Our analysis of relationships among the Protolepidodendrales, the Selaginellaceae, and the rhizomorphic lycopsid clade gave two sets of conflicting results. One set of trees supported a sister group relationship between rhizomorphic lycopsids and the Selaginellaceae based on heterospory, but a second set supported a sister group relationship between the Protolepidodendrales and the rhizomorphic forms based on sporangium morphology. A clearer understanding of the morphology of stem group Selaginellaceae will be critical to further resolution of this issue.

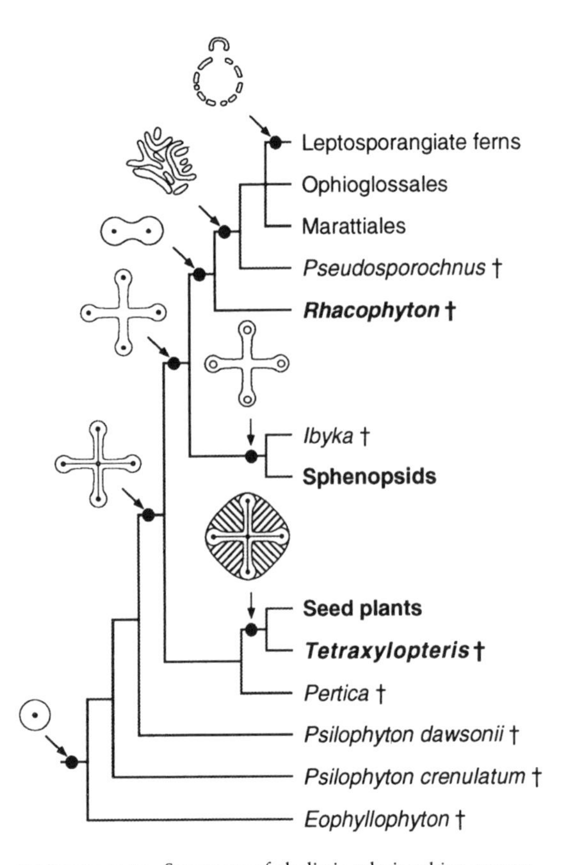

FIGURE 7.10. Summary of cladistic relationships among basal Euphyllophytina. † = extinct. Some important changes in stelar anatomy are indicated on branches. Small circles and lines within steles = position of protoxylem; open circles = protoxylem lacunae. Taxa in bold indicate independent acquisitions of secondary xylem with rays (probably underestimated; see Table 7.7). See Table 7.2 and Chapter 4 for synapomorphies.

Euphyllophytina

The Euphyllophytina are a previously unnamed group that has been widely recognized in earlier studies (see above; Figure 7.10). The Euphyllophytina is a well-supported group defined by (1) pseudomonopodial or monopodial branching, (2) basically helical arrangement of branches, (3) small, pinnulelike vegetative branches (which are nonplanated in basal taxa), (4) recurvation of branch apexes, (5) scalariform bordered pitting of metaxylem cells, (6) paired sporangia grouped into terminal trusses, (7) sporangium dehiscence along one side through a single slit, and possibly (8) radially aligned xylem in larger axes (Table 7.2). Multiflagellate spermatozoids (convergent with Isoetaceae) and a major chromosome inversion in the chloroplast genome (Raubeson and Jansen 1992) are further synapomorphies of this group, but these features are unknown in the relevant fossils (Crane 1990; Doyle and Donoghue 1986). Many of these synapomorphies are most clearly observed in early fossils and have undergone considerable alteration in extant members of this group.

The Euphyllophytina show a remarkable range of morphological diversity and divergence (compare the Equisetaceae with the Orchidaceae) in comparison to their relatively conservative sister group, the Lycophytina. Because extant ferns, seed plants, and *Equisetum* have highly divergent morphologies, some of the homology assignments employed in previous cladistic treatments are problematic and need to be reconsidered. Bremer,

Humphries, et al. (1987) recognize the Euphyllophytina (equivalent to their *Equisetum*–fern–seed plant clade) on the basis of "sporangia on leaves" (character 69). This character is problematic because early fossils in these lineages are leafless. A more appropriate comparison perhaps corresponds to our character 4.24, a specialized fertile zone, which recognizes homology in the dense terminal clusters of paired sporangia in early members of this clade (e.g., *Psilophyton dawsonii:* Figure 4.13). These structures appear to have been modified subsequently into peltate sporangiophores in the Equisetopsida, into sporangium-bearing leaves in the Filicopsida, and into pollen organs and cupules containing seeds in early seed plants.

Relationships in the basal Euphyllophytina require a more detailed treatment than has been possible in this book. Nevertheless, our inclusion of many early fossils as exemplars for major extant clades allows a preliminary assessment of relationships that should be tested by more detailed studies. The Euphyllophytina stem group contains early fossils such as *Psilophyton* (*P. dawsonii* and *P. crenulatum:* Trimerophytina) and *Eophyllophyton,* as well as a clade defined on a distinctive, lobed primary xylem strand that comprises all other Euphyllophytina (Figure 7.10). Our results support a basal dichotomy in the Euphyllophytina crown group between taxa related to lignophytes (progymnosperms and seed plants) (Radiatopses: Table 7.1) and taxa related to a fern-*Equisetum* clade (Moniliformopses: Table 7.1).

Lignophytes, represented in our analysis by the progymnosperm *Tetraxylopteris,* form a clade with the trimerophyte *Pertica* based on (1) tetrastichous branching and (2) a distinctive form of protoxylem ontogeny with multiple strands occurring along the midplanes of the primary xylem ribs (Beck and Stein 1993). This clade (Radiatopses: Tables 7.1 and 7.2) corresponds to the "radiate protoxylem" group recognized by Beck and Stein (1993) and Stein (1993). Data from the fossil record are consistent with lignophyte monophyly, but the group does not appear to be strongly supported and its status has been questioned (Beck and Stein 1993; Galtier 1992; Galtier and Beck 1992; Galtier and Rowe 1989). The bifacial cambium and other features of the secondary xylem form a character complex having traits that may be more general in the Euphyllophytina. Cambial activity occurs in putative ferns (e.g., *Ophioglossum* and the early fossils *Zygopteris* and *Rhacophyton*) and in the Equisetopsida (e.g., Calamitaceae, Archaeocalamitaceae), as well as in *Sphenophyllum.* In the basal lignophytes the cambial initials undergo radial longitudinal divisions that do not occur in the bifacial cambium of *Sphenophyllum* (Cichan 1985b). Xylem rays have also been observed in *Rhacophyton* (Dittrich, Matten, and Phillips 1983; Schultka 1978) and in the Calamitaceae (Cichan 1986a; Cichan and Taylor 1983, 1990). Incipient cambial activity may occur in such basal taxa as *Psilophyton,* in which radially aligned metaxylem elements have been reported (Banks, Leclercq, and Hueber 1975). Further evaluation of lignophyte monophyly requires a critical anatomical survey that includes the relevant closely related Paleozoic fossils.

The sphenopsid-fern clade (Moniliformopses: Tables 7.1 and 7.2) is supported by a basically mesarch protoxylem distribution, and this clade corresponds to the "permanent protoxylem" group of Beck and Stein (1993). Protoxylem strands are confined to the outer lobed ends of the xylem strand, as opposed to the condition in the *Pertica*-lignophyte clade, in which the protoxylem is also distributed along the midpoints of the lobed arms and at the center of the xylem. In our analysis the Equisetopsida were represented by *Ibyka* and the ferns by *Pseudosporochnus* and *Rhacophyton.*

The Equisetopsida (Archaeocalamitaceae, Calamitaceae, Equisetaceae) are a distinctive group defined by (1) whorled appendages, (2) a characteristic sporangiophore morphology, (3) a characteristic stelar morphology, (4) regular alternation of appendages at successive nodes, (5) "microphyllous" leaves, and possibly (6) cambium (which is lost in the Equisetaceae) and (7) the presence of a perispore (possibly more general) (Table 7.2; Stein, Wight, and Beck 1984). Additional characters noted by Bateman (1991) include (1) a medullated stele, (2) operculate strobili, and (3) columnar wall thickenings on the sporangium epidermis.

The monophyly of the Sphenophyllales *(Sphenophyllum, Bowmanites, Sphenophyllostachys, and* possibly *Cheirostrobus* and *Eviostachya)* is more problematic than that of the Equisetopsida, although character analysis suggests that a close relationship with the Equisetopsida is likely (Stein, Wight, and Beck 1984). Similarities to the Equisetopsida include (1) whorled appendages, (2) leaves, (3) secondary xylem, (4) sporangiophores, and (5) a perispore (Stein, Wight, and Beck 1984). Several of these putative homologies have been questioned, in particular that of leaves and secondary xylem (Stein, Wight, and Beck 1984). We regard the leafy appendages of *Sphenophyllum* as relatively unproblematic because of similarities to those of the Archaeocalamitaceae and Calamitaceae. Heterophylly in *Sphenophyllum* is common, with leaf morphology ranging from wedge-shaped or spatulate with a distinct lamina to deeply dissected leaves, the individual units of which resemble leaves of the Equisetales (Batenburg 1977, 1981, 1982). Secondary xylem development shows some differences to that of the Equisetopsida (Cichan 1985b, 1986a), but these features may be autapomorphic. The exarch development of the metaxylem has suggested comparisons with lycopsids (Stewart 1983; Stewart and Rothwell 1993), but other similarities with the Equisetopsida make a close relationship with lycopsids unlikely (Stein, Wight, and Beck 1984).

Filicopsida

In the analysis that we present here, we group the putative early ferns *Pseudosporochnus* and *Rhacophyton* together based on the distinctive clepsydroid xylem shape. Both taxa also have potential homologies to fern "megaphylls" in the planar orientation of the distal axes of the branching system. *Pseudosporochnus* (Stein and Hueber 1989) further resembles such basal ferns as the Marattiales and some Ophioglossales, Osmundaceae, and Schizaeaceae in the elaboration of the vascular tissues of the stem to produce a polystelic appearance caused by the vascular traces that supply the megaphylls and parenchyma inclusions within the stele. This anatomy is further developed into the elaborate siphonosteles and dictyosteles of the leptosporangiate ferns. Other potential synapomorphies for the leptosporangiate ferns plus the Marattiales and Ophioglossales include (1) webbing of megaphyll homologues to produce "fern megaphylls" (*webbing* is a feature convergent with that of some progymnosperms and seed plants); (2) circinate vernation, which is present, but less pronounced in the Ophioglossales (Tryon and Tryon 1982), possibly lost in the Psilotaceae, and convergent with that in the Zosterophyllidae; and (3) septate rhizoids on gametophyte, which are present in the Psilotaceae and in some Ophioglossaceae, Osmundaceae, Schizaeaceae, Stromatopteridaceae (Bierhorst 1971), and Marattiales (Camus 1990; Hill and Camus 1986; Table 7.2). The monophyly of leptosporangiate ferns, including the Osmundaceae, is supported by (1) a distinctive annulate dehiscence of the sporangium, (2) superficial antheridia, (3) an operculate cell in the antheridium, (4) a C-shaped leaf trace (Bierhorst 1971), and (5) a siphono-dictyostelic anatomy, which is possibly a defining feature at a more inclusive level (Table 7.2).

There have been several recent cladistic treatments of higher-level relationships in ferns based on molecular evidence (Hasebe, Ito, et al. 1993; Hasebe, Omori, et al. 1994; Manhart 1995; Stein, Conant, et al. 1992). Stein, Conant, et al. (1992) identified structural changes in the chloroplast genome that link the Adiantaceae, Cyatheaceae, Dennstaedtiaceae, and Dryopteridaceae but that are not present in the Osmundaceae. An analysis based on *rbcL* sequences of 58 species, which represent most of the major families of leptosporangiate ferns, supported a sister group relationship between the Osmundaceae and all other leptosporangiate ferns rooted on the eusporangiate ferns Marattiales and Ophioglossales (Hasebe, Omori, et al. 1994). According to these analyses, other basal groups in the leptosporangiate ferns include the Hymenophyllaceae, Gleicheniaceae, Matoniaceae, Cheiropleuriaceae, Dipteridaceae, and Schizaeaceae and a Marsileales-Salviniales clade.

More recently, a study including both molecular *(rbcL)* and morphological data from 50 extant taxa covering the major groups of ferns found much congruence in trees generated from the two independent data sets (Pryer, Smith, and Skog 1996). The most robust support was found in a "total evidence" analysis that combined the molecular and morphological data (Figure 7.11; Pryer, Smith, and Skog 1996). This analysis provided strong support for the monophyly of the leptosporangiate ferns (with *Osmunda* as basal taxon), the heterosporous ferns (see also Rothwell and Stockey 1994), *Cheiropleuria-Dipteris, Diplopteridium-Stromatopteris,* the tree ferns, the schizaeoid ferns, the pteroid ferns, and an unnamed clade containing other derived taxa excluding the dennstaedtioids and pteroids. The dennstaedtioids were found to be paraphyletic. The relationships of leptosporangiate ferns to other pteridophytes were poorly resolved. *Psilotum* grouped weakly with *Botrychium. Equisetum* was found to be more closely related to leptosporangiate ferns than to either *Angiopteris* or *Botrychium.* Relationships among extant ferns and putative early fossil ferns such as the Zygopteridales, Rhaco-

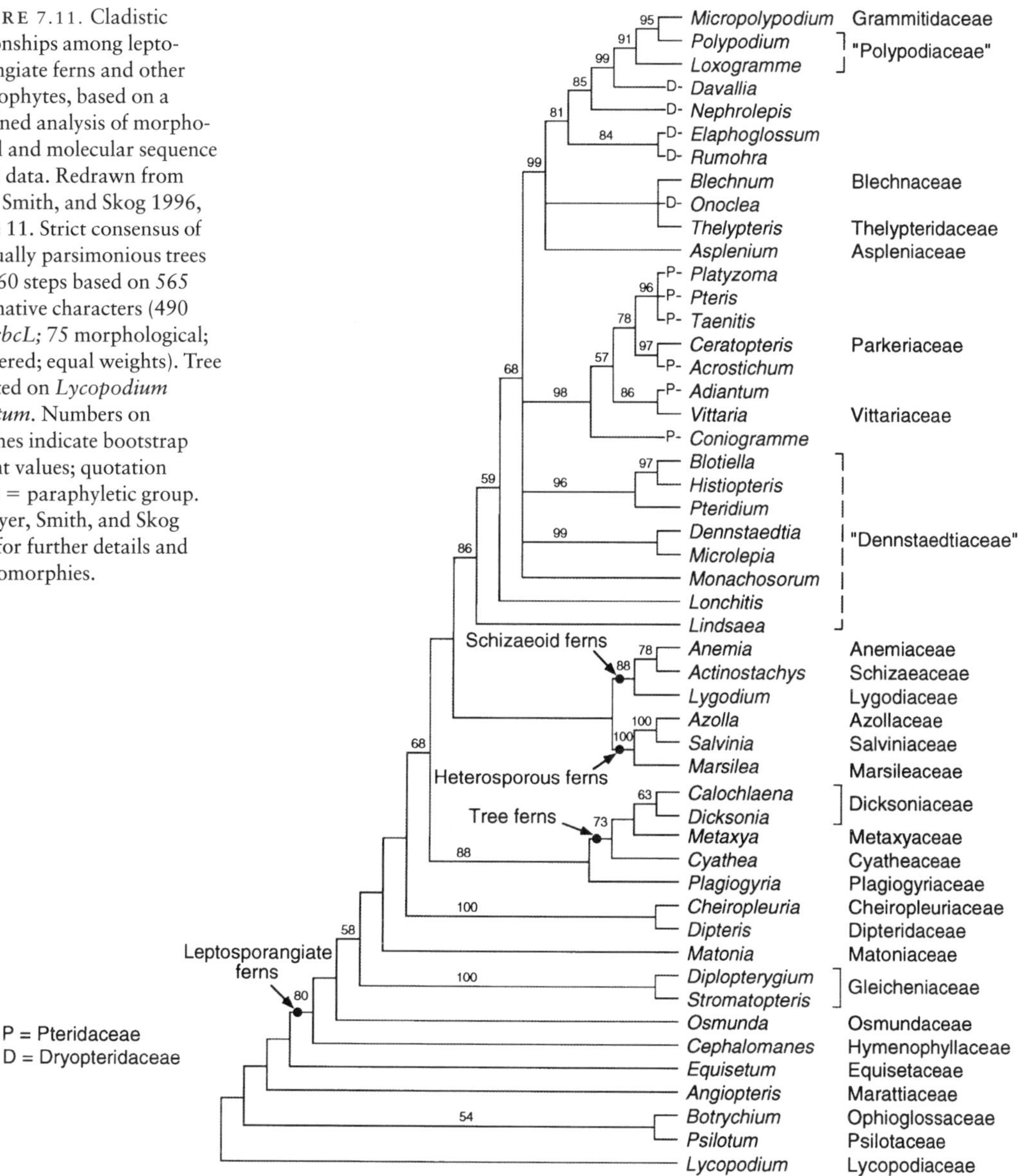

FIGURE 7.11. Cladistic relationships among leptosporangiate ferns and other pteridophytes, based on a combined analysis of morphological and molecular sequence *(rbcL)* data. Redrawn from Pryer, Smith, and Skog 1996, Figure 11. Strict consensus of 12 equally parsimonious trees of 4,360 steps based on 565 informative characters (490 from *rbcL;* 75 morphological; unordered; equal weights). Tree is rooted on *Lycopodium digitatum.* Numbers on branches indicate bootstrap percent values; quotation marks = paraphyletic group. See Pryer, Smith, and Skog 1996 for further details and synapomorphies.

phytales, and Cladoxylales are currently poorly understood. In contrast to the results of our study, Rothwell (1996) resolved these early fossils as more closely related to lignophytes than to extant ferns.

Preliminary cladistic studies on the comparative morphology of extant taxa suggested a sister group relationship between ferns and seed plants (e.g., Bremer 1985; Mishler, Lewis, et al. 1994) based on (1) megaphyllous leaves and (2) trichomes, but this hypothesis is not supported by our analysis. Inclusion of early fossils demonstrates that a simple hypothesis of homology between the megaphyllous leaves in ferns and seed plants is problematic. Lateral appendages in ferns, seed plants, and *Equisetum* are homologous at the general level of small, dichotomous, nonplanar, ultimate appendages in the basal Euphyllophytina (e.g., *Psilophyton:* Figure 4.13), but such "megaphyll precursors" are also present in aneurophytalean progymnosperms (seed-plant stem group) and such early fernlike fossils as *Rhacophyton,* clearly demonstrating that the planation and webbing characteristic of megaphylls evolved independently in ferns and seed plants (Crane 1990; Donoghue, Doyle, et al. 1989; Doyle and Donoghue 1992). The distribution of trichomes or other epidermal appendages that are commonly seen on stems and leaves and that have been used to support the fern–seed plant clade is equivocal because their potential homology in extant taxa has not been assessed in detail and their occurrence in many critical fossil taxa is currently unknown. Multicellular and unicellular epidermal appendages of this type are certainly widespread in polysporangiophytes (e.g., *Psilophyton crenulatum,* Sawdoniales) and are one of the most homoplastic features. Early cladistic treatments focusing on seed plants have left the relationships among sphenopsids, lignophytes, and ferns unresolved and have questioned the monophyly of ferns (Crane 1990; Doyle and Donoghue 1986). Our preliminary study provides tentative support for the monophyly of a leptosporangiate fern-Marattiales-Ophioglossales clade.

Other recent studies addressing general relationships among land plants have focused on extant taxa using chloroplast DNA nucleotide sequences (*rbcL* gene) (Albert, Backlund, and Bremer 1994; Albert, Backlund, Bremer, Chase, et al. 1994; Hasebe, Ito, et al. 1993; Hasebe, Omori, et al. 1994; Manhart 1994), male gamete ultrastructure (Garbary, Renzaglia, and Duckett 1993), and combined molecular sequence, general morphology, and ultrastructure (Mishler, Lewis, et al. 1994). The *Equisetum*-fern clade receives support from the molecular sequence data on *rbcL* and from the studies on male gametes and gametogenesis. A preliminary study based on characters of male gametogenesis supported an *Equisetum*-fern clade but controversially placed the Marsileales as a sister group to this clade plus seed plants (Figure 3.10; Garbary, Renzaglia, and Duckett 1993). We consider the position of *Marsilea* in this analysis to be highly unlikely given the morphological similarities to ferns in general (i.e., circinate vernation, solenostelic anatomy, megaphyllous leaves, vestigial apical annulus in the microsporangia of *Pilularia,* and C-shaped stele in leaf petiole). Furthermore, a close relationship between the Marsileales and Lygodiaceae in particular—suggested by Bierhorst (1971) on the basis of morphology—receives strong support from *rbcL* data (Manhart 1994) and from more recent molecular studies that support a close relationship between the Marsileales and Salviniales within leptosporangiate ferns (Hasebe, Omori, et al. 1994; Pryer, Smith, and Skog 1996). A sister group relationship between the Marsileales and Salviniales is also supported by the recent description of a fossil of intermediate morphology (*Hydropteris:* Rothwell and Stockey 1994; see also Bateman 1996b). In general, the chloroplast DNA study by Manhart (1994) produced obviously anomalous results for ancient cladogenic events but was more congruent with other studies of relationships among more recent clades (Chapter 4; see Albert, Backlund, and Bremer 1994; Albert, Backlund, Bremer, Chase, et al. 1994; Mishler, Lewis, et al. 1994). This study consistently grouped *Equisetum* with ferns. Total evidence analysis has not adequately addressed relationships at this level because of the absence of important groups such as ferns (Mishler, Lewis, et al. 1994).

Psilotaceae

The problematic group Psilotaceae (Appendix 1) was not explicitly included in our numerical analysis, which focused mainly on sporophyte morphology at the polysporangiophyte level. We follow Bierhorst (1971, 1977) in interpreting morphological simplicity in the sporophyte of the Psilotaceae as a derived feature because many morphological features of the gametophyte clearly indicate a close relationship to leptosporangiate ferns in general and to the Gleicheniaceae and Stromatopteridaceae in particular. In our view, the simple sporophyte morphology may be attributable to a major (probably heterochronic) developmental modification (see Bateman 1996b; Bierhorst 1977; Kaplan 1977), and those morphological features that can be observed are also indicative of reduction (e.g., synangia).

There are few unequivocal points of comparison between the Psilotaceae and early fossils in the Rhyniophytina and Trimerophytina *sensu* Banks (1975b; Gensel 1977, 1992). Previous cladistic analyses of morphological data have resolved the Psilotaceae as a sister group to all other extant vascular plants based on the absence of roots (Bremer 1985; Parenti 1980) or have given inconclusive results (Gensel 1992). However, these treatments have not provided a stringent test of the likely relationship with leptosporangiate ferns because of their emphasis on sporophyte morphology and on the simplicity of sporophyte form in the Psilotaceae (i.e., *Psilotum nudum*). An adequate cladistic treatment of the Psilotaceae must also consider details of the gametophyte generation and the morphology of *Tmesipteris,* which is considerably more complex than that of *Psilotum.* More recent cladistic studies based on extant taxa group *Psilotum* with seed plants and other fern allies to the exclusion of lycopsids. Morphological data place *Psilotum* in the Euphyllophytina either with *Equisetum* (Pryer, Smith, and Skog 1996) or in the base of an *Equisetum*–fern–seed plant clade (Stevenson and Loconte 1996). A relationship within basal euphyllophytes is strongly supported by 18S rRNA sequences (Kranz and Huss 1996) and the presence of a 30-kb inversion in the chloroplast genome (Raubeson and Jansen 1992). Combined analyses based on morphology and *rbcL* sequences favor a sister group relationship between *Psilotum* and *Botrychium* (Pryer, Smith, and Skog 1996). There is currently no cladistic support from these analyses for a close relationship with basal leptosporangiate ferns in the Gleicheniaceae and Stromatopteridaceae. Bateman (1996b) places the Psilotaceae as a sister group to the Ophioglossaceae.

Similarities between gleicheniaceous ferns and the Psilotaceae include (1) bean-shaped monolete spores with similar ontogenetic features (Lugardon 1979), (2) an axial gametophyte resembling subterranean sporophyte axes, (3) septate rhizoids on the gametophyte (found elsewhere among vascular plants in the Ophioglossaceae, Osmundaceae, Schizaeaceae, Stromatopteridaceae, and Marattiales), (4) a rootless embryo with foot and stem apex only, and (5) antheridia with lateral opercular cells (Bierhorst 1977). More general features suggesting a relationship to filicalean ferns rather than to early fossil groups include (1) superficial antheridia (Bierhorst 1977; Darnell-Smith 1917; Lawson 1917), (2) an enation ontogeny similar to the marginal meristem of fern fronds (Kaplan 1977), (3) a C-shaped xylem strand at the frond base in *Tmesipteris* (Sykes 1908), (4) the presence of a perispore (Lugardon 1979), (5) a characteristic archegonial morphology (Bierhorst 1977), and (6) indications of a rudimentary annulus at the apex of the sporangium (Bierhorst 1977). Two other more general characters suggest a closer relationship to the Euphyllophytina than to lycopsids: multiflagellate antherozoids (convergent with *Isoetes*) and a major inversion in the chloroplast genome, which is absent from extant lycopsids (Raubeson and Jansen 1992). The limited chemical evidence presented by Cooper-Driver (1977) in favor of a distant relationship between the Psilotaceae and gleicheniaceous ferns is equivocal. The Psilotaceae were shown to be distinct in only one feature, which can be interpreted as an autapomorphy.

Classification

The classification outlined in Tables 1.1 and 7.1 follows as closely as possible the cladistic classification of the Chlorobiota (green plants) subdivision Embryobiotes of Bremer (1985) and Bremer, Humphries, et al. (1987) (Table 7.3). Orders and families are listed for some of the important taxa discussed in detail in this book.

The classifications presented here use the annotated Linnean conventions recommended by Wiley (1981). All of the taxa are monophyletic, except for those enclosed in quotation marks, which are of uncertain status. Redundant categories have been avoided, but some widely used names have been retained (e.g., class names such as Bryopsida and Anthocerotopsida within superclasses in the basal Embryobiotes). Unannotated lists at the same rank form a dichotomous sequence with the first taxon being a sister group to all subsequent taxa (Nelson 1972, 1974; Wiley 1981). The term *sedis mutabilis* indicates taxa of ambiguous relationship (polytomies) with respect to other taxa of similar rank within a group (e.g., relationships among the four families within the Sawdoniales are unresolved; Table 7.1). The term *incertae sedis* indicates ambiguity regarding relationships between a low-level taxon and two or more high-level taxa (e.g., *Zosterophyllum* subgenus *Zosterophyllum;* Table 7.1). The *plesion* convention is adopted for extinct groups in order to avoid further taxonomic inflation with respect to important monotypic taxa in the stem groups of major extant clades (e.g., the Horneophytopsida and *Aglaophyton major* in the Tracheophyta stem group). Synapomorphy-based taxon definitions are given in Table 7.2, and a list of informal names for monophyletic groups and their corresponding Linnean taxa in Table 1.2. Problematic groups (paraphyletic or polyphyletic) are listed in Table 7.4.

The hierarchical structure of the general classification (Table 7.1) is broadly similar to that of Bremer, Humphries, et al. (1987; Table 7.3), but there are several important differences. First, several new ranks have been recognized to accommodate additional hierarchical levels (Table 7.1). Second, we

TABLE 7.3
Cladistic Classification of Green Plants According to Bremer, Humphries, et al.

SUBKINGDOM CHLOROBIONTA, GREEN PLANTS

Division Chlorophyta

Class "Ulvophyceae"
Class Pleurastrophyceae
Class Chlorophyceae

Division Streptophyta

SUBDIVISION CHLOROKYBOPHYTINA
 Class Chlorokybophyceae
SUBDIVISION "ZYGOPHYTINA"
 Class Zygophyceae
 Class "Klebsormidiophyceae"
SUBDIVISION CHAETOSPHAERIDIOPHYTINA
 Class Chaetosphaeridiophyceae
SUBDIVISION CHAROPHYTINA
 Class Charophyceae
SUBDIVISION "COLEOCHAETOPHYTINA"
 Class "Coleochaetophyceae"
SUBDIVISION EMBRYOPHYTINA
Superclass Marchantiatae
 Class Marchantiopsida
Superclass Anthocerotatae
 Class Anthocerotopsida
Superclass Bryatae
 Class Bryopsida
Superclass Tracheidatae
 Class Psilotopsida
 Class Lycopodiopsida
 Class Equisetopsida
 Class "Polypodiopsida"
  Subclass Ophioglossidae
  Subclass Marattiidae
  Subclass Polypodiidae
 Class Spermatopsida

Source: Bremer, Humphries, et al. 1987.
Note: Quotation marks = paraphyletic group.

TABLE 7.4

Problematic Groups and Their Corresponding Monophyletic Bases

Taxon	Status	Monophyletic base
Green algae	Paraphyletic with respect to land plants (Embryobiotes); comprises green plants that are not Embryobiotes (Chapter 2)	Chlorobiota
Charophycean algae	Paraphyletic with respect to land plants (Embryobiotes); comprises such embryophyte stem group taxa as Chlorokybophyceae, Zygnemophyceae, Klebsormidiophyceae, Chaeto-sphaeridiophyceae, Charophyceae, and Coleochaetophyceae (Chapter 2)	Streptobionta
Bryophytes	Probably paraphyletic with respect to vascular plants (Tracheophyta); comprises three monophyletic groups: Marchantiopsida, Anthocerotopsida, and Bryopsida (Chapter 3)	Embryobiotes
Protracheophytes	Paraphyletic with respect to Tracheophyta; comprises extinct, nonvascular poly-sporangiophytes such as *Aglaophyton major* and *Horneophyton lignieri* and some nonvascular *Cooksonia*-like fossils (Chapters 3 and 4)	Polysporangiomorpha
Rhyniophytina *sensu* Banks (rhyniophytes)	Paraphyletic (possibly polyphyletic) with respect to eutracheophytes; much disagreement over scope and definition (Chapter 4)	Polysporangiomorpha or Tracheophyta
Zosterophyllophytina *sensu* Banks ("zosterophylls")	Paraphyletic with respect to Lycopsida; comprises Zosterophyllopsida and basal Lycophytina (Chapter 5)	Lycophytina
Trimerophytina *sensu* Banks ("trimerophytes")	Paraphyletic with respect to Moniliformopses and Radiatopses; comprises taxa such as *Psilophyton* and *Pertica* (Chapter 4)	Euphyllophytina
Pteridophytes	Paraphyletic with respect to seed plants (Spermatophytata); comprises non–seed plant tracheophytes (Chapters 4 and 7)	Tracheophyta
Progymnosperms	Paraphyletic with respect to seed plants (Spermatophytata); comprises woody seed-plant stem group taxa such as *Tetraxylopteris* and *Archaeopteris* (Chapters 4 and 7)	Lignophytia

Continues on next page

Note: Quotation marks = paraphyletic group.

TABLE 7.4
Continued

Taxon	Status	Monophyletic base
Pteridosperms	Paraphyletic or polyphyletic assemblage of extinct basal seed plants (Spermatophytata); comprises taxa in the seed-plant stem group, such as hydraspermans and medullosans, as well as taxa that are more closely related to extant seed plants, such as *Callistophyton, Caytonia,* glossopterids, etc.	Spermatophytata
Gymnosperms	Paraphyletic with respect to angiosperms (Magnolidra); comprises all non-angiosperm seed plants	Spermatophytata

have removed the class Psilotopsida *(Psilotum* and *Tmesipteris),* which was placed as the basal clade within the Tracheophyta by Bremer, Humphries, et al. (1987). Our analysis supports the inclusion of the Psilotopsida within the order Filicopsida (ferns) as argued by Bierhorst (1971, 1977) and supported by recent molecular and morphological studies (Hasebe, Wolf, et al. 1995; Pryer, Smith, and Skog 1996). Third, we formally recognize two extant subdivisions within Tracheophyta that correspond to the two well-established major clades: the Lycophytina and the Euphyllophytina. The new subdivision name Euphyllophytina is necessary because although the group has been widely discussed in the literature, it is currently unnamed (e.g., Banks 1968, 1972; Bremer 1985; Chaloner and Sheerin 1979; Crane 1990; Doyle and Donoghue 1986; Hueber 1992; Kenrick 1994; Kenrick and Crane 1991; Raubeson and Jansen 1992; Stein, Wight, and Beck 1984). The name refers to the leaf type that is characteristic of the group and that is homologous at the level of lateral branches ("megaphyll precursors") in early fossils such as *Psilophyton* (Figure 4.13). Fourth, our analysis recognizes a sister group relationship between the Equisetopsida and the Filicopsida (but see Bateman 1996b and Rothwell 1994, 1996), which are grouped within the new subclass Moniliformopses (L. *moniliformis,* "necklace-like") on the basis of the position and ontogeny of protoxylem in the lobed primary xylem of early fossil groups ("permanent protoxylem" group of Beck and Stein [1993]). We also recognize the monophyly of the Filicopsida (ferns), which were of equivocal systematic status in the study of Bremer, Humphries, et al. (1987). Fifth, the Spermatophytata (seed plants) and early fossils such as progymnosperms and derived trimerophytes are grouped within the new subclass Radiatopses also on the basis of xylem ontogeny in early fossils. The Radiatopses (L. *radialis,* "radiating from common center") corresponds to the "radiate protoxylem" group of Beck and Stein (1993) and Stein (1993).

Several additional groups are required to accommodate taxa from the early fossil record. The Tracheophyta stem group contains two extinct groups at the division level: the plesion Horneophytopsida (two taxa) and the plesion *Aglaophyton major* (one taxon). A new definition of the Rhyniophytina *sensu* Banks is required because in its current form the group is paraphyletic by definition and content (Chapter 4). Our synapomorphy-based definition (Table 7.2) restricts unequivocal group membership to the well-known early fossil *Rhynia gwynne-vaughanii* from the Rhynie Chert and to a few other recently described taxa such as *Stockmansella langii* and *Huvenia kleui.* Other taxa previously included in the group (e.g., Banks 1975b; Edwards and Edwards 1986; Gensel and Andrews 1984; T. N. Taylor 1988a) were found to have diverse relationships in the Tracheophyta, eutracheophyte, and Lycophytina stem groups (Chapter 4). We have renamed the group Rhyniopsida to emphasize this change in scope and definition.

Our classification reflects the well-established sister group relationship between the Lycopsida and the Zosterophyllopsida by classifying both as classes within the Lycophytina. The new synapomorphy-based definition of Zosterophyllophytina *sensu* Banks (1975b) (Zosterophyllopsida: Table 7.2) excludes some equivocal basal taxa that may be more closely related to the Lycopsida. Analysis of relationships within the Lycophytina indicates that the status of the important and diverse early fossil genus *Zosterophyllum* is equivocal. *Zosterophyllum* is paraphyletic to the Sawdoniales and possibly also to the Lycopsida. Both of the previously recognized subgenera of *Zosterophyllum* are probably also paraphyletic. We classify species with marked bilateral symmetry (e.g., *Z. divaricatum*) in the Sawdoniales stem group (Table 7.1). The relationships of species with radial symmetry (e.g., *Z. myretonianum*) are unresolved with respect to both the Lycopsida and Zosterophyllopsida. These taxa are classified as Lycophytina *incertae sedis* (Table 7.1).

Finally, we include a classification of seed plants that reflects some broad areas of agreement at a basic level in recent cladistic studies, although many uncertainties remain (Crane 1985; Doyle and Donoghue 1986, 1992; Nixon, Crepet, et al. 1994; Rothwell and Serbet 1994).

A SUMMARY OF THE EARLY FOSSIL RECORD

The early fossil record of the Chlorobiota (green plants) is dominated by calcareous "algae," by the decay-resistant stages of microscopic "algae," by the spores of land plants, and by vascular plant macrofossils. Of these, the record of vascular plants has had the greatest impact on macrosystematic and evolutionary studies. The importance of the tracheophyte fossil record is attributable to two main factors. First, vascular plants are relatively complex organisms, and much useful comparative data can be obtained from the fossil record, particularly where mineralization of soft tissues preserves cellular detail. It is therefore possible to integrate paleobotanical and neobotanical data across a wide spectrum of organismal morphologies. Second, tracheophytes are of great ecological prominence, and the morphology, systematics, and diversity of extant taxa are relatively well understood. These factors have generated considerable interest in the fossil record of the group, which has facilitated synthetic comparative studies.

The following section briefly summarizes the earliest fossil evidence of some of the major basal groups of Embryobiotes (Tables 7.1 and 7.5) and evaluates the potential contribution of paleobotanical data for resolving systematic problems at various hierarchical levels. Problems concerning group recognition are discussed with reference to the early fossil record of bryophytes.

Embryobiotes

The microfossil record of the Embryobiotes (land plants) significantly predates the macrofossil record and probably reflects the number, ubiquity, and transport capabilities of spores compared to spore-producers. The appearance of obligate spore tetrads in sediments of mid-Ordovician (Llanvirn-Llandeilo) age provides the first evidence of a land (embryophyte) flora (Figure 7.12; Gray 1991, 1993), but the systematic affinities of the early spore-producers are controversial because of the generalized spore morphology and the absence of associated macrofossils. A hepatic affinity for some sporomorphs is consistent with the limited ultrastructural data on spore wall morphology (W. A. Taylor 1995a,b). Other early evidence for land plants comes from a variety of Silurian-age dispersed spores, cuticles, and other putative plant fragments (Chaloner 1967b, 1970; Gensel, Johnson, and Strother 1990; Gray 1985; Gray and Boucot 1971, 1977; Gray, Massa, and Boucot 1982; Hemsley 1994a; Nøhr-Hansen and Koppelhus 1988; Strother 1991, 1993; Strother and Traverse 1979; W. A. Taylor 1995a,b). The early spore

TABLE 7.5

Evidence for First and Last Appearances of Selected Monophyletic Higher Taxa in Basal Embryophytes

Clade	First appearance	Last appearance
Embryobiotes	Middle Ordovician (Llanvirn)—microfossils (Gray 1993) Middle Silurian—macrofossils of *Cooksonia* (Edwards, Feehan, and Smith 1983)	Extant
Marchantiopsida	[Lower Silurian—putative sphaerocarpalean microspores of *Dyadospora* (W. A. Taylor 1995a)] Upper Devonian—macrofossil of *Pallavicinites* (Hueber 1961)	Extant
Stomatophytes	[Upper Ordovician—microfossil, trilete spores (Gray 1993)] Upper Silurian (Pridoli)—microfossil, cuticles (Jeram, Selden, and Edwards 1990)	Extant
Anthocerotopsida	[Upper Silurian—microfossil, spores (Richardson 1985); Lower Cretaceous—macrofossil, *Notothylacites* (Nemejc and Pacltova 1974)] Upper Cretaceous—microfossil, spores (Jarzen 1979)	Extant
Bryopsida and Polysporangiomorpha	Middle Silurian (Wenlock)—macrofossils of *Cooksonia* (Edwards, Feehan, and Smith 1983)	Extant
Bryopsida	[Lower Devonian (Pragian)—macrofossil of *Sporogonites* (Halle 1936a)] Lower Carboniferous (Visean)—macrofossil of *Muscites* (Thomas 1972)	Extant
Polysporangiomorpha	Middle Silurian (Wenlock)—macrofossils of *Cooksonia* (Edwards, Feehan, and Smith 1983)	Extant
Horneophytopsida	Upper Silurian (Pridoli)—macrofossil of *Caia* (Fanning, Edwards, and Richardson 1990)	Lower Devonian (Pragian)—macrofossil, *Horneophyton* (Gensel and Andrews 1984; Kidston and Lang 1920a)
Tracheophyta	[Upper Silurian (Ludlow)—macrofossil, *Baragwanathia* (Garratt and Rickards 1987; Hueber 1992)] Lower Devonian (Lochkovian)—macrofossil (Edwards, Davies, and Axe 1992; Kenrick and Crane 1991)	Extant
Rhyniopsida	Lower Devonian (Pragian)—macrofossil, *Rhynia* (D. S. Edwards 1980)	(?) Upper Devonian—macrofossil, *Taeniocrada* (Banks 1980)

Notes: Taxa listed in approximate taxonomic sequence. Early records that are equivocal or disputed are reported first, in square brackets, and followed by a more secure later date.

TABLE 7.5
Continued

Clade	First appearance	Last appearance
Eutracheophytes	[Upper Silurian (Ludlow)—macrofossil, *Baragwanathia* (Garratt and Rickards 1987; Hueber 1992)] Lower Devonian (Lochkovian)—macrofossil, (Edwards, Davies, and Axe 1992; Kenrick and Crane 1991)	Extant
Lycophytina	[Upper Silurian (Ludlow)—macrofossil, *Baragwanathia* (Garratt and Rickards 1987; Hueber 1992)] Lower Devonian (Lochkovian)—macrofossil (Hueber 1992)	Extant
Lycopsida	[Upper Silurian (Ludlow)—macrofossil, *Baragwanathia* (Garratt and Rickards 1987; Hueber 1992)] Lower Devonian (Lochkovian)—macrofossil (Hueber 1992)	Extant
Drepanophycales	[Upper Silurian (Ludlow)—macrofossil, *Baragwanathia* (Garratt and Rickards 1987; Hueber 1992)] Lower Devonian (Lochkovian)—macrofossil (Hueber 1992)	Upper Devonian (Frasnian)—macrofossil, *Drepanophycus* (Gensel and Andrews 1984)
Lycopodiales	[Upper Devonian—macrofossil, *Lycopodites* (Chaloner 1967a)]	Extant
Ligulate lycopsids	Middle Devonian—macrofossil, *Leclercqia* (Grierson and Bonamo 1979)	Extant
Protolepidodendrales	Middle Devonian—macrofossil, *Leclercqia* (Grierson and Bonamo 1979)	(?) Upper Devonian—macrofossil, *Archaeosigillaria* (Chaloner 1967a)
Heterosporous lycopsids	Upper Devonian—macrofossil, *Cyclostigma* (Chaloner 1968) and *Bisporangiostrobus* (Chitaley and McGregor 1988)	Extant
Selaginellales	[Lower Carboniferous—macrofossil, *Selaginellites resimus* (Rowe 1988a)] Upper Carboniferous—macrofossil, *Selaginellites suissei* (Thomas 1992)	Extant
Isoetales	[Upper Devonian—macrofossil, *Lepidosigillaria* (Bateman, DiMichele, and Willard 1992; Grierson and Banks 1963)]	Extant
Zosterophyllopsida	[Upper Silurian (Ludlow)—macrofossils (Tims and Chambers 1984)] Lower Devonian (Lochkovian)—macrofossils (D. S. Edwards 1990)	Lower Carboniferous (Tournasian)—macrofossil, *Protobarinophyton* (Scheckler 1984)

Continues on next page

TABLE 7.5
Continued

Clade	First appearance	Last appearance
Sawdoniales	[Upper Silurian (Ludlow)—macrofossils (Tims and Chambers 1984)] Lower Devonian (Lochkovian)—macrofossils (D. S. Edwards 1990)	Lower Carboniferous (Tournasian)—macrofossil, *Protobarinophyton* (Scheckler 1984)
Sawdoniaceae	Lower Devonian (Lochkovian)—macrofossils (D. S. Edwards 1990)	Upper Devonian—macrofossil, *Serrulacaulis* (Hueber and Banks 1979)
Barinophytaceae	[Lower Devonian—macrofossil, *Protobarinophyton* (Høeg 1967)]	Lower Carboniferous (Tournasian)—macrofossil, *Protobarinophyton* (Scheckler 1984)
Gosslingiaceae	Lower Devonian (Lochkovian)—macrofossils (D. S. Edwards 1990)	Lower Devonian—macrofossil, *Oricilla* (Gensel 1982b)
Hsuaceae	Lower Devonian—macrofossil (Li 1982)	Lower Devonian—macrofossil (Li 1982)
Euphyllophytina	Lower Devonian—macrofossil, *Pertica*, *Psilophyton* (Gensel and Andrews 1984)	Extant
Moniliformopses	Middle Devonian—macrofossil, *Ibyka* (Banks 1980; Skog and Banks 1973; Stein 1982)	Extant
Equisetopsida	Middle Devonian—macrofossil, *Ibyka* (Banks 1980; Skog and Banks 1973; Stein 1982)	Extant
Polypodiidae	Lower Carboniferous—macrofossil, *Senftenbergia*-like annulate sporangium (Galtier and Scott 1985; Rothwell 1996)	Extant
Radiatopses	Middle Devonian—macrofossil, *Crossia* (Beck and Stein 1993)	Extant
Lignophytia	Middle Devonian—macrofossil, *Crossia* (Beck and Stein 1993)	Extant
Spermatophytata	Upper Devonian—macrofossil, *Elkinsia* (Rothwell, Scheckler, and Gillespie 1989; Rothwell and Serbet 1994; Rowe 1992)	Extant

record is dominated by obligate tetrads, diads, and monads (cryptospores), but a profound floral change in the early Silurian is signaled by the appearance, and apparent replacement, of many early cryptospores by single trilete spores (Burgess 1991; Burgess and Richardson 1991; Gray 1985, 1991). This change in spore flora has been tentatively interpreted as a change from a land flora dominated by plants at the hepatic level of organization to one dominated by early vascular plants or vascular plant stem group taxa (Gray 1985, 1991, 1993; Gray and Shear 1992).

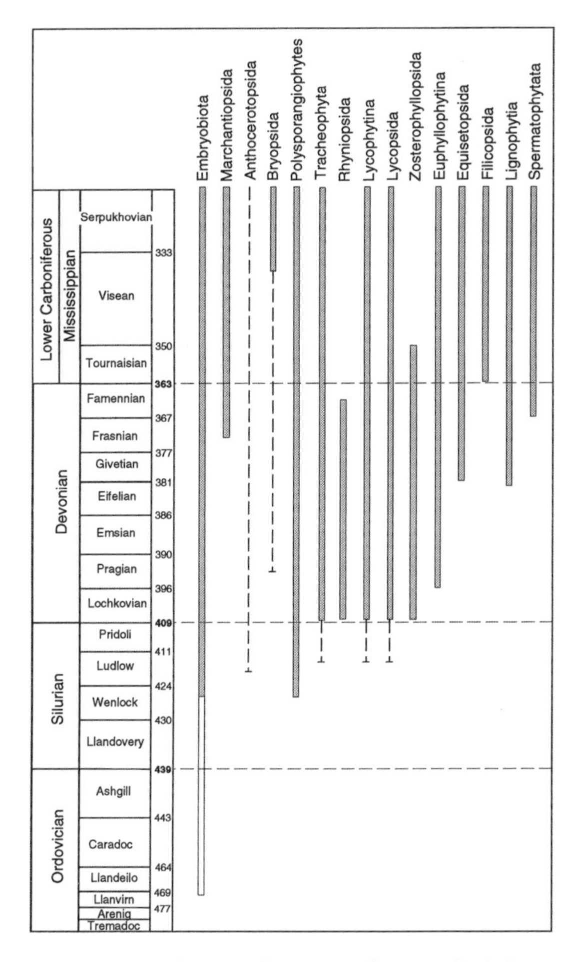

FIGURE 7.12. Stratigraphic ranges of some early clades plotted against geologic time. Stippled column = macrofossil record; open column = possible range extension based on microspore evidence; vertical dashed line = possible occurrence based on disputed or relatively incomplete evidence. The fossil evidence of minimum ages of the groups and respective sources are detailed in Table 7.5. The time scale is from Harland, Armstrong, et al. 1990.

Marchantiopsida

The fossil record of liverworts (Marchantiopsida) is poor, and in orders such as the Calobryales, Jungermanniales, Metzgeriales, Treubiales, Sphaerocarpales, and Monocleales, this situation has been attributed to the delicate nature of the thallus (Krassilov and Schuster 1984; Schuster and Janssens 1989). Recent detailed reviews of the fossil record are given by Krassilov and Schuster (1984), Oostendorp (1987), and T. N. Taylor and E. L. Taylor (1993). Most records of fossil hepatics are thalloid or leafy gametophytes, and the sporophyte generation is rarely observed. Paleozoic fossils seem to be predominantly of metzgerialean affinity. The earliest unequivocal hepatic macrofossil is the late Devonian (Frasnian) *Pallavicinites (Hepaticites)*, which consists of a branching thallus with nonseptate rhizoids (Figure 7.12; Hueber 1961). Reproductive structures are unknown. Several other unequivocal metzgerialean fossils have been recorded from the Carboniferous Period (Krassilov and Schuster 1984). The first records of the Sphaerocarpales, Marchantiales, Jungermanniales, and possibly Calobryales are from the early Mesozoic (Krassilov and Schuster 1984; Oostendorp 1987; see also *Naiadita:* Appendix 1). Recent ultrastructural work on lower Silurian spores suggests that taxa such as *Dyadospora* may be related to the Sphaerocarpales (W. A. Taylor 1995a).

Anthocerotopsida

There are no unequivocal records of hornworts (Anthocerotopsida) prior to the Cretaceous Period (Krassilov and Schuster 1984; Oostendorp 1987; T. N. Taylor and E. L. Taylor 1993). The earliest well-substantiated records are of *Phaeoceros*-like spores of late Cretaceous age (Maastrichtian, Jarzen 1979), although some late Silurian and early Devonian spore genera such as *Streelispora* and *Aneurospora* have been compared to spores from extant *Anthoceros* (Figure 7.12; Richardson 1985). Cretaceous whole plants with in situ spores reported by Nemejc and Pacltova (1974) are poorly preserved. An upper Cretaceous permineralized fossil from the Deccan Intertrappean beds, India, has been compared with the sporophyte of extant *Notothylas* (Gupta 1956).

Bryopsida

Mosses are common in Tertiary and Quaternary sediments but are comparatively rare in Mesozoic and Paleozoic fossil floras (Jovet-Ast 1967; Krassilov and Schuster 1984; Lacey 1969; Miller 1984; Oostendorp 1987; T. N. Taylor and E. L. Taylor 1993). Most fossil mosses are leafy gametophytes, some of which are difficult to distinguish from small herbaceous lycopsids (Harris 1961; Thomas 1972). Gametangia and sporophytes are extremely rare, although the recent discovery of polytrichaceous and dicranalean sporophytes and gametophytes from the late Cretaceous has dramatically expanded the Mesozoic record (A. Konopka and P. Herendeen, pers. comm. to P.R.C., 1996). Evidence from these discoveries and from Tertiary floras suggest that many modern genera had diverged by the end of the Cretaceous (Miller 1984). The most diverse Paleozoic moss floras are found in the upper Permian of Angaraland and indicate that groups such as the Bryales and Sphagnales (*Protosphagnum:* Appendix 1) had diverged by this time (Ignatov 1990; Krassilov and Schuster 1984; Neuburg 1960). Several leafy moss gametophytes have been described from the Carboniferous (Lacey 1969), and one of the earliest records of a moss is *Muscites plumatus* (lower Carboniferous), which is of possible bryalean affinity (Thomas 1972; Figure 7.12). Records of Devonian and Silurian mosses are equivocal. The sporangium of *Sporogonites* (lower Devonian: Halle 1916a, 1936a) strongly resembles sporangia of extant valvate mosses in the Andreaeopsida, but knowledge of gametophyte morphology is currently poor (Andrews 1958).

Polysporangiomorpha

The macrofossil record of Polysporangiomorpha is substantially better documented than that of bryophytes and charophycean algae. The earliest unequivocal polysporangiophyte compression fossils are Laurussian and appear in small numbers by the mid-Silurian (Figure 7.12; Edwards, Feehan, and Smith 1983). Evidence of a diverse Gondwanan flora comprising lycopsids, zosterophylls, and rhyniophytes from the upper Silurian (Ludlow) (*Baragwanathia* flora, Australia: Tims and Chambers 1984) is still regarded as problematic by some (see *Baragwanathia:* Appendix 1; Garratt and Rickards 1987; Hueber 1992). Early Chinese records include well-documented cooksonioid plants from the late Silurian (Cai, Dou, and Edwards 1993) and possibly the problematic early Silurian (late Llandovery) *Pinnatiramosus* (Cai, Shu, et al. 1996; Geng 1986; see Appendix 1 for a discussion of the affinities of *Pinnatiramosus*). Fossil evidence for polysporangiophytes is widespread by the uppermost Silurian, with records from Britain, Australia, Greenland, Czechoslovakia, Kazakhstan, Podolia, China, United States, and Libya (D. Edwards 1990).

Tracheophyta

The earliest unequivocal Tracheophyta sporophytes are of early Devonian age (Lochkovian) (Figure 7.12; Edwards, Davies, and Axe 1992; Kenrick and Crane 1991). Vascular tissues have not been documented in earlier Laurussian fossils. Earlier records from the upper Silurian of Australia and the lower Silurian of China remain problematic (see polysporangiophytes, above). By the early Devonian (Lochkovian) (Kenrick and Crane 1991, 1992), three major lineages of vascular plants were already well established. The Rhyniopsida never became diverse, but the Lycophytina and Euphyllophytina were prominent elements of the early Devonian flora (Edwards and Davies 1990; Gensel and Andrews 1984; Knoll, Niklas, et al. 1984).

Rhyniopsida

The Rhyniopsida, as defined here, have a poorly understood fossil record. The earliest unequivocal evidence of the clade comes from early Devonian (Pragian) localities in the Old Red Sandstone of northern Europe (Figure 7.12). *Rhynia gwynne-vaughanii* from the Rhynie Chert is one of several early taxa that include the recently described and more widespread *Taeniocrada*-like compression fossils such as *Stockmansella langii* (Fairon-Demaret 1985, 1986b) and *Huvenia kleui* (Hass and Remy 1991). The Rhyniopsida may extend down into the Lochkovian (lowermost lower Devonian) (Schweitzer 1983b, 1987) and probably persist into the upper Devonian (Famennian), based on the stratigraphic range of *Taeniocrada* (Figure 7.12; Banks 1980).

Lycophytina

The Lycophytina first appear unequivocally in the fossil record in the lower Devonian Period (Lochkovian), where they are represented by small homosporous forms such as *Zosterophyllum* and *Drepanophycus* (Figure 7.12; Schweitzer 1987). The earlier appearance of similar taxa in the late Silurian (Ludlow) of Australia is based on associated graptolites, but there is continued disagreement concerning age determination (see *Baragwanathia* in Appendix 1; discussion and literature in Garratt and Rickards 1987; Hueber 1992). The Lycophytina are a prominent element of Devonian plant assemblages (Edwards and Davies 1990; Gensel and Andrews 1984; Knoll, Niklas, et al. 1984; Niklas, Tiffney, and Knoll 1985). The Zosterophyllopsida are particularly diverse in the early Devonian (Edwards and Davies 1990; Kenrick 1988), and current evidence suggests that the clade became extinct in the early Carboniferous (Tournaisian), based on the last records of *Protobarinophyton* (Figure 7.12; Scheckler 1984). The Lycopsida diversified through the Devonian, and by the early Carboniferous, the group represented about half of all named vascular plant species (Knoll, Niklas, et al. 1984; Niklas, Tiffney, and Knoll 1985).

The fossil record of the Lycopodiales and Selaginellales is poorly understood, and this is partly caused by difficulties in defining the extant groups. The earliest fossil assigned to the Lycopodiales is *Lycopodites* (upper Devonian) (Chaloner 1967a; Kräusel and Weyland 1937), but there are no clear synapomorphies supporting this assignment. The earliest possible Selaginellales are Carboniferous (e.g., *Selaginellites resimus* [Rowe 1988a], Visean; see also putative upper Carboniferous taxa in Chaloner 1967a; Thomas 1992). Recent morphological (Karrfalt 1981; Rothwell and Erwin 1985) and cladistic studies (Bateman 1992; Bateman, DiMichele, and Willard 1992) show that the upper Carboniferous *Paurodendron (Selaginella) fraipontii* is more closely related to the rhizomorphic lycopsids than to the Selaginellales. Ligulate lycopsids first appear in the middle Devonian (Givetian) (Grierson and Bonamo 1979), and Isoetales (rhizomorphic lycopsids) first appear in the upper Devonian (Bateman, DiMichele, and Willard 1992). The extensive fossil record of the Isoetales has been reviewed recently by Pigg (1992).

Euphyllophytina

Euphyllophytina first appear unequivocally in the lower Devonian (Emsian, Banks 1980; Gensel and Andrews 1984), where they are represented by small, leafless homosporous taxa assigned to *Psilophyton.* Several major extant clades within the Euphyllophytina had diverged before the end of the Devonian. Early equisetaleans such as *Ibyka* appear in the middle Devonian (Givetian) (Skog and Banks 1973), and the group diversified to form an important component of upper Carboniferous coal-swamp floras (Bateman 1991; DiMichele, Hook, et al. 1992; Stein, Wight, and Beck 1984; Stewart and Rothwell 1993; T. N. Taylor and E. L. Taylor 1993). The first appearance of lignophytes in the middle Devonian is documented by petrified progymnosperm wood of *Crossia* (upper Eifelian?) (Beck and Stein 1993) and *Rellimia* (Givetian) (Bonamo 1977; Dannenhoffer and Bonamo 1989), and such taxa as *Archaeopteris* are an important component of upper Devonian floras (Banks 1980). The earliest seed plants are recorded from upper Devonian sediments of Europe (Fairon-Demaret and Scheckler 1987; Rowe 1992) and North America (Rothwell, Scheckler, and Gillespie 1989).

Filicopsida

The early record of the Filicopsida is documented by diverse reproductive structures, as well as by petrifactions of stem and leaf anatomy from the upper Devonian (Famennian) through to the lower Carboniferous (Galtier and Scott 1985). These putative stem group or basal Filicopsida belong to taxa of uncertain status such as the Cladoxylales and Coenopteridales. Clusters of exannulate, elongate sporangia of the *Musatea*-type (lower Carboniferous [Visean]) may represent the earliest evidence of the Marattiales—a group with an abundant upper Carboniferous and Permian fossil record including such taxa as the tree-fern *Psaronius* (Millay and Taylor 1979). Annulate sporangia characteristic of extant Polypodiidae first appear in the lower Carboniferous. Sporangia of *Senftenbergia*-type occur in the Tournaisian and may represent the earliest record of the Schizaeaceae, and the upper Carboniferous *Oligocarpa* is possibly the earliest Gleicheniaceae (Galtier and Scott 1985). Possible stem group Osmundales first appear in the upper Carboniferous with such taxa as *Grammatopteris* and *Catenopteris* (Miller 1971). The Ophioglossales have a poor fossil record, and the earliest specimens come from the Paleocene of western Canada (Rothwell and Stockey 1989).

TEMPORAL AND BIOGEOGRAPHIC PATTERNS IN THE EARLY FOSSIL RECORD

A Comparison of Systematic and Stratigraphic Chronologies

Comparison of the relative chronology implicit in the cladistic hierarchy with the stratigraphic appearance of land plants in the fossil record shows important points of agreement but also marked discrepancies between the predicted and actual first appearances of certain major groups. The first appearance of liverworts, hornworts, and mosses in the fossil record compared to the first appearance of vascular plants is the most obvious and interesting discrepancy. Two basic predictions from systematics are that sister groups should appear at about the same time in the fossil record and that more inclusive groups should appear before less inclusive groups. The late appearance of unequivocal bryophyte macrofossils is apparently inconsistent with both the basal position of these groups in embryophytes and the substantially earlier occurrence of dispersed spores that indicates a land flora at the bryophyte level of organization beginning in the mid-Ordovician. Macrofossils of vascular plants appear in the earliest Devonian, fully 50 million years before the first record of their bryophyte sister group (mosses) in the early Carboniferous. Similarly, the earliest appearance of liverwort macrofossils is late Devonian, and hornworts are unknown before the Cretaceous. Some authors have interpreted these stratigraphic data as favoring a polyphyletic origin of embryophytes, but in our view this is highly unlikely given the strong morphological and molecular support for embryophyte monophyly (Chapter 3).

Krassilov and Schuster (1984) discussed several possible reasons for the poor fossil record of bryophytes. First, collector bias favoring the more prominent vascular plants was considered an important factor in the general underrepresentation of bryophytes in the fossil record. Second, the delicate nature of the hepatic gametophyte is thought to be a serious obstacle to fossilization in this group. Third, because most extant mosses are considerably more robust and decay-resistant than hepatics, they explained underrepresentation on the grounds that early forms may have been less robust than their extant relatives. Although it is likely that poor preservation plays a role in the general paleobotanical underrepresentation of these groups, it is unlikely to be an important factor under exceptionally favorable conditions such as the silicified peats of the Rhynie Chert and calcium carbonate concretions (coal balls) of the Carboniferous. Under circumstances such as these, the more crucial issues probably relate to recognition problems, which lead to collector bias, and also the interesting problem of recognizing stem group taxa that frequently lack the most distinctive features of the crown group (see, for example, Crane, Friis, and Pedersen 1995, with respect to pre-Cretaceous angiosperms; Doyle and Donoghue 1993).

The history of research in paleobotany provides several examples of the difficulty of recognizing stem group taxa. Such recognition requires a readjustment of expectations that are based on extant groups so that unfamiliar character combinations in extinct taxa may be accommodated. For example, in the Bryopsida crown group, synapomorphies include multicellular rhizoids and gametophytic leaves (Figure 7.5; Mishler and Churchill 1984). Excluding the problematic Sphagnales, other potential synapomorphies are an aerial calyptra (i.e., an archegonial remnant remaining attached to the sporangium apex), an elaborate protonema, elongate antheridia, paraphyses (i.e., sterile hairs among the gametangia—also present in some Lycopodiaceae), and costate leaves (i.e., vascularization of the leaf). In a paleobotanical context, the most recognizable crown group synapomorphy is the distinctive gametophytic leaves, a feature present in the earliest record of a moss from the Carboniferous (Thomas 1972). Recognition of mosses becomes more problematic in the stem group. Nonleafy, stem group taxa may strongly resemble their leafless sister group at the protracheophyte level of organization (Figure 7.13). We think it is likely that in the floras of the late Devonian, which are dominated by vascular plants, there may be substantial biases against recognizing stem group bryophytes.

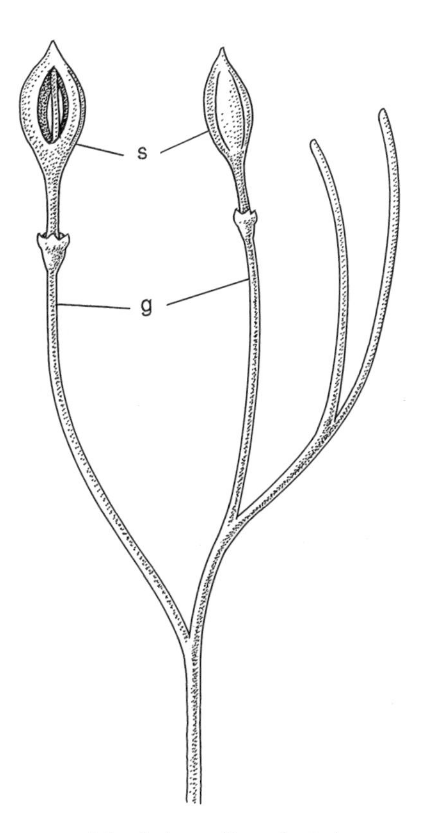

FIGURE 7.13. Morphology of hypothetical stem group moss before the evolution of gametophytic leaves. Small, unbranched, valvate sporophytes *(s)* with columella of *Andreaea* type attached terminally to naked, dichotomously branching, axial gametophyte *(g)* (×3).

Biogeographic and Macroevolutionary Patterns in Early Land Plants

Studies of the macrofossil record highlight the Devonian as perhaps the most innovative phase in the evolutionary history of land plants (e.g., DiMichele, Hook, et al. 1992; Gensel and Andrews 1984; Knoll, Niklas, et al. 1984; Niklas, Tiffney, and Knoll 1980). During this period, there was a great increase in the number of species of land plants, and a corresponding increase in morphological diversity that culminated in the evolutionary differentiation of most of the major clades of vascular plants. Temporal and biogeographic patterns in the early fossil record are documented mainly by data from the well-studied Devonian continental sediments of northern Europe and North America. These data are supplemented by records from other areas of the world, and increasingly by rapidly accumulating macrofossil data from Southeast Asia, especially China. Most significantly, however, the palynological record of land plants has been extended substantially over the last two decades by discoveries of mid-Ordovician spores that have highlighted a significant stratigraphic hiatus between the two main classes of fossil evidence (macrofossils and microfossils) and as a result have challenged traditional ideas on the mode and tempo of embryophyte evolution (Gray 1985, 1991; Gray and Boucot 1977; Gray, Massa, and Boucot 1982; Gray and Shear 1992).

Despite these recent advances and the apparent wealth of data, the interpretation of biogeographic and evolutionary patterns in the early fossil record of land plants is complex. In common with many other groups of organisms, the interpretation of patterns in the fossil record of land plants is strongly influenced by the status of the taxonomic units considered (i.e., monophyletic, paraphyletic, polyphyletic) and by biases such as relative abundance, geographical range, potential for preservation, and proximity of habitat to sites of preservation, as well as by extrinsic factors such as lithofacies variation, post-depositional processes (e.g., reworking and erosion), sampling bias, and the geographic distribution of well-studied areas (Behrensmeyer, Damuth, et al. 1992a). The effects of these factors must be evaluated and removed before biological processes can confidently be assigned as causal explanations for the pattern in the fossil record (Niklas, Tiffney, and Knoll 1980; Smith 1994). Major factors that influence patterns in the early fossil record of land plants include strong regional biases toward well-studied areas, lithofacies bias involving marine and continental sediments, differences in the relative abundance of macrofossils and microfossils, and problems with defining the taxonomic status of many early groups.

Regional Bias

Regional bias is a major problem in the early fossil record of land plants because most of the well-documented early macrofossil localities occur in northern Europe and eastern North America (Boucot and Gray 1982; Edwards 1990; Edwards and Berry 1991; Gray 1985). The proximity of these areas on the southeastern margin of the Laurussian continent in the late Silurian and early Devonian (Scotese and McKerrow 1990; Van der Voo 1993) places limits on the global generalizations that can be made from these data. Information from outside this region is often problematic because of poorly constrained age and paleogeography or because the plant fossils are of uncertain taxonomic status (D. Edwards 1990). This regional bias, which is related to the historical development of paleobotany in Europe and North America, is widely acknowledged and is currently being addressed through the detailed description of plants and floras from palaeogeographically remote areas, in particular, from China (Hao 1988, 1989a,b; 1992; Hao and Beck 1991a,b, 1993; Li 1982, 1990, 1992; Li and Edwards 1992; Li and Hsü 1987; Li, Cai, and Wang 1995; Schweitzer and Cai 1987) and from South America (Berry 1994; Berry, Casas, and Moody 1993). This bias toward well-studied areas for early land plants parallels biases in the fossil record of other groups (animal and plant) at other times, which are also generally weighted in favor of Euramerican and Chinese data (Smith 1994).

Lithofacies Bias

Land plant fossils generally are more common in continental sediments than in marine sediments, even though megafossils may be locally abundant offshore, and there is considerable variation in the abundance of fossils in continental settings, depending on sedimentary context (Behrensmeyer, Hook, et al. 1992b). In the early fossil record of land plants, a strong lithofacies bias occurs through the Silurian and into the early Devonian. All records of Silurian land plants occur in marine sediments (Edwards and Berry 1991), whereas the diverse early Devonian floras in northern Europe and eastern North America are mainly in continental deltaic, fluvial, and lacustrine facies. This lithofacies problem is related to a widespread marine regression in the Euramerican region that began in the late Silurian (Ludlow) and ended with the emergence of fully continental conditions in the early Devonian (Pragian-Emsian) (Allen 1985). This marine regression is driven by tectonic events related to closure of the Iapetus Ocean and the suturing of the Laurentia and Baltica lithospheric plates (Van der Voo 1993). Similar transitional conditions are found in many parts of the North Atlantic region and eastern and northern Canada at this time (Allen 1985).

Taxonomic Bias

Taxonomic biases relate to the status of the high-level taxa in traditional systematic treatments. The presence of monophyletic, paraphyletic, and polyphyletic groups affects the interpretation of diversification and extinction patterns in the early fossil record (Smith 1994), yet the status of many high-level plant groups is unknown. In the fossil record of fish and echinoderms, Smith (1994) estimates that only 30% of commonly recognized taxa are monophyletic, and the systematic status of the plant fossil database is unlikely to be better. Because of the problematic status of many high-level taxa and difficulties with the equivalency of such units, quantitative assessments of diversity in the plant fossil record have been based on low-level taxa (species or genera—many of which are monotypic in the Devonian), and the origination and extinction of high-level taxa has been assessed by overlaying these data on total diversity curves (e.g., Edwards and Davies 1990; Knoll, Niklas, et al. 1984).

Our systematic treatment of the early fossil record shows that several important high-level taxa are paraphyletic with respect to other high-level groups (e.g., Rhyniophytina *sensu* Banks and Trimerophytina *sensu* Banks) whereas others appear to be monophyletic or nearly so (e.g., Zosterophyllophytina *sensu* Banks). These results principally affect the interpretation of extinctions at high taxonomic levels in early land plants and of diversity at the base of major clades. The apparent extinction of manifestly paraphyletic groups such as the Rhyniophytina and Trimerophytina (Edwards and Davies 1990; Knoll, Niklas, et al. 1984) is clearly an artifact.

Temporal Patterns in the Diversification of Embryophytes

The earliest evidence for embryophytes (land plants) comes from the palynological record of dispersed spores. Small, obligate spore tetrads are characteristic of the early dispersed spore floras and first appear in the mid-Ordovician (Llanvirn–Llandeilo). Tetrads are abruptly replaced by single trilete spores in the early Silurian (late Landovery). The palynological data therefore provide evidence of an initial mid-Ordovician origin and Silurian radiation of embryophytes, but because spores present few features that are diagnostic of particular embryophyte subgroups, it is difficult to assess the relative contributions of the various major clades. In contrast, the earliest unequivocal records of embryophytes based on macrofossils are much younger and are dominated by polysporangiophyte and tracheophyte groups. These data suggest a mid-Silurian origin and a late Silurian through Devonian radiation.

Several biases may contribute to this discrepancy between macrofossil and microfossil evidence (Gray 1985, 1991). First, although the macrofossil record of vascular plants is excellent, the early record of the Marchantiopsida, Anthocerotopsida, and Bryopsida is extremely poor (see above). Systematic studies strongly suggest that these major groups are products of early cladogenic events that predate the evolution of tracheophytes, implying that these clades are present in the mid-Silurian and earlier (Chapter 3). The absence of macrofossil evidence may therefore reflect one or more sampling biases that do not seriously affect the microfossil record. Second, early palynological data have been collected over a wide geographic area, including North America (Tennessee to Nova Scotia), South America (Brazil, Paraguay), Europe (Sweden, Norway, Czechoslovakia, Belgium, United Kingdom), Africa (Libya, Egypt, Ghana, South Africa), Arabia, and Australia (Gray 1991), thus avoiding some of the problems of regional biases that affect the early macrofossil record. Third, microspores are more likely to be sampled in the fossil record than macrofossils because they are vastly more abundant and more widely dispersed.

Temporal Patterns in the Diversification of Tracheophytes

Quantitative analyses of large-scale changes in plant diversity through time are a recent innovation in paleobotany. Early analyses used similar techniques and approaches to those employed in studying the much more extensive marine invertebrate fossil record (Raup 1972, 1976a,b; Raup and Boyajian 1988; Raup, Gould, et al. 1973; Raup and Sepkoski 1986; Sepkoski 1978, 1984; Sepkoski, Bambach, et al. 1981). In the Devonian, the patterns that emerged from these analyses have been interpreted as providing evidence about the general "shape" of vascular plant diversification and the changing relative contributions of the various major clades to total land plant diversity through this time interval (Knoll, Niklas, et al. 1984; Niklas, Tiffney, and Knoll 1980). Diversity curves are typically interpreted in biological and ecological terms, but critical evaluation of contemporaneous geological changes in the late Silurian and early Devonian highlight the difficulties of distinguishing between geological and biological signals in the plant fossil record during this time period.

Numerical compilations highlight the Devonian as one of three periods in the history of land plants that are characterized by a marked increase in species diversity (Knoll, Niklas, et al. 1984; Niklas, Tiffney, and Knoll 1980, 1983, 1985; Tiffney 1981). This result is consistent with earlier qualitative studies indicating a mid-Silurian origin of vascular plants and a rapid radiation of the group that began in the early Devonian (e.g., Banks 1968, 1970; Chaloner and Sheerin 1979). Three numerical studies have focused in detail on patterns of vascular plant diversification within the critical Silurian–early Devonian interval (Edwards and Davies 1990; Knoll, Niklas, et al. 1984; Raymond and Metz 1995). Knoll, Niklas, et al. (1984) compiled data on taxonomic ranges at the generic level to investigate rates of diversification, origination, and extinction. This analysis found that the early generic diversification of vascular plants conforms to Sepkoski's kinetic model of taxon diversification as exemplified by marine invertebrates (Figure 7.14). The kinetic model predicts that the early diversification of a group will show a sigmoidal increase in taxon numbers with time (Sepkoski 1978, 1984), and the vascular plant macrofossil data conform to this pattern. Knoll, Niklas, et al. (1984) identified a lag period of low diversification that persists for some 30 million years between the first appearance of vascular plant macrofossils in the mid-Silurian and the early Devonian (Lochkovian ≈ Gedinnian). The lag phase is followed by a rapid increase in diversity from the Lochkovian to the Emsian (early middle Devonian) followed by a rapid tailing off and a plateau that persists through to the Famennian. Dispersed spore data were also interpreted to be consistent with this pattern, although the sigmoidal nature of the curve is much less pronounced. A later numerical analysis by Edwards and Davies (1990) focused in detail on Euramerican macrofossil data for the late Silurian and early Devonian and found a broadly similar early pattern of diversification. Knoll, Niklas, et al. (1984) interpreted the observed increase in taxon diversity through the late Silurian and early Devonian and the shape of the diversity curves in biological terms. Increasing taxon diversity through this period was viewed as consistent with the initial radiation of the vascular plant clade, and the sigmoidal shape of the diversity curve was interpreted as indicating the importance of density-dependent processes (e.g., competition) in governing the early phase of this diversification.

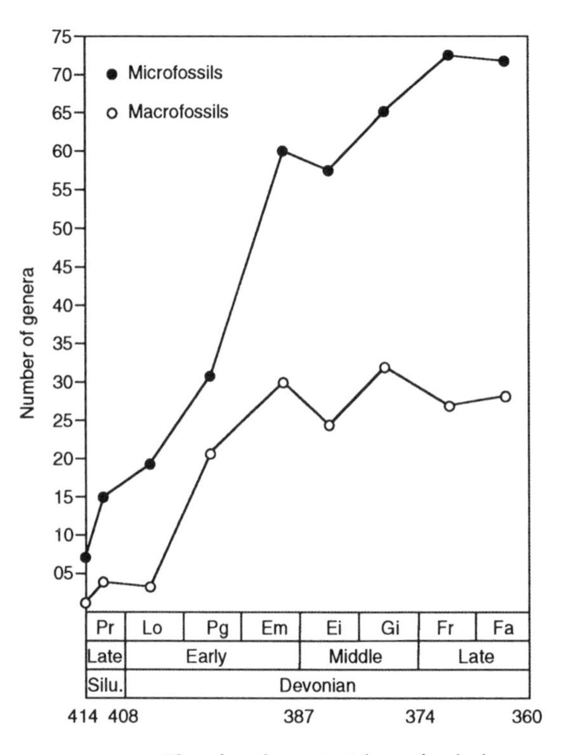

FIGURE 7.14. Plot of total generic richness for the latest Silurian and Devonian periods. Pr = Pridoli; Lo = Lochkovian; Pg = Pragian; Em = Emsian; Ei = Eifelian; Gi = Givetian; Fr = Frasnian; Fa = Famennian. Redrawn from Knoll, Niklas, et al. 1984, Figure 1; principal data sources: Banks 1980; Chaloner 1967b; Chaloner and Sheerin 1979. Time scale is that of Harland, Cox, et al. 1982; numbers along bottom axis denote millions of years before present (Ma).

The kinetic model of vascular plant diversification has been challenged recently in a study that questioned the utility of the diversity measurements employed in previous work. Raymond and Metz (1995) showed that sampling intensity—measured as the number of plant assemblages per time interval—varies widely in the early record of land plants. In the late Silurian assemblage numbers are low, with 1 assemblage in the Wenlock, 2 in the Ludlow, and 8 in the Pridoli, whereas in the lower Devonian much higher sampling values occur with 22 assemblages recorded from the Lochkovian (≈Gedinnian) to the early Pragian (≈Siegenian) and 64 from the late Pragian to the Emsian. Furthermore, there is a statistically significant, positive correlation between sampling density through the late Silurian and Devonian and taxic diversity, based on a range of measurements including (1) total diversity (number of genera per interval + number of implicit genera—those absent from the time interval but present on either side), (2) mean genus richness of floras (number of genera per interval in a formation or member), (3) median assemblage diversity, (4) most diverse assemblage, and (5) standing diversity at interval boundaries (number of genera shared between successive intervals + number of implicit genera). This result suggests that sampling intensity has an important influence on the shape of diversity curves during this time period.

Raymond and Metz (1995) found that standing diversity at interval boundaries is the measure least sensitive to sampling bias, particularly when low-diversity (late Silurian) assemblages were excluded. Standing diversity at interval boundaries measures the number of genera shared between successive time intervals and differs from measurements of total diversity principally in its exclusion of genera unique to a particular time interval. In other words, diversity curves resulting from this measurement sample common genera (genera with ranges spanning two or more time intervals) and ignore data on rare or unique taxa. For the late Silurian and early Devonian, Raymond and Metz (1995) conclude that the data support an overall increase in land plant diversity through time but that there is little support for the sigmoidal shape of the diversity curve documented by Knoll, Niklas, et al. (1984).

Although ecological and evolutionary factors undoubtedly had a major impact on taxic diversity in the late Silurian and early Devonian, the presence of profound contemporaneous geological changes may override biological signals in generating the observed pattern. Geological effects may also explain the differences in sampling intensity among time intervals. The studies by Edwards and Davies (1990) and Raymond and Metz (1995) focused on Laurussian assemblages, and the early diversification patterns documented by Knoll, Niklas, et al. (1984) are also based mainly on Euramerican data. During the late Silurian to early Devonian (Ludlow–Pragian) the North Atlantic region, notably Norway, Spitsbergen, the Anglo-Welsh area, and eastern and northern Canada, was undergoing a period of transition from the deep-water marine conditions that dominated the early Paleozoic to continental, freshwater conditions typical of the late Paleozoic (Allen 1985; see above sections on regional bias and lithofacies bias). Gray and Boucot (1977) pointed out that the progressive increase in spore abundance through the late Silurian and early Devonian of Britain and North Africa correlates with this general temporal progression from older offshore environments to younger nearshore environments. This widespread transition suggests that the major factor governing the observed patterns of taxon diversification during the late Silurian and early Devonian is the emergence of a major land mass in the main Euramerican sampling region. The low levels of diversity and low sampling intensity characteristic of the late Silurian (Ludlow) coincide with the early, deep-water phase of the marine regression, in which a poor sample of land plant diversity would be expected. The rapid increase in diversity during the early Devonian (Pragian–Emsian) coincides with the progressive shallowing and the emergence of fully continental sediments (Old Red Sandstone), providing a greater sample of land plant diversity.

Evidence from the geological record of the late Silurian and early Devonian indicates that the timing and pattern of the vascular plant diversification is strongly influenced by the nature of the important Laurussian plant assemblages. The principal effect of the geological changes through this interval is to lead to an underestimation of vascular plant diversity in the Silurian. Our analysis of the range extensions for major groups as predicted by their cladistic relationships and the maximum age of sister taxa (Figure 7.15) also points to the Silurian as a time of major cladogenesis in basal vascular plants. We suggest that the origination of the Marchantiopsida, Anthocerotopsida, Bryopsida, polysporangiophytes, Tracheophyta, eutracheophytes, Rhyniopsida, Lycophytina, and Euphyllophytina occurred no later than the late Silurian (Figure 7.15). The early appearance of *Pertica* suggests a late Pragian (early Devonian) origin for lignophytes and the *Equisetum*-fern clade. These results are more consistent with the mid-Ordovician origin of embryophytes and early Silurian diversifications that are suggested by the microfossil record (Gray 1985, 1991; Gray and Boucot 1971, 1977; Gray, Massa, and Boucot 1982). Further evaluation of the relative importance of biological and geological factors on observed patterns of taxon diversification in early land plants requires additional sampling from Silurian terrestrial sediments. Appropriate sediments are more likely to be found in paleogeographically remote areas that were not influenced by Euramerican tectonic events.

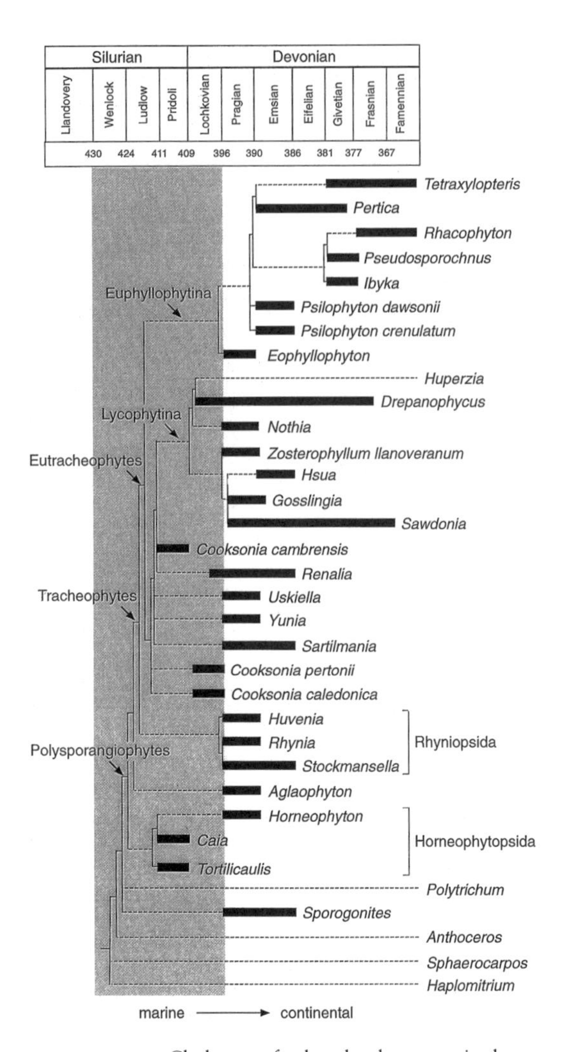

FIGURE 7.15. Cladogram for basal polysporangiophytes plotted against geologic time, showing the appearance of major clades in the late Silurian and early Devonian, during a widespread marine regression *(stippling)* in the main sampling areas (southern Laurussian supercontinent). Time scale is that of Harland, Armstrong, et al. 1990. Thick bars = approximate known stratigraphic ranges of taxa. Broken lines = implicit range extensions based on cladistic evidence. Note that the rapid appearance of taxa and major groups in the early Devonian is probably a taphonomic effect related to increasingly widespread continental conditions. Because of the relatively early appearance of such taxa as *Drepanophycus* and *Cooksonia cambrensis,* the cladogram implies that the origin and diversification of many major clades of vascular plants occurred significantly earlier (*at least* late Silurian for such groups as polysporangiophytes, tracheophytes, eutracheophytes, the Rhyniopsida, the Lycophytina, and the Euphyllophytina). An even earlier origin of major groups is implied by the early appearance of the lycopsid *Baragwanathia* in sediments dated as Upper Silurian (late Ludlow) and by the occurrence of dispersed spores resembling those of embryophytes from the mid-Ordovician.

Biogeographic Patterns

As with biostratigraphy, analysis of Silurian and Devonian phytogeography is severely constrained by poor sampling at a global level and significant variability in the intensity of study among different areas (D. Edwards 1990; Knoll 1984, 1986; Raymond 1987; Raymond and Metz 1995; Raymond, Parker, and Barrett 1985; Ziegler, Bambach, et al. 1981). Data compiled in a recent study by Raymond (1987) were from 32 localities and 170 taxon occurrences (39 taxa) and spanned a period in the early Devonian of about 10 million years from the mid-Pragian to Emsian. Over 50% of the localities were in the well-studied Euramerican region. D. Edwards (1990) focused on the late Silurian (Pridoli) to early Devonian (Lochkovian), a period of about 15 million years. She considered 35 localities and 100 taxon occurrences with a similar bias toward the Laurussian continent. The numerical treatment by Raymond (1987) recognized floristic areas and their relationships on the basis of commonality. Paired comparisons of localities were used to construct similarity indexes based on the presence and absence of taxa. Floristic areas and their relationships were then defined by the hierarchical clustering of localities. The effects of different taxon treatments in different parts of the world were evaluated by rescoring localities for plant traits (characters), and this analysis yielded broadly similar results. Raymond (1987) concluded that there are three major phytogeographic units in the early Devonian: Australia, Kazakhstan–north Gondwana, and an equatorial to middle latitude region. The equatorial–middle-latitude region is by far the largest and was subdivided into three subregions: Siberia–north Laurussia, south Laurussia, and China. A qualitative assessment of the late Silurian to early Devonian period by D. Edwards (1990) was more cautious but noted a similar pattern and split the equatorial middle latitude region into two regions: Laurussia and Siberia. Possible relationships among other areas were not specified.

The significance of the patterns recognized by Raymond (1987) and D. Edwards (1990) are difficult to evaluate, but the quality and quantity of data from regions outside Laurussia and China suggest that much of the pattern may reflect sample bias. In the Raymond study, the largest phytogeographic unit—the equatorial–middle-latitude region—contains 80% of the sampled localities, all of which are located in the most intensively studied areas (i.e., Laurussia and China). Of the other two major areas, Australia is recognized on the basis of two localities segregated from the others by two very poorly understood endemic taxa: *Yarravia* and *Hedeia.* The existence of a Kazakhstan–north Gondwanan region is also questionable because it is based on only two low-diversity localities segregated on the presence of *Cooksonia,* a poorly defined and possibly polyphyletic assemblage that also occurs in Laurussia. The identification of phytogeographic regions based on traits is also problematic because, as noted by Raymond (1987), there are almost no traits specifically associated with a single phytogeographic region or subregion.

An alternative approach to understanding biogeographic patterns is to identify areas of endemism and to approach the problem of area relationships through an analysis of cladistic relationships among endemic taxa (e.g., Nelson and Platnick 1981). The interrelationships of the areas are then specified by one or more area cladograms. This approach could, in principle, be used to investigate the relationships of such major Silurian land masses as Baltica, Laurentia, Siberia, and Gondwana and perhaps even to locate the geographic origin of major embryophyte groups. Several localities identified by D. Edwards (1990)—one in Siberia and several in Kazakhstan—are interesting because they appear to contain a relatively high proportion of endemic vascular plants. However, taxa in both areas are rather poorly understood. A cladistic approach to Paleozoic biogeography requires robust cladograms based on informative taxa from the major land masses. Current late Silurian and early Devonian paleobotanical data are far too limited for detailed examination of these interesting questions.

COMPARATIVE MORPHOLOGY

The phylogenetic treatments outlined in this book provide a cladistic perspective on early land plant morphology and constitute the basis for a specific and rigorous approach to understanding the evolution of plant form. Explicit discussion and analysis of character distributions focus attention on the key concepts of homology and homoplasy, and the inclusion of many early fossil taxa allows a more rigorous test of homology than has been possible in previous neobotanical cladistic studies. Furthermore, a clear understanding of character distributions among the major groups of land plants is essential to making generalizations from experimental studies on model organisms (e.g., Canatella and de Sá 1993), as well as to recognizing such evolutionary phenomena as convergence or loss. Evaluating morphology in terms of homology and homoplasy is also critical to developing experimental approaches for investigating the structural and functional constraints and the developmental controls that govern plant form. Patterns of character change (acquisition or loss) discovered through cladistic analysis provide the conceptual framework for investigating the evolution of plant form and development.

In this section we analyze a selection of key morphological innovations in land plants from a phylogenetic perspective, assessing hypotheses on homology, homoplasy, and putative evolutionary transformations. Ideally, this type of analysis should be built upon detailed and comprehensive cladograms, but these are still unavailable for most major groups of land plants. The discussion here focuses on basal groups within land plants and on some specific evolutionary problems that are directly relevant to the systematic treatments in this book. This section is divided into two parts. The first deals with hypotheses of taxic homology; the second considers related ideas on transformational homologies.

Taxic Homologies

The general problems associated with formulating evolutionary hypotheses of organismal morphology in terms of ecological, paleontological, functional, or structural criteria have been widely discussed and are currently the subject of much research (e.g., Bateman and DiMichele 1994a; Brooks and McLennan 1991, 1994; Coddington 1988; Donoghue, Doyle, et al. 1989; Harvey and Pagel 1991; Rieppel 1988; Smith 1994; Wanntorp, Brooks, et al. 1990). One useful perspective developed in the study of adaptation has been to view morphology in terms of unique (homology) and iterative (homoplasy) events (Coddington 1988, 1994; Pagel 1994; Sanderson and Hufford in press; Wenzel and Carpenter 1994). Iterative character distributions are interesting because they imply the presence of strong directional forces (extrinsic or intrinsic) operating independently in many clades. Investigation of iterative events may lead to the discovery of general, lawlike processes. The processes driving iterative evolutionary change can be investigated by seeking correlations with other possible causal factors (Bateman 1996b; Coddington 1988, 1994; Harvey and Pagel 1991; Pagel 1994; Wenzel and Carpenter 1994). Iterative events may be evidence of adaptation to similar selective pressures, or they may be caused by structural or developmental constraints in response to different selective regimes. In contrast, unique evolutionary events require unique explanations, but they may also be adaptations with important consequences in catalyzing major diversifications or evolutionary change. Distinguishing homology from homoplasy is therefore critical to understanding causality in the evolution of organismal form. Below we reexamine some important aspects of land plant morphology from this perspective.

HETEROSPORY Heterospory, together with other heterosporic phenomena, forms a suite of characters related to spore production, gametophyte reduction, and control of sexual expression in vascular plants (Table 7.6; Bateman and DiMichele 1994a). Heterospory has long been regarded as an iterative event, and its adaptive significance has been widely discussed, particularly in relation to the evolution of the seed. Much of the evolutionary success of seed plants has been attributed to heterospory combined with additional morphological innovations in pollination biology and other features (e.g., Chaloner and Hemsley 1991; Chaloner and Pettitt 1987; DiMichele, Davis, and Olmstead 1989; Tiffney 1981); however, other heterosporous groups are much less diverse than their homosporous antecedents. Chaloner and Pettitt (1987) highlighted this paradox by likening heterospory to a valley in the topography of success between the highlands of homospory and seed reproduction. In other words, the evolution of heterospory from homosporous ancestors would seem to imply certain negative consequences resulting in the loss of a competitive edge, which has been overcome successfully only in seed plants. An alternative view is that heterospory was a highly successful, but narrowly specialist, form of life cycle that dominated tropical aquatic and amphibious habitats through the Carboniferous Period (Bateman and DiMichele 1994a; DiMichele, Hook, et al. 1992; DiMichele, Phillips, and Peppers 1985; Phillips and Peppers 1984).

Despite widespread interest in the phenomenon

TABLE 7.6

Maximum Numbers of Heterosporic Characters Acquired by Specific Orders

Taxon	Heterospory*	Dioecism	Heterosporangy*	Endospory	Monomegaspory*	Endomegasporangy	Integumentation*	Lagenostomy*	In situ pollination	In situ fertilization	Pollen tube formation	Siphonogamy
Barinophytaceae†	+	+?	–	–?	–	–	–	–	–	–	–	–
Selaginellales	+	+	+	+	–	–	–	–	–	–	–	–
Isoetales[1]	+	+	+	+	+	+?	+	–	–	–	–	–
Equisetales	+	+?	+	+?	+	+?	–	–	–	–	–	–
Sphenophyllales†	+?	+?	–	–?	–	–	–	–	–	–	–	–
Stauropteridae†	+	+?	+	+?	–[2]	+?	–	–	–	–	–	–
Salviniales	+	+	+	+	+[3]	+	–	–[4]	–	–	–	–
Marsileales	+	+	+	+	+	–	–	–	–	–	–	–
Platyzoma	+	+	+	–	–	–	–	–	–	–	–	–
Aneurophytales†	+	+?	–[5]	–?	–	–	–	–	–	–	–	–
Archaeopteridales†	+	+?	+	–?	–	–	–	–	–	–	–	–
Protopityales†	+?	+?	+?	–?	–	–	–	–	–	–	–	–
Noeggerathiales†	+	+?	+	+?	+	+?	–	–	–	–	–	–
Cecropsidales†	+	+?	+	+?	+	+?	–	–	–	–	–	–
Spermatophytata	+	+	+	+	+	+	+	+	+	+	+	+

Source: Adapted from Bateman and DiMichele 1994a, Figure 13.

Notes: The characters are listed in approximate sequence of acquisition. + = presence. – = absence. † = extinct. * = heterosporic characters most likely to be detected in fossils.

[1] *sensu* Chapter 6.

[2] Two viable megaspores.

[3] *Salvinia** only.

[4] *Salvinia* possesses a cellular perispore that superficially resembles a pteridospermalean nucellus.

[5] Sporangia of *Chaleuria* contain spores that are dominantly but not exclusively of one kind (and presumed gender).

of heterospory and its adaptive significance (reviewed by Bateman and DiMichele 1994a), the number of iterations among major clades of land plants and the pattern of acquisition of heterosporic phenomena are poorly understood. Estimation of the number of heterosporic events is hampered by a poor understanding of cladistic relationships among early members of the Euphyllophytina, including important extinct heterosporous groups. Much greater precision is required in order to measure the diversity of heterosporous groups and to test hypotheses of adaptation. Based on recent cladistic studies and on estimates of character distributions among clades, Bateman and DiMichele (1994a) identified a minimum of 11 independent origins of heterospory in vascular plants, most having occurred during a relatively short time interval in the late Devonian and early Carboniferous (Table 7.6). The extent to which additional heterosporic phenomena subsequently developed varies greatly among clades, but based on the analysis by Bateman and DiMichele, the sequence of acquisition appears to be broadly similar in different clades, usually occurring in the order: dioecism, heterosporangy, endospory, monomegaspory, endomegasporangy, and integumentation (Table 7.6). Lagenostomy, in situ pollination, in situ fertilization, pollen tubes, and siphonogamy are all unique features of, but do not necessarily define, seed plants. No complete reversal to homospory has been documented, although loss of some heterosporic phenomena may have occurred (Bateman and DiMichele 1994a; see also for definition of above terms).

The cladistic analyses presented in this book support at least two independent origins of heterospory in the Lycophytina: one origin in the Zosterophyllopsida (Barinophytaceae) and at least one origin in the Lycopsida (Selaginellales plus Isoetales). In zosterophylls, heterospory is of a peculiar kind (megaspores and microspores are present in a single sporangium), heterosporangy has not developed, and there are no other associated heterosporic phenomena (Table 7.6). In the Lycopsida, heterospory is associated with a range of related phenomena, and multiple character acquisitions occur at some nodes (Bateman and DiMichele 1994a). Heterospory, dioecism, heterosporangy, and endospory appear simultaneously (based on current taxon sampling) as synapomorphies of heterosporous lycopsids, whereas endomegasporangy, monomegaspory, and integumentation are acquired in a stepwise fashion within the Isoetales. Bateman (1994) and Bateman and DiMichele (1994a) argue that the simultaneous appearance of developmentally related characters on a cladogram should be interpreted as evidence of saltational evolution of a character complex (null hypothesis) rather than a gradualistic stepwise acquisition. From this perspective, evolution of heterospory, dioicy, heterosporangy, and endospory in the Lycopsida could be interpreted as a saltational event, at least pending the discovery of additional early heterosporous lycopsids with some but not all of these characteristics. This view is inconsistent with gradualistic models for the evolution of heterospory (e.g., Chaloner and Hemsley 1991; Chaloner and Pettitt 1987; Haig and Westoby 1988) but may need to be modified in light of future paleobotanical discoveries.

The highly iterative nature of heterospory and the similar patterns of character acquisition in different lineages suggest that the evolution of this phenomenon was for the most part adaptively driven. Bateman and DiMichele (1994a; see also DiMichele and Bateman, in press) argue that increasing life-cycle compression results in the sporophyte playing an increasingly important role in sex determination and the control of reproductive timing. According to this interpretation, one of the main advantages of heterospory is that life histories become more "holistic," as the target for selection is reduced from two independent phases to the one integrated gametophytic-sporophytic phase. However, one problematic consequence of increasingly heterosporous life histories is that the gametophyte generation loses its ability to change sex, and hence its ability to facilitate fertilization in response to changing environmental conditions. In this respect, heterosporous taxa appear to be at a disadvantage relative to homosporous groups. Bateman and DiMichele (1994a) argue that several strategies have been developed by heterosporous taxa to alleviate this problem. Some taxa avoid sexual reproduction through reliance on apomixis (some *Selaginella*); others inbreed (self-fertilize) by dispersing megaspores with microspores attached (Selaginellaceae, Isoetaceae); some exploit stable, ecologically buffered (e.g., aquatic) environments where the ability to change sex is superfluous (many lycopsids, Salviniales, Marsileales); and seed plants use more complex pollination biologies.

The "narrow specialists" hypothesis for the evolution of heterospory in semiaquatic habitats is consistent with the early fossil record of such groups as the Equisetopsida and Isoetales. According to Bateman and DiMichele's interpretation (1994a), the evolution of heterospory was a convergent adaptation to an aquatic or semiaquatic lifestyle. More integrated holistic life cycles would allow heterosporous groups to colonize that specialized and difficult-to-exploit habitat because an adaptive response is required principally from the sporophyte. In contrast, in homosporous plants, both the sporophyte and gametophyte generations must adapt to extended growth in aquatic or semiaquatic conditions. Furthermore, the semiaquatic habitat provides a favorable and more stable environment for plants having life cycles with the fixed sex ratios characteristic of heterospory. Newly evolved heterosporous taxa would avoid direct competition with their homosporous antecedents because few homosporous groups are able to live under semiaquatic conditions. For typical fully terrestrial heterosporous groups, such as seed plants and the Selaginellaceae, DiMichele and Bateman (in press) have postulated an early aquatic phase that was followed by subsequent recolonization of and radiation in fully terrestrial habitats. There is no evidence that competition with seed plants drove heterosporous pteridophytes from ecological dominance of the tropical wetlands during the Carboniferous. Paleoecological studies suggest that their demise was caused by profound warming and drying of the regional or global climate (DiMichele, Hook, et al. 1992; DiMichele, Phillips, and Peppers 1985).

SECONDARY GROWTH Secondary growth is a prerequisite for the evolution of arboreous forms in most groups of plants. Among extant land plants, extensive secondary growth is restricted to seed plants, but the phenomenon was widespread among several major clades of tracheophytes in the late Paleozoic. Secondary growth usually results in the expansion of stem and root circumferences through the addition of tissues to the primary plant body. Additional tissues are formed from one or more lateral meristems, which typically produce xylem, phloem, or cortical sclerenchyma. Secondary growth in land plants is a complex phenomenon involving various processes related to cell growth, division, and differentiation that can be regarded as a suite of potential evolutionary innovations, some of which may be more general than others. Some of these process are recognizable as potentially independent characters or character states (Table 7.7).

In many woody plants, the distinctive alignment of cells in rows in the xylem, phloem, and periderm results from the periclinal cell division of cambial initials. Production of secondary xylem is usually associated with the activity of a well-defined meristematic layer (cambium) that is typically differentiated into elongate fusiform initials and more or less isodiametric ray initials. Fusiform initials divide periclinally to produce rows of tracheids, vessel elements, and fibers or parenchyma or both. Ray initials divide periclinally to produce radially oriented parenchymatous channels within the wood. In fossil plants the presence of a cambium is often difficult to demonstrate because the fusiform initials and their immediate derivative cells (incipient secondary tissues) may not be preserved (Cichan and Taylor 1990). Furthermore, cell alignment by itself is not sufficient to demonstrate cambial activity because alignment can be caused by other forms of regularized cell division (e.g., a primary thickening meristem). Cambial activity is consistent with patterns of secondary growth that result in a significant increase in secondary xylem toward the base of the plant. The effect of a primary thickening meristem is to produce similar amounts of secondary xylem at different levels in the plant—a pattern that would also appear in plants with a determinate cambium (Rothwell and Pryor 1991). Meristematic activity resulting in cell alignment in periderm or xylem can also be stimulated by wounding (e.g., Banks and Colthart 1993; Scott, Stephenson, and Chaloner 1992), but it is distinguishable from normal secondary growth because it is localized within the stem. Production of significant quantities of secondary xylem in all early groups is always linked to the presence of rays, implying the presence of some structural or functional constraints (Table 7.7).

In addition to periclinal cell divisions, other factors contributing to the production of secondary xylem and accommodating the expansion in girth include anticlinal divisions of fusiform initials, intrusive growth of fusiform initials, and increased cell diameter. In seed plants, occasional anticlinal divisions of the fusiform initials add to the number of cell rows as girth increases, whereas in some extinct groups, "additional" rows of tracheids arose principally through elongation and intrusive growth of the fusiform initials (Cichan 1985b; Cichan and Taylor 1982). In many taxa, tracheid diameter also increases significantly in the later-formed wood and so contributes to the increasing girth of the stem (Cichan 1985b, 1986a; Cichan and Taylor 1982, 1990).

TABLE 7.7
Characters Associated with Secondary Growth in Selected Basal Vascular Plants

Taxon	Secondary xylem	Rays	Secondary phloem	Cambium	Tracheid diameter*	Periderm	Sources
Elkinsia † (Spermatophytata)	+	+	?	?	?	–	Serbet and Rothwell 1992
Archaeopteris † (Lignophytia)	+	+	+?	+?	?	–?	Arnold 1930; Beck 1971; Beck and Wight 1988
Tetraxylopteris † (Lignophytia)	+	+	+	?	?	–	Beck 1957; Beck and Wight 1988; Scheckler and Banks 1971a
Rhacophyton † (Filicopsida stem group)	+	+	?	?	?	?	Cornet, Phillips, and Andrews 1976; Dittrich, Matten, and Phillips 1983
Zygopteris † (?Filicopsida stem group)	+	+	–	?	?	+	Dennis 1974
Botrychium (Filicopsida)	+	+	+	?	?	+	White and Turner 1995
Arthropitys † (Equisetopsida)	+	+	–	?	+	+?	Cichan 1986a; Cichan and Taylor 1982; Wilson and Eggert 1974
Metacladophyton † (Equisetopsida stem group)	+	+	?	?	+	–	Wang Zhong and Geng Bao Yin, in press
Sphenophyllum † (?Equisetopsida)	+	+	+	+	+	+	Cichan 1985b; Cichan and Taylor 1982; Eggert and Gaunt 1973
Isoetales (Lycopsida)	+	+	–	+	+	+	Eggert and Kanemoto 1977; Rothwell and Pryor 1991; Stewart and Rothwell 1993
Protopteridophyton † (basal Euphyllophytina)	+?	–	?	–?	+	–	Li and Hsü 1987
Psilophyton † (basal Euphyllophytina)	+?	–	?	–?	?	–?	Banks and Colthart 1993; Banks, Leclercq, and Hueber 1975

Notes: * = increase in tracheid diameter in later-formed wood. + = presence. – = absence. † = extinct.

In seed plants, the production of small amounts of secondary phloem usually accompanies secondary xylem formation, and both tissue types are produced from the same bifacial cambium. There is evidence for secondary phloem in aneurophytalean progymnosperms such as *Tetraxylopteris* (Beck 1957), *Triloboxylon* (Scheckler and Banks 1971a), *Proteokalon* (Scheckler and Banks 1971b), *Aneurophyton* (Schweitzer and Matten 1982), and *Rellimia* (Dannenhoffer and Bonamo 1989) and possibly in the Archaeopteridales (Arnold 1930) and *Protopitys* (Walton 1969). Secondary phloem is much rarer in other major clades of land plants (Cichan and Taylor 1990; E. L. Taylor 1990). In arborescent Lycopsida and Equisetopsida the cambium appears to be unifacial, producing only secondary xylem (Eggert and Kanemoto 1977; Wilson and Eggert 1974), but secondary phloem has been described in other taxa such as *Sphenophyllum* (Eggert and Gaunt 1973). In many taxa, one or more "cork cambia" near the periphery of the stem produce additional cortical tissues that are often lignified or suberized. Periderm development is greatest in the arboreous Lycopsida, in which it forms the principal structural support for the stem (Eggert 1961).

Cladistic analysis of basal land plants shows clearly that secondary growth in the Lycophytina and Euphyllophytina represents two independent evolutionary events, even though there are strong similarities in some aspects of secondary tissue production. In the Lycophytina, periclinal and anticlinal cell divisions in the stem produce a pattern of secondary xylem with rays that strongly resembles the wood of other vascular plants (Cichan 1985a; Eggert and Kanemoto 1977; Rothwell and Pryor 1991). Furthermore, peridermal tissue is present, although in much larger quantities than in other vascular plants. Major differences between the wood of the Lycophytina and of seed plants include the absence of secondary phloem in lycopsids (unifacial cambium, Eggert and Kanemoto 1977) and the probable presence of a primary thickening meristem in stigmarian rooting organs (Rothwell and Pryor 1991).

In extant Isoetaceae and many angiosperms, the absence of secondary growth, or presence of only limited secondary growth, clearly represents a loss, but in the Lycopodiaceae the herbaceous condition is plesiomorphic. In many extant pteridophytes (Equisetales, Filicopsida), an assessment of the evolutionary status of herbaceousness remains problematic. In basal Euphyllophytina, assessing the evolutionary acquisition of secondary growth is difficult because of the absence of a detailed cladistic hypothesis of relationship among early members of this clade. The presence of aligned metaxylem in the larger stems of *Psilophyton* (Banks, Leclercq, and Hueber 1975), and possibly of other taxa such as *Protopteridophyton* (Li and Hsü 1987), indicates that this element of secondary growth may be more general in the Euphyllophytina. In *Psilophyton,* it is unclear whether the aligned metaxylem is the product of periclinal divisions of tracheid initials or of cambial activity. Some secondary growth in the sterome of *Psilophyton* probably results from wounding (Banks and Colthart 1993).

Both tracheid alignment and rays are present in seed plants, in progymnosperms, and in many other taxa, including some probably related to the Polypodiopsida (e.g., *Rhacophyton, Zygopteris, Botrychium*), others related to the Equisetopsida *(Arthropitys, Metacladophyton),* and others that are more phylogenetically ambiguous (e.g., *Sphenophyllum*) (Table 7.7). The distribution of secondary growth in basal Euphyllophytina could reflect the homoplastic acquisition of secondary xylem with rays in many early taxa, implying that its absence in most ferns is plesiomorphic. Alternatively, acquisition of secondary xylem may be less homoplastic within the Euphyllophytina, implying a greater number of losses. According to this hypothesis, the absence of wood in many modern euphyllophytes, including the Equisetaceae, Marattiales, and Filicopsida, could represent a return to the herbaceous habit of their Devonian precursors.

STOMATES The cladistic analyses of basal embryophytes presented here are generally consistent with a single evolutionary origin of stomates but also imply many subsequent losses in mosses, hornworts, and vascular plants. The absence of stomates in liverworts can be interpreted as persistence of the plesiomorphic state. Loss of stomates is a sporadic, but widespread, phenomenon in land plants and appears to be related to ecological factors or functional requirements that render stomates nonfunctional or unnecessary. Loss or nonfunctionality of stomates is most common among taxa having a very simple sporophyte morphology, such as mosses and hornworts, and among aquatic, saprophytic, and parasitic vascular plants.

Among basal land plants, stomates have been lost or are vestigial in the hornworts *Notothylas, Dendroceros,* and *Megaceros* (Renzaglia 1978). Stomates are also absent from several plesiomorphic mosses such as *Takakia* (Smith and Davison 1993), *Andreaea,* and *Andreaeobryum* (Murray 1988), as well as from more derived taxa (e.g., many species of *Polytrichum*). In a survey of British mosses and hornworts, Paton and Pearce (1957) noted that stomates are absent from the capsules of

37 species in 21 genera. In the absence of a detailed cladistic analysis of mosses, it is difficult to estimate the number of independent losses or reductions within the group and to critically test adaptive scenarios. Nevertheless, the range of taxa that have lost stomates suggests that this phenomenon has been highly iterative.

Stomates also appear to be nonfunctional in certain mosses: they are present but probably nonfunctional in *Sphagnum* and are poorly developed in the Tetraphidales. Nonfunctionality of stomates in *Sphagnum* sporophytes may be related to the requirements of the special "air-gun" capsule dehiscence mechanism. In *Sphagnum,* spore dispersal occurs through a build up of pressure (up to five atmospheres) caused by the shrinkage of the sporangium as the columella dries and decomposes. At a critical pressure, the operculum and spores are discharged explosively. Ingold (1939) showed that this explosive dehiscence requires a sporangium wall that is impermeable to gases. In *Sphagnum,* the stomatal pore never forms a complete passage, and here the nonfunctionality of stomates is presumably related to the wall impermeability required for explosive dehiscence.

In other mosses, the loss of stomates may be related to the loss of other aspects of morphology. In some species of *Polytrichum,* stomate distribution is restricted to a groove between the apophysis—a well-defined bulge at the base of the sporangium—and the main body of the sporangium (Schofield 1985). The absence of stomates in other *Polytrichum* species may be a developmental consequence of the loss of the apophysis and its associated groove. In other cases, the absence of stomates is more difficult to link to particular functional or structural requirements. The loss of stomates in the Andreaeopsida, *Anomodon, Leucodon,* and *Campylopus* has been attributed to growth in xeric habitats, whereas in *Fontinalis, Cinclidotus,* and *Cyclodictyon,* the loss may be an adaptation to wet habitats (Paton and Pearce 1957). In the hornwort *Notothylas,* the loss of stomates or stomate function may be related to the morphology and the microenvironment of the developing sporophyte, which never fully emerges from the calyptra.

The loss of stomates has been documented in many vascular plants, including some aquatic species of the Isoetaceae (Keeley, Osmond, and Raven 1984), many submerged aquatic angiosperms such as *Ceratophyllum demersum* (Fritsch and Salisbury 1938; Metcalfe and Chalk 1950), and many monocotyledons in the Alismatiflorae (e.g., Zosteraceae, Posidoniaceae, Cymodoceaceae, Zannichelliaceae, Najadaceae, and some Hydatellaceae and Hydrocharitaceae: Dahlgren and Clifford 1982). In many partially submerged aquatic angiosperms, stomates are absent from submerged leaves and are restricted to the upper surface of floating leaves (e.g., *Hydrocharis morsus-ranae, Nymphaea*). In others, stomates are present but presumably nonfunctional (e.g., *Alisma plantago, Myosotis palustris:* Fritsch and Salisbury 1938). Stomates have also been lost (or are sparse) in some saprophytic or parasitic monocotyledons in the Triuridales and Burmanniales (Dahlgren and Clifford 1982) and in dicotyledons such as *Monotropa* and *Sarcodes* (Metcalfe and Chalk 1950). The loss of stomates in terrestrial ferns in the Hymenophyllaceae (Iwatsuki 1990) and Hymenophyllopsidaceae (Kramer and Lellinger 1990) appears to be related to leaf morphology and the ecological requirements of a group of organisms that obtain water principally by absorption through the leaf surface. The cuticle and leaves in this group are exceptionally thin (one or only a few cells thick) and the stomates are presumably functionally redundant.

Inclusion of data from the fossil record supports the earlier conclusion of Mishler and Churchill (1984, 1985b) that stomates evolved before sophisticated internal water-conducting systems (e.g., tracheids with internal thickenings). They based this conclusion on the observation that stomates are a more general feature in land plants. This result contrasts with arguments put forward by Raven (1977, 1984, 1985) favoring the origin of xylem before stomates. Arguing from ecophysiological principles, Raven concluded that low-resistance conducting systems without stomates would be functional, but that systems with stomates and only conducting parenchyma would soon become nonfunctional as plant height increased. In larger plants, water transport via parenchyma has too much inherent resistance and is not sufficient to replenish distal parts. The resulting highly negative water potentials would inhibit photosynthesis. However, both types of system occur in extant groups of low stature. Low-resistance conductance in the absence of stomates occurs in the gametophytes of many mosses and also in some Marchantiopsida (Mishler and Churchill 1984, 1985b), whereas stomates occur in the absence of conducting cells in the sporophytes of Anthocerotopsida and some Bryopsida. Since a diffusion-based system with stomates does appear to be possible in plants that are less than a few tens of centimeters in height (Raven 1984, 1993), ecophysiological data do not conflict with the "stomate first" hypothesis.

LEAVES IN VASCULAR PLANTS Leaf evolution in basal vascular plants has been widely discussed, and two main types of leaf have been recognized. Megaphyllous leaves typically are large and planated, have a complex branched vascular structure, and particularly in early groups, are pinnate. These leaves are the basic type in ferns and seed plants (e.g., Marattiales, Cycadales). Microphyllous leaves are usually small, contain a single vascular strand (however, a unique form with two strands occurs in some species of *Sigillaria:* Bateman, DiMichele, and Willard 1992), and are usually unbranched. Microphyllous leaves *sensu lato* characterize lycopsids and equisetaleans (e.g., Lycopodiaceae, Equisetaceae). Because all extant plants bear leaves or potential leaf homologues, early controversy over leaf homology in vascular plants could not easily be resolved. Bower (1908) argued that megaphyllous types could have evolved by elaboration from microphyllous ancestors resembling lycopsids. He cited two lines of evidence in favor of this hypothesis. First, a sister group relationship between lycopsids and other vascular plants was widely supported in several early phylogenetic schemes. Second, the early ontogeny of microphylls and megaphylls is similar. Since the discovery of leafless early fossils related to lycopsids, ferns, and seed plants, the microphyllous leaves of lycopsids have been interpreted as nonhomologous with megaphylls (Bower 1935; Lignier 1908). In contrast, leaf morphology in early fossil equisetaleans suggests that microphylls of modern *Equisetum* probably evolved by reduction from megaphyll-like precursors (Lignier 1908).

Cladistic analysis of basal land plants is consistent with the independent evolution of leaves, or major aspects of leaflike morphology, in several clades of vascular plants. The inclusion of early leafless fossils in this study supports the independent origin of microphylls in lycopsids. Microphylls are a synapomorphy of the Lycopsida, and putative microphyll precursors occur in many closely related early fossils (e.g., *Asteroxylon,* many Zosterophyllopsida: see below). Megaphylls can be interpreted as comprising a suite of characters or potential independent evolutionary innovations showing varying degrees of homoplasy in the Euphyllophytina (Crane 1990; Doyle and Donoghue 1986, 1992; Stein, Wight, and Beck 1984). Stem group seed plants (e.g., Aneurophytales), putative stem group ferns (e.g., *Pseudosporochnus, Rhacophyton*), and equisetaleans (e.g., *Ibyka, Archaeocalamites*) possess potential megaphyll precursors in the form of small, often nonplanar, and unwebbed lateral branches. These data suggest that megaphylls in the Euphyllophytina may be homologous at the level of lateral branches (megaphyll precursors), but it is likely that other aspects of megaphyll morphology (i.e., planation, webbing, circinate vernation in ferns) evolved independently in these groups. In other words, a lateral branch system of the type found in Euphyllophytina stem group fossils (e.g., *Psilophyton:* Figure 4.13; *Eophyllophyton:* Hao and Beck 1993) forms the basic morphological framework for "leaves" in this group. Current uncertainty regarding the sister group to seed plants (i.e., Archaeopteridales or Aneurophytales) (Crane 1985, 1990; Doyle and Donoghue 1986, 1992; Nixon, Crepet, et al. 1994; Rothwell and Serbet 1994) implies that planation and webbing may also have evolved independently in the lateral branch system of *Archaeopteris.*

ROOTING STRUCTURES The range in morphology of rooting structures in the sporophyte generation of polysporangiophytes is probably as great as the architectural diversity of aerial parts but has received much less attention. Among basal groups, major aspects of "root" morphology are strongly linked to polarity of growth. In unipolar systems (e.g., Lycopodiaceae), growth is basically unidirectional with rooting structures and aerial axes developing from the same apical meristem. In this type of system, "roots" typically develop from prostrate, stemlike axes (rhizomes). In bipolar systems (e.g., seed plants), growth is basically bidirectional, with one apical meristem producing uniquely aerial parts and a second, downward growing meristem producing "roots." Bipolar growth appears to be strongly, but not exclusively (e.g., tree ferns), linked to the tree habit because the specialized root producing meristem also facilitates penetration of the soil, resulting in an effective water-gathering and anchoring system for large, tall plants.

Rooting structures are unknown for many early vascular plants because much fossil material comprises the fragmentary remains of distal parts that have been transported by water currents away from their site of growth. Under exceptional conditions (e.g., Carboniferous coal swamps, volcanic ashfalls, Rhynie Chert), in situ fossilization occurs and rooting structures are preserved attached to aerial parts. Because of the absence of information for many fossil taxa, root characters were not included in the general polysporangiophyte analysis (Chapter 4), but certain features were included in other analyses where more data are available (e.g., Lycopsida: Chapter 6).

Unipolar systems with a rhizome characterize many early fossil members of the major vascular plant clades. Rhizomes have been described in two protracheophytes *(Horneophyton, Aglaophyton)*, the Rhyniopsida *(Rhynia)*, several Zosterophyllopsida (see character 5.21), several early Lycopsida (e.g., *Asteroxylon, Drepanophycus*), and some early members of the Euphyllophytina (e.g., *Psilophyton crenulatum, Protopteridophyton devonicum:* Li and Hsü 1987; *Hyenia:* Schweitzer 1972). Rhizoids, a general feature of extant Embryobiotes (Chapter 3), have been observed in the sporophyte generation in several fossil taxa (e.g., *Horneophyton, Aglaophyton, Rhynia, Serrulacaulis*) but appear to be absent from others (e.g., *Asteroxylon*). Rhizomes are also a feature of modern Lycopodiaceae, Equisetaceae, and many ferns. This character distribution suggests that the sporophytic rhizome may be a synapomorphy of polysporangiophytes because it is a more general feature than tracheids. Early sporophytic rooting structures in polysporangiophytes evolved as modified prostrate stems in plants with a basically unipolar growth form.

Bipolar growth forms having indeterminate meristems producing negatively geotropic stems and well-developed positively geotropic roots appear independently in several clades. One well-documented example is the pseudobipolar growth of rhizomorphic lycopsids, a clear synapomorphy of this clade (Bateman 1992, 1994; Bateman, DiMichele, and Willard 1992; Crane 1990). In rhizomorphic lycopsids the first dichotomy of the embryonic *shoot* produces an aerial stem and downward growing rooting organ—the *rhizomorph* (Phillips and DiMichele 1992; Rothwell 1984; Rothwell and Erwin 1985; Rothwell and Pryor 1991; Stubblefield and Rothwell 1981). In this group, ontogenetic studies have confirmed earlier interpretations that the "roots" of rhizomorphic lycopsids are modified stems and that the "rootlets" are modified leaves.

Among the Euphyllophytina, bipolar growth appears to have evolved independently in the Equisetopsida and in seed plants. The branched roots of arborescent Equisetopsida (e.g., Calamitales) develop endogenously as adventitious lateral branches arising in whorls from a basal rhizome (Eggert 1962). Similar growth forms have been observed in the more plesiomorphic Archaeocalamitaceae (Bateman 1991; Smoot, Taylor, and Serlin 1982) and in extant *Equisetum*. In contrast, bipolar growth in seed plants begins with the clear differentiation of root and shoot in the early embryo.

Root mantles, resembling those of basal ferns such as *Psaronius* (Marattiales), have been documented in several early fernlike plants (e.g., *Rhacophyton:* Cornet, Phillips, and Andrews 1976; *Lorophyton:* Fairon-Demaret and Li 1993; *Pseudosporochnus:* Leclercq and Banks 1962). In *Psaronius,* as in other extant Marattiales, adventitious roots arise endogenously from the vascular segments of the stem and grow basipetally. In the basal parts of the stem of *Psaronius* a substantial root mantle develops with a compact inner region and an outer zone of free roots (Ehret and Phillips 1977; Morgan 1959). Unfortunately, anatomical information on roots and their initiation within the stem is unavailable for *Rhacophyton, Lorophyton,* and *Pseudosporochnus.*

The loss of roots is a sporadic but widespread phenomenon in vascular plants and is usually associated with aquatics, epiphytes, and rupestrals. In these plants, the absorptive and anchoring function of roots has been taken over by another organ system (e.g., leaves, mycorrhizal fungi). Roots of aquatic angiosperms generally are adventitious, slender, and rarely branched and function mainly as organs of attachment. Some aquatic angiosperms are rootless (e.g., *Utricularia:* Fritsch and Salisbury 1938; *Ceratophyllum,* Les 1993), and in others, such as the free-floating *Lemna* and *Hydrocharis,* reduced root systems may be important in orienting the plant correctly (Fritsch and Salisbury 1938).

Among ferns, roots have been lost in the free-floating aquatic family Salviniaceae, where highly modified leaves may be the functional equivalent of rooting organs (Bierhorst 1971; Schneller 1990b); however, very simple roots are present in the sister group, Azollaceae (Schneller 1990a). Roots are absent from the rhizome of the mainly epiphytic or rupestral Psilotaceae (Bierhorst 1971; Kramer 1990), in which rhizomes bear rhizoids and contain a well-developed mycorrhizal fungus (Bierhorst 1971). The absence of roots in the Psilotaceae has been interpreted as a retained primitive feature (e.g., Bremer 1985), but it is more likely to represent another loss (see above). Roots are also absent from the embryonic stages of putatively related taxa such as *Stromatopteris* and *Actinostachys.* Roots may also have been lost entirely from some epiphytic Hymenophyllaceae (Bierhorst 1971), in which their nutritional function is performed by the delicate leaves.

STROBILI The acquisition of strobili (aggregations of micro- and megasporangiophores, typically around an unbranched determinate axis) is a widespread and homoplastic feature in polysporangiophytes (Bateman 1992; Bateman, DiMichele, and Willard 1992; see also Chapters 5 and 6). Strobili occur in many extant seed plants, in lycopsids, and in *Equisetum* but are less common in ferns. In most extant taxa, strobili usually comprise sporangia and the associated sporangiophores, which are clearly modified microphylls in lycopsids, are probably highly modified megaphylls in seed plants, and are reduced branches in the Equisetopsida. In the Zosterophyllopsida, sporangia are aggregated into strobili in which microphylls and sporophylls are generally absent ("naked strobili"—which is a plesiomorphic condition: characters 5.12 and 6.22, Figures 5.1 and 5.4). Several Filicopsida also possess naked strobili, a character that has apparently arisen by reduction (e.g., some Ophioglossales and Osmundales). Early fossil members of the Euphyllophytina, including taxa in the Spermatophytata, Equisetopsida, and Filicopsida stem groups, seldom possess strobili although sporangia are typically aggregated into terminal clusters (e.g., *Psilophyton:* Figure 4.13). The iterative origin, and possibly also loss, of strobili in polysporangiophytes is a phenomenon that is poorly understood but may be related to functional factors including the nutritional requirements of developing sporangia, the temporal control of sporogenesis, the control of micro- and megaspore release, the protection of spores or developing embryos, and possibly structural constraints, including a determinate apical meristem.

BRANCHING Sporophyte branching is a synapomorphy of polysporangiophytes and is one of the most important and basic components of vascular plant architecture. In the simplest polysporangiophytes, isotomous dichotomy predominates, but in many early taxa, branching is more complex and involves the activities of different types of meristem that are characterized by their position and the morphology of the shoot systems that they produce. Meristem dormancy and death also play an important role in determining plant architecture, although these phenomena are only well documented in seed plants (e.g., Bell 1994; Corner 1951). Shoot meristem abortion has been documented in the distal branching systems of *Psilophyton* (Banks, Leclercq, and Hueber 1975) and may also be responsible for the short lateral branches present in some stem group Lycophytina (e.g., *Renalia:* Gensel 1976) and in the rhizome systems of such taxa as *Zosterophyllum* (e.g., Gensel 1982a; Gerrienne 1988; Lang 1927; Lele and Walton 1961). In many Sawdoniales (Zosterophyllopsida), dormant circinate shoots are common in the axils of the main branching system but may be more widely distributed throughout the plant (e.g., *Tarella:* Edwards and Kenrick 1986), The axillary shoots may develop fully into lateral branches, which resemble the main pseudomonopodial system (e.g., *Anisophyton:* Remy, Hass, and Schultka 1986; Remy, Schultka, and Hass 1986). Dormant meristematic regions have also been documented in *Rhynia* (D. S. Edwards 1980). Adventitious branching has been described in early Lycopsida (Kidston and Lang 1920b, 1921; Li and Edwards 1995), some of which (e.g., *Drepanophycus:* Li and Edwards 1995) produce small lateral branches that resemble the bulbils of extant *Huperzia* (Øllgaard 1990).

Cladistic analysis shows that pseudomonopodial systems evolved iteratively in several groups, such as the Lycopsida, Zosterophyllopsida, and Euphyllophytina. In many Zosterophyllopsida, well-developed pseudomonopodial systems characteristically branch in one plane in a basically alternate pattern (Figures 4.22, 5.2, 5.11). Pseudomonopodial systems also occur in the Lycophytina, and in extant Selaginellaceae, these may also be planar. Cladistic analysis supports an independent origin of planar pseudomonopodial branching and rhizophore-like axillary shoots in the Selaginellaceae and Sawdoniales (Chapter 6). Plesiomorphic pseudomonopodial systems are basically helical in early Euphyllophytina, such as *Psilophyton* (Figures 4.13 and 4.22), but other patterns occur in putative stem group Spermatophytata, Filicopsida, and Equisetopsida. In the Euphyllophytina the modification of lateral branches into a range of complex structures is responsible for much of the architectural diversity in the group.

TRACHEID MORPHOLOGY Cladistic analysis of basal land plants supports a single origin of tracheids in vascular plants but at least two independent origins of tracheid pitting. The presence of water-conducting cells with conspicuous internal helical-annular thickenings and a decay-resistant (presumably lignified) cell wall is the plesiomorphic condition for vascular plants (Kenrick and Crane 1991). This cell type is characteristic of the protoxylem (first formed elements) of extant vascular plants (Bierhorst 1960) and is also the earliest well-documented tracheid type in the fossil record. In many early fossil groups (e.g., Rhyniopsida, Zosterophyllopsida, some early Lycopsida) the xylem comprises exclusively helical-annular thickened tracheids, whereas in most extant pteridophytes the bulk of the xylem comprises metaxylem tracheids with conspicuous pits (Bierhorst 1960). Character distributions in basal land plants strongly suggest that metaxylem pitting evolved independently at least twice—once in the Lycopsida and once in the Euphyllophytina. In other words, although helical-annular thickenings in protoxylem tracheids can be regarded as homologous in vascular plants, metaxylem pitting is convergent in lycopsids and the Euphyllophytina.

Loss of tracheids is a rare phenomenon and is usually associated with a highly specialized submerged aquatic habit. Xylem is absent from the aquatic angiosperms *Ceratophyllum demersum* and *Potamogeton pectinatus,* but phloem is well developed. Xylem is presumably redundant in these submerged aquatics because absorption of water and other necessary chemicals occurs directly through the plant surface. Phloem, however, is necessary for transporting photosynthetic product (Fritsch and Salisbury 1938). In some species of *Potamogeton* (angiosperm) and in

Phylloglossum drummondii (Lycopodiaceae), only helical-annular protoxylem tracheids occur. This phenomenon appears to be related to sporophyte reduction, probably through progenesis in *Phylloglossum* and redundancy associated with the aquatic habit of *Potamogeton.*

Cladistic analysis indicates that absence of tracheid thickenings in the early lycopodiopsid fossil *Nothia* is most parsimoniously interpreted as a loss, and the water-conducting cells otherwise resemble the decay-resistant cells in such early lycopsids as *Asteroxylon.* The description of *Nothia* is based on in situ fossils from the lower Devonian Rhynie Chert peat deposit, one of the earliest examples of an in situ wetland community in the fossil record. Based on the absence of tracheid thickenings and environmental interpretations of the Rhynie Chert, we suggest that *Nothia* may have been a semiaquatic plant. Reversion to a semiaquatic habitat for *Nothia* is also consistent with two important aspects of the epidermis. The cuticle is reportedly thinner than that in any other Rhynie Chert plant (<2.5 μm compared with <5.0 μm for *Aglaophyton*), and the epidermal cells lack the thick cuticular flanges typical of *Rhynia, Aglaophyton,* and *Asteroxylon* (Edwards 1993; Edwards, Edwards, and Rayner 1982). The stomates of *Nothia* are unique because each is situated at the apex of a conspicuous mound of epidermal tissue. Guard cells in *Nothia* are large, but the distinctive stomatal pore is very small (Edwards, Edwards, and Rayner 1982; El-Saadawy and Lacey 1979a). These features suggest that stomates in *Nothia* may have been nonfunctional.

SPERMATOZOID MORPHOLOGY Sperm motility is characteristic of land plant reproduction in which fertilization requires movement of the male gamete from the gametophyte to the egg cell in the archegonium. In many groups, this part of the life cycle requires the presence of free water so that the male gamete can be propelled to the archegonium by a pair of flagella. This flagellate stage provides one of the clearest indications of an aquatic ancestry for land plants. The gametes of most green plants are biflagellate, but more flagella occur in several groups (e.g., Isoetaceae, Equisetales, Filicopsida, Cycadales, Ginkgoales). Flagella are also clearly lost in several groups of green plants, including the Zygnematales (conjugating Charophyceae) and some major groups of seed plants (e.g., Coniferales, Gnetales, Angiospermales) in association with the development of siphonogamy. This loss occurs under conditions in which free movement of the male gamete is no longer necessary to ensure fertilization. For example, in the Zygnematales, sexual reproduction occurs through conjugation involving the aggregation of filaments or unicells within a gelatinous matrix and fusion of nonflagellate gametes (Graham 1993). In filamentous taxa such as *Spirogyra,* syngamy takes place via a conjugation tube produced as an outgrowth from the cell wall.

Ultrastructural studies with TEM are beginning to provide a detailed picture of spermatozoid ultrastructure in green plants, including the particulars of motile apparatus as well as of plastid structure (Carothers, Brown, and Duckett 1983; Carothers and Duckett 1980; Duckett and Renzaglia 1988a,b; Garbary, Renzaglia, and Duckett 1993; Mattox and Stewart 1984; Pickett-Heaps 1975; Pickett-Heaps and Marchant 1972; Renzaglia and Duckett 1988, 1989, 1991; Sluiman 1983, 1985; Stewart and Mattox 1975, 1978, 1980). Ultrastructural data on the male gamete has had the greatest impact on understanding relationships among plants at the green algal level of organization because many of the major groups diverged from morphologically simple unicellular ancestors. Homologies at the ultrastructural level have been assessed in several recent cladistic studies (Bremer 1985; Bremer, Humphries, et al. 1987; Garbary, Renzaglia, and Duckett 1993; Graham, Delwiche, and Mishler 1991; Mishler and Churchill 1984, 1985b; Mishler, Lewis, et al. 1994) and reviews (Graham 1993; Renzaglia and Duckett 1991).

The male gametes of embryophytes, as well as those of groups within the Streptobionta, are autapomorphic in several features (Bremer 1985; Garbary, Renzaglia, and Duckett 1993; Graham 1993; Graham, Delwiche, and Mishler 1991; Mishler and Churchill 1984, 1985b). Cladistic studies show that the highly distinctive ultrastructural features of the male gamete of the Anthocerotopsida are autapomorphic and that spermatozoid morphology does not support an independent origin of this group from other land plants, as has been suggested previously.

Biflagellate motile cells are clearly plesiomorphic for green plants (Bremer 1985; Mishler and Churchill 1984, 1985b) and are present in the male gametes of several basal land plant groups (Marchantiopsida, Anthocerotopsida, Bryopsida, Lycopodiaceae, Selaginellaceae). Cladistic analysis supports the independent origin of the multiflagellate state in male gametes within the Lycophytina and Euphyllophytina. In extant Lycophytina the multiflagellate state is clearly autapomorphic in the Isoetaceae. In the Euphyllophytina it is uncertain whether the multiflagellate condition is homologous in the Polypodiopsida (including Psilotaceae), Equisetopsida, and Spermatopsida (Chapter 4; Crane 1990). Enclosure of the female gametophyte within sporophytic tissues in seed plants overcomes the requirement for free water in fertilization, and delivery of the pollen directly to the archegonial region via the micropyle reduces the distance the flagellate gamete must travel. In angiosperms the female gametophyte is completely enclosed by sporophytic tissues, creating other barriers to fertilization. In all extant seed plants, and in the early fossil *Callistophyton* (Cordaitales, Rothwell 1972), pollen germinates to form a branched pollen tube. In cycads and *Ginkgo,* the pollen tube probably serves a nutritive function, and motile gametes bearing flagella in a distinct "ciliate band" are released directly into the pollen chamber when the basal portion of the pollen tube ruptures. In contrast, the male gametophytes of conifers, Gnetales, and angiosperms are siphonogamous. The pollen tube grows toward the egg cells, and nonflagellate sperm are transported to the archegonium through the living cytoplasm of the tube (Friedman 1993). Flagella are presumably redundant in siphonogamous taxa.

SPORE MORPHOLOGY Spore morphology in embryophytes provides a suite of useful taxonomic characters that allow the identification of many major groups on the basis of exine structure. Because spores are produced in great numbers and may be distributed over wider areas than macrofossils, the palynological record provides important information on the time of origin and patterns of diversification of the major land plant clades. The excellent fossil record of embryophyte spores is partly attributable to important innovations in the chemical characteristics of the cell wall. Embryophyte spore walls contain *sporopollenin*—a complex, probably aromatic, polymer that provides a degree of desiccation resistance, robustness, and protection from ultraviolet radiation. The basic molecular structure of sporopollenin is currently poorly understood, but its presence in extant groups can be detected using a variety of analytical techniques (Graham 1993, 1992; Hemsley, Collinson, and Brain 1992). Although sporopollenin occurs widely in green plants (e.g., dinoflagellate cysts, zygotes of charophycean algae), its presence in meiospore walls is unique to embryophytes (Graham 1990). Few attempts have been made to identify sporopollenin chemically in fossil material, and the results have been equivocal (Hemsley, Chaloner, et al. 1992; Hemsley, Collinson, and Brain 1992). Analysis for sporopollenin residues has not been attempted for the earliest spores in the fossil record, but the decay resistance of the cell wall is generally attributed to the presence of sporopollenin, and so an embryophyte relationship for these fossils has been inferred (Graham 1993; Gray and Shear 1992).

Spore exine sculpture shows remarkable morphological diversity in land plants and is widely used in systematics and biostratigraphy (e.g., Allen 1980; Boros and Járai-Komlódi 1975; Burgess 1991; Burgess and Richardson 1991; Chaloner 1967b; Chaloner and Sheerin 1979; Fanning, Richardson, and Edwards 1991; Gensel 1980; Gensel, Johnson, and Strother 1990; Hemsley 1989, 1990, 1993; Hemsley, Clayton, and Galtier 1994; Hemsley and Scott 1991; T. N. Taylor 1990; W. A. Taylor 1990; W. A. Taylor and T. N. Taylor 1990; Tryon and Lugardon 1991). Despite this diversity, there appear to be few features in spore wall morphology that unequivocally identify major basal clades of land plants. A trilete mark is one of the most distinctive features of the spore wall of basal embryophytes and forms during the tetrad stage at the contact faces of the spores. In most taxa, tetrads ultimately disassociate and the spores are dispersed singly, carrying the trilete mark as a signature of meiotic cell division. Well-defined trilete marks are characteristic of some embryophyte groups (e.g., Anthocerotopsida, polysporangiophytes, Tracheophyta) but are rarer in the Marchantiopsida and Bryopsida (Boros and Járai-Komlódi 1975). The distribution of this feature has not been assessed in a cladistic context in the Bryopsida and Marchantiopsida, and the homology among basal taxa is currently difficult to assess. It is probable that the absence of a distinct trilete mark in many basal Bryopsida is the result of phylogenetic loss.

A perine—or extra-exinous wall layer—was identified as a moss-tracheophyte synapomorphy by Mishler and Churchill (1984, 1985b). It is well established from developmental studies of extant taxa that the perine of pteridophytes is derived from tapetal material. Little is known about the origin of the perine in mosses, although it is probably also of extrasporal origin (Brown and Lemmon 1990b; Lugardon 1990). A perine has been identified in putative basal mosses such as *Sphagnum, Andreaeobryum,* and *Andreaea* (Brown and Lemmon 1988), as well as more derived taxa such as *Polytrichum* (Olesen and Mogensen 1978). Character distributions among basal embryophytes are consistent with a hypothesis of homology between the perine of mosses and tracheophytes. This feature is difficult to assess in fossil taxa and is unknown for many basal extinct tracheophytes.

Characters associated with exine micro-ornament in general appear to be subject to considerable homoplasy in land plants. In a study of relationships in the Lycopsida, Bateman (1992) and Bateman and DiMichele (1994a) showed that characters of spore morphology are relatively homoplastic in comparison to many other medium-scale morphological features. This phenomenon was attributed to the low "burden" of exine ornament characters (i.e., spore morphology appears at the end of an ontogenetic cascade and is not an integral part of a complex tissue system) and probable structural or ontogenetic constraints that limit change to a relatively small number of possibilities.

STELAR MORPHOLOGY In contrast to fossils of most animal groups, preservation of tissues at the cellular level is common in plants because of the robust and relatively decay-resistant cellulose cell walls. Further resistance to decay also occurs in some tissue types through the deposition of additional complex biopolymers such as lignin and sporopollenin. As a consequence, stelar morphology is one of several aspects of vascular plant anatomy that has an excellent fossil record extending back to the late Silurian (Duckett 1986; Edwards 1993).

As evidenced in the fossil record, the simplest vascular plant steles comprise a solid central strand of xylem. Phloem is rarely preserved, but when it is present, it surrounds the xylem (E. L. Taylor 1990). In several extinct polysporangiophytes (e.g., *Aglaophyton*), tracheid thickenings are absent and the stele bears a striking resemblance to the steles of some of the larger mosses in extant Polytrichales (D. S. Edwards 1986; Hébant 1977; Kenrick and Crane 1991). Simple steles comprising water-conducting cells (e.g., hydroids) surrounded by phloemlike tissues (leptoids) are present in many moss gametophytes and sporophytes, and hydroids have been documented in the gametophytes of a few liverworts (Hébant 1977, 1979). Leptoids in mosses are characterized by sieve element–like features, including cell elongation, oblique end walls, many enlarged plasmodesmata in the end walls, various degrees of later wall thickening, controlled autolysis of protoplast giving a "clear" appearance on electron microscopy, and the presence of refractive spherules (Scheirer 1980). Hydroids resemble tracheids in several respects: elongate shape, oblique end walls, loss of protoplast at maturity, strong phosphomonoesterase activity during differentiation, and highly permeable hydrolyzed end walls (Scheirer 1980). Most liverwort gametophytes (and all sporophytes) lack a true water-conducting strand. Hydroids have been documented in a few leafy Calobryales and thalloid Metzgeriales, but unequivocal leptoids are unknown (Hébant 1977). In mosses, hydroids are more common than leptoids and occur in the gametophytes and sporophytes of a wide range of taxa (Hébant 1977). Hydroids and leptoids are absent from the Anthocerotopsida. In many bryophytes without hydroids or leptoids, a central strand of conducting parenchyma is present. Cells of the conducting parenchyma possess some, but not all, of the characteristics of leptoids (e.g., position in the axis, cell elongation, sometimes plasmodesmatal morphology: Hébant 1977).

A hypothesis of homology between the moss "stele" and the stele of polysporangiophytes is consistent with recent cladistic studies (Chapters 3 and 4; Mishler and Churchill 1984, 1985b; Mishler, Lewis, et al. 1994) and is supported by the strong ontogenetic and morphological similarities among water-conducting cells and phloemlike cells in mosses and tracheophytes (Hébant 1977, 1979; Kenrick and Crane 1991; Scheirer 1980, 1990). Perhaps significant also is the similarity in the pattern of differentiation of stelar tissues and the regulation of differentiation by auxin. The xylem or water-conducting tissues of mosses and vascular plants are nearly always central in the stem (except *Sphagnum*) and are always central with respect to phloem or leptoids (Hébant 1977; Mishler and Churchill 1984, 1985b; Stein 1993). Furthermore, the differentiation of water-conducting tissues in mosses and vascular plants is regulated by auxin, whereas auxin is absent from liverworts with conducting tissues. Recent studies of auxin metabolism in land plants suggest that the ability to conjugate IAA with alcohols, sugars, and amino acids is important in the regulation of auxin-mediated developmental processes such as vascular tissue differentiation (Sztein, Cohen, et al. 1995). Several similar conjugation products have been observed in mosses, hornworts, and vascular plants, but these are absent from liverworts and the charophycean alga *Nitella*. Assessing the homologies of the moss stele is complicated by the absence of hydroids in basal mosses such as *Sphagnum* and the Andreaeopsida, as well as by the apparent restriction of well-developed leptoids to the Polytrichales. Further resolution of these problematic homologies would benefit from a stronger hypothesis of relationship among bryophytes and polysporangiophytes, as well as from additional comparative studies that include information from the fossil record on stem group mosses.

Among basal extant vascular plants, the position of the protoxylem with respect to the metaxylem is a relatively conservative character. Protoxylem comprises the earliest elements to differentiate within the stem, and its differentiation occurs close to the apex before cell elongation is complete in the surrounding tissues. Protoxylem cells are usually narrow, often have simple helical-annular thickenings, and are often distorted by cell elongation in the surrounding tissues (Bierhorst 1960). Metaxylem differentiation occurs later, after cell elongation is complete. Metaxylem cells are large, have thicker cell walls with more complex pitting, and are generally not distorted by cell elongation (Bierhorst 1960).

Two clearly distinct patterns in protoxylem position occur in early vascular plants. Centrarch maturation (protoxylem central and surrounded by metaxylem) appears to be the plesiomorphic state in polysporangiophytes, whereas exarch maturation (protoxylem peripheral to metaxylem) is characteristic of the Lycophytina. More detailed interpretation is complicated because the position of the protoxylem is unknown for many basal fossil taxa. Furthermore, in some early protracheophytes (e.g., *Aglaophyton, Horneophyton*) and in the Rhyniopsida (e.g., *Rhynia, Sennicaulis*), the distinction between protoxylem and metaxylem on the basis of cell size is problematic. The entire delicate vascular strand in *Rhynia* consists of a few tracheids with conspicuous helical wall thickenings that may be developmentally equivalent to protoxylem in more derived groups such as *Psilophyton*. According to this interpretation, protoxylem is more general than metaxylem in vascular plants, a hypothesis consistent with the function and ontogeny of protoxylem in extant groups. The evolution of metaxylem (i.e., a developmental "extension" of xylem differentiation beyond the cell elongation zone near the axis apex) may be an innovation of eutracheophytes.

Variations in metaxylem shape and protoxylem position are among the most conspicuous anatomical changes in the ontogeny and evolution of vascular plants (Bower 1930; Niklas 1984; Raven 1994; Stein 1993; Wight 1987). In the Euphyllophytina the plesiomorphic state for polysporangiophytes (terete, centrarch xylem) is retained in *Psilophyton,* but a diverse range of lobed, multistranded, or medullated forms appear in the equisetalean, polypodialean, and spermatophytalean stem groups. In contrast, stelar morphology and protoxylem position in the Lycophytina are relatively conservative. A polystelic condition occurs in some Selaginellaceae, and lobed or dissected steles are common in the Lycopodiaceae. Medullation also occurs in the steles of the larger extinct rhizomorphic lycopsids.

Based on detailed measurements of xylem morphology and axis size, Bower (1923b, 1930) and Wardlaw (1924, 1925, 1928, 1952) documented an increase in xylem complexity in vascular plants that was correlated with an increase in axis diameter. In these studies, much of the increase in lobing, medullation, and dissection of the stele was attributed to physiological needs that required the maintenance of an adequate ratio of surface area to volume between the stele and the other tissues of the stem. According to this general hypothesis, the evolution of more complex stelar types is favored in larger plants. This straightforward interpretation of a volumetric relationship between stelar shape and axis size is problematic for several reasons, some of which were recognized by Bower and Wardlaw. First, differences in stelar morphology among distantly related groups of similar stem size indicate that there is also a "historical" component to xylem complexity. Second, other factors that may influence stelar morphology, such as number and position of lateral appendages, also change with increasing axis size. Third, it is not clear what the most appropriate volumetric measurements are nor what kind of morphometric relationship to expect (Niklas 1984; Stein 1993). Fourth, the nature of the putative physiological restrictions imposed by stele shape is unclear (Raven 1994).

More recently, Wight (1987) and Stein (1993) proposed an alternative hypothesis that relates stelar morphology to lateral branch development and hormonal signals. In plants with multiple, closely spaced, lateral appendages in longitudinal orthostiches, hormonal signals reinforce the differentiation of vascular tissue in the direction of the lateral appendages. In other words, the Wight-Stein model suggests that xylem shape and protoxylem position co-vary with branch position. This idea is consistent with well-founded hypotheses of hormone-mediated vascular tissue differentiation based on studies of extant plants and has been used by Wight (1987) and Stein (1993) to explain the evolution of stelar morphology in basal Euphyllophytina. The explanatory power of this hypothesis was tested by Stein (1993) using a computer model of the stem apex. By changing several variables relating to branching pattern, hormonal gradients, and tissue susceptibility, Stein was able to reproduce patterns of xylem differentiation observable in several early fossil Euphyllophytina with remarkable precision. These data strongly suggest that lateral appendage type and position have a greater influence on stelar morphology than does absolute stem size alone.

We suggest that this approach to understanding stelar morphology and ontogeny in the Euphyllophytina could be extended to the Lycophytina by substituting the effect of a few, large, lateral branch appendages for the more subtle combined effects of the numerous, small, lateral microphylls and sporangia that are unique to this clade. In the Lycophytina, branching is rarely closely spaced or in well-defined orthostiches and has much less influence on stelar shape than in the Euphyllophytina. With the exception of the lobed xylem strands of the Lycopodiaceae and the highly reduced vascular tissue of *Phylloglossum* and *Isoetes,* the steles of lycopsid stems are more or less circular or elliptical in outline and all are exarch. The most conspicuous lateral appendages are sporangia and microphylls, both of which are very closely spaced in orthostiches or helical patterns or both. Both types of lateral appendage influence the differentiation of vascular tissues and are known to have their own vascular supply in basal clades.

We interpret the peripheral protoxylem patterns that characterize the stems of Lycophytina as a developmental consequence of the presence of many small lateral meristems (i.e., sporangia) acting as auxin sources (e.g., *Zosterophyllum*). The evolution of microphylls in the Lycopsida provides additional lateral appendages of comparable size and position to sporangia and would be expected to reinforce the exarch protoxylem position without a general alteration of xylem shape. The small external ribs of the xylem strand in the Protolepidodendrales and rhizomorphic lycopsids may thus be directly relatable to microphyll ontogeny through leaf trace departure. This hypothesis of the origin of exarchy in the Lycophytina is consistent with the Wight-Stein model (Stein 1993) for the Euphyllophytina.

In the Lycophytina, there are some interesting departures from the basically terete, protostelic form of xylem. In many Selaginellaceae and Zosterophyllopsida the xylem is an elliptical to strap-shaped protostele. In these groups, the flattened xylem may be related to the strongly planar, opposite, and alternate branching. By analogy with the Wight-Stein model for the Euphyllophytina, lateral branching in these groups occurs in two well-defined orthostiches and may influence xylem shape.

Medullation of the xylem strand is an iterative phenomenon in basal polysporangiophytes and clearly an independent acquisition in arboreous lycopsids. In the latter, medullation occurs in between the protostelic sporeling and solenostelic distal branch stages and appears to be related to the presence of a massive primary plant body and presumably a large apical meristem (Eggert 1961). In the Wight-Stein model for the Euphyllophytina (Stein 1993), the presence of a pith is favored when a large apical dome is surrounded by a compact phyllotaxis of hormone sources. The greater the number of influential lateral appendages, the larger the pith. If large microphylls are substituted for lateral branches as hormone sources, then the same model may also be applicable to medullation in arboreous lycopsids.

The most problematic stelar morphologies in the Lycophytina are the lobed or dissected steles of the Lycopodiaceae (and of some closely related early fossils such as *Asteroxylon, Baragwanathia,* and *Drepanophycus*) and root stele anatomy in the Lycopodiaceae and Selaginellaceae. Sporelings in all Lycopodiaceae begin their ontogeny with a small, lobed protostele that develops into a large, lobed stele or a dissected stele (Wardlaw 1924). As in all lycopsids, the xylem is basically exarch. There is no one-to-one relationship between the leaf traces and the prominent xylem lobes (Stevenson 1976), and this situation creates problems with interpreting a causal relationship between microphylls and this aspect of xylem shape. The regular formation of small, closely spaced adventitious branches (bulbils) near the stem apex in the *Huperzia selago* group (Stevenson 1976) may contribute to reinforcing xylem lobing. Very similar structures are also present in closely related early fossils with similar lobed steles, such as *Drepanophycus* (Li and Edwards 1995). However, bulbils do not occur in other Lycopodiaceae with lobed and dissected steles. This aspect of stele morphology, as well as the crescent-shaped xylem strand with phloem located on the embayed side in the root and rhizophore steles of many Lycopodiaceae and Selaginellaceae, is difficult to explain in terms of appendage type or orientation. It is these "residual" features that perhaps offer the best possibilities for limited vindication of the Bower-Wardlaw hypothesis.

GAMETANGIA Despite some structural and ontogenetic variation among groups, the unique morphologies of the archegonia and antheridia in land plants have long been regarded as strong indicators of embryophyte monophyly. In basal land plants the archegonium comprises an egg cell located in the base of a typically flask-shaped receptacle (Figure 3.33). Spermatozoids enter through the open neck of the flask and swim down the neck canal to fertilize the egg. In the Bryopsida and Marchantiopsida the flask is situated on a short stalk on the surface of the gametophyte, whereas in the Anthocerotopsida and Tracheophyta the archegonium is embedded in the surface of gametophyte with the neck protruding. The antheridia of basal land plants are more or less spherical to obovate, comprising a central mass of spermatozoids encapsulated in a "jacket" one cell thick. In the Bryopsida, Marchantiopsida, and Anthocerotopsida the antheridium is situated on a well-defined multicellular stalk. In the Bryopsida and Marchantiopsida the antheridium is superficial, whereas in the Anthocerotopsida the whole structure is located within a conspicuous chamber in the thallus. In the Tracheophyta the antheridium stalk is either absent or very short. In taxa with very short stalks, such as the leptosporangiate ferns and possibly some early stem group polysporangiophytes, the antheridium is superficial, whereas in groups lacking a stalk (e.g., Lycopodiaceae) the antheridium is embedded in the surface of the gametophyte.

With the exception of *Chara* and *Coleochaete,* structures comparable to antheridia and archegonia do not exist in green algae. In multicellular groups with oogamous life cycles, spermatozoids and oogonia differentiate directly from superficial cells of the gametophyte. *Chara* has the most complex gametangia of any green algae, but distinctive differences in morphology and ontogeny make homology with gametangia of embryophytes unlikely (see below; Fritsch 1948; Pickett-Heaps 1975). In *Chara* the archegonium is obovate and comprises a massive egg cell *(oogonium)* contained in a flask formed through the growth of five or more basal cells in a helix around the oogonium (Figure 3.4). The antheridia are spherical and comprise numerous filaments of spermatogenous cells attached to small sterile filaments *(capitula)* on a stalk *(manubrium)*. The spermatogenous cells and their stalks are encapsulated in a simple jacket ("shield").

Cladistic analyses support the monophyly of land plants, and the similarities among the gametangia of different major groups are consistent with a hypothesis of homology. This interpretation contrasts with the alternative hypothesis that resemblances among the gametangia of land plants are only superficial and may have resulted from convergent evolution early in the history of the group (e.g., Crandall-Stotler 1980, 1984, 1986; Duckett and Renzaglia 1988a; Longton 1990; Schuster 1977, 1979, 1981, 1984c). Arguments for nonhomology are based on observed differences in morphology and ontogeny among groups. From a cladistic perspective, these differences are best interpreted as autapomorphic changes that occurred after the cladogenic events that gave rise to the basic gametangium form in major land plant groups. From this viewpoint, differences among groups (e.g., see Schuster 1984a) represent interesting modifications to a basic gametangium type in embryophytes rather than morphological convergence resulting from similar strategies for sexual reproduction.

Ontogenetic studies of gametangia in embryophytes provide much detailed information on the patterns of cell division and differentiation. We have summarized some of this information in diagrammatic form to highlight the similarities and differences of the developmental pathways in the antheridia of some major groups (Figures 7.16 and 7.17). In all groups, the development of antheridia follows a similar pattern in which stalk, sperm, and jacket tissues are determined early and in a similar sequence. Antheridium ontogeny begins with a single superficial cell, which divides periclinally to produce an inner and outer cell. In the Marchantiaceae (Figure 7.16a), the inner cell produced by the division of the superficial cell (antheridium initial 1) becomes the stalk initial and initiates a cycle of divisions that ultimately give rise to the stalk (stalk pathway). The outer cell (antheridium initial 2) initiates a series of divisions that block out the main body and tissue systems of the antheridium (jacket and spermatogenous tissues). Several horizontal divisions of antheridium initial 2 produce a single-celled row in which each cell divides vertically twice at right angles to produce a filament of four cell rows. Each of these cells then divides periclinally once, initiating a further change in developmental trajectory. The inner cells begin a series of divisions that produce spermatozoids; the outer cells initiate a series of anticlinal divisions producing the jacket. The mature antheridium of the

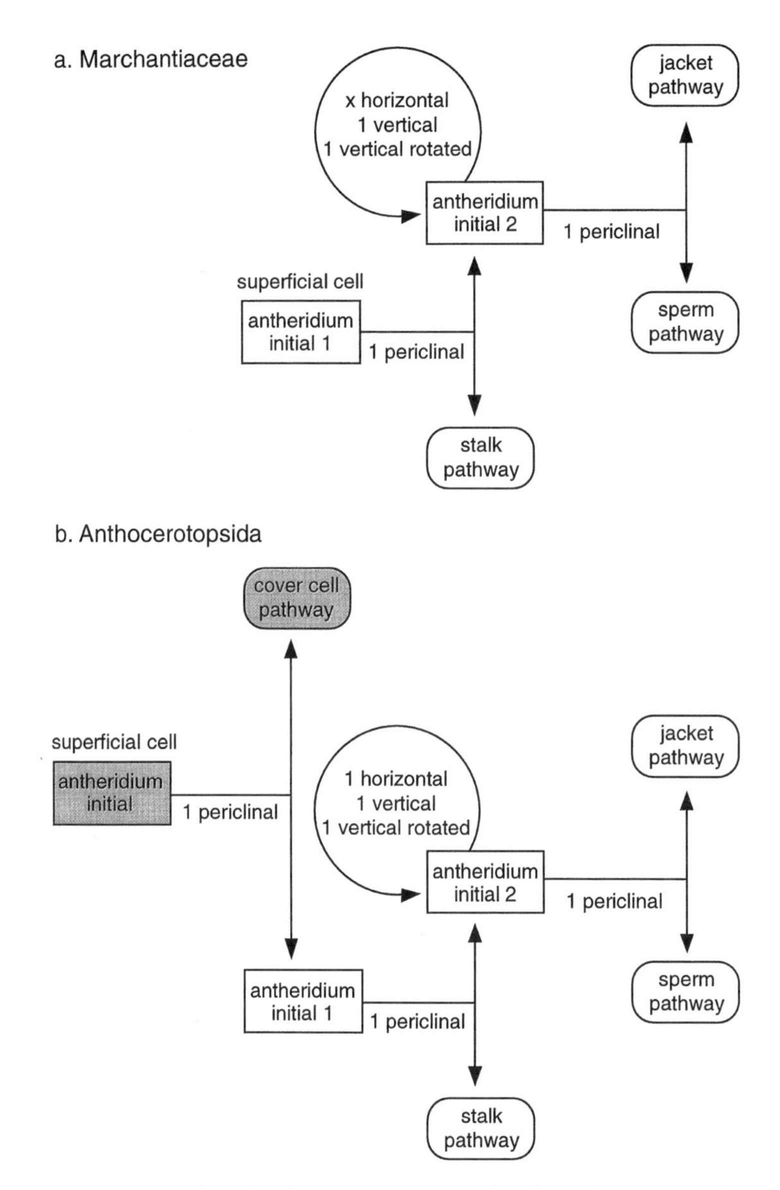

FIGURE 7.16. Early antheridium ontogeny in (**a**) selected Marchantiaceae and (**b**) selected Anthocerotopsida. Early determination of ontogenetic pathways following specific patterns of cell division, based on ontogenetic data from Schuster 1984a,b,c. In Anthocerotopsida, for example, a single superficial *antheridium initial* may divide once periclinally and the daughter cells may follow the different ontogenetic pathways shown in (b). The outer cell divides to produce a covering cell layer over the antheridium chamber *(cover cell pathway)*. The inner cell *(antheridium initial 1)* will ultimately differentiate into the antheridium. A single division of antheridium initial 1 produces an inner cell that ultimately differentiates into the antheridium stalk *(stalk pathway)* and an outer cell *(antheridium initial 2)* that will produce jacket and sperm tissues. Antheridium initial 2 undergoes a series of horizontal and vertical divisions to produce an octant of cells. Each cell of the octant divides once periclinally, marking the inception of the antheridium jacket *(jacket pathway)* and sperm *(sperm pathway)*. Antheridium ontogeny in the Marchantiaceae shows similarities to that in the Anthocerotopsida but results in superficial antheridia. Internalization of antheridia in the Anthocerotopsida is achieved through an early modification in ontogeny *(stippled boxes)*. See text for further discussion.

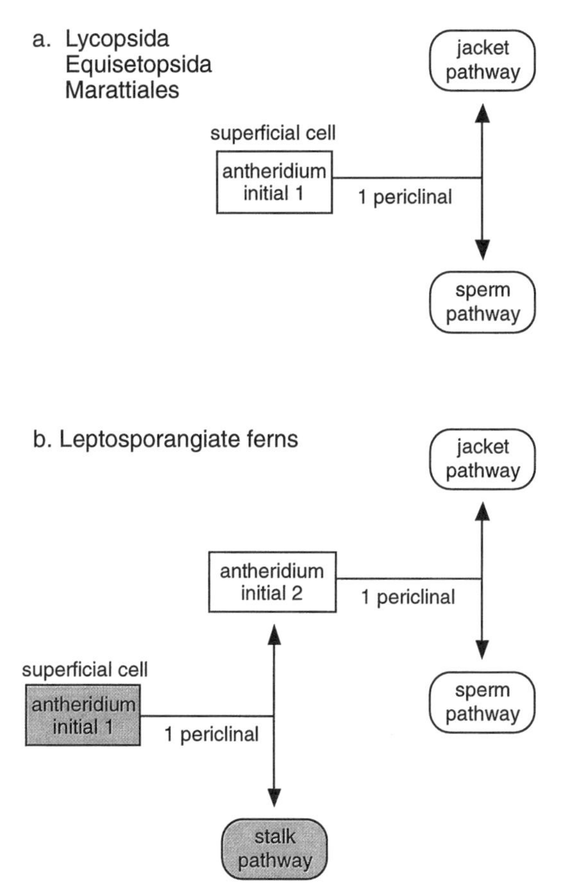

FIGURE 7.17. Antheridium ontogeny in selected vascular plants: (**a**) Lycopsida, Equisetopsida, and Marattiales, and (**b**) leptosporangiate ferns. Early determination of ontogenetic pathways following specific patterns of cell division, modified from Gifford and Foster 1989, Figures 5-2 and 5-5. In basal leptosporangiate ferns such as the Gleicheniaceae and Osmundaceae, antheridia are superficial and borne on a short stalk; the antheridium initial 1 and the stalk pathway *(stippled boxes)* are absent (lost) in other groups.

Marchantiaceae is obovate, stalked, and superficial (Figure 3.33).

In contrast to the superficial antheridia of the Marchantiopsida and Bryopsida, the antheridia of the Anthocerotopsida are located in chambers within the thallus (Figure 3.33). Cladistic analysis suggests that this positional difference is autapomorphic in the Anthocerotopsida and that the ancestors of the group had superficial antheridia. In other words, the antheridia of the Anthocerotopsida have become internalized, perhaps reflecting a degree of xeromorphism. Antheridium morphology and ontogeny in the Anthocerotopsida and Marchantiaceae is otherwise very similar (Figure 7.16; Schuster 1984a), the principal difference being the insertion of an extra cycle of cell division before the initiation of the main antheridial pathway in the Anthocerotopsida. The superficial antheridium initial divides periclinally to produce an inner and outer cell. The inner cell is developmentally equivalent to antheridium initial 1 of the Marchantiaceae and develops normally into a stalked antheridium. The outer cell divides periclinally once, and then subsequent divisions are strictly anticlinal, producing a two-cell-thick cover over the developing antheridium (jacket 1 pathway). The chamber develops as the passive consequence of continued growth of the surrounding tissues, with the developing stalked antheridium attached to the sinking chamber floor. According to this hypothesis, the remarkable positional difference between antheridia of the Anthocerotopsida and those of other bryophytes is the developmental consequence of a small addition to the beginning of antheridium ontogeny.

The antheridia of the Tracheophyta are generally simpler than those of bryophytes, but they share a broadly similar ontogeny (Figure 7.17). In the Lycophytina the superficial antheridium initial is developmentally equivalent to antheridium initial 2 of the bryophyte ontogenies (Figures 7.16 and 7.17). This cell divides periclinally to produce an inner cell that initiates a sequence of divisions leading to spermatozoid formation (sperm pathway). The outer cell divides anticlinally to produce the jacket (jacket pathway). The major difference between antheridium ontogeny in the Lycophytina, compared with that of bryophytes, is the absence of the developmental pathway leading to stalk development. We interpret this as a loss and as one of many manifestations of gametophyte reduction. The morphological consequence of this loss is that the antheridium is internalized—becoming embedded in the thallus surface. This form of internalization is clearly developmentally and morphologically different from internalization of the antheridia in the Anthocerotopsida (above), in which development of a stalked antheridium within a chamber is brought about by an additional cell division early in ontogeny.

In contrast to the Lycophytina, the Equisetales, Marattiales, Ophioglossales, and leptosporangiate ferns have superficial antheridia. Cladistic analysis suggests that this is a potential autapomorphy of the group. The antheridia of leptosporangiate ferns develop in a manner analogous to the superficial antheridia of bryophytes (Figure 7.17). The initial periclinal division produces an inner cell that develops into a weakly developed stalk or basal cell (stalk pathway). The outer cell is developmentally equivalent to antheridium initial 2 in Lycopsida ontogeny and develops normally into the antheridium. We interpret the insertion of a stalk pathway into the early ontogeny of the antheridia of leptosporangiate ferns as an evolutionary innovation, but this interpretation needs to be tested by more detailed phylogenetic analyses of the moniliform group.

With the exception of *Chara* and *Coleochaete*, spermatogenesis in charophycean algae does not involve morphologically complex structures comparable to the antheridia of land plants. In filamentous species of *Coleochaete*, spermatozoid mother cells are borne in clusters at the ends of branches. In discoid species, they are commonly produced at the margins of the disk. The antheridia of *C. scutata* comprise four small sperm-producing cells underlain by two vegetative cells (Graham 1993). The antheridia of *Chara* are morphologically and ontogenetically complex (Figures 3.4 and 7.18). A superficial antheridium initial divides periclinally to produce one or two lower cells that act as a short stalk (stalk pathway). The upper cell (antheridium initial 2) initiates a series of divisions to produce an octant of cells. A periclinal division in each cell marks the inception of the jacket-manubrium and the sperm-capitula pathways. The jacket-manubrium initials undergo one further periclinal division to produce the eight external cells of the

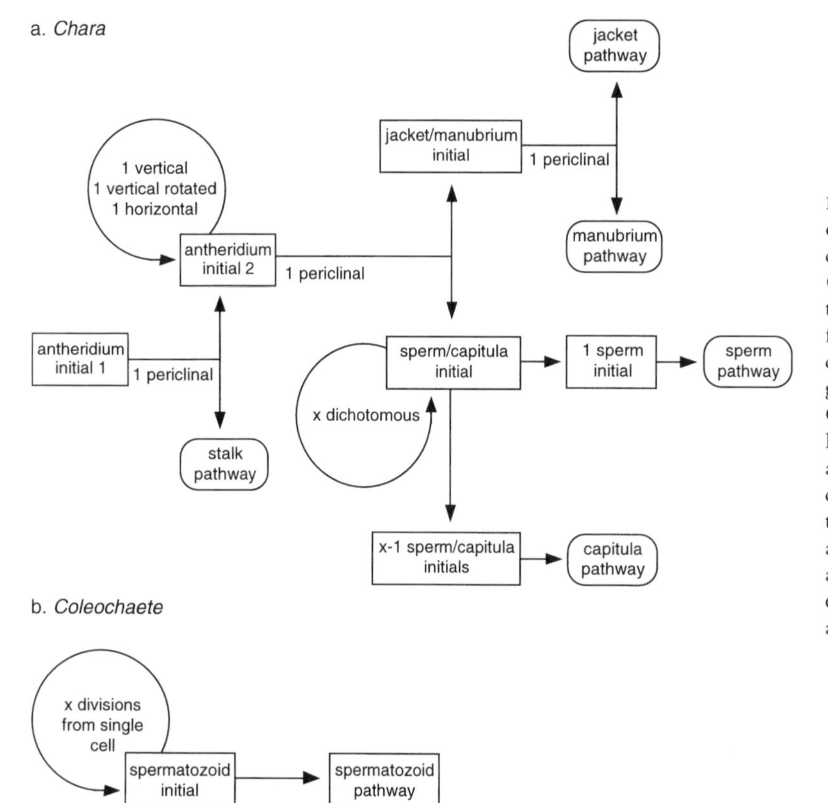

FIGURE 7.18. Antheridium ontogeny in selected charophycean algae: (**a**) *Chara* and (**b**) *Coleochaete*. Early determination of ontogenetic pathways following specific patterns of cell division, based on ontogenetic data from Fritsch 1948, Graham 1993, and Pickett-Heaps 1975. The complexity of antheridium ontogeny in *Chara* contrasts strongly with spermatozoid ontogeny in *Coleochaete* and with the relatively simple antheridium ontogenies of embryophytes (see Figures 7.16 and 7.17).

jacket ("shield") and the manubrium cell. The eight jacket cells do not divide further but enlarge and adopt a convex shield shape (jacket pathway). The manubrium cell elongates to form a rodlike structure projecting into the antheridium from the inner surface of each jacket cell and ultimately bearing the capitula and spermatozoid filaments at the other end (manubrium pathway). The sperm-capitula initial undergoes several divisions to produce a dichotomous filament (capitula pathway). The terminal cells of the filament divide horizontally numerous times to produce an elongate filament of spermatozoid mother cells (Fritsch 1948; Pickett-Heaps 1975).

Antheridium ontogeny in *Chara* shows some similarities to that of bryophytes at the stage of antheridium initials 1 and 2—in particular, the development of a quadrant phase and the differentiation of sperm- and jacket-producing pathways at this stage. Other aspects of sperm and jacket ontogeny show marked differences in the patterns of cell division and differentiation, as well as in the presence of additional structures (manubrium and capitula). Given these ontogenetic differences, any homology between the antheridia of *Chara* and of land plants could probably only relate to the inception of antheridia as far as the divergence of jacket and sperm pathways. An argument for more extensive homology at this level is also poorly supported by systematic evidence. The phylogenetic position of *Chara* and *Coleochaete* with respect to land plants is currently equivocal. Cladistic studies consistently place either one or a clade comprising both as the land plant sister group. Given these difficulties, the widely held view that the antheridia of *Chara* and land plants are not homologous seems very likely. From this perspective, the antheridia of land plants evolved de novo from very simple systems such as those in other charophycean algae (c.f. *Coleochaete:* Figure 7.18).

SPORANGIA The sporangia of land plants comprise a central spore mass encapsulated in a thin unicellular or multicellular jacket. In the Bryopsida, Anthocerotopsida, and early polysporangiophyte fossil *Horneophyton,* the spore mass surrounds a conspicuous central region of sterile tissue, the columella (Figures 3.19, 3.20, 3.21, 3.26, 3.29). In the Marchantiopsida and Anthocerotopsida, certain cells within the archesporium differentiate into sterile, often elongate, elaters that aid spore dispersal (Figures 3.30 and 3.31). Sporangia range from spherical to fusiform or reniform in basal groups and are usually borne on a short stalk (Figures 3.28 and 4.29). Sporangium ontogeny in land plants is characterized by early determination of the ontogenetic pathways leading to the stalk, spore mass, and jacket, but there are many detailed differences in ontogeny among groups. Cladistic studies suggest that sporangia are homologous in all land plants, and this homology implies that many of the differences among groups should be interpreted as modifications of a basic ancestral morphology and ontogeny. We have summarized some of the ontogenetic information on sporophyte and sporangium morphology in several basal groups to highlight similarities and differences in the developmental pathways.

The sporangium of bryophytes comprises almost the entire sporophyte generation. In all Marchantiidae (except *Monoclea,* which has an early free-nuclear stage) the first periclinal division of the zygote (sporophyte initial: Figure 7.19) results in a basal cell (a seta or foot initial) that initiates two separate series of cell divisions (the seta pathway or the foot pathway) leading to the formation of the seta and the foot. In many Marchantiidae, both cells undergo a series of vertical and transverse divisions to form an octant stage. In the sporangium pathway (sporangium initial), a further periclinal division marks the divergence of the spore-elater and jacket pathways, and a further series of cell divisions in the spore-elater mother cells leads to the differentiation of the spores and elaters.

The ontogeny of the sporophyte of Anthocerotopsida shows two conspicuous differences from that of the Marchantiidae (Figure 7.19). First, the sporophytes possess a columella, and this structure is determined before the jacket and spore pathways are initiated. Second, the equivalent developmental trajectory to the seta pathway in the Jungermanniales has been altered to a basal meristem pathway,

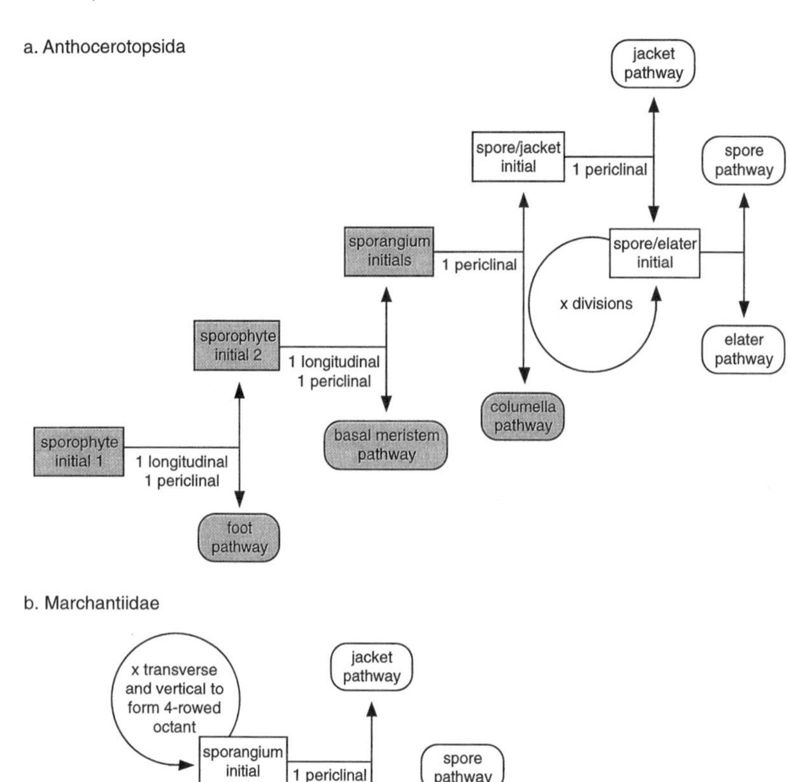

FIGURE 7.19. Summary of sporophyte ontogeny in selected (**a**) Anthocerotopsida and (**b**) Marchantiidae. Early determination of ontogenetic pathways following specific patterns of cell division, based on ontogenetic data from Schuster 1984c, 1992. Nonstippled boxes = similarities (putative homologies) in ontogeny between Anthocerotopsida and Marchantiidae.

which yields a meristematic zone that is unique to the Anthocerotopsida. This basal meristem divides continuously, producing files of cells that add to the columella, spore, and jacket pathways in an indeterminate manner resulting in a sporophyte that grows continually from the base. Note that the ontogeny of the original sporangium initial in the Anthocerotopsida is determinate, as in other bryophytes, and that it initiates a series of cell divisions that form the columella, spores, and jacket at the distal end of the sporangium.

Sporangia in basal vascular plants differentiate from epidermal initials. Initiation of sporangia in the Lycophytina is usually recognized as beginning from a single row of epidermal initials (Bower 1894). This row probably originates through a series of anticlinal divisions that can be traced to a single epidermal initial (sporangium initial 1: Figure 7.20a). The basal cell of the first periclinal division initiates the stalk pathway. The upper cell (sporangium initial 2) divides periclinally to initiate the spore and jacket pathways. Sporangia in leptosporangiate ferns clearly originate from a single initial. The first division produces the inner cell that initiates the stalk pathway. The outer cell (sporangium initial 2) undergoes four oblique di-

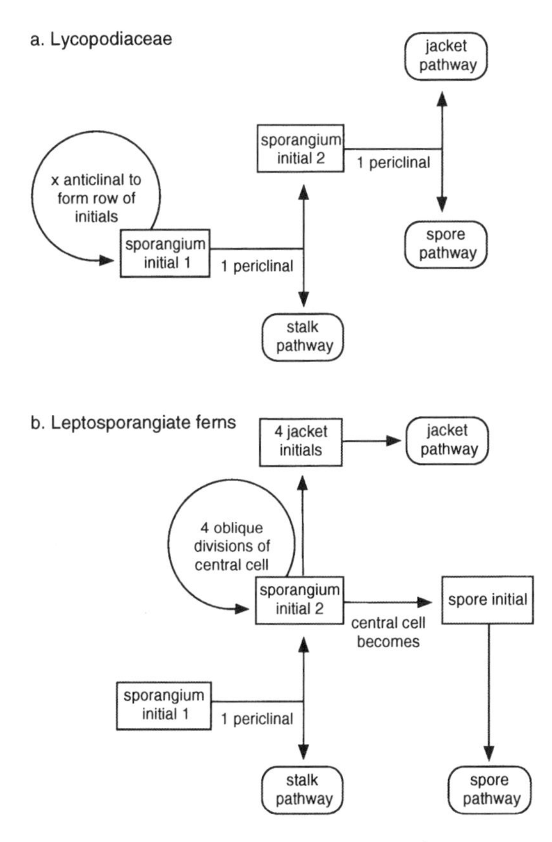

FIGURE 7.20. Sporangium ontogeny in (a) the Lycopodiaceae and (b) leptosporangiate ferns. Early determination of ontogenetic pathways following specific patterns of cell division, based on ontogenetic data from Bower 1894, 1904, and Gifford and Foster 1989.

visions (three vertical, one horizontal) producing four jacket initials and a single central spore pathway initial. Major ontogenetic differences between the Lycopsida and leptosporangiate ferns include the early formation of a row of epidermal initials in the Lycophytina and the oblique cell divisions in sporangium initial 2 that initiate the jacket and spore pathways in leptosporangiate ferns (Figure 7.20).

From an ontogenetic perspective, the shape of sporangia in basal Lycophytina (flattened and reniform) is clearly traceable to two developmental features unique to the group. The bilateral symmetry originates with the row of epidermal initials, and the reniform shape is caused by further lateral cell divisions in the spore and jacket pathways (Bower 1894). Within the Lycophytina, significant changes in sporangium shape occur in the Protolepidodendrales and Isoetales. In both groups, sporangia are dorsiventrally elongated rather than flattened as in the Lycopodiaceae, but they are still basically reniform in section. Ontogenetic studies of the Isoetales show that the number of sporangium rows at the stage of sporangium initial 1 increases from one (plesiomorphic) to many (Bower 1894). This amplification of cell rows results in a dorsiventrally elongate sporangium, and because it occurs early in ontogeny—before the initiation of the stalk pathway—the sporangium becomes fused to the leaf along most of its length. This form of sporangium is typical also of the closely related extinct rhizomorphic lycopsids. In contrast, the sporangium of the Protolepidodendrales is attached to the leaf through a small contact surface at one end. This would suggest that the ontogeny of sporangium elongation in this group is fundamentally different from that in rhizomorphic lycopsids. In the Protolepidodendrales, it is likely that elongation occurred through the amplification of tissues later in ontogeny, after the stalk pathway has been initiated.

Transformational Homologies

Many aspects of land plant morphology can be understood in terms of the evolutionary transformation of one tissue type or organ system into another. Classical ideas on transformation have various bases, including hypotheses of phylogeny, theories of homology, the sequence of appearance of fossils, and arguments built around the plausibility of different scenarios of developmental change. Usually, the principle of parsimony is invoked implicitly both in the recognition of homology (pattern) and at the level of developmental plausibility (process). In our view, the most useful hypotheses of transformation are based on patterns of taxic homology (Patterson 1982). From this perspective, hypotheses about the evolutionary transformation of one organ system into another become a posteriori hypotheses of character change occurring between nested taxic homologies. Transformational hypotheses are not supported if they are not congruent with the pattern of taxic homology (Patterson 1982). If two competing hypotheses of transformation are congruent with the pattern of taxic homology, developmental parsimony provides an additional criterion for selecting hypotheses. The following section reexamines some proposed evolutionary transformations in basal land plants in this conceptual framework and evaluates the plausibility of competing hypotheses against current cladistic schemes.

MICROPHYLL EVOLUTION IN LYCOPSIDS The evolution of microphyllous leaves in lycopsids has been widely discussed, and the competing theories are strongly influenced both by the morphology of early fossils and by ideas about the relationships among basal groups of vascular plants. We recognize three competing hypotheses of microphyll evolution—reduction, enation, and sterilization—and each is critically dependent on putative microphyll homologues and their relative position in the systematic hierarchy. Ancestral character states for the microphyll must define a clade that includes all taxa possessing that state, as well as lycopsids. If the putative microphyll homologue and the microphyll character are shown to originate independently in separate clades, then that particular theory of transformation is falsified. The following section discusses and evaluates each hypothesis on the basis of character distributions. We conclude that the least probable scenario is the *reduction hypothesis* (hypothesis 1), which is based on Zimmermann's (1938, 1952) Telome Theory. Certain scenarios involving Bower's (1935) *enation hypothesis* (hypothesis 2) are also ruled out, but we still consider this hypothesis to be plausible. Overall, we favor the *sterilization hypothesis* (hypothesis 3, developed by us here) as the simplest explanation for the origin of microphylls in lycopsids.

The reduction hypothesis (hypothesis 1) interprets the major organ systems of land plants as modifications of a simple, primitive structural unit composed of a terminal branch *(telome)*, which may be either fertile (i.e., terminating in a sporangium) or sterile. According to this theory, the microphyll and the characteristic sporangium-sporophyll association in lycopsids evolved through reduction of a lateral branch system bearing dichotomous fertile and sterile telomes (Figure 7.21). Extreme reduction results in one sterile telome (microphyll) subtending one fertile telome (sporangium), accounting for the invariable association of sporangium and leaf in extant lycopsids.

The reduction hypothesis predicts that taxa in the lycopsid stem group will have strobilar regions comprising small dichotomously branched sterile and fertile axes. Supporters of this hypothesis point to a handful of early fossils (e.g., Protolepidodendrales, *Estinnophyton*) that show putative transitional stages with branched leaves (Schweitzer 1980c; Stewart 1983; Zimmermann 1952). This fossil evidence is problematic because cladistic studies indicate that some of these taxa are only questionably related to lycopsids, whereas others are nested within the Lycopsida, implying that their branched leaves are autapomorphic. Such taxa as *Leclercqia complexa* (Protolepidodendrales) have forked microphylls (Bonamo, Banks, and Grierson 1988), but cladistic studies place this group firmly within ligulate lycopsids (Chapter 5; Bateman 1992; Crane 1990). The leaves of *Leclercqia* otherwise show typical lycopsid features such as a single, adaxial sporangium, a distal ligule, and a single well-developed vascular strand. The relationships of other plants with forked leaves, such as *Estinnophyton* (Fairon-Demaret 1978, 1979), are less secure. The anatomy of *Estinnophyton* is unknown, and the morphology of its sporangia, although poorly understood in detail, is atypical of that in lycopsids and zosterophylls. Also, unlike all other lycopsids, *Estinnophyton* bears two or four sporangia per leaf. Many of the features of this genus suggest a relationship with the Equisetopsida, but more anatomical and morphological information is required to permit a more secure systematic placement (Fairon-Demaret 1978, 1979).

Our analysis indicates that the group most closely related to the Lycopsida is the Zosterophyllopsida, and these leafless plants bear simple, cauline, lateral sporangia on short, unbranched stalks (Figure 5.1). This combination of characters is

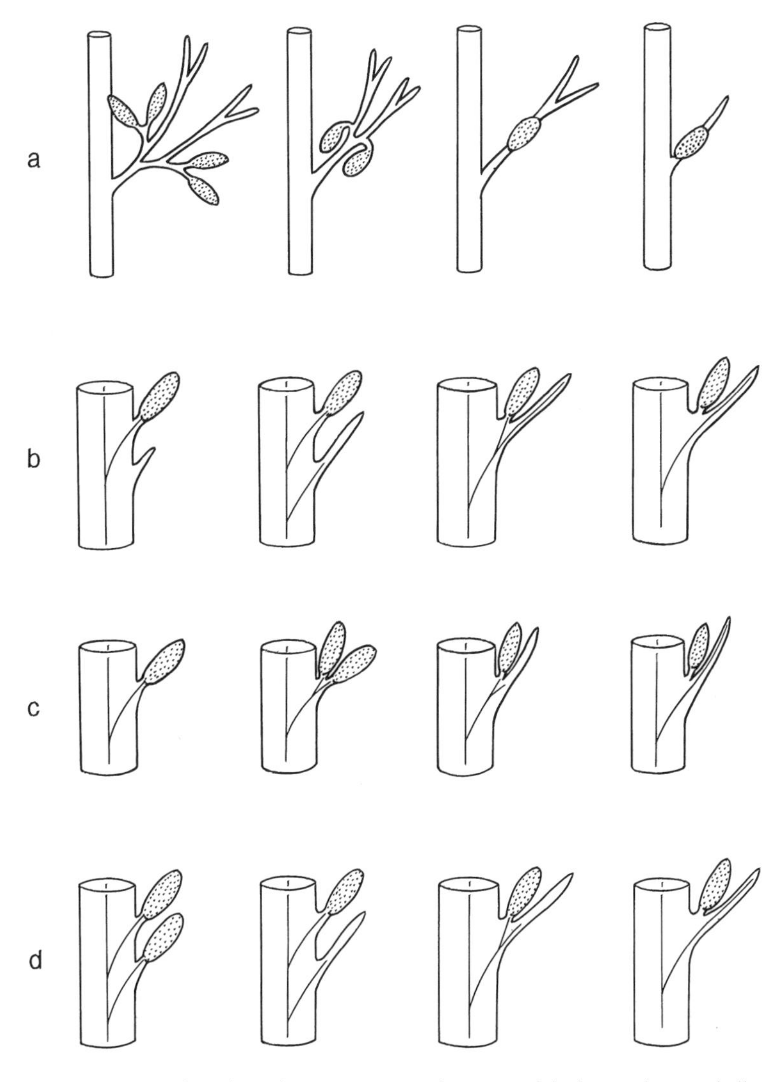

FIGURE 7.21. Three hypotheses concerning the origin of the lycopsid microphyll focusing on the sporophyll-sporangium appendage (transitions occurring from left to right). (**a**) The reduction hypothesis interprets the sporangium-sporophyll appendage in lycopsids as a highly reduced lateral branch. Sporophylls of putative extinct intermediates are characterized by branched "leaves" and multiple sporangia. Modified after Stewart and Rothwell 1993. (**b**) The enation hypothesis interprets the sporophyll as a new structure that evolved as a sterile outgrowth of the stem. Sporophylls of putative extinct intermediates are characterized by unbranched and nonvascular or partially vascular "leaves" (path of vascular tissue indicated within stem). (**c, d**) The sterilization hypothesis interprets the sporophyll as a sterilized sporangium. Sporophylls of putative extinct intermediates are characterized by unbranched and partially vascularized leaves that may be reniform or spatulate in shape (path of vascular tissue indicated within stem). In (c), the sporangium-sporophyll association arises by sporangium duplication prior to sterilization. In (d), the sporangium-sporophyll association arises by subsequent association of sporangium and microphyll.

plesiomorphic in the lycopsid-zosterophyll clade and does not support microphyll evolution by reduction. If the leaves and lateral sporangia in lycopsids evolved by reduction, then leaflessness in the Zosterophyllopsida represents a further reduction and loss of leaves. Alternatively, there may be a more distant relationship between lycopsids and zosterophylls, implying the independent origin of such features as lateral sporangia. Neither scenario is supported by this study. While the evolution of lateral sporangia in both lycopsids and zosterophylls by reduction from basal polysporangiophytes with terminal sporangia is plausible, we conclude that the evolution of microphylls in this manner is unlikely.

The enation hypothesis (hypothesis 2) proposes that the simple lycopsid microphyll evolved de novo as a lateral outgrowth of the stem (Figure 7.21; Bower 1935). Intermediate stages involve nonvascular or partially vascular, multicellular spinelike outgrowths (enations) (Bower 1935; Stewart 1983). Spinelike enations are a common feature in early land plants, but because they are small and more numerous than microphylls and are not ordered in any detectable phyllotactic pattern, the evolution of microphylls via this route requires enlargement, vascularization, reduction in number, phyllotactic regularization, and synchronization of enation phyllotaxis with sporangiotaxis. Furthermore, there are additional positional difficulties in taxa that bear numerous spines on the sporangia themselves (e.g., *Discalis*). This hypothesis accounts for the simple morphology and ontogeny of the microphyll but requires substantial developmental innovation and has difficulties explaining the association of sporangium and leaf. The enation hypothesis has received wide support in general botanical texts (e.g., Gifford and Foster 1989; Raven, Evert, and Eichhorn 1992).

The enation hypothesis predicts that the microphyll character will be nested *within* a more general clade defined by the presence of enations. Although nonvascular enations are common in many early land plants, this feature is highly homoplastic (see Chapter 3). Bower (1935) used the enations of *Psilophyton princeps* as a model for an intermediate stage in microphyll evolution, but our analysis shows that enations in *Psilophyton* evolved independently from those in the Lycophytina (Chapter 4). The most likely microphyll homologues occur in the Zosterophyllopsida and closely related taxa, and enations in these plants have been the focus of attention (Gensel 1991, 1992; Gensel, Andrews, and Forbes 1975). The leaf of the early fossil lycopsid *Asteroxylon* is often cited as evidence favoring the enation hypothesis because of the leaf's enation-like morphology. Unlike in other lycopsids, the leaf trace of the fossil terminates at the base of the leaf. *Asteroxylon* is unique in this respect, but it is equally parsimonious to interpret the absence of vasculature within the leaf as a loss (rather than as plesiomorphic), particularly given the elongated nature of the cells that can be detected in the leaf itself (Hueber 1992).

Previous cladistic studies have not stringently tested the relationship between microphylls and enations. In Gensel's cladistic study (1992), the enation hypothesis was used as the basis for the linear ordering of a "leaf" ("stem outgrowths") character consisting of four character states: emergences absent, emergences nonvascularized (i.e., enations), emergences vascularized to base (i.e., *Asteroxylon*), and emergences fully vascularized (i.e., microphyll). Character-state distributions were found to be consistent with the enation hypothesis (i.e., microphyllous taxa were nested within a clade of enation-bearing zosterophylls), but as Gensel acknowledged, treating microphylls and enations as part of an *ordered* multistate character constitutes significant weighting in favor of the enation hypothesis. In our analysis we scored microphylls and enations as two states of an unordered, multistate character. This scoring is more neutral with respect to different theories of character evolution. To further test the hypothesis of homology, we also rescored microphylls and enations as homologous in a binary character and ran the analysis a second time (Chapter 4). Although this scoring favors the enation hypothesis, microphylls again emerged as an independent acquisition in the Lycopsida, and enations appeared as independent acquisitions in the Zosterophyllopsida and in one or more plesiomorphic *Zosterophyllum*-like taxa. Although evidence from the early fossil record is equivocal and evidence from cladistic relationships is against it, the enation theory of microphyll evolution remains a plausible hypothesis because of the very scattered systematic occurrence of enations among early land plants. Firm support would come from the discovery of spine-bearing taxa in the Lycopsida stem group. It is conceivable, for example, that plesiomorphic zosterophylls such as *Discalis* (which have enations) could be placed in an appropriate position with respect to lycopsids if they were more completely understood.

The sterilization hypothesis (hypothesis 3) interprets all the appendicular structures of lycopsids (microphylls, ligules, sporangia) as iterative modifications of a single basic developmental pathway (Figure 7.21). We suggest that the lycopsid microphyll is a transformational homologue of the sporangium. According to this hypothesis the microphyll evolved through sterilization of the sporangium developmental pathway. The close association of the sporangium and sporophyll might also be explained by duplication of the sporangium developmental pathway to form sporangium pairs, followed by abortion or sterilization of the archesporium developmental pathway in the basal member of the pair to form a sterile bract or microphyll. Progressive extension of "sporangium" development in basal regions of the plant results in an extended, sterile, leafy, nonfertile zone. This hypothesis is supported by similarities in position and ontogeny between the sporangium and the leaf. Sporangium and sporophyll share similar positions on the stem, and the expression of this positional similarity is continued in the phyllotaxis of vegetative leaves. Both organs also have a similar ontogeny, developing as lateral structures from epidermal initials close to the stem apex (Bower 1894; Goebel 1887; Turner 1924). Similarity of sporangium vascularization among early members of the Lycophytina to vascularization of microphylls in Lycopsida is also significant. The sterilization hypothesis is parsimonious developmentally because it does not require the evolution of a new organ system but modifies an existing developmental trajectory possessed by the sporangium—a trajectory that already contains many of the essential microphyll attributes: a single vascular strand, lateral position, epidermal origin near the apex, and helical phyllotaxy. This hypothesis is also consistent with general character distributions in the zosterophyll-lycopsid lineage. The potential microphyll homologue—the lateral sporangium with a helical sporangiotaxis and a single vascular strand—defines a more inclusive clade that contains zosterophylls and lycopsids.

According to the sterilization hypothesis, the problematic leaf vasculature of the early lycopsid *Asteroxylon* is explained as a developmental consequence of sterilization. In *Asteroxylon,* both microphylls and sporangia possess their own vascular trace. The absence of vasculature within the microphyll of *Asteroxylon* is consistent with the absence of vasculature in equivalent areas of the sporangium. In this early taxon, absence is vestigial, perhaps representing the retention in the microphyll of a developmental feature of the sporangium that suppresses the differentiation of vascular tissue in the archesporial area.

One variation of the sterilization hypothesis predicts that stem group Lycopsida will have strobilar regions composed of sporangium pairs or sporangia with sporangium-shaped sporophylls on otherwise naked axes. Paired sporangia have not been recorded in the fossil record of this group but would be difficult to recognize in plants with relatively compact strobilate regions. However, sporangium-like sporophylls bearing functional sporangia have been documented in a plesiomorphic member of the zosterophyll-lycopsid clade recently described from China. *Adoketophyton subverticillatum* bears conspicuous fan-shaped sporophylls, each with a single, adaxial, reniform sporangium on otherwise naked axes (Li and Edwards 1992). The sporophylls of this plant were described originally as sporangia on the basis of their shape, and the plant was named originally as a new species of *Zosterophyllum.* The exact relationship of *Adoketophyton* with respect to the main Lycopsida and Zosterophyllopsida clades is uncertain because much information is still missing, but the morphology of this plant is consistent with the sterilization hypothesis.

We suggest that the ligule is a transformational homologue of the microphyll that arose through further iteration of the microphyll developmental pathway in a process that parallels the evolution of the microphyll from the sporangium. Both structures share positional and ontogenetic similarities. The ligule differentiates and matures early in leaf ontogeny and develops from a transverse strip of epidermal initials in a manner similar to sporangium and leaf to form a flattened tongue-shaped or fan-shaped structure (Harvey-Gibson 1896). The rapid growth of the ligule near the stem apex parallels that of the microphyll, but cell division ceases much earlier and the ligule remains minute.

The sterilization hypothesis is consistent with character-state distributions in the Lycopsida and Zosterophyllopsida and with the highly conservative appendicular structure of lycopsids in general. However, the sterilization hypothesis, like the enation hypothesis, lacks unequivocal support in the fossil record. Stronger support would come from the discovery of sporangium-shaped sporophylls in taxa securely assignable to the Lycopsida stem group.

CUPULE AND INTEGUMENT EVOLUTION IN SEED PLANTS Recent cladistic studies support a single origin of seed plants, and homologies in the seeds and pollen organs are traced to ancestors at the progymnosperm grade (Beck and Wight 1988; Crane 1985; Doyle and Donoghue 1986, 1992; Galtier 1988; Meyen 1987; Nixon, Crepet, et al. 1994; Rothwell and Serbet 1994; Stewart and Rothwell 1993; T. N. Taylor and E. L. Taylor 1993). Evidence from the early fossil record plays a critical role in understanding cupule and seed evolution because of putatively intermediate fossil forms and the isolated evolutionary position of the five extant seed plant groups. The most widely cited, current hypothesis for seed evolution, the Telome Theory, interprets the cupule and integument of early seed plants as reduced lateral branches of the *Psilophyton* type (Andrews 1961; Halle 1933; Stewart and Rothwell 1993; T. N. Taylor and E. L. Taylor 1993; Zimmermann 1952, 1965). This process is seen as involving more or less symmetrically arranged branches of several orders. A second, older hypothesis, the *synangial hypothesis,* views the integuments of seed plants as strictly homologous with a ring of sporangia that became sterile (Benson 1904). The following section discusses and evaluates the telome and synangial hypotheses on the basis of putative homologues in the progymnosperm stem group. Overall, we favor a modern interpretation of the synangial hypothesis that Benson (1904) originally proposed.

The synangial hypothesis interprets the integument of seed plants as homologous to a cluster of sporangia (Benson 1904). The integument evolved through sterilization and fusion of a ring of sporangia surrounding a central megasporangium. From a modern perspective, the seed-cupule precursor would resemble the "sporophyll" of an aneurophytalean progymnosperm such as *Tetraxylopteris* (Figure 4.14). In aneurophytalean progymnosperms, fusiform sporangia are borne in clusters on short stalks that are oriented toward the adaxial surface of a dichotomously branched sporophyll. This *neosynangial hypothesis* postulates the ovule as a transformation of a stalked sporangium cluster that involves the sterilization of the peripheral sporangia to form a ring of free integumentary lobes surrounding a single central megasporangium (Figure 7.22). Additional transformation results in the development of a single vascular strand within each integumentary lobe. According to this interpretation, the cupule is a transformational homologue of the aneurophytalean "sporophyll," and cupule evolution would involve few additional changes. We view the sterilization of sporangia to form free integumentary lobes as potentially saltational, with the immediate effect of focusing nutrient resources on a single developing megasporangium.

Although character distributions in basal Euphyllophytina have not yet been comprehensively examined, the neosynangial hypothesis is consistent with cladistic studies that support the evolution of seed plants from taxa at the progymnosperm grade. The hypothesis is also consistent with

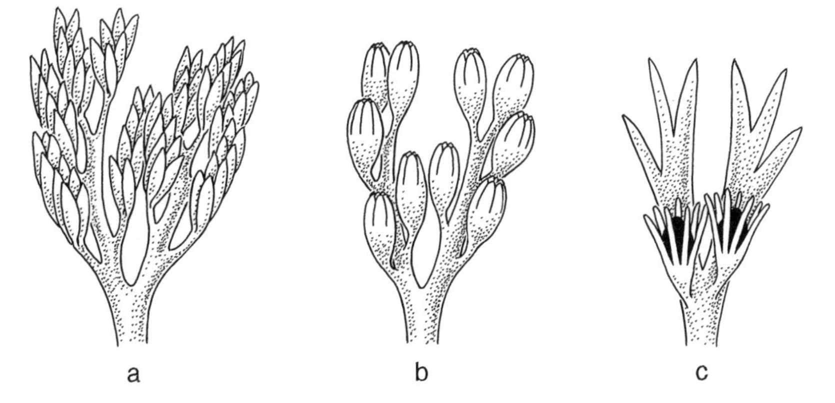

FIGURE 7.22. Homologies of pollen organs and the seed-integument-cupule appendage of basal seed plants and aneurophytalean progymnosperms. (**a**) Aneurophytalean-type sporophyll bearing clusters of fusiform sporangia on short stalks oriented on the adaxial side of the sporophyll. (**b**) Early pteridosperm-type pollen organ (e.g., *Crossotheca-, Telangium-, Codonotheca*-type) bearing stalked synangiate organs comprising rings of partially fused sporangia. Synangiate organs are homologous with the stalked sporangium clusters in the progymnosperm. (**c**) Seed-integument-cupule appendage from early pteridosperm (e.g., *Moresnetia*) showing sterile lobed cupule bearing two stalked pre-ovules comprising megasporangium with free integumentary lobes. The pre-ovule is homologous with the stalked sporangium clusters in the progymnosperm. Free integumentary lobes are sterile sporangia. The lobed cupule is basically homologous with the aneurophytalean sporophyll, which has undergone complete sterilization distally.

the morphology of the fertile branches of aneurophytalean progymnosperms in which sporangium position and morphology is similar to the free and elongate integumentary lobes of early pteridosperms. In addition, integumentary lobes in basal seed plants are never branched and contain unbranched vasculature. Striking parallels to integument formation also occur in the pollen organs of early seed plants, in which rings of fused or partially fused fusiform sporangia are common (Figure 7.22; Halle 1933; Millay and Taylor 1979; T. N. Taylor 1988b). Cupule morphology in early seed plants resembles the morphology of sporophylls in progymnosperms, and there are positional similarities between stalked sporangia and ovule position within the cupule. The asymmetric, racemose branching typical of the cupuliferous systems of basal seed plants (Rothwell and Scheckler 1988) is also present in the fertile lateral branches of aneurophytalean progymnosperms.

The Telome Theory interprets the integument of seed plants as homologous with a cluster of sterile branches (of several orders) more or less symmetrically surrounding a central megasporangium (Walton 1953) and is regarded in recent textbooks as the most probable transformation hypothesis for the evolution of the seed (Stewart and Rothwell 1993; T. N. Taylor and E. L. Taylor 1993). This theory envisages a process of reduction of branched sterile and fertile telomes (*sensu* Zimmermann [1952]) through a series of forms originally modeled on *Psilophyton,* on *Hedeia,* and on a hypothetical intermediate comprising a single central megasporangium surrounded by sterile, dichotomously branched pre-integuments (Figure 7.23; Andrews 1961). Ultimately, loss of branching in the surrounding sterile axes results in a ring of unbranched, vascularized integumentary lobes (Andrews 1961; Stewart and Rothwell 1993; T. N. Taylor and E. L. Taylor 1993). According to this interpretation, evolution of the cupule occurs in a similar manner, but the reduction of the branches does not proceed as far.

The Telome Theory seems less plausible because the morphology of the fertile branch of the hypothetical pre-ovule (Andrews 1961; Stewart and

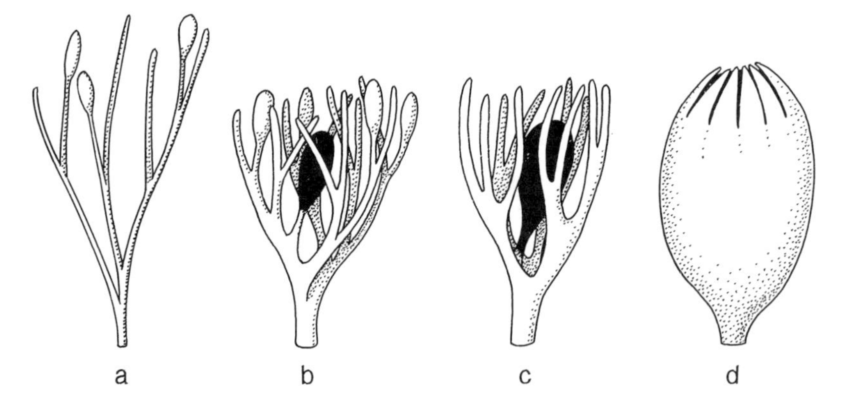

FIGURE 7.23. Suggested stages in the enclosure of a sporangium by branches of several orders to form the nucellus and integument of a pteridosperm seed. Redrawn from Andrews 1961, Figure 13-4. (a) Terminal branchlets (telomes), based on a plant of the *Rhynia* type. (b) Corymbose clustering of fertile and sterile telomes with one sporangium assuming a central location; suggested by *Hedeia corymbosa.* (c) Hypothetical stage: one central enlarged sporangium surrounded by telomes that are tending to web together. (d) A seed with sporangium (nucellus) enclosed by a lobed integument.

Rothwell 1993; T. N. Taylor and E. L. Taylor 1993) does not bear a strong resemblance to the fertile axes of progymnosperms or any other basal psilophytopsid. The plesiomorphic sporophyll morphology in lignophytes is a somewhat planar and dichotomously branched "sporophyll" bearing clusters of stalked (e.g., *Tetraxylopteris*) or more or less sessile sporangia (e.g., *Archaeopteris*) on the adaxial surface. Putative transformations from aneurophytalean or archaeopteridalean types would require a greater series of modifications than is inherent in the neosynangial hypothesis. In particular, the position and morphology of the integumentary lobes of seed plants does not fit well with a hypothesis of reduction from a heterogeneous mixture of sterile and fertile branches. An origin of cupule and integument from higher-order lateral branches in progymnosperms would also predict that early seed plants would retain some vestiges of this arrangement in ovule position and cupule lobing. However, the free integumentary lobes of basal seed plants are never branched, and both ovule position and cupule morphology bear a stronger resemblance to the sporophyll-sporangium morphology than to the higher-order branches of progymnosperms.

THE ORIGIN OF MEGAPHYLLS The term *megaphyll* describes the leafy appendages of seed plants, ferns, and basal Equisetopsida. Megaphylls are characterized by complex venation patterns, and although they are typically large with a pinnate organization, they show extreme diversity in form and size. The long-standing and widely accepted Telome Theory traces the origin of megaphyllous leaves to the dichotomously branched lateral appendages that characterize early extinct members of the Euphyllophytina. Hypothetical transformations are based on a series of modifications to a dichotomous, three-dimensional lateral branch of the *Psilophyton*-type (Figure 4.13). Evolution of the megaphyll is thought to involve a series of transformations, beginning with a change to planar branching systems *(planation)* followed by modification of the lateral branch through lateral laminar-like outgrowths *(webbing)* and eventual fusion of those outgrowths to produce a true leaf lamina with open dichotomous venation (Figure 7.24; Zimmermann 1952, 1965). Putative intermediate stages have been documented in many early fossil taxa, and it is likely that certain features of the megaphyll evolved convergently in different groups (e.g., Crane 1990; Doyle and Donoghue

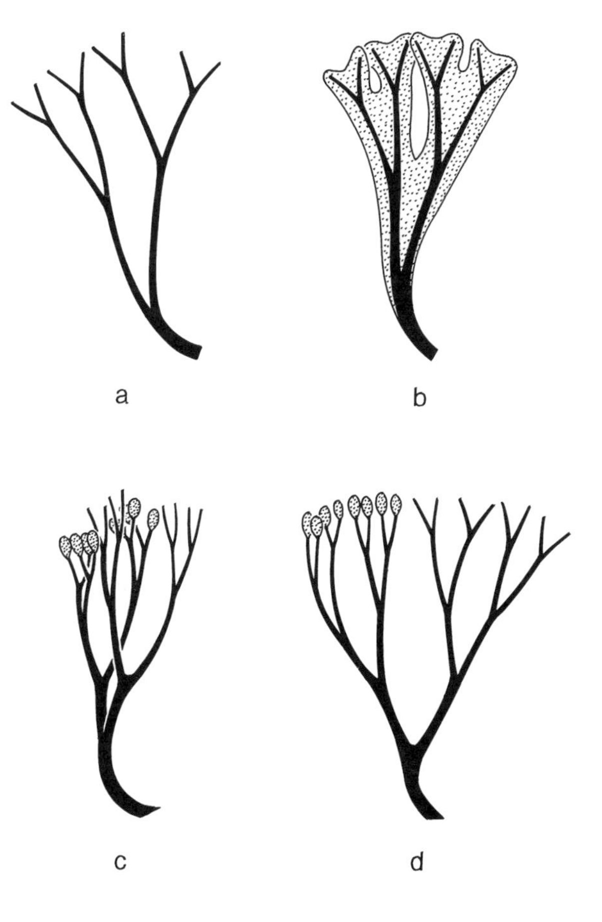

FIGURE 7.24. The Telome Theory on the origin of megaphylls from early *Psilophyton*- or *Rhynia*-like plants. The processes of webbing (lamina formation) (a, b) and planation (c, d). Drawings adapted from Stewart and Rothwell 1993, Figure 9.10. (**a**) Vegetative lateral branching in one plane. (**b**) Webbing of planar lateral to form leaf lamina. (**c**) Cluster of fertile and sterile lateral branches. (**d**) Planation of lateral branch system.

1986, 1992; Stein, Wight, and Beck 1984). Although the general outlines of this hypothesis are fully consistent with recent cladistic studies, tracing specific megaphyll homologies among early Euphyllophytina is difficult because there is no comprehensive cladistic treatment of basal clades within this group, and the details of branching in many critical taxa are unclear. The following section evaluates general homologies among megaphyllous leaves and discusses the best documented case of megaphyll evolution in the seed-plant stem group.

Our preliminary cladogram for the Euphyllophytina supports a hypothesis of homology among megaphylls in seed plants, ferns, and *Equisetum* (highly reduced) at the level of dichotomous, three-dimensional lateral branches of the *Psilophyton*-type. This result is consistent with the independent evolution of planation, webbing, and fusion in the major clades of megaphyllous plants. Thus, many of the similarities between megaphylls in ferns and in basal seed plants are convergent. Probably only the open dichotomous vasculature typical of many basal members of these clades can be viewed as homologous.

Ideas on megaphyll evolution in basal lignophytes have been strongly influenced by phylogenetic hypotheses on the origin of seed plants and by the morphology of leaves and fertile branches in basal coniferophytes and cycadophytes. The two most widely discussed hypotheses trace the origin of seed plants to progymnosperms, but putative megaphyll transformations are critically dependent on the monophyletic status of seed plants and the nature of the progymnosperm sister taxa. Beck (1971, 1981, and references cited therein) and Beck and Wight (1988) argue that seed plants are diphyletic—pteridosperms and their descendants

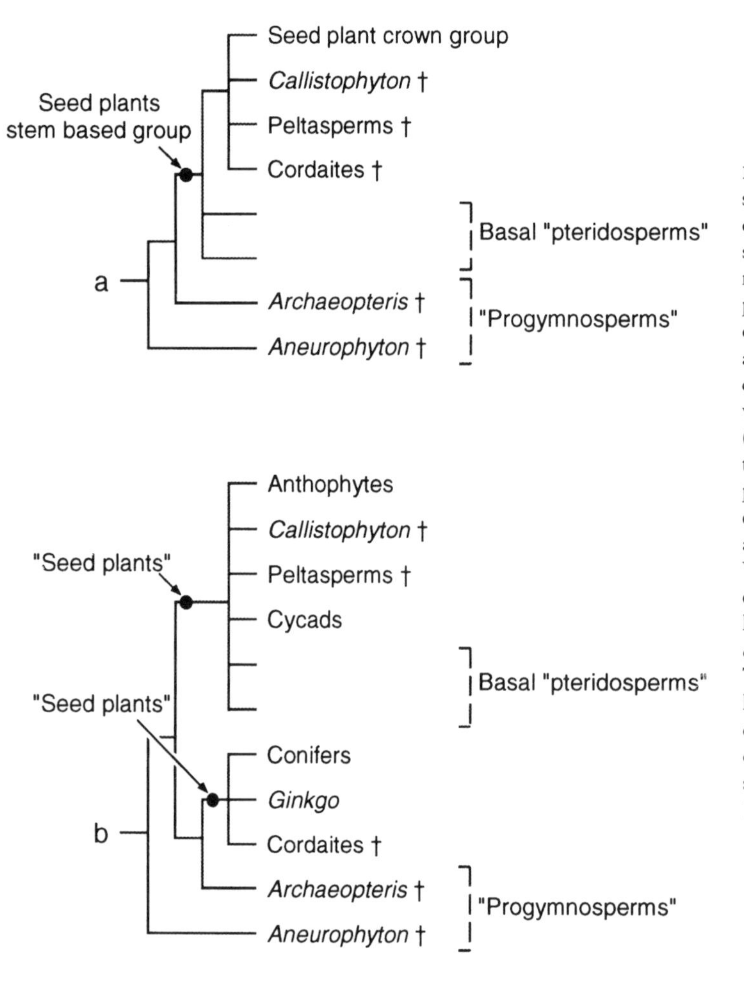

FIGURE 7.25. Diagrammatic representation of two hypotheses on the origin of seed plants from progymnosperm ancestors (simplified). Quotation marks = paraphyletic group. (**a**) Seed plants monophyletic: most parsimonious cladistic hypothesis (see Rothwell and Serbet 1994). Basal seed plants comprise a grade of extinct plants previously included in the pteridosperms. (**b**) Seed plants diphyletic: representation of the Beck hypothesis based on putative homologies between the lateral branch systems of *Archaeopteris* and coniferophytes (see Beck and Wight 1988). Conifers, *Ginkgo,* and extinct cordaites are more closely related to the progymnosperm *Archaeopteris* than to other seed plants. The Beck hypothesis (b) is 16 steps less parsimonious than the hypothesis depicted in (a). For alternative views on relationships among seed plants, see Crane 1985, Doyle and Donoghue 1986, and Nixon, Crepet, et al. 1994.

having evolved from aneurophytalean progymnosperms—whereas the origin of cordaites and conifers is traced to archaeopteridaleans (Figure 7.25). According to this hypothesis, megaphyll evolution in coniferophytes began with an *Archaeopteris*-type branch system, whereas the cycadophyte megaphyll evolved through transformation of an aneurophytalean lateral branch. Rothwell (1982) argued for seed plant monophyly and an origin from within aneurophytalean progymnosperms (Figure 7.25). According to this hypothesis, megaphylls in seed plants evolved from an aneurophytalean-type branch system, and the leaves of coniferophytes are derived from seed plants resembling *Callistophyton* at the pteridosperm grade.

We consider Rothwell's hypothesis to be the most plausible of these two alternatives because current cladistic studies, although focused mainly within seed plants, support seed plant monophyly and have all extant groups emerging from a basal grade of early, extinct pteridosperms (Crane 1985; Doyle and Donoghue 1986; 1992; Rothwell and Serbet 1994). Cladistic studies are consistent with the evolution of pinnately compound vegetative leaves in stem group seed plants (e.g., *Elkinsia, Heterangium, Lyginopteris*), and of some medullosans, from sterile, dichotomous, three-dimensional lateral branches of aneurophytalean progymnosperms. Transformations require planation, webbing, and fusion to form *Sphenopteris*-type foliage. Whereas vegetative fronds in basal seed plants are pinnately compound, fertile fronds lack pinnae except for pollen organs and cupulate structures bearing ovules. The lateral branch system of *Archaeopteris* provides a less suitable model for the derivation of megaphyllous leaves in basal pteridosperms. The vegetative pinnules are in distinctive helical orthostichies on axes with eustelic anatomy, whereas in basal pteridosperms anatomy is protostelic and pinnules are opposite and alternate. In particular, the fertile branch differs profoundly in morphology from pollen organs and cupulate systems. The sporophylls of *Archaeopteris* are arranged in dense helical orthostichies on branches that begin and terminate with vegetative pinnules. We interpret leaflike pinnules as a uniquely derived feature of *Archaeopteris* and other similarities between the lateral branch systems of *Archaeopteris* and coniferophyte seed plants as convergent.

Frond dimorphism in fertile and sterile leaves is also a feature of basal megaphyllous plants and may originate with the segregation of fertile and sterile branches in basal Euphyllophytina. This observation suggests that the transformations affecting the evolution of vegetative and fertile megaphylls in seed plants were different, at least initially. Cupulate branches and pollen organs probably evolved through modification of an aneurophytalean-like lateral fertile branch ("sporophyll"). In microsporangiate branches, most transformations involved fusion, and perhaps duplication, of sporangia to form a variety of synangiate pollen organs. In megasporangiate systems, transformations involved principally fusion and sterilization of sporangia in the formation of ovules (see above). The absence of pinnules on fertile branches in basal seed plants is probably related to the transformation of the positional equivalents to pinnules (i.e., sporangia-"sporophyll" systems) into pollen organs and cupulate seeds.

THE ORIGIN OF CORTICAL FIBERS IN SEED PLANTS Recent biomechanical studies highlight the important structural role of cortical sclerenchyma in early land plants (Mosbrugger 1990; Niklas 1990, 1992; Rowe, Speck, and Galtier 1993; Spatz, Speck, and Vogellehner 1990; Speck, Spatz, and Vogellehner 1990; Speck and Vogellehner 1988, 1991, 1994). The biomechanical properties of sclerenchyma, combined with a peripheral position in the cortex, make this tissue an important contributor to flexural stiffness of the stem. Stems with high flexural stiffness are more resistant to buckling and therefore permit the growth of taller, self-supporting, plants. In nonwoody taxa, the contribution of metaxylem to structural support is relatively low, because this tissue is located centrally within the stem and represents a relatively small fraction of the total tissue area. Other tissue types that contribute significantly to flexural stiffness include secondary xylem and periderm, but both of these tissues are poorly developed or absent in many basal Euphyllophytina.

Cortical sclerenchyma in basal Euphyllophytina, such as the relatively plesiomorphic *Psilophyton*, forms a continuous hypodermal band or sterome several cells thick (Figure 7.26). This feature is common in other basal tracheophytes such as the Zosterophyllopsida, Lycopsida, and some *Cooksonia*-like stem group vascular plants and is probably plesiomorphic in the Euphyllophytina. The stem cortex of basal seed plants is characterized by the presence of numerous bundles of cortical fibers (Figure 7.26). In some taxa the fibers are vertically aligned—a *sparganum cortex*—whereas in others they form an anastomosing netlike system—a *dictyoxylon cortex* (T. N. Taylor and E. L. Taylor 1993). The sparganum cortex is widespread in basal pteridosperms and has been documented in many stem group seed plants, including hydraspermans (e.g., *Elkinsia*), *Heterangium*, and *Lyginopteris;* in more problematic and poorly understood taxa, such as the Calamopityaceae and Buteoxylonaceae; as well as in more derived pteridosperms, such as the Medullosaceae, Callistophytaceae, and Cordaitaceae (Galtier 1988; Meyen 1987; Stewart and Rothwell 1993; T. N. Taylor and E. L. Taylor 1993). A sparganum cortex has also been documented in aneurophytalean progymnosperms such as *Tetraxylopteris* (Scheckler and Banks 1971a) and in the putative progymnosperm *Triloboxylon* (Beck and Wight 1988; Scheckler and Banks 1971a; Stein and Beck 1983). Our preliminary cladistic analysis of basal Euphyllophytina indicates that the sparganum cortex may have evolved

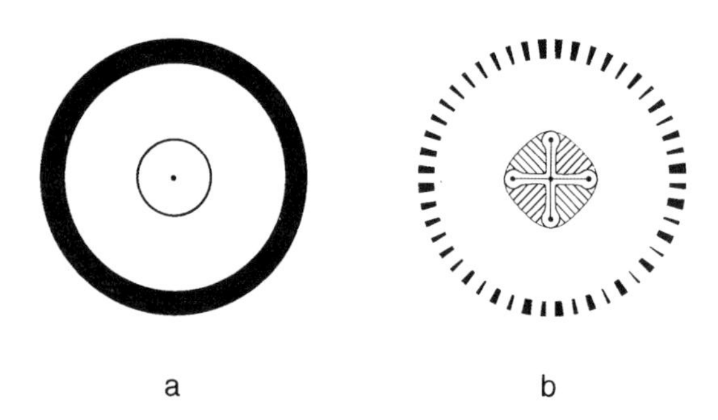

FIGURE 7.26. Origin of sparganum cortex in lignophytes through modification of a continuous sclerenchymatous sterome in basal Euphyllophytina. (**a**) Transverse section through the stem of *Psilophyton dawsonii* (basal Euphyllophytina), illustrating continuous sclerenchymatous sterome *(outer black ring)* and central terete primary xylem with centrarch protoxylem *(inner circle)*. (**b**) Transverse section through the stem of an aneurophytalean progymnosperm (basal lignophyte), illustrating sparganum cortex comprising bands of anastomosing sclerenchymatous tissues *(black)* separated by parenchyma *(white)* and central cruciate metaxylem surrounded by secondary xylem *(inner ring)*.

through modification of a continuous sclerenchymatous cortex of the *Psilophyton*-type. Transformations would include an increase in fiber cell length and a restriction of fiber differentiation within the cortex to vertically continuous or anastomosing bundles.

The transformation of a continuous hypodermis into a discontinuous ring of fibers, as is the case in the sparganum cortex, implies the evolutionary loss of sclerenchymatous tissues from parts of the stem cortex in lignophytes. Extrapolating from biomechanical studies (e.g., Rowe, Speck, and Galtier 1993), it is probable that this loss of cortical sclerenchyma means that the cortical tissues of lignophytes contribute relatively less to flexural stiffness of the stem than do similar tissues in basal Euphyllophytina. This effect may be compensated for by the production of secondary xylem to give stems with an overall level of flexural stiffness similar to that in the stems of basal Euphyllophytina. In other words, the loss of cortical sclerenchyma resulting in the sparganum cortex of early lignophytes may be related to the increasingly important structural role of secondary xylem in this group.

THE EVOLUTION OF SPOROPHYTE BRANCHING

Sporophyte branching may be the single most important evolutionary innovation of the polysporangiophyte clade. The evolution of branching in land plants had two far-reaching consequences for the sporophyte generation. First, dichotomous branching facilitated the multiplication of parts, including branches and sporangia, thereby potentially increasing photosynthetic capability and spore output. Iterative dichotomy through apical growth results in an essentially indeterminate system of repeated and developmentally equivalent morphological units. Second, branching provided a complex architectural framework for the sporophyte, parts of which were apparently modified subsequently to form a variety of structures. In polysporangiophytes, this framework has evolved into a diverse range of organ systems, including stems, branches, roots, and megaphyllous leaves. The extent to which branching systems have become modified into other structures through time is reflected in the morphological diversity among major clades. Much of the morphological diversity in the Euphyllophytina seems attributable to profound modifications to the branching systems, whereas morphological conservatism within the Lycophytina is reflected in the absence of elaborate branching and the dearth of organ systems that can be interpreted as transformed branches.

Among early polysporangiophytes, branching is relatively simple (Rothwell 1995, see above) and may contain a significant stochastic component (Niklas 1982). A series of more complex and increasingly deterministic branching systems appear early within the major clades of vascular plants(Bateman 1996a,b; Bateman and DiMichele 1994b; Niklas 1982; Rothwell 1995; Stein 1993; Wight 1987). Our analysis confirms earlier suggestions (e.g., Niklas 1982) that many aspects of branching are homoplastic, reflecting iterative evolution of features such as overtopping, apical dormancy, planation, and various other forms of branch arrangement (see above). In contrast to the polysporangiophytes, the unbranched simple sporophytes of bryophytes have a very limited range of morphological expression and show relatively little morphological divergence, even in groups as distantly related as the Marchantiopsida, Anthocerotopsida, and Bryopsida.

Cladistic studies indicate that the absence of sporophyte branching in bryophytes is the plesiomorphic state for land plants (Chapter 3). Absence of sporophyte branching at this grade is also understandable in terms of sporophyte ontogeny and ecology. In bryophytes the three main components of the sporophyte generation (foot, seta or sporophyte axis, sporangium) are determined early in ontogeny and develop in parallel. Taking the Jungermanniales (Marchantiopsida) as an example, the sporangium and foot-seta pathways are determined at the first division of the zygote (Figure 7.19). Likewise, the sporangium and seta pathways are fully determined within the next few cell divisions. One consequence of this early determination of developmental pathways is that the seta does not undergo apical growth but develops through intercalary cell division. The cell lineage that develops into the seta is constrained at the base by the developing foot and at the apex by the developing sporangium. The primary function of the seta in the Marchantiopsida is to facilitate sporangium exsertion. In most Marchantiopsida, the sporangium and the seta develop more or less in parallel, but toward the end of ontogeny, the seta undergoes rapid expansion to push the sporangium out through the calyptra in preparation for release of spores.

In the Bryopsida the sporangium and foot-seta pathways are determined at the first division of the zygote. However, it may be significant that the foot-seta initial undergoes a limited period of apical divisions to produce a short multicellular filament. Cells of the filament divide vertically to produce the quadrant stage, followed by periclinal divisions to initiate the developmental pathways producing epidermal and interior tissues. The basal cells within the filament undergo further intercalary growth to form the seta, while the sporangium develops from the distal cells. In mosses, the growth of the seta and sporangium is more or less parallel, with the seta slightly in advance. In both the Marchantiopsida and Bryopsida, the early determination of these developmental pathways, combined with limited apical growth, ensures that seta dichotomy is an exceptionally rare phenomenon.

The evolution of dichotomous branching in polysporangiophytes probably occurred in land plants at the bryophyte level of organization. Cladistic studies indicate that land plants originated from charophycean algae having a haploid, haplobiontic life cycle (Chapter 3; Graham 1993; Kenrick 1994; Mishler and Churchill 1984, 1985b). Because charophycean algae have unicellular "sporophytes" (zygote), multicellularity in the sporophyte generation is a unique evolutionary innovation of land plants. Although relationships among the major groups of bryophytes and polysporangiophytes are still poorly resolved, cladistic studies indicate that bryophytes are paraphyletic to polysporangiophytes. Furthermore, notwithstanding the generally poor fossil record of bryophytes, some of the earliest and most plesiomorphic polysporangiophyte fossils (e.g., *Sporogonites*) strongly resemble bryophyte sporophytes in size and morphology. Evidence from systematic studies therefore implies that small, unbranched sporophytes (comprising a foot, seta, and sporangium) that are "parasitic" on a generally larger, fully autotrophic, and independent gametophyte are plesiomorphic for embryophytes. Using extant bryophytes as models for investigating the evolution of dichotomous branching in embryophytes requires careful separation of autapomorphies characteristic of particular bryophyte groups from ontogenetic features that may be more general in embryophytes. General features may include differentiation into three organ systems (sporangium, seta, foot), early determination of developmental pathways, more or less parallel development of organ systems, and physiological dependency on the gametophyte.

From the results of an analysis of general ontogenetic features in embryophytes, it appears that dichotomous branching in polysporangiophytes most likely evolved from groups with ontogenies that exhibit apical growth. Duplication of apical cells followed by continued apical growth would produce a simple dichotomy. Apical growth is rare in the sporophytes of the Marchantiopsida and Anthocerotopsida—although it is present in the gametophytes of both groups—but it is more common in the early ontogeny of the Bryopsida and, of course, also in extant vascular plants. Further requirements for the evolution of dichotomous branching in polysporangiophytes are the prolongation of apical growth in early ontogeny and a significant delay in initiation of the sporangium developmental pathway. Significant postponement of sporangium formation, combined with apical growth and apical dichotomy, are the fundamental steps necessary for the evolution of large branched systems with multiple terminal sporangia from bryophyte-like plants.

THE EVOLUTION OF TRACHEID PITTING Pitted cell walls are a characteristic feature of the metaxylem tracheids of extant vascular plants. A great variety of pit types and pit distributions have been documented, and many features are taxon specific (e.g., Bailey 1953; Bierhorst 1960). Pits represent areas of the cell wall where there has been little or no deposition of secondary wall, and they are often precisely aligned with similar areas in adjacent cells. Functionally, pitting facilitates the lateral movement of water and solutes within the plant stem. In general, pits are circular to scalariform in outline and may be arranged in files, helical patterns, or groups. In addition to variations in pit morphology and distribution among taxa, there is a great variation within individuals that correlates with tracheid differentiation at different ontogenetic stages. In the earliest xylem elements that differentiate close to the apical meristem (protoxylem), secondary wall material is laid down in distinctive helical or annular-type thickenings and pitting is usually absent. In later-formed tracheids, pit morphology often shows a graded series of forms ranging from a loose reticulate network in xylem cells having relatively thin secondary walls to regular and well-defined circular or scalariform pits in the latest metaxylem tracheids.

Critical examination of tracheid morphology in early fossil vascular plants shows that, in contrast to extant groups, pitting is absent from the tracheid cell walls of many basal taxa (Kenrick and Crane 1991). In these early fossils, all tracheids have conspicuous helical or annular thickenings that strongly resemble the early-formed protoxylem elements of extant groups. Cladistic analysis indicates that pitted tracheids evolved independently at least twice (Lycopsida, Euphyllophytina) and that xylem composed entirely of tracheids with helical-annular thickenings is the plesiomorphic condition in vascular plants. The persistence of this plesiomorphic cell type in the protoxylem of extant vascular plants is probably strongly linked to functional constraints. In xylem tissues, cell elongation cannot occur by further growth after differentiation because metabolic death must occur before the cell is able to function. In vascular plants, most differentiation of primary tissues, including metaxylem, occurs after a period of cell elongation at a distance below the apex. This constraint means that the developing metaxylem tissue is unable to immediately supply the metabolically demanding growing point with xylem product (e.g., water and mineral nutrients). The role of the protoxylem is to ensure an uninterrupted supply of xylem product to the apex, and the protoxylem achieves this by early differentiation within the zone of tissue elongation immediately below the apex. Ultimately, many of the protoxylem cells become nonfunctional through severe distortion by cell elongation in the neighboring tissues, and the role of supplying water is taken over by the differentiating metaxylem with its more robust, heavily lignified, and pitted secondary walls. During the brief functional period of protoxylem, the helical-annular thickenings serve a dual role in resisting the lateral collapse (cavitation) of the cell walls while allowing axial elongation of the cell through stretching in response to growth of the surrounding tissues (Bailey 1953). The critical role of protoxylem in vascular plant ontogeny explains the persistence of the plesiomorphic helical-annular thickened cell wall and its widespread occurrence among extant groups.

Evidence from the early fossil record documents several tracheid wall features that, when taken together, are consistent with the evolution of pitting, through a series of relatively minor modifications, to a basically helical or annular thickened cell. In many early taxa, cells with annular thickenings show occasional diagonal bars connecting adjacent helical-annular rings (e.g., G-type tracheid: Figure 7.27; Kenrick and Crane 1991). The devel-

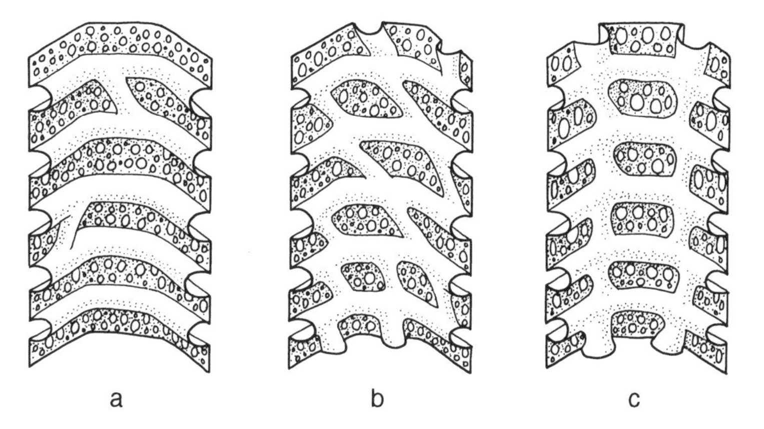

FIGURE 7.27. Evolution of tracheid pitting in vascular plants from cells with helical-annular thickenings and numerous inter-thickening perforations. Illustrations represent longitudinal sections through part of a tracheid, exposing the interior surface of the cell wall. (**a**) Early G-type tracheid with helical-annular thickenings and occasional cross-connections between adjacent thickenings. (**b**) Increase in frequency of cross-connections outlining simple irregular pits. (**c**) Regularization of cross-connections, forming thickenings in scalariform arrays. Note that perforations between thickenings in the G-type cell (a) become internalized within the pits in scalariform pitted elements (c).

opment of these diagonal bars results in the formation of a simple unbordered pit. In basal tracheophytes the frequency of diagonal bars is quite low, and their position and number varies such that the cell has a mixture of helical-annular thickenings and simple pits. This feature is more common in those taxa in which pitting may resemble a loose reticulate network. We suggest that tracheid pitting evolved through an increase in the regularity and frequency of diagonal bars to produce a more complex pattern of regular pits. According to this scenario, early pitted tracheids would have simple unbordered reticulate pits, and further organization of pits into files would result in scalariform pitting. Pit borders would develop through the accretion of additional wall material to the inner surface of the cell in the later stages of secondary wall development. As well as adding more material to the inner surface of the cell wall, this process would gradually add to the edge of the pit, causing it to overarch the underlying primary wall to form a pit chamber.

This hypothesis of pit evolution is consistent with the distribution of tracheid-wall characters in our cladistic study of basal tracheophytes. Early fossil tracheophytes clearly show that helical-annular thickenings and the diagonal bars between thickenings are more general features than pits in the tracheids of vascular plants. The transformation of helical-annular thickenings to pitting is also consistent with other aspects of the tracheid cell wall in basal Lycophytina and Euphyllophytina. In particular, the presence of a conspicuous perforate sheet of secondary wall material within pits in some basal Euphyllophytina and Lycophytina can be related directly to a similar wall feature of the plesiomorphic G-type tracheid. In G-type tracheids a sheet of secondary wall material between adjacent helical-annular thickenings contains simple perforations or pitlets of irregular shape and size. Similar features occur within the pits of *Psilophyton* (Euphyllophytina) and many basal Lycopsida. In lycopsids such as *Minarodendron* (Li 1990), the perforations within the pit are very similar to those in the G-type cell. In many rhizomorphic lycopsids the perforations are much larger, and this structure appears as strands of secondary wall material (Williamson striations) stretching across the pit aperture (Cichan, Taylor, and Smoot 1981; Wesley and Kuyper 1951). In such basal Euphyllophytina as *Psilophyton,* the perforate sheet is raised above the chamber floor of the pit and is attached to the edge of the pit border (Hartman and Banks 1980). We interpret such perforate sheets as retention of a plesiomorphic feature homologous to the perforate sheet of the G-type cell.

LIFE CYCLES

Recent discoveries of gametophytes in the early fossil record of land plants add an important new dimension to understanding the evolution of the life cycles in basal embryophytes. Exceptionally well preserved gametangiophores bearing archegonia and antheridia have been described from the Rhynie Chert (early Devonian: 380–408 Ma), and diplobiontic life cycles have been partially reconstructed for three plants (Table 7.8). Observations on the gametophyte generation have been extended to other Devonian localities on the basis of evidence from compression fossils and from morphological comparisons with the well-preserved Rhynie Chert fossils (detailed reviews by Kenrick [1994] and Remy, Gensel, and Hass [1993]). These discoveries provide the first well-substantiated evidence of diplobiontic life cycles in early vascular plants and stem group protracheophytes. The interest of these new paleobotanical data is heightened by the clear morphological similarities between the gametophyte and sporophyte generations—similarities suggesting that more or less isomorphic life cycles were a feature of basal polysporangiophytes.

Gametophytes in the Early Fossil Record

The most complete evidence regarding the gametophyte generation of early land plants comes from the silicified fossils of the Rhynie Chert. What is known of gametophytes from the Rhynie Chert is based on gametangiophores from three taxa (Remy, Gensel, and Hass 1993). All are small, probably dichotomously branched plants bearing terminal gametangiophores. Anatomically, the gametophyte axes are indistinguishable from those of the associated sporophytes and contain well-developed water-conducting tissues as well as an epidermis with stomates. For about a century, these similarities impeded the recognition of these structures as distinct from the sporophyte and still make their identification difficult. Gametangia are borne on the upper surface of distinctive cup-shaped (concave) or shield-shaped (convex) gametangiophores, which develop terminally on an axial system through expansion of the apex, probably via growth from a ring meristem as in extant Lycopodiaceae (Figure 7.28; Kenrick 1994). The position and morphology of the archegonia are similar to those of basal Tracheophyta and Anthocerotopsida. The archegonia are sunken, flask-shaped structures comprising a neck canal and egg chamber. The antheridia are spherical or club-shaped, superficial, and sessile or have a short, poorly defined stalk.

Observations on early land plant gametophytes have been extended to other Devonian localities based on evidence from compression fossils. Compressions of *Sciadophyton*-type have been known for more than 60 years, but their morphology has proved difficult to interpret and until recently their affinities remained obscure (Gensel and Andrews 1984; Høeg 1967; Remy, Hass, and Schultka 1992; T. N. Taylor and E. L. Taylor 1993). Cellular preservation is much rarer in this type of material, but the presence of distinctive S-type tracheids in some species of *Sciadophyton* indicates an affinity with

TABLE 7.8
Sporophytes and Corresponding Gametophytes from the Rhynie Chert

Sporophyte	Gametophyte	Gametangia
Aglaophyton major	*Lyonophyton rhyniensis*	Antheridia
Horneophyton lignieri	*Langiophyton mackiei*	Archegonia
Nothia aphylla	*Kidstonophyton discoides*	Antheridia

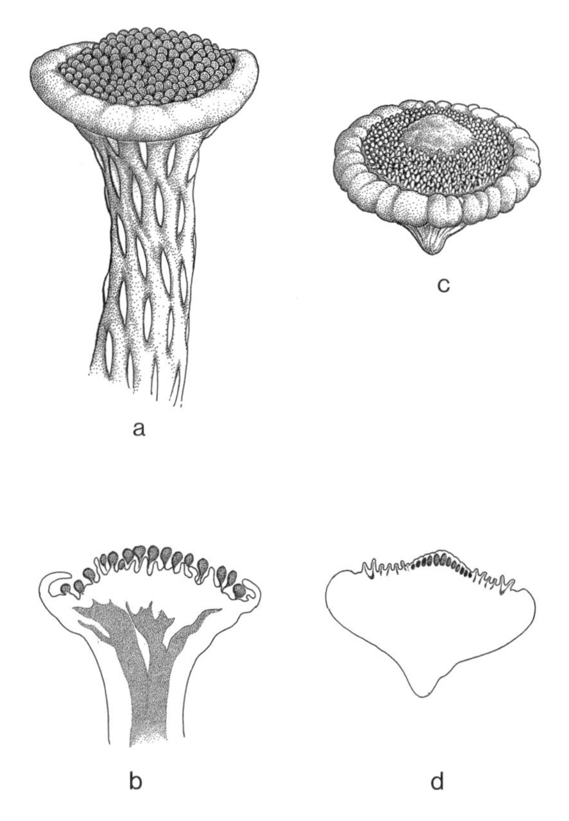

FIGURE 7.28. Comparison of gametangiophore morphology in an early fossil from the Rhynie Chert and extant Lycopodiaceae. (**a, b**) *Kidstonophyton discoides* from the Rhynie Chert, showing axis ([**a**] external; [**b**] longitudinal section) terminating in a gametangiophore with possible ring meristem (×5). Club-shaped antheridia are situated on the convex apex. Redrawn from Remy and Hass 1991c, Text-Figures 1 and 2. (**c, d**) Extant *Lycopodium clavatum*, illustrating gametangiophore with ring meristem ([**c**] external; [**d**] longitudinal section) (×5). Antheridia (central dome) surrounded archegonia. Adapted from Engler and Prantl 1902, Figure 357.

vascular plants (Kenrick and Crane 1991; Kenrick, Remy, and Crane 1991). Based on remarkable similarities between these enigmatic compressions and the well-preserved gametangiophores from the Rhynie Chert, Remy, Remy, et al. (1980) concluded that *Sciadophyton* is a vascular gametophyte. *Sciadophyton* is a simple plant with sparse dichotomous branching and terminal disk-shaped structures (gametangiophores) having irregular margins (Figure 7.29) and strongly resembling the Rhynie Chert gametangiophores in shape, size, and position on the plant. In compression, these disks bear many small, circular mounds on the upper surface that are similar in size and distribution to the unequivocal Rhynie Chert gametangia (Kenrick 1994; Remy, Gensel, and Hass 1993).

Isomorphic Life Cycles

Fossils from the Rhynie Chert provide the first direct evidence on the morphology of alternating generations in the life cycle of early land plants. At this locality, sporophytes and gametophytes were fossilized where they grew, and similarities in stem anatomy, including the morphology of internal tissue systems and epidermal features, have been used to reconstruct life cycles by linking gametophytes to associated sporophytes (Remy 1982; Remy and Hass 1991a,b,c,d; Remy and Remy 1980). Diplobiontic life cycles have been reconstructed for two stem-group vascular plants (*Aglaophyton major, Horneophyton lignieri;* Figure 3.26) and for *Nothia aphylla* (interpreted here as stem group Lycophytina; Table 7.9). The gametophytes of *Rhynia gwynne-vaughanii* and the early lycopsid *Asteroxylon mackiei* are still unknown. Studies to date have focused on the terminal, gametangia-bearing portions of the Rhynie Chert plants, so the overall habit is still poorly understood. Habit reconstructions of fossils from this locality require time-consuming serial sectioning, which is often hampered by fragmentation of the chert into small blocks. Such disintegration limits the extent to which continuity among specimens can be maintained, creating difficulties for reconstructing whole plants. Nevertheless, significant anatomical similarities with the better-known sporophytes suggest that an isomorphic alternation of generations is likely, and this conclusion is supported by direct evidence on gametophyte habit at other Devonian localities.

In general, plant habit is better understood for compression fossil gametophytes. Plants of the *Sciadophyton* type comprise 12 or more axes that diverge from a central basal point and terminate in cup-shaped gametangia (Figure 7.29). The main axes are occasionally branched with a vascular strand of simple, helically thickened elements (S-

FIGURE 7.29. Reconstruction of the Devonian gametophyte *Sciadophyton* sp. (×2). Gametangia are borne on the terminal disk-shaped gametangiophores (see Kenrick 1994, Figure 11). Based on Remy, Remy, et al. 1980, and on material in the Paleobotanical Collections at the Swedish Museum of Natural History, Stockholm.

type tracheids). Several early vascular plants from various localities have been identified as the possible sporophyte generation of *Sciadophyton*. This genus is currently poorly understood and appears to contain taxa with diverse relationships (Remy, Hass, and Schultka 1992). For some species, sporophyte associations and anatomical similarity support an affinity with the Rhyniopsida (Kenrick and Crane 1991; Kenrick, Remy, and Crane 1991). Schweitzer (1983a,b) has suggested an affinity with the extinct lycopsid sister group (zosterophylls) for other *Sciadophyton* species based on putative early developmental stages attached to gametangiophores. Similarities in branching pattern also support a zosterophyll affinity (Lycophytina) for the *Sciadophyton*-like compression fossil *Calyculiphyton* (Remy, Schultka, and Hass 1991). The general habit of *Sciadophyton* gametophytes and the presence of vascular tissue in some taxa lend further support to the hypothesis of more or less isomorphic life cycles in a range of basal polysporangiophytes.

THE ORIGIN OF LAND PLANT LIFE CYCLES The new paleobotanical data on life cycles in early fossils adds an unexpected level of complexity to the knowledge of early embryophyte life cycles. The discovery of an isomorphic alternation of generations contrasts strongly with the anisomorphic alternation typical of extant groups. This difference clearly has important implications for the evolution of land plant life cycles, as well as for the origin of the morphological differences between sporophyte and gametophyte.

The origin of alternation of generations in land plants has been widely discussed in the context of two competing theories on life cycles in ancestral embryophytes, the Antithetic and Homologous

Theories (see Chapter 1, Figure 1.2). Recent systematic studies favor the Antithetic Theory because cladistic treatments strongly support a charophycean origin of land plants, and all extant charophycean algae have haplobiontic life cycles (Figures 1.2 and 7.1, Chapter 1; Graham 1993).

At the more detailed level however, tracing life-cycle evolution among basal land plants is complex because relationships among the major basal clades (Marchantiopsida, Anthocerotopsida, Bryopsida, polysporangiophytes) are weakly supported. The hypothesis of relationship with the widest support from independent data sets indicates that bryophytes are paraphyletic to polysporangiophytes and that the Marchantiopsida basal are in embryophytes and the Bryopsida are a sister group to polysporangiophytes (Figure 7.2, Chapter 3; Mishler, Lewis, et al. 1994). The results of our cladistic study show that early fossils with more or less isomorphic life cycles are more closely related to polysporangiophytes than to any bryophyte group. The presence of isomorphic life cycles therefore does not contradict an antithetic origin for the embryophyte life cycle but rather suggests that an elaboration of life-cycle morphology occurred within embryophytes at the polysporangiophyte level (Kenrick 1994; Kenrick and Crane 1991).

GAMETOPHYTE AND SPOROPHYTE MORPHOLOGY IN VASCULAR PLANTS The new paleobotanical data on alternation of generations in early land plants is most relevant to life-cycle evolution in polysporangiophytes and tracheophytes. Gametophytes of all extant, free-sporing vascular plants are small, inconspicuous, simple, and often subterranean. However, a basic axial morphology is characteristic of many groups such as the Lycopodiaceae, Psilotaceae, Stromatopteridaceae, and Ophioglossaceae, whereas a simple thalloid morphology is typical of the Marattiales and many leptosporangiate ferns (Bierhorst 1971; Gifford and Foster 1987). The simple gametophyte morphology in extant vascular plants contrasts strongly with the large, morphologically complex gametophytes of the putative Bryopsida sister group. Similarly, the thalloid morphology of hornworts contrasts with the simple axial morphology in other vascular plants and bears only superficial resemblance to the thalloid gametophytes of leptosporangiate ferns. A range of forms that encompasses simple thalloid to complex axial gametophyte morphologies occurs within liverworts. Because of this variation within and among groups, the plesiomorphic condition in vascular plants is not easily recognized on neobotanical evidence.

Mishler and Churchill (1985b) concluded that a simple axial gametophyte morphology is probably plesiomorphic in vascular plants. This hypothesis implies that gametophytes have always been simple in vascular plants, whereas sporophytes have evolved considerable morphological complexity from simple beginnings. The new paleobotanical evidence contradicts this interpretation of gametophyte evolution. Cladistic analysis indicates that fossils with isomorphic life cycles are related to stem group tracheophytes and probably even to basal Lycophytina (e.g., *Nothia aphylla:* Table 7.8). This evidence points to a significant elaboration of *both* the gametophyte and sporophyte in early polysporangiophytes and implies that the gametophytes of extant pteridophytes are highly reduced compared to those of some of the earliest protracheophytes.

GAMETOPHYTE MORPHOLOGY IN LYCOPODIACEAE AND OPHIOGLOSSACEAE: VESTIGES OF EARLY COMPLEXITY The paleobotanical data on the gametophyte generation in early land plants shed new light on the morphology of the unique and highly distinctive gametophytes in extant Lycopodiaceae and Ophioglossaceae. During gametophyte ontogeny in the Lycopodiaceae, apical growth ceases at a very small size, and the peripheral cells on the apical flanks become meristematic, forming a ring meristem. Subsequent cell divisions in the ring meristem produce an expanded, disk-shaped apex that is typically 3–15 mm in diameter and has a smooth or convoluted margin (Figure 7.28). The archegonia and antheridia are located on the upper surface of the disk. Several distinctive gametophyte morphologies have been described in terms of variations in size and lobing of the apex (e.g., Bierhorst 1971; Bruce 1976b, 1979; Whittier 1977, 1981; Whittier and Braggins 1992; Whittier and Webster 1986), but recent work has shown that much interspecific variation is reducible to permutations of a basic body plan built around the ring meristem (Bruce 1979). Similar peripheral meristematic regions are also a feature of gametophytes in the Ophioglossaceae (Gifford and Foster 1989). From a neobotanical perspective, this unique gametophyte morphology appears to be autapomorphic in the Lycopodiaceae and Ophioglossaceae.

The disk-shaped gametangiophores in the permineralized Rhynie Chert plant *Kidstonophyton* (sporophyte = *Nothia aphylla*) and in the compression fossils of the *Sciadophyton* type strongly resemble gametangiophore morphology in extant Lycopodiaceae (Figure 7.28). Both *Kidstonophyton* and *Lycopodium clavatum* gametangiophores comprise an expanded, disk-shaped structure of similar size and shape. Both bear gametangia on the convex upper surface and have a raised marginal rim without gametangia. In *L. clavatum*, these features are known to be produced by the ring meristem. Differences between the two genera include superficial antheridia alternating with sterile tissue and the presence of a larger axial system in the fossil *Kidstonophyton*. The probable close phylogenetic relationship between these groups strengthens these comparisons and suggests that this type of gametophyte morphology may be more general in vascular plants. Further testing of potential homologies among these structures requires a more detailed understanding of the distribution of complex gametophytes within early land plant lineages. Future studies should focus on life cycles in zosterophylls (lycopsid stem group) and in early lycopsids such as *Asteroxylon mackiei* from the Rhynie Chert. Support for the hypothesis of homology would imply that ring meristems are not unique features of the Lycopodiaceae and Ophioglossaceae as they appear in a neobotanical context, but rather are one of the last vestiges of a more complex gametophyte morphology that existed early in the history of vascular plants.

CONCLUSIONS

In this book we have attempted a cladistic synthesis of the available data on the origin and early evolution of land plants, focusing on the comparative morphology of living and fossil groups. There is an enormous, and almost overwhelming, amount of relevant information from disciplines such as paleobotany, palynology, geology, and stratigraphy, as well as from neobotanical studies of comparative plant morphology, anatomy, and molecular systematics. Nevertheless, a coherent and complete hypothesis on land plant origins should be based on an evaluation of all of this varied and complementary evidence. In addition to recent empirical advances, important theoretical developments embodied in the cladistic approach provide new, and more rigorous, methods for evaluating the systematic and phylogenetic evidence. We hope that the application of these methods through the cladistic treatments in this book will provide a productive focus for future research because clearly much still remains to be done. The following section summarizes some of our main conclusions and highlights some of the major areas of uncertainty or disagreement that require further study.

Systematics

Previous cladistic studies of phylogenetic relationships in basal land plants have focused on neobotanical data. Our morphological study has integrated a significant new source of data from the fossil record, much of which we have attempted to summarize for future use. In general, the relationships recognized in this study are congruent with previous morphology-based studies, but the inclusion of fossils lends further support to some clades, clarifies relationships among important extinct groups, and contributes significantly to refuting or clarifying ideas concerning homology. In particular, our analysis provides further support for the monophyly of vascular plants and for an early divergence of lycopsids from other tracheophytes. Our results strongly support the monophyly of the Lycophytina (Lycopsida and extinct Zosterophyllopsida) and the monophyly of the widely recognized but previously unnamed Euphyllophytina clade (ferns–*Equisetum*–seed plants). But our phylogenetic results conflict principally with recent cladistic studies based on *rbcL* gene sequences (Manhart 1994) and on male gametes and gametogenesis (Garbary, Renzaglia, and Duckett 1993). The source of this conflict is not yet fully resolved but may be attributable to the divergent nature of the living groups with respect to *rbcL* sequences and gamete form and to the reinforcement of this divergence by massive extinction in the stem lineages. In other words, the major basal clades of extant vascular plants are characterized by "long branches," which increase the risk of phylogenetic signals being overprinted by homoplastic noise. Under these circumstances the inclusion of fossils is particularly informative because of the opportunities they provide for greater sampling along stem lineages.

Our preliminary analysis of relationships in the Euphyllophytina recognizes several major groups but emphasizes the need for a more comprehensive cladistic treatment at this level. Further work should include the integration of neobotanical and paleobotanical morphological data on high-level relationships in the Euphyllophytina and ferns. Although combined morphological and molecular studies are making major progress on relationships among extant ferns (e.g., Pryer, Smith, and Skog 1996), a full understanding of fern evolution will require the integration of data from the fossil record, and most particularly that on the early ferns of the Paleozoic. Ongoing morphology-based studies of fern phylogeny (Pryer, Smith, and Skog 1996) are also crucial for evaluating molecular studies, understanding the evolution of plant form, and integrating paleobotanical data. We view the phylogenetic problems in ferns and basal Euphyllophytina as analogous to the problems addressed here among early land plants and in previous cladistic treatments of seed plants. Because of the enormous amount of data on the comparative morphology of these groups, we expect that resolution of the systematic problems will again initially require substantial synthesis.

Our cladistic study of basal embryophytes is broadly consistent with embryophyte monophyly and the bryophyte paraphyly hypothesis of Mishler and Churchill (1984, 1985b). The early fossil record of land plants is heavily biased toward vascular plants, and there are no unequivocal stem group hornworts, mosses, liverworts, or embryophytes. Additional data on putative early mosses such as *Sporogonites* would be of great interest. Because of the absence of paleobotanical data at that level, the fossil record provides only indirect support for these hypotheses. The strongest evidence for embryophyte monophyly comes from neobotanical cladistic studies based on comparative morphology and molecular data. Embryophytes are one of the best-supported monophyletic groups in the Chlorobionta, and this result is not contradicted by the fossil record. Bryophyte paraphyly was relatively weakly supported but was nevertheless the most parsimonious hypothesis in our analysis. We found no evidence that bryophytes have evolved by reduction from stem group tracheophytes or basal polysporangiophytes. The discovery of well-preserved stem group bryophytes could have a profound effect on current understanding of relationships and homologies at this level of plant evolution.

Critical reevaluation of previous macrosystematic treatments of early vascular plants shows that many widely recognized groups are either paraphyletic or ambiguously defined. Our analysis focused on the innovative and influential classification of Banks (1968, 1975b). The Rhyniophytina *sensu* Banks are a strongly paraphyletic group comprising grades of taxa that fall into the vascular plant and Lycophytina stem groups. The Rhyniophytina are vascular plants, or perhaps polysporangiophytes, that are not included in the Lycophytina or Euphyllophytina. Similarly, the Trimerophytina *sensu* Banks are a paraphyletic group comprising a grade of taxa of basal Euphyllophytina. The Zosterophyllophytina *sensu* Banks are also, strictly speaking, paraphyletic but mainly comprise a large and well-supported monophyletic subgroup (Zosterophyllopsida) that is a sister group to the Lycopsida.

Morphology

The origin and early evolution of land plants was a period of great innovation in plant morphology. Although many aspects of embryophyte cell biology can be traced to aquatic green algal ancestors, other important biochemical and ultrastructural characteristics, as well as most of the supracellular features of the group, are innovations that followed colonization of the land. The early diversification of land plants involved massive morphological changes that led from simple aquatic plants at the charophycean level of organization (unicellular "sporophytes" and simple gametophytes) in the Ordovician to complex lycopsids, ferns, and seed plants by the end of the Devonian. Much of the supracellular morphology that characterizes extant land plants evolved during this interval, and these innovations provided the basic architectural framework for subsequent morphological diversification.

A clear understanding of homologies among the major organ systems of land plants is central to developing and testing theories on the evolution of form. It is also helpful in designing experiments that widen the scope of current investigations on model organisms (e.g., *Arabidopsis*, crop plants) by placing such studies in a broader evolutionary context. Critical analysis of the data of comparative plant morphology shows that much morphological diversity is reducible to transformations involving relatively few organs or organ systems. This idea was first developed, graphically represented, and popularized in Zimmermann's Telome Theory (1952, 1965) and is supported by recent studies aimed at understanding the role of regulatory genes in plant ontogeny. Reconstructing basic body plans for major extant clades and tracing homologies is fundamental to furthering knowledge on the historical and ontogenetic evolution of plant form. Because of extensive extinction in stem lineages, the fossil record provides important additional information that complements comparative studies of extant groups.

In this book we have attempted to critically examine hypotheses of homology (taxic and transformational) for selected organs and organ systems that have been the focus of previous discussions. These analyses are based on, and placed in the context of, our cladistic treatments. Many of the early innovations of embryophytes, such as spores with sporopollenin-impregnated cell walls, waxy cuticles, stomates, water-conducting tissues, archegonia, and antheridia, and general elaboration of the sporophyte generation can be viewed as responses of aquatic plants to interaction with a "new" fluid—the atmosphere. Furthermore, many of these innovations are consistently explained as unique events. The most controversial hypotheses of homology among basal embryophytes include the homologies of archegonia, antheridia, and sporangia in bryophytes and vascular plants. One view interprets these structures as convergent adaptations to the terrestrial environment because of structural and developmental differences among groups. In contrast, our work focuses on the striking similarities as a basis for hypotheses of homology that are supported by cladistic analysis. According to our interpretation the ancestor of modern embryophytes had a gametophyte bearing gametangia and a sporophyte with a sporangium. The differences in these features among modern groups are interpreted as autapomorphies that evolved after the cladogenic events that gave rise to the major clades of early embryophytes.

There are many clear examples of convergence in the early evolution of embryophytes. Cladistic analysis indicates that leaves evolved independently on at least six occasions (gametophytic leaves in the Jungermanniales and Bryopsida and sporophytic leaves in the Lycophytina, Spermatopsida, Polypodiopsida, and Equisetopsida). The independent evolution of sporophytic leaves in four groups of vascular plants is one of the clearest examples of how direct evidence from the early fossil record can be used to clarify problems of homology. All crown group vascular plants are leafy, and testing leaf homologies in extant taxa involves demonstrating substantial differences in leaf morphology, ontogeny, and position. These differences provide grounds for suggesting that microphylls in lycopsids and *Equisetum* are not homologous to megaphyllous leaves, but similarities between the megaphylls of ferns and seed plants are much greater. Inclusion of leafless extinct vascular plants provides strong evidence for the independent evolution of leaves in lycopsids and for substantial convergence in the Euphyllophytina (seed plants, ferns, *Equisetum*). Leaves in the Euphyllophytina are homologous at the level of dichotomously branched lateral appendages, but other leaflike features, such as planation, lamina formation, and circinate vernation, evolved independently in ferns and seed plants.

In addition to making predictions about the morphology of putative stem group members of these clades, these hypotheses of homology suggest possible experimental tests. It is probable that there are substantial similarities in the genetic control of ontogeny in homologous structures. Furthermore, given that much of the diversity in the Euphyllophytina appears to have evolved from the same basic structural units (e.g., Zimmermann 1952), one would expect frequent and massive development of complex gene families. Based on our cladistic study, we predict that substantial similarities will be found in the mechanisms controlling the ontogeny of such structures as gametangia and sporangia in diverse embryophytes. Similarly, the mechanisms controlling leaf initiation in ferns and seed plants are likely to show substantial similarities at the genetic level because these structures appear to be homologous at the level of lateral branches. Such similarities are unlikely to extend to other aspects of leaf development, such as lamina formation, which appears to be convergent in these groups (compare fern and angiosperm leaf development; see Doyle and Hickey 1976).

Similarities in developmental control are also expected to extend to structures that show substantial morphological differences in extant groups but that are hypothesized to be part of the same transformation series. Two of our transformational hypotheses argue in favor of sporangium sterilization as a major factor in the evolution of microphylls in lycopsids and in the evolution of integuments in seed plants. Based on these hypotheses of transformation, we would expect genetic and other similarities between the development of sporangia and leaves in lycopsids and the development of sporangia and integuments in seed plants. There are also significant analogies to structural innovations in other groups of plants, such as the iterative development of petals from stamens in angiosperm flowers and the origin of interseminal scales from ovules in the Bennettitales.

Early Land Plant Diversity

Evidence from the microfossil record and critical evaluation of macrofossil data suggest that the origin and early diversification of land plants occurred significantly earlier than is suggested from the stratigraphic record of macrofossils. Macrofossil evidence of land plants first appears in the mid-Silurian and becomes abundant in the early Devonian, but these data are almost exclusively from vascular plants and polysporangiophytes. Systematic studies indicate that the three major extant clades of bryophytes originated before vascular plants, yet there are almost no well-documented records of bryophytes in the early macrofossil record of land plants. The analysis of range extensions based on our cladistic treatment of polysporangiophytes indicates that the origin of major basal clades such as the Tracheophyta, eutracheophytes, Lycophytina, Lycopsida, Zosterophyllopsida, Rhyniopsida, and Euphyllophytina were at least late Silurian events and that other important groups, such as lignophytes, may also extend back into the early Devonian.

Regional bias in the macrofossil record is perhaps the single most important factor undermining studies aimed at understanding mode and tempo in the early radiation of vascular plants. This observation points to the need for well-documented early macrofossil data from areas outside the Euramerican region. These requirements are currently being addressed, principally through the collection of new data from south east Asia (China).

Modern Diversity

The origin of modern species diversity and the fates of the major Paleozoic clades have been traced through time by compiling macrofossil data on taxa and taxonomic ranges at species or generic level (Knoll, Niklas, et al. 1984; Niklas, Tiffney, and Knoll 1980, 1983, 1985). Species diversity curves show that many groups of low extant diversity (e.g., Lycophytina, Equisetopsida, cycads, conifers) were formerly much more diverse and more important elements of the land flora in the Paleozoic and Mesozoic (Niklas, Tiffney, and Knoll 1985). Some major groups of high extant diversity underwent major diversifications more recently. The radiation of angiosperms during the Cretaceous Period is one of the major features that has shaped the flora of the modern world (Crane, Friis, and Pedersen 1995). Angiosperms comprise more than 90% of extant vascular plants, and the radiation of this group corresponded with a major increase in total species diversity that masks a substantial numerical decline in the species of other plant groups (Crane and Lidgard 1990; Lidgard and Crane 1988; Niklas, Tiffney, and Knoll 1985). To date, however, overall measures of taxic diversity have revealed little on the origins of modern diversity in those major Paleozoic clades of "non-angiosperm" vascular plants that have persisted to the present. It seems likely, for example, that much modern species diversity at the family level and below in ferns, lycopsids, mosses, hornworts, and liverworts is a result of late Cretaceous and Cenozoic radiations in habitats dominated by angiosperms, rather than of the persistence of numerous, slowly evolving lineages from relatively high levels of Paleozoic or Mesozoic diversity. One approach to addressing this issue is through detailed systematic, paleobotanical, and biogeographic studies within major groups of extant non-angiosperm land plants. A second approach is to investigate more closely rates of sequence divergence through molecular studies within non-angiosperm groups. These and many other issues clearly provide fertile ground for further studies of land plant evolution that integrate both neobotanical and paleobotanical data.

APPENDIX 1

Summary Descriptions of Fossil† and Extant Taxa

Coleochaete

Fifteen extant species of *Coleochaete* are currently recognized (Graham 1993), and most occur as epiphytes in the littoral zone of clean freshwater lakes. The genus includes various thallus types that range from filamentous to pseudoparenchymatous to parenchymatous disks (Figure 3.3). *Coleochaete*, along with the other genus in the Coleochaetales *(Chaetosphaeridium),* is characterized by the presence of unique sheathed hairs that appear to function as an antiherbivore defense. Thallus cells in this group are also characterized by the presence of a single large, platelike chloroplast in each cell, each of which has a conspicuous pyrenoid.

Antheridium form in *Coleochaete* appears to be correlated with the morphology of the thallus (Graham 1984). In filamentous species, unicellular antheridia develop laterally on vegetative cells. In contrast, thalloid forms produce multicellular antheridia, which develop by the subdivision of a single cell to typically yield four sperm-producing cells and two underlying vegetative cells (Graham 1984, 1993). Spermatozoids are minute and colorless, and the egg cells have a prominent protuberance through which the sperm swims to effect fertilization. *Coleochaete* is unique among the charophycean algae both in retaining the zygote on the parental thallus and in producing more than four meiospores per zygote (Graham and Repavich 1989).

The late Silurian–early Devonian fossil genus *Parka* has been interpreted as morphologically and ecologically similar to *Coleochaete* (Niklas 1976), but this comparison needs to be treated with caution in view of the poor preservation of all *Parka* material that has been described to date.

Charales

The Charales (stoneworts) comprise an undoubted monophyletic group of macroscopic freshwater "green algae" with a cosmopolitan distribution. Six extant genera *(Chara, Lamprothamnium, Lychnothamnus, Nitellopsis, Nitella, Tolypella)* are usually recognized, and all members of the group are readily distinguished from other green plants by the unique structure of their sex organs and their characteristic thallus morphology (Figure 3.4). The most comprehensive systematic treatment of the group (Wood and Imahori 1965) recognizes 3 species of *Lamprothamnium,* 1 species of *Lychnothamnus,* 3 species of *Nitellopsis,* and 2 species of *Tolypella. Chara* and *Nitella* are more diverse, having 19 and 52 species respectively in Wood and Imahori's treatment, although estimates of species diversity in these genera by different authors vary considerably (Grant 1990). A cladistic analysis of the extant genera was provided by McCourt, Karol, et al. (1996), and Martín-Closas and Schudack (1991) gave an initial analysis based on fossil and recent gyrogonites.

The thallus of charophytes consists of an erect main axis up to 30 cm long, which produces regular whorls of determinate lateral branches and branched, multicellular rhizoids in the basal regions (Fritsch 1948). Frequently the thallus is covered by an encrustation of calcium carbonate. The internodes are formed entirely by a cylinder of extremely elongated cells, which may be up to 15 cm long. The thallus grows at the apex by division of an apical cell similar to that in simple land plants (Pickett-Heaps 1975). Asexual reproduction occurs by fragmentation of the thallus and also by bulbils produced on the rhizoids. Species are either monoecious or dioecious (Grant 1990; Proctor 1980).

The gametangia are produced at the nodes and attached by a single basal cell. Ontogenetic details were given by Fritsch (1948), Pickett-Heaps (1975), and Graham (1993). The antheridia are spherical, typically bright yellow or red at maturity, and may be up to 1.5 mm in diameter. The mature antheridium wall is formed from eight shield-shaped cells enclosing a space filled by branched filaments borne on an elongate basal cell. Each cell within the filament ultimately develops into an elongate, coiled, biflagellate spermatozoid. Oogonia are ellipsoidal and up to about 1 mm long and consist of a central egg cell surrounded by a jacket of five elongated cells with a marked sinistral spiral. The zygote is retained within the oogonium following fertilization, and no motile zoospore stage is produced. New gametophytes emerge from the oogonium as a filamentous protonema bearing one or two rhizoids. The mature plant develops as a lateral branch from the protonema (Fritsch 1948).

The Charales have an excellent fossil record based on the calcified remains of their oospores (Feist and Grambast-Fessard 1991). In our analysis, we use the genus *Chara* as a placeholder for the Charales.

Sphaerocarpos

The extant genus *Sphaerocarpos* contains about 12 species with a cosmopolitan distribution (Schofield 1985; Schuster 1984b). The description given below is based mainly on those of Smith (1955), Schuster (1984b), and Schofield (1985).

The sporophyte is unbranched and differentiated into a foot, a delicate seta, and a sporangium (Figure 3.18). The foot is bulbous, and placental transfer cells occur in both gametophyte and sporophyte (Ligrone and Gambardella 1988a). The poorly developed seta consists of only four vertical rows of cells and does not function in exsertion of the sporangium. The sporangium is spherical and the differentiation of spores and elaters is synchronous. The sporangium wall is one cell thick and stomates are absent. Internally, there is no columella and the sporangium is indehiscent. The spores are released by decay of the sporangium wall. The archesporium develops into spores and sterile elater cells in precisely delineated cell lineages. The number of spores equals that of the elaters because of continued mitotic cell division in the spore cell lines. The elaters are small and ovoid to spherical without helical thickenings. Spores form in tetrads that may be obligate; ornament is mostly rugulate to reticulate; no perine has been observed. Spore wall ultrastructure is typically metzgerialean (Brown and Lemmon, pers. comm. to P.K., 1992). Placental morphology was reviewed in Ligrone, Duckett, and Renzaglia (1993).

The gametophyte is a dorsiventral, prostrate thallus with almost no internal cellular differentiation (Figure 3.18). Plastid ultrastructure was reviewed by Duckett and Renzaglia (1988b). The thallus consists of a broad midrib, which is several cells thick, supporting two lateral winglike expansions that are one cell thick. This thallus is sometimes described as an axis with two rows of horizontal unistratose "leaves" (Schuster 1984b). Each gametangium is surrounded by an involucre, which begins development shortly after initiation of the sex organs. Rhizoids are unicellular and smooth-walled. The gametangia are scattered along the midrib and over a wider area on the upper surface of the thallus. Archegonia develop from a superficial cell and are surficial. The mature archegonium is flask-shaped, having an elongate neck, six vertical rows of neck cells, and a basal egg cell. The antheridia also develop from single superficial initials into a surficial, stalked, elliptical structures. Spermatogenesis and spermatid ultrastructure were described by Carothers, Brown, and Duckett (1983).

Monoclea

Monoclea is a small extant genus of two species restricted to South America, Jamaica, and New Zealand (Schofield 1985). The description given below is based on those of Schuster (1984b) and Schofield (1985).

The sporophyte is unbranched and differentiated into a foot, a seta, and a sporangium (Figure 3.16). The foot is bulbous, and placental transfer cells in the sporophyte-gametophyte junction have not been described. The seta is massive but lacks internal differentiation and generally elongates after the spores are mature and ready to be shed. The sporangium is elongate and cylindrical, and the differentiation of spores and elaters is synchronous. The sporangium wall is one cell thick, and stomates are absent. Internally, the archesporium differentiates into spores and elaters; there is no columella. Dehiscence is via a single longitudinal slit that follows one of the four cell contact lines from the quadrant stage of the embryo (Figure 3.28). The archesporium develops into spores and sterile elater cells in precisely delineated cell lineages, with the number of spores exceeding that of the elaters (32–36:1) because of continued mitotic cell division in the spore cell lines. The elaters are elongate and helically thickened (Figure 3.30). The spores are reticulate and usually form in permanent tetrads; no perine has been observed. Spore wall ultrastructure is typically metzgerialean (Brown and Lemmon, pers. comm. to P.K., 1992).

The gametophyte is a dorsiventral, prostrate, dichotomously branching thallus that lacks a distinct midrib and has almost no internal cellular differentiation (Figure 3.16). Each antheridium is embedded in a receptacle that opens onto the thallus surface, and each archegonium is surrounded by an involucre, which begins development shortly after initiation of the sex organs. Unicellular rhizoids occur on the ventral surface. The gametangia are scattered along the midline on the upper surface of the thallus. The archegonium is flask-shaped, having an elongate neck of six vertical rows of neck cells and a basal egg cell. The antheridia also develop from single superficial initials into surficial (within the receptacle), stalked elliptical structures.

Naiadita†

Naiadita is an extinct genus consisting of one well-known species, *N. lanceolata,* which is characterized mainly on the basis of compressions and permineralizations from the Rhaetic, West Midlands, England (Harris 1939). The age of the locality is given as late Triassic.

The sporophyte is unbranched and differentiated into a foot, a very short seta (possibly absent), and a sporangium. The foot is probably bulbous, and placental transfer cells have not been observed. The sporangium is spherical and the differentiation of spores is synchronous. The sporangium wall is one cell thick, and stomates were not observed. Internally, the archesporium differentiates into spores only. Because of exceptional preservation and the detail of Harris' observations, it is probable that elaters and columella were not present. The sporangium does not appear to be dehiscent *(cleistocarpic).* The spores measure about 40 μm and are alete, with small spines or tubercles, and a marked equatorial ridge. Spore ultrastructure was described by Hemsley (1989).

The gametophyte is erect, slender, and leafy. Sporophytes are enveloped initially in a delicate calyptra, which develops from the archegonium and surrounding tissue. A peculiar pseudopodium analogue is present in the gametophyte (Harris 1939). The internal anatomy of the gametophyte has not been described. Unicellular (nonseptate) rhizoids are present. The leaves are unistratose, non-costate, and helically arranged. The archegonium is sessile, flask-shaped, with an elongate neck and a basal egg cell. The archegonia are scattered along the main axis but become terminal on small, pseudopodium-like, side branches that develop toward the base. Antheridia are unknown.

Haplomitrium

The extant genus *Haplomitrium* contains approximately 12 species with a wide distribution in both hemispheres. The greatest diversity occurs in India and Southeast Asia; *H. hookeri* is widely distributed in the northern hemisphere both at sea level and in alpine areas (Schofield 1985; Schuster 1966). The description given below is based mainly on Schuster's (1966, 1984a,b).

The sporophyte is unbranched and differentiated into a foot, a massive seta, and a sporangium (Figure 3.17). The foot is bulbous, and placental morphology was reviewed by Ligrone, Duckett, and Renzaglia (1993). The seta is massive but has little internal differentiation and generally elongates after the spores are mature and ready to be shed. The sporangium is elongate and cylindrical, and the differentiation of spore and elaters is synchronous. The sporangium wall is one cell thick and stomates are absent. Internally, the archesporium differentiates into spores and elaters; there is no columella. Dehiscence is via one or two longitudinal slits that follow two of the four cell contact lines established during the quadrant stage of the embryo (Figure 3.28), and the valves of the sporangium frequently remain attached at the apex. The archesporium develops into spores and sterile elater cells in precisely delineated cell lineages. Spores equal or often substantially exceed elaters in number because of continued mitotic cell division in the spore cell lines. The elaters are elongate, usually bispirally thickened (Figure 3.30), and form a more or less distinct elaterophore at the base of the capsule. Spores form monads, diads, and tetrads and have pilate to retipilate ornament; no perine has been observed. Spore wall ultrastructure and various ultrastructural features of cell division have been described by Brown and Lemmon (1986, 1990b).

The gametophyte is erect, fleshy, and leafy and arises from a creeping rhizomatous basal region (Figure 3.17). Plastid ultrastructure was reviewed by Duckett and Renzaglia (1988b). The apical cell is tetrahedral (Schuster 1966). Sporophytes are enveloped initially in a calyptra, which develops from the archegonium and surrounding tissue. Internal differentiation of the axis results in three zones: an epidermis, a parenchymatous cortex, and a central strand of water-conducting cells. The water-conducting cells loose their protoplast late in development and have microperforate end walls (Hébant 1977). Rhizoids are absent, but slime papillae are conspicuous. The leaves are pluristratose, somewhat variable in shape, three-ranked, and non-costate. Gametangia are clustered on a dilated shoot apex or may be more scattered, with the antheridia axillary to leaves. Archegonia develop from a superficial cell and are surficial. The archegonium is flask-shaped with an elongate neck of six to ten vertical rows of neck cells and a basal egg cell. The antheridia also develop from single superficial initials into surficial, stalked elliptical structures. Spermatogenesis and spermatid ultrastructure were reviewed by Carothers and Duckett (1980), Renzaglia and Duckett (1991), and Garbary, Renzaglia, and Duckett (1993).

Anthoceros

The extant genus *Anthoceros* contains fewer than 100 species of cosmopolitan, terrestrial, and rupestral plants (Schuster 1984c, 1992). The description given below is based mainly on those of Schuster (1966, 1984c, 1992), Parihar (1977), Renzaglia (1978), and Crandall-Stotler (1981, 1984).

The sporophyte is unbranched and differentiated into a foot, a very short seta (corresponding to intercalary meristem), and a sporangium (Figure 3.19). The foot is massive and bulbous (Renzaglia 1978), with morphologically differentiated placental transfer cells present only on the gametophyte side (Ligrone and Gambardella 1988b). The seta is short and undifferentiated and consists of a persistent meristematic region that produces files of cells toward the apex, which then differentiate into the tissues of the sporangium. The sporangium is elongate and characterized by a developmental gradient from apex (most mature) to base (least mature) resulting from the continuous development of all tissues from the basal meristem. The sporangium wall is from four to six cells thick and has a distinctive epidermis with stomates consisting of two guard cells. The archesporium surrounds and overarches an elongate (square in transverse section) sterile columella that consists of 16 vertical rows of elongate cells. Dehiscence occurs along one, two, three, or four longitudinal slits (depending on the species) that follow the cell contact lines from the quadrant stage of the embryo (Figure 3.28). The valves of the dehisced sporangium usually remain attached at the apex. The archesporium develops into spores and sterile elater cells in precisely delineated cell lineages (Figure 3.30). The number of elater cells equals or exceeds the number of spores, and elaters may be multicellular and branched because of continued mitotic cell division in the elater cell lines. Spores are trilete and bear prominent furcate spines or papillae. Perine has not been observed on the spore wall. Spore wall ultrastructure and various ultrastructural features of cell division have been described by Brown and Lemmon (1990b). Placenta morphology was reviewed by Ligrone, Duckett, and Renzaglia (1993).

The gametophyte is a small, dark-green, dorsiventral, prostrate, lobed thallus with almost no internal cellular differentiation (Figure 3.19). Plastid ultrastructure was reviewed by Duckett and Renzaglia (1988b). The apical cell is cuneate (Renzaglia 1978). Sporophytes are initially enveloped in a calyptra, which develops from the archegonium and surrounding tissue. Mucilage-containing cavities develop within the thallus and may open to the exterior through a small pore that sometimes resembles a stomate (Renzaglia 1978; Schuster 1984b). Unicellular, smooth-walled rhizoids occur on the lower surface of the thallus. Archegonia develop from a superficial cell of the thallus but become embedded during development. They are initiated close to the apex and have a scattered distribution along the midline of the thallus. The archegonium is flask-shaped with six vertical rows of neck cells and a basal egg cell. The antheridia also develop from single superficial initials, but when mature, they are borne in an antheridium chamber within the thallus. The chamber usually contains more than one stalked antheridium. Spermatogenesis and spermatid ultrastructure were reviewed by Renzaglia and Duckett (1989, 1991).

Notothylas

The extant genus *Notothylas* contains from 8 to 16 species of cosmopolitan but mainly temperate and tropical plants (Schuster 1984c, 1992). The description given below is based mainly on those of Schuster (1966, 1984c), Parihar (1977), Renzaglia (1978), and Crandall-Stotler (1981, 1984).

The sporophyte is unbranched and differentiated into a foot, a very short seta (corresponding to intercalary meristem), and a sporangium. The foot is rounded and smaller than that of *Anthoceros* (Renzaglia 1978). The exact location of placental transfer cells in the sporophyte-gametophyte junction has not been described. The seta is short and not differentiated and consists of a meristematic region, which produces files of cells toward the apex that differentiate into the tissues of the sporangium. Unlike in other hornworts, there is little or no meristematic activity in this region in the mature sporophyte. The sporangium is elongate and characterized by a developmental gradient from apex (most mature) to base (least mature) resulting from the activity of the basal meristem. In *Notothylas* the developmental gradient is least marked of all the hornwort genera because intercalary meristematic activity is not persistent. The sporangium wall is usually four cells thick with a distinctive epidermis. Stomates are vestigial or absent (Schuster 1984c). The columella is absent or very poorly developed; when it is present, it is surrounded and overarched by the archesporium. The sporangia of *Notothylas* are often indehiscent but, depending on the species, may dehisce via one to two longitudinal slits that follow the four cell contact lines established during the quadrant stage of the embryo (Figure 3.28). The archesporium develops into spores and sterile elater cells in precisely delineated cell lineages. The number of elater cells equals or exceeds that of spores because of continued mitotic cell division in the elater cell lines. Elaters are spherical and similar in size to the spores. Spores are trilete and minutely granulate or smooth-walled. A perine has not been observed. Spore wall ultrastructure and various ultrastructural features of cell division have been described by Brown and Lemmon (1990b). Placenta morphology was reviewed by Ligrone, Duckett, and Renzaglia (1993).

The gametophyte is a small, light-green, dorsiventral, prostrate, lobed thallus with almost no internal cellular differentiation. Plastid ultrastructure has been reviewed by Duckett and Renzaglia (1988b). The apical cell is cuneate (Renzaglia 1978). Sporophytes are initially enveloped in a characteristic horizontal calyptra, which develops from the archegonium and surrounding tissue. Cavities containing mucilage develop within the thallus of some species and may open to the exterior through a small pore that sometimes resembles a stomate (Renzaglia 1978; Schuster 1984c). Unicellular, smooth-walled rhizoids occur on the lower surface of the thallus. Archegonia develop from a superficial cell of the thallus but become embedded during development. They are initiated close to the apex and have a scattered distribution along the midline of the thallus. The archegonium is flask-shaped with six vertical rows of neck cells and a basal egg cell. The antheridia also develop from a superficial initial, which develops into an antheridium chamber within the thallus; the chamber usually contains more than one stalked antheridium. Spermatogenesis and spermatid ultrastructure were reviewed by Renzaglia and Duckett (1989, 1991) and Garbary, Renzaglia, and Duckett (1993).

Sphagnum

The extant genus *Sphagnum* contains approximately 150 species of cosmopolitan mosses that reach their greatest abundance in the cooler temperate regions of the northern hemisphere, where they may dominate wetland vegetation (Schofield 1985).

The sporophyte is unbranched and differentiated into a foot, a very short seta, and a sporangium (Figure 3.21). The foot is large and bulbous, and morphologically differentiated transfer cells are absent in both the sporophyte and gametophyte (Ligrone and Renzaglia 1989). The seta is very reduced and undifferentiated. The sporangium is spherical and develops synchronously. The sporangium wall is usually from four to six cells thick and has a distinctive epidermis. Stomates consisting of two guard cells may be present, but their occurrence varies among species and they are probably nonfunctional (Paton and Pearce 1957). The archesporium surrounds and overarches a dome-shaped columella. Dehiscence is via a circular slit which separates a disk-shaped operculum from the capsule; there is no peristome (Figure 3.28). The archesporium develops into spores; sterile elater cells are absent. Spores are trilete and smooth to granular or papillose, and a perine has been observed. Spore wall ultrastructure and various ultrastructural features of cell division have been described by Brown and Lemmon (1990b). Placental morphology is reviewed by Ligrone, Duckett, and Renzaglia (1993).

The gametophyte is erect and leafy and has a characteristic pattern of branching (Figure 3.21). Plastid ultrastructure was reviewed by Duckett and Renzaglia (1988b). Sporophytes are enveloped in a delicate calyptra, which develops from the lower part of the archegonium. A pseudopodium is present and formed by post-fertilization intercalary growth of the upper part of the axis of an archegonial branch. Three tissue zones can be distinguished within the gametophyte axis: a cortex, which contains distinctive hyaline cells; an inner cylinder of small, narrow, thick-walled cells; and a central cylinder of thin-walled, elongate parenchymatous cells (Hébant 1977). Multicellular rhizoids are present, but only in protonemal stages. Leaves are unistratose and non-costate, contain a characteristic pattern of hyaline and chlorophyllous cells, and are initially helical but become three-ranked later in development. The gametangia are clustered at or near the shoot apex, with the antheridia axillary to leaves. Archegonia develop from a superficial cell and are surficial, apical, and surrounded by a group of enlarged leaves, the *perichaetium*. The archegonium is flask-shaped, having an elongate neck, six vertical rows of neck cells, and a basal egg cell, and is borne on a long stalk. The antheridia also develop from single superficial initials into surficial, stalked elliptical structures. They originate apically, but continued apical growth of the antheridium branch results in a distal fertile zone. Spermatogenesis and spermatid ultrastructure were described by Garbary, Renzaglia, and Duckett (1993).

Protosphagnum†

Protosphagnum is an extinct genus based on incomplete fragments of leafy axes from the upper Permian of Angara, USSR. This description is based on that of Jovet-Ast (1967). The sporophyte is unknown and the gametophyte is leafy. Leaves are costate (unlike in *Sphagnum*) and contain a pattern of hyaline and chlorophyllous cells, which is characteristic of *Sphagnum;* however, details of the leaf cell arrangement differ from the extant genus. Other features of *Protosphagnum* gametophytes are unknown. The sporophyte is unknown.

Andreaea

The extant genus *Andreaea* contains about 50 species of cosmopolitan, primarily rock-dwelling mosses that are especially abundant in the Arctic, Antarctic, and colder temperate regions. The detailed review provided by Murray (1988) forms the basis for most of the description given below.

The sporophyte is unbranched and differentiated into a foot, a very short seta, and a sporangium (Figure 3.20). The foot is tapering, and morphologically differentiated transfer cells are present in the sporophyte only (Ligrone and Gambardella 1988a). The seta is poorly developed and undifferentiated. The sporangium is fusiform and spore development is synchronous. The sporangium wall is usually from five to ten cells thick, with a distinctive, epidermis from which stomates are absent. The archesporium surrounds and overarches a massive rodlike columella, which is stellate in transverse section. Dehiscence is via four, rarely six to ten, longitudinal slits with the valves remaining attached at the apex; there is no peristome (Figure 3.28). The archesporium develops into spores; sterile elater cells are absent. Spores are trilete and papillose to reticulate-papillose. A perine has been observed. Spore wall ultrastructure and various ultrastructural features of cell division have been described by Brown and Lemmon (1990b). Placenta morphology was reviewed by Ligrone, Duckett, and Renzaglia (1993).

The gametophyte is generally erect, leafy, and branched (Figure 3.20). The apical cell is tetrahedral, having three cutting faces (Parihar 1977). The sporophytes are enveloped in a delicate calyptra, part of which, after rupturing, may be borne on the apex of the sporophyte. A pseudopodium is present and formed by postfertilization intercalary growth of the upper part of the axis bearing the archegonia. Internal axial differentiation is not marked: the cortex consists of small uniform cells, inner cells are thin-walled and collenchymatous, and there is probably an inner cylinder of conducting parenchyma. In older axes the entire inner cylinder is formed of stereids (Hébant 1977; Murray 1988). Multicellular rhizoids are present on the stems of mature gametophytes. Leaves are unistratose and non-costate in some species, but multistratose and costate in others. Leaves are helically arranged. The gametangia are clustered at or near the shoot apex. The archegonia develop from a superficial cell and are surficial. They are borne apically or lateral near the shoot apex and are surrounded by a cluster of enlarged leaves, the perichaetium. The archegonium is flask-shaped, having an elongate neck, six vertical rows of neck cells, and a basal egg cell, and is situated on a long stalk. The antheridia also develop from a superficial initial and are surficial, stalked, and elliptical. They originate apically and are apical or lateral near the apex. Data on spermatogenesis and spermatid ultrastructure were given by Garbary, Renzaglia, and Duckett (1993).

Andreaeobryum

The extant genus *Andreaeobryum* is monospecific and only relatively recently described. Plants are primarily rock-dwelling and known specimens are from the Arctic. The description given below is based mainly on the detailed review provided by Murray (1988).

The sporophyte is unbranched and differentiated into a foot, a seta, and a sporangium. The foot is long and tapering, and morphologically differentiated transfer cells are present in the sporophyte only. The seta is short, stout, and differentiated into a cortex of three to five rows of thick-walled cells and an inner zone of seven to eleven rows of thin-walled collenchyma. The sporangium is fusiform (turbinate) and spore development is synchronous. The sporangium wall is usually five to ten cells thick and has a distinctive epidermis from which stomates are absent. The archesporium surrounds and overarches a massive rodlike columella, which is stellate in transverse section. Dehiscence is via four to six main, and one to two subsidiary, longitudinal slits with the valves remaining attached at the apex; there is no peristome (Figure 3.28). The archesporium develops into spores; sterile elater cells are absent. A perine is present. Spores are trilete and papillose to reticulate-papillose. Spore wall ultrastructure and various ultrastructural features of cell division have been described by Brown and Lemmon (1990b). Placental morphology was reviewed by Ligrone, Duckett, and Renzaglia (1993).

The gametophyte is generally erect, leafy, and branched. Sporophytes are enveloped in a large, multistratose calyptra, part of which, after rupturing, may be borne on the apex of the sporophyte. There is no pseudopodium. Internal differentiation of the gametophyte axis is not marked: the cortex consists of small uniform cells, the inner cells are thin-walled and collenchymatous, and there is probably an inner cylinder of conducting parenchyma. In older axes the entire inner cylinder is formed of stereids (Hébant 1977; Murray 1988). Multicellular rhizoids are present on stems of mature gametophytes. Leaves are bistratose, costate, and helically arranged. The gametangia are clustered at or near the shoot apex. Archegonia develop from a superficial cell and are surficial at or near the apex. A perichaetium forms after fertilization. The archegonium is flask-shaped, with an elongate neck, six vertical rows of neck cells, and a basal egg cell, and is borne on a long stalk. The antheridia also develop from a superficial initial and are surficial, stalked, and elliptical. Antheridia originate apically and are apical or lateral near the apex.

Takakia

The extant genus *Takakia* contains two species—*T. lepidozioides* and *T. ceratophylla*—which are distributed throughout the Northern Hemisphere, including the Himalayas (Nepal, Sikkim, China), Japan, Borneo, the Aleutian Islands, southeastern Alaska, and British Columbia (Murray 1988; Schuster 1983, 1984b; Smith and Davison 1993). Until recently, the systematic position of *Takakia* within bryophytes was regarded as problematic because key evidence on the morphology of the antheridia and the sporophyte generation was not available (e.g., Crandall-Stotler 1986; Hattori and Mizutani 1958; Murray 1988; Schuster 1966, 1984b). Based on analyses of gametophyte morphology, *Takakia* was classified with liverworts or mosses or in a division of its own. Several discoveries have recently provided strong evidence for a close affinity of *Takakia* with the basal valvate mosses *Andreaea* and *Andreaeobryum* by several findings: the description of antheridia in *T. ceratophylla* (Davison, Smith, and McFarland 1989; Smith and Davison 1993), the documentation of male gamete ultrastructure (Garbary, Renzaglia, and Duckett 1993), the discovery of the sporophyte generation (Smith and Davison 1993), and ultrastructural analysis of the sporophyte-gametophyte junction (Ligrone, Duckett, and Renzaglia 1993). The description given below is based mainly on the above sources.

Sporophytes are unbranched, erect, and terminal on the gametophyte and are differentiated into a foot, a seta, and a sporangium (Figure 3.23). The foot is long and tapering and penetrates the conducting strand of the gametophyte. Morphologically differentiated transfer cells are present in the sporophyte only. The seta is elongate, becomes slightly twisted with age, and contains a broad zone of hydroids (leptoids absent). Sporangia are fusiform, having helically aligned epidermal cells. A columella is present and spore development is synchronous. Dehiscence is via one helical longitudinal slit that begins near the middle of the sporangium and extends to the base and apex (Figure 3.28). Suture cells are absent, as are stomates. Elaters are absent. Spores are trilete, 29–32 μm in diameter, and finely ornamented.

Gametophytes are erect, leafy, and branched and arise from a creeping rhizomatous axis (Figure 3.23). There is no distinctive apical cell, but instead there is a dome-shaped apical meristem (Crandall-Stotler and Bozzola 1988). Perichaetial structures are absent. The calyptra is mitriform, erose, and effaced at the base. A pseudopodium is absent. Internal axial differentiation comprises an outer cortex of one to two layers of thick-walled cells, one to two layers of thinner-walled parenchyma, and an inner cylinder of water-conducting cells. Water-conducting cells loose their protoplast at maturity and have microperforate end walls (Hébant 1977). Multicellular rhizoids are absent, but mucilage pads on lower shoots may be rhizoid homologues (Murray 1988). Leaves are cylindrical, pluristratose, non-costate, and irregularly arranged, although perhaps in three ranks (Hattori and Mizutani 1958). Gametangia are clustered at or near the shoot apex. Archegonia are superficial on a long stalk and flask-shaped. They have an elongate neck of six vertical rows of neck cells, and contain a basal egg cell. Antheridia are superficial on a short multiseriate stalk, are elliptical to clavate, and have a jacket that is one cell thick and apically dehiscent.

Polytrichum

The genus *Polytrichum* contains about 100 extant species of cosmopolitan but primarily cool temperate or tropical plants. The description is based mainly on Schofield 1985.

The sporophyte is unbranched and differentiated into foot, seta, and sporangium (Figure 3.22). The foot is long and tapering, and morphologically differentiated transfer cells are present in the sporophyte only (Hébant 1977; Ligrone and Gambardella 1988a). The seta is long and stout and differentiated into an outer cortex of ten rows of thick-walled cells, an inner cortex of thinner-walled cells, and a conducting strand, which consists of a cylinder of leptoids surrounding an inner strand of thin-walled hydroids (Figure 4.20; Hébant 1977; Tansley and Chick 1901). The hydroids loose their protoplast when they reach functional maturity (Hébant 1977). The sporangium is fusiform and angular in transverse section, and spore development is synchronous. Dehiscence is operculate, and multicellular peristome teeth are usually present (Figure 3.28). The sporangium wall is several cells thick and has a distinctive epidermis. Stomates are present in some species, absent from others, and usually restricted to a groove at the base of the sporangium (Paton and Pearce 1957). The archesporium surrounds but does not overarch a massive rod-like columella, which is square in transverse section. Dehiscence is via a circular slit that separates a disk-shaped operculum from the capsule, and a peristome is present. The archesporium develops into spores; sterile elater cells are absent. A perine is present (Brown and Lemmon 1990b). Spores are small (5–8 μm), trilete, and papillose to reticulate-papillose. Placenta morphology was reviewed by Ligrone, Duckett, and Renzaglia (1993).

The gametophyte is erect, leafy, and branched and arises from an underground rhizome that is morphologically distinct from the aerial axes (Figure 3.22). Plastid morphology was reviewed by Duckett and Renzaglia (1988b). Sporophytes are enveloped in a distinctive hairy calyptra, part of which, after rupturing, may be borne on the apex of the sporophyte. There is no pseudopodium. Internal differentiation of the gametophyte axes is highly developed. In aerial axes, an epidermis with a thin cuticle surrounds an outer cortex of thick-walled cells, an inner cortex of thinner-walled cells, a "rudimentary pericycle," and a central strand of leptoids and hydroids (Hébant 1977). Leaf traces are conspicuous in the stem. Multicellular rhizoids are present in the mature gametophyte. Leaves are helical, multistratose, and costate. Gametangia are clustered at the shoot apex. Archegonia develop from a superficial cell, are surficial and apical, and are surrounded by a perichaetium. The archegonium is flask-shaped with an elongate neck, six vertical rows of neck cells, and a basal egg cell and is situated on a long stalk. Antheridia also develop from a superficial initial into a surficial, stalked, elliptical antheridium. They originate close to the apex. Details of spermatogenesis were reviewed by Garbary, Renzaglia, and Duckett (1993).

Sporogonites exuberans†

Sporogonites exuberans is the type species and is based on relatively well preserved compressions and permineralizations from Røragen, Norway (Halle 1916a,b). The age of the localities was given as early Devonian (Pragian [?] to Emsian) by Edwards and Davies (1990).

Halle (1916a) originally described the genus and later provided more extensive illustrations (1916b). He also illustrated the anatomy of a second specimen from the type locality (1936a). Specimens assigned to *S. exuberans* also have been described from the lower Devonian of Belgium (Andrews 1958; Lang 1937b; Stockmans 1940) and from the lower Devonian (Pragian) of Wales (Croft and Lang 1942). Other species are based on poorly preserved material and were reviewed by Høeg (1967). Poorly preserved and fragmentary specimens assigned to the genus have been recorded from Australia, Argentina, the Falkland Islands, Turkestan, Siberia, and the United States (Maine) (Høeg 1967). We restrict our analysis to the relatively well preserved specimens of *S. exuberans,* which we treat as a monophyletic unit based on the presence of six or twelve longitudinal furrows, interpreted as the dehiscence zone, on the sporangium wall.

Main axes are naked and unbranched (Figure 4.18; Halle 1936a; Lang 1937b). Cellular preservation in pyrite indicates that thickened water-conducting cells are probably absent. Axis epidermal cells are elongate (Halle 1936a) and stomates are present (Croft and Lang 1942). The sporangium epidermis consists of isodiametric cells (Halle 1936a), and the sporangia appear to be more or less fusiform, but the basal half is composed of sterile tissue (Halle 1936a). The spore-bearing part is terminal and probably dehisces through six or twelve longitudinal slits. The sporangium wall (several cells thick) surrounds a continuous spore mass and, possibly, a central columella (Halle 1936a). Microspores in situ are more or less circular, papillate or foveolate, trilete (?), and 18–25 μm in diameter (Allen 1980).

Andrews (1958) suggested a basically thalloid gametophyte morphology for *Sporogonites* based on compression fossils from Estinnes-au-Mont, Belgium. Some specimens in this material show a strong parallel alignment of sporophyte axes that appear to be attached to a discontinuous carbonaceous film. Because the material is poorly preserved, we regard the thalloid nature of the gametophyte as poorly substantiated.

Horneophyton lignieri†

The monospecific genus *Horneophyton* is based on exceptionally well preserved permineralizations from the lower Old Red Sandstone Rhynie Chert outlier in Scotland. The age of the locality is early Devonian (Pragian) based on dispersed spores (Richardson 1967; Westoll 1977) and radiometric dates (Rice et al. 1994).

This plant was described originally by Kidston and Lang (1920a) under the generic name *Hornea,* which was later found to be a homonym of *Hornea* Baker, a flowering plant, and the name *Horneophyton* was proposed by Barghoorn and Darrah (1938). A reconstruction of *H. lignieri* was given by Kidston and Lang (1921). Subsequent work has concentrated mainly on reconstructing the morphology and anatomy of the sporangia (Bhutta 1973b; Eggert 1974; El-Saadawy and Lacey 1979b). A sporangium possibly assignable to *H. lignieri* was illustrated also by Lemoigne (1966).

The nature of the gametophyte generation of *H. lignieri* has been discussed recently by Remy and Hass (1991b,d), who concluded that the gametangiophore described under the name *Langiophyton mackiei* is part of the gametophyte generation of *H. lignieri.* Evidence supporting this conclusion includes the close association of the sporophyte and gametophyte at the same locality, the structural similarities of the water-conducting cells, and similarities in the epidermal cell patterns and stomatal shapes.

The axes branch isotomously, are nonplanar, and may bear small multicellular enations similar to those on the sporangia (Figure 3.26; Eggert 1974; also see below). Circinate vernation has not been observed. The rhizome is a lobed parenchymatous structure of considerable size and bears numerous nonseptate rhizoids. Xylem is a more or less terete, solid protostele of water-conducting cells. Well-developed helical to annular thickenings are absent. The water-conducting cells in the center are consistently smaller than those at the periphery of the xylem. Surrounding the xylem is an inner zone of thin-walled elements and an outer cortical zone showing a centripetal decrease in cell size. The epidermis is composed of elongate cells (Remy and Hass 1991d) and is covered with a well-developed cuticle. Stomates occur at the tip of small multicellular protuberances both on the axes and on the sporangia (El-Saadawy and Lacey 1979b; Remy and Hass 1991c). Sporangia are terminal on the main axes, cylindrical (barely distinguishable from main axes), and columellate (Figure 3.29). They branch in a nonplanar fashion two or three times, and bear multicellular protuberances (Bhutta 1973b; Eggert 1974; El-Saadawy and Lacey 1979b). There is a continuous spore cavity between sporangium branches. The sporangia probably dehisce through an apical slit (Eggert 1974; El-Saadawy and Lacey 1979b). In situ microspores are 39–49 μm (Eggert 1974) or 42– 71 μm (Bhutta 1973a), trilete, rounded, and apiculate (cf. *Retusotriletes* and *Apiculiretusispora;* Figure 3.29; Allen 1980; Gensel 1980).

A gametangiophore, which probably belongs to the *Horneophyton* sporophyte, was described under the name of *Langiophyton mackiei* (Figure 3.26; Remy and Hass 1991d). The gametophyte is probably surficial and autotrophic. Axes terminate in peltate structures (c. 10 mm in diameter) that bear up to 30 short, archegonia-bearing axes on the upper surface. The archegonia are characterized by a sunken egg chamber and a massive, protruding neck of about five tiers of cells. The number of vertical rows of neck cells varies between 2 or 3 that incompletely surround the neck canal at the apex, 8 rows in the midregion, and 12 rows at the base of the archegonium. Anatomically, axes are similar to those of *H. lignieri* in their conducting cells, epidermis, and stomatal structure. Vascular tissue breaks up into concentric rings of discrete bundles in the peltate apex, and each vascular bundle enters one of the short, archegonia-bearing axes.

Langiophyton mackiei†

Langiophyton mackiei is a gametophyte from the Rhynie Chert (see *Horneophyton lignieri* for details of morphology).

Caia langii†

The monospecific genus *Caia* is based on 24 compression fossils from the Rushall Beds of the Downton Group, Herefordshire, England (Fanning, Edwards, and Richardson 1990). The age of the type locality, Perton Lane, is given as late Silurian (Pridoli) (Fanning, Edwards, and Richardson 1990).

The main axes are smooth and branch dichotomously (isotomously) at least twice and in more than one plane. No internal anatomy or cuticular features were observed. The sporangia are upright, elongate cylindrical, and terminal, sometimes dichotomize, and are barely distinguishable from subtending axis (Figure 4.19). They bear fewer than ten conical emergences. No dehiscence feature was observed, nor were stomates. In situ microspores are 23–32 μm (mean = 26 μm) in diameter, trilete, and curvaturate with circular to irregular equatorial outlines *(ambs).* Spores are laevigate and thin-walled (exine less than 1 μm thick) (cf. *Retusotriletes*). The gametophyte is unknown.

Tortilicaulis offaeus†

The type species is *Tortilicaulis transwalliensis* (D. Edwards 1979), but we chose to include *T. offaeus* in our analysis because the fossil material is more informative. This species is based on fragmentary but well-preserved coalified fossils from the lower Old Red Sandstone, Ditton Series, of North Brown Clee Hill, Shropshire, England (Edwards, Fanning, and Richardson 1994). The age of the locality is given as early Devonian (Lochkovian).

The axes are naked and isotomously branched. Epidermal cells are elongate and thickened internally. Stomate-like pores have been observed. Hypodermal cells are thick-walled, but a multilayered sterome is absent. Other internal tissues have not been described and tracheids have not been observed. Sporangia are fusiform to ovoid, simple or bifurcating, and borne terminally on the axes. Sporangium epidermal cells are elongate and helically oriented. The sporangia dehisce by splitting into two, possibly more, valves (Figure 3.28). There is no substantial modification to the cell wall at the dehiscence line. A columella is unknown. Elaters are absent. Spores (in situ) are trilete, 15–19 μm (mean = 17 μm) in diameter, possibly azonate with rounded to subtriangular ambs and rays extending to the equator. Ornament is of fine, dense grana. The gametophyte is unknown.

Aglaophyton major†

Aglaophyton is a monospecific genus based on exceptionally well preserved permineralizations from the lower Old Red Sandstone Rhynie Chert outlier in Scotland. The age of the locality is early Devonian (Pragian), based on dispersed spores (Richardson 1967; Westoll 1977) and radiometric dates (Rice et al. 1994).

Specimens attributable to this species were confused originally with *Rhynia gwynne-vaughanii* (Kidston and Lang 1917) but were later described in detail as a second species, *Rhynia major* (Kidston and Lang 1920a,b). Subsequent studies have concentrated on detailed documentation of various anatomical features: spores (Bhutta 1973c), cuticle (Edwards, Edwards, and Rayner 1982), water-conducting tissue (D. S. Edwards 1986; Lemoigne and Zdebska 1980; Remy and Hass 1991a), and sporangia (D. S. Edwards 1986; Remy 1978). Phloem was reported by Satterthwait and Schopf (1972), but Lemoigne and Zdebska (1980) have pointed out that the specimens on which this study were based were more probably *R. gwynne-vaughanii*. In more comprehensive studies of the morphology and anatomy of *R. gwynne-vaughanii* and *R. major,* D. S. Edwards (1980, 1986) clarified the morphological and anatomical differences between the two species. One important difference is that the water-conducting cells of *R. major* are unlike those of *R. gwynne-vaughanii* because they lack the internal thickenings characteristic of typical tracheids. Because of this and other differences, D. S. Edwards (1986) created the new genus *Aglaophyton* to accommodate *R. major.*

The nature of the gametophyte generation of *Aglaophyton major* has been widely discussed (D. S. Edwards 1980). Some authors have suggested that *R. gwynne-vaughanii* was wholly, or in part, the gametophyte of *A. major* (Lemoigne 1968a,b, 1969a,b, 1970; Merker 1958, 1959; Pant 1962), and this idea had some validity because the original evidence for the sporophytic status of *R. gwynne-vaughanii* was based mainly on the association of sporangia with axes (Kidston and Lang 1917, 1920a, 1921). Further support for the gametophyte interpretation was provided by the identification of several putative gametangia in the "hemispherical projections" of *R. gwynne-vaughanii* (Lemoigne 1968b, 1969a,b). D. S. Edwards (1980) subsequently showed that axes of *R. gwynne-vaughanii* do bear sporangia, thus conclusively excluding the idea that this species was the gametophyte of *A. major.* More convincing evidence for the gametophyte of *A. major* was presented by Remy and Hass (1991a). In a detailed study of the putative antheridiophore *Lyonophyton rhyniensis* (Remy and Remy 1980), Remy and Hass (1991a) showed that many antheridia were attached to cup-shaped structures, which themselves were attached to axes identical to those of *A. major* (see below).

The main axes of the sporophyte are naked and branch isotomously (Figure 4.7; D. S. Edwards 1986; Kidston and Lang 1920a). The rhizome is prostrate and has localized, nonseptate rhizoids that develop from epidermal cells (Kidston and Lang 1920a). The xylem is more or less circular and composed of an inner zone of elongate thin-walled cells and an outer zone of elongate thick-walled cells. A thick layer (about six cells deep) of thin-walled, elongate cells immediately surrounds the xylem and is described as phloem based on its position. The cells of the outer cortex are thick-walled and are a distinctive feature of the axes (Figure 4.21; D. S. Edwards 1986; Kidston and Lang 1920a). The epidermis is composed of elongate cells, and the stomates consist of two guard cells surrounded by six to eight slightly modified epidermal cells (Edwards, Edwards, and Rayner 1982; D. S. Edwards 1986; Kidston and Lang 1917). Water-conducting cells are elongate and unthickened, but unequivocal protoxylem has not been identified (D. S. Edwards 1986).

The sporangia are fusiform, radially symmetrical in transverse section (unflattened), and borne terminally. The sporangium wall comprises three layers. The outer layer consists of thin-walled epidermal cells oriented at an angle to the long axis of the sporangium. Stomates are present in this layer. The inner layer is parenchymatous and similar to the inner cortex of the stem. The third layer is a "tapetal layer" surrounding the central spore mass (D. S. Edwards 1986). There is no obvious line of sporangium dehiscence; however, dehiscence may occur through several obliquely oriented slits (Figure 3.28; D. S. Edwards 1986; Remy 1978). Microspores have been observed in situ. Spores are laevigate, possibly crassitate, and 52–78 μm in diameter (cf. *Ambitisporites* sp.: Allen 1980; Bhutta 1973c).

A gametangiophore, which probably belongs to the *Aglaophyton* sporophyte, has been described under the name of *Lyonophyton rhyniensis* (Remy and Hass 1991a; Remy and Remy 1980). The link between the two plants is based on similarities in the structures of the stomates and surrounding epidermal cells, in internal anatomy, and in the wall structures of the water-conducting cells, as well as the association of the gametophytes and sporophytes at the same locality. The gametophyte was probably surficial and autotrophic. Only the distal end of the antheridiophore is known and is a naked axis (at least 16 mm long) that expands into a disk-shaped gametangiophore 2.8–9.0 mm wide having globular antheridia situated on the upper surface. Cuticular features, stomates, and internal anatomy of the axis are similar to those of the sporophyte, *A. major* (Remy and Hass 1991a; Remy and Remy 1980).

Lyonophyton rhyniensis†

Lyonophyton rhyniensis is a gametophyte from the Rhynie Chert (see *Aglaophyton major* for description of morphology).

Rhynia gwynne-vaughanii†

The monospecific genus *Rhynia* is based on exceptionally well preserved permineralizations from the lower Old Red Sandstone Rhynie Chert outlier in Scotland. The age of the locality is early Devonian (Pragian) based on dispersed spores (Richardson 1967; Westoll 1977) and radiometric dates (Rice et al. 1994).

In their original description, Kidston and Lang (1917) confused two plants that were later described as different species, *Rhynia gwynne-vaughanii* and *R. major* (Kidston and Lang 1920a). A reconstruction of *R. gwynne-vaughanii* was given by Kidston and Lang (1921). Both species were interpreted as sporophytes; however, some authors subsequently pointed out that direct evidence for the sporophytic status of *R. gwynne-vaughanii* was weak. Merker (1958, 1959) suggested that some of the horizontal axes may be gametophytes. This suggestion was taken up by Pant (1962), who interpreted most of the plant as a gametophyte—the adventitious branches being young sporophytes arising from archegonia situated inside the "hemispherical projections." Evidence of possible archegonia and antheridia was provided by Lemoigne (1968a,b, 1969a,b), but preservation was poor, and gametangia have not been observed by others in similar material. Lemoigne (1970) went on to suggest that *R. gwynne-vaughanii* is the gametophyte of *R. major* and modified the specific diagnosis of *R. gwynne-vaughanii* to incorporate this idea. In response to the uncertainty surrounding the status of *R. gwynne-vaughanii,* D. S. Edwards (1980) reinvestigated material from the type locality showing that sporangia were attached to axes typical of this species and providing detailed information on the attachment of the sporangium. Most authors follow D. S. Edwards's interpretation of *R. gwynne-vaughanii* as a sporophyte; however, the possibility of a morphologically similar gametophyte generation that is somewhat different from that conceived of by Lemoigne (1970) and Pant (1962) seems likely on the basis of recent work by Remy and Hass (1991b). D. S. Edwards (1980, 1986) identified a number of morphological and anatomical differences between *R. gwynne-vaughanii* and *R. major.* On the basis of these observations, D. S. Edwards (1986) created the new genus *Aglaophyton* to accommodate *R. major* and amended the diagnosis of *R. gwynne-vaughanii* given by Kidston and Lang (1920a).

We treat *R. gwynne-vaughanii* as monophyletic and base our analysis mainly on the descriptions given by Kidston and Lang (1917, 1920a) and D. S. Edwards (1980, 1986).

The main axes are naked, branch isotomously (Figure 4.8), and bear small hemispherical projections that give rise to lateral branches in which the vascular strand is not continuous with the main axis strand. Circinate vernation has not been observed. The rhizome is similar to the aerial axes and bears nonseptate rhizoids. The xylem strand is more or less terete and composed of helical, S-type elements (Kenrick and Crane 1991). Surrounding the xylem is a zone of thin-walled cells that are interpreted as phloem on the basis of their position and a wide cortex in which cells near the periphery tend to be larger. The epidermises of the axes and sporangia are composed of elongate, fusiform cells, and stomates are present on the axes. The sporangia are borne singly along the main axes on short, unbranched (?), subordinate vascular axes and are attached to shallow depressions or "pads of tissue." The sporangia are fusiform and dehisce possibly by splitting along several vertical slits (Figure 3.28). The sporangium wall consists of an epidermal layer that lacks stomates, an underlying poorly preserved parenchymatous layer, and a tapetal layer surrounding the spore cavity. Sterile tissue projects into the base of the sporangium, and in empty sporangia, there is a conspicuous black layer at the point where the sporangium is attached to the axis (Figure 4.27). Microspores (in situ) are circular to subcircular, azonate, finely apiculate, some curvaturate, and 35–65 μm in diameter (cf. *Granulatisporites muninensis, Cyclogranisporites* sp., *Apiculiretusispora* sp.: Allen 1980). The gametophyte is unknown.

Huvenia kleui†

The monospecific genus *Huvenia* is based on compressions from the Wahnbach Schichten of the Rheinisches Schiefergebirge, Germany (Hass and Remy 1991). The age of the type locality is given as early Devonian (Pragian).

This genus was established for plants previously

described as having *Taeniocrada*-type morphology (Hass and Remy 1991). The gametophyte generation is probably of the *Sciadophyton* sp. (Remy, Remy, et al. 1980) type. There are two lines of evidence that support this conclusion. First, *Sciadophyton* sp. and a new, unpublished species of *Huvenia* are found in close association at the type locality of *H. kleui* (Remy, pers. comm. to P.K., 1996). Second, there are significant ultrastructural similarities in the water-conducting cells of *H. kleui* and *Sciadophyton* sp. from the type locality (Kenrick, Remy, and Crane 1991).

The main axes of *H. kleui* bifurcate or trifurcate isotomously. Axes are mainly smooth, but some bear multicellular spines and others bear small, hemispherical projections. The sporangia are borne singly at various points along the main axes on small, subordinate, unbranched or bifurcate vascularized branches, which are sometimes so short that the sporangium appears almost sessile. These subordinate fertile branches may be clustered at points along the main axes. The sporangia are fusiform with a mucronate tip, are borne terminally in a shallow depression, and dehisce along numerous helical slits (Figures 3.28 and 4.27). Anatomical information is scant because of the poor preservation. The central vascular strand is composed, at least in part, of helically thickened cells with an S-type wall structure (Kenrick, Remy, and Crane 1991). Stomates are unknown. No spores were isolated.

The gametophyte generation of *Huvenia* is probably of the *Sciadophyton* sp. type (Kenrick, Remy, and Crane 1991; Remy, Remy, et al. 1980). The main axes are isotomously branched or unbranched, are naked, and arise in a dense cluster from a central area. Subordinate lateral branches may arise from the main axes, which terminate in an expanded, disk-shaped structure that bears small circular features (presumed gametangia) on the upper surface. The presumed gametangia appear larger toward the center of the disk and smaller and more numerous toward the edges; this gradation suggests a maturation sequence. The shape of the expanded disk-shaped gametangiophore, as well as the centripetal increase in size of the probable gametangia on its upper surface, suggest the activity of a "marginal" or "ring" meristem, like that of certain extant species of *Lycopodium*. Anatomical information is scant due to the poor preservation. The central vascular strand is composed, at least in part, of helically thickened cells with an S-type wall structure (Kenrick, Remy, and Crane 1991). Stomates are unknown.

Stockmansella langii†

The monospecific genus *Stockmansella* is based on compressions and permineralizations from several localities in the lower Old Red Sandstone of Belgium (Fairon-Demaret 1985, 1986b). The age of the localities is given as early Devonian (late Pragian to middle-late Emsian) (Fairon-Demaret 1985).

S. langii is based on specimens originally called *Taeniocrada langii* (Stockmans 1939, 1940). Following a thorough investigation of new collections, Fairon-Demaret (1985) created the genus *Stockmansia* because of significant morphological differences between *T. langii* and the type species, *T. decheniana* (Kräusel and Weyland 1930). *Stockmansia* was later found to be an illegitimate name (the name had previously been assigned to a late Paleozoic fern) and was renamed *Stockmansella* (Fairon-Demaret 1986b). Similar material was described as *T. langii* by Schweitzer (1980b, 1983b) from the lower Devonian (upper Pragian–lower Emsian) of Germany.

We treat *S. langii* as monophyletic based on the presence of sessile or subsessile sporangia, and the loss of vascular strand differentiation near the sporangium. Our analysis is based mainly on the description given by Fairon-Demaret (1985).

The main axes are naked and isotomously branched (Fairon-Demaret 1985; Schweitzer 1987; Stockmans 1940). The presence of circinate vernation has not been clearly established (Fairon-Demaret 1985; Schweitzer 1987). The sporangia are borne singly along the main axes and are more or less sessile, nonvascular, and attached to shallow depressions or "pads" of tissue from which they are shed (Figure 4.28). The sporangia are fusiform and dehisce along possibly one (Fairon-Demaret 1985) or many (Schweitzer 1980b) vertical slits (Figure 3.28). Anatomical information is scant due to the nature of the preservation. The central vascular strand is probably terete and composed, at least in part, of helically thickened cells with an S-type wall structure (Kenrick, Remy, and Crane 1991). A decay-resistant hypodermis has not been observed. Axis and sporangium epidermal cells are elongate, and plausible stomates have been identified. Microspores (in situ) are subtriangular, 60 μm in diameter, probably smooth, and have a rigid, dense, equatorial margin. The gametophyte is unknown.

Eogaspesiea gracilis†

The monospecific genus *Eogaspesiea* is based on tiny compression fossils from the Battery Point Formation of the Gaspé Bay, Canada (Daber 1960; Høeg 1967). The locality is dated as early Devonian.

The axes are naked and branch isotomously. Poorly preserved tracheids with thickenings were isolated from larger axes associated with the fertile specimens and were interpreted as possibly the rhizome of the *Eogaspesiea* plant. This interpretation should be treated with caution in the absence of a connection to the fertile axes. Tracheids have not been isolated from fertile specimens, and the nature of the thickenings is unclear. Stomates are unknown, and no other anatomical features have been observed. The sporangia are elongate, fusiform, and borne terminally. Microspores have been recognized in situ and are smooth-walled and probably alete. The gametophyte is unknown.

Steganotheca striata†

The species *Steganotheca striata* is based on compressions from Capel Horeb Quarry in the Grey Downtonian formation of Wales (D. Edwards 1970a). The age of the locality is given as late Silurian (Ludlow–Pridoli) by Edwards and Davies (1990). *S. striata* is probably synonymous with *Cooksonia downtonensis* (D. Edwards 1970a; Heard 1939).

The axes of *S. striata* are naked and branch isotomously at least four times (Figure 4.9). No internal anatomy was preserved. Axis and sporangium (?) epidermal cells are elongate, and the latter are oriented at an angle to the long axis of the sporangium. The sporangia are elongate with a thickened, truncated, elliptical apex and are borne terminally. The sporangium dehiscence mechanism is unknown, and no spores have been isolated. Stomates are unknown. The gametophyte is also unknown.

Dutoitea pulchra†

The genus *Dutoitea* contains three species based on fragmentary compression fossils from South Africa (Høeg 1930, 1967; Plumstead 1967; Rayner 1988). The type species, *D. pulchra,* was described originally by Høeg (1930) from a locality near Knysna, Upper Bokkeveld Series of the Cape Supergroup. Another two species, *D. alfreda* and *D. maraisia,* were characterized by Plumstead (1967), and additional material was described by Rayner (1988). The age of these localities is not well constrained, and estimates range from the early Devonian to the early Carboniferous (Rayner 1988). Rayner drew attention to the similarity between *Dutoitea* and *Cooksonia* (Lang 1937a) and suggested that the latter is a later synonym. Lang (1937a) was unaware of Høeg's paper (1930) on *Dutoitea* when he described *Cooksonia.* The main feature distinguishing *Dutoitea* from *Cooksonia* is the presence of axial spines; however, these occur only in the type species, *D. pulchra,* and the other two species described by Plumstead (1967) appear to be naked. Possible differences in sporangium shape might be used to distinguish the two genera, but poor preservation in all three species of *Dutoitea* makes comparison difficult. We base our description on the type specimen, *D. pulchra.*

The main axes of *D. pulchra* branch dichotomously (isotomously) at least once and bear multicellular spines. No internal anatomy or cuticular features have been observed. Possible sporangia are terminal and obconical with a flat top. Dehiscence is unknown, and no spores have been isolated. Stomates have not been observed, and the gametophyte is unknown.

Cooksonia hemisphaerica†

The species *Cooksonia hemisphaerica* is based on compression fossils from lower Old Red Sandstone sediments of Shropshire, England (Lang 1937a), but an amplified diagnosis was given by D. Edwards (1979). The age of the type locality, Targrove, was given as the early Devonian (Lochkovian) by Edwards and Fanning (1985). We base our analysis on specimens described by D. Edwards (1979) from Freshwater East Bay, Dyfed, Wales. The age of the locality is given as probably late Silurian (Pridoli) by Edwards and Fanning (1985).

The axes are naked and branch isotomously at least once. No internal anatomy or cuticular features have been observed. The sporangia are hemispherical to elliptical and borne terminally, and no dehiscence feature has been observed. Stomates are unknown. Microspores (in situ) are circular, smooth, and 23–35 μm in diameter. The gametophyte is unknown.

Cooksonia caledonica†

The species *Cooksonia caledonica* is based on compression fossils from several localities in the Cairnconnan Group, Angus, Scotland (D. Edwards 1970a). The age of the localities was given as early Devonian (Lochkovian) by Edwards and Fanning (1985).

The axes are naked and branch isotomously at least four times (Figure 4.9). No internal anatomy or cuticular features have been observed, and stomates are unknown. The sporangia are oval to reniform, borne terminally, and probably dehisce into two equal valves (Figure 3.28). Spores have not been described. The gametophyte is unknown.

Cooksonia pertonii†

Cooksonia pertonii is based on compression fossils from the Rushall Beds of the Downton Group, Herefordshire, England (Fanning, Richardson, and Edwards 1988; Lang 1937a). The age of the type locality, Perton Lane, was given as late Silurian (Pridoli) by Fanning, Richardson, and Edwards (1988). We base our analysis on the better-preserved specimens referred to as *C. pertonii* by Edwards, Fanning, and Richardson (1986) from Shropshire, England. The age of this material was given as early Devonian (Lochkovian).

The axes are unbranched (?) and naked and have a peripheral layer of thick-walled, decay-resistant cells. No water-conducting cells are preserved. The epidermal cells of the axes are elongate and stomates are present; sporangium epidermal cells are isodiametric. The sporangia are oval (wider than high) and borne terminally, and no dehiscence feature was observed. Microspores (in situ) are subtriangular, are trilete, and possess an equatorial crassitude and conate ornament (cf. *Streelispora newportensis*). The gametophyte is unknown.

Sartilmania jabachensis†

The monospecific genus *Sartilmania* is based on numerous compression fossils from lower Devonian (Emsian 2–Emsian 3) sediments at the University of Liège campus at Sart Tilman, Belgium, and from basal Pragian to upper Emsian sediments of the Wahnbachschichten, Germany (Fairon-Demaret 1986a). Specimens assignable to *S. jabachensis* were described originally as *Dawsonites jabachensis* (Kräusel and Weyland 1935).

The main axes of *S. jabachensis* are smooth and more or less isotomously branched and give rise to closely spaced smaller laterals arranged around the main axis, possibly helically. The laterals may be unbranched or dichotomize once, and each branch bears a terminal sporangium. No internal anatomy or cuticular features were observed. Sporangia are upright, elongate elliptical to spatulate, and partially flattened (spoon-shaped), having two tiny lobes sometimes visible either side of the point of attachment of the stalk. Dehiscence is isovalvate along a thickened rim and results in complete distal separation of the valves (Figure 3.28). Stomates are unknown. Spores were not observed. The gametophyte is unknown.

Yunia dichotoma†

The monospecific genus *Yunia* is based on compressions and permineralizations from the Posongchong Formation of Yunnan province, China (Hao and Beck 1991b). The age of the locality is given as early Devonian (late Pragian). The plant was assigned tentatively to the Trimerophytales (Hao and Beck 1991b).

The main axes branch more or less isotomously and bear prominent, multicellular spines with acute tips. Branching is said to be cruciate (i.e., successive dichotomies occur in planes at 90° to each other), but in view of the fragmentary nature of the remains, this interpretation should be treated with caution. Circinate vernation was not observed. Basal parts of the plant are unknown. The xylem is circular in transverse section and diarch: for most of its length it contains two prominent, lateral areas of parenchyma surrounded by protoxylem. Dichotomy of the xylem produces two centrarch strands, but the diarch condition returns shortly thereafter. A peripheral, hypodermal layer of thick-walled cells is preserved. Cuticular features were not observed on the axis and sporangia. The sporangia were found isolated and in association with (but not attached to) the permineralized axes or compressed spiny axes. In view of the lack of direct connection, Hao and Beck (1991b) emphasized that the description of the fertile region should be treated with caution. Associated sporangia are elongate and elliptical to broadly spatulate in outline, and dehiscence is isovalvate around the distal margin. Hao and Beck (1991b) interpreted sporangium distribution as possibly terminal—one or occasionally two sporangia were borne apically on dichotomous branches—but because of the fragmentary nature of the fossils, we adopt a more conservative interpretation and treat this feature as unknown in our analysis. Spores (in situ in associated sporangia) are more or less circular, trilete (with the arms of the trilete mark extending more than one half of the spore radius), smooth or finely ornamented, and 35–53 μm in diameter. The gametophyte is unknown.

Uskiella spargens†

The genus *Uskiella* is based on compressions and permineralized sporangia from two localities in the Senni Beds Formation of South Wales (Croft and Lang 1942; Shute and Edwards 1989). The localities are dated as early Devonian (ranging from latest Lochkovian to Pragian) based on dispersed spores (Edwards, Kenrick, and Carluccio 1989) A second species, *U. reticulata,* was recently described based on coalified sporangia (Fanning, Edwards, and Richardson 1992).

Compression fossils assignable to this genus were described originally by Croft and Lang (1942) as *Cooksonia* sp. These specimens were later renamed *Uskiella* based on a reinvestigation of compressions and permineralizations (Shute and Edwards 1989). Because there is some question concerning the identity of all the compression fossils assigned to *Uskiella* (Shute and Edwards 1989, 130), we base our analysis on the permineralized material and associated compressions.

The main axes of *U. spargens* are naked and branch isotomously (Figure 4.10). In axes preserved in limonite, a peripheral layer of thick-walled cells was present, but water-conducting tissues were not preserved. One coalified axis yielded a single thickened cell that may be a tracheid but which could also represent a "banded tube" of nematothallean affinity (Shute and Edwards 1989, Figures 64 and 65). Stomates are unknown. The sporangia are fusiform, somewhat flattened, and borne terminally. The whole sporangium appears twisted, with dehiscence progressing in a helical fashion along the sporangium wall, apparently resulting in two equal valves. The sporangium epidermis is composed of a single layer of diamond-shaped cells that are slightly elongate, with the long axis parallel to the line of dehiscence. The anticlinal outer wall, and to a lesser extent the radial walls, are thickened, except for those cells in the zone of dehiscence. Between the epidermis and the spore mass is a poorly preserved area interpreted as three to four layers of thin-walled cells. Microspores (in situ) are circular to elliptical, probably alete, and have a fine, granular ornament.

Renalia hueberi†

Renalia hueberi is the type species for the genus and is based on compressions and cuticles from the Battery Point Formation of the Gaspé Sandstone, Quebec, Canada (Gensel 1976). The age of the locality is given as early Devonian (Emsian). Two other species, *R. grabertii* and *R. major,* are based on less well preserved and less complete material from the Emsian of Germany (Schweitzer 1980a, 1983b), and *Renalia*-like fossils have been reported from the lower Devonian (Lochkovian) of Wales (Edwards and Fanning 1985). Because the other species are relatively poorly known, we base our analysis solely on *R. hueberi* and caution that other material assigned to the genus may have diverse relationships.

The main axes of *R. hueberi* are pseudomonopodially branched (isotomously distally), possibly planar in larger axes, and naked (Figure 4.11). Small, subordinate, vascular axes are observed occasionally but do not occur in a regular pattern. Internal anatomy is not well preserved. Tracheids are reported as having spiral to scalariform thickenings, but this feature has not been clearly demonstrated. In view of the misidentification of scalariform secondary wall deposition in many early land plants, this report should be treated with caution (Kenrick and Edwards 1988a). Epidermal cell outlines on the axis are elongate, and over the sporangia they are isodiametric. Sporangia are round to reniform, isovalvate (Figure 3.28), and in contrast to most other zosterophylls, are borne on subordinate, isotomously branched axes. Spores (in situ) are circular to subtriangular, trilete (with arms of the trilete mark that are two-thirds to four-fifths of the spore radius), smooth, curvaturate, and 40–70 μm in diameter (cf. *Retusotriletes:* Gensel 1980). *Retusotriletes* is a common type of Devonian and lower Carboniferous dispersed spore of general trilete, azonate, laevigate form. Spores of this kind have been recorded from rhyniophytes, zosterophylls, trimerophytes, sphenopsids, and other early plant fossils of uncertain systematic position (Traverse 1988). The gametophyte is unknown.

Cooksonia cambrensis†

The species description for *Cooksonia cambrensis* is based on tiny, fragmentary compression fossils from the lower Red Marl Group, Freshwater East Bay, Dyfed, Wales. The locality is dated as late Silurian (Pridoli) on the basis of dispersed spores (D. Edwards 1979).

The axes are naked and branch isotomously at least once. No internal anatomy or cuticular features have been observed and stomates are unknown. Presumed sporangia are circular to elliptical and borne terminally. No dehiscence features were observed. No spores were isolated. The gametophyte is unknown.

Hicklingia edwardii†

The species *Hicklingia edwardii* is based on one unique and relatively complete compression fossil from the middle Old Red Sandstone of Caithness,

Scotland (D. Edwards 1976; Høeg 1967; Kidston and Lang 1923). The age is given as middle Devonian. One other species, *H. erecta,* has been described on the basis of poorly preserved and very fragmentary material from the middle Devonian of Germany (Kräusel and Weyland 1929). D. Edwards (1976) considered that specimens assigned to *Cooksonia* cf. *hemisphaerica* by Canright (1970) from the United States (Arizona) may have a closer affinity to *Hicklingia.* Our analysis is based on the one relatively well preserved specimen from Scotland.

The main axes are naked, are isotomously and possibly pseudomonopodially branched, and because they diverge from a basal region, have a tufted appearance. Internal anatomy and cuticular features are unknown. The sporangia are hemispherical to globose, probably isovalvate, and are borne on short stalks around the axis. In contrast to most zosterophylls the junction between the sporangium and stalk is very wide. No spores have been isolated from the sporangia. The gametophyte is unknown.

Zosterophyllum myretonianum†

Zosterophyllum myretonianum is the type species of the genus *Zosterophyllum* and is based on relatively well preserved compression fossils from several localities in the early Devonian (Lochkovian) Dundee Formation of Scotland (D. Edwards 1975; Edwards and Fanning 1985). Lang (1927) and D. Edwards (1975) gave summaries of the history of the species and contributed significant new morphological and anatomical information. Other studies provided additional information on branching and cuticles (Lele and Walton 1961; Walton 1964). Edwards (1975) gave an amended specific diagnosis incorporating her own observations and those of previous workers. *Z. myretonianum* is placed in the subgenus *Zosterophyllum* (Croft and Lang 1942; D. Edwards 1975; Hueber 1972). Specimens from the Soviet Union that are attributed to *Z. myretonianum* have also been described (Høeg 1967).

Prostrate basal axes, with closely spaced dichotomies, give rise to long, possibly isotomously branched or unbranched, naked, upright axes (Figure 5.1; D. Edwards 1975; Lang 1927; Lele and Walton 1961). Prostrate basal axes have occasional small, subordinate, noncircinate laterals (Lele and Walton 1961). Internal anatomy is almost unknown. Poorly preserved, coalified tracheids having annular thickenings were illustrated by Lang (1927). Well-preserved cuticles with stomates are composed mainly of elongate cell outlines and have isodiametric cell outlines near branches (Lele and Walton 1961). Cell rosettes are absent. Sporangia are reniform, probably isovalvate (Figure 3.28), and are borne helically on short vascular stalks on upright axes to form loose spikes (D. Edwards 1975; Lang 1927). Spores (in situ) appear to be more or less circular, azonate, laevigate, and 25–35 μm in diameter (Allen 1980; Lang and Cookson 1930). The gametophyte is unknown.

Zosterophyllum deciduum†

The species *Zosterophyllum deciduum* is based on compression fossils from the Grès de Wépion Formation, Belgium. The age of the locality is given as early Devonian (Emsian) (Gerrienne 1988). *Z. deciduum* belongs to the subgenus *Zosterophyllum.*

Prostrate basal axes, with closely spaced dichotomies, give rise to long, more or less isotomously branched, naked, upright axes (Figure 5.9). The prostrate axes have small, subordinate, noncircinate laterals. No internal anatomy has been described. Axis epidermal cells are elongate, whereas the sporangium epidermis comprises isodiametric cells. The sporangia are reniform, isovalvate (Figure 3.28), and borne helically on short stalks around branched upright axes. Spores (in situ) are circular to subtriangular, trilete (with arms of the trilete mark extending four-fifths of the spore radius), smooth to minutely granulate, curvaturate, and 50–78 μm in diameter (cf. *Retusotriletes goensis*). *Retusotriletes* is a common form of Devonian and lower Carboniferous dispersed spore of general trilete, azonate, laevigate form. Spores of the same general kind have been recorded from rhyniophytes, zosterophylls, trimerophytes, sphenopsids, and a variety of other early plant fossils of uncertain systematic position (Traverse 1988). The gametophyte is unknown.

Adoketophyton subverticillatum†

Adoketophyton is a monospecific genus based on compression fossils from the Posongchong Formation, Yunnan Province, China (Li and Edwards 1992). The age of the locality is given as early Devonian (late Pragian to early Emsian).

Main axes branch more or less isotomously and are naked except for fan-shaped sporophylls (Figure 5.19). The presence of circinate vernation has not been established unequivocally. Thick-walled, elongate cortical cells are preserved in iron oxides, but other aspects of anatomy and cuticular structure are unknown. The sporangia are circular to elliptical, strongly flattened, and more or less isovalvate with distal dehiscence (Figure 3.28). Each sporangium is borne near the base and on the adaxial surface of a sporophyll. Sporophylls are fan-shaped with narrow stalks and an expanded distal lamina; some sporophylls are three to four times larger than the sporangium. The sporophylls are attached around the axis in four rows in an opposite and decussate arrangement forming an unbranched, terminal spike. Spores are unknown. The gametophyte is unknown.

Discalis longistipa†

The monospecific genus *Discalis* is based on well-preserved compression fossils from the Posongchong Formation of Yunnan Province, China (Hao 1989b). The age of the locality is given as early Devonian (late Pragian).

Prostrate basal axes, with closely spaced dichotomies, give rise to long, apparently unbranched, upright axes that terminate circinately (Figure 5.12). All axes and sporangia are covered with large, multicellular spines of somewhat variable shape (Figure 5.19). Spines terminate in expanded heads. Permineralization of soft tissues is not reported, but coalified tracheids with annular or helical thickenings, as well as some poorly preserved cuticles, are illustrated. Sporangia are flattened, reniform, isovalvate, and borne on short stalks. They are arranged around upright axes (sometimes spirally) and form loose spikes. Spores (in situ) are circular to subcircular, trilete, possibly curvaturate (lines connect extremity of laesura to trilete mark), smooth or minutely granulate, and 30–50 μm in diameter. The gametophyte is unknown.

Nothia aphylla†

The monospecific genus *Nothia* is based on exceptionally well preserved permineralizations from the lower Old Red Sandstone Rhynie Chert outlier in Scotland. The age of the locality is early Devonian (Pragian) based on dispersed spores (Richardson 1967; Westoll 1977) and radiometric dates (Rice et al. 1994).

Axes and sporangia assignable to *N. aphylla* were illustrated originally by Kidston and Lang (1920b), who cautiously suggested that they might represent the fertile parts of *Asteroxylon mackiei,* based on their association at the Rhynie locality. But Lyon (1964) showed that the fertile parts of *A. mackiei* were, in fact, quite different, implying that the axes and sporangia described by Kidston and Lang belonged to a different plant, for which the name *Nothia aphylla* was proposed. This name was validated by Høeg (1967), and El-Saadawy and Lacey (1979a) gave a detailed description of the fertile parts of the plant and an amended diagnosis. The antheridiophore of the probable gametophyte generation of the *N. aphylla* plant was described recently under the name *Kidstonophyton discoides* by Remy and Hass (1991c).

The axes of *N. aphylla* are more or less isotomously branched, nonplanar, and covered with short, nonvascular, cortical outgrowths, each of which bears a stomate at the tip (Figure 5.13). The xylem is elliptical in transverse section and maturation is described as centrarch. Because the position of the protoxylem is not clearly illustrated, this report (El-Saadawy and Lacey 1979a) should be treated with caution. The xylem is composed of thick-walled cells that have no differential annular, helical, or pitted thickenings. Cells identified by their position as phloem are not well preserved. The cortex consists of parenchymatous cells, which are slightly larger at the periphery than toward the center of the axes. There is no thick-walled, peripheral hypodermis. The epidermis of the axes consists of elongate cells, and based on illustrations in El-Saadawy and Lacey (1979a), the sporangium epidermis probably consists of isodiametric cells. The sporangia are more or less reniform and isovalvate (Figure 3.28), are borne on short vascular stalks of variable length, and are arranged around the axis, possibly in a helical pattern. Spores (in situ) are azonate, curvaturate, granulate, and 59–74 μm in diameter (cf. *Apiculiretusispora:* Allen 1980). *Apiculiretusispora* is a common dispersed spore of general trilete, azonate, echinate form, which has a wide stratigraphic range. Spores of this type have been reported from rhyniophytes, trimerophytes, and progymnosperms (Traverse 1988).

The probable gametophyte, *Kidstonophyton discoides,* is known only from five antheridiophore fragments (Remy and Hass 1991c). An unbranched axis, anatomically indistinguishable from the distinctive sterile axes of *N. aphylla,* gives rise to a terminal, disk-shaped gametangiophore. Numerous, flask-shaped antheridia are borne on the upper surface of the antheridiophore and separated by growths of sterile tissue (Remy and Hass 1991c).

Kidstonophyton discoides†

Kidstonophyton discoides is a gametophyte from the Rhynie Chert (see *Nothia aphylla* for details of morphology).

Gumuia zyzzata†

The monospecific genus *Gumuia* is based on compression fossils from the Posongchong Formation of Yunnan province, China (Hao 1989a). The age of locality is given as early Devonian (late Pragian).

The prostrate basal axes of *Gumuia zyzzata,* which have closely spaced dichotomies, give rise to long, apparently unbranched, naked, upright axes (Figure 5.14). Small, subordinate, unbranched axes ("peg-like projections") are present in basal region. The internal anatomy and cuticle were not illustrated. The sporangia are round to oval, probably isovalvate, and borne on short vascular stalks around the axis (Figure 3.28). Spores (in situ) are subcircular to subtriangular, have a simple trilete, and are 32–75 μm in diameter. The gametophyte is unknown.

Huia recurvata†

The monospecific genus *Huia* is based on compressions and permineralizations from the Posongchong Formation of Yunnan Province, China (Geng 1985). The age of the locality is given as early Devonian (Pragian). *H. recurvata* was assigned to a new family, the Taeniocradaceae, which is within the Rhyniophytina.

The main axes are pseudomonopodially branched, possibly planar, and naked. The xylem is terete to elliptical in transverse section, and its maturation is described as centrarch, but since this is not clearly illustrated, this interpretation should be treated with caution. Tracheids with annular thickenings are very obvious in the scanning electron micrographs (Geng 1985), and we interpret the wall structure as probably of the G-type (Kenrick and Crane 1991; Kenrick and Edwards 1988a). Cuticular features have not been illustrated. Sporangia appear to be oval to reniform, are borne on short stalks, and are helically arranged in a relatively compact, sometimes branched spike. Spores (in situ) are circular to subtriangular, trilete (with the arms of the trilete mark extending three-quarters to four-fifths of the spore radius), tuberculate, and 40–53 μm in diameter. The gametophyte is unknown.

Trichopherophyton teuchansii†

The monospecific genus *Trichopherophyton* is based on fragmentary but exceptionally well preserved permineralizations from the lower Old Red Sandstone Rhynie Chert outlier in Scotland (Lyon and Edwards 1991). The age of the locality is early Devonian (Pragian) based on dispersed spores (Richardson 1967; Westoll 1977) and radiometric dates (Rice et al. 1994).

The distal axes branch isotomously, have circinate vernation, and bear unicellular hairs, but the basal regions of the plant are unknown. The xylem is exarch, more or less terete, and composed of tracheids with helical or annular thickenings. The cortical cells are entirely thin-walled. The epidermis comprises elongate cells, some of which develop into unicellular hairs on axes and sporangium valves. Stomates have not been observed. The sporangia are broadly oval to reniform and are borne on short vascularized stalks in an upright orientation. Distribution of sporangia and the nature of the fertile zone are unknown. The sporangia are anisovalvate (adaxial thinner than abaxial) and dehisce along the distal edge into two valves. Spores (in situ) are smooth, trilete, and about 55 μm in diameter. The gametophyte is unknown.

Zosterophyllum fertile†

The species *Zosterophyllum fertile* is based on a poorly preserved compression fossil from Nonceveux in the Bois d'Ausse formation of the Ardenne, Belgium (Leclercq 1942). The age of the type locality is early Devonian (Lochkovian) according to Gerrienne (1993). Material of very similar type has been described from one other locality in Belgium by Gerrienne (1993) and from the lower Old Red Sandstone of Wales by Croft and Lang (1942). The following account is based on a detailed description of the Welsh material (D. Edwards 1969b).

The main axes are naked and branch isotomously. The presence of circinate vernation has not been established unequivocally. The internal anatomy is preserved in pyrite. The xylem is weakly elliptical or terete in transverse section and probably exarch. A peripheral, hypodermal layer of thick-walled cells is present. Tracheids are reported as having scalariform pitting, but this feature is not clearly illustrated. In view of the misidentification of scalariform secondary wall deposition in many early land plants (Kenrick and Edwards 1988a), the presence of this feature should be regarded as equivocal until the material can be reexamined. Cuticular features and stomates have not been observed. The sporangia are reniform, isovalvate (Figure 3.28), and borne on short *vascular* stalks and appear to be helically arranged, mostly over a compact, unbranched fertile zone. The sporangia are well preserved in pyrite. The sporangium wall consists of a hypodermal layer of thick-walled cells, similar to that of the axes, and is at least three cell layers thick. The spore mass is distinctly concave (D. Edwards 1969b, Figure 19), a feature that may indicate the presence of a sterile, central columella. In situ spores are circular to subtriangular, trilete (with the arms of the trilete mark extending one-half to two-thirds of the spore radius), curvaturate, and smooth (cf. *Retusotriletes dubius*). *Retusotriletes* is a common type of Devonian and lower Carboniferous dispersed spore of general trilete, azonate, laevigate form. Spores assigned to this form have been recorded from rhyniophytes, zosterophylls, trimerophytes, sphenopsids, and other early plant fossils of uncertain systematic position (Traverse 1988). The gametophyte is unknown.

Krithodeophyton croftii†

The monospecific genus *Krithodeophyton* is based on fragmentary compression fossils from the Senni Beds of central South Wales (D. Edwards 1968). More anatomical information, possibly from *K. croftii*, was given by D. Edwards (1980). The age of the type locality is given as early Devonian (Pragian) by Edwards, Kenrick, and Carluccio (1989), and this taxon was assigned to the Barinophytaceae (D. Edwards 1968).

The main axes are naked, branch isotomously and possibly pseudomonopodially, and are probably planar. Internal anatomy is only known from sterile axes associated with the fertile material. The assignment of this anatomically preserved material to *K. croftii* should therefore be treated with caution because no distinctive morphological features (e.g., cuticles, spines) were described that could support the link between the sterile and fertile remains (D. Edwards 1968, 1980). The xylem is reported as terete in transverse section and centrarch. In view of the poor preservation of the central region, the report of a centrarch strand should be treated with caution and requires further investigation. A peripheral layer of thick-walled hypodermal cells is preserved. Metaxylem tracheids were described originally as scalariform or reticulate, but a more thorough investigation of some of this putatively conspecific material showed basically annular thickenings (D. Edwards 1980) in a cell wall that can be interpreted as of the G-type (Kenrick and Crane 1991; Kenrick and Edwards 1988a). The cuticle from the axes is composed of elongate cell outlines.

Sporangia are described as oval in outline and sessile and occur in two rows in a compact, unbranched spike. Narrow appendages that alternate with the sporangia are reported. Experience with this type of material suggests that these "appendages" may, in fact, represent the thickened distal rims of other, partially buried sporangia, and this explanation is consistent with their appearance and distribution. Investigation of similar "sterile bracts" in *Barinophyton citrulliforme* by Brauer (1980) showed that they were merely parts of the sporangia, and so this feature of *K. croftii* also requires further investigation. Spores (in situ) are mostly circular, trilete (with the arms of the trilete mark extending the full radius of the spore), curvaturate, finely apiculate, and 55–68 μm in diameter (cf. *Apiculiretusispora brandtii:* Allen 1980; D. Edwards 1968). *Apiculiretusispora* is a common dispersed spore of general trilete, azonate, echinate form, having a wide stratigraphic range. Spores assigned to this form have been reported from rhyniophytes, trimerophytes, and progymnosperms (Traverse 1988). The gametophyte is unknown.

Zosterophyllum llanoveranum†

The species *Zosterophyllum llanoveranum* is based on fragmentary compression fossils from two localities in the Senni Beds of central South Wales (Croft and Lang 1942; D. Edwards 1969a). The ages of these localities were given as early Devonian (Pragian) by Edwards, Kenrick, and Carluccio (1989). *Z. llanoveranum* was first described by Croft and Lang (1942) based on fertile axes and spores isolated from sporangia. D. Edwards (1969a) illustrated the anatomy of the axes and sporangia and amended the specific diagnosis. This species is placed in the subgenus *Platyzosterophyllum* (Croft and Lang 1942; Hueber 1972). Specimens assigned to *Z. llanoveranum* have also been reported from the Lower Devonian of Siberia (Ananiev 1959; Høeg 1967).

The main axes are naked, probably terminate circinately (D. Edwards 1969a, Figure 7), and branch infrequently in a pseudomonopodial (Croft and Lang 1942, Figure 88) and possibly isotomous fashion. The xylem is elliptical to terete in transverse section and is reported to be exarch, although the position of the protoxylem is not clearly illustrated (D. Edwards 1969a). A peripheral layer of thick-walled hypodermal cells is present. Tracheids were reported originally as having scalariform pitting, but further investigation (D. Edwards 1980) suggested annular thickenings similar to those in *Hostinella heardii*. Cuticular features are unknown. The sporangia are reniform, isovalvate (Figure 3.28), and borne on short stalks in one or two opposite rows. A section through a permineralized sporangium showing the line of dehiscence was illustrated by D. Edwards (1969a). Spores (in situ) are circular to subtriangular, smooth, trilete (with the arms of the trilete mark extending one-half to three-quarters of the spore radius), azonate, laevigate, and 50–59 μm in diameter (cf. *Leiotriletes* sp.: Allen 1980; D. Edwards 1969a) *Leiotriletes* is a common form of dispersed spore with a generalized trilete, azonate, laevigate morphology. Spores referable to this genus have been recorded from bryophytes, rhyniophytes, ferns, and some lycopsids (Traverse 1988). The gametophyte is unknown.

Rebuchia ovata†

The genus *Rebuchia* is based on compression fossils from the Beartooth Butte Formation, in Wyoming, United States. (Dorf 1933; Hueber 1970, 1972). The age of the locality is given as early Devonian (probably Pragian to Emsian) by Hueber (1972).

Originally described under the invalid name *Bucheria ovata* (Dorf 1933), which had already been applied to an angiosperm, the plant was renamed *Rebuchia ovata* by Hueber (1970), who then amended the generic and specific diagnosis (1972) based on the description of more complete specimens, which were also collected from the Beartooth Butte Formation. Hueber (1972) considered that *R. mucronata* (Mägdefrau) Høeg was synonymous to *R. ovata,* and this equivalence would extend the geographic and stratigraphic range of the species to the lower Devonian (Emsian) of Germany. Several other questionable species are based on fragments of axis bearing sporangia, and these are reviewed by Høeg (1967) and Hueber (1972). Our analysis is based on the description of relatively complete specimens of *R. ovata* (Hueber 1972).

The main axes of *R. ovata* are naked and branch isotomously. Internal anatomy and epidermal structure are unknown. The sporangia are reniform, isovalvate (Figure 3.28), and borne on short stalks in two opposite vertical rows to form a compact, unbranched spike. In contrast to most zosterophylls, the sporangium stalks are curved in such a way as to direct the sporangia toward one side of the axis. Spores (in situ) are circular, smooth, curvaturate, and 68–75 μm in diameter (cf. *Retusotriletes:* Allen 1980). *Retusotriletes* is a common type of Devonian and lower Carboniferous dispersed spore genus of general trilete, azonate, laevigate form. No triradiate mark or aperture was observed. Spores assigned to this form have been recorded from rhyniophytes, zosterophylls, trimerophytes, sphenopsids, and other early fossils of uncertain systematic position (Traverse 1988). The gametophyte is unknown.

Zosterophyllum divaricatum†

The species *Zosterophyllum divaricatum* is based on well-preserved compressions and cuticles from the Gaspé Sandstone, New Brunswick, Canada. The age of the locality was given as early Devonian (early Emsian) by Gensel (1982a). *Z. divaricatum* is placed in the subgenus *Platyzosterophyllum.*

The main axes are naked and branch isotomously. Associated sterile axes with H-type branching, small, subordinate, branches, and circinate vernation may be part of the same plant. The internal anatomy is poorly preserved. Coalified tracheids with annular thickenings were isolated, and they have a wall structure that we interpret as probably of the G-type (Kenrick and Crane 1991; Kenrick and Edwards 1988a). Axis epidermal cells are predominantly elongate, whereas the sporangium epidermis comprises isodiametric cells. The sporangia are reniform, isovalvate (Figure 3.28), borne on short stalks, and usually in two rows oriented to one side of the axis over a relatively loose, branched fertile zone (Figure 5.10). Spores (in situ) are circular to subtriangular, trilete (with the arms of the trilete mark extending one-third to one-half of the spore radius), curvaturate, smooth or minutely granular, and 55–85 μm in diameter (cf. *Retusotriletes*). *Retusotriletes* is a common type of Devonian and lower Carboniferous dispersed spore of general trilete, azonate, laevigate form. Spores of this kind have been recorded from rhyniophytes, zosterophylls, trimerophytes, sphenopsids, and other early plant fossils of uncertain systematic position (Traverse 1988). The gametophyte is unknown.

Deheubarthia splendens†

The monospecific genus *Deheubarthia* is known only from the lower Old Red Sandstone sediments of Wales (Senni Beds and Cosheston Group) but strongly resembles other more poorly known spiny axes such as *Psilophyton goldschmidtii* (Høeg 1967) and *Margophyton goldschmidtii* (Zakharova 1981), which are widely distributed throughout Europe and Asia (Høeg 1967). The ages of the Welsh localities are given as early Devonian (ranging from latest Lochkovian to Pragian: Edwards, Kenrick, and Carluccio 1989).

The analysis given here is based on extensive collections described by Kenrick (1988) and Edwards, Kenrick and Carluccio (1989). A poorly preserved specimen assignable to *D. splendens* was illustrated by Croft and Lang (1942) under the name cf. *Psilophyton princeps.*

The main axes are pseudomonopodially branched (isotomous distally), planar, circinate at the apex, and bear large, multicellular spines with acute tips (Figures 5.2 and 5.20). Small, subordinate, circinate, isotomously branched, spiny axes occur on one side and just below dichotomies of the main axis (Figure 5.15). The exarch xylem strand is strap-shaped in cross section in large axes, elliptical in smaller ones, and terete at the base of the small, subordinate axes. A peripheral hypodermal layer of thick-walled cells is sometimes preserved. The xylem cells are of the G-type (Kenrick and Crane 1991; Kenrick and Edwards 1988a). Branching involves two closely spaced dichotomies of the xylem (Figure 5.17). Cuticles of the axes show elongate cell outlines with papillae, possible stomates, and patterns of cell rosettes. Sporangia are isovalvate, reniform, and borne on short stalks in two opposite rows over a loose, branched, fertile zone. No spores have been isolated from the sporangia. The gametophyte is unknown.

Sawdonia ornata†

Originally described by Dawson (1871) under the name *Psilophyton princeps* var. *ornatum,* this plant was renamed *Sawdonia ornata* by Hueber (1971b), and accounts of the convoluted taxonomic and nomenclatural history were given by Hueber and Banks (1967) and Banks, Leclercq, and Hueber (1975). An updated morphological description of the Canadian material and the first anatomical information was provided by Hueber and Banks (1967) and Hueber (1971b). Rayner (1983) described in detail the anatomy and morphology of specimens from Scotland and gave an amplified specific diagnosis. A detailed description of the cuticle was given by W. N. Edwards (1924) and Lang (1931), and further cuticular information has been obtained from Poland (Zdebska 1972), Scotland (Rayner 1983), and England (Chaloner, Hill, and Rogerson 1978). Spine morphology was critically examined by Gensel (1991). One other species, *S. acanthotheca,* has been described from New Brunswick, Canada (Gensel, Andrews, and Forbes 1975), but certain aspects of its morphology (particularly the branching) and anatomy are less well known than in the type species. Fossils assigned to *S. ornata* have been described from Canada (W. N. Edwards 1924; Hueber 1971b; Hueber and Banks 1967; Lang 1931), North America (Andrews, Kasper, et al. 1977; Kasper, Gensel, et al. 1988), Scotland (Rayner 1983), England (Chaloner, Hill, and Rogerson 1978), Belgium (Steemans and Gerrienne 1984), Poland (Zdebska 1972), and southwestern Siberia (Ananiev and Stepanov 1968; Hueber and Banks 1979). These occurrences range in age from Pragian to Frasnian, with most in strata dated as Pragian–Emsian. We provisionally treat *Sawdonia* as a monophyletic group. Potential synapomorphies include the globose shape of the mature sporangium and the position of the small, subordinate axes *above* the dichotomies of the main axis.

The main axes are pseudomonopodial, planar, and circinate at the apex and bear numerous, large, multicellular spines that taper to an acute tip (Hueber 1971b; Rayner 1983). Small, subordinate, branched axes occur regularly just *above* dichotomies of the main axes (Rayner 1983). Preservation of internal tissues has been recorded twice (Hueber and Banks 1967; Rayner 1983). The xylem is elliptical to strap-shaped in transverse section and probably exarch. A peripheral hypodermal layer of thick-walled cells is sometimes preserved. The xylem cells are interpreted as possibly G-type (Kenrick and Crane 1991; Kenrick and Edwards 1988a) on the basis of scanning electron micrographs provided by Rayner (1983) and the light micrograph of Hueber and Banks (1967). Cuticular features have been illustrated by several authors. The axis epidermis shows papillate, isodiametric to elongate cell outlines (Rayner 1983), scattered stomates (Lang 1927; Rayner 1983; Zdebska 1972), and cell rosettes (Edwards, Edwards, and Rayner 1982; W. N. Edwards 1924; Rayner 1983; Zdebska 1972). The sporangia are isovalvate (Figure 3.28), round to oval, and borne on short stalks in two opposite rows. Spores (in situ) described by Hueber (1971b) are circular to subtriangular, smooth, trilete (with arms extending one-third of the spore radius), and 54–64 μm in diameter (cf. *Calamospora atava:* Allen 1980), and those descibed by Rayner (1983) are trilete (with the arms of the trilete mark extending one-half to two thirds of the spore radius), noncurvaturate (cf. *Retusotriletes*), and 20–45 μm in diameter. *Retusotriletes* and *Calamospora* are common types of a Devonian and lower Carboniferous dispersed spore of general trilete, azonate, laevigate form. Spores assigned to this form have been reported from rhyniophytes, zosterophylls, trimerophytes, sphenopsids, and other early plant fossils of uncertain systematic position (Traverse 1988). The gametophyte is unknown.

Crenaticaulis verruculosus†

The monospecific genus *Crenaticaulis* is based on relatively well preserved specimens from the Battery Point Formation of the Gaspé Sandstone, Canada (Banks and Davis 1969). The age of the type locality is given as early Devonian (upper Emsian).

The main axes are pseudomonopodially branched (isotomous distally), probably planar, and circinate at the apex and bear one or two opposite vertical rows of multicellular deltoid spines. Small, subordinate axes occur on one side of the plant, just below dichotomies of the main axis, but details of the branching, morphology, and anatomy of these subordinate axes are unclear. The xylem is clearly elliptical and probably exarch. A peripheral, hypodermal layer of thick-walled cells is preserved. G-type xylem cells (Kenrick and Crane 1991; Kenrick and Edwards 1988a) have been well demonstrated with SEM (Gensel, Johnson, and Strother 1990, Figure 12c). Cuticular features of the axis are well preserved and show the outlines of small, papillate cells intermixed with elongate cells and stomates. Cuticles isolated from sporangia have isodiametric cells, some of which are papillate. The sporangia are strongly anisovalvate, reniform to nearly spherical, and borne on short stalks in two opposite rows over a loose, branched, fertile zone. No spores have been isolated from the sporangia. The gametophyte is unknown.

Serrulacaulis furcatus†

The monospecific genus *Serrulacaulis* is based on well-preserved compression fossils from Schoharie County, New York State, United States (Hueber and Banks 1979). The fossiliferous strata of the type locality are interpreted as equivalent to the lower part of the continental Oneonta Shale and Sandstone Member of the Genesee Group, and the age is given as late Devonian (lower Frasnian).

Several references to *S. furcatus* prior to valid publication were discussed by Hueber and Banks (1979). Specimens assignable to *S. furcatus* have been described also from the middle Devonian (Givetian) of the Brabant Massif of Belgium. Other material, from the Campo Chico Formation, Zulia State, Venezuela, has been described recently, and the generic diagnosis has been amended (Berry and Edwards 1994). The age of this material is given as middle or early late Devonian. We base our description and analyses on the relatively well preserved material of Hueber and Banks (1979) and supplement them with the observations of Berry and Edwards (1994).

The axes branch isotomously, terminate circinately, and bear two, opposite, rows of distinctive multicellular, deltoid or prism-shaped spines (Berry and Edwards 1994). Basal, prostrate axes bear nonseptate rhizoids but are otherwise identical to the upright axes. Details of stem anatomy are unknown, but the xylem is composed of G-type tracheids (Berry and Edwards 1994). Cuticular cell outlines of the axis and proximal half of the sporangium are predominantly elongate, but this pattern is interrupted by cell rosettes. The cuticle of the distal half of the sporangium shows isodiametric cell outlines. Sporangia are reniform, isovalvate (Figure 3.28), borne on short stalks in two rows on the axis. In contrast to most other zosterophylls, these two rows are both on the ventral side of the axis. Spores (in situ) are circular to subtriangular, smooth to slightly granulate, trilete (with the arms of the trilete mark extending one-third to one-half of the spore radius), and 42–66 μm in diameter. The gametophyte is unknown.

Bathurstia denticulata†

The monospecific genus *Bathurstia* is known only from two poorly preserved axis fragments from the Bathurst Island Formation, Bathurst Island, in the Canadian Arctic. The age of the locality is given as early Devonian (Pragian) by Hueber (1972).

The main axes, branching unknown, have circinate tips and are covered with multicellular deltoid spines. The internal anatomy and epidermal features are unknown. Sporangia are elongate but reniform in outline and are arranged in two opposite vertical rows in a compact spike. No spores have been isolated from the sporangia. The gametophyte is unknown.

Anisophyton gothanii†

The species *Anisophyton gothanii* is based on compression fossils from the southern part of the Wilbringhäuser Scholle of the Rheinisches Schiefergebirge, Germany. The age of the locality is given as early Devonian (middle or upper Emsian) (Remy, Schultka, and Hass 1986). One other poorly known species, *A. potoniei*, also from the Rheinisches Schiefergebirge, has been assigned to this genus (Remy, Hass, and Schultka 1986). Our analysis is based on the type species, *A. gothanii.*

The main axes are pseudomonopodially branched (isotomous distally), planar, circinate, and covered with large, multicellular spines that terminate in expanded heads (Figure 5.19). Smaller, subordinate axes occur on one side of the main axes just below dichotomies and develop into branching systems similar to those of the main axes. The internal anatomy and cuticular features were not described. Structures interpreted as reniform sporangia are borne on short stalks in a single row that is oriented at right angles to the plane of dichotomy of the main axis (Figure 5.24). No spores have been isolated from the sporangia. The gametophyte is unknown.

Konioria andrychoviensis†

The monospecific genus *Konioria* is based on short but well-preserved axis fragments from two boreholes in the Bielsko-Andrychów area of the Polish Western Carpathians (Zdebska 1982). The age of the locality is given as early Devonian (Emsian).

The axes dichotomize isotomously and anisotomously, apparently are nonplanar and circinate at the apex, and bear large, multicellular spines with acute tips. Basal regions of the plant are unknown. The xylem is clearly elliptical in transverse section and probably exarch. A peripheral, hypodermal layer of thick-walled cells is also preserved. Tracheids are reported as having scalariform pitting, but this feature is not clearly illustrated. In view of the frequent misidentification of scalariform secondary wall deposition in early land plants (Kenrick and Edwards 1988a), this feature should be treated with caution until it can be reexamined in more detail. Cuticular features are not well preserved but include possible stomates. Structures interpreted as oval to weakly reniform sporangia are borne on short stalks and occur singly just below dichotomies of the main axis. In contrast to most other zosterophylls, the sporangia are attached at right angles to the plane of dichotomy of the main axis. No spores have been isolated from the putative sporangia. The gametophyte is unknown.

Hsua robusta†

The species *Hsua robusta* is based on relatively well preserved compressions and permineralizations from the Xujiachong Formation of Yunnan Province, China (Li 1982, 1992). The age of the locality is given as early Devonian (Emsian), and the plant was assigned to the Cooksoniaceae of the Rhyniales.

The main axes are pseudomonopodially branched (isotomous distally), possibly planar, circinate at the apex, and naked (Figure 4.12). Small, subordinate, vascularized, branched axes occur regularly just *above* dichotomies of the main axes. The xylem is elliptical to strap-shaped in transverse section. Xylem maturation is described as centrarch, but in view of the variable appearance of the sizes of the cells in the sections illustrated, this interpretation should be treated with caution. Tracheids were described originally as scalariform, but a later SEM study (Li 1992) showed annular thickenings in the cell wall that we interpret as of the G-type (Kenrick and Crane 1991; Kenrick and Edwards 1988a). The axis cuticle bears stomates (Li 1992), and cuticular cell outlines over the axis and proximal half of the sporangium are predominantly elongate. The cuticle of the distal half of the sporangium consists of isodiametric cell outlines. The sporangia are round to reniform, are isovalvate (Figure 3.28), and in contrast to most other zosterophylls, are borne on subordinate, isotomously branched axes. Spores (in situ) are 18–36 μm in diameter. The gametophyte is unknown.

Thrinkophyton formosum†

The monospecific genus *Thrinkophyton* is based on compressions and permineralizations from lower Old Red Sandstone sediments in Dyfed, South Wales (Kenrick and Edwards 1988b). The age of the locality is given as early Devonian (uppermost Lochkovian or Pragian).

The main axes are pseudomonopodially branched (isotomous distally), planar, circinate at the apex, and naked (Figure 5.11). Small, subordinate axes occur on one side, just below dichotomies of the main axis, but details of their morphology are unknown. The xylem in main axes is elliptical in transverse section and exarch. Preservation of tracheids is poor, but annular thickenings have been observed. No peripheral hypodermal layer was preserved. Cuticular features were not well preserved, but very small hairs and papillae were observed on some axes. Sporangia are reniform, isovalvate (Figure 3.28), and borne on short stalks in one or two opposite rows over a loose, branched, fertile zone. No spores were isolated. The gametophyte is unknown.

Gosslingia breconensis†

The species *Gosslingia breconensis* is based on well-preserved compressions and permineralizations from the lower Old Red Sandstone (Senni Beds) of Wales.

The age of the type locality, Brecon Beacons Quarry, is given as early Devonian (Pragian) (Edwards, Kenrick, and Carluccio 1989).

A detailed account of the anatomy and morphology of *Gosslingia* was published by Heard (1927). Croft and Lang (1942) showed that Heard's original interpretation of the fertile region was incorrect and that sporangia occur on both the main and lateral branches. D. Edwards (1970b) described anatomical changes in the shape of the xylem strand during branching and further extended the morphological description. Edwards also amended the diagnosis to include her own observations and those of Croft and Lang (1942). Other specimens attributed to *Gosslingia* have been described from Germany (Schweitzer 1979, 1983b, 1987), Belgium (Gerrienne 1983, 1993), and Wyoming, United States (Tanner 1982), and there are poorly documented reports of *G. breconensis* from the Volyno-Podolian margin of the Russian continental platform (Petrosyan 1967, not illustrated) and from the Dnestrovsk series of the Podolian Dnieper region, Czechoslovakia (Istchenko 1974, not illustrated). The fragmentary nature of the remains, the lack of illustrations, or both, make these other species or reports of *G. breconensis* difficult to evaluate. The analysis given here is based entirely on the well-documented specimens from Wales.

The main axes are pseudomonopodially branched (isotomous distally), planar, circinate at the apex, and naked. Small, subordinate axes of unknown morphology occur on one side, just below dichotomies of the main axis. The xylem is elliptical and exarch in main axes (Kenrick and Edwards 1988a) and is terete at the base of the small, subordinate axes (axillary tubercles) (Figure 5.21). Branching involves two closely spaced dichotomies of the xylem. A peripheral, hypodermal layer of thick-walled cells is present (Figure 5.21). The xylem cells are of the G-type (Kenrick and Crane 1991; Kenrick and Edwards 1988a). Cuticular features of the axis are not well preserved but show elongate cell outlines, papillae, and possible stomates.

The sporangia are isovalvate, oval to reniform, and borne on short stalks in one or two opposite rows over a loose, branched, fertile zone (Figure 3.28). In contrast to most other zosterophylls, sporangium orientation is auricular (Figure 5.24). Spores (in situ) are mainly subtriangular, azonate, apiculate (with small spines and cones over the surface), and 36–50 μm in diameter. The trilete mark has not been observed (Allen 1980; D. Edwards 1970b). The gametophyte is unknown.

Oricilla bilinearis†

The monospecific genus *Oricilla* is based on well-preserved compression fossils from the Gaspé Sandstone, New Brunswick, Canada. The age of the locality is given as early Devonian (probably Emsian) by Gensel (1982b).

The main axes branch pseudomonopodially or isotomously and apparently are planar, circinate at the apex, and naked. No subordinate branching has been observed. The internal anatomy is not preserved, but cuticular features have been described. The cuticular cell outlines that occur over the axis are predominantly elongate, but this pattern is interrupted by occasional cell rosettes. Sporangium cuticles show isodiametric cell outlines. The sporangia are isovalvate (Figure 3.28), strongly reniform, and borne on short stalks in two opposite rows over a loose, branched, fertile zone. In contrast to most other zosterophylls, sporangium orientation is auricular (Figure 5.24). Spores (in situ) are circular, trilete (with the arms of the trilete mark extending in length from one-third to one-half spore radius), mostly curvaturate, smooth, and 68–92 μm in diameter. The gametophyte is unknown.

Tarella trowenii†

The monospecific genus *Tarella* is based on numerous but fragmentary compression fossils from the Senni Beds of central South Wales (Edwards and Kenrick 1986). The age of the type locality is given as early Devonian (Pragian). A reconstruction of *T. trowenii* was provided by Kenrick (1988) and Edwards and Davies (1990).

The main axes branch isotomously and are planar, circinate at the apex, and naked. Small, subordinate axes that are smooth, circinate, and branched are scattered over the main axes with no apparent pattern to their arrangement. The internal anatomy and cuticular features were not observed. The sporangia are reniform, isovalvate (Figure 3.28), and borne on short stalks in two opposite rows over a loose, branched, fertile zone. In contrast to most other zosterophylls, sporangium orientation is auricular (Figure 5.24). Spores (in situ) are circular to subtriangular, minutely granulate, equatorially thickened, and 31–43 μm in diameter. The trilete mark was not observed. The gametophyte is unknown.

***Barinophyton citrulliforme*†**

The genus *Barinophyton* was created by White (Smith and White 1905) with his description of *B. richardsonii.* Several species have been described subsequently (Arnold 1939; Brauer 1981; Høeg 1967; Kräusel and Weyland 1941), but the most thoroughly studied and best known is *B. citrulliforme.* Arnold (1939) briefly described the morphology of the fertile axes and isolated megaspores from the sporangia in upper Devonian material assigned to *B. citrulliforme* from New York State, United States. In a detailed reinvestigation based on better-preserved material from the Catskill Formation, Pennsylvania, Brauer (1980) described the gross morphology, anatomy, tracheid wall structure, and spores of *B. citrulliforme* using SEM.

Specimens assigned to *B. citrulliforme* have been described from North America (Arnold 1939; Brauer 1980) and Belgium (Stockmans 1948). Other species have been described from North America (Brauer 1981; Høeg 1967; Kasper, Gensel, et al. 1988), Canada (Høeg 1967; Kräusel and Weyland 1941), Siberia, and Australia (Høeg 1967). The age of many of these records is poorly delimited. North American, Australian, and Belgian material is described as late Devonian, there is one early Devonian record from Canada, and one middle Devonian record from the Soviet Union. We provisionally treat *Barinophyton* as a monophyletic group and base our analysis on the only well-known species, *B. citrulliforme.* Potential synapomorphies include the orientation and attachment of the sporangia, which appear to be similar in all species, and the unique form of heterospory that has been clearly demonstrated for *B. citrulliforme* (Brauer 1980) and probably also occurs in the type species, *B. richardsonii* (Arnold 1939; Pettitt 1965).

The main axes of *B. citrulliforme* are pseudomonopodially branched and naked (Figure 5.4). The arrangement of branches is unclear: Arnold (1939) described them as planar, but Brauer (1980) described the strobili as possibly borne spirally. A planar arrangement is supported by the illustration of vascular trace departure to adjacent strobili (Brauer 1980, Figures 12–16), which is clearly opposite. The xylem is elliptical to strap-shaped in transverse section and exarch in appearance (Brauer 1980). Xylem cells were interpreted as G-type (Kenrick and Crane 1991; Kenrick and Edwards 1988a) and have been well illustrated using SEM and polished thick sections (Brauer 1980). Cuticular features have not been described.

Sporangia are oval and borne on short stalks ("sporangiferous appendages": Brauer 1980) in two opposite rows in a compact, unbranched spike. In contrast to most other zosterophylls, sporangium orientation is clasping (Figure 5.24), there is no obvious dehiscence feature, and the plant produces spores of two distinct sizes (Arnold 1939; Brauer 1980; Taylor and Brauer 1983). Heterospory in *Barinophyton* is unique because both microspores and megaspores develop in the same sporangium. Megaspores are circular to oval, trilete (*Calamospora* sp.), and 700–900 μm in diameter; microspores are circular, trilete (with the arms of the trilete mark extending for one-half to three-quarters of the spore radius), curvaturate, minutely granulate (cf. *Calamospora* sp.), and 33–49 μm in diameter (Allen 1980; Brauer 1980; Taylor and Brauer 1983). *Calamospora* is a common type of Devonian and lower Carboniferous dispersed spore of general trilete, azonate, laevigate form. Spores of this kind have been recorded from rhyniophytes, zosterophylls, trimerophytes, sphenopsids, and other early plant fossils of uncertain systematic position (Traverse 1988). The gametophyte is unknown.

***Barinophyton obscurum*†**

Barinophyton obscurum is based on compressions from Auckland County, New South Wales, Australia (White in Smith and White 1905). The age of the locality is given as late Devonian (Frasnian) (Høeg 1967). Additional information was provided by Brauer (1981) based on specimens from the upper Devonian Catskill Formation, Pennsylvania.

The main axes are pseudomonopodially branched and naked. The arrangement of branches is unclear, but they may be planar (Brauer 1981). Anatomy is unknown. The sporangia are oval and borne on short stalks in two opposite rows in a compact, unbranched spike. Sporangium orientation is difficult to discern but is probably clasping (Figure 5.24). The dehiscence mechanism is unknown. Megaspores and microspores develop within the same sporangium. Megaspores are circular to oval, folded, and 480–600 μm in diameter; microspores are circular, trilete, curvaturate, and 47–65 μm in diameter. The gametophyte is unknown.

***Protobarinophyton obrutschevii*†**

Protobarinophyton obrutschevii is the type species for the genus and is based on compressions and permineralizations from the south eastern part of Siberia. The age of the type locality is given as early Devonian.

The genus was created by Ananiev (1955) for specimens previously assigned to *Barinophyton*. Specimens assignable to *P. obrutschevii* have also been described from the Eifelian and Givetian of the Altaï-Saïan mountains of Siberia (Lepekhina, Petrosian, and Radchenko 1962). Lepekhina, Petrosian, and Radchenko (1962) considered that *Distichophyton mucronata* (Ananiev 1959; Høeg 1967) was synonymous with *P. obrutschevii.*

The main axes are pseudomonopodially branched (isotomous distally), probably planar and naked (Figure 5.3). Small, subordinate axes of uncertain morphology occurring just below dichotomies of the main axis have been recorded in some specimens (Lepekhina, Petrosian, and Radchenko 1962). The xylem is composed of cells with annular thickenings, but other anatomical features have not been described. The sporangia are oval and borne in two opposite rows in a compact, unbranched spike. The sporangia are described as sessile, but in view of the difficulty of verifying this feature in related taxa (see *Barinophyton*), this report should be treated with caution. In contrast to most other zosterophylls, sporangium orientation is clasping (Figure 5.24). Spores (in situ) are azonate, laevigate, and 45–63 μm in diameter (cf. *Calamospora atava:* Allen 1980). *Calamospora* is a common type of Devonian and lower Carboniferous dispersed spore of general trilete, azonate, laevigate form. Spores of this kind have been recorded from rhyniophytes, zosterophylls, trimerophytes, sphenopsids, and other early plant fossils of uncertain systematic position (Traverse 1988). There is no evidence of megaspores. The gametophyte is unknown.

Protobarinophyton pennsylvanicum†

The species *Protobarinophyton pennsylvanicum* is based on compression fossils from the Catskill Formation, Pennsylvania, United States (Brauer 1981). The age of the locality is given as late Devonian (Brauer 1981).

The main axes are naked and branch pseudomonopodially, giving rise to thin, probably short, nonfertile laterals. Anatomy unknown. The sporangia are oval, are attached by short stalks in two opposite rows in a compact, terminal spike, and contain both megaspores and microspores. Spore morphology has been described in detail by Cichan, Taylor, and Brauer (1984). Microspores (in situ) are circular, trilete, smooth, and 30–42 μm in diameter. Sporoderm is 1 μm thick and homogenous except for a layer of lacunae immediately below the surface. Megaspores are 410–560 μm, alete, and smooth. Sporoderm is 6.5 μm thick, comprising a thin surface layer and a thick, two-layered alveolate layer distinguished by small variations in lacunae shape and size. The gametophyte is unknown.

Drepanophycus spinaeformis†

Drepanophycus spinaeformis is the type species for the genus and is based on compressions and permineralizations from a number of localities in the Spirifer-sandstone near Hachenburg, Nassau, Germany (Chaloner 1967a).

Drepanophycus was named by Göppert (1852), but prior to 1930, much material was described under the name *Arthrostigma,* which was established by Dawson (1871). Kräusel and Weyland (1930, 1935) reduced the two taxa to synonymy and provided new information on the fertile parts of the plant. The generic diagnosis was amended by Rayner (1984), who also included *Kaulangiophyton* (Gensel, Kasper, and Andrews 1969). In addition to *D. spinaeformis,* several other species have been proposed (Chaloner 1967a; Grierson and Banks 1963; Li and Edwards 1995; Plumstead 1967; Stockmans 1940), but these are generally very poorly circumscribed and are based on variation in the density and size of the vegetative leaves, as well as the shape of the leaf base. Fertile material is known only for specimens assigned to *D. spinaeformis,* and anatomy is known only for *D. spinaeformis* and *D. gaspianus.* Reviews of the geographic and stratigraphic ranges of species were given by Kräusel and Weyland (1930), Grierson and Banks (1963), and Chaloner (1967a).

The artificial nature of the genus *Drepanophycus* has been recognized for some time (Grierson and Banks 1963, 1983; Grierson and Hueber 1967). The main problem is that the genus is not well defined, and this has been compounded by a tendency to assign species to *Drepanophycus* on a temporary basis with the knowledge that when more fully investigated their affinities would probably be elsewhere (Grierson and Banks 1983). The diagnostic feature of *Drepanophycus* is given as the falcate leaf (Grierson and Banks 1963, 1983; Grierson and Hueber 1967; Rayner 1984), but many compression fossils of lycopodiaceous aspect and of diverse relationships also show this feature. For example, Grierson and Banks (1983) created the genus *Haskinsia* for specimens formerly called *Drepanophycus colophyllus* because further investigation showed significant differences in xylem and leaf shape. In *Haskinsia* the falcate leaf was described originally as petiolate with a simple, laminar, lanceolate blade, but it was later shown to be sagittate not lanceolate (Bonamo, Banks, and Grierson 1988). Furthermore, the generic concept of *Drepanophycus* cannot be defended by defining leaf shape as simple and more or less parallel margined (Rayner 1984) because such a broad circumscription is probably applicable to lycopsids in general. The presence of anomocytic stomates (see below) may provide a potential synapomorphy for the genus, but this feature cannot be observed in most compression fossils.

The stomates have been studied in detail by Lang (1932), Stubblefield and Banks (1978), and Rayner (1984), and there are two different interpretations. Stubblefield and Banks (1978) interpreted the stomates as *paracytic,* consisting of a pore, two guard cells, and two reniform subsidiary cells. However, Lang (1932) and Meyen (1987) interpreted the structure as *anomocytic,* consisting of a pair of guard cells (the reniform subsidiary cells of Stubblefield and Banks 1978) having a thickened ridge of cuticle surrounding the pore and with thinner cuticle continuing downward into the pore (the guard cells of Stubblefield and Banks 1978). Rayner (1984) illustrated the inner cuticular surface of the stomatal apparatus using SEM and found no evidence for a cuticular flange between the "guard cells" and "subsidiary cells" of Stubblefield and Banks, even though cuticular flanges clearly delimited other epidermal cells. However, on the outer surface of the cuticle, he found ridges around the pore that could easily be misinterpreted in light micrographs as the junction between the subsidiary cells and a guard cells. Based on this evidence and on our experience with interpreting angiosperm stomates, we favor Lang's interpretation of the subsidiary cells as anomocytic. Similar stomate morphology has also been described for *Drepanophycus qujingensis* (Li and Edwards 1995). We base our analysis only on material from which cuticle bearing the distinctive stomatal apparatus has been described.

Kräusel and Weyland (1935) and Croft and Lang (1942) both described fertile material of *D. spinaeformis* but were unable to determine satisfactorily whether the sporangia were cauline or attached to sporophylls. Schweitzer (1980c) described cauline sporangia in material from several localities in Germany, but the assignment of these specimens to *Drepanophycus* has been questioned by Gensel and Andrews (1984). Xylem shape in transverse section has been described in material from Belgium (Fairon-Demaret 1971), Canada (Grierson and Hueber 1967; Hartman 1981), and Scotland (Rayner 1984). Tracheids have been illustrated by several authors (Fairon-Demaret 1971; Grierson and Hueber 1967; Halle 1916b; Kräusel and Weyland 1935; Lang 1932), and tracheid wall structure was described in more detail by Grierson and Hueber (1967), Hartman (1981), and Rayner (1984). An amended diagnosis of *Drepanophycus spinaeformis* was given by Rayner (1984).

The main axes of *D. spinaeformis* are isotomously branched and may produce smaller, branched laterals, some of which are rootlike and others stemlike (Rayner 1984). Microphylls are simple, parallel-sided, contain a single vascular strand, and are arranged helically. The presence or absence of a ligule has not been determined. Anatomical information is relatively well preserved in pyrite in some specimens. The xylem is stellate in transverse section, and five to six lobes were observed in one axis with deep bays between adjacent lobes. A decay-resistant hypodermis was not observed (Rayner 1984). Metaxylem tracheids have an annular wall structure with a connecting pitlet sheet between the thickenings (Hartman 1981; Rayner 1984), which has been interpreted as a G-type wall structure (Kenrick and Crane 1991; Kenrick and Edwards 1988a). Epidermal cells of the axis are elongate to polygonal. Stomates are anomocytic (see above). Sporangia are oval to reniform, perhaps cauline (Schweitzer 1980c), and borne on short stalks. The line of sporangium dehiscence is probably tangential to the stem (Figure 3.28). No spores have been isolated from unequivocal *Drepanophycus* material. The gametophyte is unknown.

Drepanophycus qujingensis†

The species *Drepanophycus qujingensis* is based on compressions and permineralizations from the Xujianchong Formation, Yunnan Province, southwestern China (Li and Edwards 1995). The age of the type locality is given as early Devonian (Emsian). The species was originally described under the name of *Drepanophycus spinaeformis* (Halle 1927, 1936b; Li and Edwards 1995).

The main axes are isotomously branched and produce much narrower, unbranched leafy, bulbil-like laterals. Microphylls are simple, taper to an acute apex, contain a single vascular strand, and are arranged helically. Ligules are probably absent. Dichotomously branched adventitious roots are common along both fertile and sterile leafy stems. Some anatomical information was preserved in pyrite and limonite. The xylem is probably lobed in transverse section with deep bays between adjacent lobes. The position of the protoxylem and metaxylem has not been clearly determined, and a decay-resistant hypodermis has not been observed. Metaxylem tracheids have G-type wall structure (Kenrick and Crane 1991; Kenrick and Edwards 1988a). Epidermal cells of the axis are elongate to polygonal. Stomates are anomocytic: they comprise two guard cells surrounding a pore with thickened margins. The sporangia occur on leafy stems and are probably arranged in alternating sterile and fertile zones. The sporangia are elliptical, cauline, and borne on short vascularized stalks and do not appear to be attached to leaves or leaf bases. The sporangia dehisce into two valves of equal size, and the line of dehiscence is tangential to the stem (Figure 3.28). No spores were observed. The gametophyte is unknown.

Asteroxylon mackiei†

The monospecific genus *Asteroxylon* is based on exceptionally well preserved permineralizations from the lower Old Red Sandstone Rhynie Chert outlier in Scotland. The age of the locality is early Devonian (Pragian) based on dispersed spores (Richardson 1967; Westoll 1977) and radiometric dates (Rice et al. 1994).

We follow Høeg (1967) and Hoskins and Cross (1951) and exclude the species *A. elberfeldense* Kräusel and Weyland (equivalent to *Thursophyton elberfeldense* Høeg) and *A. setchellii* Reid and Campbell (equivalent to *Stenokoleos setchellii* Hoskins and Cross) from *Asteroxylon*. *Thursophyton elberfeldense* is an interesting but poorly understood plant of uncertain affinity originally described from the lower Devonian of the Rhine Valley (Høeg 1967). No fertile material has been recognized, and the presence of circinate vernation and pseudomonopodial branching are more suggestive of affinities with zosterophylls than with *Asteroxylon*. *Stenokoleos setchellii* is known only from a permineralized axis fragment from the New Albany Shale and probably represents part of a zygopterid fern rachis (Hoskins and Cross 1951).

Kidston and Lang (1920b) described *A. mackiei* in great detail and also presented a reconstruction (1921). Fertile material was not found attached to leafy axes, but the possible form of the fertile parts was reconstructed on the basis of associated sporangia attached to branched, naked axes. These sporangia were later shown to be part of a different plant, *Nothia aphylla* (El-Saadawy and Lacey 1979a; Høeg 1967), when Lyon (1964) found sporangia of a different type attached to leafy axes of *A. mackiei*.

The main axes of *A. mackiei* branch isotomously, but anisotomous lateral branching giving rise to small axes is more common. Microphylls are simple and arranged helically. A single vascular trace develops from the xylem strand but stops at the base of the leaf (i.e., does not enter the microphyll). The well-preserved leaves appear to be eligulate. The xylem is stellate in transverse section, and seven lobes were observed in one large axis with deep bays between adjacent lobes (Figure 6.16). The protoxylem occurs in the enlarged ends of the xylem lobes. G-type xylem cells have been identified with SEM and thin sections (Kenrick and Crane 1991). Phloem tissue has been described only on the basis of its presumed position in the bays formed by the xylem lobes. The inner cortex has large intercellular spaces formed by plates of cells (trabeculae). The outer cortex is a narrow zone, about six cells thick in transverse section. A decay-resistant hypodermis has not been observed. The epidermis of the axis is composed of elongate cells, some of which are papillate, and stomates consist of two guard cells (Edwards, Edwards, and Rayner 1982; Remy and Hass 1991c). Dichotomies of the main axis involve two types of stelar division: isotomous branching results from the division of a stellate xylem strand into two smaller stellate strands; and smaller lateral branches result from division of one lobe of the main stellate strand to produce a terete strand that goes on to develop the stellate shape distally. Sporangia are large (up to 7 mm wide), reniform, isovalvate, cauline, and borne on short, vascularized stalks (Lyon 1964). The sporangia are not axillary to the leaves, but the details of their position with respect to each other and microphylls are unknown (Hueber 1992). The leaves in the vicinity of the sporangia are of similar morphology to those on vegetative axes. Microspores with a distinctive trilete mark and little ornamentation are known in situ within the sporangia (Lyon 1964).

The rhizomes are described as naked, isotomously branched, prostrate axes. The xylem is more or less cylindrical and surrounded by tissue interpreted as phloem, an inner cortex of narrow delicate cells, an outer cortex of larger cells, and an epidermis. No root hairs were observed. The gametophyte is unknown.

Baragwanathia longifolia†

Baragwanathia longifolia is the type species of the genus, based on compressions and permineralizations from several localities in the Wilson Creek Shale, Victoria, Australia (Lang and Cookson 1935). The age of these and other new localities in Victoria has been the subject of much recent discussion, and the arguments and literature have been summarized by Garratt and Rickards (1984, 1987), Garratt, Tims, et al. (1984), Hueber (1983, 1992), and Tims and Chambers (1984).

Lang and Cookson's (1935) material was described originally as late Silurian (lower Ludlow), based on a species of graptolite found associated with the plants at two of their localities. This graptolite was later redescribed as a new species, thus invalidating the original age determination. A further two graptolite species were then identified and interpreted as indicating an early Devonian (Pragian) age for Lang and Cookson's (1935) original material. However, Garratt (1978) identified a second, stratigraphically lower, plant assemblage that also contains *Baragwanathia* (Tims and Chambers 1984). This lower assemblage was dated as late Silurian (Ludlow) based on its geological relationships in the field and on the presence of two graptolite species *(Bohemograptus bohemicus, Monograptus* aff. *unicatus unicatus)*, each from different localities containing *Baragwanathia* (Garratt and Rickards 1984; Garratt, Tims, et al. 1984). A new stratigraphically intermediate locality has recently been documented and dated as late Silurian (Pridoli) based on the graptolite *Monograptus* cf. *prognatus* (Garratt and Rickards 1987). Controversy over the late Silurian age of the lower localities persists because the *Baragwanathia* assemblage is more typical of early or early middle Devonian floras in the Northern Hemisphere. Furthermore, the age determinations have been criticized based on ambiguity in identification and poor preservation of the graptolites (Hueber 1992).

A second species of the genus, *B. abitibiensis,* has been described from the Sextant Formation, Ontario, Canada (Hueber 1983). The age of the locality was given as early Devonian (middle to late Emsian) on the basis of dispersed spores. A third species, *B. oelheyi* Hundt, based on two axis fragments, is probably not a plant but rather some form of marine organism (Chaloner 1967a; Hueber 1983) and is not considered further. We treat *Baragwanathia* as monophyletic because of the presence of distinctive long, narrow, parallel-sided microphylls in both species. Our analysis of the relationships of the genus is based on the characters of *B. longifolia* except where otherwise stated.

The main axes of *B. longifolia* branch isotomously but also produce smaller lateral branches ("lateral buds"). Microphylls are simple, long, narrow, parallel-sided, contain a single vascular strand, and are arranged helically. Well-preserved leaves of *B. abitibiensis* are eligulate (Hueber 1983). Anatomical structure was preserved in pyrite in some *B. longifolia* material. The xylem is stellate in transverse section, and 12 lobes were observed in one large axis with deep bays between adjacent lobes. A decay-resistant hypodermis has not been observed even though the cortex and leaves are well preserved. All tracheids have an annular or helical thickenings, and those illustrated for *B. abitibiensis* have a G-type appearance (Hueber 1983, Figures 47–49). Fertile material is known only for *B. longifolia* and is poorly preserved. The sporangia are reniform and interspersed among leaves of identical morphology to those on vegetative axes. Whether the sporangia are cauline or axillary is unknown. Microspores with a distinctive trilete mark and little ornamentation have been isolated from the sporangia. The gametophyte is unknown.

Huperzia selago

The taxon *Huperzia selago* is based on the extant *H. selago* species group of Øllgaard (1987) and is equivalent to *Huperzia sensu* Wagner and Beitel (1992). The group comprises ground-dwelling taxa and is widely distributed in northern and southern temperate regions, as well as in montane areas in the paleotropics. There is a single neotropical species, *H. serrata* (Øllgaard 1992).

The main axes are erect and branch isotomously, but anisotomous lateral branching gives rise to characteristic small axes termed bulbils or gemmae (Figure 3.24; Øllgaard 1987; Stevenson 1976). Microphylls are simple, eligulate, contain a single vascular strand, and are helically arranged. The xylem is stellate in transverse section (as many as seven to nine lobes) with deep bays between adjacent lobes. Protoxylem having one or more poles occurs in the enlarged ends of the xylem lobes, and the metaxylem is composed of tracheids with circular bordered pits (Bierhorst 1971). The thickenings in the protoxylem are annular and connected by thin strands of additional wall material (Bierhorst 1960). A peripheral layer of thick-walled, nonlignified cortical cells is known to occur in some species. Vascular supply to the bulbils develops from one lobe of the main stellate strand to produce a large, elongate strand that becomes stellate distally (Stevenson 1976). Cortical roots are produced endogenously near the stem apex and have a crescent-shaped xylem strand. The sporangia are reniform, isovalvate, axillary to an unmodified sporophyll and are borne on short, nonvascular stalks (Figure 3.28). Sporangia are confined to relatively loose fertile areas that may alternate with sterile regions. Microspores are triradiate and belong to the "foveolate-fossulate" group (Wilce 1972) and are characteristically triangular with truncate corners and concave sides (Øllgaard 1987; Wilce 1972).

Gametophytes are subterranean, holosaprophytic, more or less cylindrical, unbranched, and have dorsiventral symmetry. Pluricellular, uniseriate hairs occur among the gametangia (Bruchmann 1898; Øllgaard 1987).

Huperzia squarrosa

The taxon *Huperzia squarrosa* is based on the *Huperzia squarrosa* species group of Øllgaard (1987) and is equivalent to *Phlegmariurus sensu* Wagner and Beitel (1992). The group comprises epiphytes and rupestrals and has a paleotropical distribution.

The main axes are pendant and branch isotomously. Microphylls are simple and eligulate, contain a single vascular strand, and are helically arranged. The xylem is stellate in transverse section (up to seven to nine lobes) with deep bays between adjacent lobes. Protoxylem having one or more poles occurs in the enlarged ends of the xylem lobes. Metaxylem is composed of tracheids with circular bordered pits (Bierhorst 1971). The thickenings in the protoxylem are annular and connected by thin strands of additional wall material (Bierhorst 1960). A peripheral layer of thick-walled, nonlignified cortical cells is known to occur in some species. Cortical roots are produced near the stem apex and have a crescent-shaped xylem strand. Sporangia are reniform, isovalvate, axillary to an unmodified sporophyll, and borne on short, nonvascular stalks (Figure 3.28). The sporangia are confined to relatively loose fertile areas that may alternate with sterile regions. Microspores are triradiate and belong to the "foveolate-fossulate" group with proximal face unsculptured (Wagner and Beitel 1992; Wilce 1972).

Gametophytes are subterranean, holosaprophytic, more or less cylindrical, branched, with dorsiventral symmetry. Pluricellular, uniseriate hairs occur among the gametangia (Bruchmann 1898; Øllgaard 1987).

Phylloglossum drummondii

The monospecific extant genus *Phylloglossum* is found only in Australia, Tasmania, and New Zealand. The most important literature relating to *P. drummondii* was reviewed by Øllgaard (1987).

The main axis is usually unbranched and arises from a subterranean tuber to bear 1, to as many as 20, filiform (up to 3 cm long) microphylls arranged in a compressed helix (Figure 6.9; Hackney 1950). The xylem is a hollow amphiphloic cylinder at the base of stems and breaks into several isolated strands distally (Bierhorst 1971). There is no peripheral layer of thick-walled cells. The xylem is composed entirely of tracheids with annular thickenings (Bierhorst 1960). One to three adventitious roots are produced from the tuber and each contains a crescent-shaped xylem strand (Hackney 1950). Sporangia are reniform, isovalvate (Figure 3.28), and borne on short, nonvascular stalks in the axils of ephemeral, peltate sporophylls. The sporangia are confined to a compact spike on the end of an elongate peduncle. Microspores are of the "phlegmaria-type" in the "foveolate-fossulate" group (Breckon and Falk 1974; Øllgaard 1987; Wilce 1972).

Gametophytes are surficial and green with an expanded disk-shaped gametangiophore borne apically on a short stalk (Holloway 1935).

Lycopodiella

Lycopodiella is a small (c. 40 extant species) genus of vascular plants that is widespread in temperate and tropical regions. The description of *Lycopodiella* is based mainly on features listed by Øllgaard (1987, 1990) and Wagner and Beitel (1992).

The main axes branch anisotomously and bear microphylls in a helical (sections *Campylostachys* and *Lateristachys*) or whorled (sections *Lycopodiella* and *Caroliniana*) arrangement. Leaves contain a single vascular strand and are eligulate. In smaller distal axes and sporelings the xylem is stellate in transverse section (up to seven to nine lobes) with deep bays between adjacent lobes. However, in larger axes the stellate shape is fragmented by parenchyma into an interconnected meshwork of xylem strands (Wardlaw 1924). Protoxylem having one or more poles occurs in the enlarged ends of the xylem lobes. Metaxylem is composed of tracheids with circular bordered pits (Bierhorst 1971). In the protoxylem, thickenings are annular and connected by thin strands of additional wall material (Bierhorst 1960). A subperipheral layer of thick-walled, lignified, cortical cells is known to occur in some species. The sporangia are reniform, strongly anisovalvate (except for section *Caroliniana*), and borne on short, nonvascular stalks in the axils of ephemeral subpeltate sporophylls. Sporangia are confined to a compact spike, which is sessile or pedunculate. Microspores are trilete and belong to the "rugate group" of Wilce (1972).

Gametophytes are surficial, green, hemisaprophytic, tuberous, and lobed on the upper side (Øllgaard 1987).

Lycopodium

Lycopodium is a small (c. 40 extant species) genus of vascular plants that is widespread in temperate and tropical montane regions. The description of *Lycopodium* is based mainly on features listed in Øllgaard (1987, 1990) and Wagner and Beitel (1992).

The main axes branch anisotomously and bear microphylls in a helical (sections *Lycopodiastrum, Magellania, Pseudolycopodium*), whorled (sections *Lycopodium, Annotina, Obscura*), ranked (sections *Diaphasium, Pseudodiaphasium*), or decussate (section *Complanatum*) arrangement (Figure 6.8). Leaves contain a single vascular strand and are eligulate. The xylem of the main branches and main roots is parallel banded in transverse section, whereas in smaller axes it is stellate with deep bays between adjacent lobes. Metaxylem is composed of tracheids with circular bordered pits (Bierhorst 1971). The thickenings of the protoxylem are annular and connected by thin strands of additional wall material (Bierhorst 1960). A subperipheral layer of thick walled, lignified, cortical cells is known to occur in some species. Roots emerge directly from the underside of main stems. The sporangia are reniform, isovalvate or slightly anisovalvate (Figure 3.28), and borne on short, nonvascular stalks in the axils of axillary to ephemeral peltate or subpeltate sporophylls. The sporangia are confined to a compact spike, which is sessile or pedunculate. Microspores are trilete and belong to the "reticulate group" of Wilce (1972).

The gametophytes are predominantly subterranean, holosaprophytic, having an expanded terminal gametangiophore produced from a ring meristem (Bruce 1979; Whittier 1981). Pluricellular, uniseriate hairs are absent from among the gametangia (Øllgaard 1987).

Leclercqia complexa†

The monospecific genus *Leclercqia* is based on well-preserved compressions and permineralizations from the upper part of the Panther Mountain Formation, Schoharie County, New York, United States (Banks, Bonamo, and Grierson 1972). The locality is dated as middle Devonian (Givetian).

The genus was created by Banks, Bonamo, and Grierson (1972) for lycopsids having a distinctive five-forked leaf, and the generic diagnosis was amended by Bonamo, Banks, and Grierson (1988) based on their own and earlier observations (Grierson and Bonamo 1979) of leaf and sporophyll morphology. The structure of the cell walls in the metaxylem was described in detail by Grierson (1976). *Leclercqia* is homosporous but was shown to be ligulate (Grierson and Bonamo 1979), and in this feature it differs from all other known (extinct or extant) homosporous lycopsids. Microspore morphology was illustrated by Streel (1972). Other specimens assignable to *Leclercqia* have been described from the following locations: the Trout Valley Formation, Maine, United States (Kasper and Forbes 1979; Kasper, Gensel, et al. 1988); New Brunswick, Canada (Kasper 1977); the Campo Chico Formation, Venezuela (Berry 1994); the Vesdre Synclinorium and the Namur Basin, Belgium (Fairon-Demaret 1981); and the Burdekin Basin, Queensland, Australia (Fairon-Demaret 1974). Recent reinvestigations of leaf morphology in *Protolepidodendron scharyanum* by Fairon-Demaret (1974, 1978, 1979, 1980, 1981) suggest that this taxon may be synonymous with *L. complexa*. This work has been reviewed and discussed by Bonamo, Banks, and Grierson (1988), Gensel and Andrews (1984), Li (1990), and Schweitzer and Cai (1987).

The main axes of *L. complexa* branch isotomously and pseudomonopodially and bear helically arranged microphylls with a single vascular strand (Figure 6.10). The microphylls have five forks: a long, narrow, lamina gives rise to two lateral projections about two-thirds of the way along its length; each lateral projection dichotomizes once, and a central, unbranched portion of the leaf forms the fifth lobe. The lateral projections are in the plane at right angles to the lamina, and the central projection turns sharply downward (Bonamo, Banks, and Grierson 1988). Except for the projections, the margin of the leaf lamina is entire. The ligule occurs on the leaf about two-thirds of the distance to the branch point, and on sporophylls, is just distal to the sporangium (Bonamo, Banks, and Grierson 1988; Grierson and Bonamo 1979). Anatomical information is preserved in pyritized specimens. The xylem is circular in transverse section, with small, distinctive ridges of protoxylem around the edge. Fourteen to eighteen such ridges are observed in some axes. A peripheral layer of thick-walled hypodermal cells is present in the cortex but is mainly associated with the leaf bases (Banks, Bonamo, and Grierson 1972). The cells of the metaxylem have predominantly round to oval bordered pits but no interconnecting pitlet sheet. Early-formed protoxylem is annular or helical. The sporangia are globose to ellipsoidal, and the long axis is radially, rather than tangentially, aligned with respect to the stem. Sporangia are sessile, isovalvate, and borne on unmodified leaves just below the point of forking in a loose, branched, fertile zone (Banks, Bonamo, and Grierson 1972; Bonamo, Banks, and Grierson 1988). Echinate microspores have been isolated from hundreds of sporangia (Grierson and Bonamo 1979; Streel 1972). The gametophyte is unknown.

Minarodendron cathaysiense†

The monospecific genus *Minarodendron* is based on compressions and permineralizations from the Haikou Formation of Yunnan Province, China (Li 1990). The formation is dated as middle Devonian (Givetian).

The genus was created by Li (1990) for specimens formerly named *Protolepidodendron scharyanum* (Krejci) Kräusel and Weyland, but renamed *P. cathaysiense* (Schweitzer and Cai 1987). A summary of the taxonomic history was given by Li (1990). The distinguishing feature of *Minarodendron* is its very small leaf, which is distinctively trifurcate.

The main axes branch isotomously and sometimes pseudomonopodially and bear helically arranged microphylls with a single vascular strand. The leaf is trifurcate: near the tip of the lamina two narrow lateral projections depart upward and the middle projection turns sharply downward. The margin of the leaf lamina bears small spines, but the presence or absence of ligules on the leaf has not been determined. Anatomical information is preserved in limonite. The xylem is circular with small, distinctive ridges of protoxylem around the edge—15 to 18 such ridges were observed in one axis. A peripheral layer of thick-walled hypodermal cells may be present (Li 1990, Figures 4, 5, 7, 10–16). Metaxylem cells have scalariform bordered pits with interconnecting pitlet sheets and protoxylem having annular or helical thickenings. The sporangia are elliptical, and the long axis is radially, rather than tangentially, aligned with respect to the stem. They are sessile, probably isovalvate, and borne on unmodified leaves just below the point of forking in a loose, branched fertile zone. No spores have been isolated from the sporangia. The gametophyte is unknown.

Colpodexylon deatsii†

Colpodexylon deatsii is the type species for the genus and is based on compressions and permineralizations from a number of localities in the Delaware River Flags (Portage Formation) and the Bellvale Sandstones (Marcellus Formation) in New York State, United States (Banks 1944). The localities are dated as late Devonian (lower part).

Two species, *C. deatsii* and *C. trifurcatum,* were recognized initially by Banks (1944), and a review of the genus with a discussion of its stratigraphic range was given by Grierson and Banks (1963). A third species, *C.* (?) *schopfii,* was described by Lemoigne and Itschenko (1980) based on compressions and permineralizations from upper Devonian (Famennian) volcanic tuffs on the Mokraya Volnovakha River in Ukraine. A fourth species, *C. cachiriense,* was named on the basis of compression fossils from the middle to upper Devonian (Givetian [?] to Frasnian) Campo Chico Formation of northwestern Venezuela (Edwards and Benedetto 1985). A fifth species, *C. variabile,* was created by Schweitzer and Cai (1987) for compression fossils from the middle Devonian (Givetian) of the Xichong Formation, Yunnan Province, China.

We tentatively treat *Colpodexylon* as monophyletic based on leaf shape. Our analysis is based on *C. deatsii* (Banks 1944), with additional anatomical information from *C.* (?) *schopfii* (Lemoigne and Itschenko 1980).

The axes of *C. deatsii* are more or less isotomously branched and bear helically arranged microphylls containing a single vascular strand. About half way along the lamina, two narrow lateral projections depart from opposite sides at an acute angle, and the leaf therefore has a trifurcate appearance. The presence or absence of ligules on the leaf has not been determined in *C. deatsii,* but possible ligules have been illustrated at the base of leaves in *C.* (?) *schopfii* (Lemoigne and Itschenko 1980). The anatomy of *C. deatsii* is preserved in pyrite. In transverse section the xylem varies from cylindrical, with small, distinctive ridges of protoxylem around the edge (Banks 1944) to stellate, with deep bays between adjacent lobes (Banks 1944; Lemoigne and Itschenko 1980). Five to seven lobes have been observed in one such axis. When stellate, the outer margin of individual lobes bears several (up to four have been observed) protoxylem ridges (Banks 1944). A decay-resistant hypodermal layer was not observed. The tracheids were described as scalariform (Banks 1944; Lemoigne and Itschenko 1980), but this feature has not been well illustrated and it is important to establish conclusively that these are not poorly preserved annular thickenings. The sporangia are elliptical, and the long axis is radially, rather than tangentially, aligned with respect to the stem. Sporangia are probably isovalvate, borne on unmodified leaves just below the point of forking, and appear to form a loose, fertile zone. No spores have been isolated from *Colpodexylon*. The gametophyte is unknown.

Haskinsia colophylla†

The species *Haskinsia colophylla* is based on compressions and permineralizations from several localities in the Catskill clastic wedge of New York State, United States (Grierson and Banks 1983). The age of the localities is given as middle to late Devonian (Givetian to early Frasnian).

This species was originally as *Drepanophycus colophyllus* (Grierson and Banks 1963), but the discovery of xylem anatomy and an investigation of leaf morphology led Grierson and Banks (1983) to create a new genus, *Haskinsia*. Bonamo, Banks, and Grierson (1988) further modified the description of the leaf morphology. One other species, *H. sagittata*, has been described based on compressions from the Campo Chico Formation of northwestern Venezuela (Edwards and Benedetto 1985). The specimens are dated as middle to late Devonian (Givetian [?] to Frasnian).

We tentatively treat *Haskinsia* as a monophyletic group based on the distinctive leaf morphology. The leaves are petiolate with an expanded, sagittate, distal lamina (Bonamo, Banks, and Grierson 1988; Edwards and Benedetto 1985; Grierson and Banks 1983). Our analysis is based on the most completely understood species, *H. colophylla* (Bonamo, Banks, and Grierson 1988; Grierson and Banks 1983).

The axes of *H. colophylla* are isotomously branched and bear helically arranged microphylls containing a single vascular strand (Grierson and Banks 1983). The microphylls are petiolate and have an expanded, sagittate, distal lamina (Bonamo, Banks, and Grierson 1988; Grierson and Banks 1983). The presence or absence of ligules has not been determined. Some anatomical structure is preserved in pyrite. The xylem is circular in transverse section with small, distinctive ridges of protoxylem around the edge—14 to 16 such ridges were observed in one axis. There is a peripheral layer of thick-walled, decay-resistant cells in the cortex. The metaxylem tracheids have naked, uniseriate to multiseriate, round, oval, and scalariform bordered pits. Early-formed protoxylem has annular and helical thickenings. The fertile region and spores are unknown. The gametophyte is unknown.

Selaginella selaginoides

Selaginella selaginoides is one of two extant species of *Selaginella* subgenus *Selaginella* (Selaginellaceae: Jermy 1990b). *S. selaginoides* is widely distributed in the northern hemisphere but is absent from Africa, and the second species, S. *deflexa*, is endemic to the Hawaiian archipelago.

Stems are erect and new shoots arise from prostrate creeping axes (Figure 6.7). Microphylls are isophyllous and helical, but there is little differentiation of the vegetative leaves and sporophylls. Ligules are present. The roots develop endogenously from the basal hypocotylar node (Karrfalt 1981). The stems are monostelic, with a terete, exarch xylem strand suspended within a stem cavity by trabeculate endodermal cells (Harvey-Gibson 1894, 1902). Protoxylem tracheids are annular, and the metaxylem tracheids have scalariform pitting (Bierhorst 1971). Some xylem elements resemble vessels (Harvey-Gibson 1894). Microsporangia are more or less reniform in face view and spheroidal in profile. Dehiscence is transverse into two equal valves (Figure 3.28). Strobili are bisporangiate with a megasporangiate zone at the base (Horner and Arnott 1963). Four megaspores are present per megasporangium. They are trilete and tuberculate and have compact particulate meshwork in the main part of wall (Tryon and Lugardon 1991). The microspores are trilete, and the distal surface has an echinate ornament with spines that have bifurcating tips (Tryon and Lugardon 1991).

The microgametophytes are endosporic and comprise a single antheridium. The spermatozoids are biflagellate and are released on breakdown of the spore wall. The megagametophytes begin their development within the spore wall. Tufts of rhizoids and archegonia develop superficially and are exposed upon rupture of the megaspore wall (Bierhorst 1971).

Selaginella subgenus *Stachygynandrum*

Our preliminary cladistic treatment of the Selaginellaceae (Chapter 6) suggests that *Stachygynandrum* is paraphyletic to *Heterostachys sensu* Jermy (1990b). We therefore include both subgenera in our concept of *Stachygynandrum*. *Stachygynandrum* is the largest subgenus of the Selaginellaceae, comprising some 360 extant species (subgenus *Stachygynandrum* c. 300, subgenus *Heterostachys* c. 60: Jermy 1990b). *Stachygynandrum* is widespread throughout the tropics, occurring predominantly in lowland to midmontane primary and secondary rain forest (Jermy 1990b). A few species occur in temperate regions where they are able to withstand seasonal drought (e.g., "resurrection plants" such as *S. denticulata* and *S. lepidophylla*).

The stems of *Stachygynandrum* are creeping or semiprostrate, sometimes erect, pseudomonopodial with much-branched laterals, and often strongly planate (Figure 6.6). The microphylls are ligulate and tetrastichous, and the vegetative leaves are dimorphic, the ventral pair being larger than the dorsal pair. The sporophylls are uniform or dimorphic. Rooting occurs through exogenous rhizophores at branch axils, or in erect forms, at the base of the main stem. Stems are polystelic or partly bistelic, with a flattened stele in some taxa. The stele is suspended within the stem cavity by trabeculate endodermal cells (Harvey-Gibson 1894, 1902). Protoxylem tracheids are annular, and metaxylem cells have scalariform pitting (Bierhorst 1971). Xylem maturation is exarch. Larger stems have hypodermal sclerenchyma (Bierhorst 1971). The root stele is C-shaped (Bierhorst 1971).

We interpret the basic strobilus morphology of the group as bisporangiate containing a basal megasporangiate zone (Horner and Arnott 1963). This condition is modified in some taxa to monosporangiate cones (e.g., *S. viridangula, S. inaequalifolia*) and in others to a single basal megasporangium (e.g., *S. galeotii, S. kraussiana*) or megasporangia and microsporangia in two rows (e.g., *S. lepidophylla*). We interpret microsporangium morphology as basically reniform for the group, but for many taxa the shape is more or less globose (Koller and Scheckler 1986). Microsporangium dehiscence is basically transverse into two valves, and a variety of spore dispersal mechanisms have been described (Koller and Scheckler 1986). Special modifications that allow repeated flexing of the sporangium valves to eject the spores (Koller and Scheckler 1986) may be a synapomorphy of the group. Microspores are trilete, usually echinate but sometimes cristate, regulate, papillate, and striate or spherulate (Tryon and Lugardon 1991). Usually there are four megaspores per megasporangium, but the number can vary (Hemsley 1993). Megaspores are trilete and predominantly reticulate with an equatorial ridge. Various types of wall structure have been reported (e.g., Hemsley, Collinson, and Brain 1992; Minaki 1984; Taylor 1989; Tryon and Lugardon 1991).

The microgametophytes are endosporic and multicellular when dispersed. An antheridium comprising an eight-celled jacket surrounding 256 biflagellate sperm forms within the microspore in *S. kraussiana* (Robert 1973). The developmental pathways of the jacket and sperm are determined through a periclinal division at the quadrant stage. The megagametophytes begin development when the megaspore undergoes a stage of free nuclear divisions followed by cell wall formation. Growth of the gametophyte causes the megaspore to split along the triradiate mark. Paraphyses and archegonia develop on the exposed surface (Bierhorst 1971; Webster 1992).

Selaginella subgenus *Tetragonostachys*

Tetragonostachys, a distinctive subgenus of the Selaginellaceae, comprises some 50 extant species that have a pantropical distribution, mainly in seasonally dry areas. A few species occur in mesophytic heathy woodlands in northern temperate regions (Jermy 1990b).

The stems are much branched, dichotomous, prostrate, and often mat-forming. The leaves are helical, ligulate, and dimorphic on prostrate axes and have little differentiation into vegetative leaves and sporophylls. The sporophylls are tetrastichous. Exogenous rhizophores are distributed along stems and are not confined to branch axils. The stems are monostelic with an elliptical to strap-shaped exarch xylem strand suspended within stem cavity by trabeculate endodermal cells (Harvey-Gibson 1894, 1902). Protoxylem tracheids are annular, and those in the metaxylem have scalariform pitting. Vessels have been documented in some taxa (Bierhorst 1971). The strobili are bisporangiate and have a megasporangiate zone at the base (Horner and Arnott 1963). Four megaspores are present per megasporangium, and they are trilete and reticulate (Tryon and Lugardon 1991). Microspores are trilete and granulate to papillate (Tryon and Lugardon 1991).

The microgametophytes are endosporic and comprise a single antheridium. Spermatozoids are biflagellate and are released on breakdown of the spore wall. Megagametophytes begin their development within the spore wall. Tufts of rhizoids and archegonia develop superficially and are exposed through rupture of the megaspore wall (Bierhorst 1971).

Paralycopodites†

The widespread genus *Paralycopodites* contains three species and is based on a suite of cone, stem, and root compressions and permineralizations formerly called *Lepidodendron brevifolium, L. veltheimianum, Paralycopodites brevifolius, Anabathra pulcherrima* (permineralized stems), *Ulodendron* (stem compressions), and *Flemingites* (cones) (DiMichele 1980). A complete synonymy and description is given by Pearson (1986), who suggested that *Anabathra* is synonymous with *Paralycopodites;* however, others regard this synonymy as questionable because of poor preservation in the type material of *Anabathra* (DiMichele and Bateman, in press). Permineralized stems of *Paralycopodites* range in age from early Visean to the end of the Westphalian (early Carboniferous). The description given here is based on the material described by DiMichele (1980) and Pearson (1986).

Paralycopodites is an arboreous rhizomorphic lycopsid with a columnar to sparsely dichotomized main trunk bearing deciduous lateral branch systems. The axes have persistent microphylls (no foliar scars) and branch isotomously and pseudomonopodially in the lower portions of the lateral branch system. The leaves are linear, have a single vascular strand, and are arranged in a helical phyllotaxy. A ligule and *foliar parichnos* (two parallel parenchymatous strands that extend the length of the leaf) are present. The xylem is a medullated exarch strand that is protostelic distally, and the protoxylem is uniformly distributed around a smooth to slightly undulated primary xylem margin. Tracheids are scalariform with fine strands of secondary wall material traversing the pit apertures. Wood rays are narrow, generally one to two cells wide, and many cells high. Phloem (sieve cells) have not been observed. The cortex has three zones. The outer cortex comprises homogenous, thick-walled, slightly elongate cells. The two inner zones form from a bifacial *periderm* (outer tissues of the stem comprising the cork cambium and its derivatives), which produces a massive homogenous *phellem* (external derivatives of the cork cambium) and thin *phelloderm* (internal derivatives of the cork cambium). Stomates are present.

The cones are bisporangiate with a basal megasporangiate zone and more than one tetrad per megasporangium. Roots are unknown but presumed to be stigmarian. The sporophylls are helically arranged and have lateral laminae. The sporangia are radially extended and sessile on the adaxial surface of the sporophyll stalk, which extends into an upturned lamina and downturned heel. The ligule is distal to the sporangium. Megaspores are of the *Lagenicula* or *Lagenioisporites* type. Microspores are of the *Lycopospora* type, having a narrow equatorial flange, smooth proximal surface, and papillate distal surface. The gametophyte is unknown.

Isoetes

Isoetes is a small (c. 150 extant species) and distinctive extant genus of vascular plants with a cosmopolitan distribution. The description of *Isoetes* is based on features listed by Bierhorst (1971) and Jermy (1990a) and in the provisional cladistic treatments of Hickey (1986, 1990) and Taylor and Hickey (1992).

The main axes are unbranched or, more rarely, branch isotomously (Figure 6.11). Microphylls are linear, laminate, ligulate, and helically arranged and contain a single vascular strand surrounded by four air canals. Foliar gemmae are present in some taxa (Hickey 1986). Stomates are absent in some submerged aquatics (Keeley, Osmond, and Raven 1984). The roots develop endogenously from a basal meristem, which may extend into a longitudinal groove along the stem and is similar anatomically and ontogenetically to the meristem of extinct arboreous lycopsids (Karrfalt 1984b; Rothwell and Erwin 1985; Rothwell and Pryor 1991; Stewart 1947). The roots possess a single monarch vascular bundle located on the side of a central lacuna. The stele of the stem is anchor-shaped in one longitudinal plane and more or less terete in transverse section. The cambium produces secondary cortex externally and secondary xylem, phloem, and parenchyma (Bierhorst 1971). The tracheids have annular and helical thickenings and therefore resemble protoxylem (Bierhorst 1960). The sporophylls and vegetative leaves are similar and all leaves have the potential to become fertile (sporophylls). Sporangia are radially elongate, ovoid to globose, and have no dehiscence line. They are sometimes covered by a thin membrane (velum). *Isoetes* is heterosporous, with the megaspores and microspores usually in separate sporangia. Microspores are 20–55 μm in diameter, monolete, and reniform to subspheroidal and range from laevigate to echinate or tuberculate (Tryon and Lugardon 1991). Megaspores are 300–800 μm in diameter, trilete, and spheroidal with equatorial flange, contain large amounts of silica in the wall, and have a range of surface ornament (Taylor 1992; Tryon and Lugardon 1991).

Microgametophytes are endosporic and differentiate into four sterile jacket cells that encapsulate a few multiflagellate sperm (Bierhorst 1971). Megagametophytes are also endosporic, beginning their development with a series of free nuclear divisions. Continued development ruptures the megaspore wall to expose the archegonia (Bierhorst 1971).

Eophyllophyton bellum†

The monospecific genus *Eophyllophyton* is based on relatively well preserved compressions and permineralizations from the Posongchong Formation of Yunnan Province, China (Hao 1988; Hao and Beck 1993). The age of the locality is given as early Devonian (Pragian) (Hao 1989b; Hao and Beck 1991a).

The main axes are pseudomonopodially branched (isotomous distally), are nonplanar, and bear small, blunt, multicellular spines. The lateral appendages are small, forked axes that recurve at the apex or are laminar, much divided, and leaflike. The cuticle consists of elongate cells. The xylem is terete in transverse section, centrarch, and composed of G-type tracheids having annular or reticulate thickenings. The stems contain a continuous outer zone of thick-walled, decay-resistant cells. The sporangia are circular to reniform (dehiscence unclear) and are borne in clusters on the adaxial surface of leaflike appendages. Spores are unknown. The gametophyte is unknown.

Psilophyton crenulatum†

The species *Psilophyton crenulatum* is based on exceptionally well preserved material from a road-cut section in the Dalhousie Group, Atholville, New Brunswick, Canada. The age of the locality is given as early Devonian (Emsian). The description is based on Doran 1980.

The main axes are spiny and pseudomonopodially branched (isotomous distally) with helical laterals and probably developed from a prostrate rhizome. The spines are prominent, multicellular, and bifurcate or trifurcate but have no phyllotactic order (Figure 5.19). There is segregation of the fertile and sterile lateral appendages. The ultimate appendages, where fertile, comprise pairs of sporangia borne terminally in much-branched clusters on stalks that are recurved toward the axis at maturity. The sterile ultimate appendages are nonplanar and terminate in slender blunt tips that are recurved. Circinate vernation has not been demonstrated. The primary xylem is protostelic and terete; metaxylem alignment (possible incipient cambial activity) occurs in larger axes. The protoxylem is centrarch. Cortical tissues are not preserved. The tracheids are scalariform pitted but otherwise are not well preserved. Stomates are present. The sporangia are fusiform and dehisce longitudinally along one side (Figure 3.28). Spores (in situ) are trilete, circular to subcircular, curvaturate, minutely ornamented (coni 0.5–1.0 μm), of one size (48–120 μm, mean = 70 μm), and of the *Apiculiretusispora* type. The gametophyte is unknown.

Psilophyton dawsonii†

The species *Psilophyton dawsonii* is based on well-preserved mineralizations and compression fossils from several localities in the Battery Point Formation, Gaspé Bay, Quebec, Canada and the Sextant Formation, Abitibi River, Ontario, Canada. The age of both localities is given as early Devonian (Emsian) (Banks, Leclercq, and Hueber 1975). The description given here is based on Banks, Leclercq, and Hueber 1975.

The main axes are naked and pseudomonopodially branched (isotomous distally) with helical laterals (Figure 4.13). There is segregation of fertile and sterile lateral appendages. Fertile ultimate appendages comprise sporangia borne terminally in much-branched clusters and in pairs on short stalks that are recurved toward the axis at maturity. Sterile ultimate appendages are nonplanar and terminate in slender blunt tips that are recurved. Circinate vernation has not been observed. Small, more or less axillary branches sometimes occur in the distal parts of the plant. The primary xylem is protostelic and terete (lobed in areas of profuse branching); metaxylem alignment (possible incipient cambial activity) occurs in the larger axes. The protoxylem is centrarch (Figure 4.24). The cortex has two zones: the inner cortex is not preserved; the outer cortex comprises a continuous layer of thick-walled sterome. The tracheids are scalariform pitted and have a thick, decay-resistant inner wall layer and a distinctive perforate layer within the pit chamber (Hartman and Banks 1980). The epidermis comprises small, box-shaped cells and stomates. The sporangia are fusiform and dehiscence longitudinally along one side (Figure 3.28). Spores (in situ) are trilete, commissurate, curvaturate, minutely ornamented, of one size (40–75 μm), and of the *Retusotriletes* or *Apiculiretusispora* type (Allen 1980; Gensel 1980). The gametophyte is unknown.

Ibyka amphikoma†

The monospecific genus *Ibyka* is based on compressions and permineralizations from the Moscow Formation, Schoharie County, New York State, United States (Skog and Banks 1973). The age of the locality is given as late middle Devonian (Givetian).

The axes of *I. amphikoma* are smooth and branch pseudomonopodially (isotomously distally) in three dimensions (Figure 4.16). The ultimate appendages, originally described as helical, have been reinterpreted as probably whorled (Stein, Wight, and Beck 1984). Whether vegetative or fertile, these appendages are dichotomously (isotomously) branched and have recurved tips or terminal sporangia. The primary xylem forms a solid protostele possessing three main arms with lobed ends containing mesarch protoxylem. The branch traces are initially more or less terete and unlobed but become lobed distally. The protoxylem disintegrates to form conspicuous lacunae. Outer tissues of the axis are unknown but possibly comprise sclereids associated with ridges that are on the outer surface. The tracheids are probably scalariform pitted. Cuticle and stomates are unknown. The sporangia are obovoid to pyriform (more or less fusiform) and clustered in terminal pairs on short lateral branches. The mode of sporangium dehiscence is unknown. Spores are unknown. The gametophyte is unknown.

Equisetum

Equisetum is a small (c. 15 extant species), distinctive, and divergent extant genus of vascular plants with a cosmopolitan distribution with the exception of Australia, New Zealand, and Antarctica. Most species are found between 40° and 60° north latitude (Hauke 1990), where they are primarily confined to seasonally wet ground, often around rivers and lake margins. Two extant subgenera have been recognized on the basis of stomatal position: *Equisetum* (eight species, superficial stomates) and *Hippochaete* (seven species, sunken stomates) (Hauke 1990). Defining features of the genus include the well-developed lateral fusion of leaf laminae and eight or more sporangia per "operculum" (protective shield) (Bateman 1991). Fossils assigned to *Equisetum* or *Equisetites* date mostly from the Triassic Period or later, with possible records going back to the Carboniferous (T. N. Taylor and E. L. Taylor 1993). Relationships among fossil and extant taxa are poorly understood, but some fossils, such as *Equisetum fluviatoides* from the Paleocene of Canada (McIver and Basinger 1989), are clearly closely related to crown group species. Cladistic studies show that *Equisetum* is nested in the diverse sphenopsid clade (Bateman 1991). Sphenopsids first appear in the upper Devonian and include important Carboniferous groups such as the arborescent Archaeocalamitaceae and Calamitaceae (Bateman 1991; Skog and Banks 1973; Stein, Wight, and Beck 1984). The description of *Equisetum* given here is based mainly on Bierhorst 1971 and Hauke 1990.

The main axes of *Equisetum* are characterized by a jointed appearance with small leaves in whorls fused into a nodal sheath (Figure 3.25). Superficial ridges, corresponding to underlying vascular strands, alternate on successive internodes. Branches develop exogenously at internodes in some species and alternate with individual leaves. A persistent and extensive rhizome bears roots, tubers (in some species), and aerial branches. The stem anatomy is highly distinctive (Bierhorst 1971). Circling the pith is a ring of vascular bundles equal in number to the leaves. These bundles comprise an inner group of xylem cells that contain a conspicuous cavity *(lacuna* or *carinal canal)* at maturity and an outer group of phloem cells. The position and number of endodermal layers vary within the genus. Other cavities, *vallecular canals,* develop within the cortex and extend the full length of the internode. The outer cortex comprises a layer of chlorenchyma. The tracheids are distinctive: early-formed elements have annular or helical thickenings, and some later-formed elements bear circular bordered pits between adjacent annular rings (Bierhorst 1971). The stomates occur in sunken or superficial rows. The roots contain a simple cortex around a stele that comprises one large metaxylem element having two or three peripheral protoxylem points. The cones terminate the stems and comprise whorls of peltate sporophylls bearing eight or more elongate sporangia. Spores are alete and chlorophyllous and have four distinctive, straplike hygroscopic "elaters."

The gametophytes are small (up to 1 cm in diameter), surficial, chlorophyllous, and parenchymatous, possessing unicellular ventral rhizoids and dorsal plates of chlorophyllous tissue. The archegonia are sunken and flask-shaped with a short neck. The antheridia are circular to oval and superficial. Spermatozoids are helical and multiflagellate (Bierhorst

1971; Duckett 1973; Hauke 1990). Ultrastructural details of male gametes and gametogenesis were reviewed by Garbary, Renzaglia, and Duckett (1993).

Pseudosporochnus nodosus†

The species *Pseudosporochnus nodosus* is based on well-preserved compressions and permineralizations from the middle Devonian Brandt Quarry, Goé, Belgium (Gensel and Andrews 1984; Leclercq and Banks 1962; Leclercq and Lele 1968). Detailed descriptions of the anatomy of *P. hueberi* were given by Stein and Hueber (1989).

P. nodosus is a large plant comprising a main stem (trunk) and second-order axes that branch terminally and digitately. The third-order axes are pseudomonopodial and bear helical ultimate appendages, and the terminal axes are more or less isotomous (Figure 4.17). Fertile and sterile ultimate appendages are similar. The ultimate appendages dichotomize unequally in one plane, but isotomously toward the apexes. Fusiform sporangia are borne in clusters in terminal pairs. Dehiscence is probably longitudinal along one side of the sporangium (Figure 3.28). The xylem comprises many individual steles that make an anastomosing vascular cylinder and is similar between first- and third-order axes. Individual steles are elongate to strap-shaped; some are U-shaped and radially aligned in the outer part of the stem. The foliar traces become elliptical and bilaterally symmetrical and develop two protoxylem poles. Peripheral loops are present in the protoxylem. Some radial alignment of the protoxylem may occur (Leclercq and Lele 1968), but otherwise, there is no evidence of cambial activity. The cortex contains a discontinuous ring of sclereid bundles arranged in clusters ("nests"). Tracheid pitting is mainly scalariform or reticulate. Elongate, rootlike axes form an acute angle with the main stem. Spores are unknown. The gametophyte is unknown.

Rhacophyton ceratangium†

The taxon *Rhacophyton ceratangium* is based on well-preserved compressions and permineralizations from the Hampshire Formation at Valley Head, West Virginia, United States (Andrews and Phillips 1968). The age of the locality is given as late Devonian (Famennian). Two more recent studies provide additional important information on morphology (Cornet, Phillips, and Andrews 1976) and anatomy (Dittrich, Matten, and Phillips 1983).

The axes are basically quadriseriate: lateral branches are in pairs with a common base and a pair of aphlebia-like appendages. Second- and third-order branches *(pinnules)* are planar, with basal stipulelike structures *(aphlebia)* and alternate laterals in two rows (Figure 4.15). This planar appearance is interpreted as a modification of a basic quadriseriate branching pattern throughout the plant as a result of the reduction of one branch of a pair to the aphlebia-like structure (Cornet, Phillips, and Andrews 1976). The most common type of vegetative pinnules has recurved tips. Fertile second-order axes are quadriseriate and nonplanar and bear third-order sterile and fertile axes. Third-order axes on fertile branches comprise a pair of planar vegetative pinnules and a pair of nonplanar, much-branched axes bearing dense clusters of sporangia. The sporangia are borne on short stalks along one side of the ultimate axes are fusiform, with an extended distal tip, and dehisce longitudinally, probably along one side (Figure 3.28). The roots are probably adventitious with terete primary xylem and secondary wood (Cornet, Phillips, and Andrews 1976). Primary and secondary branches of the stem have a clepsydroid primary xylem surrounded by secondary xylem with uniseriate rays (Cornet, Phillips, and Andrews 1976; Dittrich, Matten, and Phillips 1983). Protoxylem lacunae (peripheral loops) are well developed, and the position of the protoxylem appears to be mesarch. Pinna trace emission is of the etapteroid type (Dittrich, Matten, and Phillips 1983). The tracheids have scalariform bordered pits (Dittrich, Matten, and Phillips 1983). *Rhacophyton* is homosporous, and the spores (in situ) are trilete, ovoid with a thin perispore, 52–70 μm in diameter, and granulate. The gametophyte is unknown.

Psilotum

Psilotum is a small extant genus (c. 2–3 species) of vascular plants that is widespread in moist tropical to warm temperate regions occurring in the southeastern United States, extreme southwestern Europe, central Japan, South Korea, and many Pacific islands (Kramer 1990). The genus has not been documented in the fossil record. The description is based mainly on Bierhorst 1971 and Kramer 1990.

The main axes branch isotomously and bear small, weakly vascular or nonvascular, awl-shaped to squamiform leaves. Branching is decussate, and the leaves are helically arranged in *P. nudum,* whereas in *P. complanatum* and the possibly conspecific epiphyte *P. flaccidum,* branching is mostly planar and the leaves are borne in rows. Aerial axes develop a stellate xylem with mesarch to exarch protoxylem. A well-developed sclerenchymatous cortex is present in the basal regions of aerial axes. The protoxylem tracheids have annular to helical thickenings; metaxylem tracheids have scalariform to circular bordered pitting. Stomates are present. The subterranean axes are branched and leafless and bear numerous multicellular rhizoids. The steles of these axes have one to several tracheids and become larger and dissected in the larger axes. An endodermis is present.

The sporangia develop in a distinct distal fertile zone. The *synangium,* usually comprising three fused sporangia, is borne on the adaxial surface of a small, two-lobed sporophyll. A vascular strand develops between the sporangium and the stele but does not extend into the sporophyll. The synangium wall is four to six cells thick, and each of the fused sporangia has a distinctive group of superficial apical cells that have been interpreted as a vestigial annulus (Bierhorst 1971, 180). Dehiscence proceeds from the apex of the synangium to the "annulus" of each sporangium. Spores are elongate-ellipsoidal ("bean-shaped"), 63–78 μm long, and monolete with a coarsely rugose to shallowly and compactly verrucate surface (Tryon and Lugardon 1991).

The gametophytes are axial, branched, subterranean, and mycorrhizal and resemble the subterranean sporophyte axes. A simple stele comprising one to three tracheids, phloem, and endodermis has been documented in some gametophytes of tetraploid sporophytes. The gametophytes bear rhizoids and a cuticle. Gemmae occur in both sporophytes and gametophytes of *P. nudum.* The antheridia are large and superficial and have a jacket that is one cell thick with an opercular cell. Spermatozoids are multiflagellate. The archegonia are flask-shaped with a tiered neck, basal egg, and ventral canal cell.

Tmesipteris

Tmesipteris is a small (c. 13 species) extant genus of mainly epiphytic vascular plants with a predominantly tropical distribution that encompasses parts of Southeast Asia, Australia, New Zealand, and several Pacific islands (Braithwaite 1988; Kramer 1990). The description below is based mainly on Bierhorst 1971 and Kramer 1990.

The main axes are usually unbranched (isotomously branched in some species) and bear leaves that range from tiny, nonvascular, and squamiform (*Psilotum*-like) at the base to larger, vascular, more or less lanceolate, obtuse to acute, often mucronate, distally. The leaf arrangement ranges from two-rowed (e.g., *T. lanceolata*) to helical (e.g., *T. vieillardi*), and the axes terminate in a leaf. Aerial axes develop a C-shaped xylem at the base and a ring of mesarch strands distally. The protoxylem tracheids have annular to helical thickenings; metaxylem tracheids have scalariform to circular bordered pitting. Stomates are present. The subterranean axes are branched and leafless and bear numerous multicellular rhizoids. The steles of these axes have one to several tracheids and become larger and dissected in the larger axes. An endodermis and cortical collenchyma are present.

The sporangia develop in a fertile zone at the base or in one or more zones along the length of the frond. The synangium, usually comprising two fused sporangia, is borne on the adaxial surface of a small, two-lobed sporophyll. A vascular strand develops between the sporangium and the stele, and both lobes of the sporophyll are vascular. The synangium wall is four to six cells thick. Spores are elongate-ellipsoidal ("bean-shaped"), 45–62 μm long, and monolete with an irregularly rugose to shallowly and compactly verrucate surface (Tryon and Lugardon 1991).

The gametophytes are axial, branched, subterranean, and mycorrhizal and resemble the subterranean sporophyte axes. The gametophytes bear rhizoids and have a cuticle, and the gametangia occur in distinct zones in some species. The antheridia are large and superficial and have a jacket that is one cell thick with an opercular cell. Spermatozoids are multiflagellate. The archegonia are flask-shaped with a tiered neck, basal egg, and ventral canal cell.

Pertica varia†

The type species is *Pertica quadrifaria* (Kasper and Andrews 1972). We chose *P. varia* as the basis for our analysis because more morphological details are available. This species is based on compression fossils from the Battery Point Formation, Gaspé Bay, Quebec, Canada (Granoff, Gensel, and Andrews 1976). The age of the type locality is given as early Devonian (Emsian).

The main axes are papillate and pseudomonopodially branched (isotomous distally) with a decussate (four-rowed) arrangement of laterals. The papillae are less than 0.4 mm long. There is segregation of the fertile and sterile lateral appendages. The fertile ultimate appendages comprise sporangia borne terminally and erect in branched clusters and in pairs on short stalks. The sterile ultimate appendages are nonplanar and terminate in slender blunt tips that are recurved toward the axis. Circinate vernation has not been demonstrated. Stem anatomy is unknown. The sporangia are oval to fusiform, and dehiscence occurs longitudinally along one side of sporangium (Figure 3.28). Spores (in situ) are subcircular to subtriangular, azonate, curvaturate, trilete with minute ornament (coni) less than 1 μm high, of one size (c. 56–90 μm), and of the *Apiculiretusispora* type. The gametophyte is unknown.

Tetraxylopteris schmidtii†

The monospecific genus *Tetraxylopteris* is based on compressions and permineralizations from localities in New York State, United States (Beck 1957; Bonamo and Banks 1967; Scheckler and Banks 1971a). The age of the localities is given as late Devonian (Gensel and Andrews 1984).

The main axes are naked, monopodial, and decussately branched (four-rowed), isotomous distally (Figure 4.14; Bonamo and Banks 1967; Scheckler and Banks 1971a). There is segregation of the fertile and sterile lateral appendages. The fertile ultimate appendages comprise pairs of sporangia borne terminally in much-branched clusters that are oriented adaxially (Bonamo and Banks 1967). The sterile ultimate appendages are decussately branched and not planated (Scheckler and Banks 1971a). Circinate vernation and marked recurvation of ultimate branches has not been observed. The primary xylem is protostelic and has four lobes at all levels, except in the ultimate branches, where it is terete. The protoxylem "radiates" from the center of the stele to a mesarch position in the stelar lobes (Beck 1957; Gensel and Andrews 1984; Wight 1987) and contains some thin-walled parenchyma (Scheckler and Banks 1971a). The cambium is bifacial and produces secondary xylem and phloem in radial rows (Beck 1957). Xylem rays are up to 100 cells high and are mostly uniseriate (Scheckler and Banks 1971a). The cortex has two zones: the inner cortex is wide and contains large parenchymatous cells with some fibers; the outer cortex comprises alternating bundles of thick-walled fibers and parenchyma (Beck 1957; Scheckler and Banks 1971a). Tracheids have scalariform and multiseriate circular bordered pitting. The epidermis has elongate cells with stomates (Scheckler and Banks 1971a). The sporangia are oval to fusiform, have acute apexes, and dehisce longitudinally along one side (Figure 3.28; Bonamo and Banks 1967). Spores (in situ) are spherical, trilete, pseudosaccate (the exine is expanded into a much-folded bladder), of one size (73–176 μm; mean = 130 μm), and of the *Rhabdosporites* type. The ornament consists of fine, dense, granular rods that are less than 1 μm high (Bonamo and Banks 1967). The gametophyte is unknown.

Pinnatiramosus qianensis†

The problematic monospecific genus *Pinnatiramosus* is based on compression fossils and permineralizations from the Xiushan Formation, which is in the northern part of Guizhou, China (Geng 1986). The age of the locality is given as early Silurian (late Llandovery) because of the presence of brachiopods, chitinozoans, and bivalves (Cai, Shu, et al. 1996). This description is based on Geng (1986) and on personal examination of the holotype by one of us (P.K.).

The plant is at least 60 cm long and has main axes that branch anisotomously and give rise to much-branched, planar, "pinnate" laterals. The "pinnae" comprise long, narrow axes that are slightly constricted at the base and are characteristically curved at their point of departure. The extent of the curvature may result in pinnae running parallel to the "midrib" on some laterals. Several orders of pinnate branching occur on some axes. The axes contain tracheid-like cells with scalariform and circular pitting. Stomates have not been observed. Reproductive structures are unknown.

The presence of tracheid-like cells that have scalariform pitting suggests that *P. qianensis* is a vascular plant, but the anatomical preservation is poor and the details of cell wall structure are difficult to interpret. In the absence of further information on axis anatomy and critical data on the reproductive structures, the systematic position of *P. qianensis* remains obscure.

A further problem with interpreting the affinities and age of *P. qianensis* is that the fossils lie 2.2 m below a major unconformity between Silurian and overlying Permian strata (Cai, Shu, et al. 1996). The sediments containing *P. qianensis* are clearly Silurian (probably upper Llandovery: Cai, Shu, et al. 1996) but because of the proximity to the unconformity, it is possible that the plant fossils were derived from younger strata. *P. qianensis* could be interpreted as part of a root system from a late Paleozoic (e.g., Permian) vascular plant that grew on or close to the present surface of the unconformity. The roots may have penetrated loosely consolidated or weathered Silurian sediments and grown along bedding planes to produce the planar "pinnate" morphology. Further lithifaction of the Silurian sediments and fossilization of the roots would have followed the deposition of the overlying Permian strata. This hypothesis was rejected by Cai, Shu, et al. (1996) because putative roots have not been found in the 2.2 m of Silurian sediments overlying the *P. qianensis* fossils and because the Silurian sediments appear to be unweathered. Documentation of reproductive structures (currently unknown) would provide conclusive evidence that *P. qianensis* is not a root.

APPENDIX 2

Data for Analyses of Embryobiota

TABLE A2.1

Characters and Character States for Taxa Used in the Cladistic Analysis in Chapter 3

Taxon	Sporophyte cellularity (character 3.1)	Sporophyte independence or dependence (character 3.2)	Well-developed sporangiophore (character 3.3)	Sporophyte branching (character 3.4)	Intercalary meristem (character 3.5)
Andreaea	1 multicellular	0 dependent	0 absent	0 absent	0 absent
Andreaeobryum	1 multicellular	0 dependent	1 present	0 absent	0 absent
Anthoceros	1 multicellular	0 dependent	0 absent	0 absent	1 present
Chara	0 unicellular	0 dependent	? NA	? NA	? NA
Coleochaete	0 unicellular	0 dependent	? NA	? NA	? NA
Equisetum	1 multicellular	1 independent	1 present	1 present	0 absent
Haplomitrium	1 multicellular	0 dependent	0 absent	0 absent	0 absent
Horneophyton	1 multicellular	1 independent	1 present	1 present	0 absent
Huperzia	1 multicellular	1 independent	1 present	1 present	0 absent
Monoclea	1 multicellular	0 dependent	0 absent	0 absent	0 absent
Notothylas	1 multicellular	0 dependent	0 absent	0 absent	1 present
Polytrichum	1 multicellular	0 dependent	1 present	0 absent	0 absent
Sphaerocarpos	1 multicellular	0 dependent	0 absent	0 absent	0 absent
Sphagnum	1 multicellular	0 dependent	0 absent	0 absent	0 absent
Takakia	1 multicellular	0 dependent	1 present	0 absent	0 absent

TABLE A2.1
Continued

Taxon	Foot form (character 3.6)	Xylem-cell thickenings (character 3.7)	Sporangium (character 3.8)	Sporangium dehiscence (character 3.9)	Columella (character 3.10)
Andreaea	1 tapering	0 absent	1 present	0 linear	1 present
Andreaeobryum	1 tapering	0 absent	1 present	0 linear	1 present
Anthoceros	0 bulbous	0 absent	1 present	0 linear	1 present
Chara	? NA	? NA	? NA	? NA	? NA
Coleochaete	? NA	? NA	? NA	? NA	? NA
Equisetum	0 bulbous	1 present	1 present	0 linear	0 absent
Haplomitrium	0 bulbous	0 absent	1 present	0 linear	0 absent
Horneophyton	? unknown	0 absent	1 present	? unknown	1 present
Huperzia	0 bulbous	1 present	1 present	0 linear	0 absent
Monoclea	0 bulbous	0 absent	1 present	0 linear	0 absent
Notothylas	0 bulbous	0 absent	1 present	0 linear	1 present
Polytrichum	1 tapering	0 absent	1 present	1 operculate	1 present
Sphaerocarpos	0 bulbous	0 absent	1 present	0 linear	0 absent
Sphagnum	0 bulbous	0 absent	1 present	1 operculate	1 present
Takakia	1 tapering	0 absent	1 present	0 linear	1 present

TABLE A2.1
Continued

Taxon	Elaters and pseudoelaters (character 3.11)	Zoospore flagella (character 3.12)	Perine (character 3.13)	Cuticle (character 3.14)	Stomates (character 3.15)
Andreaea	0 absent	1 absent	1 present	1 present	0 absent
Andreaeobryum	0 absent	1 absent	1 present	1 present	0 absent
Anthoceros	1 present	1 absent	0 absent	1 present	1 present
Chara	0 absent	1 absent	? unknown	0 absent	? NA
Coleochaete	0 absent	0 present	0 absent	0 absent	? NA
Equisetum	0 absent	1 absent	1 present	1 present	1 present
Haplomitrium	1 present	1 absent	0 absent	1 present	0 absent
Horneophyton	0 absent	1 absent	? unknown	1 present	1 present
Huperzia	0 absent	1 absent	1 present	1 present	1 present
Monoclea	1 present	1 absent	0 absent	1 present	0 absent
Notothylas	1 present	1 absent	0 absent	1 present	1 present
Polytrichum	0 absent	1 absent	1 present	1 present	1 present
Sphaerocarpos	1 present	1 absent	0 absent	1 present	0 absent
Sphagnum	0 absent	1 absent	1 present	1 present	1 present
Takakia	0 absent	1 absent	? unknown	1 present	0 absent

TABLE A2.1
Continued

Taxon	Gametophyte form (character 3.16)	Gametophyte leaves (character 3.17)	Archegonium (character 3.18)	Antheridium morphology (character 3.19)	Antheridium development (character 3.20)
Andreaea	1 axial	1 present	1 present	1 jacket	0 exogenous
Andreaeobryum	1 axial	1 present	1 present	1 jacket	0 exogenous
Anthoceros	0 thalloid	0 absent	1 present	1 jacket	1 endogenous
Chara	1 axial	0 absent	0 absent	0 naked	0 exogenous
Coleochaete	0 thalloid	0 absent	0 absent	0 naked	0 exogenous
Equisetum	0 thalloid	0 absent	1 present	1 jacket	0 exogenous
Haplomitrium	1 axial	1 present	1 present	1 jacket	0 exogenous
Horneophyton	1 axial	0 absent	1 present	? unknown	? unknown
Huperzia	1 axial	0 absent	1 present	1 jacket	0 exogenous
Monoclea	0 thalloid	0 absent	1 present	1 jacket	0 exogenous
Notothylas	0 thalloid	0 absent	1 present	1 jacket	1 endogenous
Polytrichum	1 axial	1 present	1 present	1 jacket	0 exogenous
Sphaerocarpos	0 thalloid	0 absent	1 present	1 jacket	0 exogenous
Sphagnum	1 axial	1 present	1 present	1 jacket	0 exogenous
Takakia	1 axial	1 present	1 present	1 jacket	0 exogenous

TABLE A2.1
Continued

Taxon	Archegonium position (character 3.21)	Gametangia distribution (character 3.22)	Paraphyses (character 3.23)	Rhizoid cellularity (character 3.24)	Protonema type (character 3.25)
Andreaea	0 superficial	1 terminal	1 present	1 multicellular	1 filamentous
Andreaeobryum	0 superficial	1 terminal	1 present	1 multicellular	1 filamentous
Anthoceros	1 sunken	0 nonterminal	0 absent	0 unicellular	0 nonfilamentous
Chara	0 superficial	0 nonterminal	0 absent	0 unicellular	0 nonfilamentous
Coleochaete	0 superficial	0 nonterminal	0 absent	0 unicellular	0 nonfilamentous
Equisetum	1 sunken	0 nonterminal	0 absent	0 unicellular	0 nonfilamentous
Haplomitrium	0 superficial	1 terminal	0 absent	? NA	0 nonfilamentous
Horneophyton	1 sunken	1 terminal	0 absent	0 unicellular	? unknown
Huperzia	1 sunken	1 terminal	1 present	0 unicellular	0 nonfilamentous
Monoclea	0 superficial	0 nonterminal	0 absent	0 unicellular	0 nonfilamentous
Notothylas	1 sunken	0 nonterminal	0 absent	0 unicellular	0 nonfilamentous
Polytrichum	0 superficial	1 terminal	1 present	1 multicellular	1 filamentous
Sphaerocarpos	0 superficial	0 nonterminal	0 absent	0 unicellular	0 nonfilamentous
Sphagnum	0 superficial	1 terminal	0 absent	1 multicellular	0 nonfilamentous
Takakia	0 superficial	1 terminal	0 absent	? NA	? unknown

TABLE A2.1
Continued

Taxon	Oil bodies (character 3.26)	Mucilage clefts (character 3.27)	Pseudopodium (character 3.28)	Pyrenoids (character 3.29)	Bicentriolar centrosomes (character 3.30)
Andreaea	0 absent	0 absent	1 present	? unknown	1 bicentriolar
Andreaeobryum	0 absent	0 absent	0 absent	? unknown	? unknown
Anthoceros	0 absent	1 present	0 absent	0 present	1 bicentriolar
Chara	0 absent	0 absent	0 absent	? unknown	0 orthogonal
Coleochaete	0 absent	0 absent	0 absent	0 present	0 orthogonal
Equisetum	0 absent	0 absent	0 absent	? unknown	2 peripheral
Haplomitrium	1 present	0 absent	0 absent	1 absent	1 bicentriolar
Horneophyton	? unknown	0 absent	0 absent	? unknown	? unknown
Huperzia	0 absent	0 absent	0 absent	? unknown	? unknown
Monoclea	1 present	0 absent	0 absent	? unknown	? unknown
Notothylas	0 absent	1 present	0 absent	0 present	1 bicentriolar
Polytrichum	0 absent	0 absent	0 absent	? unknown	1 bicentriolar
Sphaerocarpos	1 present	0 absent	0 absent	1 absent	1 bicentriolar
Sphagnum	0 absent	0 absent	1 present	1 absent	1 bicentriolar
Takakia	0 absent	0 absent	0 absent	1 absent	1 bicentriolar

TABLE A2.1
Continued

Taxon	Orientation of lamellae in multi-layered structure (character 3.31)	Preprophase bands in mitosis (character 3.32)	Lunularic acid (character 3.33)	D-Methionine (character 3.34)
Andreaea	1 40–45°	? unknown	? unknown	1 distinguished
Andreaeobryum	? unknown	? unknown	? unknown	? unknown
Anthoceros	? unknown	? unknown	0 absent	1 distinguished
Chara	0 90°	0 absent	? unknown	? unknown
Coleochaete	0 90°	0 absent	? unknown	? unknown
Equisetum	1 40–45°	? unknown	0 absent	1 distinguished
Haplomitrium	1 40–45°	? unknown	1 present	? unknown
Horneophyton	? unknown	? unknown	? unknown	? unknown
Huperzia	? unknown	? unknown	? unknown	1 distinguished
Monoclea	? unknown	? unknown	1 present	? unknown
Notothylas	1 40–45°	1 present	? unknown	? unknown
Polytrichum	1 40–45°	? unknown	0 absent	1 distinguished
Sphaerocarpos	1 40–45°	? unknown	1 present	0 not distinguished
Sphagnum	1 40–45°	1 present	0 absent	1 distinguished
Takakia	1 40–45°	? unknown	? unknown	? unknown

TABLE A2.2

Data Matrix for Chapter 3 in Numeric Format

Andreaea	1000010101011101111001111001?11??1
Andreaeobryum	1010010101011101111001111000??????
Anthoceros	1000100101110110011110000001001??01
Chara	00????????01?0?1000000000000?000??
Coleochaete	00????????0000?00000000000000000??
Equisetum	1111001100011110011010000000?21?01
Haplomitrium	10000001001101011110010?0100111?1?
Horneophyton	11110?01?101?11101??1100??00??????
Huperzia	1111001100011111011011100000?????1
Monoclea	1000000100110100011000000100????1?
Notothylas	10001001011101100111100000100111??
Polytrichum	1010010111011111111001111000?11?01
Sphaerocarpos	1000000100110100011000000100111?10
Sphagnum	1000000111011111111001010001111101
Takakia	1010010101 01?1011110010??000111???

TABLE A2.3

Completeness of Data for Taxa Used in the Cladistic Analysis in Chapter 3

Taxon	Total inapplicable characters	Total unknown characters	Total characters scored	Percentage complete
Andreaea	0	3	31	91
Andreaeobryum	0	6	28	82
Anthoceros	0	2	32	94
Chara	9	4	21	62
Coleochaete	9	2	23	68
Equisetum	0	2	32	94
Haplomitrium	1	2	31	91
Horneophyton	0	13	21	62
Huperzia	0	5	29	85
Monoclea	0	5	29	85
Notothylas	0	2	32	94
Polytrichum	0	2	32	94
Sphaerocarpos	0	1	33	97
Sphagnum	0	0	34	100
Takakia	1	5	28	82

TABLE A2.4

Completeness of Data for Characters Used in Cladistic Analysis in Chapter 3

Character		Total taxa with inapplicable conditions	Total taxa with no information	Total taxa scored
3.1	Sporophyte cellularity	0	0	15
3.2	Sporophyte independence or dependence	0	0	15
3.3	Well-developed sporangiophore	2	0	13
3.4	Sporophyte branching	2	0	13
3.5	Intercalary meristem	2	0	13
3.6	Foot form	2	1	12
3.7	Xylem-cell thickenings	2	0	13
3.8	Sporangium	2	0	13
3.9	Sporangium dehiscence	2	1	12
3.10	Columella	2	0	13
3.11	Elaters and pseudoelaters	0	0	15
3.12	Zoospore flagella	0	0	15
3.13	Perine	0	3	12
3.14	Cuticle	0	0	15
3.15	Stomates	2	0	13
3.16	Gametophyte form	0	0	15
3.17	Gametophyte leaves	0	0	15
3.18	Archegonium	0	0	15
3.19	Antheridium morphology	0	1	14
3.20	Antheridium development	0	1	14
3.21	Archegonium position	0	0	15
3.22	Gametangia distribution	0	0	15
3.23	Paraphyses	0	0	15
3.24	Rhizoid cellularity	2	0	13
3.25	Protonema type	0	2	13
3.26	Oil bodies	0	1	14
3.27	Mucilage clefts	0	0	15
3.28	Pseudopodium	0	0	15
3.29	Pyrenoids	0	8	7
3.30	Bicentriolar centrosomes	0	4	11
3.31	Orientation of multilayered structures	0	5	10
3.32	Preprophase bands in mitosis	0	11	4
3.33	Lunularic acid	0	8	7
3.34	D-Methionine	0	8	7

APPENDIX 3

Data for Analyses of Polysporangiophytes

TABLE A3.1

Characters and Character States for Taxa Used in the Cladistic Analysis in Chapter 4

Taxon	Sporophyte independence or dependence (character 4.1)		Well-developed sporangiophore (character 4.2)		Sporophyte branching (character 4.3)		Branching type (character 4.4)	
Aglaophyton	1	independent	1	present	1	present	0	±isotomous
Anthoceros	0	dependent	0	absent	0	absent	?	NA
Asteroxylon	1	independent	1	present	1	present	0	±isotomous
Caia	?	unknown	1	present	1	present	0	±isotomous
Cooksonia caledonica	?	unknown	1	present	1	present	0	±isotomous
Cooksonia cambrensis	?	unknown	1	present	1	present	0	±isotomous
Cooksonia pertonii	?	unknown	1	present	1	present	0	±isotomous
Eophyllophyton	1	independent	1	present	1	present	1	pseudomonopodial
Gosslingia	1	independent	1	present	1	present	1	pseudomonopodial
Haplomitrium	0	dependent	0	absent	0	absent	?	NA
Horneophyton	1	independent	1	present	1	present	0	±isotomous
Hsua	1	independent	1	present	1	present	1	pseudomonopodial
Huperzia	1	independent	1	present	1	present	0	±isotomous
Huvenia	1	independent	1	present	1	present	0	±isotomous
Ibyka	1	independent	1	present	1	present	1	pseudomonopodial
Nothia	1	independent	1	present	1	present	0	±isotomous
Psilophyton crenulatum	1	independent	1	present	1	present	1	pseudomonopodial
Psilophyton dawsonii	1	independent	1	present	1	present	1	pseudomonopodial
Pertica	1	independent	1	present	1	present	1	pseudomonopodial
Polytrichum	0	dependent	1	present	0	absent	?	NA
Pseudosporochnus	1	independent	1	present	1	present	1	pseudomonopodial
Renalia	1	independent	1	present	1	present	1	pseudomonopodial
Rhacophyton	1	independent	1	present	1	present	1	pseudomonopodial
Rhynia	1	independent	1	present	1	present	0	±isotomous
Sartilmania	?	unknown	1	present	1	present	0	±isotomous
Sawdonia	1	independent	1	present	1	present	1	pseudomonopodial
Sphaerocarpos	0	dependent	0	absent	0	absent	?	NA
Sporogonites	?	unknown	1	present	0	absent	?	NA
Stockmansella	1	independent	1	present	1	present	0	±isotomous
Tetraxylopteris	1	independent	1	present	1	present	1	pseudomonopodial
Tortilicaulis	?	unknown	1	present	1	present	0	±isotomous
Uskiella	?	unknown	1	present	1	present	0	±isotomous
Yunia	1	independent	1	present	1	present	0	±isotomous
Zosterophyllum llanoveranum	1	independent	1	present	1	present	0	±isotomous

TABLE A3.1
Continued

Taxon	Branching pattern (character 4.5)	Subordinate branching (character 4.6)	*Rhynia*-type adventitious branching (character 4.7)	Circinate vernation (character 4.8)
Aglaophyton	0 absent	0 absent	0 absent	0 absent
Anthoceros	? NA	? NA	? NA	0 absent
Asteroxylon	0 absent	0 absent	0 absent	0 absent
Caia	0 absent	0 absent	0 absent	0 absent
Cooksonia caledonica	0 absent	0 absent	0 absent	0 absent
Cooksonia cambrensis	0 absent	0 absent	0 absent	0 absent
Cooksonia pertonii	0 absent	0 absent	0 absent	0 absent
Eophyllophyton	? unknown	0 absent	0 absent	2 recurvation
Gosslingia	2 planar-alter.	1 ±axillary	0 absent	1 present
Haplomitrium	? NA	? NA	? NA	0 absent
Horneophyton	0 absent	0 absent	0 absent	0 absent
Hsua	2 planar-alter.	1 ±axillary	0 absent	1 present
Huperzia	0 absent	0 absent	0 absent	0 absent
Huvenia	0 absent	0 absent	1 present	0 absent
Ibyka	1 helical	0 absent	0 absent	2 recurvation
Nothia	0 absent	0 absent	0 absent	0 absent
Psilophyton crenulatum	1 helical	0 absent	0 absent	2 recurvation
Psilophyton dawsonii	1 helical	2 abortion	0 absent	2 recurvation
Pertica	3 tetrastichous	0 absent	0 absent	2 recurvation
Polytrichum	? NA	? NA	? NA	0 absent
Pseudosporochnus	1 helical	0 absent	0 absent	0 absent
Renalia	? unknown	0 absent	0 absent	0 absent
Rhacophyton	4 quadriseriate	0 absent	0 absent	2 recurvation
Rhynia	0 absent	0 absent	1 present	0 absent
Sartilmania	0 absent	0 absent	0 absent	0 absent
Sawdonia	2 planar-alter.	1 ±axillary	0 absent	1 present
Sphaerocarpos	? NA	? NA	? NA	0 absent
Sporogonites	? NA	? NA	? NA	0 absent
Stockmansella	0 absent	0 absent	? unknown	0 absent
Tetraxylopteris	3 tetrastichous	0 absent	0 absent	2 recurvation
Tortilicaulis	? unknown	? unknown	? unknown	0 absent
Uskiella	0 absent	0 absent	0 absent	0 absent
Yunia	0 absent	0 absent	0 absent	0 absent
Zosterophyllum llanoveranum	0 absent	0 absent	0 absent	1 present

TABLE A3.1
Continued

Taxon	Multicellular appendages (character 4.9)	Dichotomous pinnulelike appendages (character 4.10)	Conical sporangium emergences (character 4.11)	Stomates (character 4.12)
Aglaophyton	0 absent	0 absent	0 absent	1 present
Anthoceros	0 absent	? NA	0 absent	1 present
Asteroxylon	2 microphylls	0 absent	0 absent	1 present
Caia	0 absent	0 absent	1 present	? unknown
Cooksonia caledonica	0 absent	0 absent	0 absent	? unknown
Cooksonia cambrensis	0 absent	0 absent	0 absent	? unknown
Cooksonia pertonii	0 absent	0 absent	0 absent	1 present
Eophyllophyton	1 spines	1 present	0 absent	? unknown
Gosslingia	0 absent	0 absent	0 absent	1 present
Haplomitrium	0 absent	? NA	0 absent	0 absent
Horneophyton	0 absent	0 absent	1 present	1 present
Hsua	0 absent	0 absent	0 absent	1 present
Huperzia	2 microphylls	0 absent	0 absent	1 present
Huvenia	1 spines	0 absent	0 absent	? unknown
Ibyka	0 absent	1 present	0 absent	? unknown
Nothia	0 absent	0 absent	0 absent	1 present
Psilophyton crenulatum	1 spines	? unknown	0 absent	1 present
Psilophyton dawsonii	0 absent	1 present	0 absent	1 present
Pertica	0 absent	1 present	0 absent	? unknown
Polytrichum	0 absent	? NA	0 absent	1 present
Pseudosporochnus	0 absent	1 present	0 absent	? unknown
Renalia	0 absent	0 absent	0 absent	? unknown
Rhacophyton	0 absent	1 present	0 absent	? unknown
Rhynia	0 absent	0 absent	0 absent	1 present
Sartilmania	0 absent	0 absent	0 absent	? unknown
Sawdonia	1 spines	0 absent	0 absent	1 present
Sphaerocarpos	0 absent	? NA	0 absent	0 absent
Sporogonites	0 absent	? NA	0 absent	1 present
Stockmansella	0 absent	0 absent	0 absent	1 present
Tetraxylopteris	0 absent	1 present	0 absent	1 present
Tortilicaulis	0 absent	0 absent	0 absent	? unknown
Uskiella	0 absent	0 absent	0 absent	? unknown
Yunia	1 spines	0 absent	0 absent	? unknown
Zosterophyllum llanoveranum	0 absent	0 absent	0 absent	? unknown

TABLE A3.1
Continued

Taxon	Sterome (character 4.13)	Xylem strand shape (character 4.14)	Protoxylem (character 4.15)	Decay-resistance of tracheid cell wall (character 4.16)
Aglaophyton	0 absent	0 ±terete	? unknown	0 limited
Anthoceros	? NA	? NA	? NA	? NA
Asteroxylon	0 absent	2 stellate	2 exarch	1 extensive
Caia	? unknown	? unknown	? unknown	? unknown
Cooksonia caledonica	? unknown	? unknown	? unknown	? unknown
Cooksonia cambrensis	? unknown	? unknown	? unknown	? unknown
Cooksonia pertonii	1 continuous	? unknown	? unknown	? unknown
Eophyllophyton	1 continuous	0 ±terete	1 centrarch	1 extensive
Gosslingia	1 continuous	1 elliptical	2 exarch	1 extensive
Haplomitrium	? NA	? NA	? NA	? NA
Horneophyton	0 absent	0 ±terete	1 centrarch	0 limited
Hsua	? unknown	1 elliptical	? unknown	1 extensive
Huperzia	0 absent	2 stellate	2 exarch	1 extensive
Huvenia	? unknown	? unknown	? unknown	0 limited
Ibyka	2 discontinuous	3 lobed	3 mesarch	1 extensive
Nothia	0 absent	0 ±terete	? unknown	1 extensive
Psilophyton crenulatum	? unknown	0 ±terete	1 centrarch	1 extensive
Psilophyton dawsonii	1 continuous	0 ±terete	1 centrarch	1 extensive
Pertica	? unknown	? unknown	? unknown	? unknown
Polytrichum	0 absent	0 ±terete	0 absent	0 limited
Pseudosporochnus	2 discontinuous	4 clepsydroid	3 mesarch	1 extensive
Renalia	? unknown	? unknown	? unknown	1 extensive
Rhacophyton	? unknown	4 clepsydroid	3 mesarch	1 extensive
Rhynia	0 absent	0 ±terete	1 centrarch	0 limited
Sartilmania	? unknown	? unknown	? unknown	? unknown
Sawdonia	1 continuous	1 elliptical	2 exarch	1 extensive
Sphaerocarpos	? NA	? NA	? NA	? NA
Sporogonites	? unknown	? unknown	0 absent	0 limited
Stockmansella	? unknown	0 ±terete	0 absent	0 limited
Tetraxylopteris	2 discontinuous	3 lobed	1 centrarch	1 extensive
Tortilicaulis	0 absent	? unknown	? unknown	? unknown
Uskiella	1 continuous	? unknown	? unknown	? unknown
Yunia	1 continuous	0 ±terete	1 centrarch	1 extensive
Zosterophyllum llanoveranum	1 continuous	1 elliptical	2 exarch	1 extensive

TABLE A3.1
Continued

Taxon	Tracheid thickenings (character 4.17)	Metaxylem tracheid pitting (character 4.18)	Simple pitlets in tracheid wall (character 4.19)	Aligned xylem (character 4.20)
Aglaophyton	0 absent	0 absent	0 absent	0 absent
Anthoceros	? NA	? NA	? NA	? NA
Asteroxylon	1 present	0 absent	1 present	0 absent
Caia	? unknown	? unknown	? unknown	? unknown
Cooksonia caledonica	? unknown	? unknown	? unknown	? unknown
Cooksonia cambrensis	? unknown	? unknown	? unknown	? unknown
Cooksonia pertonii	1 present	0 absent	? unknown	? unknown
Eophyllophyton	1 present	0 absent	1 present	0 absent
Gosslingia	1 present	0 absent	1 present	0 absent
Haplomitrium	? NA	? NA	? NA	? NA
Horneophyton	0 absent	0 absent	0 absent	0 absent
Hsua	1 present	0 absent	1 present	0 absent
Huperzia	1 present	1 present	0 absent	0 absent
Huvenia	1 present	0 absent	0 absent	0 absent
Ibyka	1 present	1 present	? unknown	? unknown
Nothia	0 absent	0 absent	? unknown	0 absent
Psilophyton crenulatum	1 present	1 present	1 present	1 present
Psilophyton dawsonii	1 present	1 present	1 present	1 present
Pertica	? unknown	? unknown	? unknown	? unknown
Polytrichum	0 absent	0 absent	0 absent	0 absent
Pseudosporochnus	1 present	1 present	? unknown	1 present
Renalia	1 present	? unknown	? unknown	? unknown
Rhacophyton	1 present	1 present	? unknown	1 present
Rhynia	1 present	0 absent	0 absent	0 absent
Sartilmania	? unknown	? unknown	? unknown	? unknown
Sawdonia	1 present	0 absent	1 present	0 absent
Sphaerocarpos	? NA	? NA	? NA	? NA
Sporogonites	0 absent	0 absent	? unknown	0 absent
Stockmansella	1 present	0 absent	0 absent	0 absent
Tetraxylopteris	1 present	1 present	? unknown	1 present
Tortilicaulis	0 absent	0 absent	? unknown	0 absent
Uskiella	? unknown	? unknown	? unknown	? unknown
Yunia	1 present	0 absent	1 present	0 absent
Zosterophyllum llanoveranum	1 present	0 absent	? unknown	0 absent

TABLE A3.1
Continued

Taxon	Xylem rays (character 4.21)	Sporangiotaxis (character 4.22)	Sporangium attachment (character 4.23)	Specialized fertile zone (character 4.24)
Aglaophyton	0 absent	? NA	0 terminal	0 absent
Anthoceros	? NA	? NA	0 terminal	? NA
Asteroxylon	0 absent	2 helical	1 lateral-stalked	0 absent
Caia	? unknown	? NA	0 terminal	0 absent
Cooksonia caledonica	? unknown	? NA	0 terminal	0 absent
Cooksonia cambrensis	? unknown	? NA	0 terminal	0 absent
Cooksonia pertonii	0 absent	? NA	0 terminal	0 absent
Eophyllophyton	0 absent	0 none	1 lateral-stalked	1 present
Gosslingia	0 absent	1 rowed	1 lateral-stalked	0 absent
Haplomitrium	? NA	? NA	0 terminal	? NA
Horneophyton	0 absent	? NA	0 terminal	0 absent
Hsua	0 absent	? NA	0 terminal	0 absent
Huperzia	0 absent	2 helical	1 lateral-stalked	0 absent
Huvenia	0 absent	? NA	0 terminal	0 absent
Ibyka	0 absent	? NA	0 terminal	1 present
Nothia	0 absent	0 none	1 lateral-stalked	0 absent
Psilophyton crenulatum	0 absent	? NA	0 terminal	1 present
Psilophyton dawsonii	0 absent	? NA	0 terminal	1 present
Pertica	? unknown	? NA	0 terminal	1 present
Polytrichum	0 absent	? NA	0 terminal	? NA
Pseudosporochnus	0 absent	? NA	0 terminal	1 present
Renalia	? unknown	? NA	0 terminal	0 absent
Rhacophyton	1 present	? NA	0 terminal	1 present
Rhynia	0 absent	? NA	2 lateral-sessile	0 absent
Sartilmania	0 absent	? NA	0 terminal	0 absent
Sawdonia	0 absent	1 rowed	1 lateral-stalked	0 absent
Sphaerocarpos	? NA	? NA	0 terminal	? NA
Sporogonites	0 absent	? NA	0 terminal	? NA
Stockmansella	0 absent	? NA	2 lateral-sessile	0 absent
Tetraxylopteris	1 present	? NA	0 terminal	1 present
Tortilicaulis	0 absent	? NA	0 terminal	? NA
Uskiella	? unknown	? NA	0 terminal	0 absent
Yunia	0 absent	? NA	0 terminal	0 absent
Zosterophyllum llanoveranum	0 absent	1 rowed	1 lateral-stalked	0 absent

TABLE A3.1
Continued

Taxon	Specialized sporangium-axis junction (character 4.25)	Sporangium abscission (character 4.26)	Sporangium shape (character 4.27)	Sporangium symmetry (character 4.28)
Aglaophyton	0 absent	0 absent	0 ±fusiform	0 radial
Anthoceros	0 absent	0 absent	0 ±fusiform	0 radial
Asteroxylon	0 absent	0 absent	1 ±reniform	1 bilateral
Caia	0 absent	0 absent	4 axislike	0 radial
Cooksonia caledonica	0 absent	0 absent	5 ±elliptical	1 bilateral
Cooksonia cambrensis	0 absent	0 absent	1 ±reniform	? unknown
Cooksonia pertonii	0 absent	0 absent	5 ±elliptical	? unknown
Eophyllophyton	0 absent	0 absent	5 ±elliptical	1 bilateral
Gosslingia	0 absent	0 absent	1 ±reniform	1 bilateral
Haplomitrium	0 absent	0 absent	0 ±fusiform	0 radial
Horneophyton	0 absent	0 absent	4 axislike	0 radial
Hsua	0 absent	0 absent	1 ±reniform	1 bilateral
Huperzia	0 absent	0 absent	1 ±reniform	1 bilateral
Huvenia	1 present	? unknown	0 ±fusiform	0 radial
Ibyka	0 absent	0 absent	0 ±fusiform	0 radial
Nothia	0 absent	0 absent	1 ±reniform	1 bilateral
Psilophyton crenulatum	0 absent	0 absent	0 ±fusiform	0 radial
Psilophyton dawsonii	0 absent	0 absent	0 ±fusiform	0 radial
Pertica	0 absent	0 absent	0 ±fusiform	0 radial
Polytrichum	0 absent	0 absent	0 ±fusiform	0 radial
Pseudosporochnus	0 absent	0 absent	0 ±fusiform	0 radial
Renalia	0 absent	0 absent	1 ±reniform	1 bilateral
Rhacophyton	0 absent	0 absent	0 ±fusiform	0 radial
Rhynia	1 present	1 present	0 ±fusiform	0 radial
Sartilmania	0 absent	0 absent	2 ±spatulate	1 bilateral
Sawdonia	0 absent	0 absent	1 ±reniform	1 bilateral
Sphaerocarpos	0 absent	0 absent	3 ±spherical	0 radial
Sporogonites	0 absent	0 absent	0 ±fusiform	0 radial
Stockmansella	1 present	1 present	0 ±fusiform	0 radial
Tetraxylopteris	0 absent	0 absent	0 ±fusiform	0 radial
Tortilicaulis	0 absent	0 absent	0 ±fusiform	0 radial
Uskiella	0 absent	0 absent	2 ±spatulate	1 bilateral
Yunia	0 absent	0 absent	2 ±spatulate	1 bilateral
Zosterophyllum llanoveranum	0 absent	0 absent	1 ±reniform	1 bilateral

TABLE A3.1
Continued

Taxon	Sporangium dehiscence (character 4.29)	Thickened sporangium valve rim (character 4.30)	Sporangium branching (character 4.31)	Columella (character 4.32)	Elaters and pseudoelaters (character 4.33)
Aglaophyton	? unknown	0 absent	0 absent	0 absent	0 absent
Anthoceros	0 multislit	0 absent	0 absent	1 present	1 present
Asteroxylon	2 single apical	? unknown	0 absent	0 absent	0 absent
Caia	? unknown	0 absent	1 present	? unknown	? unknown
Cooksonia caledonica	2 single apical	0 absent	0 absent	? unknown	? unknown
Cooksonia cambrensis	? unknown	1 present	0 absent	? unknown	? unknown
Cooksonia pertonii	? unknown	0 absent	0 absent	0 absent	0 absent
Eophyllophyton	? unknown	0 absent	0 absent	? unknown	? unknown
Gosslingia	2 single apical	1 present	0 absent	0 absent	0 absent
Haplomitrium	0 multislit	0 absent	0 absent	0 absent	1 present
Horneophyton	? unknown	0 absent	1 present	1 present	0 absent
Hsua	2 single apical	1 present	0 absent	0 absent	0 absent
Huperzia	2 single apical	1 present	0 absent	0 absent	0 absent
Huvenia	0 multislit	0 absent	0 absent	0 absent	0 absent
Ibyka	? unknown	0 absent	0 absent	0 absent	? unknown
Nothia	2 single apical	1 present	0 absent	0 absent	0 absent
Psilophyton crenulatum	1 single, one side	0 absent	0 absent	0 absent	0 absent
Psilophyton dawsonii	1 single, one side	0 absent	0 absent	0 absent	0 absent
Pertica	? unknown	0 absent	0 absent	0 absent	0 absent
Polytrichum	3 operculate	0 absent	0 absent	1 present	0 absent
Pseudosporochnus	1 single, one side	0 absent	0 absent	0 absent	? unknown
Renalia	2 single apical	1 present	0 absent	0 absent	0 absent
Rhacophyton	1 single, one side	0 absent	0 absent	0 absent	0 absent
Rhynia	? unknown	0 absent	0 absent	0 absent	0 absent
Sartilmania	2 single apical	1 present	0 absent	? unknown	? unknown
Sawdonia	2 single apical	1 present	0 absent	0 absent	0 absent
Sphaerocarpos	4 indehiscent	0 absent	0 absent	0 absent	1 present
Sporogonites	0 multislit	0 absent	0 absent	1 present	0 absent
Stockmansella	0 multislit	0 absent	0 absent	0 absent	0 absent
Tetraxylopteris	1 single, one side	0 absent	0 absent	0 absent	0 absent
Tortilicaulis	? unknown	0 absent	1 present	? unknown	? unknown
Uskiella	2 single apical	1 present	0 absent	0 absent	0 absent
Yunia	2 single apical	1 present	0 absent	? unknown	0 absent
Zosterophyllum llanoveranum	2 single apical	1 present	0 absent	0 absent	0 absent

TABLE A3.2

Data Matrix for Chapter 4 in Numeric Format

Aglaophyton	11100000000100?000000?000000?0000
Anthoceros	000???00?01???????????0?000000011
Asteroxylon	111000002001022110100210001122?000
Caia	?1100000001??????????000040?01??
Cooksonia caledonica	?1100000000???????????000051200??
Cooksonia cambrensis	?1100000000???????????00001??10??
Cooksonia pertonii	?110000000011???10??0?00005??0000
Eophyllophyton	1111?002110?1011101000110051?00??
Gosslingia	111121010001112110100110001121000
Haplomitrium	000???00?00???????????0?000000001
Horneophyton	111000000011001000000?000040?0110
Hsua	11112101000 1?1?110100?00001121000
Huperzia	111000002001022111000210001121000
Huvenia	11100010100????010000?001?0000000
Ibyka	111110020 10?233111??0?010000?000?
Nothia	11100000000100?100?00010001121000
Psilophyton crenulatum	111110021?01?0111111 0?01000010000
Psilophyton dawsonii	111112020101101111110?01000010000
Pertica	11113002010???????????010000?0000
Polytrichum	010???00?010000000000?0?000030010
Pseudosporochnus	11111000010?243111?10?0100001000?
Renalia	1111?000000????11?????00001121000
Rhacophyton	1111400201 0??43111?11?01000010000
Rhynia	111000100001001010000?201100?0000
Sartilmania	?1100000000?????????0?00002121 0??
Sawdonia	111121011001112110100110001121000
Sphaerocarpos	000???00?00???????????0?003040001
Sporogonites	?10???00?01??0000?00?0?000000010
Stockmansella	111000?00001?00010000?20110000000
Tetraxylopteris	111130020101231111?11?01000010000
Tortilicaulis	?110???0000?0???00?00?0?0000?01??
Uskiella	?1100000000?1?????????00002121000
Yunia	11100000100?101110100?00002121 0?0
Zosterophyllum llanoveranum	11100001000?112110?00110001121000

TABLE A3.3

Completeness of Data for Taxa Used in the Cladistic Analysis in Chapter 4

Taxon	Total inapplicable characters	Total unknown characters	Total characters scored	Percentage complete
Aglaophyton	1	2	30	91
Anthoceros	16	0	17	52
Asteroxylon	0	1	32	97
Caia	1	14	18	55
Cooksonia caledonica	1	13	19	58
Cooksonia cambrensis	1	15	17	52
Cooksonia pertonii	1	8	24	73
Eophyllophyton	0	5	28	85
Gosslingia	0	0	33	100
Haplomitrium	16	0	17	52
Horneophyton	1	1	31	94
Hsua	1	2	30	91
Huperzia	0	0	33	100
Huvenia	1	5	27	82
Ibyka	1	5	27	82
Nothia	0	2	31	94
Psilophyton crenulatum	1	2	30	91
Psilophyton dawsonii	1	0	32	97
Pertica	1	11	21	64
Polytrichum	7	0	26	79
Pseudosporochnus	1	3	29	88
Renalia	1	9	23	70
Rhacophyton	1	3	29	88
Rhynia	1	1	31	94
Sartilmania	1	12	20	61
Sawdonia	0	0	33	100
Sphaerocarpos	16	0	17	52
Sporogonites	7	4	22	67
Stockmansella	1	2	30	91
Tetraxylopteris	1	1	31	94
Tortilicaulis	2	12	19	58
Uskiella	1	10	22	67
Yunia	1	2	30	91
Zosterophyllum llanoveranum	0	2	31	94

TABLE A3.4

Completeness of Data for Characters Used in the Cladistic Analysis in Chapter 4

Character		Total taxa with inapplicable conditions	Total taxa with no information	Total taxa scored
4.1	Sporophyte independence or dependence	0	8	26
4.2	Well-developed sporangiophore	0	0	34
4.3	Sporophyte branching	0	0	34
4.4	Branching type	5	0	29
4.5	Branching pattern	5	3	26
4.6	Subordinate branching	5	1	28
4.7	*Rhynia*-type adventitious branching	5	2	27
4.8	Circinate vernation	0	0	34
4.9	Multicellular appendages	0	0	34
4.10	Dichotomous pinnulelike appendages	5	1	28
4.11	Conical sporangium emergences	0	0	34
4.12	Stomates	0	15	19
4.13	Sterome	3	12	19
4.14	Xylem strand shape	3	11	20
4.15	Protoxylem	3	13	18
4.16	Decay-resistance of the tracheid cell wall	3	8	23
4.17	Tracheid thickenings	3	6	25
4.18	Metaxylem tracheid pitting	3	7	24
4.19	Simple pitlets in the tracheid wall	3	16	15
4.20	Aligned xylem	3	9	22
4.21	Xylem rays	3	6	25
4.22	Sporangiotaxis	27	0	7
4.23	Sporangium attachment	0	0	34
4.24	Specialized fertile zone	6	0	28
4.25	Specialized sporangium-axis junction	0	0	34
4.26	Sporangium abscission	0	1	33
4.27	Sporangium shape	0	0	34
4.28	Sporangium symmetry	0	2	32
4.29	Sporangium dehiscence	0	10	24
4.30	Thickened sporangium valve rim	0	1	33
4.31	Sporangium branching	0	0	34
4.32	Columella	0	7	27
4.33	Elaters and pseudoelaters	0	8	26

APPENDIX 4

Data for Analyses of Zosterophyllopsida

TABLE A4.1

Characters and Character States for Taxa Used in the Cladistic Analysis in Chapter 5

Taxon	Branching type (character 5.1)		Branching pattern (character 5.2)		Subordinate branching (character 5.3)		Circinate vernation (character 5.4)	
Adoketophyton	0	isotomous	?	unknown	0	absent	?	unknown
Anisophyton	1	pseudomonopodial	1	planated	1	±axillary	1	present
Asteroxylon	0	isotomous	0	nonplanar	0	absent	0	absent
Baragwanathia	0	isotomous	0	nonplanar	0	absent	0	absent
Barinophyton citrulliforme	1	pseudomonopodial	?	unknown	?	unknown	?	unknown
Barinophyton obscurum	1	pseudomonopodial	1	planated	?	unknown	?	unknown
Bathurstia	?	unknown	?	unknown	?	unknown	1	present
Crenaticaulis	1	pseudomonopodial	1	planated	1	±axillary	1	present
Deheubarthia	1	pseudomonopodial	1	planated	1	±axillary	1	present
Discalis	0	isotomous	0	nonplanar	0	absent	1	present
Drepanophycus	0	isotomous	0	nonplanar	0	absent	0	absent
Gosslingia	1	pseudomonopodial	1	planated	1	±axillary	1	present
Gumuia	0	isotomous	0	nonplanar	0	absent	?	unknown
Hicklingia	?	unknown	0	nonplanar	?	unknown	0	absent
Hsua	1	pseudomonopodial	1	planated	1	±axillary	1	present
Huia	1	pseudomonopodial	1	planated	0	absent	0	absent
Huperzia	0	isotomous	0	nonplanar	0	absent	0	absent
Konioria	1	pseudomonopodial	0	nonplanar	?	NA	1	present
Krithodeophyton	?	unknown	1	planated	0	absent	?	unknown
Nothia	0	isotomous	0	nonplanar	0	absent	?	unknown
Oricilla	1	pseudomonopodial	1	planated	0	absent	1	present
Protobarinophyton	1	pseudomonopodial	1	planated	1	±axillary	?	unknown
Psilophyton	1	pseudomonopodial	0	nonplanar	0	absent	0	absent
Rebuchia	0	isotomous	0	nonplanar	0	absent	0	absent
Rhynia	0	isotomous	0	nonplanar	0	absent	0	absent
Sawdonia	1	pseudomonopodial	1	planated	1	±axillary	1	present
Serrulacaulis	?	unknown	?	unknown	?	unknown	1	present
Tarella	?	unknown	1	planated	2	unconfined	1	present
Thrinkophyton	1	pseudomonopodial	1	planated	1	±axillary	1	present
Trichopherophyton	?	unknown	?	unknown	?	unknown	1	present
Zosterophyllum deciduum	0	isotomous	0	nonplanar	0	absent	0	absent
Zosterophyllum divaricatum	0	isotomous	0	nonplanar	0	absent	1	present
Zosterophyllum fertile	0	isotomous	0	nonplanar	0	absent	?	unknown
Zosterophyllum llanoveranum	0	isotomous	0	nonplanar	0	absent	1	present
Zosterophyllum myretonianum	0	isotomous	0	nonplanar	0	absent	0	absent

TABLE A4.1
Continued

Taxon	Multicellular appendages (character 5.5)		Appendage phyllotaxy (character 5.6)		Xylem strand shape (character 5.7)		Protoxylem (character 5.8)	
Adoketophyton	3	"bracts"	3	4-rowed	?	unknown	?	unknown
Anisophyton	2	spines	0	absent	?	unknown	?	unknown
Asteroxylon	1	microphylls	1	helical	2	stellate	1	exarch
Baragwanathia	1	microphylls	1	helical	2	stellate	?	unknown
Barinophyton citrulliforme	0	absent	?	NA	1	elliptical	1	exarch
Barinophyton obscurum	0	absent	?	NA	?	unknown	?	unknown
Bathurstia	2	spines	0	absent	?	unknown	?	unknown
Crenaticaulis	2	spines	2	2-rowed	1	elliptical	1	exarch
Deheubarthia	2	spines	0	absent	1	elliptical	1	exarch
Discalis	2	spines	0	absent	?	unknown	?	unknown
Drepanophycus	1	microphylls	1	helical	2	stellate	1	exarch
Gosslingia	0	absent	?	NA	1	elliptical	1	exarch
Gumuia	0	absent	?	NA	?	unknown	?	unknown
Hicklingia	0	absent	?	NA	?	unknown	?	unknown
Hsua	0	absent	?	NA	1	elliptical	?	unknown
Huia	0	absent	?	NA	1	elliptical	?	unknown
Huperzia	1	microphylls	1	helical	2	stellate	1	exarch
Konioria	2	spines	0	absent	1	elliptical	1	exarch
Krithodeophyton	0	absent	?	NA	?	unknown	?	unknown
Nothia	0	absent	?	NA	1	elliptical	?	unknown
Oricilla	0	absent	?	NA	?	unknown	?	unknown
Protobarinophyton	0	absent	?	NA	?	unknown	?	unknown
Psilophyton	0	absent	?	NA	0	terete	0	centrarch
Rebuchia	0	absent	?	NA	?	unknown	?	unknown
Rhynia	0	absent	?	NA	0	terete	0	centrarch
Sawdonia	2	spines	0	absent	1	elliptical	1	exarch
Serrulacaulis	2	spines	2	2-rowed	?	unknown	?	unknown
Tarella	0	absent	?	NA	?	unknown	?	unknown
Thrinkophyton	0	absent	?	NA	1	elliptical	1	exarch
Trichopherophyton	0	absent	?	NA	0	terete	1	exarch
Zosterophyllum deciduum	0	absent	?	NA	?	unknown	?	unknown
Zosterophyllum divaricatum	0	absent	?	NA	?	unknown	?	unknown
Zosterophyllum fertile	0	absent	?	NA	1	elliptical	1	exarch
Zosterophyllum llanoveranum	0	absent	?	NA	1	elliptical	1	exarch
Zosterophyllum myretonianum	0	absent	?	NA	?	unknown	?	unknown

TABLE A4.1
Continued

Taxon	Tracheid thickenings (character 5.9)		Sporangiotaxis (character 5.10)		Sporangium rows (character 5.11)		Sporangium distribution (character 5.12)	
Adoketophyton	?	unknown	2	4-rowed	?	NA	0	±unbranched
Anisophyton	?	unknown	1	1–2-rowed	1	dorsiventral	1	2+ branches
Asteroxylon	1	present	?	unknown	?	NA	?	unknown
Baragwanathia	1	present	0	helical	?	NA	?	unknown
Barinophyton citrulliforme	1	present	1	1–2-rowed	2	lateral	2	compact
Barinophyton obscurum	?	unknown	1	1–2-rowed	2	lateral	2	compact
Bathurstia	?	unknown	1	1–2-rowed	2	lateral	?	unknown
Crenaticaulis	1	present	1	1–2-rowed	2	lateral	1	2+ branches
Deheubarthia	1	present	1	1–2-rowed	2	lateral	1	2+ branches
Discalis	?	unknown	0	helical	?	NA	0	±unbranched
Drepanophycus	1	present	?	unknown	?	NA	?	unknown
Gosslingia	1	present	1	1–2-rowed	2	lateral	1	2+ branches
Gumuia	?	unknown	0	helical	?	NA	0	±unbranched
Hicklingia	?	unknown	0	helical	?	NA	0	±unbranched
Hsua	1	present	?	NA	?	NA	?	NA
Huia	1	present	0	helical	?	NA	0	±unbranched
Huperzia	1	present	0	helical	?	NA	1	2+ branches
Konioria	?	unknown	1	1–2-rowed	1	dorsiventral	?	unknown
Krithodeophyton	?	unknown	1	1–2-rowed	2	lateral	0	±unbranched
Nothia	0	absent	0	helical	?	NA	1	2+ branches
Oricilla	?	unknown	1	1–2-rowed	2	lateral	1	2+ branches
Protobarinophyton	?	unknown	1	1–2-rowed	2	lateral	2	compact
Psilophyton	1	present	?	NA	?	NA	?	NA
Rebuchia	?	unknown	1	1–2-rowed	2	lateral	0	±unbranched
Rhynia	1	present	?	NA	?	NA	?	NA
Sawdonia	1	present	1	1–2-rowed	2	lateral	1	2+ branches
Serrulacaulis	?	unknown	1	1–2-rowed	0	ventral	?	unknown
Tarella	?	unknown	1	1–2-rowed	2	lateral	1	2+ branches
Thrinkophyton	1	present	1	1–2-rowed	2	lateral	1	2+ branches
Trichopherophyton	1	present	?	unknown	?	unknown	?	unknown
Zosterophyllum deciduum	?	unknown	0	helical	?	NA	1	2+ branches
Zosterophyllum divaricatum	1	present	1	1–2-rowed	2	lateral	1	2+ branches
Zosterophyllum fertile	?	unknown	1	1–2-rowed	2	lateral	0	±unbranched
Zosterophyllum llanoveranum	?	unknown	1	1–2-rowed	2	lateral	0	±unbranched
Zosterophyllum myretonianum	?	unknown	0	helical	?	NA	0	±unbranched

TABLE A4.1
Continued

Taxon	Sporangium shape (character 5.13)	Sporangium symmetry (character 5.14)	Relative valve size (character 5.15)	Sporangium dehiscence (character 5.16)
Adoketophyton	1 ±reniform	1 bilateral	1 isovalvate	1 two
Anisophyton	1 ±reniform	1 bilateral	? unknown	? unknown
Asteroxylon	1 ±reniform	1 bilateral	1 isovalvate	1 two
Baragwanathia	1 ±reniform	1 bilateral	? unknown	? unknown
Barinophyton citrulliforme	1 ±reniform	1 bilateral	? unknown	2 none
Barinophyton obscurum	1 ±reniform	1 bilateral	? unknown	? unknown
Bathurstia	1 ±reniform	1 bilateral	? unknown	? unknown
Crenaticaulis	1 ±reniform	1 bilateral	2 anisovalvate	1 two
Deheubarthia	1 ±reniform	1 bilateral	1 isovalvate	1 two
Discalis	1 ±reniform	1 bilateral	1 isovalvate	1 two
Drepanophycus	1 ±reniform	1 bilateral	? unknown	? unknown
Gosslingia	1 ±reniform	1 bilateral	1 isovalvate	1 two
Gumuia	1 ±reniform	1 bilateral	1 isovalvate	1 two
Hicklingia	2 ±spatulate	1 bilateral	1 isovalvate	1 two
Hsua	1 ±reniform	1 bilateral	1 isovalvate	1 two
Huia	1 ±reniform	1 bilateral	1 isovalvate	1 two
Huperzia	1 ±reniform	1 bilateral	1 isovalvate	1 two
Konioria	1 ±reniform	1 bilateral	1 isovalvate	1 two
Krithodeophyton	1 ±reniform	1 bilateral	? unknown	? unknown
Nothia	1 ±reniform	1 bilateral	1 isovalvate	1 two
Oricilla	1 ±reniform	1 bilateral	1 isovalvate	1 two
Protobarinophyton	1 ±reniform	1 bilateral	? unknown	? unknown
Psilophyton	0 fusiform	0 radial	? NA	0 one
Rebuchia	1 ±reniform	1 bilateral	1 isovalvate	1 two
Rhynia	0 fusiform	0 radial	? NA	2 none
Sawdonia	1 ±reniform	1 bilateral	1 isovalvate	1 two
Serrulacaulis	1 ±reniform	1 bilateral	1 isovalvate	1 two
Tarella	1 ±reniform	1 bilateral	1 isovalvate	1 two
Thrinkophyton	1 ±reniform	1 bilateral	1 isovalvate	1 two
Trichopherophyton	1 ±reniform	1 bilateral	2 anisovalvate	1 two
Zosterophyllum deciduum	1 ±reniform	1 bilateral	1 isovalvate	1 two
Zosterophyllum divaricatum	1 ±reniform	1 bilateral	1 isovalvate	1 two
Zosterophyllum fertile	1 ±reniform	1 bilateral	1 isovalvate	1 two
Zosterophyllum llanoveranum	1 ±reniform	1 bilateral	1 isovalvate	1 two
Zosterophyllum myretonianum	1 ±reniform	1 bilateral	1 isovalvate	1 two

TABLE A4.1
Continued

Taxon	Thickened sporangium valve rim (character 5.17)	Sporangium attachment (character 5.18)	Sporangium orientation (character 5.19)	Spore size (character 5.20)
Adoketophyton	1 present	1 lateral-stalked	0 upright	? unknown
Anisophyton	? unknown	1 lateral-stalked	0 upright	? unknown
Asteroxylon	1 present	1 lateral-stalked	0 upright	0 homosporous
Baragwanathia	? unknown	? unknown	0 upright	0 homosporous
Barinophyton citrulliforme	0 absent	1 lateral-stalked	2 clasping	1 heterosporous
Barinophyton obscurum	? unknown	1 lateral-stalked	2 clasping	1 heterosporous
Bathurstia	? unknown	? unknown	? unknown	? unknown
Crenaticaulis	1 present	1 lateral-stalked	0 upright	? unknown
Deheubarthia	1 present	1 lateral-stalked	0 upright	? unknown
Discalis	1 present	1 lateral-stalked	0 upright	0 homosporous
Drepanophycus	? unknown	1 lateral-stalked	0 upright	? unknown
Gosslingia	1 present	1 lateral-stalked	1 auricular	0 homosporous
Gumuia	1 present	1 lateral-stalked	0 upright	0 homosporous
Hicklingia	0 absent	1 lateral-stalked	0 upright	? unknown
Hsua	1 present	0 terminal	0 upright	0 homosporous
Huia	1 present	2 lateral-sessile	0 upright	0 homosporous
Huperzia	1 present	1 lateral-stalked	0 upright	0 homosporous
Konioria	? unknown	1 lateral-stalked	0 upright	? unknown
Krithodeophyton	? unknown	? unknown	0 upright	0 homosporous
Nothia	1 present	1 lateral-stalked	0 upright	0 homosporous
Oricilla	1 present	1 lateral-stalked	1 auricular	0 homosporous
Protobarinophyton	? unknown	1 lateral-stalked	2 clasping	1 heterosporous
Psilophyton	0 absent	0 terminal	0 upright	0 homosporous
Rebuchia	1 present	1 lateral-stalked	0 upright	0 homosporous
Rhynia	0 absent	3 tissue pad	0 upright	0 homosporous
Sawdonia	1 present	1 lateral-stalked	0 upright	0 homosporous
Serrulacaulis	1 present	1 lateral-stalked	0 upright	0 homosporous
Tarella	1 present	1 lateral-stalked	1 auricular	0 homosporous
Thrinkophyton	1 present	1 lateral-stalked	0 upright	? unknown
Trichopherophyton	1 present	2 lateral-sessile	0 upright	0 homosporous
Zosterophyllum deciduum	1 present	1 lateral-stalked	0 upright	0 homosporous
Zosterophyllum divaricatum	1 present	1 lateral-stalked	0 upright	0 homosporous
Zosterophyllum fertile	1 present	1 lateral-stalked	0 upright	0 homosporous
Zosterophyllum llanoveranum	1 present	1 lateral-stalked	0 upright	0 homosporous
Zosterophyllum myretonianum	1 present	1 lateral-stalked	0 upright	0 homosporous

TABLE A4.2

Data Matrix for Chapter 5 in Numeric Format

Adoketophyton	0?0?33????2?01111110?
Anisophyton	111120???11111???10?
Asteroxylon	000011211???11111100
Baragwanathia	0000112?10??11????00
Barinophyton citrulliforme	1???0?11112211?20121
Barinophyton obscurum	11??0????12211???121
Bathurstia	???120???12?11??????
Crenaticaulis	1111221111211121110?
Deheubarthia	1111201111211111110?
Discalis	000120???0?011111100
Drepanophycus	000011211???11???10?
Gosslingia	11110?11112111111110
Gumuia	000?0????0?011111100
Hicklingia	?0?00????0?02111010?
Hsua	11110?1?1???11111000
Huia	11000?1?10?011111200
Huperzia	0000112110?111111100
Konioria	10?12011?11?1111?10?
Krithodeophyton	?10?0????12011????00
Nothia	000?0?1?00?111111100
Oricilla	11010????12111111110
Protobarinophyton	111?0????12211???121
Psilophyton	10000?001???00?00000
Rebuchia	00000????12011111100
Rhynia	00000?001???00?20300
Sawdonia	11112011112111111100
Serrulacaulis	???122???10?11111100
Tarella	?1210????12111111110
Thrinkophyton	11110?1111211111110?
Trichopherophyton	???10?011???11211200
Zosterophyllum deciduum	00000????0?111111100
Zosterophyllum divaricatum	00010???112111111100
Zosterophyllum fertile	000?0?11?12011111100
Zosterophyllum llanoveranum	00010?11?12011111100
Zosterophyllum myretonianum	00000????0?011111100

TABLE A4.3

Completeness of Data for Taxa Used in the Cladistic Analysis in Chapter 5

Taxon	Total inapplicable characters	Total unknown characters	Total characters scored	Percentage complete
Adoketophyton	1	6	13	65
Anisophyton	0	7	13	65
Asteroxylon	1	2	17	85
Baragwanathia	1	6	13	65
Barinophyton citrulliforme	1	4	15	75
Barinophyton obscurum	1	8	11	55
Bathurstia	0	13	7	35
Crenaticaulis	0	1	19	95
Deheubarthia	0	1	19	95
Discalis	1	3	16	80
Drepanophycus	1	6	13	65
Gosslingia	1	0	19	95
Gumuia	2	4	14	70
Hicklingia	2	6	12	60
Hsua	4	1	15	75
Huia	2	1	17	85
Huperzia	1	0	19	95
Konioria	1	4	15	75
Krithodeophyton	1	9	10	50
Nothia	2	2	16	80
Oricilla	1	3	16	80
Protobarinophyton	1	7	12	60
Psilophyton	5	0	15	75
Rebuchia	1	3	16	80
Rhynia	5	0	15	75
Sawdonia	0	0	20	100
Serrulacaulis	0	7	13	65
Tarella	1	4	15	75
Thrinkophyton	1	1	18	90
Trichopherophyton	1	6	13	65
Zosterophyllum deciduum	2	3	15	75
Zosterophyllum divaricatum	1	2	17	85
Zosterophyllum fertile	1	2	17	85
Zosterophyllum llanoveranum	1	1	18	90
Zosterophyllum myretonianum	2	3	15	75

TABLE A4.4

Completeness of Data for Characters Used in the Cladistic Analysis in Chapter 5

Character		Total taxa with inapplicable conditions	Total taxa with no information	Total taxa scored
5.1	Branching type	0	6	29
5.2	Branching pattern	0	5	30
5.3	Subordinate branching	1	6	28
5.4	Circinate vernation	0	8	27
5.5	Multicellular appendages	0	0	35
5.6	Appendage phyllotaxy	22	0	13
5.7	Xylem strand shape	0	16	19
5.8	Protoxylem	0	20	15
5.9	Tracheid thickenings	0	18	17
5.10	Sporangiotaxis	3	3	29
5.11	Sporangium rows	15	1	19
5.12	Sporangium distribution	3	7	25
5.13	Sporangium shape	0	0	35
5.14	Sporangium symmetry	0	0	35
5.15	Relative valve size	2	8	25
5.16	Sporangium dehiscence	0	7	28
5.17	Thickened sporangium valve rim	0	8	27
5.18	Sporangium attachment	0	3	32
5.19	Sporangium orientation	0	1	34
5.20	Spore size	0	9	26

APPENDIX 5
Data for Analyses of Lycopsida

TABLE A5.1
Characters and Character States for Taxa Used in the Cladistic Analysis in Chapter 6

Taxon	Circinate vernation (character 6.1)		Branching type (character 6.2)		Branching pattern (character 6.3)		Subordinate branching (character 6.4)	
Asteroxylon	0	absent	0	±isotomous	0	nonplanar	0	absent
Baragwanathia	0	absent	0	±isotomous	0	nonplanar	0	absent
Drepanophycus	0	absent	0	±isotomous	0	nonplanar	0	absent
Gosslingia	1	present	1	pseudomonopodial	1	planated	1	present
Huperzia selago	0	absent	0	±isotomous	0	nonplanar	0	absent
Huperzia squarrosa	0	absent	0	±isotomous	0	nonplanar	0	absent
Isoetes	0	absent	0	±isotomous	0	nonplanar	0	absent
Leclercqia	0	absent	0	±isotomous	0	nonplanar	0	absent
Lycopodiella	0	absent	0	±isotomous	0	nonplanar	0	absent
Lycopodium	0	absent	0	±isotomous	0	nonplanar	0	absent
Minarodendron	0	absent	1	pseudomonopodial	0	nonplanar	0	absent
Paralycopodites	0	absent	0	±isotomous	0	nonplanar	0	absent
Phylloglossum	0	absent	0	±isotomous	0	nonplanar	0	absent
Rhynia gwynne-vaughanii	0	absent	0	±isotomous	0	nonplanar	0	absent
Sawdonia	1	present	1	pseudomonopodial	1	planated	1	present
Selaginella (Stachygynandrum)	0	absent	1	pseudomonopodial	1	planated	1	present
Selaginella (Tetragonostachys)	0	absent	1	pseudomonopodial	1	planated	1	present
Selaginella selaginoides	0	absent	0	±isotomous	0	nonplanar	0	absent
Zosterophyllum llanoveranum	?	unknown	0	±isotomous	0	nonplanar	0	absent
Zosterophyllum myretonianum	0	absent	0	±isotomous	0	nonplanar	0	absent

TABLE A5.1
Continued

Taxon	Bulbils (character 6.5)		Microphylls (character 6.6)		Leaf vasculature (character 6.7)		Phyllotaxy (character 6.8)	
Asteroxylon	1	present	1	present	0	partial	0	helical
Baragwanathia	1	present	1	present	1	full	0	helical
Drepanophycus	1	present	1	present	1	full	0	helical
Gosslingia	0	absent	0	absent	?	NA	?	NA
Huperzia selago	1	present	1	present	1	full	0	helical
Huperzia squarrosa	0	absent	1	present	1	full	0	helical
Isoetes	0	absent	1	present	1	full	0	helical
Leclercqia	0	absent	1	present	1	full	0	helical
Lycopodiella	0	absent	1	present	1	full	0	helical
Lycopodium	0	absent	1	present	1	full	0	helical
Minarodendron	0	absent	1	present	1	full	0	helical
Paralycopodites	0	absent	1	present	1	full	0	helical
Phylloglossum	0	absent	1	present	1	full	0	helical
Rhynia gwynne-vaughanii	0	absent	0	absent	?	NA	?	NA
Sawdonia	0	absent	0	absent	?	NA	?	NA
Selaginella (Stachygynandrum)	0	absent	1	present	1	full	1	4-rowed
Selaginella (Tetragonostachys)	0	absent	1	present	1	full	0	helical
Selaginella selaginoides	0	absent	1	present	1	full	0	helical
Zosterophyllum llanoveranum	0	absent	0	absent	?	NA	?	NA
Zosterophyllum myretonianum	0	absent	0	absent	?	NA	?	NA

TABLE A5.1
Continued

Taxon	Leaf shape (character 6.9)		Anisophylly (character 6.10)		Ligule (character 6.11)		Mucilage canals (character 6.12)	
Asteroxylon	0	simple	0	absent	0	absent	0	absent
Baragwanathia	0	simple	0	absent	0	absent	0	absent
Drepanophycus	0	simple	0	absent	?	unknown	?	unknown
Gosslingia	?	NA	?	NA	0	absent	?	NA
Huperzia selago	0	simple	0	absent	0	absent	0	absent
Huperzia squarrosa	0	simple	0	absent	0	absent	0	absent
Isoetes	0	simple	0	absent	1	present	0	absent
Leclercqia	1	forked	0	absent	1	present	0	absent
Lycopodiella	0	simple	0	absent	0	absent	1	present
Lycopodium	0	simple	0	absent	0	absent	1	present
Minarodendron	1	forked	0	absent	?	unknown	?	unknown
Paralycopodites	0	simple	0	absent	1	present	0	absent
Phylloglossum	0	simple	0	absent	0	absent	0	absent
Rhynia gwynne-vaughanii	?	NA	?	NA	0	absent	?	NA
Sawdonia	?	NA	?	NA	0	absent	?	NA
Selaginella (Stachygynandrum)	0	simple	1	present	1	present	0	absent
Selaginella (Tetragonostachys)	0	simple	1	present	1	present	0	absent
Selaginella selaginoides	0	simple	0	absent	1	present	0	absent
Zosterophyllum llanoveranum	?	NA	?	NA	0	absent	?	NA
Zosterophyllum myretonianum	?	NA	?	NA	0	absent	?	NA

TABLE A5.1
Continued

Taxon	Sporangium shape (character 6.13)	Sporangium attachment (character 6.14)	Sporophyll (character 6.15)	Sporophyll shape (character 6.16)
Asteroxylon	1 ±reniform	1 lateral-stalked	0 absent	0 unmodified
Baragwanathia	1 ±reniform	1 lateral-stalked	? unknown	0 unmodified
Drepanophycus	1 ±reniform	1 lateral-stalked	0 absent	0 unmodified
Gosslingia	1 ±reniform	1 lateral-stalked	0 absent	? NA
Huperzia selago	1 ±reniform	1 lateral-stalked	1 present	0 unmodified
Huperzia squarrosa	1 ±reniform	1 lateral-stalked	1 present	0 unmodified
Isoetes	1 ±reniform	1 lateral-stalked	1 present	0 unmodified
Leclercqia	1 ±reniform	1 lateral-stalked	1 present	0 unmodified
Lycopodiella	1 ±reniform	1 lateral-stalked	1 present	1 peltate
Lycopodium	1 ±reniform	1 lateral-stalked	1 present	1 peltate
Minarodendron	1 ±reniform	1 lateral-stalked	1 present	0 unmodified
Paralycopodites	1 ±reniform	1 lateral-stalked	1 present	1 peltate
Phylloglossum	1 ±reniform	1 lateral-stalked	1 present	1 peltate
Rhynia gwynne-vaughanii	0 fusiform	0 terminal	0 absent	? NA
Sawdonia	1 ±reniform	1 lateral-stalked	0 absent	? NA
Selaginella (Stachygynandrum)	1 ±reniform	1 lateral-stalked	1 present	0 unmodified
Selaginella (Tetragonostachys)	1 ±reniform	1 lateral-stalked	1 present	0 unmodified
Selaginella selaginoides	1 ±reniform	1 lateral-stalked	1 present	0 unmodified
Zosterophyllum llanoveranum	1 ±reniform	1 lateral-stalked	0 absent	? NA
Zosterophyllum myretonianum	1 ±reniform	1 lateral-stalked	0 absent	? NA

TABLE A5.1
Continued

Taxon	Sporangiotaxis (character 6.17)	Sporangium vasculature (character 6.18)	Sporophyll structure (character 6.19)	Sporangium symmetry (character 6.20)
Asteroxylon	? unknown	1 present	0 simple	1 flattened
Baragwanathia	0 helical	? unknown	0 simple	1 flattened
Drepanophycus	? unknown	1 present	0 simple	1 flattened
Gosslingia	1 2-rowed	? unknown	? NA	1 flattened
Huperzia selago	0 helical	0 absent	0 simple	1 flattened
Huperzia squarrosa	0 helical	0 absent	0 simple	1 flattened
Isoetes	0 helical	0 absent	0 simple	2 elongated
Leclercqia	0 helical	0 absent	0 simple	2 elongated
Lycopodiella	0 helical	0 absent	1 ephemeral	1 flattened
Lycopodium	0 helical	0 absent	1 ephemeral	1 flattened
Minarodendron	0 helical	? unknown	0 simple	2 elongated
Paralycopodites	0 helical	0 absent	2 complex	2 elongated
Phylloglossum	0 helical	0 absent	1 ephemeral	1 flattened
Rhynia gwynne-vaughanii	? NA	1 present	? NA	0 radial
Sawdonia	1 2-rowed	? unknown	? NA	1 flattened
Selaginella (Stachygynandrum)	2 4-rowed	0 absent	0 simple	0 radial
Selaginella (Tetragonostachys)	0 helical	0 absent	0 simple	0 radial
Selaginella selaginoides	0 helical	0 absent	0 simple	0 radial
Zosterophyllum llanoveranum	1 2-rowed	? unknown	? NA	1 flattened
Zosterophyllum myretonianum	0 helical	1 present	? NA	1 flattened

TABLE A5.1
Continued

Taxon	Sporangium dehiscence (character 6.21)	Sporangium distribution (character 6.22)	Xylem strand shape (character 6.23)	Protoxylem (character 6.24)
Asteroxylon	0 tangential	? unknown	2 stellate	1 exarch
Baragwanathia	0 tangential	? unknown	2 stellate	? unknown
Drepanophycus	0 tangential	? unknown	2 stellate	? unknown
Gosslingia	0 tangential	0 extended	1 elliptical	1 exarch
Huperzia selago	0 tangential	0 extended	2 stellate	1 exarch
Huperzia squarrosa	0 tangential	0 extended	2 stellate	1 exarch
Isoetes	2 absent	1 compact	4 I-type	? unknown
Leclercqia	1 radial	0 extended	0 ±terete	1 exarch
Lycopodiella	0 tangential	1 compact	2 stellate	1 exarch
Lycopodium	0 tangential	1 compact	2 stellate	1 exarch
Minarodendron	1 radial	0 extended	0 ±terete	1 exarch
Paralycopodites	1 radial	1 compact	0 ±terete	1 exarch
Phylloglossum	0 tangential	1 compact	3 P-type	1 exarch
Rhynia gwynne-vaughanii	? unknown	0 extended	0 ±terete	0 centrarch
Sawdonia	0 tangential	0 extended	1 elliptical	1 exarch
Selaginella (Stachygynandrum)	0 tangential	1 compact	1 elliptical	1 exarch
Selaginella (Tetragonostachys)	0 tangential	1 compact	1 elliptical	1 exarch
Selaginella selaginoides	0 tangential	1 compact	0 ±terete	1 exarch
Zosterophyllum llanoveranum	0 tangential	0 extended	0 ±terete	1 exarch
Zosterophyllum myretonianum	0 tangential	1 compact	? unknown	? unknown

TABLE A5.1
Continued

Taxon	Stelar suspension (character 6.25)	Cortical sclerenchyma (character 6.26)	Sclerenchyma position (character 6.27)	Secondary growth (character 6.28)
Asteroxylon	2 partial cavity	? unknown	? unknown	0 absent
Baragwanathia	? unknown	? unknown	? unknown	0 absent
Drepanophycus	? unknown	? unknown	? unknown	0 absent
Gosslingia	? unknown	1 lignified	0 peripheral	0 absent
Huperzia selago	0 absent	2 nonlignified	0 peripheral	0 absent
Huperzia squarrosa	0 absent	2 nonlignified	0 peripheral	0 absent
Isoetes	0 absent	1 lignified	0 peripheral	1 present
Leclercqia	? unknown	1 lignified	0 peripheral	0 absent
Lycopodiella	0 absent	1 lignified	1 subperipheral	0 absent
Lycopodium	0 absent	1 lignified	1 subperipheral	0 absent
Minarodendron	? unknown	? unknown	? unknown	0 absent
Paralycopodites	? unknown	1 lignified	0 peripheral	1 present
Phylloglossum	0 absent	? unknown	? unknown	0 absent
Rhynia gwynne-vaughanii	0 absent	0 absent	? NA	0 absent
Sawdonia	? unknown	1 lignified	0 peripheral	0 absent
Selaginella (Stachygynandrum)	1 stelar cavity	1 lignified	0 peripheral	0 absent
Selaginella (Tetragonostachys)	1 stelar cavity	1 lignified	0 peripheral	0 absent
Selaginella selaginoides	1 stelar cavity	1 lignified	0 peripheral	0 absent
Zosterophyllum llanoveranum	? unknown	1 lignified	0 peripheral	0 absent
Zosterophyllum myretonianum	? unknown	? unknown	? unknown	? unknown

TABLE A5.1
Continued

Taxon	Metaxylem tracheid pitting (character 6.29)	Pseudobipolar growth (character 6.30)	Rootlet anatomy (character 6.31)	Root stele symmetry (character 6.32)
Asteroxylon	0 absent	0 monopolar	0 simple	0 radial
Baragwanathia	0 absent	0 monopolar	? unknown	? unknown
Drepanophycus	0 absent	0 monopolar	? unknown	? unknown
Gosslingia	0 absent	0 monopolar	? unknown	? unknown
Huperzia selago	1 present	0 monopolar	0 simple	1 bilateral
Huperzia squarrosa	1 present	0 monopolar	0 simple	1 bilateral
Isoetes	1 present	1 pseudobipolar	1 rhizomorphic	1 bilateral
Leclercqia	1 present	0 monopolar	? unknown	? unknown
Lycopodiella	1 present	0 monopolar	0 simple	1 bilateral
Lycopodium	1 present	0 monopolar	0 simple	1 bilateral
Minarodendron	1 present	0 monopolar	? unknown	? unknown
Paralycopodites	1 present	1 pseudobipolar	1 rhizomorphic	1 bilateral
Phylloglossum	0 absent	0 monopolar	0 simple	1 bilateral
Rhynia gwynne-vaughanii	0 absent	0 monopolar	0 simple	? NA
Sawdonia	0 absent	0 monopolar	? unknown	? unknown
Selaginella (Stachygynandrum)	1 present	0 monopolar	0 simple	1 bilateral
Selaginella (Tetragonostachys)	1 present	0 monopolar	0 simple	1 bilateral
Selaginella selaginoides	1 present	0 monopolar	0 simple	1 bilateral
Zosterophyllum llanoveranum	0 absent	0 monopolar	? unknown	? unknown
Zosterophyllum myretonianum	? unknown	0 monopolar	? unknown	? unknown

TABLE A5.1
Continued

Taxon	Root xylem shape (character 6.33)	Cortical roots (character 6.34)	Spore size (character 6.35)	Microspore ornamentation (character 6.36)
Asteroxylon	0 ±circular	0 absent	0 homosporous	0 absent
Baragwanathia	? unknown	0 absent	0 homosporous	0 absent
Drepanophycus	? unknown	0 absent	? unknown	? unknown
Gosslingia	? unknown	0 absent	0 homosporous	0 absent
Huperzia selago	1 crescent-shaped	1 present	0 homosporous	1 foveolate-fossulate
Huperzia squarrosa	1 crescent-shaped	1 present	0 homosporous	1 foveolate-fossulate
Isoetes	0 ±circular	0 absent	1 heterosporous	2 echinate
Leclercqia	? unknown	0 absent	0 homosporous	2 echinate
Lycopodiella	1 crescent-shaped	0 absent	0 homosporous	1 foveolate-fossulate
Lycopodium	1 crescent-shaped	0 absent	0 homosporous	1 foveolate-fossulate
Minarodendron	? unknown	0 absent	? unknown	? unknown
Paralycopodites	0 ±circular	0 absent	1 heterosporous	3 densely granulate
Phylloglossum	1 crescent-shaped	0 absent	0 homosporous	1 foveolate-fossulate
Rhynia gwynne-vaughanii	? NA	0 absent	0 homosporous	0 absent
Sawdonia	? unknown	0 absent	0 homosporous	0 absent
Selaginella (Stachygynandrum)	1 crescent-shaped	0 absent	1 heterosporous	2 echinate
Selaginella (Tetragonostachys)	1 crescent-shaped	0 absent	1 heterosporous	4 papillate
Selaginella selaginoides	1 crescent-shaped	0 absent	1 heterosporous	2 echinate
Zosterophyllum llanoveranum	? unknown	0 absent	0 homosporous	0 absent
Zosterophyllum myretonianum	? unknown	? unknown	0 homosporous	0 absent

TABLE A5.1
Continued

Taxon	Microgametophyte development (character 6.37)		Megagametophyte development (character 6.38)		Gametophyte habit (character 6.39)	
Asteroxylon	0	exosporic	?	unknown	?	unknown
Baragwanathia	?	unknown	?	unknown	?	unknown
Drepanophycus	?	unknown	?	unknown	?	unknown
Gosslingia	?	unknown	?	unknown	?	unknown
Huperzia selago	0	exosporic	0	cellular	1	subterranean
Huperzia squarrosa	0	exosporic	0	cellular	1	subterranean
Isoetes	1	endosporic	1	free nuclear	0	superficial
Leclercqia	?	unknown	?	unknown	?	unknown
Lycopodiella	0	exosporic	0	cellular	0	superficial
Lycopodium	0	exosporic	0	cellular	1	subterranean
Minarodendron	?	unknown	?	unknown	?	unknown
Paralycopodites	?	unknown	?	unknown	0	superficial
Phylloglossum	0	exosporic	0	cellular	0	superficial
Rhynia gwynne-vaughanii	?	unknown	?	unknown	?	unknown
Sawdonia	?	unknown	?	unknown	?	unknown
Selaginella (Stachygynandrum)	1	endosporic	1	free nuclear	0	superficial
Selaginella (Tetragonostachys)	1	endosporic	1	free nuclear	0	superficial
Selaginella selaginoides	1	endosporic	1	free nuclear	0	superficial
Zosterophyllum llanoveranum	?	unknown	?	unknown	?	unknown
Zosterophyllum myretonianum	?	unknown	?	unknown	?	unknown

TABLE A5.2
Data Matrix for Chapter 6 in Numeric Format

```
Asteroxylon                       00001100000001100?1010?212??0000000000??
Baragwanathia                     0000111000000011?00?010?2????000???000???
Drepanophycus                     0000111000??1100?1010?2????000???0?????
Gosslingia                        111100????0?110?1??10011?10000???000???
Huperzia selago                   0000111000000111000010021020010011101001
Huperzia squarrosa                0000011000000111000010021020010011101001
Isoetes                           000001100010111000002214?010111110012110
Leclercqia                        00000110101010111000021001?10010???002???
Lycopodiella                      0000011000001111100110121011010011001000
Lycopodium                        0000011000001111100110121011010011001001
Minarodendron                     0100011010??11100?021001???010???0?????
Paralycopodites                   00000110001011110022 1101?10111110013??0
Phylloglossum                     000001100000111100110131 0??000011001000
Rhynia gwynne-vaughanii           000000????0?000??1?0?00000?0000??000???
Sawdonia                          111100????0?110?1??10011?10000???000???
Selaginella (Stachygynandrum)     0111011101101110200001111100100110121 10
Selaginella (Tetragonostachys)    0111011001101110000001111100100110141 10
Selaginella selaginoides          0000011000101110000000101110010011012110
Zosterophyllum llanoveranum       ?00000????0?110?1??10001?10000???000???
Zosterophyllum myretonianum       000000????0?110?01?101???????0????00???
```

TABLE A5.3

Completeness of Data for Taxa Used in the Cladistic Analysis in Chapter 6

Taxon	Total inapplicable characters	Total unknown characters	Total characters scored	Percentage complete
Asteroxylon	0	6	33	85
Baragwanathia	0	13	26	67
Drepanophycus	0	16	23	59
Gosslingia	7	8	24	62
Huperzia selago	0	0	39	100
Huperzia squarrosa	0	0	39	100
Isoetes	0	1	38	97
Leclercqia	0	7	32	82
Lycopodiella	0	0	39	100
Lycopodium	0	0	39	100
Minarodendron	0	14	25	64
Paralycopodites	0	3	36	92
Phylloglossum	0	2	37	95
Rhynia gwynne-vaughanii	11	4	24	62
Sawdonia	7	8	24	62
Selaginella (Stachygynandrum)	0	0	39	100
Selaginella (Tetragonostachys)	0	0	39	100
Selaginella selaginoides	0	0	39	100
Zosterophyllum llanoveranum	7	9	23	59
Zosterophyllum myretonianum	7	14	18	46

TABLE A5.4

Completeness of Data for Characters Used in the Cladistic Analysis in Chapter 6

Character		Total taxa with inapplicable conditions	Total taxa with no information	Total taxa scored
6.1	Circinate vernation	0	1	19
6.2	Branching type	0	0	20
6.3	Branching pattern	0	0	20
6.4	Subordinate branching	0	0	20
6.5	Bulbils	0	0	20
6.6	Microphylls	0	0	20
6.7	Leaf vasculature	5	0	15
6.8	Phyllotaxy	5	0	15
6.9	Leaf shape	5	0	15
6.10	Anisophylly	5	0	15
6.11	Ligule	0	2	18
6.12	Mucilage canals	5	2	13
6.13	Sporangium shape	0	0	20
6.14	Sporangium attachment	0	0	20
6.15	Sporophyll	0	1	19
6.16	Sporophyll shape	5	0	15
6.17	Sporangiotaxis	1	2	17
6.18	Sporangium vasculature	0	5	15
6.19	Sporophyll structure	5	0	15
6.20	Sporangium symmetry	0	0	20
6.21	Sporangium dehiscence	0	1	19
6.22	Sporangium distribution	0	3	17
6.23	Xylem strand shape	0	1	19
6.24	Protoxylem	0	4	16
6.25	Stelar suspension	0	9	11
6.26	Cortical sclerenchyma	0	6	14
6.27	Sclerenchyma position	1	6	13
6.28	Secondary growth	0	1	19
6.29	Metaxylem tracheid pitting	0	1	19
6.30	Pseudobipolar growth	0	0	20
6.31	Rootlet anatomy	0	8	12
6.32	Root stele symmetry	1	8	11
6.33	Root xylem shape	1	8	11
6.34	Cortical roots	0	1	19
6.35	Spore size	0	2	18
6.36	Microspore ornamentation	0	2	18
6.37	Microgametophyte development	0	10	10
6.38	Megagametophyte development	0	11	9
6.39	Gametophyte habit	0	10	10

Bibliography

Albert, V. A., A. Backlund, and K. Bremer. 1994. DNA characters and cladistics: the optimization of functional history. In *Models in phylogeny reconstruction,* edited by R. W. Scotland, D. J. Siebert, and D. M. Williams. Oxford: Clarendon Press.

Albert, V. A., A. Backlund, K. Bremer, M. W. Chase, J. R. Manhart, B. D. Mishler, and K. C. Nixon. 1994. Functional constraints and *rbcL* evidence for land plant phylogeny. *Annals of the Missouri Botanical Garden* 81:534–567.

Allen, J. R. L. 1985. Marine to fresh water: the sedimentology of the interrupted environmental transition (Ludlow-Siegenian) in the Anglo-Welsh region. *Philosophical Transactions of the Royal Society of London* B309:85–104.

Allen, K. C. 1980. A review of *in situ* late Silurian and Devonian spores. *Review of Palaeobotany and Palynology* 29:253–270.

Amstutz, E. 1957. *Stylites,* a new genus of Isoetaceae. *Annals of the Missouri Botanical Garden* 44:121–123.

Ananiev, A. R. 1955. Rasteniia. *Atlas rukovod. form iskopaemykh fauny i flory zapadnoï Sibiri* 1:279–296.

———. 1959. Vazhneichie mestonakhozhdeniia devonskikh flor v Saïano-Altaïskoi gornoï oblasti. *Tomskij gosudarstvennij universitet,* pp. 1–99.

———. 1960. Biostratigrafiia paleozoia Saiano-Altaiskoi gornoi oblasti. *Srednii paleozoi* 2: 301–318.

Ananiev, A. R., and S. A. Stepanov. 1968. Finds of sporangiferous organs in *Psilophyton princeps* Dawson emend. Halle in the Lower Devonian of the southern Minusinsk Trough, western Siberia. In *New materials on the stratigraphy and paleontology of the Lower and Middle Paleozoic of western Siberia. Tomsk State University Publication* 202:30–46.

———. 1969. The first finding of the *Psilophyton* flora in Lower Devonian Salairsky Ridge (western Siberia). *Tomsk State University Publication* 203:13–28.

Andrews, H. N. 1947. *Ancient plants and the world they lived in.* New York: Comstock.

———. 1958. Notes on Belgian specimens of *Sporogonites. The Palaeobotanist* 7:85–89.

———. 1959. Evolutionary trends in early vascular plants. *Cold Spring Harbor Symposia on Quantitative Biology* 24:217–234.

———. 1961. *Studies in paleobotany.* 1st ed. New York: Wiley.

Andrews, H. N., and K. S. Alt. 1956. A new fossil plant from the New Albany Shale, with some comments on the origin of land vascular plants. *Annals of the Missouri Botanical Garden* 43:355–378.

Andrews, H. N., C. A. Arnold, E. Boureau, J. Doubinger, and S. Leclercq, eds. 1970. *Filicophyta.* Vol. 4, Part 1 of *Traité de paléobotanique,* edited by E. Boureau. Paris: Masson et Cie.

Andrews, H. N., A. E. Kasper, W. H. Forbes, P. G.

Gensel, and W. G. Chaloner. 1977. Early Devonian flora of the Trout Valley Formation of northern Maine. *Review of Palaeobotany and Palynology* 23:255–285.

Andrews, H. N., and T. L. Phillips. 1968. *Rhacophyton* from the Upper Devonian of West Virginia. *Botanical Journal of the Linnean Society* 61:37–64.

Arber, E. A. N. 1921. *Devonian floras: a study of the origin of Cormophyta.* Cambridge: Cambridge University Press.

Arnold, C. A. 1930. Bark structure of *Callixylon. Botanical Gazette* 90:427–431.

———. 1939. Observations on fossil plants from the Devonian of eastern North America. IV. Plant remains from the Catskill Delta deposits of northern Pennsylvania and southern New York. *Contributions from the Museum of Paleontology, The University of Michigan* 5:271–314.

———. 1947. *An introduction to paleobotany.* 1st ed. New York: McGraw-Hill.

Awramik, S. M. 1992. The oldest records of photosynthesis. *Photosynthesis Research* 33:75–89.

Ax, P. 1987. *The phylogenetic system.* Chichester: Wiley.

———. 1989. Basic phylogenetic systematization of the Metazoa. In *The hierarchy of life: molecules and morphology in phylogenetic analysis,* edited by B. Fernholm, K. Bremer, and H. Jörnvall. Amsterdam: Elsevier.

Axelrod, D. I. 1952. A theory of angiosperm evolution. *Evolution* 6:29–60.

———. 1959. Evolution of the psilophyte paleoflora. *Evolution* 13:264–275.

Bailey, I. W. 1953. Evolution of the tracheary tissue of land plants. *American Journal of Botany* 40:4–8.

Baker, R. G. E. 1988. The morphology and distribution of pits in the cell walls of *Sphagnum. Journal of the Hattori Botanical Laboratory* 64:359–365.

Baldauf, S. L., J. R. Manhart, and J. D. Palmer. 1990. Different fates of the chloroplast *tufA* gene following its transfer to the nucleus in green algae. *Proceedings of the National Academy of Sciences* 87: 5317–5321.

Baldauf, S. L., and J. D. Palmer. 1990. Evolutionary transfer of the chloroplast *tufA* gene to the nucleus. *Nature* 344:262–265.

———. 1993. Animals and fungi are each other's closest relatives: congruent evidence from multiple proteins. *Proceedings of the National Academy of Sciences* 90:11558–11562.

Banks, H. P. 1944. A new Devonian lycopod genus from southeastern New York. *American Journal of Botany* 31:649–659.

———. 1968. The early history of land plants. In *Evolution and environment,* edited by E. T. Drake. New Haven: Yale University Press.

———. 1970. *Evolution and plants of the past.* Belmont, Calif.: Wadsworth.

———. 1975a. Palaeogeographic implications of some Silurian–Early Devonian floras. In *Gondwana geology,* edited by K. S. W. Campbell. Canberra: Australian National University Press.

———. 1975b. Reclassification of Psilophyta. *Taxon* 24:401–413.

———. 1980. Floral assemblages in the Siluro-Devonian. In *Biostratigraphy of fossil plants,* edited by D. L. Dilcher and T. N. Taylor. Stroudsburg, Pa.: Hutchinson and Ross.

———. 1992. The classification of early land plants—revisited. *Geophytology* 22:49–64.

Banks, H. P., P. M. Bonamo, and J. D. Grierson. 1972. *Leclercqia complexa* gen. et sp. nov., a new lycopod from the late Middle Devonian of eastern New York. *Review of Palaeobotany and Palynology* 14:19–40.

Banks, H. P., and B. J. Colthart. 1993. Plant-animal-fungal interactions in early Devonian trimerophytes from Gaspé, Canada. *American Journal of Botany* 80:992–1001.

Banks, H. P., and M. R. Davis. 1969. *Crenaticaulis,* a new genus of Devonian plants allied to *Zosterophyllum,* and its bearing on the classification of early land plants. *American Journal of Botany* 56:436–449.

Banks, H. P., S. Leclercq, and F. M. Hueber. 1975. Anatomy and morphology of *Psilophyton dawsonii,* sp.n. from the late Lower Devonian of Quebec (Gaspé), and Ontario, Canada. *Palaeontographica Americana* 48:77–127.

Barghoorn, E. S., and W. C. Darrah. 1938. *Horneophyton,* a necessary change of name for *Hornea. Harvard University Botanical Museum Leaflets* 6:142–144.

Baroin, A., R. R. Perasso, L. H. Qu, G. Brugerolle, J. P. Bachellerie, and A. Andoutte. 1988. Partial phylogeny of the unicellular eukaryotes based on rapid sequencing of a portion of 28S ribosomal RNA. *Proceedings of the National Academy of Sciences* 85:3474–3478.

Barr, D. 1992. Evolution of the kingdoms of organ-

isms from the perspective of a mycologist. *Mycologia* 84:1–11.
Baschnagel, R. A. 1966. New fossil algae from the Middle Devonian of New York. *Transactions of the American Microscopical Society* 85:297–302.
Bateman, R. M. 1991. Palaeobiological and phylogenetic implications of anatomically-preserved *Archaeocalamites* from the Dinantian of Oxroad Bay and Loch Humphrey Burn, southern Scotland. *Palaeontographica* B223:1–59.
———. 1992. Morphogenetic reconstruction, palaeoecology, and phylogeny of *Oxroadia gracilis* Alvin emend., and *O. coniferata* sp. nov.: anatomically preserved lycopods from Oxroad Bay, SE Scotland. *Palaeontographica* B228:29–103.
———. 1994. Evolutionary-developmental change in the growth architecture of fossil rhizomorphic lycopods: scenarios constructed on cladistic foundations. *Biological Reviews of the Cambridge Philosophical Society* 69:527–598.
———. 1996a. Architectural radiations cannot be optimally interpreted without morphological and molecular phylogenies. In *The evolution of plant architecture,* edited by M. H. Kurmann and A. R. Hemsley. London: Royal Botanic Gardens, Kew.
———. 1996b. Non-floral homoplasy and evolutionary scenarios in living and fossil land plants. In *Homoplasy and the evolutionary process,* edited by M. J. Sanderson and L. Hufford. London: Academic Press.
———. 1996c. An overview of lycophyte phylogeny. In *Pteridology in perspective,* edited by J. M. Camus and R. J. Johns. London: Royal Botanic Gardens, Kew.
Bateman, R. M., and W. A. DiMichele. 1994a. Heterospory: the most iterative key innovation in the evolutionary history of the plant kingdom. *Biological Reviews* 69:345–417.
———. 1994b. Saltational evolution of form in vascular plants: a neoGoldschmidtian synthesis. In *Shape and form in plants and fungi,* edited by D. S. Ingram and A. Hudson. London: Academic Press.
Bateman, R. M., W. A. DiMichele, and D. A. Willard. 1992. Experimental cladistic analysis of anatomically preserved lycoposids from the Carboniferous of Euramerica: an essay on paleobotanical phylogenetics. *Annals of the Missouri Botanical Garden* 79:500–559.
Batenburg, L. H. 1977. The *Sphenophyllum* species in the Carboniferous flora of Holtz (Westphalian D, Saar Basin, Germany). *Review of Palaeobotany and Palynology* 24:69–99.
———. 1981. Vegetative anatomy and ecology of *Sphenophyllum zwickaviense, S. emarginatum,* and other "compression species" of *Sphenophyllum. Review of Palaeobotany and Palynology* 32: 275–313.
———. 1982. "Compression species" and "petrifaction species" of *Sphenophyllum* compared. *Review of Palaeobotany and Palynology* 36:335–339.
Beck, C. B. 1957. *Tetraxylopteris schmidtii* gen. et sp. nov., a probable pteridosperm precursor from the Devonian of New York. *American Journal of Botany* 44:350–367.
———. 1971. On the anatomy and morphology of lateral branch systems of *Archaeopteris. American Journal of Botany* 58:758–784.
———. 1981. *Archaeopteris* and its role in vascular plant evolution. In *Paleobotany, paleoecology, and evolution,* edited by K. J. Niklas. New York: Praeger.
Beck, C. B., R. Schmid, and G. R. Rothwell. 1982. Stelar morphology of the primary vascular system of seed plants. *Botanical Review* 48:691–815.
Beck, C. B., and W. E. Stein. 1993. *Crossia virginiana* gen. et sp. nov., a new member of the Stenokoleales from the Middle Devonian of southwestern Virginia. *Palaeontographica* B229:115–134.
Beck, C. B., and D. C. Wight. 1988. Progymnosperms. In *Origin and evolution of gymnosperms,* edited by C. B. Beck. New York: Columbia University Press.
Behnke, H.-D., and R. D. Sjolund, eds. 1990. *Sieve elements: comparative structure, induction, and development.* Berlin: Springer-Verlag.
Behrensmeyer, A. K., J. D. Damuth, W. A. DiMichele, R. Potts, H.-D. Sues, and S. L. Wing, eds. 1992a. *Terrestrial ecosystems through time: evolutionary paleoecology of terrestrial plants and animals.* Chicago: University of Chicago Press.
Behrensmeyer, A. K., R. W. Hook, C. E. Badgley, J. A. Boy, R. E. Chapman, P. Dodson, R. A. Gastaldo, R. W. Graham, L. D. Martin, P. E. Olsen, R. A. Spicer, R. E. Taggart, and M. V. H. Wilson. 1992b. Paleoenvironmental contexts and taphonomic modes. In *Terrestrial ecosystems through time: evolutionary paleoecology of terrestrial plants and animals,* edited by A. K. Behrensmeyer, J. D. Damuth, W. A. DiMichele, R. Potts, H.-D. Sues,

and S. L. Wing. Chicago: University of Chicago Press.

Bell, A. D. 1994. A summary of the branching process in plants. In *Shape and form in plants and fungi,* edited by D. S. Ingram and A. Hudson. London: Academic Press.

Bell, P. R., and C. L. F. Woodcock. 1971. *The diversity of green plants.* 2nd ed. London: Clowes.

Benson, M. 1904. *Telangium scotti,* a new species of *Telangium (Calymmatotheca)* showing structure. *Annals of Botany* 18:161–177.

Berger, S., and M. Kaever. 1992. *Dasycladales: an illustrated monograph of a fascinating algal order.* Stuttgart: Thieme.

Bernhardi, J. J. 1801. *Tentamen alterum filices in genera redigendi. Journal für die Botanik,* pp. 121–136.

Berry, C. M. 1994. First record of the Devonian lycophyte *Leclercqia* from South America. *Geological Magazine* 131:269–272.

Berry, C. M., J. E. Casas, and J. M. Moody. 1993. Diverse Devonian plant assemblages from Venezuela. *Documents des laboratoires de géologie, Lyon* 125:29–42.

Berry, C. M., and D. Edwards. 1994. New data on the morphology and anatomy of the Devonian zosterophyll *Serrulacaulis* Hueber and Banks from Venezuela. *Review of Palaeobotany and Palynology* 81:141–150.

Bhutta, A. A. 1973a. Observations on the sporangia of *Horneophyton lignieri* (Kidston and Lang) Barghoorn and Darrah 1938. *Pakistan Journal of Botany* 4:27–34.

———. 1973b. On the spores (including germinating spores) of *Horneophyton (Hornea) lignieri* (Kidston and Lang) Barghoorn and Darrah. *Pakistan Journal of Botany* 5:45–55.

———. 1973c. On the spores (including germinating spores) of *Rhynia major* Kidston and Lang. *Biologia* 19:47–57.

Bierhorst, D. W. 1960. Observations on tracheary elements. *Phytomorphology* 10:249–305.

———. 1971. *Morphology of vascular plants.* New York: Macmillan.

———. 1977. The systematic position of *Psilotum* and *Tmesipteris. Brittonia* 29:3–13.

Blackman, F. E. 1900. The primitive algae and the flagellate. An account of modern work bearing on the evolution of the algae. *Annals of Botany* 14:647–688.

Bold, H. C. 1957. *Morphology of plants.* New York: Harper.

Bold, H. C., and M. J. Wynne. 1985. *Introduction to the algae: structure and reproduction.* Englewood Cliffs: Prentice-Hall.

Bonamo, P. M. 1977. *Rellimia thomsonii* (Progymnospermopsida) from the Middle Devonian of New York State. *American Journal of Botany* 64:1272–1285.

Bonamo, P. M., and H. P. Banks. 1967. *Tetraxylopteris schmidtii:* its fertile parts and its relationships within the Aneurophytales. *American Journal of Botany* 54:755–768.

Bonamo, P. M., H. P. Banks, and J. D. Grierson. 1988. *Leclercqia, Haskinsia,* and the role of leaves in the delineation of Devonian lycopod genera. *Botanical Gazette* 149:222–239.

Bopp, M., and I. Capesius. 1996. New aspects of bryophyte taxonomy provided by a molecular approach. *Botanica Acta* 109:1–5.

Boros, Á., and M. Járai-Komlódi. 1975. *An atlas of European moss spores.* Budapest: Akadémiai Kiadó.

Boucot, A. J., and J. Gray. 1982. *Geologic correlates of early land plant evolution.* In Vol. 1, *Proceedings of the Third North American Paleontological Convention,* edited by B. Mamet and M. J. Copeland. Toronto: Business and Economic Service Ltd.

Boureau, E., ed. 1964. *Sphenophyta, Noeggerathiophyta.* Vol. 3 of *Traité de paléobotanique,* edited by E. Boureau. Paris: Masson et Cie.

Boureau, E., and J. Doubinger, eds. 1975. *Pteridophylla.* Vol. 4, Part 2 of *Traité de paléobotanique,* edited by E. Boureau. Paris: Masson et Cie.

Boureau, E., S. Jovet-Ast, O. A. Høeg, and W. G. Chaloner, eds. 1967. *Bryophyta, Psilophyta, Lycophyta.* Vol. 2 of *Traité de paléobotanique,* edited by E. Boureau. Paris: Masson et Cie.

Bower, F. O. 1890. On antithetic as distinct from homologous alternation of generations in plants. *Annals of Botany* 4:347–370.

———. 1891. Is the eusporangiate or leptosporangiate the more primitive type in the ferns? *Annals of Botany* 5:109–134.

———. 1894. Studies in the morphology of spore-producing members: Equisetineae and Lycopodineae. *Philosophical Transactions of the Royal Society of London* B185:473–572.

———. 1904. Studies in the morphology of spore-producing members. V. General comparisons and

conclusions. *Philosophical Transactions of the Royal Society of London* B196:191–257.
———. 1908. *The origin of a land flora.* London: Macmillan.
———. 1920. The earliest known land flora. Paper read at weekly evening meeting of the Royal Institution of Great Britain, April 30, 1920.
———. 1923a. *The ferns (Filicales) treated comparatively with a view to their natural classification.* Vol. 1. Cambridge: Cambridge University Press.
———. 1923b. The relation of size to the elaboration of form and structure of the vascular tracts in primitive plants. *Proceedings of the Royal Society of Edinburgh* 43:117–126.
———. 1930. *Size and form in plants, with special reference to conducting tracts.* London: Macmillan.
———. 1935. *Primitive land plants.* London: Macmillan.
Brack-Hanes, S. D. 1978. On the megagametophytes of two lepidodendracean cones. *Botanical Gazette* 139:140–146.
Brack-Hanes, S. D., and J. C. Vaughn. 1978. Evidence of Paleozoic chromosomes from lycopod microgametophytes. *Science* 200:1383–1385.
Braithwaite, A. F. 1988. Cytological and anatomical observations on *Tmesipteris* (Tmesipteridaceae: Pteridophyta) species from New Caledonia. *Fern Gazette* 13:199–208.
Brauer, D. F. 1980. *Barinophyton citrulliforme* (Barinophytales *incertae sedis,* Barinophytaceae) from the Upper Devonian of Pennsylvania. *American Journal of Botany* 67:1186–1206.
———. 1981. Heterosporous, barinophytacean plants from the Upper Devonian of North America and a discussion of the possible affinities of the Barinophytaceae. *Review of Palaeobotany and Palynology* 33:347–362.
Breckon, G. J., and R. H. Falk. 1974. External spore morphology and taxonomic affinities of *Phylloglossum drummondii* Kunze (Lycopodiaceae). *American Journal of Botany* 61:481–485.
Bremer, K. 1985. Summary of green plant phylogeny and classification. *Cladistics* 1:369–385.
Bremer, K., C. J. Humphries, B. D. Mishler, and S. P. Churchill. 1987. On cladistic relationships in green plants. *Taxon* 36:339–349.
Bremer, K., and H. E. Wanntorp. 1981a. The cladistic approach to plant classification. Paper read at the first meeting of the Willi Hennig Society.
———. 1981b. A cladistic classification of green plants. *Nordic Journal of Botany* 1:1–3.
Brooks, D. R., and D. A. McLennan. 1991. *Phylogeny, ecology, and behavior: a research program in comparative biology.* Chicago: University of Chicago Press.
———. 1994. Historical ecology as a research programme: scope, limitations, and the future. In *Phylogenetics and ecology,* edited by R. I. Vane-Wright and P. Eggleton. London: Academic Press.
Brown, R. C., and B. E. Lemmon. 1986. Spore wall development in the liverwort, *Haplomitrium hookeri. Canadian Journal of Botany* 64:1174–1182.
———. 1988. Sporogenesis in bryophytes. *Advances in Bryology* 3:159–223.
———. 1990a. Monoplastidic cell division in lower land plants. *American Journal of Botany* 77: 559–571.
———. 1990b. Sporogenesis in bryophytes. In *Microspores: evolution and ontogeny,* edited by S. Blackmore and R. B. Knox. London: Academic Press.
———. 1991a. The cytokinetic apparatus in meiosis: control of division plane in the absence of a preprophase band of microtubules. In *The cytoskeletal basis of plant growth and form,* edited by C. Lloyd. London: Academic Press.
———. 1991b. Sporogenesis in simple land plants. In *Pollen and spores,* edited by S. Blackmore and S. H. Barnes. Oxford: Clarendon Press.
Bruce, J. G. 1976a. Development and distribution of mucilage canals in *Lycopodium. American Journal of Botany* 63:481–491.
———. 1976b. Gametophytes and subgeneric concepts in *Lycopodium. American Journal of Botany* 63:919–924.
———. 1979. Gametophyte of *Lycopodium digitatum. American Journal of Botany* 66:1138–1150.
Bruchmann, H. 1898. *Über die Prothallien und Keimpflanzen mehrerer Europäischer Lycopodien.* Gotha.
Burgess, N. D. 1991. Silurian cryptospores and miospores from the type Llandovery area, south-west Wales. *Palaeontology* 34:575–599.
Burgess, N. D., and D. Edwards. 1988. A new Palaeozoic plant closely allied to *Prototaxites* Dawson. *Botanical Journal of the Linnean Society* 97: 189–203.
———. 1991. Classification of uppermost Ordovician to Lower Devonian tubular and filamentous

macerals from the Anglo-Welsh Basin. *Botanical Journal of the Linnean Society* 106:41–66.

Burgess, N. D., and J. D. Richardson. 1991. Silurian cryptospores and miospores from the type Wenlock area, Shropshire, England. *Palaeontology* 34: 601–628.

Butterfield, N. J., A. H. Knoll, and K. Swett. 1988. Exceptional preservation of fossils in an Upper Proterozoic shale. *Nature* 334:424–427.

———. 1990. A bangiophyte red alga from the Proterozoic of arctic Canada. *Science* 250:104–107.

Cai, C., and H.-J. Schweitzer. 1983. Über *Zosterophyllum yunnanicum* Hsü aus dem unterdevon Südchinas. *Palaeontographica* B185:1–10.

Cai, Chong-Yang, Dou Ya-Wei, and D. Edwards. 1993. New observations on a Pridoli plant assemblage from north Xinjiang, northwest China, with comments on its evolutionary and palaeogeographical significance. *Geological Magazine* 130: 155–170.

Cai, Chongyang, Shu Ouyang, Yi Wang, Zongjie Fang, Jiayu Rong, Liangyu Geng, and Xingxue Li. 1996. An early Silurian vascular plant. *Nature* 379:592.

Campbell, D. H. 1895. *The structure and development of the mosses and ferns.* London: Macmillan.

———. 1940. *The evolution of the land plants (Embryophyta).* Stanford: Stanford University Press.

Camus, J. M. 1990. Marattiaceae. In *Pteridophytes and gymnosperms,* edited by K. U. Kramer and P. S. Green. Vol. 1 of *The families and genera of vascular plants,* edited by K. Kubitzki. Berlin: Springer-Verlag.

Canatella, D. C., and R. O. de Sá. 1993. *Xenopus laevis* as a model organism. *Systematic Biology* 42:476–507.

Canright, J. E. 1970. Spores and associated macrofossils from the Devonian of Arizona. *Geoscience and Man* 1:83–88.

Carothers, Z. B., R. C. Brown, and J. G. Duckett. 1983. Comparative spermatogenesis in the Sphaerocarpales. 1. Blepharoplast structure in *Sphaerocarpos* and *Riella. Bryologist* 86:97–106.

Carothers, Z. B., and J. G. Duckett. 1980. The bryophyte spermatozoid: a source of new phylogenetic information. *Bulletin of the Torrey Botanical Club* 107:281–297.

Cavalier-Smith, T. 1978. The evolutionary origin and phylogeny of microtubules, mitotic spindles, and eukaryote flagella. *BioSystems* 10:93–114.

———. 1981. Eukaryote kingdoms: seven or nine? *BioSystems* 14:461–481.

———. 1987. The origin of fungi and pseudofungi. In *Evolutionary biology of the fungi,* edited by A. D. M. Rayner, C. M. Brasier, and D. Moore. Cambridge: Cambridge University Press.

Cedergren, R., M. W. Gray, Y. Abel, and D. Sankoff. 1988. The evolutionary relationships among known life forms. *Journal of Molecular Evolution* 28:98–112.

Celakovsky, L. 1874. Über die verschiedenen Formen und die Bedeutung des Generationswechsels der Pflanzen. *Sitzungsberichte der K. Böhmischen Gesellschaft der Wissenschafte,* n.p.

Chaloner, W. G. 1967a. Lycophyta. In *Bryophyta, Psilophyta, Lycophyta,* edited by E. Boureau, S. Jovet-Ast, O. A. Høeg, and W. G. Chaloner. Vol. 2 of *Traité de paléobotanique,* edited by E. Boureau. Paris: Masson et Cie.

———. 1967b. Spores and land-plant evolution. *Review of Palaeobotany and Palynology* 1:83–93.

———. 1968. The cone of *Cyclostigma kiltorkense* Houghton, from the Upper Devonian of Ireland. *Journal of the Linnean Society (Botany)* 61:25–36.

———. 1970. The rise of the first land plants. *Biological Reviews* 45:353–377.

Chaloner, W. G., and A. R. Hemsley. 1991. Heterospory: cul-de-sac or pathway to the seed? In *Pollen and spores,* edited by S. Blackmore and S. H. Barnes. Oxford: Clarendon Press.

Chaloner, W. G., A. Hill, and E. C. W. Rogerson. 1978. Early Devonian plant fossils from a southern England borehole. *Palaeontology* 21:693–707.

Chaloner, W. G., and P. Macdonald. 1980. *Plants invade the land.* Edinburgh: H.M.S.O., The Royal Scottish Museum.

Chaloner, W. G., M. K. Mensah, and M. D. Crane. 1974. Non-vascular land plants from the Devonian of Ghana. *Palaeontology* 17:925–947.

Chaloner, W. G., and J. M. Pettitt. 1987. The inevitable seed. *Bulletin de la Société botanique de France* 134:39–49.

Chaloner, W. G., and A. Sheerin. 1979. Devonian macrofloras. In *The Devonian system,* edited by M. R. House, C. T. Scrutton, and M. G. Bassett. Special Papers in Palaeontology, vol. 23. London: The Palaeontological Association.

Chapman, R. L., and M. A. Buchheim. 1991. Ribosomal RNA gene sequences: analysis and significance in the phylogeny and taxonomy of green algae. *Critical Reviews in Plant Science* 10:343–368.

Chatton, E. 1925. *Pansporella perplexa:* réflexions sur la biologie et la phylogenie des protozoaires. In *Annales des sciences naturelles, zoologie: l'anatomie, le physiologie, la classification, et l'histoire naturelle des animaux,* edited by M. E. L. Bouvier. Paris: Masson et Cie.

Chitaley, S., and D. C. McGregor. 1988. *Bisporangiostrobus harrisii* gen. et sp. nov., an eligulate lycopsid cone with *Duosporites* megaspores and *Geminospora* microspores from the Upper Devonian of Pennsylvania, U.S.A. *Palaeontographica* B210:127–149.

Church, A. H. 1919. Thalassiophyta and the subaerial transmigration. *Botanical Memoires, Oxford* 3:1–95.

Cichan, M. A. 1985a. Vascular cambium and wood development in Carboniferous plants. 1. Lepidodendrales. *American Journal of Botany* 72: 1163–1176.

———. 1985b. Vascular cambium and wood development in Carboniferous plants. 2. *Sphenophyllum plurifoliatum* Williamson and Scott (Sphenophyllales). *Botanical Gazette* 146:395–403.

———. 1986a. Vascular cambium and wood development in Carboniferous plants. 3. *Arthropitys* (Equisetales; Calamitaceae). *Canadian Journal of Botany* 64:688–695.

———. 1986b. Vascular cambium and wood development in Carboniferous plants. IV. Seed plants. *Botanical Gazette* 174:227–235.

Cichan, M. A., and T. N. Taylor. 1982. Vascular cambium development in *Sphenophyllum:* a Carboniferous arthrophyte. *IAWA (International Association of Wood Anatomists) Bulletin* 3:155–160.

———. 1983. A systematic and developmental analysis of *Arthropitys deltoides* sp. nov. *Botanical Gazette* 144:285–294.

———. 1990. Evolution of cambium in geologic time: a reappraisal. In *The vascular cambium,* edited by M. Iqbal. Somerset: John Wiley.

Cichan, M. A., T. N. Taylor, and D. F. Brauer. 1984. Ultrastructural studies of *in situ* Devonian spores: *Protobarinophyton pennsylvanicum* Brauer. *Review of Palaeobotany and Palynology* 41:167–175.

Cichan, M. A., T. N. Taylor, and E. L. Smoot. 1981. The application of scanning electron microscopy in the characterization of Carboniferous lycopod wood. *Scanning Electron Microscopy* 3:197–201.

Clark, C. G., and G. A. M. Cross. 1988. Small-subunit ribosomal RNA sequence from *Naegleria gruberi* supports the polyphyletic origin of amoebas. *Molecular Biology and Evolution* 5:512–515.

Coddington, J. A. 1988. Cladistic tests of adaptational hypotheses. *Cladistics* 4:3–22.

———. 1994. The roles of homology and convergence in studies of adaptation. In *Phylogenetics and ecology,* edited by R. I. Vane-Wright and P. Eggleton. London: Academic Press.

Collinson, M. E. 1991. Diversification of modern heterosporous pteridophytes. In *Pollen and spores,* edited by S. Blackmore and S. H. Barnes. Oxford: Clarendon Press.

Collinson, M. E., D. J. Batten, A. C. Scott, and S. N. Ayonghe. 1985. Palaeozoic, Mesozoic, and contemporaneous megaspores from the Tertiary of southern England: indicators of sedimentary provenance and ancient vegetation. *Journal of the Geological Society of London* 142:375–395.

Cooper-Driver, G. 1977. Chemical evidence for separating the Psilotaceae from the Filicales. *Science* 198:1260–1262.

Copeland, H. F. 1956. *The classification of lower organisms.* Palo Alto, Calif.: Pacific Books.

Corner, E. J. H. 1951. *Wayside trees of Malaya.* Vol. 1. Singapore: The Government Printer.

Cornet, B. C., T. L. Phillips, and H. N. Andrews. 1976. The morphology and variation in *Rhacophyton ceratangium* from the Upper Devonian and its bearing on frond evolution. *Palaeontographica* B158:105–129.

Crandall-Stotler, B. 1980. Morphogenetic designs and a theory of bryophyte origins and divergence. *BioScience* 30:580–585.

———. 1981. Morphology and anatomy of hepatics and anthocerotes. *Advances in Bryology* 1:315–376.

———. 1984. Musci, hepatics, and anthocerotes: an essay on analogues. In *New manual of bryology,* edited by R. M. Schuster. Miyazaki-ken, Japan: Hattori Botanical Laboratory.

———. 1986. Morphogenesis, developmental anatomy, and bryophyte phylogenetics: contraindications of monophyly. *Journal of Bryology* 14:1–23.

Crandall-Stotler, B., and J. J. Bozzola. 1988. Fine structure of the meristematic cells of *Takakia lepidozioides* Hatt. et H. Inoue (Takakiophyta). *Journal of the Hattori Botanical Laboratory* 64: 197–218.

Crane, P. R. 1985. Phylogenetic analysis of seed plants

and the origin of angiosperms. *Annals of the Missouri Botanical Garden* 72:716–793.

———. 1989. Patterns of evolution and extinction in vascular plants. In *Evolution and the fossil record,* edited by K. C. Allen and D. E. G. Briggs. London: Belhaven Press.

———. 1990. The phylogenetic context of microsporogenesis. In *Microspores: evolution and ontogeny,* edited by S. Blackmore and R. B. Knox. London: Academic Press.

Crane, P. R., E. M. Friis, and K. R. Pedersen. 1995. The origin and early diversification of angiosperms. *Nature* 374:27–33.

Crane, P. R., and S. Lidgard. 1990. Angiosperm radiation and patterns of Cretaceous palynological diversity. In *Major evolutionary radiations,* edited by P. D. Taylor and G. P. Larwood. Oxford: Clarendon Press.

Croft, W. N., and W. H. Lang. 1942. The Lower Devonian flora of the Senni Beds of Monmouthshire and Breconshire. *Philosophical Transactions of the Royal Society of London* B231:131–164.

Cronquist, A., A. Takhtajan, and W. Zimmermann. 1966. On the higher taxa of Embryobionta. *Taxon* 15:129–134.

Crundwell, A. C. 1979. Rhizoids and moss taxonomy. In *Bryophyte systematics,* edited by G. C. S. Clarke and J. G. Duckett. London: Academic Press.

Daber, R. 1960. *Eogaspesiea gracilis* n.g. n.sp. *Geologie* 9:418–425.

Dahlgren, R. M. T., and H. T. Clifford. 1982. *The monocotyledons: a comparative study.* London: Academic Press.

Dannenhoffer, J. E., and P. M. Bonamo. 1989. *Rellimia thomsonii* from the Givetian of New York: secondary growth in three orders of branching. *American Journal of Botany* 76:1312–1325.

Darnell-Smith, G. P. 1917. The gametophyte of *Psilotum. Transactions of the Royal Society of Edinburgh* 52:79–91.

Darrah, W. C. 1939. *Textbook of paleobotany.* New York: Appleton-Century.

———. 1960. *Principles of paleobotany.* 2nd ed. New York: Ronald Press Co.

Davis, P. H., and V. H. Heywood. 1973. *Principles of angiosperm taxonomy.* New York: Krieger.

Davison, P. G., D. K. Smith, and K. D. McFarland. 1989. The discovery of antheridia in *Takakia. ASB (Association of Southeastern Biologists) Bulletin* 36:65.

Dawson, J. W. 1859. On fossil plants from the Devonian rocks of Canada. *Quarterly Journal of the Geological Society of London* 15:477–488.

———. 1861. On the pre-Carboniferous flora of New Brunswick, Maine, and eastern Canada. *Canadian Naturalist* 6:161–180.

———. 1862. On the flora of the Devonian period of northeastern America. *Quarterly Journal of the Geological Society of London* 18:296–330.

———. 1870. The primitive vegetation of the earth. *Nature* 2:85–88.

———. 1871. *The fossil plants of the Devonian and Upper Silurian formations of Canada.* Geological Survey of Canada Paper.

De Bary, A. 1877. *Vergleichende Anatomie der Vegetationsorgane der Phanerogamen und Farne.* Leipzig: Engelmann.

Delwiche, C. F., L. E. Graham, and N. Thomson. 1989. Lignin-like compounds and sporopollenin in *Coleochaete,* an algal model for land plant ancestry. *Science* 245:399–401.

Delwiche, C. F., M. Kuhsel, and J. D. Palmer. 1995. Phylogenetic analysis of *tufA* sequences indicates a cyanobacterial origin of all plastids. *Molecular Phylogenetics and Evolution* 4:110–128.

Delwiche, C. F., and J. D. Palmer. 1996. Rampant horizontal transfer and duplication of Rubisco genes in Eubacteria and plastids. *Molecular Biology and Evolution* 13:873–882.

Dennis, R. L. 1974. Studies of Paleozoic ferns: *Zygopteris* from the Middle and Upper Pennsylvanian of the United States. *Palaeontographica* 148B:95–136.

De Saporta, G., and A. F. Marion. 1881. *L'évolution du règne végétal: les cryptogames.* Paris: Librairie Germer Baillière et Cie.

DiMichele, W. A. 1980. *Paralycopodites* Morey & Morey, from the Carboniferous of Euramerica: a reassessment of generic affinities and evolution of "*Lepidodendron*" *brevifolium* Williamson. *American Journal of Botany* 67:1466–1476.

DiMichele, W. A., and R. M. Bateman. 1992. Diaphorodendraceae, fam. nov. (Lycopsida: Carboniferous): systematics and evolutionary relationships of *Diaphorodendron* and *Synchysidendron,* gen. nov. *American Journal of Botany* 79:605–617.

———. In press. The rhizomorphic lycopsids: a case study in paleobotanical classification. *Systematic Botany.*

DiMichele, W. A., J. I. Davis, and R. G. Olmstead.

1989. Origins of heterospory and the seed habit: The role of heterochrony. *Taxon* 38:1–11.

DiMichele, W. A., R. W. Hook, R. Beerbower, J. A. Boy, R. A. Gastaldo, N. Hotton III, T. M. Phillips, S. E. Scheckler, W. A. Shear, and H.-D. Sues. 1992. Paleozoic terrestrial ecosystems. In *Terrestrial ecosystems through time: evolutionary paleoecology of terrestrial plants and animals,* edited by A. K. Behrensmeyer, J. D. Damuth, W. A. DiMichele, R. Potts, H.-D. Sues, and S. L. Wing. Chicago: University of Chicago Press.

DiMichele, W. A., and T. L. Phillips. 1985. Arborescent lycopod reproduction and paleoecology in a coal-swamp environment of late Middle Pennsylvanian age (Herrin Coal, Illinois, U.S.A.). *Review of Palaeobotany and Palynology* 44:1–26.

DiMichele, W. A., T. L. Phillips, and R. A. Peppers. 1985. The influence of climate and depositional environment on the distribution and evolution of Pennsylvanian coal-swamp plants. In *Geological factors and the evolution of plants,* edited by B. H. Tiffney. New Haven: Yale University Press.

DiMichele, W. A., and J. E. Skog. 1992. The Lycopsida: a symposium. *Annals of the Missouri Botanical Garden* 79:447–449.

Dittrich, H. S., L. C. Matten, and T. L. Phillips. 1983. Anatomy of *Rhacophyton ceratangium* from the Upper Devonian (Famennian) of West Virginia. *Review of Palaeobotany and Palynology* 40:127–147.

Donoghue, M. J. 1994. Progress and prospects in reconstructing plant phylogeny. *Annals of the Missouri Botanical Garden* 81:405–418.

Donoghue, M. J., J. A. Doyle, J. Gauthier, A. G. Kluge, and T. Rowe. 1989. The importance of fossils in phylogeny reconstruction. *Annual Review of Ecology and Systematics* 20:431–460.

Donoghue, M. J., R. G. Olmstead, J. F. Smith, and J. D. Palmer. 1992. Phylogenetic relationships of Dipsacales based on *rbcL* sequences. *Annals of the Missouri Botanical Garden* 79:333–345.

Doran, J. B. 1980. A new species of *Psilophyton* from the Lower Devonian of northern New Brunswick, Canada. *Canadian Journal of Botany* 58: 2241–2262.

Doran, J. B., P. G. Gensel, and H. N. Andrews. 1978. New occurrences of trimerophytes from the Devonian of eastern Canada. *Canadian Journal of Botany* 56:3052–3068.

Dorf, E. 1933. A new occurrence of the oldest known terrestrial vegetation, from Beartooth Butte, Wyoming. *Botanical Gazette* 95:240–257.

Douglas, S. E., and D. G. Durnford. 1989. The small subunit of ribulose-1,5-bisphosphate carboxylase is plastid encoded in the chlorophyll c-containing alga *Cryptomonas* F. *Plant Molecular Biology* 13:13–20.

Doyle, J. A., and M. J. Donoghue. 1986. Seed plant phylogeny and the origin of angiosperms: an experimental cladistic approach. *The Botanical Review* 52:321–431.

———. 1992. Fossils and seed plant phylogeny reanalyzed. *Brittonia* 44:89–106.

———. 1993. Phylogenies and angiosperm diversification. *Paleobiology* 19:141–167.

Doyle, J. A., M. J. Donoghue, and E. A. Zimmer. 1994. Integration of morphological and ribosomal RNA data on the origin of angiosperms. *Annals of the Missouri Botanical Garden* 81:419–450.

Doyle, J. A., and L. J. Hickey. 1976. Pollen and leaves from the mid-Cretaceous Potomac Group and their bearing on the early angiosperm evolution. In *Origin and early evolution of angiosperms,* edited by C. B. Beck. New York: Columbia University Press.

Doyle, W. T. 1970. *The biology of higher cryptogams.* London: Macmillan.

Duckett, J. G. 1973. Comparative morphology of the gametophytes of the genus *Equisetum,* subgenus *Equisetum. Botanical Journal of the Linnean Society* 66:1–22.

———. 1986. Ultrastructure in bryophyte systematics and evolution: an evaluation. *Journal of Bryology* 14:25–42.

Duckett, J. G., and K. S. Renzaglia. 1988a. Cell and molecular biology of bryophytes: ultimate limits to the resolution of phylogenetic problems. *Botanical Journal of the Linnean Society* 98:225–246.

———. 1988b. Ultrastructure and development of plastids in the bryophytes. *Advances in Bryology* 3:33–93.

Duerden, H. 1929. Variations in megaspore number in *Selaginella. Annals of Botany* 43:451–457.

Dufka, P. 1995. Upper Wenlock miospores and cryptospores derived from a Silurian volcanic island in the Prague Basin (Barrandian area, Bohemia). *Journal of Micropalaeontology* 14:67–79.

Eames, A. J. 1936. *Morphology of vascular plants: lower groups (Psilophytales to Filicales).* 1st ed. London: McGraw-Hill.

Edwards, D. 1968. A new plant from the Lower Old Red Sandstone of South Wales. *Palaeontology* 11:683–690.

———. 1969a. Further observations on *Zosterophyllum llanoveranum* from the Lower Devonian of South Wales. *American Journal of Botany* 56:201–210.

———. 1969b. *Zosterophyllum* from the Lower Old Red Sandstone of South Wales. *New Phytologist* 68:923–931.

———. 1970a. Fertile Rhyniophytina from the Lower Devonian of Britain. *Palaeontology* 13:451–461.

———. 1970b. Further observations on the Lower Devonian plant, *Gosslingia breconensis* Heard. *Philosophical Transactions of the Royal Society of London* B258:225–243.

———. 1975. Some observations on the fertile parts of *Zosterophyllum myretonianum* Penhallow from the Lower Old Red Sandstone of Scotland. *Transactions of the Royal Society of Edinburgh* 69: 251–265.

———. 1976. The systematic position of *Hicklingia edwardii* Kidston and Lang. *New Phytologist* 76:173–181.

———. 1979. A late Silurian flora from the Lower Old Red Sandstone of south-west Dyfed. *Palaeontology* 22:23–52.

———. 1980. Studies on Lower Devonian petrifactions from Britain. 1. Pyritised axes of *Hostinella* from the Brecon Beacons Quarry, Powys, South Wales. *Review of Palaeobotany and Palynology* 29:189–200.

———. 1982. Fragmentary non-vascular plant microfossils from the late Silurian of Wales. *Botanical Journal of the Linnean Society* 84:223–256.

———. 1986. Dispersed cuticles of putative non-vascular plants from the Lower Devonian of Britain. *Botanical Journal of the Linnean Society* 93: 259–275.

———. 1990. Constraints on Silurian and early Devonian phytogeographic analysis based on megafossils. In *Palaeozoic palaeogeography and biogeography,* edited by W. S. McKerrow and C. R. Scotese. London: The Geological Society.

———. 1993. Cells and tissues in the vegetative sporophytes of early land plants. *New Phytologist* 125:225–247.

Edwards, D., and J. L. Benedetto. 1985. Two new species of herbaceous lycopods from the Devonian of Venezuela with comments on their taphonomy. *Palaeontology* 28:599–618.

Edwards, D., and C. Berry. 1991. Silurian and Devonian. In *Plant fossils in geological investigation,* edited by C. J. Cleal. New York: Ellis Horwood.

Edwards, D., K. L. Davies, and L. Axe. 1992. A vascular conducting strand in the early land plant *Cooksonia. Nature* 357:683–685.

Edwards, D., and M. S. Davies. 1990. Interpretations of early land plant radiations: "facile adaptationist guesswork" or reasoned speculation? In *Major evolutionary radiations,* edited by P. D. Taylor and G. P. Larwood. Oxford: Clarendon Press.

Edwards, D., and D. S. Edwards. 1986. A reconsideration of the Rhyniophytina Banks. In *Systematic and taxonomic approaches in palaeobotany,* edited by R. A. Spicer and B. A. Thomas. Oxford: Clarendon Press.

Edwards, D., D. S. Edwards, and R. Rayner. 1982. The cuticle of early vascular plants and its evolutionary significance. In *The plant cuticle,* edited by D. F. Cutler, K. L. Alvin, and C. E. Price. London: Academic Press.

Edwards, D., and U. Fanning. 1985. Evolution and environment in the late Silurian–early Devonian: the rise of the pteridophytes. *Philosophical Transactions of the Royal Society of London* B309: 147–165.

Edwards, D., U. Fanning, and J. B. Richardson. 1986. Stomata and sterome in early land plants. *Nature* 323:438–440.

———. 1994. Lower Devonian coalified sporangia from Shropshire: *Salopella* Edwards & Richardson and *Tortilicaulis* Edwards. *Botanical Journal of the Linnean Society* 116:89–110.

Edwards, D., J. Feehan, and D. G. Smith. 1983. A late Wenlock flora from Co. Tipperary, Ireland. *Botanical Journal of the Linnean Society* 86:19–36.

Edwards, D., and P. Kenrick. 1986. A new zosterophyll from the Lower Devonian of Wales. *Botanical Journal of the Linnean Society* 92:269–283.

Edwards, D., P. Kenrick, and L. M. Carluccio. 1989. A reconsideration of cf. *Psilophyton princeps* (Croft and Lang, 1942), a zosterophyll widespread in the Lower Old Red Sandstone of South Wales. *Botanical Journal of the Linnean Society* 100:293–318.

Edwards, D., and J. B. Richardson. 1974. Lower Devonian (Dittonian) plants from the Welsh Borderland. *Palaeontology* 17:311–324.

Edwards, D. S. 1980. Evidence for the sporophytic status of the Lower Devonian plant *Rhynia gwynne-vaughanii* Kidston and Lang. *Review of Palaeobotany and Palynology* 29:177–188.

———. 1986. *Aglaophyton major,* a non-vascular land-plant from the Devonian Rhynie Chert. *Botanical Journal of the Linnean Society* 93:173–204.

Edwards, D. S., and A. G. Lyon. 1983. Algae from the Rhynie Chert. *Botanical Journal of the Linnean Society* 86:37–55.

Edwards, W. N. 1924. On the cuticular structure of the Devonian plant *Psilophyton. Journal of the Linnean Society of London (Botany)* 46:377–385.

Eggert, D. A. 1961. The ontogeny of Carboniferous arborescent Lycopsida. *Palaeontographica* B108:43–92.

———. 1962. The ontogeny of Carboniferous arborescent Sphenopsida. *Palaeontographica* B110:99–127.

———. 1974. The sporangium of *Horneophyton lignieri* (Rhyniophytina). *American Journal of Botany* 61:405–413.

Eggert, D. A., and D. D. Gaunt. 1973. Phloem of *Sphenophyllum. American Journal of Botany* 60:755–770.

Eggert, D. A., and N. Y. Kanemoto. 1977. Stem phloem of a Middle Pennsylvanian *Lepidodendron. Botanical Gazette* 138:102–111.

Ehret, D. L., and T. L. Phillips. 1977. *Psaronius* root systems: morphology and development. *Palaeontographica* B161:147–164.

El-Saadawy, W., and W. S. Lacey. 1979a. Observations on *Nothia aphylla* Lyon ex Høeg. *Review of Palaeobotany and Palynology* 27:119–147.

———. 1979b. The sporangia of *Horneophyton lignieri* (Kidston and Lang) Barghoorn and Darrah. *Review of Palaeobotany and Palynology* 28:137–144.

Elliott, G. F. 1984. Modern developments in the classification of some fossil green algae. In *Systematics of the green algae,* edited by D. E. G. Irvine and D. M. John. London: Academic Press.

Emberger, L. 1944. *Les plantes fossiles dans leurs rapports avec les végétaux vivants.* Paris: Masson et Cie.

———. 1960. *Les végétaux vasculaires.* Vol. 2, Part 1 of *Traité de botanique (systématique),* edited by M. Chadefaud and L. Emberger. Paris: Masson et Cie.

———. 1968. *Les plantes fossiles dans leurs rapports avec les végétaux vivants.* 2nd ed. Paris: Masson et Cie.

Endlicher, S. 1841. *Enchiridion Botanicum.* Vienna: Engelmann.

Engler, A., and K. Prantl. 1898–1900. *Die natürlichen Pflanzenfamilien.* Leipzig: Engelmann.

———. 1902. *Die natürlichen Pflanzenfamilien.* Vol. 1(4). Leipzig: Engelmann.

Evert, R. F. 1990. Seedless vascular plants. In *Sieve elements: comparative structure, induction, and development,* edited by H.-D. Behnke and R. D. Sjolund. Berlin: Springer-Verlag.

Fairon-Demaret, M. 1971. Quelques caractères anatomiques du *Drepanophycus spinaeformis* Göppert. *Comptes rendus des séances de l'Académie des sciences. Sciences naturelles* D273:933–935.

———. 1974. Nouveaux specimens du genre *Leclercqia* Banks, H. P., Bonamo, P. M., et Grierson, J. D., 1972 du Givetien (?) du Queensland (Australie). *Bulletin de l'Institut royal des sciences naturelles de Belgique. Sciences de la terre* 50:1–4.

———. 1978. *Estinnophyton gracile* gen. et sp. nov., a new name for specimens previously determined *Protolepidodendron wahnbachense* Kräusel and Weyland from the Siegenian of Belgium. *Bulletin de l'Académie royale de Belgique. Classe des sciences* 64(5):597–609.

———. 1979. *Estinnophyton wahnbachense* (Kräusel et Weyland) comb. nov., une plante remarquable du Siegenien d'Allemagne. *Review of Palaeobotany and Palynology* 28:145–160.

———. 1980. A propos des spécimens déterminés *Protolepidodendron scharianum* par Kräusel et Weyland, 1932. *Review of Palaeobotany and Palynology* 29:201–220.

———. 1981. Le genre *Leclercqia* Banks, H. P., Bonamo, P. M., et Grierson, J. D., 1972 dans le Dévonien Moyen de Belgique. *Bulletin de l'Institut royal des sciences naturelles de Belgique. Sciences de la terre* 53:1–10.

———. 1985. Les plantes fossiles de l'Emsien du Sart Tilman, Belgique. 1. *Stockmansia langii* (Stockmans) comb. nov. *Review of Palaeobotany and Palynology* 44:243–260.

———. 1986a. Les plantes Emsiennes du Sart Tilman (Belgique). 2. *Sartilmania jabachensis* (Kräusel et Weyland) comb. nov. *Review of Palaeobotany and Palynology* 47:225–239.

———. 1986b. *Stockmansella,* a new name for *Stockmansia* Fairon-Demaret (fossil). *Taxon* 35:334.

Fairon-Demaret, M., and Cheng-Sen Li. 1993. *Lorophyton goense* gen. et sp. nov. from the Lower Givetian of Belgium and a discussion of the Middle Devonian Cladoxylopsida. *Review of Palaeobotany and Palynology* 77:1–22.

Fairon-Demaret, M., and S. E. Scheckler. 1987. Typification and redescription of *Moresnetia zalesskyi* Stockmans, 1948, an early seed plant from the Upper Famennian of Belgium. *Bulletin de l'Institut royal des sciences naturelles de Belgique. Sciences de la terre* 57:183–199.

Fanning, U., D. Edwards, and J. B. Richardson. 1990. Further evidence for diversity in late Silurian land vegetation. *Journal of the Geological Society, London* 147:725–728.

———. 1991. A new rhyniophytoid from the late Silurian of the Welsh Borderland. *Neues Jahrbuch für Geologie und Paläontologie* 183:37–47.

———. 1992. A diverse assemblage of early land plants from the Lower Devonian of the Welsh Borderland. *Botanical Journal of the Linnean Society* 109:161–188.

Fanning, U., J. B. Richardson, and D. Edwards. 1988. Cryptic evolution in an early land plant. *Evolutionary Trends in Plants* 2:13–24.

———. 1991. A review of *in situ* spores in Silurian land plants. In *Pollen and spores,* edited by S. Blackmore and S. H. Barnes. Oxford: Clarendon Press.

Farris, J. S. 1970. Methods of computing Wagner trees. *Systematic Zoology* 19:83–92.

———. 1989. The retention index and rescaled consistency index. *Cladistics* 5:417–419.

Feist, M., and N. Grambast-Fessard. 1991. The genus concept in Charophyta: evidence from the Palaeozoic to Recent. In *Calcareous algae and stromatolites,* edited by R. Riding. Berlin: Springer-Verlag.

Fisher, D. C. 1992. Stratigraphic parsimony. In *McClade 3: analysis of phylogeny and character evolution,* edited by W. P. Maddison and D. R. Maddison. Sunderland, Mass.: Sinauer.

Forey, P. L., C. J. Humphries, I. J. Kitching, R. W. Scotland, D. J. Siebert, and D. M. Williams. 1992. *Cladistics: a practical course in systematics.* Oxford: Oxford University Press.

Frey, W., E. O. Campbell, and H. H. Hilger. 1994a. The sporophyte-gametophyte junction in *Tmesipteris* (Psilotaceae, Psilotopsida): phylogenetic implications of transfer cell arrangement. *Beiträge zur Biologie der Pflanzen* 68:105–111.

———. 1994b. Structure of the sporophyte-gametophyte junction in *Tmesipteris elongata* P. A. Dangeard (Psilotaceae, Psilotopsida) and its phylogenetic implications: a SEM analysis. *Nova Hedwigia* 59:21–32.

Friedman, W. E. 1993. The evolutionary history of the seed plant male gametophyte. *Trends in Ecology and Evolution* 8:15–20.

Fries, T. 1897. *Lärobok i systematisk botanik. 2. De kryptogama växterna.* Stockholm: Beijers.

Fritsch, F. E. 1916. The algal ancestry of the higher plants. *The New Phytologist* 15:233–250.

———. 1935. *The structure and reproduction of the algae.* Vol. 1. London: Cambridge University Press.

———. 1945. Studies in the comparative morphology of the algae. 4. Algae and archegoniate plants. *Annals of Botany* 9:1–29.

———. 1948. *The structure and reproduction of the algae.* Vol. 1. Cambridge: Cambridge University Press.

Fritsch, F. E., and E. J. Salisbury. 1938. *Plant form and function.* London: Bell.

Galtier, J. 1988. Morphology and phylogenetic relationships of early pteridosperms. In *Origin and evolution of gymnosperms,* edited by C. B. Beck. New York: Columbia University Press.

———. 1992. On the earliest arborescent gymnosperms. *Courier Forschungsinstitut Senckenberg* 147:119–125.

Galtier, J., and C. B. Beck. 1992. *Triichnia,* a new eustelic calamopityacean from the Lower Carboniferous of France. *Palaeontographica* B224:1–16.

Galtier, J., and N. Rowe. 1989. A primitive seed-like structure and its implications for early gymnosperm evolution. *Nature* 340:225–227.

Galtier, J., and A. C. Scott. 1985. Diversification of early ferns. *Proceedings of the Royal Society of Edinburgh* B86:289–301.

Garbary, D. J., K. S. Renzaglia, and J. G. Duckett. 1993. The phylogeny of land plants: a cladistic analysis based on male gametogenesis. *Plant Systematics and Evolution* 188:237–269.

Garratt, M. J. 1978. New evidence for a Silurian (Ludlow) age for the earliest *Baragwanathia* flora. *Alcheringa* 2:217–224.

Garratt, M. J., and R. B. Rickards. 1984. Graptolite biostratigraphy of early land plants from Victoria,

Australia. *Proceedings of the Yorkshire Geological Society* 44:377–384.

———. 1987. Pridoli (Silurian) graptolites in association with *Baragwanathia* (Lycophytina). *Bulletin of the Geological Society of Denmark* 35:135–139.

Garratt, M. J., J. D. Tims, R. B. Rickards, T. C. Chambers, and J. G. Douglas. 1984. The appearance of *Baragwanathia* (Lycophytina) in the Silurian. *Botanical Journal of the Linnean Society* 89:355–358.

Gauthier, J., A. G. Kluge, and T. Rowe. 1988. Amniote phylogeny and the importance of fossils. *Cladistics* 4:105–209.

Geng, Bao-Yin. 1985. *Huia recurvata:* a new plant from the Lower Devonian of southeastern Yunnan, China. *Acta Botanica Sinica* 27:419–426.

———. 1986. Anatomy and morphology of *Pinnatiramosus,* a new plant from the Middle Silurian (Wenlockian) of China. *Acta Botanica Sinica* 28:664–670.

———. 1990. New observations of *Leptophloeum rhombicum* from the Upper Devonian of Hubei, China. *Acta Botanica Sinica* 32:951–956.

Gensel, P. G. 1976. *Renalia hueberi,* a new plant from the Lower Devonian of Gaspé. *Review of Palaeobotany and Palynology* 22:19–37.

———. 1977. Morphologic and taxonomic relationships of the Psilotaceae relative to evolutionary lines in early land vascular plants. *Brittonia* 29:14–29.

———. 1979. Two *Psilophyton* species from the Lower Devonian of eastern Canada with a discussion on morphological variation within the genus. *Palaeontographica* B168:81–99.

———. 1980. Devonian *in situ* spores: a survey and discussion. *Review of Palaeobotany and Palynology* 30:101–132.

———. 1982a. A new species of *Zosterophyllum* from the early Devonian of New Brunswick. *American Journal of Botany* 69:651–669.

———. 1982b. *Oricilla,* a new genus referable to the zosterophyllophytes from the late Early Devonian of northern New Brunswick. *Review of Palaeobotany and Palynology* 37:345–359.

———. 1984. A new Lower Devonian plant and the early evolution of leaves. *Nature* 309:785–787.

———. 1991. Notes on the cuticular morphology of *Sawdonia acanthotheca,* particularly in regard to emergences. *Neues Jahrbuch für Geologie und Paläontologie* 183:49–59.

———. 1992. Phylogenetic relationships of the zosterophylls and lycopsids: evidence from morphology, paleoecology and cladistic methods of inference. *Annals of the Missouri Botanical Garden* 79:450–473.

Gensel, P. G., and H. N. Andrews. 1984. *Plant life in the Devonian.* New York: Praeger.

Gensel, P. G., H. N. Andrews, and W. H. Forbes. 1975. A new species of *Sawdonia* with notes on the origin of microphylls and lateral sporangia. *Botanical Gazette* 136:50–62.

Gensel, P. G., W. G. Chaloner, and W. H. Forbes. 1991. *Spongiophyton* from the late Lower Devonian of New Brunswick and Quebec, Canada. *Palaeontology* 34:149–168.

Gensel, P. G., N. G. Johnson, and P. K. Strother. 1990. Early land plant debris (Hooker's "Waifs and Strays"?). *Palaios* 5:520–547.

Gensel, P. G., A. Kasper, and H. N. Andrews. 1969. *Kaulangiophyton,* a new genus of plants from the Devonian of main. *Bulletin of the Torrey Botanical Club* 96:265–276.

Gensel, P. G., and A. R. White. 1983. The morphology and ultrastructure of spores of the early Devonian trimerophyte *Psilophyton* (Dawson) Hueber and Banks. *Palynology* 7:221–223.

Gerrienne, P. 1983. Les plantes émsiennes de Marchin (Vallée du Hoyoux, Belgique). *Annales de la Société géologique de Belgique* 106:19–35.

———. 1988. Early Devonian plant remains from Marchin (north of Dinant Synclinorium, Belgium), 1. *Zosterophyllum deciduum* sp. nov. *Review of Palaeobotany and Palynology* 55:317–335.

———. 1990. Les *Pachytheca* de la Gileppe et de Nonceveux (Dévonien Inférieur de Belgique). *Annales de la Société géologique de Belgique* 113: 267–285.

———. 1993. Inventaire des végétaux éodévoniens de Belgique. *Annales de la Société géologique de Belgique* 116:105–117.

Gess, R. W., and N. Hiller. 1995. Late Devonian charophytes from the Witteberg Group, South Africa. *Review of Palaeobotany and Palynology* 89:417–428.

Gifford, E. M., and A. S. Foster. 1989. *Morphology and evolution of vascular plants.* 3rd ed. New York: Freeman.

Goebel, K. 1887. *Outlines of classification and special morphology of plants.* Translated by H. E. F.

Garnsey, and I. A. Balfour. Oxford: Clarendon Press.

Gogarten, J. P. 1995. The early evolution of cellular life. *Trends in Ecology and Evolution* 10:147–151.

Göppert, H. R. 1852. Fossile flora des Übergangsgebirges. *Nova Acta Leopoldiana* 22:1–299.

Gorham, J. 1977. Lunularic acid and related compounds in liverworts, algae, and *Hydrangea*. *Phytochemistry* 16:249–253.

Gould, S. J. 1989. *Wonderful life*. New York: Norton.

Graham, L. E. 1984. *Coleochaete* and the origin of land plants. *American Journal of Botany* 71: 603–608.

———. 1990. Meiospore formation in charophycean algae. In *Microspores: evolution and ontogeny*, edited by S. Blackmore and R. B. Knox. London: Academic Press.

———. 1993. *Origin of land plants*. New York: Wiley.

Graham, L. E., C. F. Delwiche, and B. D. Mishler. 1991. Phylogenetic connections between the "green algae" and the "bryophytes." *Advances in Bryology* 4:213–244.

Graham, L. E., and W. M. Repavich. 1989. Spermatogenesis in *Coleochaete pulvinata* (Charophyceae): early blepharoplast development. *American Journal of Botany* 76:1266–1278.

Graham, L. E., and L. W. Wilcox. 1983. The occurrence and phylogenetic significance of putative placental transfer cells in the green alga *Coleochaete*. *American Journal of Botany* 70:113–120.

Granoff, J. A., P. G. Gensel, and H. N. Andrews. 1976. A new species of *Pertica* from the Devonian of eastern Canada. *Palaeontographica* B155:119–128.

Grant, M. C. 1990. Phylum Chlorophyta, Class Charophyceae, Order Charales. In *Handbook of Protoctista*, edited by L. Margulis, J. O. Corliss, M. Melkonian, and D. J. Chapman. Boston: Jones and Bartlett.

Gray, J. 1985. The microfossil record of early land plants: advances in understanding of early terrestrialization, 1970–1984. *Philosophical Transactions of the Royal Society of London* B309:167–195.

———. 1991. *Tetrahedraletes, Nodospora*, and the "cross" tetrad: an accretion of myth. In *Pollen and spores*, edited by S. Blackmore and S. H. Barnes. Oxford: Clarendon Press.

———. 1993. Major Paleozoic land plant evolutionary bio-events. *Palaeogeography, Palaeoclimatology, Palaeoecology* 104:153–169.

Gray, J., and A. J. Boucot. 1971. Early Silurian spore tetrads from New York: earliest New World evidence for vascular plants? *Science* 173:918–921.

———. 1977. Early vascular land plants: proof and conjecture. *Lethaia* 10:145–174.

Gray, J., D. Massa, and A. J. Boucot. 1982. Caradocian land plant microfossils from Libya. *Geology* 10:197–201.

Gray, J., and W. Shear. 1992. Early life on land. *American Scientist* 80:444–456.

Greguss, P. 1955. *Identification of living gymnosperms on the basis of xylotomy*. Budapest: Akadémiai Kiadó.

Grierson, J. D. 1976. *Leclercqia complexa* (Lycopsida, Middle Devonian): its anatomy and the interpretation of pyrite petrifactions. *American Journal of Botany* 63:1184–1202.

Grierson, J. D., and H. P. Banks. 1963. Lycopods of the Devonian of New York State. *Palaeontographica Americana* 4:217–295.

———. 1983. A new genus of lycopods from the Devonian of New York State. *Botanical Journal of the Linnean Society* 86:81–101.

Grierson, J. D., and P. M. Bonamo. 1979. *Leclercqia complexa:* earliest ligulate lycopod (Middle Devonian). *American Journal of Botany* 66:474–476.

Grierson, J. D., and F. M. Hueber. 1967. Devonian lycopods from northern New Brunswick. Paper read at the international symposium on the Devonian system held in Calgary by the Alberta Society of Petroleum Geologists.

Gunderson, J. H., H. Elwood, A. Inglod, K. Kindle, and M. L. Sogin. 1987. Phylogenetic relationships between chlorophytes, chrysophytes, and oomycetes. *Proceedings of the National Academy of Sciences* 84:5823–5827.

Gunning, B. E. S., and J. S. Pate. 1969. Cells with wall ingrowths (transfer cells) in the placenta of ferns. *Planta* 87:271–274.

———. 1974. Transfer cells. In *Dynamic aspects of plant ultrastructure*, edited by A. W. Robards. London: McGraw-Hill.

Gupta, K. M. 1956. Fossil plants from the Deccan Intertrappean Series. 1. A bryophytic type of sporogonium. *Science and Culture* 21:540–541.

Hackney, F. M. V. 1950. A review of and contribution to the knowledge of *Phylloglossum drummondii* Kunze. *Proceedings of the Linnean Society of New South Wales* 75:133–152.

Haeckel, E. 1866. *Allgemeine Entwicklungsgeschichte der Organismen*. Berlin: Reimer.

———. 1876. *The history of creation.* Vol. 2. New York: Appleton and Company.

———. 1894. *Systematische Phylogenie.* Berlin: Georg Reimer.

Haig, D., and M. Westoby. 1988. A model for the origin of heterospory. *Journal of Theoretical Biology* 134:257–272.

Halanych, K. M. 1991. 5S ribosomal RNA sequences inappropriate for phylogenetic reconstruction. *Molecular Biology and Evolution* 8:249–253.

Halle, T. G. 1916a. A fossil sporogonium from the Lower Devonian of Röragen in Norway. *Botaniska Notiser,* pp. 79–81.

———. 1916b. Lower Devonian plants from Röragen in Norway. *Kungliga Svenska Vetenskapsakademiens Handlingar* 57:1–46.

———. 1927. Fossil plants from south-western China. *Palaeontologica Sinica* A1:1–26.

———. 1933. The structure of certain fossil spore-bearing organs believed to belong to pteridosperms. *Kungliga Svenska Vetenskapsakademiens Handlingar* 12:1–103.

———. 1936a. Notes on the Devonian genus *Sporogonites. Svensk Botanisk Tidskrift* 30:613–623.

———. 1936b. On *Drepanophycus, Protolepidodendron,* and *Protopteridium,* with notes on the Palaeozoic flora of Yunnan. *Palaeontologica Sinica* A1:1–38.

Hallier, H. 1902. *Beiträge zur Morphogenie der Sporophylle und des Trophophylls, Beziehung zur Phylogenie der Kormophyten.* Hamburg: Gräfe und Sillem.

Han, T. M., and B. Runnegar. 1992. Megascopic eukaryotic algae from the 21-billion-year-old Negaunee Iron-Formation, Michigan. *Science* 257:232–235.

Hao, Shou-Gang. 1988. A new Lower Devonian genus from Yunnan, with notes on the origin of leaves. *Acta Botanica Sinica* 30:441–448.

———. 1989a. *Gumuia zyzzata:* a new plant from the Lower Devonian of Yunnan, China. *Acta Botanica Sinica* 31:954–961.

———. 1989b. A new zosterophyll from the Lower Devonian (Siegenian) of Yunnan, China. *Review of Palaeobotany and Palynology* 57:155–171.

———. 1992. Some observations on *Zosterophyllum australianum* Lang & Cookson from the Lower Devonian of Yunnan, China. *Botanical Journal of the Linnean Society* 109:189–202.

Hao, Shou-Gang, and C. B. Beck. 1991a. *Catenalis digitata* gen. et sp. nov., a plant from the Lower Devonian of Yunnan, China. *Canadian Journal of Botany* 69:873–882.

———. 1991b. *Yunia dichotoma,* a Lower Devonian plant from Yunnan, China. *Review of Palaeobotany and Palynology* 68:181–195.

———. 1993. Further observations on *Eophyllophyton bellum* from the Lower Devonian (Siegenian) of Yunnan, China. *Palaeontographica* B230:27–41.

Harland, W. B., R. L. Armstrong, A. V. Cox, L. E. Craig, A. G. Smith, and D. G. Smith. 1990. *A geologic time scale 1989.* Cambridge: Cambridge University Press.

Harland, W. B., A. V. Cox, P. G. Llewellyn, C. A. G. Pickton, A. G. Smith, and R. Walters. 1982. *A geologic time scale.* Cambridge: Cambridge University Press.

Harris, T. M. 1939. *Naiadita,* a fossil bryophyte with reproductive organs. *Annales Bryologici* 12:57–70.

———. 1961. *The Yorkshire Jurassic flora.* Vol. 1. *Thallophyta to Pteridophyta.* British Museum of Natural History Publication. London: Clowes and Sons.

Harris, T. M., W. Millington, and J. Miller. 1974. *The Yorkshire Jurassic flora.* IV. *Ginkgoales and Czekanowskiales.* British Museum of Natural History Publication 724. Kettering: Staples.

Hartman, C. M. 1981. The effect of pyrite on the tracheid structure of *Drepanophycus spinaeformis,* a long-ranging Devonian lycopod. *Review of Palaeobotany and Palynology* 32:239–255.

Hartman, C. M., and H. P. Banks. 1980. Pitting in *Psilophyton dawsonii,* an early Devonian trimerophyte. *American Journal of Botany* 67:400–412.

Harvey, P. H., and M. D. Pagel. 1991. *The comparative method of evolutionary biology.* Oxford: Oxford University Press.

Harvey, W. H. 1836. Algae. In *Flora Hibernica,* edited by J. T. Mackay. Dublin: William Curry.

Harvey-Gibson, R. J. 1894. Contributions towards a knowledge of the anatomy of the genus *Selaginella.* Part 1. The stem. *Annals of Botany* 8:133–206.

———. 1896. Contributions towards a knowledge of the anatomy of the genus *Selaginella,* Spr. *Annals of Botany* 10:77–88.

———. 1902. Contributions towards a knowledge of the anatomy of the genus *Selaginella. Annals of Botany* 16:449–466.

Hasebe, M., M. Ito, R. Kofuji, K. Ueda, and K. Iwatsuki. 1993. Phylogenetic relationships of ferns deduced from *rbcL* gene sequence. *Journal of Molecular Evolution* 37:476–482.

Hasebe, M., T. Omori, M. Nakazawa, T. Sano, M. Kato, and K. Iwatsuki. 1994. *rbcL* gene sequences provide evidence for the evolutionary lineages of leptosporangiate ferns. *Proceedings of the National Academy of Sciences* 91:5730–5734.

Hasebe, M., P. G. Wolf, K. M. Pryer, K. Ueda, M. Ito, R. Sano, G. J. Gastony, J. Yokoyama, J. R. Manhart, N. Murakami, E. H. Crane, C. H. Haufler, and W. D. Hauk. 1995. Fern phylogeny based on *rbcL* nucleotide sequences. *American Fern Journal* 85:134–181.

Hasegawa, J. 1988. A proposal for a new system of the Anthocerotae, with a revision of the genera. *Journal of the Hattori Botanical Laboratory* 64:87–95.

Hass, H., and W. Remy. 1991. *Huvenia kleui* nov. gen., nov. spec.: ein Vertreter der Rhyniaceae aus dem Höheren Siegen des Rheinischen Schiefergebirges. *Argumenta Palaeobotanica* 8:141–168.

Hässel de Menéndez, G. G. 1988. A proposal for a new classification of the genera within the Anthocerotophyta. *Journal of the Hattori Botanical Laboratory* 64:71–86.

Hattori, S., and M. Mizutani. 1958. What is *Takakia lepidoziodes? Journal of the Hattori Botanical Laboratory* 20:295–303.

Hauke, R. L. 1990. Equisetatae. In *Pteridophytes and gymnosperms,* edited by K. U. Kramer and P. S. Green. Vol. 1 of *The families and genera of vascular plants,* edited by K. Kubitzki. Berlin: Springer-Verlag.

Heard, A. 1927. On Old Red Sandstone plants showing structure from Brecon (South Wales). *Quarterly Journal of the Geological Society of London* 83:195–209.

———. 1939. Further notes on Lower Devonian plants from South Wales. *Quarterly Journal of the Geological Society of London* 95:223–229.

Hébant, C. 1977. *The conducting tissues of bryophytes.* Bryophytorum Bibliotheca, vol. 10. Vaduz, Germany: Cramer.

———. 1979. Conducting tissues in bryophyte systematics. In *Bryophyte systematics,* edited by G. C. S. Clarke and J. G. Duckett. London: Academic Press.

Hemsley, A. R. 1989. The ultrastructure of the spore wall of the Triassic bryophyte *Naiadita lanceolata. Review of Palaeobotany and Palynology* 61:89–99.

———. 1990. *Parka decipiens* and land plant spore evolution. *Historical Biology* 4:39–50.

———. 1993. A review of Palaeozoic seed-megaspores. *Palaeontographica* B229:135–166.

———. 1994a. Exine ultrastructure in the spores of enigmatic Devonian plants: its bearing on the interpretation of relationships and on the origin of the sporophyte. In *Ultrastructure of fossil spores and pollen,* edited by M. H. Kurmann and J. A. Doyle. London: Royal Botanic Gardens, Kew.

———. 1994b. The origin of the land plant sporophyte: an interpolational scenario. *Biological Reviews* 69:263–274.

Hemsley, A. R., W. G. Chaloner, A. C. Scott, and C. J. Groombridge. 1992. Carbon-13 solid-state nuclear magnetic resonance of sporopollenins from modern and fossil plants. *Annals of Botany* 69:545–549.

Hemsley, A. R., G. Clayton, and J. Galtier. 1994. Further studies on a late Tournasian (Lower Carboniferous) flora from Loch Humphrey Burn, Scotland: spore taxonomy and ultrastructure. *Review of Palaeobotany and Palynology* 81:213–231.

Hemsley, A. R., M. E. Collinson, and A. P. R. Brain. 1992. Colloidal crystal-like structure of sporopollenin in the megaspore walls of Recent *Selaginella* and similar fossil spores. *Botanical Journal of the Linnean Society* 108:307–320.

Hemsley, A. R., and A. C. Scott. 1991. Ultrastructure and relationships of Upper Carboniferous spores from Thorpe Brickworks, West Yorkshire, UK. *Review of Palaeobotany and Palynology* 69:337–351.

Hennig, W. 1965. Phylogenetic systematics. *Annual Review of Entomology* 10:97–116.

———. 1966. *Phylogenetic systematics.* Urbana: University of Illinois Press.

Hickey, R. J. 1986. The early evolution and morphological diversity of *Isoëtes,* with descriptions of two new neotropical species. *Systematic Botany* 11:309–321.

———. 1990. Studies of neotropical *Isoetes* L. 1. The subgenus *Euphyllum* Hickey, subg. nov. *Annals of the Missouri Botanical Garden* 77:239–245.

Hieronymus, G. 1901. Selaginellaceae. In *Die natürlichen Pflanzenfamilien,* edited by A. Engler and K. Prantl. Leipzig: Engelmann.

Hill, C. R., and J. M. Camus. 1986. Evolutionary cladistics of marattialean ferns. *Bulletin of the*

British Museum of Natural History (Botany) 14: 219–300.

Hill, C. R., and P. R. Crane. 1982. Evolutionary cladistics and the origin of angiosperms. In *Problems of phylogenetic reconstruction,* edited by K. A. Joysey and A. E. Friday. New York: Academic Press.

Hirmer, M. 1927. *Handbuch der Palaeobotanik. Band I: Thallophyta, Bryophyta, Pteridophyta.* Munich: Oldenbourg.

Høeg, O. A. 1930. A psilophyte in South Africa. *Det Kongelige norske videnskabers selskabs Forhandlinger* 3:92–94.

———. 1935. Further contributions to the middle Devonian flora of western Norway. *Norsk geologisk tidsskrift* 15:1–18.

———. 1967. Psilophyta. In *Bryophyta, Psilophyta, Lycophyta,* edited by E. Boureau, S. Jovet-Ast, O. A. Høeg, and W. G. Chaloner. Vol. 2 of *Traité de paléobotanique,* edited by E. Boureau. Paris: Masson et Cie.

Hofmeister, W. 1869. *On the germination, development, and fructification of the higher Cryptogamia and on the fructification of the Coniferae.* London: Robert Hardwicke.

Hogg, J. 1860. On the distinctions of a plant and an animal, and on a fourth kingdom of nature. *The Edinburgh New Philosophical Journal* 12: 216–225.

Holloway, J. E. 1935. The gametophyte of *Phylloglossum drummondii. Annals of Botany* 49:513–519.

Hori, H., B.-L. Lim, and S. Osawa. 1985. Evolution of green plants as deduced from 5S rRNA sequences. *Proceedings of the National Academy of Sciences* 82:820–823.

Horner, T. H., and H. J. Arnott. 1963. Sporangial arrangement in North American species of *Selaginella. Botanical Gazette* 124:371–383.

Hoskins, J. H., and A. T. Cross. 1951. The structure and classification of four plants from the New Albany Shale. *The American Midland Naturalist* 46:684–716.

Hueber, F. M. 1961. *Hepaticites devonicus,* a new fossil liverwort from the Devonian of New York. *Annals of the Missouri Botanical Garden* 48: 125–132.

———. 1964. The psilophytes and their relationship to the origin of ferns. *Memoirs of the Torrey Botanical Club* 21:5–9.

———. 1967. *Psilophyton:* the genus and the concept. In *International symposium on the Devonian system,* edited by D. H. Oswald. Calgary: Alberta Society of Petroleum Geologists.

———. 1970. *Rebuchia:* a new name for *Bucheria* Dorf. *Taxon* 19:822.

———. 1971a. *Early Devonian plants from Bathurst Island, District of Franklin.* Geological Survey of Canada Paper 71-28.

———. 1971b. *Sawdonia ornata:* a new name for *Psilophyton princeps* var. *ornatum. Taxon* 20: 641–642.

———. 1972. *Rebuchia ovata,* its vegetative morphology and classification with the Zosterophyllophytina. *Review of Palaeobotany and Palynology* 14:113–127.

———. 1983. A new species of *Baragwanathia* from the Sextant Formation (Emsian), northern Ontario, Canada. *Botanical Journal of the Linnean Society* 86:57–79.

———. 1992. Thoughts on the early lycopsids and zosterophylls. *Annals of the Missouri Botanical Garden* 79:474–499.

Hueber, F. M., and H. P. Banks. 1967. *Psilophyton princeps:* the search for organic connection. *Taxon* 16:81–85.

———. 1979. *Serrulacaulis furcatus* gen. et sp. nov., a new zosterophyll from the lower Upper Devonian of New York State. *Review of Palaeobotany and Palynology* 28:169–189.

Huelsenbeck, J. P. 1994. Measuring and testing the fit of the stratigraphic record to phylogenetic trees. *Paleobiology* 20:470–484.

Hughes, N. F. 1976. *Palaeobiology of angiosperm origins.* Cambridge: Cambridge University Press.

———. 1994. *The enigma of angiosperm origins.* Cambridge: Cambridge University Press.

Ignatov, M. S. 1990. Upper Permian mosses from the Russian Platform. *Palaeontographica* B217: 147–189.

Imaichi, R., and M. Kato. 1991. Developmental study of branched rhizophores in three *Selaginella* species. *American Journal of Botany* 78:1694–1703.

Ingold, C. T. 1939. *Spore discharge in land plants.* Oxford: Oxford University Press.

Istchenko, T. A. 1974. *Tirasophyton,* a new early Devonian plant genus from Podolia. *Paleontological Journal* 8:104–108.

Iwatsuki, K. 1990. Hymenophyllaceae. In *Pteridophytes and gymnosperms,* edited by K. U. Kramer and P. S. Green. Vol. 1 of *The families and genera of vascular plants,* edited by K. Kubitzki. Berlin: Springer-Verlag.

Jarzen, D. M. 1979. Spore morphology of some Anthocerotaceae and the occurrence of *Phaeoceros* spores in the Cretaceous of North America. *Pollen et Spores* 21:211–232.

Jeffrey, C. 1962. The origin and differentiation of the archegoniate land-plants. *Botaniska Notiser* 115:446–454.

Jeffrey, E. C. 1902. The structure and development of the stem in the Pteridophyta and gymnosperms. *Philosophical Transactions of the Royal Society of London* B195:119–146.

Jeram, A. J., P. A. Selden, and D. Edwards. 1990. Land animals in the Silurian: arachnids and myriapods from Shropshire, England. *Science* 250:658–661.

Jermy, A. C. 1990a. Isoetaceae. In *Pteridophytes and gymnosperms,* edited by K. U. Kramer and P. S. Green. Vol. 1 of *The families and genera of vascular plants,* edited by K. Kubitzki. Berlin: Springer-Verlag.

———. 1990b. Selaginellaceae. In *Pteridophytes and gymnosperms,* edited by K. U. Kramer and P. S. Green. Vol. 1 of *The families and genera of vascular plants,* edited by K. Kubitzki. Berlin: Springer-Verlag.

Jernstedt, J. A., E. G. Cutter, E. M. Gifford, and P. Lu. 1992. Angle meristem origin and development in *Selaginella martensii. Annals of Botany* 69:351–363.

John, D. M. 1994. Alternation of generations in algae: its complexity, maintenance, and evolution. *Biological Reviews* 69:275–291.

Jovet-Ast, S. 1967. Bryophyta. In *Bryophyta, Psilophyta, Lycophyta,* edited by E. Boureau, S. Jovet-Ast, O. A. Høeg, and W. G. Chaloner. Vol. 2 of *Traité de paléobotanique,* edited by E. Boureau. Paris: Masson et Cie.

Källersjö, M., J. S. Farris, A. G. Kluge, and C. Bult. 1992. Skewness and permutation. *Cladistics* 8: 275–287.

Kaplan, D. R. 1977. Morphological status of the shoot systems of Psilotaceae. *Brittonia* 29:30–53.

Karrfalt, E. E. 1981. The comparative and developmental morphology of the root system of *Selaginella* (L.) Link. *American Journal of Botany* 68:244–253.

———. 1982. Secondary development in the cortex of *Isoetes. Botanical Gazette* 143:439–445.

———. 1984a. Further observations on *Nathorstiana* (Isoetaceae). *American Journal of Botany* 71:1023–1030.

———. 1984b. The origin and early development of the root-producing meristem of *Isoetes andicola* L. D. Gomez. *Botanical Gazette* 138:357–368.

Kasper, A. E. 1977. A new species of the Devonian lycopod genus *Leclercqia* from New Brunswick, Canada. *American Journal of Botany, Miscellaneous Series (Abstract)* 154:39.

Kasper, A. E., and H. N. Andrews. 1972. *Pertica,* a new genus of Devonian plants from northern Maine. *American Journal of Botany* 59:897–911.

Kasper, A. E., H. N. Andrews, and W. H. Forbes. 1974. New fertile species of *Psilophyton* from the Devonian of Maine. *American Journal of Botany* 61:339–359.

Kasper, A. E., and W. H. Forbes. 1979. The Devonian lycopod *Leclercqia* from the Trout Valley Formation of Maine. *Geological Society of Maine, Maine Geology Bulletin* 1:49–59.

Kasper, A. E., P. G. Gensel, W. H. Forbes, and H. N. Andrews. 1988. Plant paleontology in the state of Maine: a review. *Maine Geological Survey, Studies in Maine Geology* 1:109–128.

Keeley, J. E., C. B. Osmond, and J. A. Raven. 1984. *Stylites,* a vascular land plant without stomata absorbs carbon dioxide via its roots. *Nature* 310:694–695.

Kendrick, B. 1985. *The fifth kingdom.* Waterloo, Ontario: Mycologue Publications.

Kenrick, P. 1988. Studies on Lower Devonian plants from South Wales. Ph.D. dissertation, University of Wales, Cardiff.

———. 1994. Alternation of generations in land plants: new phylogenetic and morphological evidence. *Biological Reviews* 69:293–330.

Kenrick, P., and P. R. Crane. 1991. Water-conducting cells in early fossil land plants: implications for the early evolution of tracheophytes. *Botanical Gazette* 152:335–356.

———. 1992. Numerical cladistic analysis of relationships in early land plants: lycopods and zosterophylls. Paper presented at the fourth conference of the Organisation internationale de paléobotanique, Paris, August 30–September 3.

Kenrick, P., and D. Edwards. 1988a. The anatomy of Lower Devonian *Gosslingia breconensis* Heard based on pyritized axes, with some comments on the permineralization process. *Botanical Journal of the Linnean Society* 97:95–123.

———. 1988b. A new zosterophyll from a recently discovered exposure of the Lower Devonian Senni Beds in Dyfed, Wales. *Botanical Journal of the Linnean Society* 98:97–115.

Kenrick, P., D. Edwards, and R. C. Dales. 1991. Novel ultrastructure in water-conducting cells of the Lower Devonian plant *Sennicaulis hippocrepiformis. Palaeontology* 34:751–766.

Kenrick, P., W. Remy, and P. R. Crane. 1991. The structure of water-conducting cells in the enigmatic early land plants *Stockmansella langii* Fairon-Demaret, *Huvenia kleui* Hass et Remy, and *Sciadophyton* sp. Remy et al. 1980. *Argumenta Palaeobotanica* 8:179–191.

Kidston, R., and W. H. Lang. 1917. On Old Red Sandstone plants showing structure, from the Rhynie Chert Bed, Aberdeenshire. Part 1. *Rhynia gwynne-vaughani* Kidston and Lang. *Transactions of the Royal Society of Edinburgh* 51:761–784.

———. 1920a. On Old Red Sandstone plants showing structure, from the Rhynie Chert Bed, Aberdeenshire. Part 2. Additional notes on *Rhynia gwynne-vaughani,* Kidston and Lang; with descriptions of *Rhynia major,* n.sp., and *Hornea lignieri,* n.g., n.sp. *Transactions of the Royal Society of Edinburgh* 52:603–627.

———. 1920b. On Old Red Sandstone plants showing structure, from the Rhynie Chert Bed, Aberdeenshire. Part 3. *Asteroxylon mackiei,* Kidston and Lang. *Transactions of the Royal Society of Edinburgh* 52:643–680.

———. 1921. On Old Red Sandstone plants showing structure, from the Rhynie Chert Bed, Aberdeenshire. Part 4. Restorations of the vascular cryptogams and discussion on their bearing on the general morphology of the Pteridophyta and the origin of the organization of land-plants. *Transactions of the Royal Society of Edinburgh* 52:831–854.

———. 1923. Notes on fossil plants from the Old Red Sandstone of Scotland. 1. *Hicklingia edwardii,* K. and L. *Transactions of the Royal Society of Edinburgh* 53:405–407.

Kiss, J. Z., A. C. Vasconcelos, and R. E. Triemer. 1986. Paramylon synthesis and chloroplast structure associated with nutrient levels in *Euglena* (Euglenophyceae). *Journal of Phycology* 22:327–333.

Kluge, A. G., and J. S. Farris. 1969. Quantitative phyletics and the evolution of anurans. *Systematic Zoology* 18:1–32.

Knoll, A. H. 1984. Patterns of extinction in the fossil record of vascular plants. In *Extinctions,* edited by M. Nitecki. Chicago: University of Chicago Press.

———. 1986. Patterns of change in plant communities through geological time. In *Community ecology,* edited by J. Diamond and T. J. Chase. New York: Harper and Row.

———. 1992. The early evolution of eukaryotes: a geological perspective. *Science* 256:622–627.

Knoll, A. H., K. J. Niklas, P. G. Gensel, and B. H. Tiffney. 1984. Character diversification and patterns of evolution in early vascular plants. *Paleobiology* 10:34–47.

Knoll, A. H., K. J. Niklas, and B. H. Tiffney. 1979. Phanerozoic land plant diversity in North America. *Science* 206:1400–1402.

Knoll, A. H., and M. R. Walter. 1992. Latest Proterozoic stratigraphy and Earth history. *Nature* 356:673–678.

Kolattukudy, P. E. 1980. Biopolyester membranes of plants: cutin and suberin. *Science* 208:990–1000.

Koller, A. L., and S. E. Scheckler. 1986. Variations in microsporangia and microspore dispersal in *Selaginella. American Journal of Botany* 73:1274–1288.

Kovach, W. L. 1989. Quantitative methods for the study of lycopod megaspore structure. *Review of Palaeobotany and Palynology* 57:233–246.

Kovach, W. L., and D. J. Batten. 1989. Worldwide stratigraphic occurrences of Mesozoic and Tertiary megaspores. *Palynology* 13:247–277.

———. 1993. Diversity changes in lycopsid and aquatic fern megaspores through geologic time. *Paleobiology* 19:28–42.

Kramer, K. U. 1990. Psilotaceae. In *Pteridophytes and gymnosperms,* edited by K. U. Kramer and P. S. Green. Vol. 1 of *The families and genera of vascular plants,* edited by K. Kubitzki. Berlin: Springer-Verlag.

Kramer, K. U., and D. B. Lellinger. 1990. Hymenophyllopsidaceae. In *Pteridophytes and gymnosperms,* edited by K. U. Kramer and P. S. Green. Vol. 1 of *The families and genera of vascular plants,* edited by K. Kubitzki. Berlin: Springer-Verlag.

Kranz, H. D., and V. A. R. Huss. 1996. Molecular evolution of pteridophytes and their relationships to seed plants: evidence from complete 18S rRNA gene sequences. *Plant Systematics and Evolution* 202:1–11.

Kranz, H. D., D. Miks, M.-L. Siegler, I. Capesius, W. Sensen, and V. A. R. Huss. 1995. The origin of land plants: phylogenetic relationships among charophytes, bryophytes, and vascular plants inferred from complete small-subunit ribosomal RNA gene sequences. *Journal of Molecular Evolution* 41:74–84.

Krassilov, V. 1981. *Orestovia* and the origin of vascular plants. *Lethaia* 14:235–250.

Krassilov, V. A., and R. M. Schuster. 1984. Paleozoic and Mesozoic fossils. In *New manual of bryology,* edited by R. M. Schuster. Miyazaki-ken, Japan: Hattori Botanical Laboratory.

Kräusel, R. 1938. Psilophytinae. In *Manual of pteridology,* edited by F. Veerdorn. The Hague: Martinus Nijhoff.

———. 1950. *Versunkene floren eine Einführung in die Paläobotanik.* Frankfurt am Main: Waldemar Kramer.

Kräusel, R., and H. Weyland. 1929. Beiträge zur Kenntnis der Devonflora. *Abhandlungen Senckenbergische Naturforschende Gesellschaft* 41: 315–360.

———. 1930. Die Flora des Deutschen Unterdevons. *Abhandlungen der Preussischen Geologischen Landesanstalt, Neue Folge* 131:1–92.

———. 1935. Neue Pflanzenfunde im Rheinischen Unterdevon. *Palaeontographica* B80:171–190.

———. 1937. Pflanzenreste aus dem Devon. X. Zwei Pflanzenfunde im Oberdevon der Eifel. *Senckenbergiana* 19:338–355.

———. 1941. Pflanzenreste aus dem Devon von Nord-Amerika. *Palaeontographica* 86B:1–78.

———. 1961. Über *Psilophyton robustius* Dawson. *Palaeontographica* B108:11–21.

Kremer, B. P. 1980. Taxonomic implications of algal photoassimilate patterns. *British Phycological Journal* 15:399–409.

Lacey, W. S. 1969. Fossil bryophytes. *Biological Reviews* 44:189–205.

Lam, H. J. 1948. Classification and the new morphology. *Acta Biotheoretica* 8:107–154.

———. 1950. Stachyospory and phyllospory as factors in the natural system of the Cormophyta. *Svensk Botanisk Tidskrift* 44:517–534.

Lampa, E. 1903. Exogene Entstehung der Antheridien von *Anthoceros. Oesterreichisches Botanisches Wochenblatt* 53:436–438.

Lang, W. H. 1915. Presidential address, Botanical Section. Paper read at a meeting of the British Association for the Advancement of Science, Manchester.

———. 1927. Contributions to the study of the Old Red Sandstone flora of Scotland. 6. On *Zosterophyllum myretonianum,* Penh., and some other plant-remains from the Carmyllie Beds of the Lower Old Red Sandstone. 7. On a specimen of *Pseudosporochnus* from the Stromness Beds. *Transactions of the Royal Society of Edinburgh* 55:443–456.

———. 1931. On the spines, sporangia, and spores of *Psilophyton princeps,* Dawson, shown in specimens from Gaspé. *Philosophical Transactions of the Royal Society of London* B219:421–442.

———. 1932. Contributions to the study of the Old Red Sandstone flora of Scotland. 8. On *Arthrostigma, Psilophyton,* and some associated plant-remains from the Strathmore Beds of the Caledonian Lower Old Red Sandstone. *Transactions of the Royal Society of Edinburgh* 57:491–521.

———. 1937a. On the plant-remains from the Downtonian of England and Wales. *Philosophical Transactions of the Royal Society of London* B227:245–292.

———. 1937b. A specimen of *Sporogonites* from the "Grès de Wépion" (Lower Devonian, Belgium). *Bulletin du Musée royal d'histoire naturelle de Belgique* 13:1–7.

Lang, W. H., and I. C. Cookson. 1930. Some fossil plants of early Devonian type from the Walhalla Series, Victoria, Australia. *Philosophical Transactions of the Royal Society of London* B219:133–164.

———. 1935. On a flora, including vascular land plants, associated with *Monograptus,* in rocks of Silurian age, from Victoria, Australia. *Philosophical Transactions of the Royal Society of London* B224:421–449.

Lankester, E. R. 1870. On the use of the term homology in modern zoology, and the distinction between homogenetic and homoplastic agreements. *The Annals and Magazine of Natural History* 6:34–43.

Larsen, N., G. J. Olsen, B. L. Maidak, M. J. McCaughey, R. Overbeek, T. J. Macke, T. L. March, and C. R. Woese. 1993. The ribosomal database project. *Nucleic Acids Research* 21:3021–3023.

Lawson, A. A. 1917. The prothallus of *Tmesipteris tannensis. Transactions of the Royal Society of Edinburgh* 51:785–794.

Leclercq, S. 1942. Quelques plantes fossiles recueillies dans le Dévonien inférieur des environs de Nonceveux (bordure orientale du bassin de Dinant). *Annales de la Société géologique de Belgique* 65:193–211.

———. 1954. Are the Psilophytales a starting or a resulting point? *Svensk Botanisk Tidskrift* 48:103–315.

———. 1956. Evidence of vascular plants in the Cambrian. *Evolution* 10:109–114.

———. 1957. Etude d'une fructification de Sphenopside à structure conservée du Dévonien supérieur. *Mémoires de l'Académie royale de Belgique. Classe des sciences* 14(2):1–39.

Leclercq, S., and H. P. Banks. 1962. *Pseudosporochnus nodosus* sp. nov., a middle Devonian plant with cladoxylalean affinities. *Palaeontographica* B110:1–34.

Leclercq, S., and P. M. Bonamo. 1971. A study of the fructification of *Milleria (Protopteridium) thomsonii* Lang from the Middle Devonian of Belgium. *Palaeontographica* B136:83–114.

Leclercq, S., and K. M. Lele. 1968. Further investigation of the vascular system of *Pseudosporochnus nodosus* Leclercq et Banks. *Palaeontographica* B123:97–112.

Lee, B., and J. H. Priestley. 1924. The plant cuticle. 1. Its structure, distribution, and function. *Annals of Botany* 38:525–545.

Lele, K. M., and J. Walton. 1961. Contributions to the knowledge of *"Zosterophyllum myretonianum"* Penhallow from the Lower Old Red Sandstone of Angus. *Transactions of the Royal Society of Edinburgh* 64:469–475.

Lemoigne, Y. 1966. Sur un sporogone de bryale d'âge Dévonien. *Bulletin mensuel de la Société Linnéenne de Lyon* 35:13–16.

———. 1968a. Les genres *Rhynia* Kidston et Lang du Dévonien et *Psilotum* Seward actuel appartiennent-ils au même phylum? *Bulletin de la Société botanique de France* 115:425–440.

———. 1968b. Observations d'archégones portés par des axes du type *Rhynia gwynne-vaughanii* Kidston et Lang. Existence de gamétophytes vascularisés au Dévonien. *Comptes rendus des séances de l'Académie des sciences, Paris. Sciences naturelles* 266:1655–1657.

———. 1969a. Contribution à la connaissance du gamétophyte *Rhynia gwynne-vaughanii* Kidston et Lang; problème des protrubérances et processus de ramification. *Bulletin mensuel de la Société Linnéenne de Lyon* 4:94–102.

———. 1969b. Organe assimilable à une anthéridie et stomates épidermiques portés par des axes rampants du type *Rhynia gwynne-vaughanii* Kidston et Lang. *Comptes rendus des séances de l'Académie des sciences, Paris. Sciences naturelles* 269:1393–1395.

———. 1970. Nouvelles diagnoses du genre *Rhynia* et de l'espèce *Rhynia gwynne-vaughanii. Bulletin de la Société botanique de France* 117:307–320.

Lemoigne, Y., and T. Itschenko. 1980. Deux lycophytes avec structures conservées du Dévonien supérieur d'Ukraine (U.R.S.S.). *Géobios* 13:671–681.

Lemoigne, Y., and D. Zdebska. 1980. Structures problématiques observées dans des axes provenant du chert Dévonien de Rhynie. *Acta palaeobot* 21:3–8.

Lepekhina, V. G., N. M. Petrosian, and G. P. Radchenko. 1962. Main Devonian plants of the Altaï-Saïan mountain region. *Vsesoyuznuy nauchno-issledovatelsky Geologichesky institut* 70:61–189.

Les, D. H. 1993. Ceratophyllaceae. In *Flowering plants. Dicotyledons: magnoliid, hamamelid, and caryophyllid families,* edited by K. Kubitzki, J. G. Rohwer, and V. Bittrich. Berlin: Springer-Verlag.

Li, Cheng-Sen. 1982. *Hsüa robusta,* a new land plant from the Lower Devonian of Yunnan, China. *Acta Phytotaxonomica Sinica* 20:331–342.

———. 1990. *Minarodendron cathaysiense* (gen. et comb. nov.), a lycopod from the late Middle Devonian of Yunnan, China. *Palaeontographica* B220:97–117.

———. 1992. *Hsüa robusta,* an early Devonian plant from Yunnan Province, China and its bearing on some structures of early land plants. *Review of Palaeobotany and Palynology* 71:121–147.

Li, Cheng-Sen, and D. Edwards. 1992. A new genus of early land plants with novel strobilar construction from the Lower Devonian Posongchong Formation, Yunnan Province, China. *Palaeontology* 35:257–272.

———. 1995. A re-investigation of Halle's *Drepanophycus spinaeformis* Göpp. from the Lower Devonian of Yunnan Province, southern China. *Botanical Journal of the Linnean Society* 118:163–192.

Li, Cheng-Sen, and Jen Hsü. 1987. Studies on a new Devonian plant *Protopteridophyton devonicum* assigned to primitive fern from south China. *Palaeontographica* B207:111–131.

Li, Xing-Xue, and Chong-Yang Cai. 1978. A type-section of Lower Devonian strata in southwest China with brief notes on the succession and correlation of its plant assemblages. *Acta Geologica Sinica* 52:1–12.

Li, Xingxue, Chongyang Cai, and Yi Wang. 1995.

Hamatophyton verticillatum (Gu & Zhi) emend. a primitive plant of Sphenopsida from the Upper Devonian–Lower Carboniferous in China. *Palaeontographica* B235:1–22.

Lidgard, S., and P. R. Crane. 1988. Quantitative analyses of the early angiosperm radiation. *Nature* 331:344–346.

Lignier, O. 1903. Equisétales et Sphénophyllales. Leur origine filicinéenne commune. *Bulletin de la Société Linnéenne de Normandie* 7(5): 93–137.

———. 1908. Essai sur l'évolution morphologique du règne végétal. *Bulletin de la Société Linnéenne de Normandie* 3(6): 35–62.

Ligrone, R., J. G. Duckett, and K. S. Renzaglia. 1993. The gametophyte-sporophyte junction in land plants. *Advances in Botanical Research* 19:231–317.

Ligrone, R., and R. Gambardella. 1988a. The sporophyte-gametophyte junction in bryophytes. *Advances in Bryology* 3:225–274.

———. 1988b. The ultrastructure of the sporophyte-gametophyte junction and its relationship to bryophyte evolution. *Journal of the Hattori Botanical Laboratory* 64:187–196.

Ligrone, R., and K. S. Renzaglia. 1989. The ultrastructure of the placenta in *Sphagnum*. *New Phytologist* 111:197–201.

———. 1990. The sporophyte-gametophyte junction in the hornwort, *Dendroceros tubercularis* Hatt. (Anthocerotophyta). *New Phytologist* 114:497–505.

Linné, C. 1753. *Species plantarum*. Holmiae: L. Salvi.

Lipscomb, D. L. 1989. Relationships among the eukaryotes. In *The hierarchy of life: molecules and morphology in phylogenetic analysis,* edited by B. Fernholm, K. Bremer, and H. Jörnvall. Amsterdam: Elsevier.

Logan, K. J., and B. A. Thomas. 1985. Distribution of lignin derivatives in plants. *New Phytologist* 99:571–585.

Longton, R. E. 1990. Mosses: form and function as a foundation for phytochemical studies. In *Bryophytes: their chemistry and chemical taxonomy,* edited by H. D. Zinsmeister and R. Mues. Oxford: Clarendon Press.

Lotsy, J. P. 1916. *Evolution by means of hybridization.* The Hague: Nijhoff.

Lugardon, B. 1979. Sur la formation du sporoderme chez *Psilotum triquetrum* Sw. (Psilotaceae). *Grana* 18:145–165.

———, ed. 1990. Pteridophyte sporogenesis: a survey of spore wall ontogeny and fine structure in a polyphyletic plant group. In *Microspores: evolution and ontogeny,* edited by S. Blackmore and R. B. Knox. London: Academic Press.

Lyon, A. G. 1964. Probable fertile region of *Asteroxylon mackiei* K. and L. *Nature* 203:1082–1083.

Lyon, A. G., and D. Edwards. 1991. The first zosterophyll from the Lower Devonian Rhynie Chert, Aberdeenshire. *Transactions of the Royal Society of Edinburgh. Earth Sciences* 82:323–332.

Mackie, W. 1914. The rock series of Craigbeg and Ord Hill, Rhynie, Aberdeenshire. *Transactions of the Edinburgh Geological Society* 10:205–236.

Maddison, W. P. 1993. Missing data versus missing characters on phylogenetic analysis. *Systematic Biology* 42:576–581.

Mägdefrau, K. 1932. Über *Nathorstiana,* eine Isoetacee aus dem Neokom von Quedlinburg a. Harz. *Botanisches Centralblatt Beihefte* 49:706–718.

———. 1942. *Palaeobiologie der Pflanzen.* Jena: Fischer.

Manhart, J. R. 1994. Phylogenetic analysis of green plant *rbcL* sequences. *Molecular Phylogenetics and Evolution* 3:114–127.

———. 1995. Chloroplast 16S rDNA sequences and phylogenetic relationships of fern allies and ferns. *American Fern Journal* 85:182–192.

Manhart, J. R., R. W. Hoshaw, and J. D. Palmer. 1990. Unique chloroplast genome in *Spirogyra maxima* (Chlorophyta) revealed by physical gene mapping. *Journal of Phycology* 26:490–494.

Manhart, J. R., and J. D. Palmer. 1990. The gain of two chloroplast tRNA introns marks the green algal ancestors of land plants. *Nature* 345:268–270.

Manton, I., and M. Parke. 1960. Further observations on small green flagellates with special reference to possible relatives of *Chromulina pusila* Butcher. *Journal of the Marine Biological Association of the United Kingdom* 39:275–298.

Margulis, L. 1974. Five-kingdom classification and the origin and evolution of cells. *Evolutionary Biology* 7:45–78.

———. 1996. Archael-eubacterial mergers in the origin of Eukarya: phylogenetic classification of life. *Proceedings of the National Academy of Sciences* 93:1071–1076.

Margulis, L., J. O. Corliss, M. Melkonian, and D. J. Chapman, eds. 1990. *Handbook of Protoctista.* Boston: Jones and Bartlett.

Margulis, L., and K. Schwartz. 1988. *Five kingdoms:*

an illustrated guide to the phyla of life on earth. 2nd ed. New York: Freeman.

Markham, K. R., and L. J. Porter. 1978. Chemical constituents of the bryophytes. *Progress in Phytochemistry* 5:181–272.

Martín-Closas, C., and M. E. Schudack. 1991. Phylogenetic analysis and systematization of post-paleozoic charophytes. *Bulletin de la Société botanique de France* 138:53–71.

Mattox, K. R., and K. D. Stewart. 1984. Classification of the green algae: a concept based on comparative cytology. In *Systematics of the green algae,* edited by D. E. G. Irvine and D. M. John. London: Academic Press.

Mayr, E. 1990. A natural system of organisms. *Nature* 348:491.

———. 1991. More natural classification. *Nature* 353:122.

McCourt, R. M. 1995. Green algal phylogeny. *Trends in Ecology and Evolution* 10:159–163.

McCourt, R. M., K. G. Karol, M. Guerlesquin, and M. Feist. 1996. Phylogeny of extant genera in the family Characeae (Charales, Charophyceae) based on *rbcL* sequences and morphology. *American Journal of Botany* 83:125–131.

McFadden, G., and P. Gilson. 1995. Something borrowed, something green: lateral transfer of chloroplasts by secondary endosymbiosis. *Trends in Ecology and Evolution* 10:12–17.

McIver, E. E., and J. F. Basinger. 1989. The morphology and relationships of *Equisetum fluviatoides* sp. nov. from the Paleocene Ravenscrag Formation of Saskatchewan, Canada. *Canadian Journal of Botany* 67:2937–2943.

McKay, R. M. L., and S. P. Gibbs. 1989. Phycoerythrin is absent from the pyrenoid of *Porphyridium cruenatum:* photosynthetic implications. *Plant Physiology Supplement (Abstract)* 89:82.

Melkonian, M. 1990. Phylum Chlorophyta: introduction to the Chlorophyta. In *Handbook of Protoctista,* edited by L. Margulis, J. O. Corliss, M. Melkonian, and D. J. Chapman. Boston: Jones and Bartlett.

Merker, H. 1958. Zum fehlenden Gliede der Rhynienflora. *Botaniska Notiser* 111:608–618.

———. 1959. Analyse der Rhynien-Basis und Nachweis des Gametophyten. *Botaniska Notiser* 112: 441–452.

Metcalfe, C. R., and L. Chalk. 1950. *Anatomy of the dicotyledons.* 2 vols. Oxford: Clarendon Press.

Meyen, S. V. 1987. *Fundamentals of palaeobotany.* London: Chapman and Hall.

Millay, M. A., and T. N. Taylor. 1979. Paleozoic seed fern pollen organs. *Botanical Review* 45: 301–375.

Miller, C. N. 1971. Evolution of the fern family Osmundaceae based on anatomical studies. *Contributions from the Museum of Palaeontology, the University of Michigan* 23:105–169.

Miller, H. 1859. *The Old Red Sandstone.* London: Hamilton and Adams.

Miller, N. G. 1984. Tertiary and Quaternary fossils. In *New manual of bryology,* edited by R. M. Schuster. Miyazaki-ken, Japan: Hattori Botanical Laboratory.

Minaki, M. 1984. Macrospore morphology and taxonomy of *Selaginella* (Selaginellaceae). *Pollen et Spores* 26:421–480.

Mishler, B. D., and S. P. Churchill. 1984. A cladistic approach to the phylogeny of the "bryophytes." *Brittonia* 36:406–424.

———. 1985a. Cladistics and the land plants: a response to Robinson. *Brittonia* 37:282–285.

———. 1985b. Transition to a land flora: phylogenetic relationships of the green algae and bryophytes. *Cladistics* 1:305–328.

———. 1987. Transition to a land flora: a reply. *Cladistics* 3:65–71.

Mishler, B. D., L. A. Lewis, M. A. Buchheim, K. S. Renzaglia, D. J. Garbary, C. F. Delwiche, F. W. Zechman, T. S. Kantz, and R. L. Chapman. 1994. Phylogenetic relationships of the "green algae" and "bryophytes." *Annals of the Missouri Botanical Garden* 81:451–483.

Mishler, B. D., P. H. Thrall, J. S. Hopple Jr., E. De Luna, and R. Vilgalys. 1992. A molecular approach to the phylogeny of bryophytes: cladistic analysis of chloroplast-encoded 16S and 23S ribosomal RNA genes. *The Bryologist* 95:172–180.

Mitchell, G. 1910. Contributions towards a knowledge of the anatomy of the genus *Selaginella,* Spr. Part V. The stobilus. *Annals of Botany* 24:19–33.

Moczyolowska, M. 1991. Acritarch biostratigraphy in southeastern Poland. *Fossils and Strata* 29:1–127.

Moestrup, Ø. 1974. Ultrastructure of the scale-covered zoospores of the green alga *Chaetosphaeridium,* a possible ancestor of the higher plants and bryophytes. *Biological Journal of the Linnean Society* 6:111–125.

Mogensen, G. S. 1983. The spore. In *New manual of*

bryology, edited by R. M. Schuster. Miyazaki-ken, Japan: Hattori Botanical Laboratory.

Morbelli, M. A., and J. R. Rowley. 1993. Megaspore development in *Selaginella.* 1. "Wicks," their presence, ultrastructure, and presumed function. *Sexual Plant Reproduction* 6:98–107.

Morgan, J. 1959. The morphology and anatomy of American species of the genus *Psaronius. Illinois Biological Monographs* 27:1–107.

Mosbrugger, V. 1990. The tree habit in land plants. *Lecture Notes in Earth Science* 28:1–161.

Murray, B. M. 1988. Systematics of the Andreaeopsida (Bryophyta): two orders with links to *Takakia. Beiheft zur Nova Hedwigia* 90:289–336.

Nathorst, A. G. 1913. Die Pflanzenreste der Röragen-Ablagerung. In *Das Devongebiet am Röragen bei Röros,* edited by V. M. Goldschmidt. *Videnskapsselskapets Skrifter, Mathematische-naturwissenschaftliche Klasse* 9:1–27.

———. 1915. Zur Devonflora des westlichen Norwegens. *Bergens Museums Aarbok* 9:1–34.

Nehira, K. 1983. Spore germination, protonema development, and sporeling development. In *New manual of bryology,* edited by R. M. Schuster. Miyazaki-ken, Japan: Hattori Botanical Laboratory.

Nelson, G. J. 1972. Phylogenetic relationship and classification. *Systematic Zoology* 21:227–231.

———. 1974. Classification as an expression of phylogenetic relationships. *Systematic Zoology* 22:344–359.

Nelson, G. J., and N. Platnick. 1981. *Systematics and biogeography.* New York: Columbia University Press.

Nemejc, F., and B. Pacltova. 1974. Hepaticae in the Senonian of South Bohemia. *The Palaeobotanist* 21:23–26.

Neuburg, M. F. 1960. Mosses from the Permian deposits of Angaraland. *Trudy Geologicheskogo Instituta. Akademiya nauk SSSR* 19:1–104.

Newman, I. V. 1961. Pattern in the meristems of vascular plants. 2. A review of shoot apical meristems of gymnosperms, with comments on apical biology and taxonomy, and a statement of some fundamental concepts. *Proceedings of the Linnean Society of New South Wales* 86:9–59.

———. 1965. Pattern in the meristems of vascular plants. 3. Pursuing the patterns in the apical meristem where no cell is a permanent cell. *Journal of the Linnean Society of London* 59:185–214.

Newman, S. M., and R. A. Cattolico. 1990. Ribulose bisphosphate carboxylase in algae: synthesis, enzymology, and evolution. *Photosynthesis Research* 26:69–85.

Niklas, K. J. 1976. Morphological and ontogenetic reconstruction of *Parka decipiens* Fleming and *Pachytheca* Hooker from the Lower Old Red Sandstone, Scotland. *Transactions of the Royal Society of Edinburgh* 69:483–499.

———. 1982. Computer simulations of early land plant branching morphologies: canalization of patterns during evolution? *Paleobiology* 8:196–210.

———. 1984. Size-related change in the primary xylem anatomy of some early tracheophytes. *Paleobiology* 10:487–506.

———. 1990. Biomechanics of *Psilotum nudum* and some early Paleozoic vascular sporophytes. *American Journal of Botany* 77:590–606.

———. 1992. *Plant biomechanics: an engineering approach to plant form and function.* Chicago: University of Chicago Press.

Niklas, K. J., and H. P. Banks. 1990. A reevaluation of the Zosterophyllophytina with comments on the origin of lycopods. *American Journal of Botany* 77:274–283.

Niklas, K. J., and T. L. Phillips. 1976. Morphology of *Protosalvinia* from the Upper Devonian of Ohio and Kentucky. *American Journal of Botany* 63:9–29.

Niklas, K. J., B. H. Tiffney, and A. H. Knoll. 1980. Apparent changes in the diversity of fossil plants. *Evolutionary Biology* 12:1–89.

———. 1983. Patterns in vascular land plant diversification. *Nature* 303:614–616.

———. 1985. Patterns in vascular land plant diversification: an analysis at the species level. In *Phanerozoic diversity patterns: profiles in macroevolution,* edited by J. W. Valentine. Princeton: Princeton University Press.

Nitecki, M. H., and D. F. Toomey. 1979. Nature and classification of receptaculitids. *Bull. Cent. Rech. Explor.-Prod. Elf-Aquitaine* 3:725–732.

Nixon, K. C., W. L. Crepet, D. Stevenson, and E. M. Friis. 1994. A reevaluation of seed plant phylogeny. *Annals of the Missouri Botanical Garden* 81:484–533.

Nixon, K. C., and J. I. Davis. 1991. Polymorphic taxa, missing values, and cladistic analysis. *Cladistics* 7:233–241.

Nøhr-Hansen, H., and E. B. Koppelhus. 1988. Ordovician spores with trilete rays from Washington Land, north Greenland. *Review of Palaeobotany and Palynology* 56:305–311.

Olesen, P., and G. S. Mogensen. 1978. Ultrastructure, histochemistry, and notes on germination stages of spores in selected mosses. *The Bryologist* 81:493–516.

Oliver, F. W., and D. H. Scott. 1904. On the structure of the Palaeozoic seed *Lagenostoma lomaxi*, with a statement of the evidence upon which it is referred to *Lyginodendron*. *Philosophical Transactions of the Royal Society of London* B197:193–247.

Øllgaard, B. 1987. A revised classification of the Lycopodiaceae s. lat. *Opera Botanica* 92:153–178.

———. 1990. Lycopodiaceae. In *Pteridophytes and gymnosperms*, edited by K. U. Kramer and P. S. Green. Vol. 1 of *The families and genera of vascular plants*, edited by K. Kubitzki. Berlin: Springer-Verlag.

———. 1992. Neotropical Lycopodiaceae: an overview. *Annals of the Missouri Botanical Garden* 79:687–717.

Oostendorp, C. 1987. *The bryophytes of the Palaeozoic and the Mesozoic*. Bryophytorum Bibliotheca, vol. 34. Berlin: Cramer.

Pagel, M. D. 1994. The adaptationist wager. In *Phylogenetics and ecology*, edited by R. I. Vane-Wright and P. Eggleton. London: Academic Press.

Pant, D. D. 1962. The gametophyte of the Psilophytales. In *Proceedings of the Summer School of Botany*, edited by P. Maheshwari, B. M. Johri, and I. K. Vasil. New Delhi: Ministry of Scientific Research and Cultural Affairs.

Parenti, L. R. 1980. A phylogenetic analysis of the land plants. *Biological Journal of the Linnean Society* 13:225–242.

———. 1982. Cladistics and conventions. *Taxon* 31:96–101.

Parihar, N. S. 1977. *An introduction to Embryophyta. 1. Bryophyta.* 5th ed. Allahabad, India: Halcyon Press.

Pascher, A. 1914. Über Flagellaten und Algen. *Berichte der Deutschen Botanischen Gesellschaft* 32:136–160.

———. 1931. Systematische Übersicht über die mit Flagellaten in Zusammenhang stehenden Algenreihen und Versuch einer Einreihung dieser Algenstämme in die Stämme des Pflanzenreichs. *Beihefte zum Botanischen Centralblatt* 48:317–332.

Paton, J. A., and J. V. Pearce. 1957. The occurrence, structure, and functions of the stomata in British bryophytes. *Transactions of the British Bryological Society* 3:228–259.

Patterson, C. 1982. Morphological characters and homology. In *Problems of phylogenetic reconstruction*, edited by K. A. Joysey and A. E. Friday. London: Academic Press.

Patterson, D. J., and M. L. Sogin. 1993. Eukaryote origin and protistan diversity. In *The origin and evolution of prokaryotic and eukaryotic cells*, edited by H. Hartman and K. Matsuno. New Jersey: Wald Scientific.

Pearson, H. L. 1986. Structure and taxonomy of the Carboniferous lycopsid *Anabathra*. *Bulletin of the British Museum of Natural History (Geology)* 40:265–292.

Penhallow, D. P. 1892. Additional notes on Devonian plants from Scotland. *Canadian Record of Science* 5:1–13.

Perasso, R., A. Baroin, L. H. Qu, J. P. Bachellerie, and A. Andoutte. 1989. Origin of the algae. *Nature* 339:142–144.

Petrosyan, N. M. 1967. Stratigraphic importance of the Devonian flora of the USSR. In *International symposium on the Devonian system*, edited by D. H. Oswald. Calgary: Alberta Society of Petroleum Geologists.

Pettitt, J. M. 1965. Two heterosporous plants from the Upper Devonian of North America. *Bulletin of the British Museum Natural History (Geology)* 10:81–92.

Philipson, W. R. 1990. The significance of apical meristems in the phylogeny of land plants. *Plant Systematics and Evolution* 173:17–38.

———. 1991. A new approach to the origins of vascular plants. *Botanische Jahrbücher für Systematik* 113:443–460.

Phillips, T. L. 1979. Reproduction of heterosporous arborescent lycopods in the Mississippian-Pennsylvanian of Euramerica. *Review of Palaeobotany and Palynology* 27:239–289.

Phillips, T. L., and W. A. DiMichele. 1992. Comparative ecology and life-history biology of arborescent lycopsids in late Carboniferous swamps of Euramerica. *Annals of the Missouri Botanical Garden* 79:560–588.

Phillips, T. L., and G. A. Leisman. 1966. *Paurodendron*, a rhizomorphic lycopod. *American Journal of Botany* 53:1086–1100.

Phillips, T. L., and R. A. Peppers. 1984. Changing

patterns of Pennsylvanian coal swamp vegetation and implications of climatic control on coal occurrence. *International Journal of Coal Geology* 3:205–255.

Pichi-Sermolli, R. E. G. 1959. Pteridophyta. In *Vistas in botany,* edited by W. B. Turrill. London: Pergamon Press.

Pickett-Heaps, J. D. 1969. The evolution of the mitotic apparatus: an attempt at comparative ultrastructural cytology in dividing plant cells. *Cytobios* 1:259–280.

———. 1975. *Green algae: structure, reproduction, and evolution in selected genera.* Sunderland, Mass.: Sinauer Associates.

Pickett-Heaps, J. D., and H. J. Marchant. 1972. The phylogeny of green algae: a new proposal. *Cytobios* 6:255–264.

Pigg, K. B. 1992. Evolution of Isoetalean lycopsids. *Annals of the Missouri Botanical Garden* 79:589–612.

Platnick, N. I., C. E. Griswold, and J. A. Coddington. 1991. On missing entries in cladistic analysis. *Cladistics* 7:337–343.

Plumstead, E. P. 1967. A general review of the Devonian fossil plants found in the Cape System of South Africa. *Palaeontologia Africana* 10:1–83.

Pokorny, M. 1974. D-Methionine metabolic pathways in Bryophyta: a chemotaxonomic evaluation. *Phytochemistry* 13:965–971.

Pokorny, M., E. Marcenko, and D. Keglevic. 1970. Comparative studies of L- and D-methionine metabolism in lower and higher plants. *Phytochemistry* 9:2175–2188.

Potonié, H. 1899. *Lehrbuch der Pflanzenpalaeontologie.* Berlin: Dümmlers.

Potonié, H., and C. Bernard. 1904. *Flore dévonienne de l'étage H de Barrande.* Supplement to *Système Silurien du centre de la Bohème.* Prague: Barrande.

Pringsheim, N. 1878. Über die Sprossung der Moosfrüchte. *Jahrbuch für Wissenschaftliche Botanik* 11:1–46.

Proctor, V. W. 1980. Historical biogeography of *Chara* (Charophyta): an appraisal of the Braun-Wood classification plus a falsifiable alternative for future consideration. *Journal of Phycology* 16:218–233.

Proskauer, J. 1962. On *Takakia,* especially its mucilage hairs. *Journal of the Hattori Botanical Laboratory* 25:217–223.

Pryce, R. J. 1972. The occurrence of lunularic and abscisic acids in plants. *Phytochemistry* 11: 1759–1761.

Pryer, K. M., A. R. Smith, and J. E. Skog. 1996. Phylogenetic relationships of extant ferns based on evidence from morphology and *rbcL* sequences. *American Fern Journal* 85:205–282.

Quansah, N., and B. A. Thomas. 1985. "Sporophyllpteryx" in African and American *Selaginella. Fern Gazette* 13:49–52.

Ragan, M. A. 1989. Biochemical pathways and the phylogeny of the eukaryotes. In *The hierarchy of life: molecules and morphology in phylogenetic analysis,* edited by B. Fernholm, K. Bremer, and H. Jörnvall. Amsterdam: Elsevier.

Raubeson, L. A., and R. K. Jansen. 1992. Chloroplast DNA evidence on the ancient evolutionary split in vascular land plants. *Science* 255:1697–1699.

Rauh, W., and H. Falk. 1959. *Stylites* E. Amstutz, eine neue Isoetaceae aus den Hochanden Perus. 1. Teil: Morphologie, Anatomie, und Entwicklungsgeschichte der Vegetationsorgane. *Heidelberger Akademie der Wissenschaften, Mathematisch-naturwissenschaftliche Klasse:* 1–83.

Raup, D. M. 1972. Taxonomic diversity during the Phanerozoic. *Science* 177:1065–1071.

———. 1976a. Species diversity in the Phanerozoic: a tabulation. *Paleobiology* 2:279–288.

———. 1976b. Species diversity in the Phanerozoic: an interpretation. *Paleobiology* 2:289–297.

Raup, D. M., and G. E. Boyajian. 1988. Patterns of generic extinction in the fossil record. *Paleobiology* 14:109–125.

Raup, D. M., S. J. Gould, T. J. M. Schopf, and D. J. Simberloff. 1973. Stochastic models of phylogeny and the evolution of diversity. *Journal of Geology* 81:525–542.

Raup, D. M., and J. J. Sepkoski. 1986. Periodic extinction of families and genera. *Science* 231: 833–836.

Rausch, H., N. Larsen, and R. Schmitt. 1989. Phylogenetic relationships of the green alga *Volvox carteri* deduced from small-subunit ribosomal RNA comparisons. *Journal of Molecular Evolution* 29:255–265.

Raven, J. A. 1977. The evolution of vascular plants in relation to supracellular transport processes. *Advances in Botanical Research* 5:153–219.

———. 1984. Physiological correlates of the morphology of early vascular plants. *Botanical Journal of the Linnean Society* 88:105–126.

———. 1985. Comparative physiology of plant and arthropod land adaptation. *Philosophical Transactions of the Royal Society of London* B309:273–288.

———. 1993. The evolution of vascular plants in relation to quantitative functioning of dead water-conducting cells and stomata. *Biological Reviews* 68:337–363.

———. 1994. The significance of the distance from photosynthesizing cells to vascular tissue in extant and early vascular plants. *Botanical Journal of Scotland* 47:65–81.

Raven, P. H., R. F. Evert, and S. E. Eichhorn. 1992. *Biology of plants.* 5th ed. New York: Worth.

Raymond, A. 1987. Paleogeographic distribution of early Devonian plant traits. *Palaios* 2:113–132.

Raymond, A., and C. Metz. 1995. Laurussian land-plant diversity during the Silurian and Devonian: mass extinction, sampling bias, or both? *Paleobiology* 21:74–91.

Raymond, A., W. C. Parker, and S. F. Barrett. 1985. Early Devonian phytogeography. In *Geological factors and the evolution of plants,* edited by B. H. Tiffney. New Haven: Yale University Press.

Rayner, R. J. 1983. New observations on *Sawdonia ornata* from Scotland. *Transactions of the Royal Society of Edinburgh. Earth Sciences* 74:79–94.

———. 1984. New finds of *Drepanophycus spinaeformis* Göppert from the Lower Devonian of Scotland. *Transactions of the Royal Society of Edinburgh. Earth Sciences* 75:353–363.

———. 1988. Early land plants from South Africa. *Botanical Journal of the Linnean Society* 97: 229–237.

Remy, W. 1978. Der Dehiszenmechanismus der Sporangien von *Rhynia. Argumenta Palaeobotanica* 5:23–30.

———. 1982. Lower Devonian gametophytes: relation to the phylogeny of land plants. *Science* 215:1625–1627.

Remy, W., P. G. Gensel, and H. Hass. 1993. The gametophyte generation of some early Devonian land plants. *International Journal of Plant Sciences* 154:35–58.

Remy, W., and H. Hass. 1991a. Ergänzende Beobachtungen an *Lyonophyton rhyniensis. Argumenta Palaeobotanica* 8:1–27.

———. 1991b. Gametophyten und Sporophyten im Unterdevon: Fakten und Spekulationen. *Argumenta Palaeobotanica* 8:193–223.

———. 1991c. *Kidstonophyton discoides* nov. gen., nov. spec., ein Gametophyt aus dem Chert von Rhynie (Unterdevon, Schottland). *Argumenta Palaeobotanica* 8:29–45.

———. 1991d. *Langiophyton mackiei* nov. gen., nov. spec., ein Gametophyt mit Archegoniophoren aus dem Chert von Rhynie (Unterdevon, Schottland). *Argumenta Palaeobotanica* 8:69–117.

Remy, W., H. Hass, and S. Schultka. 1986. *Anisophyton potoniei* nov. spec. aus den Kühlbacher Schichten (Emsian) vom Steinbruch Ufersmühle, Wiehltalsperre. *Argumenta Palaeobotanica* 7:123–138.

———. 1992. *Sciadophyton* Steinmann emend. Kräusel et Weyland (1930): der einzige Vertreter eines Unterdevonischen Bauplanes? *Courier Forschungsinstitut Senckenberg* 147:87–91.

Remy, W., and R. Remy. 1980. *Lyonophyton rhyniensis* nov. gen. et nov. spec., ein Gametophyt aus dem Chert von Rhynie (Unterdevon, Schottland). *Argumenta Palaeobotanica* 6:37–72.

Remy, W., R. Remy, H. Hass, S. Schultka, and F. Franzmeyer. 1980. *Sciadophyton* Steinmann: ein Gametophyt aus dem Siegen. *Argumenta Palaeobotanica* 6:73–94.

Remy, W., S. Schultka, and H. Hass. 1986. *Anisophyton gothani* nov. gen., nov. spec. und Hinweise zur Stratigraphie der südlichen Wilbringhäuser Scholle. *Argumenta Palaeobotanica* 7:79–107.

———. 1991. *Calyculiphyton blanai* nov. gen., nov. spec., ein Gametophyt aus dem Ems. *Argumenta Palaeobotanica* 8:119–140.

Renault, B. 1879. Structure comparée de quelques tiges de la flore Carbonifère. *Nouvelles archives du Muséum, Paris* 2:213–348.

———. 1888. *Les plantes fossiles.* Paris: Baillière.

Renzaglia, K. S. 1978. A comparative morphology and developmental anatomy of the Anthocerotophyta. *Journal of the Hattori Botanical Laboratory* 44:31–90.

Renzaglia, K. S., and J. G. Duckett. 1988. Different developmental processes underlie similar spermatozoid architecture in mosses, liverworts, and hornworts. *Journal of the Hattori Botanical Laboratory* 64:219–236.

———. 1989. Ultrastructural studies of spermatogenesis in Anthocerotophyta. V. Nuclear metamorphosis and the posterior mitochondrion of *Notothylas orbicularis* and *Phaeoceros laevis. Protoplasma* 151:137–150.

———. 1991. Towards an understanding of the differences between the blepharoplasts of mosses and liverworts, and comparisons with hornworts, biflagellate lycopods, and charophytes: a numerical analysis. *New Phytologist* 117:187–208.

Renzaglia, K. S., D. K. Smith, K. D. McFarland, and P. G. Davison. 1991. Ultrastructure of the gametophyte/sporophyte junction in *Takakia ceratophylla. American Journal of Botany (Supplement)* 78:8–9.

Rice, C. M., et al. 1994. A Devonian auriferous hot spring system, Rhynie, Scotland. *Journal of the Geological Society of London* 152:229–250.

Richardson, J. B. 1967. Some British Lower Devonian spore assemblages and their stratigraphic significance. *Review of Palaeobotany and Palynology* 1:111–129.

———. 1985. Lower Palaeozoic sporomorphs: their stratigraphical distribution and possible affinities. *Philosophical Transactions of the Royal Society of London* B309:201–205.

Richardson, J. B., and D. C. McGregor. 1986. *Silurian and Devonian spore zones of the Old Red Sandstone Continent and adjacent regions.* Geological Survey of Canada Bulletin 364:1–79.

Riding, R., ed. 1991. *Calcareous algae and stromatolites.* Berlin: Springer-Verlag.

Riding, R., and L. Voronova. 1985. Morphological groups and series in Cambrian calcareous algae. In *Paleoalgology: contemporary research and applications,* edited by D. F. Toomey and M. H. Nitecki. Berlin: Springer.

Rieppel, O. 1988. *Fundamentals of comparative biology.* Basel: Birkhäuser.

Rivera, M. C., and J. A. Lake. 1992. Evidence that eukaryotes and oocyte prokaryotes are immediate relatives. *Science* 257:74–76.

Robert, D. 1973. Le gamétophyte mâle de *Selaginella kraussiana* (Kunze) A. Br. Organisation et développement. Etude en microscope électronique. *Annales des sciences naturelles. Botanique et biologie végétale* 14:465–504.

Robinson, H. 1985. Comments on the cladistic approach to the phylogeny of the "bryophytes" by Mishler and Churchill. *Brittonia* 37:279–281.

Rothwell, G. W. 1972. Evidence of pollen tubes in Paleozoic pteridosperms. *Science* 175:772–774.

———. 1982. New interpretations of the earliest conifers. *Review of Palaeobotany and Palynology* 37:7–28.

———. 1984. The apex of *Stigmaria* (Lycopsida), rooting organ of Lepidodendrales. *American Journal of Botany* 71:1031–1034.

———. 1994. Phylogenetic relationships among ferns: an overview. *Journal of Plant Research* 107:411–416.

———. 1995. The fossil history of branching: implications for the phylogeny of land plants. In *Experimental and molecular approaches to plant biosystematics,* edited by P. C. Hoch and A. G. Stephenson. St. Louis: Missouri Botanical Garden.

———. 1996. Phylogenetic relationships of ferns: a paleobotanical perspective. In *Pteridology in perspective,* edited by J. M. Camus, M. Gibby, and R. J. Johns. London: Royal Botanic Gardens, Kew.

Rothwell, G. W., and D. M. Erwin. 1985. The rhizomorph apex of *Paurodendron:* implications for homologies among the rooting organs of lycopsida. *American Journal of Botany* 72:86–98.

Rothwell, G. W., and J. S. Pryor. 1991. Developmental dynamics of arborescent lycophytes: apical and lateral growth in *Stigmaria ficoides. American Journal of Botany* 78:1740–1745.

Rothwell, G. W., and S. E. Scheckler. 1988. Biology of ancestral gymnosperms. In *Origin and evolution of gymnosperms,* edited by C. B. Beck. New York: Columbia University Press.

Rothwell, G. W., S. E. Scheckler, and W. H. Gillespie. 1989. *Elkinsia* gen. nov., a late Devonian gymnosperm with cupulate ovules. *Botanical Gazette* 150:170–189.

Rothwell, G. W., and R. Serbet. 1994. Lignophyte phylogeny and the evolution of spermatophytes: a numerical cladistic analysis. *Systematic Botany* 19:443–482.

Rothwell, G. W., and R. A. Stockey. 1989. Fossil Ophioglossales in the Paleocene of western North America. *American Journal of Botany* 76:637–644.

———. 1994. The role of *Hydropteris pinnata* gen. et sp. nov. in reconstructing the cladistics of heterosporous ferns. *American Journal of Botany* 81:479–492.

Round, F. E. 1984. The systematics of the Chlorophyta: an historical review leading to some modern concepts. In *Systematics of green algae,* edited by D. E. G. Irvine and D. M. John. London: Academic Press.

Rowe, N. P. 1988a. A herbaceous lycophyte from the Lower Carboniferous Drybrook Sandstone of the Forest of Dean, Gloucestershire. *Palaeontology* 31:69–83.

———. 1988b. Two species of the lycophyte genus *Eskdalia* Kidston from the Drybrook Sandstone (Visean) of Great Britain. *Palaeontographica* B208: 81–103.

———. 1992. Winged late Devonian seeds. *Nature* 359:682.

Rowe, N. P., T. Speck, and J. Galtier. 1993. Biomechanical analysis of a Palaeozoic gymnosperm stem. *Proceedings of the Royal Society of London* 252:19–28.

Sachs, J. 1868. *Lehrbuch der Botanik.* Leipzig: Engelmann.

Sanderson, M. J., and L. Hufford, eds. In press. *Homoplasy and the evolutionary process.* London: Academic Press.

Satterthwait, D. F., and J. W. Schopf. 1972. Structurally preserved phloem zone tissue in *Rhynia. American Journal of Botany* 59:373–376.

Scheckler, S. E. 1975. *Rhymokalon,* a new plant with cladoxylalean anatomy from the Upper Devonian of New York State. *Canadian Journal of Botany* 53:25–38.

———. 1976. Ontogeny of progymnosperms. 1. Shoots of Upper Devonian Aneurophytales. *Canadian Journal of Botany* 54:202–219.

———. 1984. Persistence of the Devonian plant group Barinophytaceae into the basal Carboniferous of Virginia. In *International congress on the stratigraphy and geology of the Carboniferous,* edited by P. K. Sutherland and W. L. Manger. Carbondale: Southern Illinois University Press.

Scheckler, S. E., and H. P. Banks. 1971a. Anatomy and relationships of some Devonian progymnosperms from New York. *American Journal of Botany* 58:737–751.

———. 1971b. *Proteokalon,* a new genus of progymnosperms from the Devonian of New York State and its bearing on phylogenetic trends in the group. *American Journal of Botany* 58:874–884.

———. 1972. Periderm in some Devonian plants. In *Advances in plant morphology,* edited by Y. S. Murty, B. M. Johri, H. Y. Mohan Ram, and T. M. Varghese. Meerut, India: Sarita Prakashan.

Scheirer, D. C. 1980. Differentiation of bryophyte conducting tissues: structure and histochemistry. *Bulletin of the Torrey Botanical Club* 107: 298–307.

———. 1990. Mosses. In *Sieve elements: comparative structure, induction, and development,* edited by H.-D. Behnke and R. D. Sjolund. Berlin: Springer-Verlag.

Schlanker, C. M., and G. A. Leisman. 1969. The herbaceous Carboniferous lycopod *Selaginella fraipontii* comb. nov. *American Journal of Botany* 130:35–41.

Schmid, R. 1976. Septal pores in *Prototaxites,* an enigmatic Devonian plant. *Science* 191:287–288.

———. 1982. The terminology and classification of steles: historical perspective and outlines of a system. *Botanical Review* 48:817–931.

Schneller, J. J. 1990a. Azollaceae. In *Pteridophytes and gymnosperms,* edited by K. U. Kramer and P. S. Green. Vol. 1 of *The families and genera of vascular plants,* edited by K. Kubitzki. Berlin: Springer-Verlag.

———. 1990b. Salviniaceae. In *Pteridophytes and gymnosperms,* edited by K. U. Kramer and P. S. Green. Vol. 1 of *The families and genera of vascular plants,* edited by K. Kubitzki. Berlin: Springer-Verlag.

Schofield, W. B. 1985. *Introduction to bryology.* New York: Macmillan.

Schofield, W. B., and C. Hébant. 1984. The morphology and anatomy of the moss gametophore. In *New manual of bryology,* edited by R. M. Schuster. Miyazaki-ken, Japan: Hattori Botanical Laboratory.

Schopf, J. W. 1968. Microflora of the Bitter Springs Formation, late Precambrian, central Australia. *Journal of Paleontology* 42:651–688.

———. 1993. Macrofossils of the early Archean Apex Chert: new evidence of the antiquity of life. *Science* 260:640–646.

Schopf, J. W., and J. M. Blacic. 1971. New microorganisms from the Bitter Springs Formation (late Precambrian) of the north-central Amadeus Basin, Australia. *Journal of Paleontology* 45:925–960.

Schultka, S. 1978. Beiträge zur Anatomie von *Rhacophyton condrusorum* Crépin. *Argumenta Palaeobotanica* 5:11–22.

Schuster, R. M. 1966. *The Hepaticae and Anthocerotae of North America.* Vol. 1. New York: Columbia University Press.

———. 1977. The evolution and early diversification of the Hepaticae and Anthocerotae. In *Beiträge zur Biologie der niederen Pflanzen,* edited by W. Frey, H. Hurka, and F. Oberwinkler. Stuttgart: Fischer Verlag.

———. 1979. The phylogeny of the Hepaticae. In

Bryophyte Systematics, edited by G. C. S. Clarke and J. G. Duckett. London: Academic Press.
———. 1981. Paleoecology, origin, distribution through time, and evolution of Hepaticae and Anthocerotae. In *Paleobotany, paleoecology, and evolution,* edited by K. J. Niklas. New York: Praeger.
———. 1983. Phytogeography of the Bryophyta. In *New manual of bryology,* edited by R. M. Schuster. Miyazaki-ken, Japan: Hattori Botanical Laboratory.
———. 1984a. Comparative anatomy and morphology of the Hepaticae. In *New manual of bryology,* edited by R. M. Schuster. Miyazaki-ken, Japan: Hattori Botanical Laboratory.
———. 1984b. Evolution, phylogeny, and classification of the Hepaticae. In *New manual of bryology,* edited by R. M. Schuster. Miyazaki-ken, Japan: Hattori Botanical Laboratory.
———. 1984c. Morphology, phylogeny, and classification of the Anthocerotae. In *New manual of bryology,* edited by R. M. Schuster. Miyazaki-ken, Japan: Hattori Botanical Laboratory.
———. 1992. *The Hepaticae and Anthocerotae of North America.* Vol. 6. Chicago: Field Museum of Natural History.
Schuster, R. M., and J. A. Janssens. 1989. On *Diettertia,* an isolated Mesozoic member of the Jungermanniales. *Review of Palaeobotany and Palynology* 57:277–287.
Schweitzer, H.-J. 1972. Die Mitteldevonflora von Lindlar (Rheinland). 3. Filicinae: *Hyenia elegans* Kräusel & Weyland. *Palaeontographica* B137: 154–175.
———. 1979. Die Zosterophyllaceae des Rheinischen Unterdevons. *Bonner Paläobotanische Mitteilungen* 3:1–32.
———. 1980a. Die Gattungen *Renalia* Gensel und *Psilophyton* Dawson im Unterdevon des Rheinlandes. *Bonner Paläobotanische Mitteilungen* 6:1–34.
———. 1980b. Die Gattungen *Taeniocrada* White und *Sciadophyton* Steinmann in Unterdevon des Rheinlandes. *Bonner Paläobotanische Mitteilungen* 5:1–38.
———. 1980c. Über *Drepanophycus spinaeformis* Goeppert. *Bonner Paläobotanische Mitteilungen* 7:1–29.
———. 1983a. Der Generationswechsel der Psilophyten. *Deutsche Botanische Gesellschaft* 96:483–496.
———. 1983b. Die Unterdevonflora des Rheinlandes. *Palaeontographica* B189:1–138.
———. 1987. Introduction to the plant bearing beds and the flora of the Lower Devonian of the Rhineland. *Bonner Paläobotanische Mitteilungen* 13:1–94.
Schweitzer, H.-J., and C. Cai. 1987. Beiträge zur Mitteldevon-Flora Südchinas. *Palaeontographica* B207:111–131.
Schweitzer, H.-J., and L. C. Matten. 1982. *Aneurophyton germanicum* and *Protopteridium thomsonii* from the Middle Devonian of Germany. *Palaeontographica* 184B:65–106.
Scotese, C. R., and W. S. McKerrow. 1990. Revised world maps and introduction. In *Palaeozoic palaeogeography and biogeography,* edited by W. S. McKerrow and C. R. Scotese. London: Geological Society of London.
Scott, A. C., J. Stephenson, and W. G. Chaloner. 1992. Interaction and coevolution of plants and arthropods during the Palaeozoic and Mesozoic. *Philosophical Transactions of the Royal Society of London* B336:129–165.
Scott, D. H. 1900. *Studies in fossil botany.* London: Black.
———. 1920. *Studies in fossil botany.* Vol. 1, *Pteridophyta.* 3rd ed. London: Black.
———. 1924. *Extinct plants and problems of evolution.* London: Macmillan.
Selden, P. A., and D. Edwards. 1989. Colonisation of the land. In *Evolution and the fossil record,* edited by K. C. Allen and D. E. G. Briggs. London: Belhaven Press.
Sepkoski, J. J. 1978. A kinetic model of Phanerozoic taxonomic diversity. 1. Post-Paleozoic families and mass extinctions. *Paleobiology* 10:246–267.
———. 1984. A kinetic model of Phanerozoic taxonomic diversity. 3. Analysis of marine orders. *Paleobiology* 4:223–251.
Sepkoski, J. J., R. K. Bambach, D. M. Raup, and J. W. Valentine. 1981. Phanerozoic marine diversity and the fossil record. *Nature* 239:435–437.
Serbet, R., and G. W. Rothwell. 1992. Characterizing the most primitive seed ferns. 1. A reconstruction of *Elkinsia polymorpha. International Journal of Plant Science* 153:602–621.
Seward, A. C., and S. O. Ford. 1906. The Araucariae, recent and extinct. *Philosophical Transactions of the Royal Society of London* B198:307–398.
Shute, C. H., and D. Edwards. 1989. A new rhyniopsid with novel sporangium organization from the Lower Devonian of South Wales. *Botanical Journal of the Linnean Society* 100:111–137.

Skog, J. E., and H. P. Banks. 1973. *Ibyka amphikoma,* gen. et sp. n., a new protoarticulate precursor from the late Middle Devonian of New York State. *American Journal of Botany* 60:366–380.

Skog, J. E., and C. R. Hill. 1992. The Mesozoic herbaceous lycopsids. *Annals of the Missouri Botanical Garden* 79:648–675.

Sluiman, H. J. 1983. The flagellar apparatus of the zoospore of the filamentous green algae *Coleochaete pulvinata:* absolute configuration and phylogenetic significance. *Protoplasma* 115:160–175.

———. 1985. A cladistic evaluation of the lower and higher green plants (Viridiplantae). *Plant Systematics and Evolution* 149:217–232.

Smith, A. B. 1994. *Systematics and the fossil record: documenting evolutionary patterns.* Oxford: Blackwell.

Smith, D. K., and P. G. Davison. 1993. Antheridia and sporophytes in *Takakia ceratophylla* (Mitt.) Grolle: evidence for reclassification among the mosses. *Journal of the Hattori Botanical Laboratory* 73:263–271.

Smith, G. M. 1938. *Cryptogamic botany. 2. Bryophytes and pteridophytes.* 1st ed. New York: McGraw-Hill.

———. 1955. *Cryptogamic botany. 2. Bryophytes and pteridophytes.* 2nd ed. New York: McGraw-Hill.

Smith, G. O., and D. White. 1905. *The geology of the Perry Basin in southeastern Maine.* U.S. Geological Survey Professional Paper 35.

Smoot, E. L., R. K. Jansen, and T. N. Taylor. 1981. A phylogenetic analysis of the land plants: a botanical commentary. *Taxon* 30:65–67.

Smoot, E. L., T. N. Taylor, and B. S. Serlin. 1982. *Archaeocalamites* from the Upper Mississippian of Arkansas. *Review of Palaeobotany and Palynology* 36:325–334.

Snigirevskaya, N. S., and Y. S. Nadler. 1994. Habit and relationships of *Orestovia* (Middle Devonian). *Palaeontographica* 233B:11–18.

Sogin, M. L., U. Edman, and H. Elwood. 1989. A single kingdom of eukaryotes. In *The hierarchy of life: molecules and morphology in phylogenetic analysis,* edited by B. Fernholm, K. Bremer, and H. Jörnvall. Amsterdam: Elsevier.

Sogin, M., J. H. Gunderson, H. J. Elwood, R. A. Alonso, and D. A. Peattie. 1989. Phylogenetic significance of the kingdom concept: an unusual eukaryotic 16S-like ribosomal RNA from *Giardia lamblia. Science* 243:75–77.

Solms-Laubach, H. G. 1891. *Fossil botany.* Translated by H. E. F. Garnsey and revised by I. B. Balfour. Oxford: Clarendon Press.

Spatz, H.-C., T. Speck, and D. Vogellehner. 1990. Contributions to the biomechanics of plants. 2. Stability against local buckling in hollow stems. *Botanica Acta* 103:123–130.

Speck, T., H.-C. Spatz, and D. Vogellehner. 1990. Contributions to the biomechanics of plants. 1. Stabilities of plant stems with strengthening elements of different cross-sections against weight and wind forces. *Botanica Acta* 103:111–122.

Speck, T., and D. Vogellehner. 1988. Biophysical examinations of the bending stability of various stele types and the upright axes of early "vascular" land plants. *Botanica Acta* 101:262–268.

———. 1991. Biomechanics and maximum height of some Devonian land plants. In *Palaeovegetational development in Europe and regions relevant to its palaeofloristic evolution* [proceedings of the Pan-European Palaeobotanical Conference, Vienna, September 19–23], edited by J. Kovar-Eder. Vienna: Museum of Natural History.

———. 1994. Devonische Landpflanzen mit und ohne hypodermales Sterom: eine biomechanische Analyse mit Überlegungen zur Frühevolution des Leit- und Festigungssystems. *Palaeontographica* B233:157–227.

Sporne, K. R. 1970. *The morphology of pteridophytes: the structure of ferns and allied plants.* London: Hutchinson.

Stafford, P. J. 1991. The northwestern European pollen flora. 44. Selaginellaceae. *Review of Palaeobotany and Palynology* 69:1–22.

Steele, K. P., K. E. Holsinger, R. K. Jansen, and D. W. Taylor. 1991. Assessing the reliability of 5S rRNA sequence data for phylogenetic analysis in green plants. *Molecular Biology and Evolution* 8:240–248.

Steemans, P., and P. Gerrienne. 1984. La micro- et macroflore du Gedinnien de la Gileppe, Synclinorium de la Vesdre, Belgique. *Annales de la Société géologique de Belgique* 107:51–71.

Stein, D. B., D. S. Conant, M. E. Ahearn, E. T. Jordan, S. A. Kirch, M. Hasebe, K. Iwatsuki, M. K. Tan, and J. A. Thomson. 1992. Structural rearrangements of the chloroplast genome provide an important phylogenetic link in ferns. *Proceedings of the National Academy of Sciences* 89:1856–1860.

Stein, W. E. 1982. *Iridopteris eriensis* from the Middle

Devonian of North America, with systematics of apparently related taxa. *Botanical Gazette* 143: 401–416.

———. 1993. Modeling the evolution of stelar architecture in vascular plants. *International Journal of Plant Science* 154:229–263.

Stein, W. E., and C. B. Beck. 1983. *Triloboxylon arnoldii* from the Middle Devonian of western New York. *Contributions to the Museum of Palaeontology, University of Michigan* 26:257–288.

Stein, W. E., G. D. Harmon, and F. M. Hueber. 1994. *Spongiophyton* from the Lower Devonian of North America reinterpreted as a lichen. Paper read at international workshop on the biology and evolutionary implications of early Devonian plants, at Westfälische Wilhelms-Universität, Münster, Germany, September 14–17, 1994.

Stein, W. E., and F. M. Hueber. 1989. The anatomy of *Pseudosporochnus: P. hueberi* from the Devonian of New York. *Review of Palaeobotany and Palynology* 60:311–359.

Stein, W. E., D. C. Wight, and C. B. Beck. 1984. Possible alternatives for the origin of Sphenopsida. *Systematic Botany* 9:102–118.

Stevenson, D. W. 1976. Observations on phyllotaxis, stelar morphology, the shoot apex, and gemmae of *Lycopodium lucidulum* Michaux (Lycopodiaceae). *Botanical Journal of the Linnean Society* 72:81–100.

Stevenson, D. W., and H. Loconte. 1996. Ordinal and familial relationships of pteridophyte genera. In *Pteridology in perspective,* edited by J. M. Camus, M. Gibby, and R. J. Johns. London: Royal Botanic Gardens, Kew.

Stewart, K. D., and K. R. Mattox. 1975. Comparative cytology, evolution, and classification of the green algae with some consideration of the origin of other organisms with chlorophylls a and b. *Botanical Review* 41:104–135.

———. 1978. Comparative evolution in the flagellated cells of green algae and land plants. *BioSystems* 10:145–152.

———. 1980. Phylogeny of phytoflagellates. In *Phytoflagellates,* edited by E. R. Cox. Amsterdam: Elsevier.

Stewart, W. N. 1947. A comparative study of stigmarian appendages and *Isoetes* roots. *American Journal of Botany* 34:315–324.

———. 1964. An upward outlook in plant morphology. *Phytomorphology* 14:120–134.

———. 1983. *Paleobotany and the evolution of plants.* Cambridge: Cambridge University Press.

Stewart, W. N., and G. W. Rothwell. 1993. *Paleobotany and the evolution of plants.* 2nd ed. Cambridge: Cambridge University Press.

Stockmans, F. 1939. Note préliminaire sur des empreintes végétales du Dévonien Inférieur de la Belgique. *Bulletin du Musée royal d'histoire naturelle de Belgique* 15:1–6.

———. 1940. Végétaux Eodévoniens de la Belgique. *Mémoires du Musée royal d'histoire naturelle de Belgique* 93:1–90.

———. 1948. Végétaux Dévoniens Supérieur de la Belgique. *Mémoires du Musée royal d'histoire naturelle de Belgique* 110:1–85.

Stokey, A. G. 1951. The contribution by the gametophyte to the classification of homosporous ferns. *Phytomorphology* 1:39–58.

Streel, M. 1972. Dispersed spores associated with *Leclercqia complexa* Banks, Bonamo and Grierson from the late Middle Devonian of eastern New York State (U.S.A.). *Review of Palaeobotany and Palynology* 14:205–215.

Strother, P. K. 1991. A classification schema for the cryptospores. *Palynology* 15:219–236.

———. 1993. Clarification of the genus *Nematothallus* Lang. *Journal of Paleontology* 67:1090–1094.

Strother, P. K., and A. Traverse. 1979. Plant microfossils from the Llandoverian and Wenlockian rocks of Pennsylvania. *Palynology* 3:1–22.

Stubblefield, S. P., and H. P. Banks. 1978. The cuticle of *Drepanophycus spinaeformis,* a long-ranging Devonian lycopod from New York and eastern Canada. *American Journal of Botany* 65:110–118.

Stubblefield, S. P., and G. W. Rothwell. 1981. Embryogeny and reproductive biology of *Bothrodendrostrobus mundus* (Lycopsida). *American Journal of Botany* 68:625–634.

Suire, C., and Y. Asakawa. 1979. Chemotaxonomy of bryophytes: a survey. In *Bryophyte systematics,* edited by G. C. S. Clarke and J. G. Duckett. London: Academic Press.

Summans, R. E., and M. R. Walter. 1990. Molecular fossils and microfossils of prokaryotes and protists from Proterozoic sediments. *American Journal of Science* 290A:212–244.

Swartz, O. 1801. *Genera et species filicum ordine systematico redactarum adiectis synonymis et iconibus selectis, nec non speciebus recenter detectis, et denum plurimis dubiosis, ulterius investigandis. Journal für die Botanik,* pp. 1–120.

Swofford, D. L. 1990. *PAUP: phylogenetic analysis using parsimony.* 3.1.1 ed. Champaign: Illinois Natural History Survey.

Sykes, M. G. 1908. The anatomy and morphology of *Tmesipteris. Annals of Botany* 22:63–89.

Sztein, A. E., J. D. Cohen, J. P. Slovin, and T. J. Cooke. 1995. Auxin metabolism in representative land plants. *American Journal of Botany* 82: 1514–1521.

Takhtajan, A. L. 1953. Phylogenetic principles of the system of higher plants. *Botanical Review* 19:1–45.

Tanner, W. R. 1982. A new species of *Gosslingia* (Zosterophyllophytina) from the Lower Devonian Beartooth Butte Formation of northern Wyoming. In Vol. 2 of *Proceedings of the Third North American Paleontological Convention,* edited by B. Mamet and M. J. Copeland. Toronto: Business and Economic Service Ltd.

Tansley, A. G. 1907. Lectures on the evolution of the filicinean vascular system. *New Phytologist* 6:25–35.

———. 1923. Some aspects of the present position of botany. Paper read at a meeting of the British Association for the Advancement of Science (section k).

Tansley, A. G., and E. C. Chick. 1901. Notes on the conducting tissue-system in Bryophyta. *Annals of Botany* 25:1–38.

Tappan, H. N. 1980. *The paleobiology of plant protists.* San Francisco: Freeman.

Taylor, E. L. 1990. Phloem evolution: an appraisal based on the fossil record. In *Sieve elements: comparative structure, induction, and development,* edited by H.-D. Behnke and R. D. Sjolund. Berlin: Springer.

Taylor, T. N. 1981. *Paleobotany: an introduction to fossil plant biology.* New York: McGraw-Hill.

———. 1988a. The origin of land plants: some answers, more questions. *Taxon* 37:805–833.

———. 1988b. Pollen and pollen organs of fossil gymnosperms: phylogeny and reproductive biology. In *Origin and evolution of gymnosperms,* edited by C. B. Beck. New York: Columbia University Press.

———. 1990. Microsporogenesis in fossil plants. In *Microspores: evolution and ontogeny,* edited by S. Blackmore and R. B. Knox. London: Academic Press.

Taylor, T. N., and D. F. Brauer. 1983. Ultrastructural studies of *in situ* Devonian spores *Barinophyton citrulliforme. American Journal of Botany* 70:106–112.

Taylor, T. N., H. Hass, and W. Remy. 1992. Devonian fungi: interactions with the green alga *Palaeonitella. Mycologia* 84:901–910.

Taylor, T. N., W. Remy, and H. Hass. 1992. Parasitism in a 400-million-year-old green alga. *Nature* 357:493–494.

Taylor, T. N., and E. L. Taylor. 1993. *The biology and evolution of fossil plants.* New Jersey: Prentice Hall.

Taylor, W. A. 1989. Megaspore wall ultrastructure in *Selaginella. Pollen et Spores* 31:251–288.

———. 1990. Comparative analysis of megaspore ultrastructure in Pennsylvanian lycophytes. *Review of Palaeobotany and Palynology* 62:65–78.

———. 1992. Megaspore wall development in *Isoetes melanopoda:* morphogenetic post-initiation changes accompanying spore enlargement. *Review of Palaeobotany and Palynology* 72:61–72.

———. 1995a. Spores in earliest land plants. *Nature* 373:391–392.

———. 1995b. Ultrastructure of *Tetrahedraletes medinensis* (Strother and Traverse) Wellman and Richardson from the Upper Ordovician of southern Ohio. *Review of Palaeobotany and Palynology* 85:183–187.

Taylor, W. A., and T. N. Taylor. 1987. Spore wall ultrastructure of *Protosalvinia. American Journal of Botany* 74:437–443.

———. 1990. Persistent ultrastructural features in microspores of heterosporous lycophytes. *Grana* 29:219–228.

Taylor, W. C., and R. J. Hickey. 1992. Habitat, evolution, and speciation in *Isoetes. Annals of the Missouri Botanical Garden* 79:613–622.

Theriot, E. 1988. A review of Sluiman's cladistic classification of green plants with particular reference to flagellar data and to land plant origins. *Taxon* 37:913–919.

Thomas, B. A. 1972. A probable moss from the Lower Carboniferous of the Forest of Dean, Gloucestershire. *Annals of Botany* 36:155–161.

———. 1986. The biochemical analysis of fossil plants and its use in taxonomy and systematics. In *Systematic and taxonomic approaches in palaeobotany,* edited by R. A. Spicer and B. A. Thomas. Oxford: Clarendon Press.

———. 1992. Paleozoic herbaceous lycopsids and the beginnings of extant *Lycopodium* sens. lat. and *Selaginella* sens. lat. *Annals of the Missouri Botanical Garden* 79:623–631.

Thomas, B. A., and S. D. Brack-Hanes. 1984. A new approach to family groupings in the lycophytes. *Taxon* 33:247–255.

Thomas, B. A., and N. Quansah. 1991. The palaeobotanical case for dividing *Selaginella* (Selaginellaceae: Pteridophyta). *Fern Gazette* 14:59–64.

Thomas, H. H. 1932. The old morphology and the new. *Proceedings of the Linnean Society of London* 176:17–44.

———. 1951. A hundred years of plant morphology. *Nature* 168:312.

———. 1958. Fossil plants and evolution. *Journal of the Linnean Society of London (Botany)* 56: 123–135.

Thompson, R. H. 1969. Sexual reproduction in *Chaetosphaeridium globosum* (Nordst.) Klebahn (Chlorophyceae) and description of a species new to science. *Journal of Phycology* 5:285–292.

Tiffney, B. H. 1981. Diversity and major events in the evolution of land plants. In *Paleobotany, paleoecology, and evolution,* edited by K. J. Niklas. New York: Praeger.

Tims, J. D., and T. C. Chambers. 1984. Rhyniophytina and Trimerophytina from the early land flora of Victoria, Australia. *Palaeontology* 27:265–279.

Towe, K. M. 1994. Earth's early atmosphere: constraints and opportunities for early evolution. In *Early life on earth,* edited by S. Bengtson. New York: Columbia University Press.

Trant, C. A., and P. G. Gensel. 1985. Branching in *Psilophyton:* a new species from the Lower Devonian of New Brunswick, Canada. *American Journal of Botany* 72:1256–1273.

Traverse, A. 1988. *Paleopalynology.* Boston: Unwin Hyman.

Tryon, A. F., and B. Lugardon. 1991. *Spores of the Pteridophyta: surface, wall structure, and diversity based on electron microscope studies.* New York: Springer.

Tryon, R. M., and A. F. Tryon. 1982. *Ferns and allied plants with special reference to tropical America.* New York: Springer.

Turner, J. J. 1924. Origin and development of vascular system of *Lycopodium lucidulum. Botanical Gazette* 78:215–225.

Uehara, K., and S. Kurita. 1991. Ultrastructural study on spore wall morphogenesis in *Lycopodium clavatum* (Lycopodiaceae). *American Journal of Botany* 78:24–36.

Valdespino, I. A. 1992. New species of *Selaginella* (Selaginellaceae) from the Guayana Highland of Venezuela. *Brittonia* 44:199–297.

Van de Peer, Y., R. De Baere, J. Cauwenberghs, and R. De Wachter. 1990. Evolution of green plants and their relationship with other photosynthetic eukaryotes as deduced from 5S ribosomal RNA sequences. *Plant Systematics and Evolution* 170:85–96.

Van der Voo, R. 1993. *Paleomagnetism of the Atlantic, Tethys, and Iapetus oceans.* Cambridge: Cambridge University Press.

Van Tieghem, P. 1891. *Traité de botanique.* Paris: Savy.

Vaughn, K. C., E. O. Campbell, T. Hasegawa, H. A. Owen, and K. S. Renzaglia. 1990. The pyrenoid is the site of ribulose 1,5-bisphosphate carboxylase/oxygenase accumulation in the hornwort (Bryophyta: Anthocerotae) chloroplast. *Protoplasma* 156:117–129.

Wagner, W. H., and M. J. Beitel. 1992. Generic classification of modern North American Lycopodiaceae. *Annals of the Missouri Botanical Garden* 79:676–686.

Wagner, W. H., M. J. Beitel, and F. S. Wagner. 1982. Complex venation patterns in the leaves of *Selaginella:* megaphyll-like leaves in lycophytes. *Science* 218:793–794.

Wainright, P. O., G. Hinkle, M. L. Sogin, and S. K. Stickel. 1993. Monophyletic origin of the Metazoa: an evolutionary link with fungi. *Science* 260:340–342.

Walne, P. L., and P. A. Kivic. 1990. Phylum Euglenida. In *Handbook of Protoctista,* edited by L. Margulis, J. O. Corliss, M. Melkonian, and D. J. Chapman. Boston: Jones and Bartlett.

Walton, J. 1940. *An introduction to the study of fossil plants.* London: Black.

———. 1953. The evolution of the ovule in the pteridosperms. *The Advancement of Science* 38:1–8.

———. 1964. On the morphology of *Zosterophyllum* and some other early Devonian plants. *Phytomorphology* 14:155–160.

———. 1969. On the structure of a silicified stem of *Protopitys* and roots associated with it from the Carboniferous Limestone, Lower Carboniferous (Mississippian) of Yorkshire, England. *American Journal of Botany* 56:808–813.

Wang, Zhong, and Bao-Yin Geng. In press. A new

Middle Devonian plant: *Metacladophyton tetrapetalum* gen. et sp. nov. *Palaeontographica.*

Wanntorp, H. E., D. R. Brooks, T. Nilsson, S. Nylin, F. Ronquist, S. C. Stearns, and N. Wedell. 1990. Phylogenetic approaches in ecology. *Oikos* 57:119–132.

Wardlaw, C. W. 1924. Size in relation to internal morphology. 1. Distribution of the xylem in the vascular system of *Psilotum, Tmesipteris,* and *Lycopodium. Transactions of the Royal Society of Edinburgh* 53:503–532.

———. 1925. Size in relation to internal morphology. 2. The vascular system of *Selaginella. Transactions of the Royal Society of Edinburgh* 54:281–308.

———. 1928. Size in relation to internal morphology. 3. The vascular system of roots. *Transactions of the Royal Society of Edinburgh* 56:19–55.

———. 1952. *Phylogeny and morphogenesis.* London: Macmillan.

Warming, E. 1891. *Den systematiske Botanik.* Copenhagen: Philipsens.

Waters, D. A., M. A. Buchheim, R. A. Dewey, and R. L. Chapman. 1992. Preliminary inferences of the phylogeny of bryophytes from nuclear-encoded ribosomal RNA sequences. *American Journal of Botany* 79:459–466.

Watson, E. V. 1971. *The structure and life of bryophytes.* London: Hutchinson.

Webster, T. R. 1992. Developmental problems in *Selaginella* (Selaginellaceae) in an evolutionary context. *Annals of the Missouri Botanical Garden* 79:632–647.

Wellman, C. H. 1993a. A land plant microfossil assemblage of Mid Silurian age from the Stonehaven Group, Scotland. *Journal of Micropalaeontology* 12:47–66.

———. 1993b. A Lower Devonian sporomorph assemblage from the Midland Valley of Scotland. *Transactions of the Royal Society of Edinburgh. Earth Sciences* 84:117–136.

———. 1995. "Phytodebris" from Scottish Silurian and Lower Devonian continental deposits. *Review of Palaeobotany and Palynology* 84:255–279.

Wellman, C. H., and J. B. Richardson. 1993. Terrestrial plant microfossils from the Silurian Inliers of the Midland Valley of Scotland. *Palaeontology* 36:155–193.

Wenzel, J. W., and J. M. Carpenter. 1994. Comparing methods: adaptive traits and tests of adaptation. In *Phylogenetics and ecology,* edited by R. I. Vane-Wright and P. Eggleton. London: Academic Press.

Wesley, A., and B. Kuyper. 1951. Electron-microscopic observations on the xylem elements of a fossil plant. *Nature* 168:137–140.

Westoll, T. S. 1977. Northern Britain. In *A correlation of Devonian rocks in the British Isles,* edited by M. R. House, J. B. Richardson, W. G. Chaloner, J. R. L. Allen, C. H. Holland, and T. S. Westoll. Geological Society of London Special Report 7.

White, R. A., and M. D. Turner. 1995. Anatomy and development of the fern sporophyte. *The Botanical Review* 61:281–305.

Whittaker, R. H. 1959. On the broad classification of organisms. *Quarterly Review of Biology* 34:210–226.

———. 1969. New concepts of the kingdoms of organisms. *Science* 163:150–160.

Whittaker, R. H., and L. Margulis. 1978. Protist classification and the kingdoms of organisms. *BioSystems* 10:3–18.

Whittemore, A. T. 1987. Transition to a land flora: a critique. *Cladistics* 3:60–65.

Whittier, D. P. 1977. Gametophytes of *Lycopodium obscurum* as grown in axenic culture. *Canadian Journal of Botany* 55:563–567.

———. 1981. Gametophytes of *Lycopodium digitatum* (formerly *L. complanatum* var. *flabelliforme*) as grown in axenic culture. *Botanical Gazette* 142:519–524.

Whittier, D. P., and J. E. Braggins. 1992. The young gametophyte of *Phylloglossum* (Lycopodiaceae). *Annals of the Missouri Botanical Garden* 79:730–736.

Whittier, D. P., and T. R. Webster. 1986. Gametophytes of *Lycopodium lucidulum* from axenic culture. *American Fern Journal* 76:48–55.

Wight, D. C. 1987. Non-adaptive change in early land plant evolution. *Paleobiology* 13:208–214.

Wilce, J. H. 1972. Lycopod spores. 1. General spore patterns and the generic segregates of *Lycopodium. American Fern Journal* 62:65–79.

Wilcox, L. W., P. A. Fuerst, and G. L. Floyd. 1993. Phylogenetic relationship of four charophycean green algae inferred from complete nuclear-encoded small subunit rRNA gene sequences. *American Journal of Botany* 80:1028–1033.

Wiley, E. O. 1981. *Phylogenetics: the theory and practice of phylogenetic systematics.* New York: Wiley.

Wiley, E. O., D. Siegel-Causey, D. R. Brooks, and V. A. Funk. 1991. *The complete cladist.* The University of Kansas Museum of Natural History Special Publication 19.

Williamson, W. C. 1871. On the organization of the fossil plants of the coal measures. Part 1. *Calamites. Philosophical Transactions of the Royal Society of London* B161:447–510.

———. 1872a. On the organization of the fossil plants of the coal measures. Part 2. Lycopodiaceae: Lepidodendra and Sigillariae. *Philosophical Transactions of the Royal Society of London* B162:197–240.

———. 1872b. On the organization of the fossil plants of the coal measures. Part 3. Lycopodiaceae (continued). *Philosophical Transactions of the Royal Society of London* B162:283–318.

———. 1881. On the organization of the fossil plants of the coal measures. Part 11. *Philosophical Transactions of the Royal Society of London* B172: 283–318.

———. 1887. On the organisation of the fossil plants of the coal measures. Part 13. *Heterangium tiliaeoides* (Williamson) and *Kaloxylon hookeri. Philosophical Transactions of the Royal Society of London* B178:289–304.

Wilson, M. L., and D. A. Eggert. 1974. Root phloem of fossil tree-sized arthrophytes. *Botanical Gazette* 135:319–328.

Woese, C. R. 1987. Bacterial evolution. *Microbiological Reviews* 51:221–271.

———. 1989. Archaebacteria and the nature of their evolution. In *The hierarchy of life: molecules and morphology in phylogenetic analysis,* edited by B. Fernholm, K. Bremer, and H. Jörnvall. Amsterdam: Elsevier.

———. 1991. A natural classification. *Nature* 351: 528–529.

Woese, C. R., O. Kandler, and M. L. Wheelis. 1990. Towards a natural system of organisms: proposal for the domains Archaea, Bacteria, and Eucarya. *Proceedings of the National Academy of Sciences* 87:4576–4579.

Wolters, J. 1991. The troublesome parasites: molecular and morphological evidence that Apicomplexa belong to the dinoflagellate-ciliate clade. *BioSystems* 25:75–83.

Wood, R. D., and K. Imahori. 1965. *A revision of the Characeae.* Vol. 1. Weinheim: Cramer.

Young, D. A., and P. M. Richardson. 1982. A phylogenetic analysis of extant seed plants: the need to utilize homologous characters. *Taxon* 31: 250–254.

Zakharova, T. V. 1981. On the systematic position of the species *"Psilophyton" goldschmidtii* from the Lower Devonian of Eurasia. *Paleontological Journal* 15(3):109–118.

Zdebska, D. 1972. *Sawdonia ornata* (=*Psilophyton princeps* var. *ornatum*) from Poland. *Acta Palaeobotanica* 13:77–100.

———. 1982. A new zosterophyll from the Lower Devonian of Poland. *Palaeontology* 25:247–263.

———. 1986. *Psilophyton szaferi* sp. nov., from the Lower Devonian of the Holy Cross Mountains, Poland. *Acta Societatis Botanicorum Poloniae* 55:315–324.

Ziegler, A. M., R. K. Bambach, J. T. Parrish, S. F. Barrett, E. H. Gierlowski, W. C. Parker, A. Raymond, and J. J. Sepkoski. 1981. Paleozoic biogeography and climatology. In *Paleobotany, paleoecology, and evolution,* edited by K. J. Niklas. New York: Praeger.

Zimmermann, W. 1930. *Die Phylogenie der Pflanzen.* Jena: Fischer.

———. 1938. Die Telometheorie. *Biologe* 7: 385–391.

———. 1952. Main results of the "Telome Theory." *The Palaeobotanist* 1:456–470.

———. 1965. *Die Telomtheorie.* Stuttgart: Fischer.

Taxonomic Index

Page-number citations followed by **t** refer to tables, and those followed by *f* refer to figures.

Subject Index

Page-number citations followed by **t** refer to tables, and those followed by *f* refer to figures.